Birkhäuser

Progress in Mathematics

Volume 358

Progress in Mathematics is a series of books intended for professional mathematicians and scientists, encompassing all areas of pure mathematics. This distinguished series, which began in 1979, includes research level monographs, polished notes arising from seminars or lecture series, graduate level textbooks, and proceedings of focused and refereed conferences. It is designed as a vehicle for reporting ongoing research as well as expositions of particular subject areas.

Michael Pevzner • Hideko Sekiguchi
Editors

Symmetry in Geometry and Analysis, Volume 2

Festschrift in Honor of Toshiyuki Kobayashi

Editors
Michael Pevzner
Mathematical Laboratory
University of Reims Champagne-Ardenne
Reims, France

Hideko Sekiguchi
Department of Mathematics
University of Tokyo
Tokyo, Japan

ISSN 0743-1643 ISSN 2296-505X (electronic)
Progress in Mathematics
ISBN 978-981-97-7661-0 ISBN 978-981-97-7662-7 (eBook)
https://doi.org/10.1007/978-981-97-7662-7

Mathematics Subject Classification: 00B30, 20G20, 17B10, 22E45, 22E46, 22E50, 33C05, 33C80, 34M03, 35H10, 37C79, 43A85, 53C35, 81R12

This book is published under the imprint Birkhäuser, www.birkhauser-science.com by the registered company Springer Nature Singapore Pte Ltd.
The registered company address is: 152 Beach Road, #21-01/04 Gateway East, Singapore 189721, Singapore

Contents

Contents of Volume 1

Contents of Volume 3

The Source Operator Method: An Overview

Salem Ben Said, Jean-Louis Clerc, and Khalid Koufany

To Toshiyuki Kobayashi for his commitment to the mathematical community.

Abstract This is an overview on the source operator method, which leads to the construction of symmetry breaking differential operators (SBDO) in the context of tensor product of two principals series representations for the conformal group of a simple real Jordan algebra. This method can be applied to other geometric contexts: in the construction of SBDO for differential forms and for spinors and also for the construction of Juhl's operators corresponding to the restriction from the sphere S^n to S^{n-1}.

1 Introduction

T. Kobayashi's article [25] entitled
A program for branching problems in the representation theory of reductive group is concerned with the following general problem : for a real reductive group G and a reductive subgroup G', how does an irreducible representation π of G behave when restricted to G'? His program consists in three stages:

Stage A: Abstract features of the restriction $\pi_{|G'}$.
Stage B: Branching laws.
Stage C: Construction of symmetry breaking operators.

S. Ben Said (✉)
Department of Mathematical Sciences, College of Science, United Arab Emirates University, Al Ain, UAE
e-mail: salem.bensaid@uaeu.ac.ae

J.-L. Clerc · K. Koufany
CNRS, IECL, Université de Lorraine, Nancy, France
e-mail: jean-louis.clerc@univ-lorraine.fr; khalid.koufany@univ-lorraine.fr

M. Pevzner, H. Sekiguchi (eds.), *Symmetry in Geometry and Analysis, Volume 2*, Progress in Mathematics 358, https://doi.org/10.1007/978-981-97-7662-7_1

Given a representation ρ of G', a *symmetry breaking operator* D is an operator from the representation space of π into the representation space of ρ, which satisfies the intertwining property, valid for any $g' \in G'$,

$$D \circ \pi(g') = \rho(g') \circ D. \tag{1}$$

For the history of the subject, examples of symmetry breaking operators came earlier than Kobayashi's program, and they gave inspiration for the first two stages. An early example is the *Rankin–Cohen brackets* [13]. In this case, the group G is the product $\mathrm{SL}(2,\mathbb{R}) \times \mathrm{SL}(2,\mathbb{R})$ and G' is the diagonal subgroup of G naturally identified with $\mathrm{SL}(2,\mathbb{R})$. Let $\mathcal{O}(\Pi)$ be the space of holomorphic functions on the upper half-plane $\Pi = \{z = x + iy : y > 0\}$, and for any $k \in \mathbb{N}$, let π_k be the representation of $\mathrm{SL}(2,\mathbb{R})$ on $\mathcal{O}(\Pi)$ given as follows : for $g \in \mathrm{SL}(2,\mathbb{R})$ such that $g^{-1} = \begin{pmatrix} a & b \\ c & d \end{pmatrix}$ and $F \in \mathcal{O}(\Pi)$, set

$$\pi_k(g)F(z) = (cz+d)^{-k} F\left(\frac{az+b}{cz+d}\right). \tag{2}$$

This formula defines a representation π_k of $\mathrm{SL}(2,\mathbb{R})$ on $\mathcal{O}(\Pi)$. Let ℓ, m be two nonnegative integers. The tensor product (in fact an appropriate completion) $\pi_\ell \otimes \pi_m$ is a representation of G, which has a natural realization on $\mathcal{O}(\Pi \times \Pi)$. Let further $k \in \mathbb{N}$. Consider the following constant coefficients bi-differential operator

$$B^{(k)}_{\ell,m} : \mathcal{O}(\Pi \times \Pi) \longrightarrow \mathcal{O}(\Pi)$$

defined by

$$B^{(k)}_{\ell,m} F(z) = \sum_{r+s=k} (-1)^r \binom{\ell+k-1}{s} \binom{m+k-1}{r} \left(\frac{\partial}{\partial z}\right)^r \left(\frac{\partial}{\partial w}\right)^s F(z,w)_{|z=w} \, . \tag{3}$$

Then, for any $g \in \mathrm{SL}(2,\mathbb{R})$, we have

$$B^{(k)}_{\ell,m} \circ (\pi_\ell(g) \otimes \pi_m(g)) = \pi_{\ell+m+2k}(g) \circ B^{(k)}_{\ell,m} \tag{4}$$

or otherwise stated, $B^{(k)}_{\ell,m}$ is a symmetry breaking operator for the pair $(\rho_\ell \otimes \rho_m, \rho_{\ell+m+2k})$. See [11] for more information and proofs on this case.

This chapter presents a method, called the *source operator method* to construct such symmetry breaking differential operators (SBDO). It turned out to be sucessfull in many situations.

We concentrate on one example (first exposed in [5]), which covers a large example of geometric situations and for which a rather explicit formula (see [10]) for the symmetry breaking operators is now known. In Sect. 2, a short presentation of the geometric situation was given. An overview of the construction of the source operator is given in Sect. 4. This comes after a functional equation, which has been established in Sect. 3, which is an important result on its own. The corresponding SBDOs are constructed in Sect. 5. Finally, Sect. 6 collects other examples where the source operator method is used.

It is not clear at the moment to which cases the source operator method may be applied, both in terms of geometric situations and in terms of representations. Notice first that we were only interested in cases where the source operator is a *differential* operator, but the first step of the construction (i.e., the use of a multiplication operator M and the Knapp–Stein operators) is quite general. It would be interesting to describe all cases where this construction actually leads to a differential operator. In the case of a tensor product representation of the conformal group $O(1, n+1)$, the answer is positive for the case of differential forms and also for the spinor case, but the question remains open for other vector-valued representations. Another question is to characterize the source operator among all possible G'-covariant operators. In the tensor product case, it is not difficult to see that there are many $\text{diag}(G \times G)$-covariant differential operators between $\mathcal{H}_\lambda \otimes \mathcal{H}_\mu$ and $\mathcal{H}_{\lambda+1} \otimes \mathcal{H}_{\mu+1}$, and our guess is that the one we have constructed is the one that has the lowest possible degree.

Finally, let us mention two connections of the source operator method with other questions. The source operator corresponding to the (Euclidean) conformal case already appeared in [2] in connection with the study of singular invariant trilinear forms on the sphere $S = S^n$. It is connected to a *Bernstein–Sato identity* for the generic kernel of the trilinear form viewed as a distribution on $S \times S \times S$. In a quite different direction, operators similar to the source operator appeared in [9], in connection with conformal analysis on hypersurfaces of manifolds, under the name of *shift operators* (see [22, 29]) or *degenerate Laplacians* (see [19]).

2 Background: Real Simple Jordan Algebra

Let V be a real Jordan algebra of dimension n, with a bilinear multiplication $V \times V \to V, (x, y) \mapsto x \cdot y$, and a unit element $\mathbf{1}$. To every $x \in V$, we associate a generic minimal polynomial (see [15, Section II.2])

$$f_x(\lambda) = \lambda^r - a_1(x)\lambda^{r-1} + \cdots + (-1)^r a_r(x).$$

Its degree r is the so-called rank of the Jordan algebra V. For $1 \leq j \leq r$, the coefficient $a_j(x)$ are homogeneous polynomials of degree j and invariant under automorphisms of V. In particular, the Jordan trace $\text{tr}(x) := a_1(x)$ and the Jordan determinant $\mathbf{det}(x) := a_r(x)$ are invariant under automorphisms of V.

The symmetric bilinear form

$$\tau(x, y) := \text{tr}(xy), \qquad x, y \in V \tag{5}$$

is called the trace form of V. The real Jordan algebra V is said to be semi-simple if τ is non-degenerate, and V is called Euclidean if τ is positive definite. Moreover, a semi-simple Jordan algebra with no nontrivial ideal is called simple. From now on we will assume that V is a simple real Jordan algebra.

An involutive automorphism α of V such that

$$(x \mid y) := \tau(x, \alpha y) \tag{6}$$

is positive definite, is called Cartan involution of V. For a semi-simple Jordan algebra, such a Cartan involution always exists and two Cartan involutions are conjugate by an automorphism of V (see [20, Satz 4.1, Satz 5.2]). We have the decomposition

$$V = V^+ \oplus V^-$$

into ± 1 eigenspaces of V. One can see that

$$V^\pm \cdot V^\pm \subseteq V^+,$$
$$V^+ \cdot V^- \subseteq V^-.$$

Hence, the $+1$ eigenspace V^+ is an Euclidean Jordan subalgebra of V with the same identity element $\mathbf{1}$. Note that if V itself is already Euclidean, then $\alpha = \mathrm{id}_V$ is a Cartan involution, and indeed the only Cartan involution. In this case, $V^+ = V$ and $V^- = 0$.

We denote by n_+ and r_+ dimension and rank of $V^+$and call r_+ the *split rank* of V. The constants n_+ and r_+ only depend on the isomophism class of the Jordan algebra V, not on the choice of the involution α.

An element $c \in V$ is said to be *idempotent* if $c \cdot c = c$. Two idempotents c_1 and c_2 are called *orthogonal* if $c_1 \cdot c_2 = 0$. A non-zero idempotent is called *primitive* if it cannot be written as the sum of two non-zero orthogonal idempotents.

Every set $\{c_1, \dots, c_k\}$ of orthogonal primitive idempotents in V_+ with the additional condition $c_1 + \cdots + c_k = \mathbf{1}$ is called a *Jordan frame* in V_+. By [15, Theorem III.1.2] the cardinal of a Jordan frame equals to the rank r_+ of V_+ and every two Jordan frames are conjugate by an automorphism of V_+ (see [15, Corollary IV.2.7]).

Fix a Jordan frame $\{c_1, \dots, c_{r_+}\}$ in V_+. By [15, Proposition III.1.3], the spectrum of the multiplication operator $L(c_k)$ by c_k is $\{0, \frac{1}{2}, 1\}$. Further, the operators $L(c_1), \dots, L(c_{r_+})$ commute and therefore are simultaneously diagonalizable. This gives the following Peirce decomposition

$$V = \bigoplus_{1 \le i \le j \le r_+} V_{ij},$$

where

$$\begin{aligned} V_{ii} &:= V(c_i, 1), && 1 \leq i \leq r_+, \\ V_{ij} &:= V(c_i, \tfrac{1}{2}) \cap V(c_j, \tfrac{1}{2}), && 1 \leq i < j \leq r_+. \end{aligned}$$

Here $V(c, \lambda)$ denotes the eigenspace of $L(c)$ corresponding to the eigenvalue λ. Since the operators $L(c_k)$, for $0 \leq k \leq r_+$, are symmetric with respect to the inner product (6), then the aforementioned direct sum is orthogonal.

Denote by d the common dimension of the subspaces V_{ij} $(i < j)$ and by $e + 1$ the common dimension of the subalgebras V_{ii}. Then, the dimension n of V satisfies

$$n = r_+(e + 1) + \frac{d}{2} r_+(r_+ - 1).$$

The Jordan algebra V is called *split* if $V_{ii} = \mathbb{R}c_i$ for every $1 \leq i \leq r_+$ (equivalently $e = 0$). Otherwise, V is called *non-split*. By [20], if V is split then $r = r_+$, otherwise $r = 2r_+$. We pin down that every Euclidean Jordan algebra is split. There is a classification of simple real Jordan algebra given in [20] and [27]. We refer the reader to [5, Appendix A] for the complete list. More precisely, we have four types of simple real Jordan algebras :

Type I	:	V is Euclidean.
Type II	:	V is split non-Euclidean.
Type III	:	V is non-split with no complex structure.
Type IV	:	V is non-split with a complex structure.

Notice that every simple real Jordan algebra is either a real form of a simple complex Jordan algebra (type I, II, III), or a simple complex Jordan algebra viewed as a real one (type IV).

The structure group $\mathrm{Str}(V)$ of V can be thought of as the subgroup of transformations $g \in \mathrm{GL}(V)$ such that for every invertible $x \in V^\times$, there exists a nonzero real number $\chi(g)$ for which

$$\mathbf{det}(gx) = \chi(g)\mathbf{det}(x).$$

The map $g \mapsto \chi(g)$ is a character of $\mathrm{Str}(V)$, which is a reductive Lie group.

For $a \in V$, denote by n_a the translation $x \mapsto x + a$. Let

$$N := \{n_a : a \in V\}$$

be the abelian Lie group of all translations. The conformal group $\mathrm{Co}(V)$ of V is, by definition, the group of rational transforms of V generated by $\mathrm{Str}(V)$, N and the inversion $\imath : x \mapsto -x^{-1}$. It can be shown that $\mathrm{Co}(V)$ is a simple Lie group (see [3, 26]). A transformation $g \in \mathrm{Co}(V)$ is conformal in the sense that, at each point x, where g is well defined, its differential $(Dg)_x$ belongs to the structure group $\mathrm{Str}(V)$.

In our context, it turned out to be more convenient to work with a group G, which is locally isomorphic to $\mathrm{Co}(V)$, instead of $\mathrm{Co}(V)$ itself. More precisely, we will denote by $\mathrm{Str}(V)^+$ the subgroup of $\mathrm{Str}(V)$ defined by

$$\mathrm{Str}(V)^+ = \{\ell \in \mathrm{Str}(V) : \chi(\ell) > 0\}.$$

Define the proper conformal group $\mathrm{Co}(V)^+$ to be the group generated by $\mathrm{Str}(V)^+$, N and the inversion ι.

For any $g \in \mathrm{Co}(V)^+$ and $x \in V_g$ (an open dense subset of V where g is defined), the differential $Dg(x) \in \mathrm{Str}(V)^+$, and therefore we may define $c(g, x) = \chi(Dg(x))^{-1}$. In particular,

(i) For any $\ell \in \mathrm{Str}(V)^+$ and $x \in V$

$$c(\ell, x) = \chi(\ell)^{-1}.$$

(ii) For any $v \in V$ and $x \in V$

$$c(n_v, x) = 1.$$

(iii) For any $x \in V^\times$,

$$c(\iota, x) = \mathbf{det}(x)^2.$$

In [5, Proposition 6.1], we proved the existence of a polynomial $p_g \in \mathcal{P}(V)$ such that for every $x \in V_g$, $c(g, x) = p_g(x)^2$ where p_g is unique, up to a sign $\{\pm\}$. This allows us to construct the twofold covering group

$$G = \Big\{(g, p_g) \in \mathrm{Co}(V)^+ \times \mathcal{P}(V) : p_g(x)^2 = c(g, x) \text{ for all } x \in V\Big\}. \quad (7)$$

The group G is a simple Lie group which acts rationally on V. The subgroup P of affine transformations in G, which is a twofold covering of $\mathrm{Str}(V)^+ \ltimes N$, is in fact a parabolic subgroup of G. Let $\bar{P}$ be the opposite parabolic subgroup to P. Then the flag variety $\mathcal{X} := G/\bar{P}$, which is compact, is the conformal compactification of V and the map $V \simeq N \longrightarrow G/\bar{P}$ gives the embedding of V in its compactification.

For a given $\widetilde{g} = (g, p_g) \in G$ and $x \in V_g$, we set $\widetilde{g}(x) := g(x)$. Further, we will use the following notation

$$a(\widetilde{g}, x) := p_g(x), \quad x \in V. \quad (8)$$

By abuse of notation, we will denote elements of G without tilde.

3 Functional Equation of Zeta Functions

Let V be a simple real Jordan algebra with Jordan determinant **det**. We mention that if V is of type I or II, then **det** takes both positive and negative values, while if V is of type III or IV, then **det** takes only positive values. For $x \in V^\times$ and $s \in \mathbb{C}$, we introduce $\mathbf{det}(x)^{s,\varepsilon}$ by

$$\mathbf{det}(x)^{s,\varepsilon} = \begin{cases} |\mathbf{det}(x)|^s & \text{for } \varepsilon = + \\ \operatorname{sign}(\mathbf{det}(x))|\mathbf{det}(x)|^s & \text{for } \varepsilon = -. \end{cases}$$

Let $\mathcal{S}(V)$ be the Schwartz space of V, and let $\mathcal{S}'(V)$ be its dual, the space of tempered distributions on V. For $f \in \mathcal{S}(V)$ and $(s, \varepsilon) \in \mathbb{C} \times \{\pm\}$, we set

$$Z_\varepsilon(f, s) = \int_V f(x)\mathbf{det}(x)^{s,\varepsilon} dx, \tag{9}$$

where dx is the Lebesgue measure on V with respect to the non-degenerate bilinear form (5). The integral in (9) is convergent for $\Re(s) > 0$, the analytic function $Z_\varepsilon(f, s)$ has a meromorphic continuation with respect to s to the whole plane $\mathbb{C}$, and the map $f \mapsto Z_\varepsilon(f, s)$ is a tempered distribution on V, called a Zeta distribution. For the details, we refer the reader to [1, 7, 23, 5, 4] when $\varepsilon = +$ and to [5] when $\varepsilon = -$.

The purpose of this section is to give an explicit expression for the Fourier transform of the Zeta distribution for the four types of simple real Jordan algebras. Further, the Fourier transform of $f \in \mathcal{S}(V)$ is defined by

$$\widehat{f}(x) = \int_V e^{2i\pi\langle x,y\rangle} f(y)dy,$$

which we extend it by duality to the space of tempered distributions in the standard way.

For $s \in \mathbb{C}$,, one defines the gamma function associated to V by

$$\Gamma_V(s) := \prod_{k=1}^{r_+} \Gamma\left(\frac{s}{2} - (k-1)\frac{d}{4}\right),$$

where r_+ denotes the split rank of V. Now we are ready to state the functional equation for the Zeta distribution when V is of type II $\not\simeq \mathbb{R}^{p,q}$, type III and type IV.

Theorem 1 *The tempered distribution $f \mapsto Z_\varepsilon(f, s)$ satisfies the following functional equation*

$$Z_\varepsilon(\widehat{f}, s) = c(s, \varepsilon) Z_\varepsilon\big(f, -s - \frac{n}{r}\big),$$

where

$$c(s,\varepsilon)=\begin{cases}\pi^{-rs-\frac{n}{2}}\dfrac{\Gamma_V\left(s+\frac{n}{r}\right)}{\Gamma_V(-s)} & \textit{for } \varepsilon=+ \textit{ and } V \textit{ of type II } \not\simeq \mathbb{R}^{p,q}\\ i^r\pi^{-rs-\frac{n}{2}}\dfrac{\Gamma_V(s+1+\frac{n}{r})}{\Gamma_V(-s+1)} & \textit{for } \varepsilon=- \textit{ and } V \textit{ of type II } \not\simeq \mathbb{R}^{p,q}\\ \pi^{-rs-\frac{n}{2}}\dfrac{\Gamma_V\left(2s+\frac{2n}{r}\right)}{\Gamma_V(-2s)} & \textit{for } \varepsilon=\pm \textit{ and } V \textit{ of type III or IV.}\end{cases}$$

When $\varepsilon=+$, the aforementioned theorem goes back to [1, Theorem 4.4] (recall that if V is of type III or IV, then $\mathbf{det}(x)^{s,+}$ and $\mathbf{det}(x)^{s,-}$ coincide). In the notation of [1], our $|\mathbf{det}|$ is equal to ∇ when V is of type II, and is equal to ∇^2 when V is of type III or IV. For the case $\varepsilon=-$ and V is of type II, the aforementioned functional equation was established by the three authors in [5, Theorem 5.2], where we were able to link the Fourier transforms of Z_- and Z_+ by means of the following Bernestein identity

$$\det\left(\frac{\partial}{\partial x}\right)\mathbf{det}(x)^{s+1,+}=\prod_{k=0}^{r-1}\left(s+1+k\frac{d}{2}\right)\mathbf{det}(x)^{s,-}$$

(see [5, Proposition 3.4]).

To close the case of real simple Jordan algebras of type II, let us consider the case $V=\mathbb{R}^{p,q}$, for $p\geq 2$ and $q\geq 1$. For $x=(x_1,\dots,x_p,x_{p+1},\dots,x_n)\in V$, with $n=p+q$, the determinant is given by $\mathbf{det}(x)=P(x):=x_1^2+\cdots+x_p^2-x_{p+1}^2-\cdots-x_n^2$. Here it is customary to define the Fourier transform of $f\in\mathcal{S}(\mathbb{R}^{p,q})$ as the integral of f against the kernel $e^{i\langle x,y\rangle}$.

Introduce the polynomial functions $P_+(x)=P(x)\chi_{\{P>0\}}(x)$ and $P_-(x)=-P(x)\chi_{\{P<0\}}(x)$. Their Fourier transforms as tempered distributions are given in [18, (2.8) and (2.9)]. Since $\mathbf{det}(x)^{s,+}=P_+(x)^s+P_-(x)^s$ and $\mathbf{det}(x)^{s,-}=P_+(x)^s-P_-(x)^s$, one then can easily deduce the statement belwo:

Theorem 2 *Assume that $V=\mathbb{R}^{p,q}$ with $p\geq 2$ and $q\geq 1$. For every $s\in\mathbb{C}$, the following system of functional equations holds*

$$\begin{bmatrix}Z_+(\widehat{f},s)\\ Z_-(\widehat{f},s)\end{bmatrix}=\gamma(s)\mathbf{A}(s)\begin{bmatrix}Z_+\left(f,-s-\frac{n}{2}\right)\\ Z_-\left(f,-s-\frac{n}{2}\right)\end{bmatrix}$$

where

$$\mathbf{A}(s) = \begin{bmatrix} \cos\frac{(p-q)\pi}{4}\left(\sin\frac{n\pi}{4} - \sin(s+\frac{n}{4})\pi\right) & \sin\frac{(p-q)\pi}{4}\left(\cos(s+\frac{n}{4})\pi - \cos\frac{n\pi}{4}\right) \\ \sin\frac{(p-q)\pi}{4}\left(\cos(s+\frac{n}{4})\pi + \cos\frac{n\pi}{4}\right) & -\cos\frac{(p-q)\pi}{4}\left(\sin\frac{n\pi}{4} + \sin(s+\frac{n}{4})\pi\right) \end{bmatrix}$$

and

$$\gamma(s) = 2^{2s+n}\pi^{\frac{n}{2}-1}\Gamma(s+1)\Gamma\big(s+\frac{n}{2}\big).$$

Next we turn our attention to the case where V is an Euclidean Jordan algebra, i.e., V is of type I. Our main result depends heavily on the paper [28] by Satake and Faraut, where the authors proved a system of functional equations for Zeta distributions of type Z_+ defined on orbits Ω_ℓ consist of elements in $V^\times$ of signature $(r-\ell,\ell)$, where r is the rank of V. We may think of our result as a far reaching generalization of Satake–Faraut's result by considering the Zeta distributions Z_+ and Z_- defined over the entire $V^\times$.

The set of invertible elements $V^\times$ in V decomposes into the disjoint union of $r+1$ open $G_\circ$-orbits

$$V^\times = \bigcup_{\ell=0}^{r}\Omega_\ell,$$

where Ω_ℓ is the set of elements of signature $(r-\ell,\ell)$. Here $G_\circ$ is the identity connected component of $G = \mathrm{Str}(V)$. In particular, Ω_0, the $G_\circ$-orbit of the unit element $\mathbf{1}$, is a self-dual homogeneous cone. We shall simply denote Ω_0 by Ω.

For $f \in \mathcal{S}(V)$, in [28], the authors defined the following system of Zeta integrals

$$Z_\ell(f,s) = \int_{\Omega_\ell} f(x)|\mathbf{det}(x)|^s dx, \qquad 1 \le \ell \le r.$$

The aforementioned integral converges for $\Re(s) > 0$ and has a meromorphic continuation to the whole plane $\mathbb{C}$. Their main result is the following system of functional equations

$$Z_\ell\big(\widehat{f}, s-\frac{n}{r}\big) = (2\pi)^{-rs}e^{i\pi\frac{rs}{2}}\Gamma_\Omega(s)\sum_{\kappa=0}^{r}u_{\ell,\kappa}(s)Z_\kappa(f,-s), \tag{10}$$

where $u_{\ell,\kappa}(s)$ are polynomials in $e^{-i\pi s}$ of degree at most r (see [28, (23)]). Above, Γ_Ω is the so-called Gindikin gamma function,

$$\Gamma_\Omega(s) = (2\pi)^{\frac{n-r}{2}}\prod_{j=1}^{r}\Gamma\left(s-(j-1)\frac{d}{2}\right).$$

For $x \in \Omega_{\ell,}$, one has $\mathbf{det}(x) = (-1)^{\ell}|\mathbf{det}(x)|$. Thus, we may express the local Zeta integrals $Z_{\varepsilon}(f, s)$ defined by (9) as following

$$\begin{aligned} Z_+(f, s) &= \int_V f(x)|\mathbf{det}(x)|^s dx = \sum_{\ell=0}^{r} Z_\ell(f, s), \\ Z_-(f, s) &= \int_V f(x)\mathrm{sgn}(\mathbf{det}(x))|\mathbf{det}(x)|^s dx = \sum_{\ell=0}^{r}(-1)^{\ell} Z_\ell(f, s). \end{aligned} \tag{11}$$

Using the system (10) of functional equations, we obtain

$$Z_+\big(\widehat{f}, s - \frac{n}{r}\big) = (2\pi)^{-rs} \mathrm{e}^{i\pi\frac{rs}{2}} \Gamma_\Omega(s) \sum_{\kappa=0}^{r} \left(\sum_{\ell=0}^{r} u_{\ell,\kappa}(s) \right) Z_\kappa(f, -s), \tag{12}$$

and

$$Z_-\big(\widehat{f}, s - \frac{n}{r}\big) = (2\pi)^{-rs} \mathrm{e}^{i\pi\frac{rs}{2}} \Gamma_\Omega(s) \sum_{\kappa=0}^{r} \left(\sum_{\ell=0}^{r} (-1)^{\ell} u_{\ell,\kappa}(s) \right) Z_\kappa(f, -s). \tag{13}$$

Next we shall compute the finite sums $\sum_\ell u_{\ell,\kappa}(s)$ and $\sum_\ell (-1)^\ell u_{\ell,\kappa}(s)$. In [28], the authors proved that for any $y \in \mathbb{R}$,

$$\sum_{\ell=0}^{r} y^{\ell} u_{\ell,\kappa}(s) = \xi^{-(r-\kappa)} P_\kappa(\xi \mathrm{e}^{-i\pi s}, y) P_{r-\kappa}(1, \xi \mathrm{e}^{-i\pi s} y), \tag{14}$$

where $\xi := \mathrm{e}^{i\frac{\pi}{2}d(r+1)}$ and

$$P_\kappa(a, b) = \begin{cases} (a+b)^\kappa & \text{if } d \text{ is even}, \\ (a+b)^{\lfloor \frac{\kappa}{2} \rfloor}(b-a)^{\kappa - \lfloor \frac{\kappa}{2} \rfloor} & \text{if } d \text{ is odd}. \end{cases} \tag{15}$$

First, we consider the case where d is even. We distinguish two cases:

Case (a): $d \equiv 0 \pmod 4$ or $d \equiv 2 \pmod 4$ and r odd,
Case (a'): $d \equiv 2 \pmod 4$ and $r$even

Then one has

$$\sum_{\ell=0}^{r} y^{\ell} u_{\ell,\kappa}(s) = \begin{cases} (y + e^{-i\pi s})^{\kappa}(ye^{-i\pi s} + 1)^{r-\kappa} & \text{in Case (a)} \\ (y - e^{-i\pi s})^{\kappa}(ye^{-i\pi s} - 1)^{r-\kappa} & \text{in Case (a').} \end{cases}$$

In Case (a), we deduce that

$$\sum_{\ell=0}^{r} u_{\ell,\kappa}(s) = 2^r e^{-i\frac{\pi}{2}rs} \cos^r\left(\frac{\pi s}{2}\right),$$

and

$$\sum_{\ell=0}^{r} (-1)^{\ell} u_{\ell,\kappa}(s) = (2i)^r (-1)^{\kappa} e^{-i\frac{\pi}{2}rs} \sin^r\left(\frac{\pi s}{2}\right).$$

Hence, in Case (a), the functional equations (12) and (13) become

$$Z_+(\widehat{f}, s - \frac{n}{r}) = 2^r (2\pi)^{-rs} \Gamma_{\Omega}(s) \cos^r\left(\frac{\pi s}{2}\right) Z_+(f, -s), \tag{16}$$

and

$$Z_-(\widehat{f}, s - \frac{n}{r}) = (2i)^r (2\pi)^{-rs} \Gamma_{\Omega}(s) \sin^r\left(\frac{\pi s}{2}\right) Z_-(f, -s). \tag{17}$$

Similarly, in Case (a') one obtains

$$\sum_{\ell=0}^{r} u_{\ell,\kappa}(s) = (-1)^{\kappa} (2i)^r e^{-i\frac{\pi}{2}rs} \sin^r\left(\frac{\pi s}{2}\right),$$

and

$$\sum_{\ell=0}^{r} (-1)^{\ell} u_{\ell,\kappa}(s) = 2^r e^{-i\frac{\pi}{2}rs} \cos^r\left(\frac{\pi s}{2}\right),$$

which yield to

$$Z_+(\widehat{f}, s - \frac{n}{r}) = (2i)^r (2\pi)^{-rs} \Gamma_{\Omega}(s) \sin^r\left(\frac{\pi s}{2}\right) Z_-(f, -s),$$

and

$$Z_-(\widehat{f}, s - \frac{n}{r}) = 2^r (2\pi)^{-rs} \Gamma_{\Omega}(s) \cos^r\left(\frac{\pi s}{2}\right) Z_+(f, -s).$$

Hence, we have the following statement:

Theorem 3 *Assume that V is an Euclidean Jordan algebra of type Case (a) or Case (a'). The Zeta integrals* (9) *satisfy the following system of functional equations:*

- *Case (a):*

$$\begin{bmatrix} Z_+(\widehat{f}, s) \\ Z_-(\widehat{f}, s) \end{bmatrix} = 2^r (2\pi)^{-r(s+\frac{n}{r})} \Gamma_\Omega\Big(s + \frac{n}{r}\Big) \mathbf{A}(s) \begin{bmatrix} Z_+\big(f, -s - \frac{n}{r}\big) \\ Z_-\big(f, -s - \frac{n}{r}\big) \end{bmatrix},$$

where

$$\mathbf{A}(s) = \begin{bmatrix} \cos^r\Big(\frac{\pi}{2}(s + \frac{n}{r})\Big) & 0 \\ 0 & i^r \sin^r\Big(\frac{\pi}{2}(s + \frac{n}{r})\Big) \end{bmatrix}.$$

- *Case (a'):*

$$\begin{bmatrix} Z_+(\widehat{f}, s) \\ Z_-(\widehat{f}, s) \end{bmatrix} = 2^r (2\pi)^{-r(s+\frac{n}{r})} \Gamma_\Omega\Big(s + \frac{n}{r}\Big) \mathbf{A}(s) \begin{bmatrix} Z_+\big(f, -s - \frac{n}{r}\big) \\ Z_-\big(f, -s - \frac{n}{r}\big) \end{bmatrix},$$

where

$$\mathbf{A}(s) = \begin{bmatrix} i^r \sin^r\Big(\frac{\pi}{2}(s + \frac{n}{r})\Big) & 0 \\ 0 & \cos^r\Big(\frac{\pi}{2}(s + \frac{n}{r})\Big) \end{bmatrix}.$$

Next we consider the case where d is odd. According to the classification theory of Euclidean Jordan algebras, we have the following two possibilities:

Case (b): d odd and $r = 2$,
Case (c): $d = 1$ and arbitrary r.

In Case (b), by (14) we have

$$\sum_{\ell=0}^{2} y^\ell u_{\ell,\kappa}(s) = \begin{cases} 1 + e^{-2\pi i s} y^2 & \text{for } \kappa = 0 \\ e^{-\pi i s} - i^d(1 - e^{-2\pi i s})y + e^{-\pi i s} y^2 & \text{for } \kappa = 1 \\ e^{-2\pi i s} + y^2, & \text{for } \kappa = 2 \end{cases}$$

for any real number y. Therefore, for $d \equiv 1 \pmod 4$, we get

$$\sum_{\ell=0}^{2} (\pm 1)^\ell u_{\ell,\kappa}(s) = \begin{cases} 2e^{-\pi i s}\cos(\pi s) & \text{for } \kappa = 0 \\ 2e^{-\pi i s} \pm 2e^{-\pi i s}\sin(\pi s) & \text{for } \kappa = 1 \\ 2e^{-\pi i s}\cos(\pi s) & \text{for } \kappa = 2. \end{cases}$$

The functional equations (12) and (13) become

$$Z_{\pm}(\widehat{f}, s - \frac{n}{2}) = 2(2\pi)^{-2s}\Gamma_{\Omega}(s)\Big\{\cos(\pi s)\big(Z_0(f, -s) + Z_2(f, -s)\big) + (1 \pm \sin(\pi s))Z_1(f, -s)\Big\}$$

On the other hand, by (11), we have $Z_0(f, -s) + Z_1(f, -s) + Z_2(f, -s) = Z_+(f, -s)$ and $Z_0(f, -s) - Z_1(f, -s) + Z_2(f, -s) = Z_-(f, -s)$. That is

$$Z_0(f, -s) + Z_2(f, -s) = \frac{1}{2}\big(Z_+(f, -s) + Z_-(f, -s)\big),$$

and

$$Z_1(f, -s) = \frac{1}{2}\big(Z_+(f, -s) - Z_-(f, -s)\big).$$

By putting pieces together, we arrive at

$$Z_+(\widehat{f}, s - \frac{n}{2}) = 2\sqrt{2}(2\pi)^{-2s}\Gamma_{\Omega}(s)\sin\Big(\frac{\pi s}{2} + \frac{\pi}{4}\Big)\Big\{\cos\Big(\frac{\pi s}{2}\Big)Z_+(f, -s) - \sin\Big(\frac{\pi s}{2}\Big)Z_-(f, -s)\Big\}$$

and

$$Z_-(\widehat{f}, s - \frac{n}{2}) = 2\sqrt{2}(2\pi)^{-2s}\Gamma_{\Omega}(s)\cos\Big(\frac{\pi s}{2} + \frac{\pi}{4}\Big)\Big\{\cos\Big(\frac{\pi s}{2}\Big)Z_+(f, -s) + \sin\Big(\frac{\pi s}{2}\Big)Z_-(f, -s)\Big\}.$$

Similarly, for $d \equiv 3 \pmod 4$, we have

$$\sum_{\ell=0}^{2}(\pm 1)^{\ell}u_{\ell,\kappa}(s) = \begin{cases} 2e^{-\pi i s}\cos(\pi s) & \text{for } \kappa = 0 \\ 2e^{-\pi i s} \mp 2e^{-\pi i s}\sin(\pi s) & \text{for } \kappa = 1 \\ 2e^{-\pi i s}\cos(\pi s) & \text{for } \kappa = 2, \end{cases}$$

which lead to the following functional equations

$$Z_+(\widehat{f}, s - \frac{n}{2}) = 2\sqrt{2}(2\pi)^{-2s}\Gamma_{\Omega}(s)\cos\Big(\frac{\pi s}{2} + \frac{\pi}{4}\Big)\Big\{\cos\Big(\frac{\pi s}{2}\Big)Z_+(f, -s) + \sin\Big(\frac{\pi s}{2}\Big)Z_-(f, -s)\Big\}$$

and

$$Z_-\left(\widehat{f}, s-\frac{n}{2}\right) = 2\sqrt{2}(2\pi)^{-2s}\Gamma_\Omega(s)\sin\left(\frac{\pi s}{2}+\frac{\pi}{4}\right)\left\{\cos\left(\frac{\pi s}{2}\right)Z_+(f,-s) - \sin\left(\frac{\pi s}{2}\right)Z_-(f,-s)\right\}.$$

Theorem 4 *Assume that V is an Euclidean Jordan algebra of type Case (b). The Zeta integrals* (9) *satisfy the following system of functional equations:*

- *If* $d \equiv 1 \pmod 4$, *then*

$$\begin{bmatrix} Z_+(\widehat{f}, s) \\ Z_-(\widehat{f}, s) \end{bmatrix} = 2\sqrt{2}(2\pi)^{-2(s+\frac{n}{2})}\Gamma_\Omega\left(s+\frac{n}{2}\right)\mathbf{A}(s)\begin{bmatrix} Z_+(f,-s-\frac{n}{2}) \\ Z_-(f,-s-\frac{n}{2}) \end{bmatrix},$$

where

$$\mathbf{A}(s) = \begin{bmatrix} \sin\left(\frac{\pi}{2}(s+\frac{n+1}{2})\right)\cos\left(\frac{\pi}{2}(s+\frac{n}{2})\right) & -\sin\left(\frac{\pi}{2}(s+\frac{n+1}{2})\right)\sin\left(\frac{\pi}{2}(s+\frac{n}{2})\right) \\ \cos\left(\frac{\pi}{2}(s+\frac{n+1}{2})\right)\cos\left(\frac{\pi}{2}(s+\frac{n}{2})\right) & \cos\left(\frac{\pi}{2}(s+\frac{n+1}{2})\right)\sin\left(\frac{\pi}{2}(s+\frac{n}{2})\right) \end{bmatrix}.$$

- *If* $d \equiv 3 \pmod 4$, *then*

$$\begin{bmatrix} Z_+(\widehat{f}, s) \\ Z_-(\widehat{f}, s) \end{bmatrix} = 2\sqrt{2}(2\pi)^{-2(s+\frac{n}{2})}\Gamma_\Omega\left(s+\frac{n}{2}\right)\mathbf{A}(s)\begin{bmatrix} Z_+(f,-s-\frac{n}{2}) \\ Z_-(f,-s-\frac{n}{2}) \end{bmatrix},$$

where

$$\mathbf{A}(s) = \begin{bmatrix} \cos\left(\frac{\pi}{2}(s+\frac{n+1}{2})\right)\cos\left(\frac{\pi}{2}(s+\frac{n}{2})\right) & \cos\left(\frac{\pi}{2}(s+\frac{n+1}{2})\right)\sin\left(\frac{\pi}{2}(s+\frac{n}{2})\right) \\ \sin\left(\frac{\pi}{2}(s+\frac{n+1}{2})\right)\cos\left(\frac{\pi}{2}(s+\frac{n}{2})\right) & -\sin\left(\frac{\pi}{2}(s+\frac{n+1}{2})\right)\sin\left(\frac{\pi}{2}(s+\frac{n}{2})\right) \end{bmatrix}.$$

We close this section by the aforementioned Case (c), that is, $d = 1$ and r arbitrary. In this case one needs to distinguish the four cases $r \equiv 0, 1, 2$ and $3 \pmod 4$. We shall present briefly the case $r \equiv 0 \pmod 4$. For the three remaining cases, we refer the reader to [5] for a detailed exposition.

In view of (14) and the fact that $r \equiv 0 \pmod 4$, we have

$$\sum_{\ell=0}^{r}(\pm 1)^{\ell} u_{\ell,\kappa}(s) = \begin{cases} (1+e^{-2\pi i s})^{r/2} = 2^{r/2}\cos^{r/2}(\pi s)e^{-i\pi\frac{rs}{2}} & \text{for even } \kappa \\ \mp i(1+e^{-2\pi i s})^{r/2} = \mp i 2^{r/2}\cos^{r/2}(\pi s)e^{-i\pi\frac{rs}{2}} & \text{for odd } \kappa. \end{cases}$$

Therefore,

$$\begin{aligned} & Z_{\pm}(\widehat{f}, s - \frac{n}{r}) \\ &= 2^{r/2}(2\pi)^{-rs}\Gamma_{\Omega}(s)\cos^{r/2}(\pi s)\Big\{ \sum_{\kappa \text{ even}} Z_{\kappa}(f,-s) \mp i \sum_{\kappa \text{ odd}} Z_{\kappa}(f,-s)\Big\} \\ &= 2^{r/2-1}(2\pi)^{-rs}\Gamma_{\Omega}(s)\cos^{r/2}(\pi s)\Big\{ Z_{+}(f,-s) + Z_{-}(f,-s) \mp i(Z_{+}(f,-s) \\ &\quad - Z_{-}(f,-s))\Big\} \\ &= 2^{r/2-1}(2\pi)^{-rs}\Gamma_{\Omega}(s)\cos^{r/2}(\pi s)\Big\{ (1 \mp i)Z_{+}(f,-s) + (1 \pm i)Z_{-}(f,-s)\Big\} \\ &= 2^{r/2-1/2}(2\pi)^{-rs}\Gamma_{\Omega}(s)\cos^{r/2}(\pi s)\Big\{ e^{\mp i\frac{\pi}{4}} Z_{+}(f,-s) + e^{\pm i\frac{\pi}{4}} Z_{-}(f,-s)\Big\}. \end{aligned}$$

The remaining cases $r \equiv 1, 2$ and $3 \pmod 4$ can be treated in a similar way (see [5, pages 2314–2317] for an exhaustive exposition). To state the complete result for Case (c), we need to introduce the following tempered distributions,

$$Z^{e}(f,s) = \sum_{\kappa=0}^{\lfloor \frac{r}{2} \rfloor}(-1)^{\kappa} Z_{2\kappa}(f,s) \quad \text{and} \quad Z^{o}(f,s) = \sum_{\kappa=0}^{\lfloor \frac{r-1}{2} \rfloor}(-1)^{\kappa} Z_{2\kappa+1}(f,s). \tag{18}$$

Theorem 5 *Assume that V is an Euclidean Jordan algebra with $d = 1$ and of arbitrary rank r. The Zeta integrals (9) satisfy the following system of functional equations:*

- *If $r \equiv 0 \pmod 4$, then*

$$\begin{aligned} \begin{bmatrix} Z_{+}(\widehat{f}, s) \\ Z_{-}(\widehat{f}, s) \end{bmatrix} &= e^{i\frac{\pi}{4}} 2^{\frac{r-1}{2}} (2\pi)^{-r(s+\frac{n}{r})}\Gamma_{\Omega}\Big(s+\frac{n}{r}\Big) \\ &\times \cos^{\frac{r}{2}}\Big(\pi(s+\frac{n}{r})\Big)\mathbf{A}(s)\begin{bmatrix} Z_{+}(f,-s-\frac{n}{r}) \\ Z_{-}(f,-s-\frac{n}{r}) \end{bmatrix}, \end{aligned}$$

where

$$\mathbf{A}(s) = \begin{bmatrix} -i & 1 \\ 1 & -i \end{bmatrix}.$$

- *If* $r \equiv 2 \pmod 4$, *then*

$$\begin{bmatrix} Z_+(\widehat{f}, s) \\ Z_-(\widehat{f}, s) \end{bmatrix} = e^{i\frac{\pi}{4}} 2^{\frac{r-1}{2}} (2\pi)^{-r(s+\frac{n}{r})} \Gamma_\Omega\Big(s + \frac{n}{r}\Big)$$

$$\times \cos^{\frac{r}{2}} \Big(\pi(s + \frac{n}{r})\Big) \mathbf{A}(s) \begin{bmatrix} Z_+\big(f, -s - \frac{n}{r}\big) \\ Z_-\big(f, -s - \frac{n}{r}\big) \end{bmatrix},$$

where

$$\mathbf{A}(s) = \begin{bmatrix} 1 & -i \\ -i & 1 \end{bmatrix}.$$

- *If* $r \equiv 1 \pmod 4$, *then*

$$\begin{bmatrix} Z_+(\widehat{f}, s) \\ Z_-(\widehat{f}, s) \end{bmatrix} = (2i)^{\lfloor \frac{r}{2} \rfloor} (2\pi)^{-r(s+\frac{n}{r})} \Gamma_\Omega\Big(s + \frac{n}{r}\Big)$$

$$\times \sin^{\lfloor \frac{r}{2} \rfloor} \Big(\pi(s + \frac{n}{r})\Big) \mathbf{A}(s) \begin{bmatrix} Z^{\mathrm{e}}\big(f, -s - \frac{n}{r}\big) \\ Z^{\mathrm{o}}\big(f, -s - \frac{n}{r}\big) \end{bmatrix},$$

where

$$\mathbf{A}(s) = \begin{bmatrix} \cos\Big(\frac{\pi}{2}(s + \frac{n}{r})\Big) & \cos\Big(\frac{\pi}{2}(s + \frac{n}{r})\Big) \\ i \sin\Big(\frac{\pi}{2}(s + \frac{n}{r})\Big) & -i \sin\Big(\frac{\pi}{2}(s + \frac{n}{r})\Big) \end{bmatrix}.$$

- *If* $r \equiv 3 \pmod 4$, *then*

$$\begin{bmatrix} Z_+(\widehat{f}, s) \\ Z_-(\widehat{f}, s) \end{bmatrix} = (2i)^{\lfloor \frac{r}{2} \rfloor} (2\pi)^{-r(s+\frac{n}{r})} \Gamma_\Omega\Big(s + \frac{n}{r}\Big)$$

$$\times \sin^{\lfloor \frac{r}{2} \rfloor} \Big(\pi(s + \frac{n}{r})\Big) \mathbf{A}(s) \begin{bmatrix} Z^{\mathrm{e}}\big(f, -s - \frac{n}{r}\big) \\ Z^{\mathrm{o}}\big(f, -s - \frac{n}{r}\big) \end{bmatrix},$$

where

$$\mathbf{A}(s) = \begin{bmatrix} i \sin\Big(\frac{\pi}{2}(s + \frac{n}{r})\Big) & -i \sin\Big(\frac{\pi}{2}(s + \frac{n}{r})\Big) \\ \cos\Big(\frac{\pi}{2}(s + \frac{n}{r})\Big) & \cos\Big(\frac{\pi}{2}(s + \frac{n}{r})\Big) \end{bmatrix}.$$

4 Construction of the Source Operator

Recall that $\mathcal{X} = G/\bar{P}$ be the conformal completion of V, where G is the double covering (7) of $\mathrm{Co}(V)^+$. The characters of $\bar{P}$ are parametrized by $(\lambda, \varepsilon) \in \mathbb{C} \times \{\pm\}$. To a character $\chi_{\lambda,\varepsilon}$, there corresponds a line bundle over $\mathcal{X}$ and let $\mathcal{H}_{\lambda,\varepsilon}$ be the corresponding space of smooth sections of this bundle, which can be realized as the space of smooth functions $f : G \longrightarrow \mathbb{C}$ satisfying

$$f(g\bar{p}) = \chi_{\lambda,\varepsilon}(\bar{p})^{-1} f(g),$$

for each $g \in G$ and $\bar{p} \in \bar{P}$. The formula

$$\big(\pi_{\lambda,\varepsilon}(g)f\big)(x) = a(g^{-1}, x)^{-\lambda,\varepsilon} f\big(g^{-1}(x)\big)$$

defines a smooth representation $\pi_{\lambda,\varepsilon}$ of G on $\mathcal{H}_{\lambda,\varepsilon}$, where $a(g, x)$ is as in (8).

Consider two pairs (λ, ε), $(\mu, \eta) \in \mathbb{C} \times \{\pm\}$. The main idea of the *source operator method*, inspired by the Ω-process (see [8, 11]), is to construct a differential operator

$$E_{(\lambda,\varepsilon),\,(\mu,\eta)} : \mathcal{H}_{\lambda,\,\varepsilon} \otimes \mathcal{H}_{\mu,\,\eta} \longrightarrow \mathcal{H}_{\lambda+1,\,-\varepsilon} \otimes \mathcal{H}_{\mu+1,\,-\eta}$$

which is G-covariant with respect to $(\pi_{\lambda,\varepsilon} \otimes \pi_{\mu,\eta}\,,\ \pi_{\lambda+1,-\varepsilon} \otimes \pi_{\mu+1,-\eta})$.

The source operator is obtained by combining a natural *multiplication operator M* and the classical *Knapp-Stein intertwining operators* [24].

To describe the operator M, it is convenient to use the noncompact picture. Recall that the injective map $V \ni v \mapsto n_v$ gives a chart on an open set of $\mathcal{X}$, which is dense and of full measure. The Jordan determinant **det** of V satisfies the following covariance relation

$$\mathbf{det}(g(x) - g(y)) = a(g, x)^{-1}\,\mathbf{det}(x - y)\,a(g, y)^{-1}, \tag{19}$$

whenever g is defined at x and y, with $x, y \in V$ and $g \in G$. For $g = \iota$, this is the famous *Hua formula*, and the general formula is obtained by verifying it on the generators of G and then by using the cocycle property of $a(g, x)$.

Let $f \in \mathcal{H}_{\lambda,\varepsilon} \otimes \mathcal{H}_{\mu,\eta}$ and let $f(x, y)$ be its corresponding local expression on $V \times V$. Then, thanks to the covariance relation (19), the function

$$\mathbf{det}(x - y) f(x, y)$$

can be interpreted as the local expression of some $Mf \in \mathcal{H}_{\lambda-1,-\varepsilon} \otimes \mathcal{H}_{\mu-1,-\eta}$, and the operator M so defined satisfies the intertwining relation

$$M \circ \big(\pi_{\lambda,\varepsilon}(g) \otimes \pi_{\mu,\eta}(g)\big) = \big(\pi_{\lambda-1,-\varepsilon}(g) \otimes \pi_{\mu-1,-\eta}(g)\big) \circ M.$$

The *Knapp–Stein* operators are a meromorphic family of operators

$$I_{\lambda,\varepsilon} : \mathcal{H}_{\lambda,\,\varepsilon} \longrightarrow \mathcal{H}_{\frac{2n}{r}-\lambda,\,\varepsilon}$$

which satisfy the intertwining relation

$$I_{\lambda,\,\varepsilon} \circ \pi_{\lambda,\,\varepsilon}(g) = \pi_{\frac{2n}{r}-\lambda,\,\varepsilon}(g) \circ I_{\lambda,\,\varepsilon} \qquad \text{for any } g \in G.$$

Now consider the following diagram

$$\begin{array}{ccc}
\mathcal{H}_{(\lambda,\varepsilon),(\mu,\eta)} & \xrightarrow{F_{(\lambda,\varepsilon),(\mu,\eta)}} & \mathcal{H}_{(\lambda+1,-\varepsilon),(\mu+1,-\eta)} \\
{\scriptstyle I_{\lambda,\varepsilon}\otimes I_{\mu,\eta}}\big\downarrow & & \big\uparrow{\scriptstyle I_{\frac{2n}{r}-\lambda-1,-\varepsilon}\otimes I_{\frac{2n}{r}-\mu-1,-\eta}} \\
\mathcal{H}_{(\frac{2n}{r}-\lambda,\varepsilon),(\frac{2n}{r}-\mu,\eta)} & \xrightarrow{M} & \mathcal{H}_{(\frac{2n}{r}-\lambda-1,-\varepsilon),(\frac{2n}{r}-\mu-1,-\eta)}
\end{array}$$

That is,

$$F_{(\lambda,\varepsilon),(\mu,\eta)} = (I_{\frac{2n}{r}-\lambda-1,-\varepsilon} \otimes I_{\frac{2n}{r}-\mu-1,-\eta}) \circ M \circ (I_{\lambda,\varepsilon} \otimes I_{\mu,\eta}). \tag{20}$$

By construction, $F_{(\lambda,\varepsilon),(\mu,\eta)}$ is covariant with respect to $(\pi_{\lambda,\varepsilon} \otimes \pi_{\mu,\eta}\,,\ \pi_{\lambda+1,-\varepsilon} \otimes \pi_{\mu+1,-\eta})$ and depends meromorphically on (λ, μ).

The main theorem can now be stated.

Theorem 6 *The operator $F_{(\lambda,\varepsilon),(\mu,\eta)}$ is a* differential operator *on $\mathcal{X} \times \mathcal{X}$. Moreover, after a normalization by a meromorphic function in (λ, μ), the operator depends polynomially on (λ, μ).*

The proof that $F_{(\lambda,\varepsilon),(\mu,\eta)}$ is a differential operator is done in the noncompact picture. We keep the same notation for the local expression of the operators involved in the definition of $F_{(\lambda,\varepsilon),(\mu,\eta)}$. The proof uses as an essential ingredient the *Fourier transform* on $V \times V$. In the local chart $V \times V$, the operator M is the multiplication by a polynomial, and a Knapp–Stein operator is a convolution by a tempered distribution on V. Both operators have nice correspondents on the Fourier side. However, the Knapp–Stein operators have singular kernels. So, in order to compose the operators, there are some technical difficulties. To circumvent these difficulties, we use the Schwartz spaces $\mathcal{S}(V)$, $\mathcal{S}(V \times V) \simeq \mathcal{S}(V) \otimes \mathcal{S}(V)$ and their dual spaces, namely the spaces of tempered distributions $\mathcal{S}'(V)$ and $\mathcal{S}'(V \times V)$. The Schwartz spaces are stable under the infinitesimal action of the representations $\pi_{\lambda,\varepsilon}$ and by the multiplication operator M, while the Knapp–Stein operators map $\mathcal{S}(V)$ into $\mathcal{S}'(V)$. So it is possible to consider the composition of M and of *one* Knapp–Stein operator, getting an operator from $\mathcal{S}(V)$ into $\mathcal{S}'(V)$. A (seemingly weaker) form of the identity used for the definition of $F_{(\lambda,\varepsilon),(\mu,\eta)}$ can be used. In fact, the Knapp–Stein operators are generically invertible (i.e., outside of a denumerable set of values of λ) and the inverse of $I_{\lambda,\varepsilon}$ is (up to a meromorphic factor) the operator

$I_{\frac{2n}{r}-\lambda,\varepsilon}$ so that (20) can be replaced by

$$M \circ (I_{\lambda,\varepsilon} \otimes I_{\mu,\eta}) = \kappa(\lambda, \mu)(I_{\lambda+1,-\varepsilon} \otimes I_{\mu+1,-\eta}) \circ F_{(\lambda,\varepsilon),(\mu,\eta)}, \tag{21}$$

where $\kappa(\lambda, \mu)$ is a meromorphic function on $\mathbb{C} \times \mathbb{C}$. As observed in the Ω-process [11], in the noncompact picture the source operator is a differential operator with *polynomial coefficients*. We also pin down that the coefficients do not depend on (ε, η), and hence, we omit these indices in the notation. The corresponding operator on the Fourier transform side is also a differential operator with polynomial coefficients. It turns out that the same properties hold in general. From these observations, it follows that (21) is clearly equivalent to an operator identity on the Fourier transform side, which we will now describe.

To the multiplication operator M corresponds on the Fourier side transform the constant coefficients differential operator $\mathbf{det}\left(\frac{\partial}{\partial \xi} - \frac{\partial}{\partial \zeta}\right)$. The kernel of the Knapp–Stein operator is (up to a normalizing factor)

$$\mathbf{det}(x)^{-\frac{2n}{r}+\lambda,\varepsilon},$$

and its Fourier transform was studied in Sect. 3. After a change of parameters and up to normalizing factors, the identity (21) turns out to be equivalent to an identity on the Fourier side, which we state as a theorem.

Theorem 7 *There exists a differential operator $D_{s,t}$ with polynomial coefficients on $V \times V$ such that*

$$\mathbf{det}\left(\frac{\partial}{\partial \xi} - \frac{\partial}{\partial \zeta}\right) \circ \mathbf{det}(\xi)^{s,\varepsilon}\, \mathbf{det}(\zeta)^{t,\eta} = \mathbf{det}(\xi)^{s-1,-\varepsilon}\mathbf{det}(\zeta)^{t-1,-\eta} \circ D_{s,t}. \tag{22}$$

Moreover, the coefficients of $D_{s,t}$ are polynomial in (s, t).

Let us sketch a proof of this result, which is simpler than the one given in [5].

Step 1. The first ingredient in the proof is a version of the Taylor formula and of the Leibniz formula, with reference to the *Fischer inner product* on the space of polynomials. Let $(E, \langle\cdot, \cdot\rangle)$ be a finite dimensional Euclidean vector space. Let $\mathcal{P}$ be the space of all (real-valued) polynomials on E. For any $p \in \mathcal{P}$, we let $p\left(\frac{\partial}{\partial x}\right)$ to be the unique constant coefficients differential operator on E characterized by the following property, valid for any $y \in E$

$$p\left(\frac{\partial}{\partial x}\right) e^{\langle x,y\rangle} = p(y)e^{\langle x,y\rangle}.$$

The same operator will occasionally be denoted by $\partial(p)$.

Recall that the Fischer inner product on $\mathcal{P}$ is given by

$$(p, q)_F = \partial(p)q(0).$$

The main property of the Fischer product is

$$(p, qr)_F = (\partial(r)p, q)_F, \qquad p, q, r \in \mathcal{P}. \tag{23}$$

For $k \in \mathbb{N}$, denote by $\mathcal{P}_k$ the subspace of all polynomials, which are homogeneous of degree k. The corresponding decomposition

$$\mathcal{P} = \bigoplus_{k=0}^{\infty} \mathcal{P}_k$$

is orthogonal with respect to the Fischer inner product.

Let $p \in \mathcal{P}$. For any $a \in E$, denote by p_a the polynomial given by

$$p_a(x) = p(x + a) \ .$$

Let $\mathcal{R}$ be the subspace of $\mathcal{P}$ spanned by all $(p_a)_{a\in E}$. It is also the subspace spanned by all partial derivatives of p. Choose an orthonormal basis (for the Fischer inner product) $(p_i)_{i\in I}$ of $\mathcal{R}$. To any element p_i of the basis, associate the polynomial $\breve{p}_i$ defined by

$$\breve{p}_i = \partial(p_i)p.$$

By the definition of $\mathcal{R}$, $\breve{p}_i$ belongs to $\mathcal{R}$.

Proposition 1

1. *For all $x, y \in E$*

$$p(x + y) = \sum_{i\in I} p_i(x)\, \breve{p}_i(y), \tag{24}$$

2. *For all smooth functions f, g on E*

$$\partial(p)(fg) = \sum_{i\in I} \partial(\breve{p}_i)f\ \partial(p_i)g. \tag{25}$$

Proof Fix $y \in E$, and let $p_y(x) = p(x + y)$. As $\mathcal{R}$ is stable by translations, p_y belongs to $\mathcal{R}$. As $(p_i)_{i\in I}$ is an orthonormal basis,

$$p_y(x) = \sum_{i\in I} (p_y, p_i)_F\ p_i(x) \ .$$

Next,

$$(p_y, p_i)_F = \partial(p_i)p_y(0) = \big(\partial(p_i)p(\cdot + y)\big)(0) = \big(\partial(p_i)p\big)(0+y) = \big(\partial(p_i)p\big)(y),$$

and (24) follows.

It is enough to prove (25) for $f, g \in \mathcal{P}$. As the differential operators $\partial(p), \partial(p_i), \partial(\breve{p}_i)$ have constant coefficients, they commute with translations. It is enough to prove the equality at $x = 0$. Next,

$$\partial(p)(fg)\,(0) = (p, fg)_F = (\partial(f)p, g)_F.$$

As $\partial(f)p$ belongs to $\mathcal{R}$, $\partial(f)p = \sum_{i\in I} a_i\, p_i$ where

$$a_i = (\partial(f)p, p_i)_F = (p, fp_i) = (\partial(p_i)p, f) = (\breve{p}_i, f)_F = \partial(\breve{p}_i)f\,(0).$$

Hence,

$$\partial(p)(fg)\,(0) = \sum_{i\in I} a_i(p_i, g)_F = \sum_{i\in I} \partial(\breve{p}_i)f\,(0)\,\partial(p_i)g\,(0).$$

□

Assume moreover that p is homogeneous of degree $k \in \mathbb{N}$. Then $\mathcal{R}$ is contained in the space of polynomials of degree less than or equal to k, and more precisely,

$$\mathcal{R} = \bigoplus_{\ell=0}^{k} \mathcal{R}_\ell, \qquad \mathcal{R}_\ell = \mathcal{R} \cap \mathcal{P}_\ell.$$

Set $d_\ell = \dim \mathcal{R}_\ell$. As the direct sum is orthogonal w.r.t. the Fischer inner product, the orthogonal basis $(p_i)_{i\in I}$ can (and will always) be chosen in the form

$$\bigcup_{\ell=0}^{k} (p_{\ell,i})_{1\le i\le d_\ell},$$

where $(p_{\ell,i})_{1\le i\le d_\ell}$ is an orthonormal basis of $\mathcal{R}_\ell$.

Step 2. Let us apply these ideas to our context. Let $\mathcal{W}$ be the subspace of $\mathcal{P}(V)$ generated by all translates of **det**, which is a L-submodule of the space $\mathcal{P} = \mathcal{P}(V)$.

Let us recall the decomposition of $\mathcal{W}$ into simple L-submodules (see [14]). Fix a Jordan frame $\{c_1, c_2, \dots, c_r\}$, and for each $1 \le k \le r$, set $e_k = c_1 + \cdots + c_k$ and let

$$V^{(k)} = V(e_k, 1) = \{x \in V : e_k x = x\}.$$

Further, let $P^{(k)}$ be the orthogonal projection on $V^{(k)}$, and let $\mathbf{det}^{(k)}$ be the determinant of the Euclidean subalgebra $V^{(k)}$. Finally, define the *principal minor* Δ_k by

$$\Delta_k(x) = \mathbf{det}^{(k)}\left(P^{(k)}x\right).$$

Lemma 1 *The decomposition of $\mathcal{W}$ into simple L-modules is given by*

$$\mathcal{W} = \bigoplus_{k=0}^{r} \mathcal{W}_k \tag{26}$$

where $\mathcal{W}_k = \mathcal{P}_k \cap \mathcal{W}$. Moreover, for $1 \leq k \leq r$, $\mathcal{W}_k$ is the L-submodule generated by Δ_k and $\mathcal{W}_0 = \mathbb{R}$.

The remarkable fact is that the primary decomposition of $\mathcal{W}$ in homogeneous components coincides with the L-decomposition in irreducibles. It has very important consequences.

Let us apply Proposition 1 to the polynomial **det** and to $\mathcal{W}$. For each $0 \leq k \leq r$, let $d_k = \dim \mathcal{W}_k$ and choose an orthonormal basis $\left(p_{k,i}\right)_{1\leq i\leq d_k}$ of $\mathcal{W}_k$ for the Fischer inner product. Then

$$\bigcup_{k=0}^{r} \left(p_{k,i}\right)_{1\leq i\leq d_k} \tag{27}$$

is an orthonormal basis of $\mathcal{W}$.

More generally, let $p \in \mathcal{W}_k$ for some $1 \leq k \leq r$, and let $\mathcal{R}(p)$ be the subspace generated by all translates of p. As p is homogeneous, $\mathcal{W}(p)$ splits as

$$\mathcal{W}(p) = \bigoplus_{\ell=0}^{k} \left(\mathcal{W}(p) \cap \mathcal{P}_\ell\right).$$

As $\mathcal{W}(p) \cap \mathcal{P}_\ell \subset \mathcal{W} \cap \mathcal{P}_\ell = \mathcal{W}_\ell$, it is possible to choose an orthonormal basis of $\mathcal{W}(p)$ of the following form

$$\bigcup_{\ell=0}^{k} \left(q_{\ell,i}\right)_{1\leq i\leq \dim(\mathcal{W}(p)\cap\mathcal{P}_\ell)} \tag{28}$$

where $q_{\ell,i} \in \mathcal{W}_\ell$.

Step 3. We are now ready for the proof of Theorem 7. The last key ingredient that is needed is a consequence (not explicitly stated in [5]) of results obtained by Faraut and Korányi (see Proposition VII.1.5 and Proposition XI.5.1 in [15]).

Proposition 2 *Let $p \in \mathcal{W}_k$. We have*

1. *The expression*

$$p^\sharp(x) = p(x^{-1})\mathbf{det}(x), \tag{29}$$

a priori defined for $x \in V^\times$, extends to V as a polynomial, which belongs to $\mathcal{W}_{r-k}$.

2. *For any $\lambda \in \mathbb{R}$, and for $x \in V^\times$,*

$$p\left(\frac{\partial}{\partial x}\right)\mathbf{det}(x)^{\lambda+1} = \prod_{j=1}^{k}\left((\lambda+1+\frac{d}{2}(j-1)\right)p^\sharp(x)\mathbf{det}(x)^\lambda. \tag{30}$$

By letting $\lambda = 0$ in (30), notice that $p^\sharp$ is proportional to $\breve{p} = \partial(p)\mathbf{det}$.

To motivate the reader, let us formulate and prove a close but simpler result, which will be used later.

Proposition 3 *There exists a polynomial $c_{\lambda,\mu}$ on $V \times V$ such that, for any $x, y \in \Omega$*

$$\mathbf{det}\left(\frac{\partial}{\partial x} - \frac{\partial}{\partial y}\right)\mathbf{det}(x)^{s+1}\mathbf{det}(y)^{t+1} = c_{\lambda,\mu}(x, y)\mathbf{det}(x)^s\mathbf{det}(y)^t\ . \tag{31}$$

The polynomial $c_{\lambda,\mu}$ is homogeneous of degree r. Its coefficents depend polynomially on s, t.

Proof By Proposition 1, and for the basis $\left(p_{k,i}\right)_{0\le k\le r, 1\le i\le d_k}$ of $\mathcal{W}$ presented in (27), we have

$$\mathbf{det}(x-y) = \sum_{k=0}^{r}(-1)^{r-k}\sum_{i=1}^{d_k} p_{k,i}(x)\breve{p}_{k,i}(y). \tag{32}$$

Hence,

$$\mathbf{det}\left(\frac{\partial}{\partial x} - \frac{\partial}{\partial y}\right)\mathbf{det}(x)^{s+1}\mathbf{det}(y)^{t+1}$$

$$\sum_{k=1}^{r}(-1)^{r-k}\sum_{i=1}^{d_k} p_{k,i}\left(\frac{\partial}{\partial x}\right)\mathbf{det}(x)^{s+1}\breve{p}_{k,i}\left(\frac{\partial}{\partial y}\right)\mathbf{det}(y)^{t+1}.$$

Use (30) to compute $p_{k,i}\left(\frac{\partial}{\partial x}\right)\mathbf{det}(x)^{s+1}$ and $\breve{p}_{k,i}\left(\frac{\partial}{\partial y}\right)\mathbf{det}(y)^{t+1}$. The conclusion follows easily. □

The proof of Theorem 7 follows similar lines. Let f be a smooth function on $V \times V$. Using again (32), we are reduced to calculate, for $0 \leq k \leq r$ and $1 \leq i \leq d_k$,

$$p_{k,i}\left(\frac{\partial}{\partial x}\right)\left(\mathbf{det}(x)^s\left(\breve{p}_{k,i}\left(\frac{\partial}{\partial y}\right)\mathbf{det}(y)^t f(x,y)\right)\right). \tag{33}$$

Compute first the inner expression

$$\breve{p}_{k,i}\left(\frac{\partial}{\partial y}\right)\mathbf{det}(y)^t f(x,y), \tag{34}$$

for which we use Proposition 1.2, applied to $\breve{p}_{k,i} \in \mathcal{W}_{r-k}$ and the basis of $\mathcal{R}(\breve{p}_{k,i})$ presented in (28). Hence, the inner term (34) is equal to

$$\mathbf{det}(y)^{t-1} d_{k,i,t}\left(y, \frac{\partial}{\partial y}\right) f(x,y)$$

where $d_{k,i,t}$ is a polynomial on $V \times V$. It remains to perform the derivation with respect to the variable x, which is treated in a similar manner, showing that the term (33) is equal to

$$\mathbf{det}(x)^{s-1}\mathbf{det}(y)^{t-1} d_{k,i,s,t}\left(x, y, \frac{\partial}{\partial x}, \frac{\partial}{\partial y}\right) f(x,y),$$

where $d_{k,i,s,t}$ is a polynomial on $V \times V \times V \times V$. It remains to sum the terms to finish the proof of Theorem 7.

The results obtained for Euclidean Jordan algebras (i.e., type I) are easily extended to other cases, first to type IV by an appropriate complexification of the main identity (22), and then by elementary tricks to yield the result for type II and type III. We again refer to [5] for details.

Example 1 Recall from Sect. 3 the example $V = \mathbb{R}^{p,q}$, where

$$\mathbf{det}(x) = P(x) = x_1^2 + \cdots + x_p^2 - x_{p+1}^2 - \cdots - x_n^2.$$

Denote by $P(x,y)$ the corresponding symmetric bilinear form on $V \times V$. In this example, the differential operator $D_{s,t}$ is given explicitly by

$$\begin{aligned} D_{s,t} = {} & P(x)P(y)\, P\left(\frac{\partial}{\partial x} - \frac{\partial}{\partial y}\right) \\ & + 4s\, P(y)\sum_{i=1}^{n} x_j\left(\frac{\partial}{\partial x_j} - \frac{\partial}{\partial y_j}\right) + 4t\, P(x)\sum_{j=1}^{n} y_j\left(\frac{\partial}{\partial y_j} - \frac{\partial}{\partial x_j}\right) \\ & + 2t(2t-2+n)P(x) - 8st\,P(x,y) + 2s(2s-2+n)P(y). \end{aligned} \tag{35}$$

Let us state the (slightly more precise) version of Theorem 6 in the noncompact picture.

Theorem 8 *The differential operators $F_{\lambda,\mu}$ have polynomials coefficients on $V\times V$, and they depend polynomially on $(\lambda,\mu)\in\mathbb{C}\times\mathbb{C}$. For any $g\in G$,*

$$F_{\lambda,\mu}\circ\big(\pi_{\lambda,\varepsilon}\otimes\pi_{\mu,\eta}\big)(g)=\big(\pi_{\lambda+1,-\varepsilon}\otimes\pi_{\mu+1,-\eta}\big)(g)\circ F_{\lambda,\mu}. \tag{36}$$

5 Construction of the Generalized Rankin–Cohen Operators

Let res : $\mathcal{S}(V\times V)\longrightarrow\mathcal{S}(V)$ be the restriction operator from $V\times V$ to $\operatorname{diag}(V)=\{(x,x):x\in V\}\simeq V$ given by

$$\operatorname{res}(f)(x)=f(x,x),\qquad\text{for } x\in V.$$

One can easily prove that for $(\lambda,\varepsilon),(\mu,\eta)\in\mathbb{C}\times\{\pm\}$, and for any $g\in G$

$$\operatorname{res}\circ\left(\pi_{\lambda,\varepsilon}\otimes\pi_{\mu,\eta}\right)(g)=\pi_{\lambda+\mu,\varepsilon\eta}(g)\circ\operatorname{res} \tag{37}$$

For any positive integer k, let

$$F^{(k)}_{\lambda,\mu}=F_{\lambda+k-1,\mu+k-1}\circ\cdots\circ F_{\lambda,\mu}$$

and

$$B^{(k)}_{\lambda,\mu}=\operatorname{res}\circ F^{(k)}_{\lambda,\mu}$$

The covariance property of the operators $F_{\lambda,\mu}$ and of res imply the following statement.

Proposition 4 *For all (λ,ε) and (μ,η) in $\mathbb{C}\times\{\pm\}$ and for all k in $\mathbb{N}^*$, the operators $B^{(k)}_{\lambda,\mu}$ are covariant with respect to $\left(\pi_{\lambda,\varepsilon}\otimes\pi_{\mu,\eta},\pi_{\lambda+\mu+2k,\varepsilon\eta}\right)$, i.e. for any $g\in G$,*

$$B^{(k)}_{\lambda,\mu}\circ\left(\pi_{\lambda,\varepsilon}\otimes\pi_{\mu,\eta}\right)(g)=\pi_{\lambda+\mu+2k,\varepsilon\eta}(g)\circ B^{(k)}_{\lambda,\mu}. \tag{38}$$

The operators $B^{(k)}_{\lambda,\mu}$ generalize the classical Rankin–Cohen operators, or more precisely their real-valued version on $\mathbb{R}\times\mathbb{R}$.

Example 2 Assume that $V = \mathbb{R}^{p,q}$. Using the same notation as in Example 1, the source operator is given by

$$\begin{aligned}F_{\lambda,\mu} &= -P(x-y)P\left(\frac{\partial}{\partial x}\right)P\left(\frac{\partial}{\partial y}\right)\\ &+4(-\lambda+\frac{n}{2}-1)\sum_{j=1}^{n}(x_j-y_j)\frac{\partial}{\partial x_j}P\left(\frac{\partial}{\partial y}\right)\\ &+4(-\mu+\frac{n}{2}-1)\sum_{j=1}^{n}(y_j-x_j)\frac{\partial}{\partial x_j}P\left(\frac{\partial}{\partial x}\right)\\ &+4\lambda(-\lambda+\frac{n}{2}-1)P\left(\frac{\partial}{\partial y}\right)+4\mu(-\mu+\frac{n}{2}-1)P\left(\frac{\partial}{\partial x}\right)\\ &+8(-\lambda+\frac{n}{2}-1)(-\mu+\frac{n}{2}-1)P\left(\frac{\partial}{\partial x},\frac{\partial}{\partial y}\right).\end{aligned}$$

In particular,

$$\begin{aligned}B^{(1)}_{\lambda,\mu} = 4\ \mathrm{res}\Big\{&\mu(-\mu+\frac{n}{2}-1)P\left(\frac{\partial}{\partial x}\right)+\lambda(-\lambda+\frac{n}{2}-1)P\left(\frac{\partial}{\partial y}\right)\\ &+2(-\lambda+\frac{n}{2}-1)(-\mu+\frac{n}{2}-1)P\left(\frac{\partial}{\partial x},\frac{\partial}{\partial y}\right)\Big\}.\end{aligned}$$

See [5, Section 9] for more details.

To get a more explicit expression of these operators, we need to extend the classical notion of symbols to bi-differential operators. For the present situation, it will be enough to consider differential operators with polynomial coefficients. Let $B\colon C^\infty(V\times V)\longrightarrow C^\infty(V)$ be a bi-differential operator with polynomial coefficients. Define its symbol $b(x,\xi,\zeta)$ for $x\in V$, $(\xi,\zeta)\in V\times V$ by the identity

$$B\left(e^{i(x\xi+y\zeta)}\right)_{x=y} = b(x,\xi,\zeta)e^{i(x\xi+x\zeta)}.$$

If B is a bi-differential operator with constant coefficients, its symbol does not depend on x and is simply denoted by $b(\xi,\zeta)$.

By repeated applications of Proposition 3, for each $k\in\mathbb{N}$, there exists a polynomial $c^{(k)}_{s,t}(\xi,\zeta)$ such that

$$\mathbf{det}\left(\frac{\partial}{\partial\xi}-\frac{\partial}{\partial\zeta}\right)^k \mathbf{det}(\xi)^{\lambda+k}\mathbf{det}(\zeta)^{\mu+k} = c^{(k)}_{\lambda,\mu}(\xi,\zeta)\mathbf{det}(\xi)^{\lambda}\mathbf{det}(\zeta)^{\mu}\,. \tag{39}$$

Theorem 9 *The symbol of the bi-differential operator* $B^{(k)}_{\lambda,\mu}$ *is given by*

$$b^{(k)}_{\lambda,\mu}(\xi,\zeta) = c^{(k)}_{\lambda-\frac{n}{r},\,\mu-\frac{n}{r}}(\xi,\zeta)\,. \tag{40}$$

Proof It is possible to compose a differential operator F on $V \times V$ followed by a bi-differential operator B on $V \times V$, yielding a bi-differential operator $B \circ F$ on $V \times V$. There is a corresponding composition formula expressing the symbol $b(x,\xi,\zeta)\#f(x,y,\xi,\zeta)$ of $B \circ F$, in terms of the symbols of B and F, similar to the classical composition formula for symbols of differential operators. Now observe that

$$\begin{aligned} B^{(k)}_{\lambda,\mu} &= \operatorname{res} \circ \left(F_{(\lambda+k-1,\,\mu+k-1)} \circ \cdots \circ F_{\lambda+1,\,\mu+1}\right) \circ F_{\lambda,\,\mu} \\ &= \operatorname{res} \circ F^{(k-1)}_{\lambda+1,\,\mu+1} \circ F_{\lambda,\,\mu} \\ &= B^{(k-1)}_{\lambda+1,\mu+1} \circ F_{\lambda,\mu} \end{aligned}$$

which implies in terms of symbols

$$\begin{aligned} b^{(k)}_{\lambda,\,\mu}(\xi,\zeta) &= \left(b^{(k-1)}_{\lambda+1,\mu+1}\# f_{\lambda,\,\mu}\right)(\xi,\zeta) \\ &= D_{\lambda-\frac{n}{r}+1,\,\mu-\frac{n}{r}+1}\left(b^{(k-1)}_{\lambda+1,\,\mu+1}\right)(\xi,\zeta), \end{aligned}$$

see the proof of Proposition 3.4 in [10] for the last equality. Hence,

$$\begin{aligned} &\mathbf{det}(\xi)^{\lambda-\frac{n}{r}}\mathbf{det}(\zeta)^{\mu-\frac{n}{r}}\, b^{(k)}_{\lambda,\mu}(\xi,\zeta) \\ &= \mathbf{det}(\xi)^{\lambda-\frac{n}{r}}\mathbf{det}(\zeta)^{\mu-\frac{n}{r}}\left(D_{\lambda-\frac{n}{r}+1,\,\mu-\frac{n}{r}+1}\, b^{(k-1)}_{\lambda+1,\,\mu+1}\right)(\xi,\eta) \\ &= \mathbf{det}\left(\frac{\partial}{\partial\xi}-\frac{\partial}{\partial\zeta}\right)\left(\mathbf{det}(\xi)^{\lambda-\frac{n}{r}+1}\mathbf{det}(\zeta)^{\mu-\frac{n}{r}+1} b^{(k-1)}_{\lambda+1,\mu+1}(\xi,\zeta)\right), \end{aligned}$$

and this is how we come up with the following result.

Lemma 2 *The symbols* $b^{(k)}_{\lambda,\mu}$ *satisfy the following recurrence relation*

$$\begin{aligned} &\mathbf{det}(\xi)^{\lambda-\frac{n}{r}}\mathbf{det}(\zeta)^{\mu-\frac{n}{r}} b^{(k)}_{\lambda,\mu}(\xi,\zeta) = \\ &\mathbf{det}\left(\frac{\partial}{\partial\xi}-\frac{\partial}{\partial\zeta}\right)\left(\mathbf{det}(\xi)^{\lambda-\frac{n}{r}+1}\mathbf{det}(\zeta)^{\mu-\frac{n}{r}+1}\, b^{(k-1)}_{\lambda+1,\,\mu+1}(\xi,\zeta)\right). \end{aligned} \tag{41}$$

To conclude, it remains to calculate the solution of the recurrence relation. Now the family $c^{(k)}_{\lambda,\mu}$ satisfies the recurrence relation

$$\begin{aligned} & c_{\lambda,\mu}(\xi,\zeta)\mathbf{det}(\xi)^{\lambda}\mathbf{det}(\zeta)^{\mu} \\ &= \mathbf{det}\left(\frac{\partial}{\partial\xi}-\frac{\partial}{\partial\zeta}\right)\left(\mathbf{det}\left(\frac{\partial}{\partial\xi}-\frac{\partial}{\partial\zeta}\right)^{k-1}\mathbf{det}(\xi)^{(\lambda+1)+(k-1)}\mathbf{det}(\zeta)^{(\mu+1)+(k-1)}\right) \\ &= \mathbf{det}\left(\frac{\partial}{\partial\xi}-\frac{\partial}{\partial\zeta}\right)\left(c^{(k-1)}_{\lambda+1,\mu+1}(\xi,\zeta)\mathbf{det}(\xi)^{\lambda+1}\mathbf{det}(\zeta)^{\mu+1}\right). \end{aligned}$$

Hence, the family $c^{(k)}_{\lambda-\frac{n}{r},\mu-\frac{n}{r}}$ satisfies the same recurrence relation as the family $b^{(k)}_{\lambda,\mu}$ and are both equal to 1 for $k=0$ and hence, they coincide for every $k\in\mathbb{N}$.

□

Remark 1 The definition of the polynomials $c^{(k)}_{\lambda,\mu}$ reminds of the definition of the Jacobi polynomials using the *Rodrigues formula*, see [11] Section 5 for more details.

6 Other Applications

Further, we shall assure that our new approach for constructing symmetry breaking differential operators is efficient to produce new examples of such operators in many different geometric contexts.

6.1 The Case of Differential Forms

In [6], we were able to built bi-differential operators acting on spaces of differential forms, which are covariant for the Lie group $G=\mathrm{SO}_0(1,n+1)$, the conformal group of $\mathbb{R}^n$. Further, we will briefly summarize the main results in [6]. All the detailed proofs can be found in [loc.cit.].

Let $\mathbb{R}^{1,n+1}$ be the $n+2$-dimensional real vector space equipped with the Lorentzian quadratic form

$$[\mathbf{x},\mathbf{x}]=x_0^2-(x_1^2+\cdots+x_{n+1}^2),\qquad \mathbf{x}=(x_0,x_1,\ldots,x_{n+1}).$$

Let $G=\mathrm{SO}_0(1,n+1)$ be the connected component of the identity in the group of isometries for the Lorentzian form on $\mathbb{R}^{1,n+1}$. The action of G on the unit sphere S^n

of $\mathbb{R}^{n+1}$ is given as follows: For $x' = (x'_1, \dots, x'_{n+1}) \in S^n$ and $g \in G$, observe that the component $(g(1, x'))_0 > 0$, which allows to define the element $g(x') \in S^n$ by

$$(1, g(x')) = (g(1, x'))_0^{-1}\, g(1, x').$$

By the inverse map of the stereographic projection, the action of G on S^n can be transferred to a rational action (not everywhere defined) on $\mathbb{R}^n$, for which we still use the notation $G \times \mathbb{R}^n \ni (g, x) \longmapsto g(x) \in \mathbb{R}^n$.

Further, we shall use the following standard notation for the Iwasawa decomposition of G,

$$G \ni g = \overline{n}(g) m(g) a_{t(g)} n(g) \in \overline{N} \times M \times A \times N.$$

We will identify the elements $n_x \in N$ with $x \in \mathbb{R}^n$, and, for $\lambda \in \mathbb{C}$, we write $a^\lambda_{t(g)} = e^{\lambda t(g)}$.

For $0 \le k \le n$, we denote by Λ^k the vector space of complex-valued alternating multilinear k-forms on $\mathbb{R}^n$. Let $\mathcal{E}^k(\mathbb{R}^n) = C^\infty(\mathbb{R}^n, \Lambda^k)$ be the space of smooth complex-valued differential k-forms on $\mathbb{R}^n$. For $g \in G$ and $\omega \in \mathcal{E}^k(\mathbb{R}^n)$ let

$$\pi^k_\lambda(g)\omega(x) = e^{-(\lambda+k)t(g^{-1}\overline{n}_x)} \sigma_k\big(m(g^{-1}\overline{n}_x)\big)^{-1} \omega(g^{-1}(x)), \tag{42}$$

where σ_k is the representation of $M = \mathrm{SO}(n)$ on Λ^k. That is,

$$\pi^k_\lambda \simeq \mathrm{Ind}^G_{MAN}(\sigma_k \otimes \chi_{\lambda+k} \otimes 1), \tag{43}$$

where, for $\lambda \in \mathbb{C}$, we denote by χ_λ the character of A given by $\chi_\lambda(a_t) = e^{\lambda t}$.

In the present framework, the Knapp–Stein intertwining operators take the form

$$I^k_\lambda \omega(x) = \int_{\mathbb{R}^n} \mathcal{R}^k_{-2n+2\lambda}(x - y)\, \omega(y) dy, \qquad \omega \in \mathcal{E}^k(\mathbb{R}^n)$$

where $\mathcal{R}^k_{-2n+2\lambda}$ is the tempered distribution defined by

$$\langle \mathcal{R}^k_s, \omega\rangle = \pi^{-\frac{n}{2}} 2^{-s-n+1} \frac{\Gamma\left(-\frac{s}{2}+1\right)}{\Gamma\left(\frac{s+n}{2}\right)} \int_{\mathbb{R}^n} \|x\|^{s-2} (\boldsymbol{\iota}_x \boldsymbol{\varepsilon}_x - \boldsymbol{\varepsilon}_x \boldsymbol{\iota}_x)\omega(x) dx, \tag{44}$$

with $\boldsymbol{\iota}_x$ and $\boldsymbol{\varepsilon}_x$ are, respectively, the interior and the exterior products of k-forms. For $-n < \Re s < 0$, $\mathcal{R}^k_s$ defines a tempered distribution depending holomorphically on s. By [17, §3.2], $\mathcal{R}^k_s$ can be analytically continued to $\mathbb{C}$, giving a meromorphic family of tempered distributions. When $k = 0$, the identity $\boldsymbol{\iota}_x \boldsymbol{\varepsilon}_x + \boldsymbol{\varepsilon}_x \boldsymbol{\iota}_x = \|x\|^2\, \mathrm{Id}_{\mathcal{E}}$

implies immediately that $\mathcal{R}_s^0$ is nothing but the classical Riesz distribution, up to the multiplication by $-s$. In particular, we have

$$I_\lambda^k \circ \pi_\lambda^k(g) = \pi_{n-\lambda}^k(g) \circ I_\lambda^k .$$

For $0 \leq k, \ell \leq n$,, let $\mathcal{E}^{k,\ell}(\mathbb{R}^n \times \mathbb{R}^n) = C^\infty(\mathbb{R}^n \times \mathbb{R}^n, \Lambda^k \otimes \Lambda^\ell)$ be the space of differential forms of bidegree (k, ℓ). The source operator $F_{\lambda,\mu}^{k,\ell}$ defined on $\mathcal{E}^{k,\ell}(\mathbb{R}^n \times \mathbb{R}^n)$ and satisfies

$$F_{\lambda,\mu}^{k,\ell} \circ \left(\pi_\lambda^k \otimes \pi_\mu^\ell\right)(g)\omega = \left(\pi_{\lambda+1}^k \otimes \pi_{\mu+1}^\ell\right)(g) \circ F_{\lambda,\mu}^{k,\ell}\, \omega$$

is given explicitly by

$$\begin{aligned}
F_{\lambda,\mu}^{k,\ell} = 16\Big\{ & \|x-y\|^2 \widetilde{\Box}_{k,\lambda} \otimes \widetilde{\Box}_{\ell,\mu} \\
& -2\sum_{j=1}^n (x_j - y_j) \\
& \quad \times \left((2\lambda - n + 2)\widetilde{\nabla}_{k,\lambda,j} \otimes \widetilde{\Box}_{\ell,\mu} - (2\mu - n + 2)\widetilde{\Box}_{k,\lambda} \otimes \widetilde{\nabla}_{\ell,\mu,j}\right) \\
& -2(2\lambda - n + 2)(2\mu - n + 2)\sum_{j=1}^n \widetilde{\nabla}_{k,\lambda,j} \otimes \widetilde{\nabla}_{\ell,\mu,j} \\
& -2(2\mu - n + 2)(\mu + 1)(\mu - \ell)(\mu - n + \ell)\widetilde{\Box}_{k,\lambda} \otimes \mathrm{Id}_{\mathcal{E}^\ell} \\
& -2(2\lambda - n + 2)(\lambda + 1)(\lambda - k)(\lambda - n + k)\, \mathrm{Id}_{\mathcal{E}^k} \otimes \widetilde{\Box}_{\ell,\mu}\Big\},
\end{aligned}$$

where

$$\begin{aligned}
\widetilde{\Box}_{p,q} &= (q - n + p)(q - p + 1)\boldsymbol{\delta d} + (q - n + p + 1)(q - p)\boldsymbol{d\delta} \\
\widetilde{\nabla}_{p,q,j} &= (q - n + p)(q - p + 1)\partial_{x_j} - (q - p)\boldsymbol{\varepsilon}_{e_j}\boldsymbol{\delta} - (q - n + p)\boldsymbol{\delta}\boldsymbol{\iota}_{e_j}.
\end{aligned}$$

Here $\boldsymbol{d}$ is the exterior differential $\boldsymbol{d} : \mathcal{E}^k(\mathbb{R}^n) \longrightarrow \mathcal{E}^{k+1}(\mathbb{R}^n)$ defined by $\boldsymbol{d} = \sum_{j=1}^n \boldsymbol{\varepsilon}_{e_j}\partial_{e_j}$, while $\boldsymbol{\delta}$ is the co-differential $\boldsymbol{\delta} : \mathcal{E}^{k+1}(\mathbb{R}^n) \longrightarrow \mathcal{E}^k(\mathbb{R}^n)$ given by $\boldsymbol{\delta} = -\sum_{j=1}^n \boldsymbol{\iota}_{e_j}\partial_{e_j}$. More generally, for any integer $m \geq 1$, the operator

$$F_{\lambda,\mu;m}^{k,\ell} := F_{\lambda+m-1,\mu+m-1}^{k,\ell} \circ \cdots \circ F_{\lambda+1,\mu+1}^{k,\ell} \circ F_{\lambda,\mu}^{k,\ell}$$

intertwines the representations $\pi_\lambda^k \otimes \pi_\mu^\ell$ and $\pi_{\lambda+m}^k \otimes \pi_{\mu+m}^\ell$.

Let res : $\mathcal{E}^{k,\ell}(\mathbb{R}^n \times \mathbb{R}^n) \longrightarrow C^\infty(\mathbb{R}^n, \Lambda^k \otimes \Lambda^\ell)$ be the restriction map defined by

$$(\mathrm{res}\,\omega)(x) = \omega(x, x).$$

The map res intertwines the representations $\pi_\lambda^k \otimes \pi_\mu^\ell$ and $\mathrm{Ind}_{MAN}^G\big((\sigma_k \otimes \sigma_\ell) \otimes \chi_{\lambda+\mu+k+\ell} \otimes 1\big)$. However, as a representation of $M = \mathrm{SO}(n)$, the tensor product $\sigma_k \otimes \sigma_\ell$ is in general not irreducible. Let Γ be a minimal invariant subspace of $\Lambda^k \otimes \Lambda^\ell$ under the action of $\mathrm{SO}(n)$. Let σ_Γ be the corresponding irreducible representation of $\mathrm{SO}(n)$ on Γ, and let p_Γ be the orthogonal projection on Γ. Then, the map $\mathrm{res}_\Gamma := p_\Gamma \circ \mathrm{res}$ intertwines the representations $\pi_\lambda^k \otimes \pi_\mu^\ell$ and $\mathrm{Ind}_P^G\big(\sigma_\Gamma \otimes \chi_{\lambda+\mu+k+\ell} \otimes 1\big)$.

Finally, from putting all pieces together, the operator $B_{\lambda,\mu;m}^{k,\ell;\Gamma} : \mathcal{E}^{k,\ell}(\mathbb{R}^n \times \mathbb{R}^n) \longrightarrow C^\infty(\mathbb{R}^n, \Gamma)$ defined by

$$B_{\lambda,\mu;m}^{k,\ell;\Gamma} := \mathrm{res}_\Gamma \circ F_{\lambda,\mu;m}^{k,\ell} ,$$

is a bi-differential operator and covariant with respect to $\pi_\lambda^k \otimes \pi_\mu^\ell$ and $\mathrm{Ind}_P^G(\sigma_\Gamma \otimes \chi_{\lambda+\mu+k+\ell+2m} \otimes 1)$.

In some cases, it is possible to give an explicit expression for $B_{\lambda,\mu;m}^{k,\ell;\Gamma}$. For instance, if $0 \leq k + \ell \leq n$, then the representation $\Lambda^{k+\ell}$ appears in the decomposition of the tensor product $\Lambda^k \otimes \Lambda^\ell$ with multiplicity one, and the projection (up to a normalization factor) is given by

$$p_{\Lambda^{k+\ell}}(\omega \otimes \eta) = \omega \wedge \eta .$$

Thus, for $m = 1$, the bi-differential operator $B_{\lambda,\mu;1}^{k,\ell;\Lambda^{k+\ell}}$ is given by

$$\begin{aligned} &B_{\lambda,\mu;1}^{k,\ell;\Lambda^{k+\ell}}(\omega \otimes \eta)(x) \\ &\quad = -32\Big\{(2\mu - n + 2)(\mu + 1)(\mu - \ell)(\mu - n + \ell)\widetilde{\Box}_{k,\lambda}\omega(x) \wedge \eta(x) \\ &\qquad +(2\lambda - n + 2)(2\mu - n + 2)\sum_{j=1}^{n} \widetilde{\nabla}_{k,\lambda,j}\omega(x) \wedge \widetilde{\nabla}_{\ell,\mu,j}\eta(x) \\ &\qquad +(2\lambda - n + 2)(\lambda + 1)(\lambda - k)(\lambda - n + k)\omega(x) \wedge \widetilde{\Box}_{\ell,\mu}\eta(x)\Big\} \end{aligned}$$

If in addition $k = \ell = 0$, i.e. $\omega, \eta \in C^\infty(\mathbb{R}^n)$, then

$$\begin{aligned} B_{\lambda,\mu;1}^{0,0;\mathbb{C}}(\omega \otimes \eta)(x) = -64(\lambda + 1)(\lambda - n)(\mu + 1)(\mu - n) \\ \Big\{\mu(\mu - \frac{n}{2} + 1)\left(\Delta\left(\partial_x\right)\omega\right)(x)\eta(x) \\ + 2(\lambda - \frac{n}{2} + 1)(\mu - \frac{n}{2} + 1)\sum_{j=1}^{n}\left(\partial_{x_j}\omega\right)(x)\left(\partial_{x_j}\eta\right)(x) \\ + \lambda(\lambda - \frac{n}{2} + 1)\omega(x)\left(\Delta\left(\partial_x\right)\eta\right)(x)\Big\}, \end{aligned}$$

which coincides, up to a scalar factor, with the multidimensional Rankin–Cohen operators in Sect. 5.

6.2 The Case of Spinors

The construction of covariant bi-differential operators acting on the spinor bundle has been developed in [12]. We will give further a brief overview of the main results. For more details, see [12].

The conformal spin group $G = \mathrm{Spin}_0(1, n+1)$, a double cover of the conformal group $G = \mathrm{SO}_0(1, n+1)$, acts by conformal rational transformations on $\mathbb{R}^n$. This action gives rise to the principal series representation $\pi_{\rho,\lambda} = \mathrm{Ind}_P^G(\rho \otimes \chi_\lambda \otimes 1)$, parameterized by a spinor representation $(\rho, \mathbb{S})$ of $M = \mathrm{Spin}(n)$ and a character χ_λ of A for $\lambda \in \mathfrak{a}_{\mathbb{C}} \simeq \mathbb{C}$. Above $P = MAN$ is a maximal parabolic subgroup of G where $A \simeq \mathbb{R}$, $N \simeq \mathbb{R}^n$ and $M = \mathrm{Spin}(n)$ is the centralizer of A in the maximal compact subgroup $K = \mathrm{Spin}(n+1)$ of G. Let M' be the normalizer of A in K. Then the Weyl group M'/M has two elements. As a representative of the non-trivial Weyl group element choose $w = e_1 e_{n+1}$, where $\{e_j\}_{j=0}^{n+1}$ is the standard basis of $\mathbb{R}^{1,n+1}$.

The non-compact realization of the representation $\pi_{\rho,\lambda}$, which will be used, turns out to be more appropriate for Clifford algebra calculus. It is given by

$$\pi_{\rho,\lambda}(g) f(\overline{n}) = \chi_\lambda\big(a(g^{-1}\overline{n})\big)^{-1} \rho\big(m(g^{-1}\overline{n})\big)^{-1} f\big(g^{-1}(\overline{n})\big),$$

where f is a smooth $\mathbb{S}$-valued function on $\overline{N} \simeq \mathbb{R}^n$.

The Knapp-Stein operators $J_{\rho,\lambda}$ are intertwining operators between $\pi_{\rho,\lambda}$ and $\pi_{w\rho,n-\lambda}$. However, $w\rho$ and ρ are equivalent (namely $\rho(-e_1) \circ w\rho(m) = \rho(m) \circ \rho(-e_1)$), thus we let

$$I_{\rho,\lambda} = \rho(-e_1) \circ J_{\rho,\lambda}.$$

which in fact are intertwining operators between $\pi_{\rho,\lambda}$ and $\pi_{\rho,n-\lambda}$.

Now consider simultaneously a spinor representation $(\rho, \mathbb{S})$ and its dual $(\rho', \mathbb{S}')$. Consider for $\lambda, \mu \in \mathbb{C}$, the corresponding induced representations are $\pi_{\rho,\lambda}$ and $\pi_{\rho',\mu}$. The tensor product $\pi_{\rho,\lambda} \otimes \pi_{\rho',\mu}$ is a representation of G by the diagonal action. This representation is not irreducible, and one is interested in constructing symmetry breaking differential operators $\pi_{\rho,\lambda} \otimes \pi_{\rho',\mu} \to \pi$ into irreducible representations π of G, for instance into another principal series representation $\pi = \pi_{\tau_k^*,\nu}$ where τ_k^* is an exterior power representation of $M = \mathrm{Spin}(n)$ on $\Lambda_k^*(\mathbb{C}^n)$.

To this aim, we consider for $s, t \in \mathbb{C}$, the following *Clifford-Riesz operators* $\mathcal{R}_s$ and $\mathcal{R}'_t$ given by

$$\mathcal{R}_s f = \int_{\mathbb{R}^n} r_s(x-y) f(y) dy, \qquad \mathcal{R}'_t f = \int_{\mathbb{R}^n} r'_t(x-y) f(y) dy$$

where r_s and r'_t are the *Clifford-Riesz distributions*

$$r_s(x) = \|x\|^s \rho\left(\frac{x}{\|x\|}\right), \qquad r'_t(x) = \|x\|^t \rho'\left(\frac{x}{\|x\|}\right), \tag{45}$$

associated respectively to the Clifford modules $(\mathbb{S}, \rho)$ and $(\mathbb{S}', \rho')$.

The main observation in this framework is that up to a shift in the parameters, and up to a constant multiple, the Knapp–Stein operators for $\pi_{\rho,\lambda}$ and $\pi_{\rho',\mu}$ are essentially the Clifford–Riesz operators. More precisely,

$$I_{\rho,\lambda} = \mathcal{R}_{2\lambda-2n}, \qquad I_{\rho',\mu} = \mathcal{R}'_{2\mu-2n}.$$

Further, let M to be the operator on $C^\infty(\mathbb{R}^n \times \mathbb{R}^n, \mathbb{S} \otimes \mathbb{S}')$ defined for a smooth function f on $\mathbb{R}^n \times \mathbb{R}^n$ with values in $\mathbb{S} \otimes \mathbb{S}'$ by

$$Mf(x, y) = \|x - y\|^2 f(x, y).$$

The operator M intertwines $\pi_{\rho,\lambda} \otimes \pi_{\rho',\mu}$ and $\pi_{\rho,\lambda-1} \otimes \pi_{\rho',\mu-1}$.

Applying the general construction of the source operator $F_{\lambda,\mu}$, which satisfies $F_{\lambda,\mu} \circ \left(\pi_{\rho,\lambda} \otimes \pi_{\rho',\mu}\right)(g) = \left(\pi_{\rho,\lambda+1} \otimes \pi_{\rho',\mu+1}\right)(g) \circ F_{\lambda,\mu}$, gives

$$\begin{aligned}
F_{\lambda,\mu} &= \|x-y\|^2 \Delta_x \otimes \Delta_y \\
&\quad + 2(2\lambda - n + 1) \sum_{j=1}^{n} (x_j - y_j) \frac{\partial}{\partial x_j} \otimes \Delta_y \\
&\quad + 2(2\mu - n + 1) \sum_{j=1}^{n} (y_j - x_j) \Delta_x \otimes \frac{\partial}{\partial y_j} \\
&\quad - 2\rho(x-y) \not{D}_x \otimes \Delta_y - 2\Delta_x \otimes \rho'(x-y) \not{D}_y \\
&\quad + (2\mu - n + 1)(2\mu + 1) \Delta_x \otimes \mathrm{Id} + (2\lambda - n + 1)(2\lambda + 1) \, \mathrm{Id} \otimes \Delta_y \\
&\quad - 2(2\lambda - n + 1)(2\mu - n + 1) \sum_{j=1}^{n} \frac{\partial}{\partial x_j} \otimes \frac{\partial}{\partial y_j}
\end{aligned} \tag{46}$$

$$+ 2(2\lambda - n + 1)\sum_{j=1}^{n}\frac{\partial}{\partial x_j}\otimes \rho'(e_j)\not{D}'_y + 2(2\mu - n + 1)\sum_{j=1}^{n}\rho(e_j)\not{D}_x\otimes\frac{\partial}{\partial y_j}$$

$$- 2(\sum_{j=1}^{n}\rho(e_j)\not{D}_x\otimes\rho'(e_j)\not{D}'_y).$$

Above, $\not{D}$ and $\not{D}'$ are respectively the Dirac operators associated to the Clifford modules $(\rho, \mathbb{S})$ and $(\rho', \mathbb{S}')$.

For $1 \le k \le n$, let

$$\widetilde{\Psi}^{(k)} : C^\infty\left(\mathbb{R}^n \times \mathbb{R}^n, \mathbb{S}\otimes\mathbb{S}'\right) \to C^\infty\left(\mathbb{R}^n, \Lambda_k^*\left(\mathbb{R}^n\right)\otimes\mathbb{C}\right), f \mapsto \left(\Psi^{(k)} f(x,y)\right)_{|x=y}$$

where $\Psi^{(k)} : \mathbb{S}\otimes\mathbb{S}' \to \Lambda_k^*(\mathbb{R}^n)\otimes\mathbb{C}$ is the usual projection. Then one can prove that $\widetilde{\Psi}^{(k)}$ intertwines $\pi_{\rho,\lambda}\otimes\pi_{\rho',\mu}$ and $\pi_{\tau_k^*,\lambda+\mu}$.

For $m \in \mathbb{N}$, define the operator $F_{\lambda,\mu}^{(m)} : C^\infty\left(\mathbb{R}^n\times\mathbb{R}^n, \mathbb{S}\otimes\mathbb{S}'\right) \to C^\infty\left(\mathbb{R}^n\times\mathbb{R}^n, \mathbb{S}\otimes\mathbb{S}'\right)$ by

$$F_{\lambda,\mu}^{(m)} = F_{\lambda+m-1,\mu+m-1}\circ\cdots\circ F_{\lambda,\mu}.$$

It is an intertwining operator of $\pi_{\rho,\lambda}\otimes\pi_{\rho',\mu}$ and $\pi_{\rho,\lambda+m}\otimes\pi_{\rho',\mu+m}$. Finlay, define

$$B_{k;\lambda,\mu}^{(m)} = \widetilde{\Psi}^{(k)}\circ F_{\lambda,\mu}^{(m)}.$$

Then the operators $B_{k;\lambda,\mu}^{(m)} : C^\infty\left(\mathbb{R}^n\times\mathbb{R}^n, \mathbb{S}\otimes\mathbb{S}'\right) \to C^\infty\left(\mathbb{R}^n, \Lambda_k^*\left(\mathbb{R}^n\right)\otimes\mathbb{C}\right)$ are constant coefficient bi-differential operators and homogeneous of degree $2m$, and for any $g \in G$,

$$B_{k;\lambda,\mu}^{(m)}\circ\left(\pi_{\rho,\lambda}\otimes\pi_{\rho',\mu}\right)(g) = \pi_{\tau_k^*;\lambda+\mu+2m}(g)\circ B_{k;\lambda,\mu}^{(m)}.$$

In particular, for $k = 0$ and $m = 1$, the operator $B_{0;\lambda,\mu}^{(1)} : C^\infty\left(\mathbb{R}^n\times\mathbb{R}^n, \mathbb{S}\otimes\mathbb{S}'\right) \to C^\infty\left(\mathbb{R}^n\right)$ is given by

$$\begin{aligned}
&B_{0;\lambda,\mu}^{(1)}\left(v(\cdot)\otimes w'(\cdot)\right)(x)\\
&\quad = (2\mu - n + 1)(2\mu + 1)\left(\Delta v(x), w'(x)\right)\\
&\qquad + (2\lambda - n + 1)(2\lambda + 1)\left(v(x), \Delta w'(x)\right)\\
&\qquad - 2(2\lambda - n + 1)(2\mu - n + 1) + \sum_{j=1}^{n}\left(\frac{\partial}{\partial x_j}v(x), \frac{\partial}{\partial y_j}w'(x)\right)\\
&\qquad - 2(2\lambda + 2\mu - n + 2)\left(\not{D}v(x), \not{D}'w'(x)\right).
\end{aligned}$$

If in addition the dimension $n = 1$, using the realization of $\mathbb{S}$ as a left ideal in the exterior algebra, one obtains

$$B^{(1)}_{0;\lambda,\mu} = 2\mu(2\mu+1)\frac{\partial^2}{\partial x^2} + 2\lambda(2\lambda+1)\frac{\partial^2}{\partial y^2} - 2(2\lambda+1)(2\mu+1)\frac{\partial^2}{\partial x\partial y}.$$

This coincides (up to a constant multiple) to the degree two Rankin–Cohen operator (3) for the group $SL(2,\mathbb{R})$ which is isomorphic to $\mathrm{Spin}_0(1,2)$.

6.3 Other Examples

Although the focus in this article is on the tensor product cases, the source operator method has been applied successfully in other geometric contexts. Let us briefly mention these applications. *Juhl operators* (see [21]) are conformal symmetry breaking differential operators corresponding to the restriction from the sphere S^n to S^{n-1}, more precisely from the conformal group $O(1, n+1)$ to $O(1, n)$ acting on densities. A construction of these operators has been presented by the second author in [9]. In the same context, a similar construction has been worked out for the case of differential forms and for the case of spinors [16]. There is a similar approach in conformal analysis of curved spades, and an alternative construction of the GJMS operators was proposed by B. Ørsted and A. Juhl along similar lines (see [22]). It is likely that these ideas can be used in other Cartan geometries.

Acknowledgments We thank the anonymous reviewers for their careful reading of our manuscript and their insightful comments. S. Ben Said is thankful to UAEU for the financial support through the UPAR grant 12S121.

References

1. Barchini, L., Sepanski, M., Zierau, R.: Positivity of zeta distributions and small unitary representations. In: The ubiquitous heat kernel. Contemp. Math. vol. 398, pp. 1–46. Amer. Math. Soc., Providence (2006)
2. Beckman, R., Clerc, J.-L.: Singular invariant trilinear forms and covariant (bi-)differential operators under the conformal group. J. Funct. Anal. **262**(10), 4341–4376 (2012)
3. Bertram, W.: The Geometry of Jordan and Lie Structures. Lecture Notes in Mathematics, vol. 1754. Springer, Berlin (2000)
4. Ben Saïd, S.: On the integrability of a representation of sl (2, R). J. Funct. Anal. **250**(2), 249–264 (2007)
5. Ben Saïd, S., Clerc, J.-L., Koufany, K.: Conformally covariant bi-differential operators on a simple real Jordan algebra. Int. Math. Res. Not. IMRN **8**, 2287–2351 (2020)
6. Ben Saïd, S., Clerc, J.-L., Koufany, K.: Conformally covariant bi-differential operators for differential forms. Commun. Math. Phys. **373**(2), 739–761 (2020)
7. Bopp, N., Rubenthaler, H.: Local zeta functions attached to the minimal spherical series for a class of symmetric spaces. Mem. Am. Math. Soc. **174**(821) (2005)

8. Clerc, J.-L.: Covariant bi-differential operators on matrix space. Ann. Inst. Fourier (Grenoble) **67**(4), 1427–1455 (2017)
9. Clerc, J.-L.: Another approach to Juhl's conformally covariant differential operators from S^n to S^{n-1}. SIGMA **13**(026), 1–18 (2017)
10. Clerc, J.-L.: Symmetry breaking differential operators, the source operator and Rodrigues formulæ. Pac. J. Math. **307**(1), 79–107 (2020)
11. Clerc, J.-L.: Four variations on the Rankin-Cohen brackets (2023), this volume
12. Clerc, J.-L., Koufany, K.: Symmetry breaking differential operators for tensor products of spinorial representations. SIGMA Symmetry Integrability Geom. Methods Appl. **17** Paper No. 049, 23 pp (2021)
13. Cohen, H.: Sums involving the values at negative integers of L-functions of quadratic characters. Math. Ann. **217**, 271–385 (1975)
14. Faraut, J., Gindikin, S.: Pseudo-Hermitian symmetric spaces of tube type, in Topics in Geometry, In Memory of J. D'Atri (S. Gindikin, Ed.), Birkhauser, Boston, 1996, pp. 123–154.
15. Faraut, J., Korányi, A.: Analysis on Symmetric Cones. Oxford University Press, New York (1994)
16. Fischmann, M., Ørsted, B., Somberg, P.: Bernstein-Sato identities and conformal symmetry breaking operators. J. Funct. Anal. **277**(11), 108219, 1–36 (2019)
17. Fischmann, M., Ørsted, B.: A family of Riesz distributions for differential forms on Euclidean space. Int. Math. Res. Not. **2021**(13), 9746–9768 (2021)
18. Gel'fand, I.M., Shilov, G.E.: Generalized Functions, vol. 1. Academic, New York/London (1964)
19. Gover, R., Waldron, A.: Boundary calculus for conformally compact manifolds. Indiana Univ. Math. J. **63**, 119–163 (2014)
20. Helwig, K.-H.: Halbeinfache reelle Jordan-Algebren. Math. Z. **109**, 1–28 (1969)
21. Juhl, A.: Families of conformally covariant differential operators. In: Q-Curvature and Holography. Progress in Mathematics, vol. 275. Birkhäuser, Basel (2009)
22. Juhl, A., Ørsted, B.: Shift operators, residue families and degenerate Laplacians. Pac. J. Math. **308**(1), 103–160 (2020)
23. Kayoya, J.B.: Zeta functional equation on Jordan algebras of type II. J. Math. Anal. Appl. **302**, 425–440 (2005)
24. Knapp, A.W.: Representation Theory of Semisimple Lie Groups, an Overview Based on Examples. Princeton Mathematical Series, vol. 36. Princeton University Press, Princeton (1986)
25. Kobayashi, T.: A program for branching problems in the representation theory of real reductive groups. In: Representations of Reductive Groups. Progress in Mathemmatics, vol. 312, pp. 277–322. Birkhäuser/Springer, Cham (2015)
26. Koecher, M.: The Minnesota Notes on Jordan Algebras and Their Applications, Edited, annotated and with a preface by Aloys Krieg and Sebastian Walcher. In Lecture Notes in Mathematics, vol. 1710. Springer, Berlin (1999)
27. Loos, O.: Bounded Symmetric Domains and Jordan Pairs. University of California, Irvine (1977)
28. Satake, I., Faraut, J.: The functional equation of zeta distributions associated with formally real Jordan algebras. Tohoku Math. J. **36**, 469–482 (1984)
29. Saïd, S.B., Ørsted, B.: Analysis on flat symmetric spaces. Journal des Mathematiques Pures et Appliquees **84**(10), 1393–1426 (2005)

Some Mixed Norm Bounds for the Spectral Projections of the Heisenberg Sublaplacian

Valentina Casarino and Paolo Ciatti

Dedicated to Toshi Kobayashi on the occasion of his 60th birthday

Abstract This chapter concerns the boundedness of the generalized spectral projections $\mathcal{P}_\lambda$ associated to the sublaplacian in the Heisenberg group, seen as operators from $L^1_t L^p_z$ to $L^\infty_t L^q_z$. We formulate a conjecture about the range of p, q for which the projections $\mathcal{P}_\lambda$ are bounded and provide some motivations.

1 Introduction

About 30 years ago D. Müller in [6] considered the operators arising in the spectral decomposition of the standard sublaplacian $\mathcal{L}$ on the Heisenberg group $\mathbb{H}_n$ (we refer to Sect. 3.1 for a precise definition) and proved some $L^p - L^{p'}$ estimates for them.

More precisely, for any Schwartz function f on the Heisenberg group, there is a decomposition

$$f = \int_0^\infty \mathcal{P}_\lambda f \, d\lambda;$$

the latter integral converges in the sense of distributions, and the distributions $\mathcal{P}_\lambda f$, defined for all $\lambda > 0$, satisfy

$$\mathcal{L}\mathcal{P}_\lambda f = \lambda \mathcal{P}_\lambda f. \tag{1}$$

V. Casarino
DTG, Università degli Studi di Padova, Vicenza, Italy
e-mail: valentina.casarino@unipd.it

P. Ciatti (✉)
Dipartimento di Matematica "Tullio Levi Civita", Università degli Studi di Padova, Padova, Italy
e-mail: paolo.ciatti@unipd.it

M. Pevzner, H. Sekiguchi (eds.), *Symmetry in Geometry and Analysis, Volume 2*,
Progress in Mathematics 358, https://doi.org/10.1007/978-981-97-7662-7_2

The operators $\mathcal{P}_\lambda$ have been named generalized spectral projections by R. Strichartz [15]. In [6] Müller proved the bounds

$$\|\mathcal{P}_\lambda f\|_{L_t^\infty L_z^{p'}} \lesssim_{n,\lambda} \|f\|_{L_t^1 L_z^p},$$

holding for $1 \leq p < 2$ and $\frac{1}{p} + \frac{1}{p'} = 1$. The mixed norms appearing in these inequalities are defined by

$$\|f\|_{L_t^r L_z^p} = \left(\int_{\mathbb{C}^n} \left(\int_{\mathbb{R}} |f(z,t)|^r \, dt \right)^{\frac{p}{r}} dz \right)^{\frac{1}{p}}, \qquad 1 \leq p, r < \infty, \tag{2}$$

(with the usual modifications if p or r is ∞); here $t \in \mathbb{R}$ is the coordinate in the centre and $z \in \mathbb{C}^n$.

We remark that estimates of this type with internal exponents different from 1 and ∞ cannot be expected (see Proposition 3.1 in [6] and also [18, p. 2]). This is due to the fact that the operator $\mathcal{P}_\lambda$ acts on f through the one dimensional Fourier transform in the central variable and, in dimension one, only the $L^1 \to L^\infty$ restriction estimate is available.

Some years ago, the authors improved on the result of Müller [1], obtaining estimates from $L_t^1 L_z^p$ to $L_t^\infty L_z^q$ holding in the range $1 \leq p \leq 2 \leq q$. Indeed, they replaced in the proof a bound used by Müller with a new and stronger one due to H. Koch and F. Ricci [4].

The description of the complete range of p, q for which $\mathcal{P}_\lambda$ is bounded from $L_t^1 L_z^p$ to $L_t^\infty L_z^q$ seems to be quite involved. In a forthcoming chapter we shall provide an example showing that in $\mathbb{H}_n$ there are no estimates of the type

$$\|\mathcal{P}_\lambda f\|_{L_t^\infty L_z^q} \lesssim \|f\|_{L_t^1 L_z^p}$$

when $q < \tilde{p}(2n)$ or when $p = q$ and $\tilde{p}(2n) \leq p \leq \tilde{p}'(2n)$; here

$$\tilde{p}(2n) = \tilde{p} := \frac{4n}{2n+1}. \tag{3}$$

There we shall also prove the optimal estimates on the three-dimensional Heisenberg group.

In this note we introduce the problem, fix some notation, and prove some preliminary results. In particular, we show that using Hölder's inequality and the Koch–Ricci bounds, one can easily prove that $\mathcal{P}_\lambda$ is bounded from $L_t^1 L_z^p$ to $L_t^\infty L_z^q$ in the range $1 \leq p \leq p_*(2n)$ and $\tilde{p}(2n) < q$ (so that q may be less than 2); here

$$p_*(2n) = p_* := 2\frac{2n+1}{2n+3}. \tag{4}$$

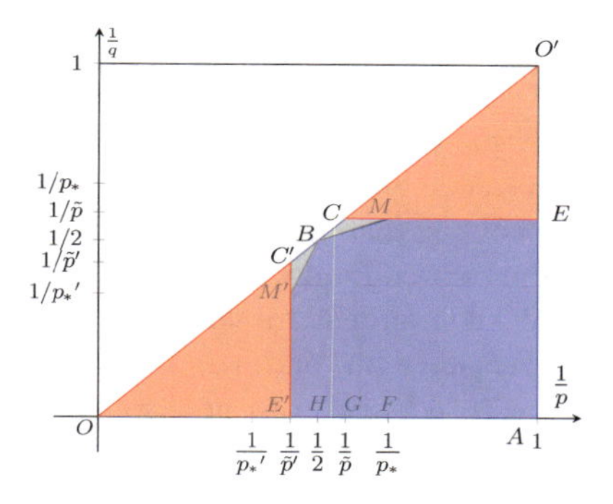

Fig. 1 On the range of boundedness of $\mathcal{P}_\lambda$. In this picture $\tilde{p} = \frac{4n}{2n+1}$, $p_* = 2\frac{2n+1}{2n+3}$

To simplify notation, from now on we shall write p_* and $\tilde{p}$ instead of $p_*(2n)$ and $\tilde{p}(2n)$.

For the Riesz diagram showing what we know see Fig. 1 in Sect. 7.

We call $O = (0, 0)$, $O' = (1, 1)$, $A = (1, 0)$, $B = (1/2, 1/2)$, $C = (1/\tilde{p}, 1/\tilde{p})$, $C' = (1/\tilde{p}', 1/\tilde{p}')$, $E = (1, 1/\tilde{p})$, $M = (1/p_*, 1/\tilde{p})$, $M' = (1/\tilde{p}', 1/p_*')$, $E' = (1/\tilde{p}', 0)$, $F = (1/p_*, 0)$, $G = (1/\tilde{p}, 0)$, $H = (1/2, 0)$.

The spectral projections of the sublaplacian are bounded inside the polygon of vertices $AEMBM'E'$ (the blue region). They are also bounded in A and on the open sides AE and $E'A$. In a forthcoming paper we will show that they are unbounded in the closed triangles $OE'C'$ and $O'EC$ (the red region). In the triangles CBM and $C'BM'$ (the grey region) we are able to prove the estimates only for $n = 1$, and this will be the main task in our upcoming article.

In light of what happens in dimension one, we are led to formulate the following conjecture.

Conjecture 1 The estimate

$$\|\mathcal{P}_\lambda f\|_{L_t^\infty L_z^q} \lesssim_{n,\lambda} \|f\|_{L_t^1 L_z^p}$$

holds for $(1/p, 1/q)$ in the interior of the polygon with vertices $AECC'E'$, in the half open segments AE and $E'A$ (closed in A and open in E and E').

The chapter is structured as follows.

In Sect. 2, we contextualize our estimates, relating them to some relevant restriction bounds. In Sect. 3, we introduce the setting of Heisenberg group and define two strictly related operators, the sublaplacian $\mathcal{L}$ on $\mathbb{H}_n$ and the λ-twisted Laplacian L_λ on $\mathbb{C}^n$. A few estimates for Laguerre functions, which are essential in our approach, are proved as well. In Sect. 4 we illustrate the spectral resolution of L_λ and of $\mathcal{L}$, and define in a rigorous way the spectral projections $\mathcal{P}_\lambda$ and Λ_k^λ, associated to the sublaplacian $\mathcal{L}$ and to the λ-twisted Laplacian L_λ, respectively. Then we prove a first general result about the boundedness of $\mathcal{P}_\lambda$. Following the approach introduced by Müller, in Sect. 5 we reduce the proof of the conjectured estimate to a bound concerning the spectral projections of the twisted Laplacian. We also establish a conjecture about the boundedness of Λ_k^λ. This conjecture is proved in Sect. 6 for functions given by the product of a radial function and a bigraded spherical harmonic. Finally, in Sect. 7, we first prove the boundedness of $\mathcal{P}_\lambda$ on the open rectangle $AEMF$ (and on the open edges MF and AE). Then we conclude by showing that the conjecture formulated in Sect. 5 for the spectral projections Λ_k^λ of the twisted Laplacian would imply Conjecture 1.

We will write C for various constants, which are not necessarily equal at different occurrences. If a and b are positive quantities, $a \lesssim b$ or equivalently $b \gtrsim a$ means $a \leq Cb$. When $a \lesssim b$ and also $b \lesssim a$, we write $a \simeq b$. By $\mathbb{N}$, we mean the set of all nonnegative integers.

2 Some Context

In this section, we give some context to the estimates we shall discuss in the chapter. The relation between the spectral theory of the Laplacian in Euclidean spaces and the restriction theory of the Fourier transform is well known since the works of Strichartz and Sogge [15, 12]. Indeed, writing the inversion formula of the Fourier transform in polar coordinates, one gets

$$f(x) = \frac{1}{(2\pi)^{d/2}} \int_0^\infty \left(\int_{\mathbb{S}^{d-1}} \hat{f}(r\omega) e^{ir\omega\cdot x} d\sigma(\omega) \right) r^{d-1} dr, \tag{5}$$

for, say, a Schwartz function f in $\mathbb{R}^d$; in this formula $\mathbb{S}^{d-1}$ is the unit sphere, $d\sigma(\omega)$ denotes the measure induced on it by the Lebesgue measure and $\hat{f}$ is the Fourier transform of f, given by

$$\hat{f}(\xi) = \frac{1}{(2\pi)^{d/2}} \int_{\mathbb{R}^d} f(x) e^{-ix\cdot\xi} dx.$$

Since

$$\Delta \int_{\mathbb{S}^{d-1}} \hat{f}(r\omega) e^{ir\omega\cdot x} d\sigma(\omega) = -r^2 \int_{\mathbb{S}^{d-1}} \hat{f}(r\omega) e^{ir\omega\cdot x} d\sigma(\omega),$$

the functions

$$\int_{\mathbb{S}^{d-1}} \hat{f}(r\omega) e^{ir\omega\cdot x} d\sigma(\omega)$$

are called generalized eigenfunctions of Δ (generalized since they are not in L^2). In this perspective, (5) yields the spectral decomposition of the function f with respect to Δ, and the map

$$P_{-r^2} f(x) = \frac{1}{(2\pi)^{d/2}} \int_{\mathbb{S}^{d-1}} \hat{f}(r\omega) e^{ir\omega\cdot x} d\sigma(\omega)$$

is the spectral projection of Δ corresponding to the (generalized) eigenvalue $-r^2$.

The celebrated Stein–Tomas restriction theorem, asserting the possibility to restrict the Fourier transform of a function in the L^2 sense, is equivalent to the estimate

$$\|P_{-r^2} f\|_{L^{p'}(\mathbb{R}^d)} \lesssim r^{-2\frac{d}{p'}} \|f\|_{L^p(\mathbb{R}^d)},$$

holding for $1 \leq p \leq 2\frac{d+1}{d+3}$, where $\frac{1}{p} + \frac{1}{p'} = 1$ (see, for instance, [13]). In this perspective, the Stein–Tomas theorem may be thought of as concerning the mapping properties of the spectral projectors of Δ.

The Stein–Tomas theorem completely settles the problem of establishing the range of exponents p for which it makes sense to consider the restriction of the Fourier transform to the sphere in the L^2 sense. Stein posed the problem of determining for which q the restriction phenomenon holds in the L^q sense, formulating the well-known restriction conjecture [13] (for an extensive review on this topic see [16]). In this respect, note, however, that for $q \neq 2$ there is no clear relation between the estimates for the spectral projectors and the restriction problem.

In the Heisenberg group setting, the Laplacian is replaced by the sublaplacian $\mathcal{L}$. Due to computational difficulties, to obtain sharp estimates, one is often lead to replace the Fourier decomposition of a function on $\mathbb{H}_n$ with its spectral decomposition with respect to the sublaplacian (for an example, see [7]). In view of these considerations Müller in [6] proved an $L^p \to L^{p'}$ estimate for the norms of $\mathcal{P}_\lambda$, calling it restriction theorem for the Heisenberg group.

Analogous results were obtained by Sogge, who proved sharp $L^p \to L^2$ estimates for the spectral projectors of the Laplace-Beltrami operators on spheres [11] and later, in collaboration with Seeger, on compact Riemannian manifolds [10]. In particular, in [11] Sogge proved an estimate holding in the optimal range ($1 \leq p < 4/3$) for the spectral projectors of the Laplace-Beltrami operator on the two dimensional sphere. This result provided the inspiration for our forthcoming article concerning the spectral projections of the sublaplacian on the three-dimensional Heisenberg group.

3 Notation and Preliminaries

3.1 The Heisenberg Group

The Heisenberg group $\mathbb{H}_n$ topologically coincides with the space $\mathbb{R}^n \times \mathbb{R}^n \times \mathbb{R}$. The product is given by the law

$$(x, y, t) \cdot (x', y', t') = \left(x + x', y + y', t + t' + \frac{1}{2}(x \cdot y' - y \cdot x')\right), \tag{6}$$

where $x \cdot y' = \sum_{j=1}^n x_j y'_j$, if $x = (x_1, \dots, x_n)$ and $y' = (y'_1, \cdots, y'_n)$.

3.2 The Sublaplacian $\mathcal{L}$ and the Twisted Laplacian L

We first introduce the left invariant vector fields defined by

$$X_j f(x,y,t) = \frac{d}{ds} f\left((x,y,t)\cdot(s\mathbf{e}_j,0,0)\right)\Big|_{s=0} = \partial_{x_j} f(x,y,t) - \frac{1}{2} y_j \partial_t f(x,y,t),$$

$$Y_j f(x,y,t) = \frac{d}{ds} f\left((x,y,t)\cdot(0,s\mathbf{e}_j,0)\right)\Big|_{s=0} = \partial_{y_j} f(x,y,t) + \frac{1}{2} x_j \partial_t f(x,y,t),$$

for $j = 1,\dots,n$, and by

$$Tf(x,y,t) = \frac{d}{ds} f\left((x,y,t)\cdot(0,0,s)\right)\Big|_{s=0} = \partial_t f(x,y,t).$$

Here f is a smooth function on $\mathbb{H}_n$ and $\{\mathbf{e}_j\}_{j=1,\cdots n}$ is an orthonormal basis of $\mathbb{R}^n$. When we identify the Lie algebra $\mathfrak{h}_n$ of $\mathbb{H}_n$ with the space of left-invariant vector fields on the group, the vector field T generates the centre of $\mathfrak{h}_n$ and the family $\{X_1,\cdots,X_n,Y_1,\cdots,Y_n\}$ is a basis of the first layer of $\mathfrak{h}_n$.

The sublaplacian $\mathcal{L}$ associated to $X_1,\dots,X_n,Y_1,\dots,Y_n$ is defined by

$$\mathcal{L} = -X_1^2 - \cdots - X_n^2 - Y_1^2 - \cdots - Y_n^2.$$

This is a second order, left-invariant, hypoelliptic differential operator. In the coordinates (x,y,t) it takes the form

$$\mathcal{L} = -\sum_{j=1}^{n}\left(\partial_{x_j}^2 + \partial_{y_j}^2\right) + \sum_{j=1}^{n}\left(x_j\partial_{y_j} - y_j\partial_{x_j}\right)\partial_t - \frac{1}{4}\sum_{j=1}^{n}\left(x_j^2+y_j^2\right)\partial_t^2.$$

Next we consider the partial Fourier transform of f in the central variable

$$f^{(\lambda)}(z) = \int_{\mathbb{R}} e^{i\lambda t} f(z,t)\,dt, \qquad \lambda\in\mathbb{R}, \quad f\in\mathcal{S}(\mathbb{H}_n), \tag{7}$$

and, if $\lambda \in \mathbb{R}\setminus\{0\}$, we introduce a differential operator L_λ on $\mathbb{R}^n\times\mathbb{R}^n$, which is called λ-twisted laplacian and is given by

$$L_\lambda f^{(\lambda)}(x,y) = (\mathcal{L}f)^{(\lambda)}(x,y) = \int_{\mathbb{R}} e^{i\lambda t}\mathcal{L}f(x,y,t)\,dt,$$

for any Schwartz function f on $\mathbb{H}_n$. In our coordinates, it takes the form

$$L_\lambda = -\sum_{i=1}^{n}\left(\frac{\partial^2}{\partial x_i^2} + \frac{\partial^2}{\partial y_i^2}\right) - i\lambda\sum_{i=1}^{n}\left(x_i\frac{\partial}{\partial y_i} - y_i\frac{\partial}{\partial x_i}\right) + \frac{1}{4}\lambda^2\sum_{i=1}^{n}\left(x_i^2+y_i^2\right).$$

The λ-twisted Laplacian L_λ admits a complete orthonormal system of eigenfunctions, given by matrix coefficients of the Schrödinger representation [19, Section 1.3]. Hence, $L^2(\mathbb{R}^{2n})$ decomposes into eigenspaces of L_λ. The orthogonal projections onto the eigenspace corresponding to the eigenvalue $(2k+n)\lambda$, $k \in \mathbb{N}$, are usually denoted by Λ_k^λ.

For $\lambda = 1$, the λ-twisted laplacian is simply called twisted laplacian and denoted by L. This is a second-order elliptic operator on $\mathbb{C}^n$, with point spectrum, and eigenvalues given by $\lambda_k = 2k+n$, $k \in \mathbb{N}$.

3.3 *Laguerre Functions*

A central role in the spectral decomposition of $\mathcal{L}$ and L_λ is played by Laguerre functions.

Recall that the Laguerre polynomials of type m and degree k, $L_k^{(m)}(t)$, are defined for $t \geq 0$ by

$$L_k^m(t) = e^t\, t^{-m} \frac{1}{k!} \frac{d^k}{dt^k}(e^{-t} t^{k+m}), \qquad k = 0, 1, 2, \cdots . \tag{8}$$

Then the Laguerre functions of order m on $\mathbb{C}^n$, $m = 0, 1, 2, \cdots$, are given by

$$\varphi_k^{(m)}(z) = L_k^{(m)}\left(\frac{|z|^2}{2}\right) e^{-\frac{|z|^2}{4}}. \tag{9}$$

In the following when $m = n-1$, we shall often write $L_k(t)$ and $\varphi_k(z)$ instead of $L_k^{n-1}(t)$ and $\varphi_k^{(n-1)}(z)$.

In order to prove a lower bound for the norms of the spectral projections, we shall need some estimates for the norms of the Laguerre functions.

Lemma 1 *Let $\ell \in \mathbb{N}$ be fixed. Then for k sufficiently large one has*

$$\left\| |\cdot|^\ell \varphi_k^{(n+\ell-1)} \right\|_{L^p(\mathbb{C}^n)} \simeq \begin{cases} 2^{\frac{\ell}{2}} k^{\frac{\ell}{2}+\frac{n}{p}-\frac{1}{2}} & \text{if } 1 \leq p < \tilde{p}', \\ 2^{\frac{\ell}{2}} k^{\frac{\ell}{2}+\frac{n}{\tilde{p}'}-\frac{1}{2}} (\log k)^{\frac{1}{\tilde{p}'}} & \text{if } p = \tilde{p}', \\ 2^{\frac{\ell}{2}} k^{\frac{\ell}{2}+\frac{n}{p'}-1} & \text{if } \tilde{p}' < p \leq \infty, \end{cases} \tag{10}$$

where $\tilde{p}' = \tilde{p}/(\tilde{p}-1)$, and $\tilde{p}$ is defined by (3).

Proof For the sake of completeness, we recall that Markett proved the following estimates [5]:

$$\left(\int_0^\infty \left| L_k^{\alpha+\beta}(s)\, e^{-\frac{s}{2}} s^{\frac{\alpha}{2}} \right|^p ds \right)^{\frac{1}{p}} \simeq \begin{cases} k^{\frac{\alpha}{2}-\frac{1}{2}+\frac{1}{p}} & \text{if } \beta < \frac{2}{p}-\frac{1}{2}, \\ k^{\frac{\alpha}{2}-\frac{1}{2}+\frac{1}{p}} (\log k)^{\frac{1}{p}} & \text{if } \beta = \frac{2}{p}-\frac{1}{2}, \\ k^{\frac{\alpha}{2}+\beta-\frac{1}{p}} & \text{if } \beta > \frac{2}{p}-\frac{1}{2}, \end{cases}$$

where $\alpha+\beta>-1$, $1\le p\le 4$, and $\alpha>-\frac{2}{p}$;

$$\left(\int_0^\infty \left|L_k^{\alpha+\beta}(s)\,e^{-\frac{s}{2}}s^{\frac{\alpha}{2}}\right|^p ds\right)^{\frac{1}{p}} \simeq \begin{cases} k^{\frac{\alpha}{2}-\frac{1}{3}+\frac{1}{3p}} & \text{if } \beta\le\frac{4}{3p}-\frac{1}{3}, \\ k^{\frac{\alpha}{2}+\beta-\frac{1}{p}} & \text{if } \beta>\frac{4}{3p}-\frac{1}{3}, \end{cases}$$

where $\alpha+\beta>-1$, $4<p\le\infty$, and $\alpha>-\frac{2}{p}$ if $p<\infty$ or $\alpha\ge 0$ if $p=\infty$.

Thus, we have

$$\begin{aligned}\left\||\cdot|^\ell\varphi_k^{(n+\ell-1)}\right\|_{L^p(\mathbb{C}^n)} &\simeq 2^{\frac{\ell}{2}}\left(\int_0^\infty\left|L_k^{n+\ell-1}(s)\,e^{-\frac{s}{2}}s^{\frac{\ell}{2}}\right|^p s^{n-1}ds\right)^{\frac{1}{p}} \\ &\simeq 2^{\frac{\ell}{2}}\left(\int_0^\infty\left|L_k^{n+\ell-1}(s)\,e^{-\frac{s}{2}}s^{\frac{\ell}{2}+\frac{n-1}{p}}\right|^p ds\right)^{\frac{1}{p}}.\end{aligned}$$

Take $\alpha=\ell+2\frac{(n-1)}{p}$ and $\beta=(n-1)\left(1-\frac{2}{p}\right)$. We first observe that $\tilde{p}'=\frac{4n}{2n-1}\le 4$. When $1\le p<\frac{4n}{2n-1}=\tilde{p}'$, then $1-\frac{2}{p}<\frac{1}{2n}$, whence $\beta<\frac{1}{2}-\frac{1}{2n}$. Since, moreover, $\frac{2}{p}-\frac{1}{2}>\frac{1}{2}-\frac{1}{2n}$, one has $\beta<\frac{2}{p}-\frac{1}{2}$, which implies

$$\left\||\cdot|^\ell\varphi_k^{(n+\ell-1)}\right\|_{L^p(\mathbb{C}^n)} \simeq 2^{\frac{\ell}{2}}k^{\frac{\ell}{2}+\frac{n-1}{p}-\frac{1}{2}+\frac{1}{p}} \simeq 2^{\frac{\ell}{2}}k^{\frac{\ell}{2}+\frac{n}{p}-\frac{1}{2}}.$$

Similarly, when $p=\frac{4n}{2n-1}=\tilde{p}'$, we have $\beta=\frac{2}{p}-\frac{1}{2}$, whence

$$\left\||\cdot|^\ell\varphi_k^{(n+\ell-1)}\right\|_{L^{\tilde{p}'}(\mathbb{C}^n)} \simeq 2^{\frac{\ell}{2}}k^{\frac{\ell}{2}+\frac{n}{\tilde{p}'}-\frac{1}{2}}(\log k)^{\frac{1}{\tilde{p}'}}.$$

Finally, we consider the case $p>\tilde{p}'$. From the earlier discussion it follows that, for $\frac{4n}{2n-1}=\tilde{p}'<p\le 4$, one has $\beta>\frac{2}{p}-\frac{1}{2}$, so that

$$\left\||\cdot|^\ell\varphi_k^{(n+\ell-1)}\right\|_{L^p(\mathbb{C}^n)} \simeq 2^{\frac{\ell}{2}}k^{\frac{\ell}{2}+\frac{n}{p'}-1}.$$

On the other hand, for $4<p$, it is easy to check that $\beta>\frac{4}{3p}-\frac{1}{3}$. Hence,

$$\left\||\cdot|^\ell\varphi_k^{(n+\ell-1)}\right\|_{L^p(\mathbb{C}^n)} \simeq 2^{\frac{\ell}{2}}k^{\frac{\alpha}{2}+\beta-\frac{1}{p}} \simeq 2^{\frac{\ell}{2}}k^{\frac{\ell}{2}+\frac{n-1}{p}+n-1-2\frac{n-1}{p}-\frac{1}{p}} \simeq 2^{\frac{\ell}{2}}k^{\frac{\ell}{2}+\frac{n}{p'}-1}.$$

□

For future convenience, we collect the estimates corresponding to $\ell = 0$ in the following result.

Corollary 1 *One has*

$$\|\varphi_k\|_{L^r(\mathbb{C}^n)} \simeq k^{\frac{n}{r}-\frac{1}{2}} \qquad \text{if } 1 \le r < \tilde{p}', \tag{11}$$

$$\|\varphi_k\|_{L^{\tilde{p}'}(\mathbb{C}^n)} \simeq k^{\frac{n}{\tilde{p}'}-\frac{1}{2}}(\log k)^{\frac{1}{\tilde{p}'}} = k^{\frac{n}{2}-\frac{3}{4}}(\log k)^{\frac{1}{2}-\frac{1}{4n}}, \tag{12}$$

$$\|\varphi_k\|_{L^r(\mathbb{C}^n)} \simeq k^{\frac{n}{r'}-1} = k^{n-\frac{n}{r}-1} \qquad \text{if } \tilde{p}' < r \le \infty, \tag{13}$$

as $k \to \infty$. Here $\tilde{p} = \tilde{p}(2n) = \frac{4n}{2n+1}$ and $\tilde{p}' = \frac{4n}{2n-1}$.

4 The Spectral Decomposition of $\mathcal{L}$

In this section, given a Schwartz function f on $\mathbb{H}_n$, we will write its spectral decomposition with respect to $\mathcal{L}$. This decomposition is due to Strichartz [15].

We start by describing the spectral theory of L_λ (for more details on this subject we refer to [19, Section 2.1 and Section 2.2]). First of all, we need the Laguerre functions of order $n-1$ on $\mathbb{C}^n$, defined in (9), since

$$L_\lambda\, \varphi_k(|\lambda|^{\frac{1}{2}} z) = |\lambda|(2k+n)\, \varphi_k(|\lambda|^{\frac{1}{2}} z).$$

We also introduce the λ-twisted convolution, given for $\lambda > 0$ by

$$f \times_\lambda g(z) = \int_{\mathbb{C}^n} f(z-z')g(z')e^{\frac{i}{2}\lambda \mathrm{Im}(z\cdot\overline{z'})}dz',$$

for any couple of Schwartz functions f, g on $\mathbb{C}^n$. When $\lambda = 1$, we write $\times$ instead of $\times_1$.

It is well-known that

$$f = \sum_{k=0}^{\infty} \Lambda_k^\lambda f, \tag{14}$$

for all Schwartz functions f on $\mathbb{C}^n$, where Λ_k^λ are orthogonal projections in $L^2(\mathbb{C}^n, dz)$, given by

$$\Lambda_k^\lambda f(z) := f \times_\lambda \varphi_k^\lambda(z), \tag{15}$$

with $\varphi_k^\lambda(z) = |\lambda|^n \varphi_k(|\lambda|^{\frac{1}{2}} z)$. Notice that we adopt here the notation from [8], which differs from that in [1, Section 2] by a factor $\lambda^{n/2}$.

Since from

$$L_\lambda(f \times_\lambda g) = f \times_\lambda L_\lambda g$$

we deduce that

$$L_\lambda(\Lambda_k^\lambda g) = |\lambda|(2k+n)\Lambda_k^\lambda g,$$

(14) provides the spectral decomposition of g with respect to L_λ.

Exploiting the homogeneity of the projections, we are reduced to prove estimates for Λ_k^1, which we will denote by Λ_k for short.

Next, given a Schwartz function f on $\mathbb{H}_n$, we spectrally decompose $f^\lambda(z)$ with respect to the twisted laplacian, as in (14), and then take the inverse Fourier transform in the central variable, obtaining

$$f(z,t) = \int_0^\infty \mathcal{P}_\lambda f(z,t) d\lambda, \tag{16}$$

where the integral converges in the sense of distributions and

$$\mathcal{P}_\lambda f(z,t) = \frac{1}{(2\pi)^{n+1}} \sum_{k=0}^{\infty} \frac{1}{2k+n} \left(e^{-i\lambda_k t} \Lambda_k^{\lambda_k} f^{(\lambda_k)}(z) + e^{i\lambda_k t} \Lambda_k^{-\lambda_k} f^{(-\lambda_k)}(z) \right), \tag{17}$$

with $\lambda_k = \frac{\lambda}{2k+n}$. This formula should be compared with [1, (2.11)], where the factor λ^n is now incorporated in $\Lambda_k^{\lambda_k}$ and μ_k is replaced by λ_k. The generalized spectral projections $\mathcal{P}_\lambda$ satisfy (1) and the decomposition (16) gives the spectral resolution of the sublaplacian $\mathcal{L}$.

Our main concern in this chapter consists of discussing some mapping properties of the generalized spectral projections $\mathcal{P}_\lambda$. We will express the bounds in terms of the mixed Lebesgue norms (2).

The estimates are based on the following result, for an earlier version of which we refer to [1, Proposition 2.4].

Proposition 1 *Let $\lambda > 0$ and $1 \leq p \leq q \leq \infty$. Then*

$$\|\mathcal{P}_\lambda f\|_{L_t^\infty L_z^q} \lesssim \lambda^{n(\frac{1}{p}-\frac{1}{q})} \left(\sum_{k=0}^{\infty} (2k+n)^{-1-n\left(\frac{1}{p}-\frac{1}{q}\right)} \|\Lambda_k\|_{L^p(\mathbb{C}^n) \to L^q(\mathbb{C}^n)} \right) \|f\|_{L_t^1 L_z^p}, \tag{18}$$

for all Schwartz functions f on $\mathbb{H}_n$.

Proof We first observe that, in the light of (15), the main estimate in Lemma 2.2 in [1] may be written as

$$\|\Lambda_k^{\lambda} g\|_{L^q(\mathbb{C}^n)} \leq \lambda^{n(\frac{1}{p}-\frac{1}{q})} \|\Lambda_k\|_{L^p(\mathbb{C}^n)\to L^q(\mathbb{C}^n)} \|g\|_{L^p(\mathbb{C}^n)} \tag{19}$$

for all $\lambda > 0$.

From (17), using the triangular inequality, we deduce

$$|\mathcal{P}_\lambda f(z,t)| \lesssim \sum_{k=0}^{\infty} \frac{1}{2k+n} \left(|\Lambda_k^{\lambda_k} f^{(\lambda_k)}(z)| + |\Lambda_k^{-\lambda_k} f^{(-\lambda_k)}(z)| \right),$$

which implies

$$\begin{aligned} &\left(\int_{\mathbb{C}^n} |\mathcal{P}_\lambda f(z,t)|^q dz \right)^{\frac{1}{q}} \\ &\lesssim \sum_{k=0}^{\infty} \frac{1}{2k+n} \left(\left(\int_{\mathbb{C}^n} |\Lambda_k^{\lambda_k} f^{(\lambda_k)}(z)|^q dz \right)^{\frac{1}{q}} + \left(\int_{\mathbb{C}^n} |\Lambda_k^{-\lambda_k} f^{(-\lambda_k)}(z)|^q dz \right)^{\frac{1}{q}} \right) \\ &\qquad \lesssim \sum_{k=0}^{\infty} \frac{1}{2k+n} \left(\|\Lambda_k^{\lambda_k} f^{(\lambda_k)}\|_{L^q(\mathbb{C}^n)} + \|\Lambda_k^{-\lambda_k} f^{(-\lambda_k)}\|_{L^q(\mathbb{C}^n)} \right). \end{aligned}$$

This estimate, combined with (19), yields

$$\begin{aligned} &\left(\int_{\mathbb{C}^n} |\mathcal{P}_\lambda f(z,t)|^q dz \right)^{\frac{1}{q}} \\ &\lesssim \sum_{k=0}^{\infty} \frac{1}{2k+n} \lambda_k^{n(\frac{1}{p}-\frac{1}{q})} \|\Lambda_k\|_{L^p(\mathbb{C}^n)\to L^q(\mathbb{C}^n)} \left(\|f^{(\lambda_k)}\|_{L^p(\mathbb{C}^n)} + \|f^{(-\lambda_k)}\|_{L^p(\mathbb{C}^n)} \right) \\ &\leq \sum_{k=0}^{\infty} \frac{1}{2k+n} \lambda_k^{n(\frac{1}{p}-\frac{1}{q})} \|\Lambda_k\|_{L^p(\mathbb{C}^n)\to L^q(\mathbb{C}^n)} \\ &\quad \times \left(\left(\int \left| \int e^{i\lambda_k t} f(z,t) dt \right|^p dz \right)^{1/p} + \left(\int \left| \int e^{-i\lambda_k t} f(z,t) dt \right|^p dz \right)^{1/p} \right) \end{aligned}$$

$$\leq \sum_{k=0}^{\infty} \frac{1}{2k+n} \lambda_k^{n(\frac{1}{p}-\frac{1}{q})} \|\Lambda_k\|_{L^p(\mathbb{C}^n)\to L^q(\mathbb{C}^n)} \Big(\int \Big(\int |f(z,t)|dt\Big)^p dz\Big)^{1/p}$$

$$= \|f\|_{L_t^1 L_z^p} \sum_{k=0}^{\infty} \frac{1}{2k+n} \lambda_k^{n(\frac{1}{p}-\frac{1}{q})} \|\Lambda_k\|_{L^p(\mathbb{C}^n)\to L^q(\mathbb{C}^n)},$$

from which (18) follows. □

To show that $\mathcal{P}_\lambda$ is bounded from $L_t^1 L_z^p$ to $L_t^\infty L_z^q$, one is therefore reduced to prove that the series in (18) converges; hence, the main concern in the following sections will be to obtain bounds for the norms of the spectral projections of the twisted Laplacian.

Remark 2 Notice that the inequality

$$\|\mathcal{P}_1 f\|_{L_t^\infty L_z^q} \lesssim \|f\|_{L_t^1 L_z^p}$$

holds if and only if the inequality

$$\|\mathcal{P}_1 f\|_{L_t^\infty L_z^{p'}} \lesssim \|f\|_{L_t^1 L_z^{q'}}$$

is satisfied, with $\frac{1}{p} + \frac{1}{p'} = \frac{1}{q} + \frac{1}{q'} = 1$. This follows from Proposition 1 since $\|\Lambda_k\|_{L^p(\mathbb{C}^n)\to L^q(\mathbb{C}^n)} = \|\Lambda_k\|_{L^{q'}(\mathbb{C}^n)\to L^{p'}(\mathbb{C}^n)}$ by duality.

5 Bounding the Spectral Projections of the Twisted Laplacian

5.1 *Koch–Ricci Estimates*

In this section, we begin to discuss the mapping properties of the spectral projection operators $\Lambda_k = \Lambda_k^1$ for the twisted Laplacian, obtaining some bounds from below for their norms.

The main results in the field are the sharp estimates for the norms $\|\Lambda_k\|_{L^p(\mathbb{C}^n)\to L^2(\mathbb{C}^n)}$, $1 \leq p \leq 2$, proved by H. Koch and F. Ricci in [4]. By improving some earlier results in [9, 14], they showed that

$$\|\Lambda_k\|_{L^p(\mathbb{C}^n)\to L^2(\mathbb{C}^n)} = \|\Lambda_k\|_{L^2(\mathbb{C}^n)\to L^{p'}(\mathbb{C}^n)} \simeq (2k+n)^{\gamma(1/p,n)}, \quad 1 \leq p \leq 2, \tag{20}$$

where γ is the piecewise affine function on $[1/2, 1]$ defined by

$$\gamma(1/p, n) := \begin{cases} n\left(\frac{1}{p} - \frac{1}{2}\right) - \frac{1}{2} & \text{if } 1 \leq p \leq p_*, \\ \frac{1}{2}(\frac{1}{2} - \frac{1}{p}) & \text{if } p_* \leq p \leq 2, \end{cases} \tag{21}$$

with critical point $p_* = p_*(2n) := 2\frac{2n+1}{2n+3}$. Observe that $p_*(m)$ coincides with the critical exponent found by E. M. Stein and P. Tomas in the restriction theorem for the Fourier transform in $\mathbb{R}^m$ (see [17]). For simplicity of notation, we shall write $\gamma(1/p)$ instead of $\gamma(1/p, n)$.

Finally, it is worth mentioning that recently, in [3], the authors proved optimal bounds for the $L^p - L^q$ norm of Λ_k when $1 \leq p \leq 2 \leq q \leq \infty$.

5.2 *Necessary Conditions*

We start with a simple lemma, from which, using the classical estimates for the norms of the Laguerre functions collected in Corollary 1, we will deduce a lower bound for the operator norms of Λ_k.

Lemma 2 *Let* $1 \leq p \leq q \leq \infty$ *and* $\frac{1}{p} + \frac{1}{p'} = 1$. *Then*

$$\|\Lambda_k\|_{L^p(\mathbb{C}^n)\to L^q(\mathbb{C}^n)} \gtrsim \frac{k!}{(k+n-1)!}\|\varphi_k\|_{L^{p'}(\mathbb{C}^n)}\|\varphi_k\|_{L^q(\mathbb{C}^n)}. \tag{22}$$

Proof We consider only the case $p > 1$ (the case $p = 1$ is similar and easier).

We begin by recalling how the spectral projections Λ_k act on radial functions. If $g(z) = G(|z|)$ is a radial function, one has

$$\Lambda_k g(z) = (2\pi)^n R_k(g)\varphi_k(z) \tag{23}$$

(see [19, formula (1.4.31)]), where

$$R_k(g) = \frac{2^{1-n}k!}{(k+n-1)!}\int_0^\infty G(s)\varphi_k(s)s^{2n-1}ds.$$

Set

$$G(t) = |\varphi_k(t)|^{\frac{p'}{p}} \operatorname{sgn}\varphi_k(t)$$

and define $g(z) = G(|z|)$. Then $\|g\|_{L^p(\mathbb{C}^n)} = \|\varphi_k\|_{L^{p'}(\mathbb{C}^n)}^{\frac{p'}{p}} \neq 0$ and

$$\begin{aligned} R_k(g) &= \frac{2^{1-n}k!}{(k+n-1)!} \int_0^\infty |\varphi_k(s)|^{\frac{p'}{p}} |\varphi_k(s)| s^{2n-1} ds \\ &= \frac{2^{1-n}k!}{(k+n-1)!} \int_0^\infty |\varphi_k(s)|^{p'} s^{2n-1} ds \\ &= C_n \frac{k!}{(k+n-1)!} \|\varphi_k\|_{L^{p'}(\mathbb{C}^n)}^{p'}. \end{aligned}$$

We deduce therefore from (23)

$$\|\Lambda_k g\|_{L^q(\mathbb{C}^n)} = |R_k(g)| \|\varphi_k\|_{L^q(\mathbb{C}^n)} = C_n \frac{k!}{(k+n-1)!} \|\varphi_k\|_{L^{p'}(\mathbb{C}^n)}^{p'} \|\varphi_k\|_{L^q(\mathbb{C}^n)}.$$

Hence,

$$\begin{aligned} \|\Lambda_k\|_{L^p(\mathbb{C}^n) \to L^q(\mathbb{C}^n)} &\geq \frac{\|\Lambda_k g\|_{L^q(\mathbb{C}^n)}}{\|g\|_{L^p(\mathbb{C}^n)}} \\ &= C_n \frac{k!}{(k+n-1)!} \frac{\|\varphi_k\|_{L^{p'}(\mathbb{C}^n)}^{p'} \|\varphi_k\|_{L^q(\mathbb{C}^n)}}{\|\varphi_k\|_{L^{p'}(\mathbb{C}^n)}^{\frac{p'}{p}}} \\ &= C_n \frac{k!}{(k+n-1)!} \|\varphi_k\|_{L^{p'}(\mathbb{C}^n)} \|\varphi_k\|_{L^q(\mathbb{C}^n)}, \end{aligned}$$

proving the assertion. □

Remark 3 It is worth noticing that the lemma does not always give the sharp estimate. In particular, the bounds proved by Koch and Ricci are better than those provided by the lemma for some values of p. Indeed, from (22) we get

$$\begin{aligned} &\|\Lambda_k\|_{L^p(C^n) \to L^2(\mathbb{C}^n)} \\ &\quad \gtrsim k^{1-n} \|\varphi_k\|_{L^{p'}(\mathbb{C}^n)} \|\varphi_k\|_{L^2(\mathbb{C}^n)} \approx k^{\frac{1}{2}-\frac{n}{2}} \|\varphi_k\|_{L^{p'}(\mathbb{C}^n)} \\ &\quad \approx \begin{cases} k^{\frac{1}{2}-\frac{n}{2}} k^{\frac{n}{p}-1} = k^{n(\frac{1}{p}-\frac{1}{2})-\frac{1}{2}} & \text{for } 1 \leq p < \tilde{p} \iff \tilde{p}' < p' \leq \infty \\ k^{\frac{1}{2}-\frac{n}{2}} k^{\frac{n}{p'}-\frac{1}{2}} = k^{\frac{n}{2}-\frac{n}{p}} & \text{for } \tilde{p} < p \leq 2 \iff 2 \leq p < \tilde{p}'. \end{cases} \end{aligned}$$

Since $p_* < \tilde{p}$, for $1 \le p \le p^*$ the two bounds coincide; for $p^* < p < \tilde{p}$ we have

$$n\left(\frac{1}{p}-\frac{1}{2}\right)-\frac{1}{2}<\frac{1}{2}\left(\frac{1}{2}-\frac{1}{p}\right),$$

and for $\tilde{p} < p \le 2$

$$n\left(\frac{1}{2}-\frac{1}{p}\right)\le\frac{1}{2}\left(\frac{1}{2}-\frac{1}{p}\right).$$

As a consequence of Lemma 2, one obtains a lower bound for the norms of the projections Λ_k. For the sake of simplicity, we will only discuss the diagonal case.

Let

$$\Phi(k;1/p,n)=\begin{cases} k^{2n\left(\frac{1}{p}-\frac{1}{2}\right)-\frac{1}{2}} & \text{if } 1\le p<\tilde{p},\\ (\log k)^{\frac{1}{\tilde{p}'}} & \text{if } p=\tilde{p},\\ 1 & \text{if } \tilde{p}<p\le 2 \end{cases}$$
$$=\begin{cases} k^{\alpha(1/p)} & \text{if } 1\le p\le 2 \text{ and } p\ne\tilde{p},\\ (\log k)^{\frac{1}{\tilde{p}'}} & \text{if } p=\tilde{p}, \end{cases} \tag{24}$$

where α is the piecewise affine function on $[\frac{1}{2},1]$ defined by

$$\alpha(1/p,n):=\begin{cases} 2n\left(\frac{1}{p}-\frac{1}{2}\right)-\frac{1}{2} & \text{if } 1\le p<\tilde{p},\\ 0 & \text{if } \tilde{p}<p\le 2. \end{cases} \tag{25}$$

To simplify notation, we shall write $\alpha(1/p)$ instead of $\alpha(1/p,n)$.

Proposition 2 *For k sufficiently large, we have*

$$\|\Lambda_k\|_{L^p(\mathbb{C}^n)\to L^p(\mathbb{C}^n)} \gtrsim \Phi(k;1/p,n), \qquad 1\le p\le 2,$$

where Φ is given by (24).

The bounds in Proposition 2 are attained in the range $1 \le p \le p_*$, as the following Proposition shows.

Proposition 3 *For $1 \le p \le q \le 2$, we have*

$$\|\Lambda_k\|_{L^p(\mathbb{C}^n)\to L^q(\mathbb{C}^n)} \lesssim \begin{cases} (2k+n)^{\frac{n}{p}+\frac{n}{q}-n-\frac{1}{2}} & \text{if } 1\le p\le p_*,\\ (2k+n)^{n\left(\frac{1}{q}-\frac{1}{2}\right)+\frac{1}{2}\left(\frac{1}{2}-\frac{1}{p}\right)} & \text{if } p_*\le p\le 2, \end{cases} \tag{26}$$

for sufficiently large k.

Remark 4 In particular, if $1 \le p \le q \le p_*$, since $p_* < \tilde{p}$ one can write

$$(2k+n)^{\frac{n}{p}+\frac{n}{q}-n-\frac{1}{2}} = (2k+n)^{\frac{1}{2}[\alpha(\frac{1}{p})+\alpha(\frac{1}{q})]},$$

with α defined by (25).

Proof Recall that there are constants $C, \delta > 0$ independent of k such that

$$|\varphi_k(z)| \le C e^{-\delta |z|^2}$$

for $|z| \ge 4\sqrt{k}$ (this follows, for instance, from [14], by combining the fourth line at page 96 with the last bound in (2.3) therein).

Let χ_k denote a smooth function with compact support that is identically 1 on the ball

$$B_k = \{z \in \mathbb{C}^n : |z| \le 12\sqrt{k}\}$$

and vanishes for $|z| \ge 15\sqrt{k}$. For any Schwartz function f we write $\Lambda_k f$ as

$$\Lambda_k f = f \times \big((1-\chi_k)\varphi_k\big) + f \times (\varphi_k \chi_k). \tag{27}$$

The first addendum in (27) may be controlled by means of Young's inequality. Indeed, for any $1 \le p \le q \le 2$, let r be such that $1/r = 1/q - 1/p + 1$. Then

$$\begin{aligned}
\|f \times \big((1-\chi_k)\varphi_k\big)\|_{L^q(\mathbb{C}^n)} &\le \left(\int \Big(\big(|f| * \big|(1-\chi_k)\varphi_k\big|\big)(z)\Big)^q dz\right)^{\frac{1}{q}} \\
&= \big\| |f| * \big|(1-\chi_k)\varphi_k\big| \big\|_{L^q(\mathbb{C}^n)} \\
&\le \|f\|_{L^p(\mathbb{C}^n)} \|(1-\chi_k)\varphi_k\|_{L^r(\mathbb{C}^n)} \\
&\simeq \|f\|_{L^p(\mathbb{C}^n)} \|\varphi_k\|_{L^r(\mathbb{C}^n \setminus B_k)} \\
&\lesssim \|f\|_{L^p(\mathbb{C}^n)} \|e^{-\delta|\cdot|^2}\|_{L^r(\mathbb{C}^n \setminus B_k)} \\
&\lesssim \|f\|_{L^p(\mathbb{C}^n)} e^{-\delta' k} \|e^{-\delta''|\cdot|^2}\|_{L^r(\mathbb{C}^n)} \\
&\lesssim e^{-\delta' k} \|f\|_{L^p(\mathbb{C}^n)},
\end{aligned} \tag{28}$$

for some positive δ', δ''.

Then we consider the latter term in (27), that is, $\tilde{\Lambda}_k f := f \times (\varphi_k \chi_k)$.

In light of (28), to prove the assertion it suffices to control the norm of $\tilde{\Lambda}_k$. This operator is local in the sense that, if g is compactly supported, then the support of $\tilde{\Lambda}_k g$ always remains within $15\sqrt{k}$ of the support of g. Since $\tilde{\Lambda}_k$ is also invariant under twisted translations, by a well-known localization principle (see [16, Lecture 3, Lemma 1.6]) it suffices to show that $\bar{\chi}_k \tilde{\Lambda}_k$ satisfies the estimates as in the statement for a function $\bar{\chi}_k$, which is 1 on the support of χ_k and has support contained in the ball of radius, say, $300\sqrt{k}$. To prove this we first apply Hölder's inequality with exponents $2/q$ and $2/(2-q)$, obtaining

$$
\begin{aligned}
\|\bar{\chi}_k \tilde{\Lambda}_k f\|_{L^q(\mathbb{C}^n)} &= \left(\int \left|\bar{\chi}_k \tilde{\Lambda}_k f(z)\right|^q dz\right)^{\frac{1}{q}} \\
&\le \left(\int \left|\tilde{\Lambda}_k f(z)\right|^2 dz\right)^{\frac{1}{2}} \left(\int \bar{\chi}_k(z) dz\right)^{\frac{1}{q}-\frac{1}{2}} \\
&\lesssim k^{n\left(\frac{1}{q}-\frac{1}{2}\right)} \|\tilde{\Lambda}_k f\|_{L^2(\mathbb{C}^n)}.
\end{aligned}
$$

Next,

$$
\begin{aligned}
\|\tilde{\Lambda}_k f\|_{L^2(\mathbb{C}^n)} &= \|f \times (\varphi_k \chi_k)\|_{L^2(\mathbb{C}^n)} \\
&\lesssim \|f \times \varphi_k\|_{L^2(\mathbb{C}^n)} + \|f \times (\varphi_k(1-\chi_k))\|_{L^2(\mathbb{C}^n)} \\
&\simeq \|\Lambda_k f\|_{L^2(\mathbb{C}^n)} + \|f \times (\varphi_k(1-\chi_k))\|_{L^2(\mathbb{C}^n)} \\
&\lesssim k^{\gamma(1/p)} \|f\|_{L^p(\mathbb{C}^n)} + \| |f| * |(\varphi_k(1-\chi_k))| \|_{L^2(\mathbb{C}^n)},
\end{aligned}
$$

where the last estimate follows from (20).

Then (28), applied with $q = 2$, yields

$$
\| |f| * |(\varphi_k(1-\chi_k))| \|_{L^2(\mathbb{C}^n)} \lesssim e^{-\delta' k} \|f\|_{L^p(\mathbb{C}^n)},
$$

whence

$$
\begin{aligned}
\|\tilde{\Lambda}_k f\|_{L^2(\mathbb{C}^n)} &\lesssim k^{\gamma(1/p)} \|f\|_{L^p(\mathbb{C}^n)} + e^{-\delta' k} \|f\|_{L^p(\mathbb{C}^n)} \\
&\lesssim k^{\gamma(1/p)} \|f\|_{L^p(\mathbb{C}^n)}
\end{aligned} \tag{29}
$$

Finally, combining (27), (28) and (29) yields

$$
\|\Lambda_k f\|_{L^q(\mathbb{C}^n)} \le k^{n\left(\frac{1}{q}-\frac{1}{2}\right)+\gamma\left(\frac{1}{p}\right)} \|f\|_{L^p(\mathbb{C}^n)},
$$

proving (26). □

In the light of Proposition 3 (in the range $1 \le p < p_*$) and of the one-dimensional results, that will appear in a forthcoming paper, we are led to formulate the following conjecture.

Conjecture 2 The bounds

$$
\|\Lambda_k\|_{L^p(\mathbb{C}^n) \to L^p(\mathbb{C}^n)} \simeq \Phi(k; 1/p, n) \tag{30}
$$

are attained for $1 \le p \le 2$ and all $n \in \mathbb{N}$.

6 An Example Involving Bigraded Spherical Harmonics

In this section we shall show that the key estimate

$$\|\Lambda_k\|_{L^{\tilde{p}}(\mathbb{C}^n)\to L^{\tilde{p}}(\mathbb{C}^n)} \simeq \Phi(k; 1/\tilde{p}, n), \tag{31}$$

where $\Phi(k; 1/\tilde{p}, n) = (\log k)^{\frac{1}{2}-\frac{1}{4n}}$, conjectured in (30), holds for functions given by the product of a radial function and a certain bigraded spherical harmonic, providing further evidence to support the estimates in Conjecture 2. Notice that the lower bound in (31) is guaranteed by Proposition 2.

In order to prove the upper bound, we use a Hecke–Bochner identity for the spectral projection of the twisted Laplacian. This identity was proved by D. Geller in [2] (see also [20, Section 2.6]), and shows that, when f is the product of a radial function and a bigraded spherical harmonic P, then $\Lambda_k f$ is given by a radial function times the same P.

Recall that a solid bigraded spherical harmonic of bidegree (ℓ, ℓ'), $\ell, \ell' = 0, 1, 2, \cdots$, is a polynomial P on $\mathbb{C}^n$, which is homogeneous of degree ℓ in z and of degree ℓ' in $\bar{z}$ and is annihilated by the Laplacian $\Delta = 4\sum_{j=1}^n \frac{\partial^2}{\partial_{z_j}\partial_{\bar{z}_j}}$ (for more details see [20, Section 2.5]). We denote by $\mathcal{H}_{\ell,\ell'}$ the space of the restrictions to S^{2n-1} of solid spherical harmonics of bidegree (ℓ, ℓ'). This space equipped with the inner product inherited from $L^2(\mathbb{C}^n)$ becomes a Hilbert space of dimension

$$d_{\ell,\ell'} = \binom{\ell+n-1}{n-1}\binom{\ell'+n-1}{n-1} - \binom{\ell+n-2}{n-1}\binom{\ell'+n-2}{n-1}. \tag{32}$$

In the proof of the upper bound, it will be crucial to estimate the Lebesgue norm of some weighted Laguerre functions of the form $|\cdot|^{\ell+\ell'}\varphi_{k-\ell}^{(n+\ell+\ell'-1)}$; this will be done by means of Lemma 1. We fix, from now on, two positive integers ℓ and ℓ'.

From (10) we readily deduce, as $k \to +\infty$, the following result.

Lemma 3 *Let $p \geq 1$. For some fixed $\ell, \ell' \in \mathbb{N}$, when $k \to +\infty$ one has*

$$2^{-\ell-\ell'}k^{1-n-\ell-\ell'} \left\| |\cdot|^{\ell+\ell'}\varphi_{k-\ell}^{(n+\ell+\ell'-1)} \right\|_{L^p(\mathbb{C}^n)} \left\| |\cdot|^{\ell+\ell'}\varphi_{k-\ell}^{(n+\ell+\ell'-1)} \right\|_{L^{p'}(\mathbb{C}^n)} \lesssim \Phi(k; 1/p, n). \tag{33}$$

Proof Let $1 \leq p < \tilde{p}'$. Set

$$\mathcal{N} := \left\| |\cdot|^{\ell+\ell'}\varphi_{k-\ell}^{(n+\ell+\ell'-1)} \right\|_{L^p(\mathbb{C}^n)} \left\| |\cdot|^{\ell+\ell'}\varphi_{k-\ell}^{(n+\ell+\ell'-1)} \right\|_{L^{p'}(\mathbb{C}^n)}.$$

Then

$$\mathcal{N} \lesssim 2^{\frac{\ell+\ell'}{2}}(k-\ell)^{\frac{\ell+\ell'}{2}+\frac{n}{p}-\frac{1}{2}} \times \Psi(\ell, \ell', k),$$

where

$$\Psi(\ell,\ell',k)=\begin{cases}2^{\frac{\ell+\ell'}{2}}(k-\ell)^{\frac{\ell+\ell'}{2}+\frac{n}{p'}-\frac{1}{2}} & \text{if } p'<\tilde{p}'\\ 2^{\frac{\ell+\ell'}{2}}(k-\ell)^{\frac{\ell+\ell'}{2}+\frac{n}{p}-1} & \text{if } p'>\tilde{p}'.\end{cases}$$

Straightforward computations lead to

$$\mathcal{N}\lesssim\begin{cases}2^{(\ell+\ell')}(k-\ell)^{(\ell+\ell')+n-1} & \text{if } p'<\tilde{p}'\\ 2^{(\ell+\ell')}(k-\ell)^{(\ell+\ell')+\frac{2n}{p}-\frac{3}{2}} & \text{if } p'>\tilde{p}',\end{cases}$$

whence

$$2^{-(\ell+\ell')}(k-\ell)^{-(\ell+\ell')-n+1}\mathcal{N}\lesssim\begin{cases}1 & \text{if } 2\geq p>\tilde{p}\\ (k-\ell)^{2n(\frac{1}{p}-\frac{1}{2})-\frac{1}{2}} & \text{if } 1\leq p<\tilde{p},\end{cases},$$

that is, one obtains (33), since $k-\ell\simeq k$.

Let now $p=\tilde{p}'$. Then

$$\begin{aligned}\mathcal{N}&\lesssim 2^{\frac{(\ell+\ell')}{2}}k^{\frac{\ell+\ell'}{2}+\frac{n}{\tilde{p}'}-\frac{1}{2}}(\log k)^{\frac{1}{\tilde{p}'}}2^{\frac{(\ell+\ell')}{2}}k^{\frac{\ell+\ell'}{2}+\frac{n}{\tilde{p}}-\frac{1}{2}}\\ &\simeq 2^{(\ell+\ell')}k^{(\ell+\ell')+n-1}(\log k)^{\frac{1}{\tilde{p}'}},\end{aligned}$$

and this implies (33).

Finally, for $p>\tilde{p}'$, we have $p'<\tilde{p}<\tilde{p}'$ and the estimate is easier, so we omit the computations. □

Now, according to the Bochner–Hecke formula, if f is a Schwartz function on $\mathbb{C}^n$ of the form $f=gP$, where $g(z)=G(|z|)$ is radial and P is a bigraded spherical harmonic in $\mathcal{H}_{\ell,\ell'}$, then

$$\begin{aligned}\Lambda_k f(z)&=f\times\varphi_k^{(n-1)}(z)\\ &=\frac{1}{2^{n+\ell+\ell'-1}}\frac{\Gamma(k-\ell+1)}{\Gamma(k+\ell'+n)}P\left(\frac{z}{|z|}\right)|z|^{l+l'}\varphi_{k-\ell}^{(n+\ell+\ell'-1)}(z)\\ &\quad\times\left(\int_0^\infty G(s)\varphi_{k-\ell}^{(n+\ell+\ell'-1)}(s)\,s^{2(n+\ell+\ell')-1}ds\right)\end{aligned}\tag{34}$$

if $k\geq\ell$ (see [20, p.73]); here we abuse notation writing $\varphi_{k-\ell}^{(n+\ell+\ell'-1)}(s)$ for $L_{k-\ell}^{n+\ell+\ell'-1}(s^2/2)\,e^{-s^2/4}$.

Proposition 4 *Let $f = gP$ be a Schwartz function on $\mathbb{C}^n$, where $g(z)$ is radial and P is a spherical harmonic of bidegree (ℓ, ℓ'), for some fixed $\ell, \ell' \in \mathbb{N}$. Then for k sufficiently large one has*

$$\|\Lambda_k f\|_{L^p(\mathbb{C}^n)} \lesssim \Phi\,(k;\,1/p, n)\,\|f\|_{L^p(\mathbb{C}^n)}, \quad 1 \le p \le 2, \tag{35}$$

where Φ is defined by (24). *The implicit constants in this formula are independent of P.*

Proof Let $g(z) = G(|z|)$. Observe that

$$\|f\|_{L^p(\mathbb{C}^n)} \simeq \|P\|_{L^p(S^{2n-1})} \left(\int_0^\infty \left| G(r) r^{\ell+\ell'} \right|^p r^{2n-1} dr \right)^{\frac{1}{p}}.$$

Using Holder's inequality, we deduce from (34)

$$\begin{aligned}
\|\Lambda_k f\|_{L^p(\mathbb{C}^n)} &\lesssim \frac{(k-\ell)^{1-n-\ell-\ell'}}{2^{\ell+\ell'}} \|P\|_{L^p(S^{2n-1})} \left\| |\cdot|^{\ell+\ell'} \varphi_{k-\ell}^{(n+\ell+\ell'-1)} \right\|_{L^p(\mathbb{C}^n)} \\
&\quad \times \left| \int_0^\infty G(s) s^{\ell+\ell'} \varphi_{k-\ell}^{(n+\ell+\ell'-1)}(s)\, s^{\ell+\ell'}\, s^{2n-1} ds \right| \\
&\lesssim \frac{(k-\ell)^{1-n-\ell-\ell'}}{2^{\ell+\ell'}} \|P\|_{L^p(S^{2n-1})} \left\| |\cdot|^{\ell+\ell'} \varphi_{k-\ell}^{(n+\ell+\ell'-1)} \right\|_{L^p(\mathbb{C}^n)} \\
&\quad \times \left\| |\cdot|^{\ell+\ell'} \varphi_{k-\ell}^{(n+\ell+\ell'-1)} \right\|_{L^{p'}(\mathbb{C}^n)} \left(\int_0^\infty \left| G(r) r^{\ell+\ell'} \right|^p r^{2n-1} dr \right)^{\frac{1}{p}} \\
&\lesssim \frac{(k-\ell)^{1-n-\ell-\ell'}}{2^{\ell+\ell'}} \\
&\quad \times \left\| |\cdot|^{\ell+\ell'} \varphi_{k-\ell}^{(n+\ell+\ell'-1)} \right\|_{L^p(\mathbb{C}^n)} \left\| |\cdot|^{\ell+\ell'} \varphi_{k-\ell}^{(n+\ell+\ell'-1)} \right\|_{L^{p'}(\mathbb{C}^n)} \|f\|_{L^p(\mathbb{C}^n)} \\
&\lesssim \frac{k^{1-n-\ell-\ell'}}{2^{\ell+\ell'}} \left\| |\cdot|^{\ell+\ell'} \varphi_{k-\ell}^{(n+\ell+\ell'-1)} \right\|_{L^p(\mathbb{C}^n)} \left\| |\cdot|^{\ell+\ell'} \varphi_{k-\ell}^{(n+\ell+\ell'-1)} \right\|_{L^{p'}(\mathbb{C}^n)} \|f\|_{L^p(\mathbb{C}^n)} \\
&\lesssim \Phi(k;\,1/p, n) \|f\|_{L^p(\mathbb{C}^n)},
\end{aligned}$$

where we used the fact that $k - \ell \simeq k$ (recall that ℓ is fixed and $k \to +\infty$). □

7 Bounding the Spectral Projections of the Sublaplacian

We first deduce some estimates for the norm of $\mathcal{P}_\lambda$ from the bounds for the norm of Λ_k obtained in Sect. 5.

In fact, with reference to Fig. 1, Proposition 3 yields the boundedness of $\mathcal{P}_\lambda$ on the open rectangle $AEMF$ (and on the open edges MF and AE).

Proposition 5 *For $1 \leq p \leq p_*$ and $q > \tilde{p}$ we have*

$$\|\mathcal{P}_\lambda f\|_{L_t^\infty L_z^q} \lesssim_{n,\lambda} \|f\|_{L_t^1 L_z^p}.$$

Proof By plugging the first of (26) in (18), one gets at the right-hand side the series

$$\sum_{k=0}^{n} (2k+n)^{-\frac{3}{2}+\frac{2n}{q}-n},$$

and this series is convergent if $q > \tilde{p}$. □

Anyway, the proof of the boundedness of $\mathcal{P}_\lambda$ on the entire blue and gray regions in Fig. 1 seems to be much more difficult. At the moment, we are able to prove that estimates conjectured in (30) would imply that the operators $\mathcal{P}_\lambda$ are bounded on the rectangle $AECG$ (including the vertical edges AE and GC, with the exception of the vertices C and E) and on the trapezoid $HGCB$ (including the open edge HB and the vertex H).

Proposition 6 *If Conjecture 2 is true, then the operators $\mathcal{P}_\lambda$ are bounded from $L_t^1 L_z^p$ to $L_t^\infty L_z^q$ for $1 \leq p \leq \tilde{p}$ and $q > \tilde{p} = \frac{4n}{2n+1}$ (that is, in $AECG$), and for $\tilde{p} < p \leq 2$ and $p < q$ (that is, in $HGCB$).*

Proof We start by noticing that, since the boundedness of $\mathcal{P}_\lambda$ on the open rectangle $AEMF$ (and on the open edges MF and AE) has already been proved in Proposition 5, it is enough to consider the smaller rectangle $GFMC$ (including the vertical edge GC, with the exception of the vertex C) instead of the rectangle $AECG$.

By hypothesis,

$$\|\Lambda_k\|_{L^r(\mathbb{C}^n)\to L^r(\mathbb{C}^n)} \lesssim k^{\alpha(1/r)}, \quad 1 \leq r \leq 2, \quad r \neq \tilde{p}, \tag{36}$$

where α was defined in (25).

From this, interpolating between (36) for some $p_* \leq r < \tilde{p}$ and the Koch-Ricci estimate (L^{p_*}, L^2)

$$\|\Lambda_k\|_{L^{p_*}(\mathbb{C}^n)\to L^2(\mathbb{C}^n)} \lesssim k^{\frac{1}{2}\left(\frac{1}{2}-\frac{1}{p_*}\right)},$$

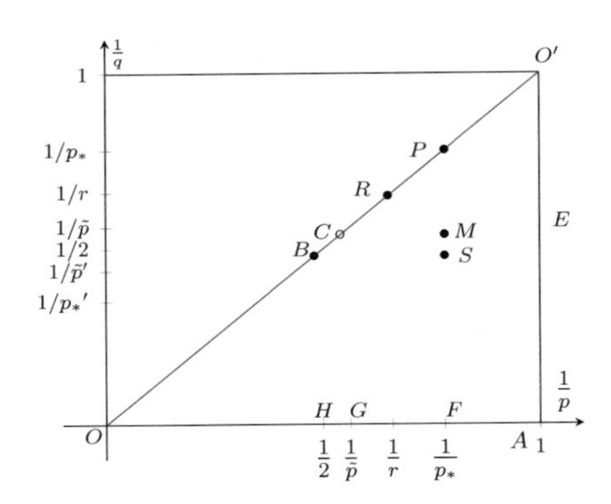

Fig. 2 In this picture $p_* \leq r < \tilde{p}$, $R = (1/r, 1/r)$, $S = (1/p_*, 1/2)$, $P = (1/p_*, 1/p_*)$, $M = (1/p_*, 1/\tilde{p})$, $C = (1/\tilde{p}, 1/\tilde{p})$, $B = (1/2, 1/2)$

that is, interpolating between the points R and S in Fig. 2, we get a bound for $\|\Lambda_k\|_{L^p(\mathbb{C}^n)\to L^q(\mathbb{C}^n)}$, with $p_* \leq p < \tilde{p}$ and $r \leq q \leq 2$ (notice that we shall use this estimate only for $q > \tilde{p}$). Indeed, setting

$$\frac{1}{p} = \frac{\theta}{r} + \frac{1-\theta}{p_*} \quad \text{and} \quad \frac{1}{q} = \frac{\theta}{r} + \frac{1-\theta}{2}, \tag{37}$$

if $p_* \leq p \leq r < \tilde{p}$ we obtain

$$\begin{aligned} \|\Lambda_k\|_{L^p(\mathbb{C}^n)\to L^q(\mathbb{C}^n)} &\lesssim k^{\theta\alpha(1/r)+(1-\theta)\frac{1}{2}\left(\frac{1}{2}-\frac{1}{p_*}\right)} \\ &= k^{-\frac{\theta}{2}+2n\left(\frac{1}{q}-\frac{1}{2}\right)+\frac{1}{2}(1-\theta)\left(\frac{1}{2}-\frac{1}{p_*}\right)}, \end{aligned} \tag{38}$$

where we used also the fact that

$$\theta\left(\frac{1}{r}-\frac{1}{2}\right) = \frac{1}{q} - \frac{1}{2}.$$

Taking the difference of the relations (37), we get

$$\frac{1}{q} - \frac{1}{p} = (1-\theta)\left(\frac{1}{2} - \frac{1}{p_*}\right) = -\frac{1-\theta}{2n+1},$$

yielding

$$\theta = (2n+1)\left(\frac{1}{q} - \frac{1}{p}\right) + 1.$$

Plugging these expressions in (38) we deduce

$$\|\Lambda_k\|_{L^p(\mathbb{C}^n)\to L^q(\mathbb{C}^n)} \lesssim k^{\frac{n}{p}+\frac{n}{q}-\frac{1}{2}-n}, \tag{39}$$

for $p_* \leq p < \tilde{p}$ and $p \leq q$. Let T_1 be the triangle with vertices $S = (1/p_*, 1/2)$, $C = (1/\tilde{p}, 1/\tilde{p})$ and $P = (1/p_*, 1/p_*)$. Then the estimate (39) holds in $T_1 \setminus \{C\}$.

Denoting by T_2 the triangle with vertices $B = (1/2, 1/2)$, $S = (1/p_*, 1/2)$ and $C = (1/\tilde{p}, 1/\tilde{p})$, by similar computations we obtain

$$\|\Lambda_k\|_{L^p(\mathbb{C}^n)\to L^q(\mathbb{C}^n)} \lesssim k^{\frac{1}{2}(1-\theta)\left(\frac{1}{2}-\frac{1}{p_*}\right)} = k^{\frac{1}{2}\left(\frac{1}{q}-\frac{1}{p}\right)} \tag{40}$$

in $T_2 \setminus \{C\}$.

Collecting together (39) and (40), we may write

$$\|\Lambda_k\|_{L^p(\mathbb{C}^n)\to L^q(\mathbb{C}^n)} \lesssim k^{\beta(\frac{1}{p},\frac{1}{q})}, \qquad p_* \leq p \leq 2,\ p \neq \tilde{p},\ p \leq q \leq 2,$$

where β is the piecewise affine function defined by

$$\beta\left(\frac{1}{p}, \frac{1}{q}\right) := \begin{cases} \frac{n}{p} + \frac{n}{q} - \frac{1}{2} - n & \text{if } (p,q) \in T_1 \setminus \{C\}, \\ \frac{1}{2}\left(\frac{1}{q} - \frac{1}{p}\right) & \text{if } (p,q) \in T_2 \setminus \{C\}. \end{cases}$$

According to Proposition 1, the operators $\mathcal{P}_\lambda$ are bounded from $L^p(\mathbb{H}_n)$ to $L^q(\mathbb{H}_n)$ if the series

$$\sum_{k=0}^{\infty} (2k+n)^{-1-n\left(\frac{1}{p}-\frac{1}{q}\right)} \|\Lambda_k\|_{L^p(\mathbb{C}^n)\to L^q(\mathbb{C}^n)} \lesssim \sum_{k=1}^{\infty} k^{-1+\frac{n}{q}-\frac{n}{p}+\beta\left(\frac{1}{p},\frac{1}{q}\right)}$$

converges. In the following, we shall prove that this is indeed the case.

First, in order to prove the boundedness of $\mathcal{P}_\lambda$ for $p_* \leq p < \tilde{p}$ and $q > \tilde{p} = \frac{4n}{2n+1}$, we consider the segment $CM \cup \{M\}$, which is contained in T_1. Along this segment one has

$$\begin{aligned} -1 + \frac{n}{q} - \frac{n}{p} + \beta\left(\frac{1}{p}, \frac{1}{q}\right) &= \frac{2n}{q} - \frac{3}{2} - n \\ &< \frac{2n+1}{2} - \frac{3}{2} - n = -1. \end{aligned}$$

Thus, the the operators $\mathcal{P}_\lambda$ are bounded along $CM \cup \{M\}$, and the boundedness on $GFMC$ follows by interpolating with the segment GF.

Then, in order to prove the boundedness of $\mathcal{P}_\lambda$ for $\tilde{p} < p \leq 2$ and $p < q$, we consider the segment BS, which is contained in T_2. Along this segment one has

$$\begin{aligned} -1 + \frac{n}{q} - \frac{n}{p} + \beta\left(\frac{1}{p}, \frac{1}{q}\right) &= -1 + \frac{n}{q} - \frac{n}{p} + \frac{1}{2}\left(\frac{1}{q} - \frac{1}{p}\right) \\ &= -1 + \frac{2n+1}{2}\left(\frac{1}{q} - \frac{1}{p}\right) < -1, \end{aligned}$$

that is, the operators $\mathcal{P}_\lambda$ are bounded along BS. By interpolating with the segments HG and $CS \setminus \{C\}$, one finally proves the boundedness on $HGCB$. □

From Proposition 6 and Lemma 4, one obtains the following result.

Corollary 2 *Let gP be a Schwartz function on $\mathbb{C}^n$, where $g(z)$ is radial and P is a solid spherical harmonics of bidegree (ℓ, ℓ') and let $f(z,t) = e^{i\lambda t} g(z)P(z)$. Then*

$$\|\mathcal{P}_\lambda f\|_{L_t^\infty L_z^q(\mathbb{C}^n)} \lesssim \lambda^{\frac{n}{p}-\frac{n}{q}} \|f\|_{L_t^1 L_z^p(\mathbb{C}^n)}$$

for $1 \le p \le \tilde{p}$ and $\tilde{p} = \frac{4n}{2n+1} < q$ or $\tilde{p} < p \le 2$ and $p < q$.

Acknowledgments The authors would like to thank the anonymous referees for a number of very useful suggestions to improve the presentation. The authors are members of Gruppo Nazionale per l'Analisi Matematica, la Probabilità e le loro Applicazioni (GNAMPA) of Istituto Nazionale di Alta Matematica (INdAM). The authors were partially supported by GNAMPA (Project 2022 "Temi di Analisi Armonica Subellittica")

References

1. Casarino, V., Ciatti, P.: A restriction theorem for Métivier groups. Adv. Math. **245**, 52–77 (2013)
2. Geller, D.: Spherical harmonics, the Weyl transform and the Fourier transform on the Heisenberg group. Can. J. Math. **36**, 615–684 (1984)
3. Jeong, E., Lee, S., Ryu, J.: Sharp $L^p - L^q$ estimate for the spectral projection associated with the twisted Laplacian. Publ. Mat. **66**(2), 831–855 (2022)
4. Koch, H., Ricci, F.: Spectral projections for the twisted Laplacian, Stud. Math. **180**, 103–110 (2007)
5. Markett, C.: Mean Cesàro summability of Laguerre expansions and norms estimates with shifted parameter. Anal. Math. **8**, 19–37 (1982)
6. Müller, D.: A restriction theorem for the Heisenberg group. Ann. Math. **131**, 567–587 (1990)
7. Müller, D., Stein, E.M.: On spectral multipliers for Heisenberg and related groups. J. Math. Pures Appl. **9**(73), 413–440 (1994)
8. Niedorf, L.: An L^p-spectral multiplier theorem with sharp p-specific regularity bound on Heisenberg type groups. J. Fourier Anal. Appl. **30**, 22 (2024)
9. Ratnakumar, P.K., Rawat, R., Thangavelu, S.: A restriction theorem for the Heisenberg motion group. Stud. Math. **126**, 1–12 (1997)
10. Seeger, A., Sogge, C.D.: Bounds for eigenfunctions of differential operators. Indiana Univ. Math. J. **38**(3), 669–682 (1989)
11. Sogge, C.D.: Oscillatory integrals and spherical harmonics. Duke Math. J. **53**(1), 43–65 (1986)
12. Sogge, C.D.: Fourier Integrals in Classical Analysis. Cambridge Tracts in Mathematics, vol. 105. Cambridge University Press, Cambridge (1993)
13. Stein, E.M.: Oscillatory integrals in Fourier analysis. In: Beijing Lectures in Harmonic Analysis, vol. 112. Princeton University Press, Princeton (1986)
14. Stempak, K., Zienkiewicz, J.: Twisted convolution and Riesz means. J. Anal. Math. **76**, 93–107 (1998)
15. Strichartz, R.: Harmonic analysis as spectral theory of Laplacians. J. Funct. Anal. **87**(1), 51–148 (1989)

16. Tao, T.: Restriction theorems and applications, Math 254B, Spring 1999. https://www.math.ucla.edu/~tao/254b.1.99s/
17. Tomas, P.: A restriction theorem for the Fourier transform. Bull. Am. Math. Soc. **81**, 477–478 (1975)
18. Thangavelu, S.: Restriction theorems for the Heisenberg group. J. reine angew. Math. **414**, 51–65 (1991)
19. Thangavelu, S.: Harmonic Analysis on the Heisenberg Group. Progress in Mathematics, vol. 105. Birkhäuser, Boston (1998)
20. Thangavelu, S.: An Introduction to the Uncertainty Principle, 105. Progress in Mathematics, vol. 217. Birkhäuser, Boston (2004)

Four Variations on the Rankin–Cohen Brackets

Jean-Louis Clerc

To Toshiyuki Kobayashi, on the occasion of his 60th birthday, for his commitment to the mathematical community

Abstract After recalling the original definition of the Rankin–Cohen brackets, four different constructions are presented, using different approaches corresponding to various possible generalizations to other geometric contexts.

1 Introduction

Let $\Pi = \{z = x + iy, y > 0\}$ be the complex upper half-plane, Let $g = \begin{pmatrix} a & b \\ c & d \end{pmatrix}$, $ad - bc = 1$ be an element of the group $G = SL(2, \mathbb{R})$. The formula

$$g(z) = \frac{az + b}{cz + d}$$

defines an action of G on Π, and in fact G is a twofold covering of the group $PSL(2, \mathbb{R})$ of holomorphic diffeomorphisms of Π. Introduce the notation

$$c(g, z) = cz + d\ .$$

The following *cocycle identity*

$$g_1, g_2 \in G, z \in \Pi, \qquad c(g_1 g_2, z) = c\big(g_1, g_2(z)\big)\, c(g_2, z) \tag{1}$$

J.-L. Clerc (✉)
Institut Elie Cartan, Université de Lorraine, CNRS, Vandœuvre-lès-Nancy, France
e-mail: jean-louis.clerc@univ-lorraine.fr

M. Pevzner, H. Sekiguchi (eds.), *Symmetry in Geometry and Analysis, Volume 2*,
Progress in Mathematics 358, https://doi.org/10.1007/978-981-97-7662-7_3

is easily verified by a direct computation. Also notice for further reference the following result, for $g \in G$ and $z \in \Pi$

$$\frac{dg(z)}{dz} = j(g, z) = c(g, z)^{-2} . \tag{2}$$

Let ℓ be a nonnegative integer. Then for $g \in G$ and F, a holomorphic function on Π, the formula

$$\rho_\ell(g)F(z) = c(g^{-1}, z)^{-\ell} F(g^{-1}(z)) \tag{3}$$

defines a representation ρ_ℓ of G on the space $\mathcal{H}(\Pi)$ of holomorphic functions on Π. A more explicit expression is given, for $g \in G$ such that $g^{-1} = \begin{pmatrix} a & b \\ c & d \end{pmatrix}$, by

$$\rho_\ell(g)f\,(z) = (cz + d)^{-\ell} f\left(\frac{az + b}{cz + d}\right) . \tag{4}$$

Let ℓ, m be two nonnegative integers, and let $\mathcal{H}(\Pi \times \Pi)$ be the space of holomorphic functions on the product space $\Pi \times \Pi$. Now the (completion of the) tensor product representation $\rho_\ell \otimes \rho_m$ can be realized on $\mathcal{H}(\Pi \times \Pi)$ by the formula

$$\rho_\ell \otimes \rho_m)(g)F(z, w) = c(g^{-1}, z)^{-\ell} c(g^{-1}, w)^{-m} F\big(g^{-1}(z), g^{-1}(w)\big) .$$

Let further $k \in \mathbb{N}$. Consider the following constant coefficients bi-differential operator

$$\mathbf{B}_{\ell,m}^{(k)} : \mathcal{H}(\Pi \times \Pi) \longrightarrow \mathcal{H}(\Pi)$$

defined by

$$\mathbf{B}_{\ell,m}^{(k)} F(z) =$$

$$\sum_{r+s=k} (-1)^r \binom{l + k - 1}{s} \binom{m + k - 1}{r} \left(\frac{\partial}{\partial z}\right)^r \left(\frac{\partial}{\partial w}\right)^s F\,(z, z) . \tag{5}$$

Notice in particular that

$$\mathbf{B}_{\ell,m}^{(0)} F(z) = F(z, z), \qquad \mathbf{B}_{\ell,m}^{(1)} F(z) = \left(-m\frac{\partial F}{\partial z} + \ell\frac{\partial F}{\partial w}\right)(z, z)$$

The following result was obtained by H. Cohen in [7].

Theorem 1.1 *For* $g \in G$,

$$\mathbf{B}^{(k)}_{\ell,m} \circ \big(\rho_\ell(g) \otimes \rho_m(g)\big) = \rho_{\ell+m+2k}(g) \circ \mathbf{B}^{(k)}_{\ell,m} \, .$$

The operators $\mathbf{B}^{(k)}_{\ell,m}$ are known as the *Rankin–Cohen brackets* and were much used in number theory for the study of modular forms.

The Rankin–Cohen brackets are a famous example of what is called *symmetry breaking differential operators*, an area in which T. Kobayashi made major contributions, in particular the *F method* (see [13]). The Rankin–Cohen brackets are symmetry-breaking differential operators for the restriction of the exterior tensor product $\rho_\ell \boxtimes \rho_m$, viewed as a representation of $G \times G$, to its diagonal subgroup $\{(g, g), g \in G\} \sim G$.

Our motivation in this chapter is to describe four different approaches of the construction of the Rankin–Cohen brackets, with only marginal claim of originality, considering the present geometric situation as a toy model for the general restriction problem.

The first construction is inspired by the Ω-*process*, going back to Cayley (see [3]) and further used by algebraists to produce the *transvectants*, a powerful tool in the theory of invariants (see [16] for a more detailed approach). The transvectants are analogous to the Rankin–Cohen brackets, but for the *finite dimensional representations* of G. The present construction covers the principal series of G, with the transvectants and the Rankin–Cohen brackets as special cases.

The second construction is based on the article [17] by V. Ovsienko and P. Redou, where they study symmetry-breaking differential operators for the tensor product of two scalar principal series of the group $O(p, q)$. The present situation is even an easier case, leading to a more explicit expression for the resulting differential operators.

The third construction is based on the *source operator method*. The source operator already appears in our presentation of the Ω-process, but, with generalization in mind, it is here constructed by using more sophisticated tools of harmonic analysis on G (Knapp–Stein intertwining operators namely), and indeed, this method has been adapted to more geometric situations (see e.g. [2]).

Whereas the three first variations deal with real harmonic analysis of $SL(2, \mathbb{R})$ and the *principal series of representations*, the last approach, inspired by a recent paper by T. Kobayashi and M. Pevzner [15], deals specifically with the *holomorphic discrete series representations*, uses their unitarity explicitly, with the Laplace transform playing an important role. Among many other works in this area, we wish to mention the paper by L. Peng and G. Zhang [18] and the work of Ibukiyama, e.g. [11] (see also [6]).

In Sect. 5, the relation between the symbols of the Rankin–Cohen brackets and the classical *Jacobi polynomials* is presented. In fact, this is due to the fact that both are obtained by a (kind of) *Rodrigues formula*.

2 The Ω-Process and the Generalized Construction of Transvectants

Our first construction of the Rankin–Cohen brackets is based on the Ω-*process*, first introduced by Cayley and used at the end of the XIXth centruy by algebraists for the construction of the *transvectants*, a key tool in the theory of invariants.

Consider the space $\mathbb{R}^2$, equipped with the symplectic form ω defined by

$$\mathbf{x} = \begin{pmatrix} x_1 \\ x_2 \end{pmatrix}, \mathbf{y} = \begin{pmatrix} y_1 \\ y_2 \end{pmatrix} \in \mathbb{R}^2, \qquad \omega(\mathbf{x}, \mathbf{y}) = x_1 y_2 - x_2 y_1$$

which can be thought as the determinant of the matrix $\begin{pmatrix} x_1 & x_2 \\ y_1 & y_2 \end{pmatrix}$ so that ω is clearly invariant under the natural action of $G = SL(2, \mathbb{R})$ on $\mathbb{R}^2$, i.e.

$$g \in G \qquad \mathbf{x}, \mathbf{y} \in \mathbb{R}^2 \qquad\qquad \omega(g\mathbf{x}, g\mathbf{y}) = \omega(\mathbf{x}, \mathbf{y}) \,.$$

For $\lambda \in \mathbb{C}$ and $\epsilon \in \{+1, -1\}$, consider the character of $\mathbb{R}^*$

$$t \longmapsto t^{\lambda,\epsilon} \quad = \quad \begin{cases} |t|^\lambda & \text{if } \epsilon = +1 \\ |t|^\lambda \operatorname{sgn}(t) & \text{if } \epsilon = -1 \end{cases} \qquad .$$

Let $\mathcal{H}_{\lambda,\epsilon}$ be the space of all smooth functions F defined on $\mathbb{R}^2 \setminus \{0\}$ such that for $s \in \mathbb{R}^*$

$$F(s\mathbf{x}) = s^{-\lambda,\epsilon} F(\mathbf{x}) \tag{6}$$

The space $\mathcal{H}_{\lambda,\epsilon}$ is stable by the action of G given by $F \longmapsto F \circ g^{-1}$, and this action of G induces a representation $\pi_{\lambda,\epsilon}$ of G on $\mathcal{H}_{\lambda,\epsilon}$.

Let $F \in \mathcal{H}_{\lambda,\epsilon}$. The formula

$$f(x) = F(x, 1) \tag{7}$$

defines a smooth function f is on $\mathbb{R}$. Conversely, let f be a smooth function on $\mathbb{R}$. For $x_1 \in \mathbb{R}$ and $x_2 \neq 0$, define

$$F(x_1, x_2) = x_2^{-\lambda,\epsilon} f\left(\frac{x_1}{x_2}\right) \,. \tag{8}$$

Now F is a smooth function on $\mathbb{R}^2 \setminus \{x_2 = 0\}$ and satisfies the homogeneity relation (6). This allows to identify the space $\mathcal{H}_{\lambda,\epsilon}$ with the subspace of smooth functions f on $\mathbb{R}$ such that the function F a priori defined by (8) on $\mathbb{R}^2 \setminus \{x_2 = 0\}$ has a smooth extension to $\mathbb{R}^2 \setminus \{0\}$. Denote by $\widetilde{\mathcal{H}}_{\lambda,\epsilon}$ the corresponding subspace of $C^\infty(\mathbb{R})$. As it will appear later, this space is not well fitted for analysis, but it can be shown that the Schwartz space $\mathcal{S}(\mathbb{R})$ is contained in $\widetilde{\mathcal{H}}_{\lambda,\epsilon}$. More precisely a function $F \in \mathcal{H}_{\lambda,\epsilon}$ corresponds through the map $F \longmapsto f$ to a function $f \in \mathcal{S}(\mathbb{R})$ if and only if the function $x_2 \longmapsto F(x_2, 1)$ has all its derivatives vanishing for $x_2 = 0$. The space $\mathcal{S}(\mathbb{R})$ turns out to be a good substitute to $\widetilde{\mathcal{H}}_{\lambda,\epsilon}$ in many situations.

Let $\widetilde{\pi}_\lambda$ be the corresponding representation of G on $\widetilde{\mathcal{H}}_{\lambda,\epsilon}$, which is classically referred to as the *noncompact picture*. An elementary calculation yields the following formula, valid for $g^{-1} = \begin{pmatrix} a & b \\ c & d \end{pmatrix}$ and $cx + d \neq 0$

$$\widetilde{\pi}_{\lambda,\epsilon}(g) f(x) = (cx+d)^{-\lambda,\epsilon} f\left(\frac{ax+b}{cx+d}\right) . \tag{9}$$

Consider now the space $\mathbb{R}^2 \times \mathbb{R}^2$. For $(\lambda,\epsilon), (\mu,\eta) \in \mathbb{C} \times \{-1,+1\}$, let $\mathcal{H}_{(\lambda,\epsilon),(\mu,\eta)}$ be the space of all functions $\big\{F \in C^\infty\big(\mathbb{R}^2 \setminus \{0\} \times \mathbb{R}^2 \setminus \{0\}\big)$ such that

$$\text{for all } s \in \mathbb{R}^*, \quad F(s\mathbf{x}, t\mathbf{y}) = s^{-\lambda,\epsilon}\, t^{-\mu,\eta}\, F(\mathbf{x},\mathbf{y}) ,$$

which we view as (a completion of) the tensor product of $\mathcal{H}_{\lambda,\epsilon} \otimes \mathcal{H}_{\mu,\eta}$. The group G acts on $\mathcal{H}_{(\lambda,\epsilon),(\mu,\eta)}$ by the natural diagonal action

$$F(\mathbf{x},\mathbf{y}) \longmapsto F(g^{-1}\mathbf{x}, g^{-1}\mathbf{y}) .$$

To the sympletic form ω introduced previously, we may associate the differential operator on $\mathbb{R}^2$

$$\Omega = \frac{\partial^2}{\partial x_1 \partial y_2} - \frac{\partial^2}{\partial x_2 \partial y_1} .$$

As Ω is a homogeneous differential operator, it is easily seen that Ω maps $\mathcal{H}_{(\lambda,\epsilon),(\mu,\eta)}$ into $\mathcal{H}_{(\lambda+1,-\epsilon),(\mu+1,-\eta)}$, thus defining an operator

$$\Omega_{(\lambda,\epsilon),(\mu,\eta)} : \mathcal{H}_{(\lambda,\epsilon),(\mu,\eta)} \longrightarrow \mathcal{H}_{(\lambda+1,-\epsilon),(\mu,-\eta)} .$$

As ω is invariant under the action of G, Ω commutes with the action of G on functions on $\mathbb{R}^2 \times \mathbb{R}^2$. Hence, $\Omega_{(\lambda,\epsilon),(\mu,\eta)}$ satisfies the intertwining relation

$$\Omega_{(\lambda,\epsilon),(\mu,\eta)} \circ \big(\pi_{\lambda,\,\epsilon}(g) \otimes \pi_{\mu,\,\eta}(g)\big) = \big(\pi_{\lambda+1,\,-\epsilon}(g) \otimes \pi_{\mu+1,\,-\eta}(g)\big) \circ \Omega_{(\lambda,\epsilon),(\mu,\eta)} .$$

Let res be the restriction map

$$\text{res} : C^\infty(\mathbb{R}^2 \setminus \{0\} \times \mathbb{R}^2 \setminus \{0\}) \longrightarrow C^\infty(\mathbb{R}^2 \setminus \{0\})$$

$$(\text{res}\, F)(\mathbf{x}) = F(\mathbf{x},\mathbf{x}) .$$

Then res maps $\mathcal{H}_{(\lambda,\epsilon),(\mu,\eta)}$ into $\mathcal{H}_{\lambda+\mu,\epsilon\eta}$, and satisfies the intertwining relation

$$g \in G \qquad \text{res} \circ \big(\pi_{\lambda,\epsilon}(g) \otimes \pi_{\mu,\eta}(g)\big) = \pi_{\lambda+\mu,\epsilon\eta}(g) \circ \text{res}\,,$$

so that the operator

$$\mathbf{B}_{(\lambda,\epsilon),(\mu,\eta)} = \text{res} \circ \Omega_{(\lambda,\epsilon),(\mu,\eta)} \quad : \quad \mathcal{H}_{(\lambda,\epsilon),(\mu,\eta)} \longrightarrow \mathcal{H}_{\lambda+\mu+2,\epsilon\eta}$$

satisfies the intertwining relation, valid for $g \in G$

$$\mathbf{B}_{(\lambda,\epsilon),(\mu,\eta)} \circ \big(\pi_{\lambda,\epsilon}(g) \otimes \pi_{\mu,\eta}(g)\big) = \pi_{(\lambda+\mu+2,\epsilon\eta}(g) \circ \mathbf{B}_{(\lambda,\epsilon),(\mu,\eta)}\,. \tag{10}$$

The same process can also be applied to the differential operator Ω^k where $k \in \mathbb{N}$, yielding an operator

$$\Omega^{(k)}_{(\lambda,\epsilon),(\mu,\eta)} : \mathcal{H}_{(\lambda,\epsilon),(\mu,\eta)} \longrightarrow \mathcal{H}_{(\lambda+k,(-1)^k\epsilon),(\mu+k,(-1)^k\eta)}\,,$$

and further an operator

$$\mathbf{B}^{(k)}_{(\lambda,\epsilon),(\mu,\eta)} = \text{res} \circ \Omega^{(k)}_{(\lambda,\epsilon),(\mu,\eta)} \quad : \quad H_{(\lambda,\epsilon),(\mu,\eta)} \longrightarrow \mathcal{H}_{\lambda+\mu+2k,\epsilon\eta}$$

which satisfies the intertwining property valid for $\in G$

$$\mathbf{B}^{(k)}_{(\lambda,\epsilon),(\mu,\eta)} \circ \big(\pi_{\lambda,\epsilon}(g) \otimes \pi_{\mu,\eta}(g)\big) = \pi_{\lambda+\mu+2k,\epsilon\eta}(g) \circ \mathbf{B}^{(k)}_{(\lambda,\epsilon),(\mu,\eta)}\,. \tag{11}$$

This construction may be transferred to the spaces $\widetilde{\mathcal{H}}_{\lambda,\epsilon}$ and $\widetilde{\mathcal{H}}_{\mu,\eta}$. The definition of the space $\widetilde{\mathcal{H}}_{(\lambda,\epsilon),(\mu,\eta)}$ follows the definition of $\widetilde{\mathcal{H}}_{\lambda,\epsilon}$ (see (7) and the discussion thereafter). The following result is obtained by a routine calculation, which is left to the reader.[1]

Proposition 2.1 *Let F be a function in $H_{(\lambda,\epsilon),(\mu,\eta)}$ and let f be the associated function in $\widetilde{\mathcal{H}}_{(\lambda,\epsilon),(\mu,\eta)}$. To the function $\Omega F \in \mathcal{H}_{(\lambda+1,-\epsilon),(\mu+1,-\eta)}$ is associated the function $\widetilde{\Omega}_{\lambda,\mu} f \in \mathcal{H}_{(\lambda+1,-\epsilon),(\mu+1,-\eta)}$ given by*

$$\widetilde{\Omega}_{\lambda,\mu} f = -\mu \frac{\partial}{\partial x} f + \lambda \frac{\partial}{\partial y} f + (x-y) \frac{\partial^2}{\partial x \partial y} f\,. \tag{12}$$

The next result is a reformulation of the property (10).

[1] The expression of the resulting operator does not depend on ϵ, η so that we suppress theses indices in the notation.

Theorem 2.2 *The bi-differential operator* $\widetilde{\mathbf{B}}_{\lambda,\mu}$ *given by*

$$\widetilde{\mathbf{B}}_{\lambda,\mu} f(x) = \left(-\mu \frac{\partial}{\partial x} f + \lambda \frac{\partial}{\partial y} f\right)(x, x)$$

satisfies the intertwining relation

$$\widetilde{\mathbf{B}}_{\lambda,\mu} \circ \left(\widetilde{\pi}_{\lambda,\epsilon}(g) \otimes \widetilde{\pi}_{\mu,\eta}(g)\right) = \widetilde{\pi}_{\lambda+\mu+2,\epsilon\eta}(g) \circ \widetilde{\mathbf{B}}_{\lambda,\mu} \,. \tag{13}$$

The same process can be used for the k-th power Ω^k, yielding a bi-differential operator $\widetilde{\mathbf{B}}^{(k)}_{\lambda,\mu}$, which satisfies the intertwining relation

$$\widetilde{\mathbf{B}}^{(k)}_{\lambda,\mu} \circ \left(\widetilde{\pi}_{\lambda,\epsilon}(g) \otimes \widetilde{\pi}_{\mu,\eta}(g)\right) = \widetilde{\pi}_{\lambda+\mu+2k,\epsilon\eta}(g) \circ \widetilde{\mathbf{B}}_{\lambda,\mu} \,. \tag{14}$$

The next computations are best presented in the context of the *Weyl algebra* of $\mathbb{R}^2$. Although we only need the case $n = 2$, the presentation is made for $\mathbb{R}^n$. The Weyl algebra $\mathcal{W} = \mathcal{W}(\mathbb{R}^n)$ consists of all differential operators with polynomial coefficients on $\mathbb{R}^n$. We use freely the multiindices notation (see e.g. [10] for more information). For D an element of $\mathcal{W}$, say

$$D = \sum_\alpha a_\alpha(\boldsymbol{x}) \frac{\partial^{|\alpha|}}{\partial \boldsymbol{x}^\alpha}$$

its *algebraic symbol* $d(\boldsymbol{x}, \boldsymbol{\xi})$ is defined by

$$d(\boldsymbol{x}, \boldsymbol{\xi}) = \sum_\alpha a_\alpha(\boldsymbol{x}) \boldsymbol{\xi}^\alpha \,.$$

There is a well-known formula for the algebraic symbol of the composition of two elements in the Weyl algebra. Let A (resp. B) be in $\mathcal{W}$ with symbol $a(\boldsymbol{x}, \boldsymbol{\xi})$ (resp. $b(\boldsymbol{x}, \boldsymbol{\xi})$). Then the symbol $c = a \sharp b$ of $C = A \circ B$ is given by

$$c(\boldsymbol{x}, \boldsymbol{\xi}) = \sum_\alpha \frac{1}{\alpha!} \left(\frac{\partial}{\partial \boldsymbol{\xi}}\right)^\alpha a(\boldsymbol{x}, \boldsymbol{\xi}) \left(\frac{\partial}{\partial \boldsymbol{x}}\right)^\alpha b(\boldsymbol{x}, \boldsymbol{\xi}) \,. \tag{15}$$

The computation of the coefficients of $\widetilde{\mathbf{B}}^{(k)}_{\lambda,\mu}$ is obtained through a recurrence on k. The trivial relation $\Omega^k = \Omega^{k-1} \circ \Omega$ implies

$$\widetilde{\Omega}^{(k)}_{\lambda,\mu} = \widetilde{\Omega}^{(k-1)}_{\lambda+1,\mu+1} \circ \widetilde{\Omega}_{\lambda,\mu} \,. \tag{16}$$

Let $\mathcal{D}$ be the subalgebra of constant coefficients differential operators on $\mathbb{R}^2$. The explicit expression of $\widetilde{\Omega}^{(k)}_{\lambda,\mu}$ can be seen by induction on k to be of the form

$$\widetilde{\Omega}^{(k)}_{\lambda,\mu} = \widetilde{E}^{(k)}_{\lambda,\mu} \mod (x-y)\mathcal{D}\,,$$

where $\widetilde{E}^{(k)}_{\lambda,\mu}$ belongs to $\mathcal{D}$, and consequently

$$\widetilde{\mathbf{B}}^{(k)}_{\lambda,\mu} = \mathrm{res} \circ \widetilde{E}^{(k)}_{\lambda,\mu},$$

so that computing $\widetilde{E}^{(k)}_{\lambda,\mu}$ is enough. Now, (16) implies

$$\widetilde{E}^{(k)}_{\lambda,\mu} = \widetilde{E}^{(k-1)}_{\lambda+1,\mu+1} \circ \widetilde{\Omega}_{\lambda,\mu} \mod (x-y)\mathcal{D}\,. \tag{17}$$

Let $e^{(k)}_{\lambda,\mu}$ (resp. $e^{(k-1)}_{\lambda+1,\mu+1}$) be the algebraic symbol of $\widetilde{E}^{(k)}_{\lambda,\mu}$ (resp. $\widetilde{E}^{(k-1)}_{\lambda+1,\mu+1}$), and let

$$\omega_{\lambda,\mu}(x,y,\xi,\eta) = -\mu\xi + \lambda\eta + (x-y)\xi\eta$$

be the algebraic symbol of $\widetilde{\Omega}_{\lambda,\mu}$. The composition formula (15) applied to (17) and after ignoring terms belonging to $(x-y)\mathcal{D}$ yields

$$e^{(k)}_{\lambda,\mu}(\xi,\eta) = e^{(k-1)}_{\lambda+1,\mu+1}(\xi,\eta)(-\mu\xi+\lambda\eta) + \xi\eta\left(\frac{\partial}{\partial\xi} - \frac{\partial}{\partial\eta}\right) e^{(k-1)}_{\lambda+1,\,\mu+1}(\xi,\eta), \tag{18}$$

Proposition 2.3 *Let $\lambda, \mu \in \mathbb{C}$.*

(i) For any $k \in \mathbb{N}$, there exists a well-determined polynomial $b^{(k)}_{\lambda,\mu}(\xi,\eta)$, homogeneous of degree k such that, for $\xi,\eta > 0$

$$\left(\frac{\partial}{\partial\xi} - \frac{\partial}{\partial\eta}\right)^k \xi^{\lambda+k-1}\eta^{\mu+k-1} = b^{(k)}_{\lambda,\mu}(\xi,\eta)\,\xi^{\lambda-1}\,\eta^{\mu-1}\,. \tag{19}$$

(ii) The polynomials $b^{(k)}_{\lambda,\mu}(\xi,\eta)$, $k \in \mathbb{N}$ satisfy

(a) $b^{(0)}_{\lambda,\mu}(\xi,\eta) \equiv 1$

(b)

$$b^{(k)}_{\lambda,\mu} = (-\mu\xi+\lambda\eta)\,b^{(k-1)}_{\lambda+1,\mu+1} + \xi\eta\left(\frac{\partial}{\partial\xi} - \frac{\partial}{\partial\eta}\right) b^{(k-1)}_{\lambda+1,\mu+1}\,. \tag{20}$$

Proof The proof is by induction on k. The case $k = 0$ being trivial, assume that the result is valid for $k - 1$. Hence

$$\left(\frac{\partial}{\partial \xi} - \frac{\partial}{\partial \eta}\right)^{k-1} \xi^{\lambda+k-1} \eta^{\mu+k-1}$$

$$= \left(\frac{\partial}{\partial \xi} - \frac{\partial}{\partial \eta}\right)^{k-1} \xi^{\left(\lambda+1)+(k-1)-1\right)} \eta^{(\mu+1)+(k-1)-1}$$

$$= b^{(k-1)}_{\lambda+1\,,\,\mu+1}(\xi,\,,\eta)\xi^{\lambda}\eta^{\mu}\ .$$

Next,

$$\left(\frac{\partial}{\partial \xi} - \frac{\partial}{\partial \eta}\right) b^{(k-1)}_{\lambda+1,\,\mu+1}(\xi,\eta)\,\xi^{\lambda}\eta^{\mu}$$

$$= \left((-\mu\xi + \lambda\eta)b^{(k-1)}_{\lambda+1,\mu+1}(\xi,\eta) + \xi\eta\left(\frac{\partial}{\partial \xi} - \frac{\partial}{\partial \eta}\right) b^{(k-1)}_{\lambda+1\,,\,\mu+1}(\xi,\eta)\right)\xi^{\lambda-1}\eta^{\mu-1}$$

so that (20) follows. □

The main result is obtained by comparing (18) and (20).

Theorem 2.4

$$\widetilde{\mathbf{B}}^{(k)}_{\lambda,\mu} f(x) = b^{(k)}_{\lambda,\mu}\left(\frac{\partial}{\partial x}, \frac{\partial}{\partial y}\right) f\,(x,x)\ .$$

An explicit formula for the coefficients of the polynomials $b^{(k)}_{\lambda,\mu}$ can easily be obtained from this recurrence formula (see Sect. 5).

3 A Straightforward Construction (after V. Ovsienko an P. Redou)

The second construction is based on the article [17] by V. Ovsienko and P. Redou, valid for the scalar principal series of group $O(p,q)$). The group $SL(2,\mathbb{R})$ is locally isomorphic to $(O(1,2)$, there are, however, some differences in our presentation. The group $SL(2,\mathbb{R})$ is a *twofold covering* of $SO_0(1,2)$, and the representation $\pi_{\lambda,-}$ (see definition thereafter) does not descent to a representation of $PSL(2,\mathbb{R}) \simeq SO_0(1,2)$. The present case is, however, computationally simpler and leads to more explicit formulas.

Proposition 3.1 *Let $B : C^\infty(\mathbb{R}^2) \longrightarrow C^\infty(\mathbb{R})$ be a bi-differential operator on $\mathbb{R}$ given as a finite sum*

$$Bf(x) = \sum_{j,i} b_{j,i}(x) \left(\frac{\partial}{\partial x}\right)^j \left(\frac{\partial}{\partial y}\right)^i f\ (x,x)\ ,$$

where $b_{j,i} \in C^\infty(\mathbb{R})$. Then B satisfies the intertwining property (11) if and only if

(i) $b_{j,i} \equiv 0$ *for* $i + j \neq k$
(ii) $b_{j,k-j}(x)$, *for* $0 \le j \le k$, *is a constant polynomial equal to (say)* $b_j \in \mathbb{C}$ *and the* (b_j) *satisfy the condition*

$$(j+1)(\lambda + j)\, b_{j+1} + (k-j)(\mu + k - j - 1)\, b_j = 0\ . \tag{21}$$

Proof First, recall that G is generated by the three subgroups[2] N, A^*, $\overline{N}$, where

$$N = \left\{\begin{pmatrix} 1 & b \\ 0 & 1\end{pmatrix}, b \in \mathbb{R}\right\},\quad A^* = \left\{\begin{pmatrix} t & 0 \\ 0 & \frac{1}{t},\end{pmatrix} t \in \mathbb{R}^*\right\},\quad \overline{N} = \left\{\begin{pmatrix} 1 & 0 \\ c & 1\end{pmatrix}, c \in \mathbb{R}\right\}.$$

For elements in N, A^*, $\overline{N}$, the representation $\widetilde{\pi}_\lambda$ is given by

$$\widetilde{\pi}_{\lambda,\epsilon}\begin{pmatrix} 1 & b \\ 0 & 1\end{pmatrix} f(x) = f(x-b)$$

$$\widetilde{\pi}_{\lambda,\epsilon}\begin{pmatrix} t & 0 \\ 0 & \frac{1}{t}\end{pmatrix} f(x) = t^{-\lambda,\epsilon} f(t^{-2}x)$$

$$\widetilde{\pi}_{\lambda,\epsilon}\left(\begin{pmatrix} 1 & 0 \\ c & 1\end{pmatrix}\right) f(x) = (-cx+1)^{-\lambda,\epsilon} f\left(\frac{x}{-cx+1}\right),$$

The intertwining property (11) for $g \in \overline{N}$ implies that B has constant coefficients, i.e. $b_{j,i}(x) = b_{j,i} \in \mathbb{C}$. For $g \in A^*$, (11) forces B to be homogeneous of degree k, i.e. B must be of the form

$$Bf(x) = \sum_{j=0}^k b_j \left(\frac{\partial}{\partial x}\right)^j \left(\frac{\partial}{\partial y}\right)^{k-j} f\ (x,x)\ ,$$

[2] For *aficionados* of harmonic analysis on general semi-simple groups, our group A^* corresponds to MA, where M is the centralizer of A in G.

where $b_j \in \mathbb{C}$ for $0 \le j \le k$. So in order to fulfill the condition (11), it remains to test it for $g \in \overline{N}$. As G is connected, it is equivalent to test the intertwining property for the derived representation. The element $X = \begin{pmatrix} 0 & 0 \\ 1 & 0 \end{pmatrix}$ generates $\overline{\mathfrak{n}} = \mathrm{Lie}(\overline{N})$. Hence, the intertwining property (11) will be satisfied if and only if

$$B \circ \left(d\pi_{\lambda,\epsilon}(X) \otimes \mathrm{id} + \mathrm{id} \otimes d\pi_{\mu,\eta}(X)\right) = d\pi_{\lambda+\mu+2k,\epsilon\eta}(X) \circ B \,. \tag{22}$$

For $c \in \mathbb{R}$

$$\pi_{\lambda,\epsilon}\left(\begin{pmatrix} 1 & 0 \\ c & 1 \end{pmatrix}\right) f(x) = (-cx+1)^{-\lambda,\epsilon} f\left(\frac{x}{-cx+1}\right),$$

so that

$$d\pi_{\lambda,\epsilon}\left(\begin{pmatrix} 0 & 0 \\ 1 & 0 \end{pmatrix}\right) f = \lambda x f + x^2 \frac{\partial f}{\partial x}\,.$$

Hence,

$$\left(d\pi_{\lambda,\epsilon}(X) \otimes \mathrm{id} + \mathrm{id} \otimes d\pi_{\mu,\eta}(X)\right) f = (\lambda x + \mu y) f + x^2 \frac{\partial f}{\partial x} + y^2 \frac{\partial f}{\partial y}\,.$$

The condition (22) can be studied using the formalism of the Weyl algebra. For computing the composition $\dfrac{\partial^k}{\partial x^j \partial y^{k-j}} \circ d\left(\pi_{\lambda,\epsilon} \otimes \pi_{\mu,\eta}\right)(X)$, use the composition formula (15) to obtain

$$\begin{aligned}
&\xi^j \eta^{k-j} \,\sharp\, (\lambda x + \mu y + x^2 \xi + y^2 \eta) \\
&= (\lambda x + \mu y)\xi^j \eta^{k-j} + x^2 \xi^{j+1} \eta^{k-j} + y^2 \xi^j \eta^{k-j+1} \\
&\quad + j\lambda \xi^{j-1} \eta^{k-j} + (k-j)\mu \xi^j \eta^{k-j-1} \\
&\quad + 2jx\xi^j \eta^{k-j} + 2(k-j)y\xi^j \eta^{k-j} \\
&\quad + \frac{1}{2}\left(2j(j-1)\xi^{j-1}\eta^{k-j} + 2(k-j)(k-j-1)\xi^j \eta^{k-j-1}\right).
\end{aligned}$$

Going back from the symbol to the differential operator, we obtain

$$\begin{aligned}
&\operatorname{res}\circ\left(\frac{\partial}{\partial x}\right)^{j}\left(\frac{\partial}{\partial y}\right)^{k-j}\circ\left(\lambda x+\mu y+x^{2}\frac{\partial}{\partial x}+y^{2}\frac{\partial}{\partial y}\right)=\\
&\quad j(\lambda+j-1)\operatorname{res}\circ\left(\frac{\partial}{\partial x}\right)^{j-1}\left(\frac{\partial}{\partial y}\right)^{k-j}+\cdots\\
&\quad+(k-j)(\mu+k-j-1)\operatorname{res}\circ\left(\frac{\partial}{\partial x}\right)^{j}\left(\frac{\partial}{\partial y}\right)^{k-j-1}\\
&\quad+(\lambda+\mu+2k)\,x\,\frac{\partial^{k}}{\partial x^{j}\partial y^{k-j}}\left(\frac{\partial}{\partial x}\right)^{j}\left(\frac{\partial}{\partial y}\right)^{k-j-1}\\
&\quad+x^{2}\operatorname{res}\circ\left(\left(\frac{\partial}{\partial x}\right)^{j+1}\left(\frac{\partial}{\partial y}\right)^{k-j}+\left(\frac{\partial}{\partial x}\right)^{j}\left(\frac{\partial}{\partial y}\right)^{k-j+1}\right).
\end{aligned}$$

Hence,

$$\begin{aligned}
&B\circ\left(d\pi_{\lambda,\epsilon}(X)\otimes\mathrm{id}+\mathrm{id}\otimes d\pi_{\mu,\eta}(X)\right)\\
&=\sum_{j}\Big((j+1)(\lambda+j)\,b_{j+1}+(k-j)(\mu+k-j-1)\,b_{j}\Big)\left(\frac{\partial}{\partial x}\right)^{j}\left(\frac{\partial}{\partial y}\right)^{k-j-1}\\
&\quad+(\lambda+\mu+2k)\,x\sum_{j}b_{j}\left(\frac{\partial}{\partial x}\right)^{j}\left(\frac{\partial}{\partial y}\right)^{k-j}\\
&\quad+x^{2}\sum_{j}b_{j}\left(\left(\frac{\partial}{\partial x}\right)^{j-1}\left(\frac{\partial}{\partial y}\right)^{k-j}+\left(\frac{\partial}{\partial x}\right)^{j}\left(\frac{\partial}{\partial y}\right)^{k-j+1}\right).
\end{aligned}$$

On the other hand,

$$\begin{aligned}
&\left(d\pi_{\lambda+\mu+2k,\epsilon\eta}(X)\circ B\right)f\,(x)=\\
&=\left(\left((\lambda+\mu+2k)x+x^{2}\frac{\partial}{\partial x}\right)\circ\sum_{j}b_{j}\left(\frac{\partial}{\partial x}\right)^{j}\left(\frac{\partial}{\partial y}\right)^{k-j}\right)f\,(x,x)\\
&\quad=(\lambda+\mu+2k)\,x\sum_{j}b_{j}\left(\frac{\partial}{\partial x}\right)^{j}\left(\frac{\partial}{\partial y}\right)^{k-j}\\
&\quad+x^{2}\sum_{j}b_{j}\left(\left(\frac{\partial}{\partial x}\right)^{j-1}\left(\frac{\partial}{\partial y}\right)^{k-j}+\left(\frac{\partial}{\partial x}\right)^{j}\left(\frac{\partial}{\partial y}\right)^{k-j+1}\right).
\end{aligned}$$

The proposition follows by comparing the two handsides of (22). □

Theorem 3.2 *Let $\lambda, \mu \in \mathbb{C}$.*

(i) The bi-differential operator

$$\mathbf{B}^{(k)}_{\lambda,\mu} = \operatorname{res} \circ \sum_{j=0}^{k} (-1)^j \binom{\lambda+k-1}{k-j} \binom{\mu+k-1}{j} \left(\frac{\partial}{\partial x}\right)^j \left(\frac{\partial}{\partial y}\right)^{k-j} \tag{23}$$

satisfies the intertwining property (14).

(ii) Assume that either $\lambda \notin -\mathbb{N}$ or $\mu \notin -\mathbb{N}$. Up to a scalar, $\mathbf{B}^{(k)}_{\lambda,\mu}$ is the unique non vanishing bi-differential operator, which satisfies (14).

Proof Assume that $\lambda \notin -\mathbb{N}$. Then for $j \geq 1$

$$b_j = (-1)^j \frac{(k-j+1)(\mu+k-j)}{(\lambda+j-1)\, j}\, b_{j-1}$$

and by induction

$$b_j = (-1)^j \frac{(k-j+1)\dots k\,(\mu+k-j)\dots(\mu+k-1)}{1\dots j\,(\lambda+j-1)\dots\lambda}\, b_0\,.$$

The uniqueness is clear and by choosing $b_0 = \dfrac{(\lambda+k-1)\dots\lambda}{1\dots k}$, formula (23) follows. In case $\lambda \in -\mathbb{N}$, then by assumption $\mu \notin -\mathbb{N}$ and just exchange the role of x and y.

For a discussion of the cases where $\lambda, \mu \in -\mathbb{N}$, see [14] Section 9. □

4 The Source Operator Method

The Ω process construction has no obvious generalization to other geometric situations. A lesser ambition is to try to generalize the operator $\widetilde{\Omega}_{\lambda,\mu}$, as it is the key element for the construction construct of the Rankin–Cohen brackets described in Sect. 1. I proposed to call it the *source operator* (see [4]). The construction is based on more sophisticated tools of harmonic analysis on semi-simple Lie groups, but does work in rather many situations (see e.g. [2]). The construction is done using the noncompact picture, see (9).

Proposition 4.1 *Let $x, y \in \mathbb{R}$ and let $g \in G$ be defined at x and y. Then*

$$g(x) - g(y) = c(g, x)^{-1}\,(x-y)\,c(g, y)^{-1}\,. \tag{24}$$

In more concrete terms, the statement amounts to the identity, valid for $a, b, c, d, \in \mathbb{R}$ such that $ad - bc = 1$:

$$\frac{ax+b}{cx+d} - \frac{ay+b}{cy+d} = \frac{1}{cx+d}(x-y)\frac{1}{cy+d} .$$

Let us introduce the operator M defined for $f \in C^\infty(\mathbb{R} \times \mathbb{R})$ by

$$Mf(x, y) = (x-y)f(x, y) .$$

Proposition 4.2 *The operator M satisfies the intertwining property, valid for $X \in \mathfrak{g}$*

$$M \circ d\big(\widetilde{\pi}_{\lambda,\epsilon} \otimes \widetilde{\pi}_{\mu,\eta}\big)(X) = d\big(\widetilde{\pi}_{\lambda-1,-\epsilon} \otimes \widetilde{\pi}_{\mu-1,-\eta}\big)(X) \circ M . \tag{25}$$

Proof Let $(x, y) \in \mathbb{R}^2$ be given, and let $g \in G$ such that g^{-1} is defined at x and y. Let f be a smooth function on $\mathbb{R}^2$. Now, using (24)

$$(x-y) = c(g^{-1}, x)\big(g^{-1}(x) - g^{-1}(y)\big)c(g^{-1}, y)$$

so that

$$M \circ \big(\widetilde{\pi}_{\lambda,\epsilon}(g) \otimes \widetilde{\pi}_{\mu,\eta}(g)\big) f(x, y)$$

$$= (x-y)\, c(g^{-1}, x)^{-\lambda,\epsilon}\, c(g^{-1}, y)^{-\mu,\eta} f\big(g^{-1}(x), g^{-1}(y)\big)$$

$$= c(g^{-1}, x)^{-\lambda+1,-\epsilon}\, c(g^{-1}, y)^{-\mu+1,-\eta} \big(g^{-1}(x) - g^{-1}(y)\big)\, f\big(g^{-1}(x), g^{-1}(y)\big)$$

$$= \Big(\big(\widetilde{\pi}_{\lambda-1,-\epsilon}(g) \otimes \widetilde{\pi}_{\mu-1,-\eta}(g)\big) \circ M\Big) f(x, y) .$$

Observe that for g in a (small enough) neighborhood of the neutral element in G, g^{-1} is defined in neighborhoods of x and of y. This allows to differentiate the last identity and obtain (25). □

The covariance property of the operator M goes in the "opposite direction" of the source operator we are trying to construct : it decreases the parameters λ, μ by 1 (and changes ϵ and η to their opposite), whereas our source operator should increase them by 1.

The next idea is to use the *Knapp-Stein intertwining operators*, a very basic tool in harmonic analysis on semi-simple Lie groups (see [12] for general information).

In the non-compact picture we are using, the Knapp–Stein intertwining operators are convolution operators by tempered distributions. So instead of using the space $C^\infty(\mathbb{R})$, it is wise to use the Schwartz space $\mathcal{S}(\mathbb{R})$ and its dual, the space of tempered distributions $\mathcal{S}'(\mathbb{R})$. Notice that the operator M acts on both spaces. For $G = SL(2, \mathbb{R})$, the theory of Knapp–Stein intertwining operators is based on the following fact (see e.g. [10] for a proof).

Proposition 4.3 *For $\Re(\alpha) > -1$, the function $x^{\alpha,\epsilon}$ is locally integrable and of moderate growth at infinity, hence defines a tempered distribution. The map*

$$\alpha \longmapsto \frac{1}{\Gamma(\alpha+1)}\, x^{\alpha,\epsilon}$$

can be analytically continued to $\mathbb{C}$ as a holomorphic $\mathcal{S}'(\mathbb{R})$-valued function.

Now define the Knapp–Stein intertwining operators by the formula

$$J_{\lambda,\epsilon} f(x) = \frac{1}{\Gamma(\lambda-1)} \int_{\mathbb{R}} (x-y)^{-2+\lambda,\epsilon} f(y)dy\,. \tag{26}$$

The Knapp–Stein operators, being convolution operators by tempered distributions map $\mathcal{S}(\mathbb{R})$ continuously into $\mathcal{S}'(\mathbb{R})$.

Proposition 4.4 *For any $X \in \mathfrak{g}$*

$$J_{\lambda,\epsilon} \circ d\widetilde{\pi}_{\lambda,\epsilon}(X) = d\widetilde{\pi}_{2-\lambda,\epsilon}(X) \circ J_{\lambda,\epsilon} \tag{27}$$

Proof (Sketch) Assume first that $\Re(\lambda) > 1$. Let f a smooth function with compact support on $\mathbb{R}$ and let $g \in G$ be such that g^{-1} is defined on the support of f. Then

$$J_{\lambda,\epsilon} \circ \widetilde{\pi}_{\lambda,\epsilon}(g) f(x) = \widetilde{\pi}_{2-\lambda,\epsilon}(g) \circ J_{\lambda,\epsilon} f(x)$$

is easily obtained by a change of variable, taking into account the formula for the Jacobian of g at x (cf (2)). By differentiation, (27) follows. By continuity of $J_{\lambda,\epsilon}$, this is also valid for $f \in \mathcal{S}(\mathbb{R})$. Finally use analytic continuation to get rid of the condition $\Re(\lambda) > 1$. □

Let $(\lambda, \epsilon), (\mu, \eta) \in \mathbb{C}\times\{\pm\}$. Then consider the following composition of operators:

$$E_{\lambda,\mu} = (J_{1-\lambda,-\epsilon} \otimes J_{1-\mu,-\eta}) \circ M \circ (J_{\lambda,\epsilon} \otimes J_{\mu,\eta})\,. \tag{28}$$

Thanks to the intertwining properties (25) and (27), $E_{\lambda,\mu}$ intertwines the representations $d(\widetilde{\pi}_{\lambda,\epsilon} \otimes \widetilde{\pi}_{\mu,\eta})$ and $d(\widetilde{\pi}_{\lambda+1,-\epsilon} \otimes \widetilde{\pi}_{\mu+1,-\eta})$. This is the correct intertwining property for a potential source operator. However, because of the singularities of the kernels of the Knapp–Stein operators, it is not possible to directly consider the composition of the three operators. And another drawback is that there seems to be no reason to believe that $E_{\lambda,\mu}$ is a *differential* operator as the Knapp–Stein operators are (generically) not local (see however comments at the end of this Section).

As $\mathcal{S}(\mathbb{R})\widehat{\otimes}\mathcal{S}(\mathbb{R}) \simeq \mathcal{S}(\mathbb{R}^2)$, the (completed) tensor product operator $J_{\lambda,\epsilon}\widehat{\otimes}J_{\mu,\eta}$ can be viewed as a convolution operator on $\mathbb{R}^2$, thus mappping $\mathcal{S}(\mathbb{R}^2)$ into $\mathcal{S}'(\mathbb{R}^2)$. So it is possible to compose M and *one* tensor product of Knapp–Stein operators. These remarks lead to consider the following diagram as a substitute for the composition (28). The lower index in the notation $\mathcal{S}(\mathbb{R}^2)_{(\lambda,\epsilon),(\mu,\eta)}$ is meant for the representation of $\mathfrak{g}$ that is acting on the space $\mathcal{S}(\mathbb{R}^2)$ or its dual.

$$\begin{array}{ccc} \mathcal{S}'(\mathbb{R}^2)_{(\lambda,\epsilon)\,(\mu,\eta)} & \xrightarrow{?} & \mathcal{S}'(\mathbb{R}^2)_{(\lambda+1,-\epsilon)\,(\mu+1,-\eta)} \\ J_{2-\lambda,\,\epsilon}\widehat{\otimes}J_{2-\mu,\,\eta}\uparrow & & \uparrow J_{1-\lambda,\,-\epsilon}\widehat{\otimes}J_{1-\mu,\,-\eta} \\ \mathcal{S}(\mathbb{R}^2)_{(2-\lambda,\epsilon),(2-\mu,\eta)} & \xrightarrow{M} & \mathcal{S}(\mathbb{R}^2)_{(1-\lambda,-\epsilon),\,(1-\mu,-\eta)} \end{array}$$

We want to complete the diagram by an operator on the upper side, so that it becomes a commutative diagram, and this should (and indeed will) produce the desired source operator.

Theorem 4.5 *There exists a differential operator $E_{\lambda,\mu}$ with polynomial coefficients on $\mathbb{R}^2$ such that*

$$(J_{1-\lambda,\,-\epsilon}\widehat{\otimes}J_{1-\mu,\,-\eta})\circ M = E_{\lambda,\mu}\circ(J_{2-\lambda,\,\epsilon}\widehat{\otimes}J_{2-\mu,\,\eta})\,, \tag{29}$$

Moreover,

$$E_{\lambda,\mu} = \left((x-y)\frac{\partial^2}{\partial x\partial y} - \mu\frac{\partial}{\partial x} + \lambda\frac{\partial}{\partial y}\right)$$

coincides with $\widetilde{\Omega}_{\lambda,\mu}$.

Our strategy for the (long) proof is to use the Fourier transform on $\mathbb{R}\times\mathbb{R}\simeq\mathbb{R}^2$. The operator $J_{\lambda,\,\epsilon}\widehat{\otimes}J_{\mu,\,\eta}$ is a convolution operator, hence the corresponding operator on the Fourier transform side is a multiplication operator by some tempered distribution. The operator corresponding to M is a partial differential operator with constant coefficients. Hence, it is possible to prove (29) by some form of symbolic calculus, reminiscent of the pseudo-differential calculus (and also of the symbolic calculus for the Weyl algebra used in Sect. 1). We present this calculus for $\mathbb{R}^n$ although we will use it only for $n = 2$.

The Fourier transform of a function f (say in $\mathcal{S}(\mathbb{R}^n)$) is defined by the formula

$$\mathcal{F}f(\xi) = \widehat{f}(\xi) = \int_{\mathbb{R}^n} f(x)e^{-i\xi.x}dx\,.$$

and the inverse Fourier transform is given by

$$f(x) = \left(\mathcal{F}^{-1}\widehat{f}\right)(x) = \left(\frac{1}{2\pi}\right)^n\int_{\mathbb{R}}\widehat{f}(\xi)e^{ix.\xi}d\xi\,,$$

where $x.\xi = \xi.x = \sum_{j=1}^n \xi_j x_j$.

The operators involved in the computation are sums of operators, which can be written as $Ef(x) = p(x)\,(k \star f)(x)$, where $p(x)$ is a polynomial, k is a tempered distribution, and f is a test function in $\mathcal{S}(\mathbb{R}^n)$. Denote by $Op\,(\mathbb{R}^n)$ the space of such operators.

Using the Fourier tranform, operators in $Op(\mathbb{R}^n)$ can be written as

$$Ef(x) = \frac{1}{(2\pi)^n}\int_{\mathbb{R}^n} p(x)\,\widehat{k}(\xi)\,\widehat{f}(\xi)\,e^{ix.\xi}d\xi\ .$$

Define the *symbol* of the operator $E = p(x)\circ(k\ \star\cdot)$ to be

$$\sigma_E(x,\xi) = e(x,\ \xi) = p(x)\,\widehat{k}(\xi)\ .$$

A differential operator D with constant coefficients can be interpreted as a convolution with the distribution $D\delta_0$. Hence, a differential operator with polynomial coefficients (i.e. an element of the Weyl algebra) belongs to our class of operators, and its symbol coincides with its (analytic) symbol σ_E defined by

$$E(e^{ix.\xi}) = \sigma_E(x,\xi)e^{ix.\xi}$$

Although our class of operators is not an algebra, some products are allowed. Here is the important result for our computation.

Proposition 4.6 *Let E be an operator in $Op(\mathbb{R}^n)$ and let D be an element of the Weyl algebra $\mathcal{W}(\mathbb{R}^n)$. Then the composed operator $E\circ D$ belongs to $Op(\mathbb{R}^n)$. Its symbol is given by*

$$\sigma_{E\circ D} = \sigma_E\,\sharp\,\sigma_D = \sum_\alpha \frac{1}{\alpha!}\left(\frac{1}{i}\partial_\xi\right)^\alpha \sigma_E\ \ \partial_x^\alpha\sigma_D\ . \tag{30}$$

The proof is a variant of the proof of the similar statement in the pseudo-differential calculus (see [4]).

Let us now sketch the computation leading to (29). The Fourier transform of the tempered distribution $\frac{1}{\Gamma(\alpha+1)}\,x^{\alpha,\epsilon}$ is given respectively by

$$2\cos\left((\alpha+1)\frac{\pi}{2}\right)\,\xi^{-\alpha-1,+}\qquad \text{for } \epsilon=+1$$

$$-2i\,\sin\left((\alpha+1)\frac{\pi}{2}\right)\,\xi^{-\alpha-1,-}\qquad \text{for } \epsilon=-1\ .$$

See e.g. [10, p. 167] or [9, p. 1120], formulas 25^7 and 26^7 (take care of a different definition of the Fourier transform!). Hence, the symbol $j_{\lambda,\epsilon}$ of the Knapp–Stein operator $J_{\lambda,\epsilon}$ is given by

$$j_{\lambda,\epsilon}(\xi)\ = \gamma(\lambda,\epsilon)\,\xi^{-\lambda+1,\epsilon} \tag{31}$$

where

$$\gamma(\lambda,+) = 2\sin\left(\lambda\frac{\pi}{2}\right), \quad \gamma(\lambda,-) = 2i\cos\left(\lambda\frac{\pi}{2}\right).$$

Notice the two formulas, valid for $\lambda \notin \mathbb{Z}$

$$\frac{d}{d\xi}\, j_{\lambda,\epsilon}(\xi) = i\,(-\lambda+1)\, j_{\lambda+1,-\epsilon}(\xi)\,, \quad \xi\, j_{\lambda,\epsilon}(\xi) = -i\, j_{\lambda-1,-\epsilon}(\xi)\,. \tag{32}$$

The symbol of the left-hand side of (29) is equal to

$$j_{1-\lambda,-\epsilon}(\xi)\, j_{1-\mu,-\eta}(\zeta) \,\sharp\, (x-y)\,,$$

which by formula (30) is equal to

$$(x-y) j_{1-\lambda,-\epsilon}(\xi)\, j_{1-\mu,-\eta}(\zeta)$$
$$+\frac{1}{i}\left(\frac{\partial}{\partial\xi} j_{1-\lambda,-\epsilon}(\xi) j_{1-\mu,-\eta}(\zeta) - j_{1-\lambda,-\epsilon}(\xi)\frac{\partial}{\partial\zeta} j_{1-\mu,-\eta}(\zeta)\right)$$

and using (32)

$$j_{1-\lambda,-\epsilon}(\xi) = i\,\xi\, j_{2-\lambda,\epsilon}(\xi)\,, \qquad j_{1-\mu,-\eta}(\zeta) = i\,\zeta\, j_{2-\mu,\eta}(\zeta)$$

$$\frac{\partial}{\partial\xi} j_{1-\lambda,-\epsilon} = -\,i\,\lambda\, j_{2-\lambda,\epsilon}\,, \qquad \frac{\partial}{\partial\zeta} j_{1-\mu,-\eta} = -i\mu\, j_{2-\mu,\eta}$$

so that the symbol of the left hand side of (29) is equal to

$$\Big(-(x-y)\,\xi\zeta + i(\lambda\zeta - \mu\xi)\Big) \,\sharp\, j_{2-\lambda,\epsilon}(\xi)\, j_{2-\mu,\eta}(\zeta)\,.$$

This expression is the symbol of $E_{\lambda,\mu} \circ \left(J_{2-\lambda,\epsilon} \otimes J_{2-\mu,\eta}\right)$, where

$$E_{\lambda,\mu} = (x-y)\frac{\partial^2}{\partial x\partial y} + \lambda\frac{\partial}{\partial y} - \mu\frac{\partial}{\partial x}\,,$$

and the main statement of Theorem 4.5 follows. However, in this approach, that is to say using (29) instead of (28) as definition of $E_{\lambda,\mu}$, it is necessary to justify that $E_{\lambda,\mu}$ is an intertwining operator. The following (slightly weaker) property is almost obvious from the definition of $E_{\lambda,\mu}$.

Proposition 4.7 *For any* $X \in \mathfrak{g}$

$$\Big[d(\widetilde{\pi}_{1+\lambda,-\epsilon}\otimes\widetilde{\pi}_{1+\mu,-\eta})(X)\circ E_{\lambda,\mu},\; E_{\lambda,\mu}\circ d(\widetilde{\pi}_{\lambda,\epsilon}\otimes\widetilde{\pi}_{\mu,\eta})(X)\Big]\circ(J_{2-\lambda,\epsilon}\otimes J_{2-\mu,\eta}) = 0.$$

Proof Let $X \in \mathfrak{g}$. Thanks to the intertwining properties of M and of the Knapp–Stein operators,

$$E_{\lambda,\mu} \circ d(\widetilde{\pi}_{\lambda,\epsilon} \otimes \widetilde{\pi}_{\mu,\eta})(X) \circ (J_{2-\lambda,\epsilon} \otimes J_{2-\mu,\eta})$$

$$= E_{\lambda,\mu} \circ (J_{2-\lambda,\epsilon} \otimes J_{2-\mu,\eta}) \circ d(\widetilde{\pi}_{2-\lambda,\epsilon} \otimes \widetilde{\pi}_{2-\mu,\eta})(X)$$

$$= (J_{1-\lambda,-\epsilon} \otimes J_{1-\mu,-\eta}) \circ M \circ d(\widetilde{\pi}_{2-\lambda,\epsilon} \otimes \widetilde{\pi}_{2-\mu,\eta})(X)$$

$$(J_{1-\lambda,-\epsilon} \otimes J_{1-\mu,-\eta}) \circ d(\widetilde{\pi}_{1-\lambda,-\epsilon} \otimes \widetilde{\pi}_{1-\mu,-\eta})(X) \circ M$$

$$= d(\widetilde{\pi}_{1+\lambda,-\epsilon} \otimes \widetilde{\pi}_{1+\mu,-\eta})(X) \circ (J_{1-\lambda,-\epsilon} \otimes J_{1-\mu,-\eta}) \circ M$$

$$= d(\widetilde{\pi}_{1+\lambda,-\epsilon} \otimes \widetilde{\pi}_{1+\mu,-\eta})(X) \circ E_{\lambda,\mu} \circ (J_{2-\lambda,\epsilon} \otimes J_{2-\mu,\eta}) \,.$$

□

To finish the proof of the intertwining property for $E_{\lambda,\mu}$, it remains to show that Proposition 4.7 implies

$$\Big[d(\widetilde{\pi}_{1+\lambda,-\epsilon} \otimes \widetilde{\pi}_{1+\mu,-\eta})(X) \circ E_{\lambda,\mu}\ ,\ E_{\lambda,\mu} \circ d(\widetilde{\pi}_{\lambda,\epsilon} \otimes \widetilde{\pi}_{\mu,\eta})(X)\Big] = 0\,. \tag{33}$$

Assume first that $\lambda = 1 + iu$, $u \in \mathbb{R}$ and $\mu = 1 + iv$, $v \in \mathbb{R}$. Then

$$j_{1+iu,+}(\xi) = 2\cosh(u\frac{\pi}{2})\,\xi^{-iu}, \qquad j_{1+iu,-}(\xi) = 2\sinh(u\frac{\pi}{2})\,\xi^{-iu}\,,$$

so that $J_{1+iu,+}$ (resp. $J_{1+iv,+}$) is an invertible operator on $L^2(\mathbb{R})$. A similar remark applies to $J_{1+iu,-}$ (resp. $J_{1+iv,-}$) except we have to assume in this case that $u \neq 0$ (resp. $v \neq 0$). Under these assumptions, the commutator in (33) has to be 0 on $\mathcal{S}(\mathbb{R}^2)$. But if a differential operator D is 0 on $\mathbb{S}(\mathbb{R}^2)$, it has to be identically 0. Hence we have proved that (33) is valid for $\lambda = 1+iu$, $\mu = 1+iv$ and $u \neq 0$ (resp. $v \neq 0$) in case $\epsilon = -$ (resp. $\eta = -$). The general statement follows as both sides of (33) are holomorphic and polynomial in (λ, μ).

Knowing the source operator $E_{\lambda,\mu}$, the symmetry-breaking bi-differential operators $\boldsymbol{B}^{(k)}_{\lambda,\mu}$ are constructed as in Sect. 1.

Remark 4.8 The reference [1] contains the first approach to intertwining operators similar to $E_{\lambda,\mu}$ for the Lorentz group. It is based on the study of *trilinear invariant forms* with respect to three representations of the principal series, a dual approach to the study of bilinear intertwining operators. Some of *singular* trilinear forms do correspond to covariant bi-differential operators, thus entlightening their *local* character.

5 The Laplace Transform, the Holographic Transform, and the Rankin–Cohen Brackets

This last construction goes back to the initial context of the introduction. It uses some facts on the holomorphic discrete series and in particular its unitarity. This construction was influenced by T. Kobayashi and M. Pevzner's recent work [15]. A generalization to Hermitian symmetric spaces of tube type was obtained in [5].

It is worth to use the universal covering $\widetilde{G}$ of $G = SL(2,\mathbb{R})$. The space Π is simply connected and for $g \in G, c(g,z) \neq 0$ for $z \in \Pi$, so that it is possible to define a holomorphic function $\psi_g(z)$ such that $e^{\psi_g}(z) = c(g,z)$, and two such functions differ by a multiple of $2i\pi$. The universal covering group is then defined by

$$\widetilde{G} = \left\{(g,\psi_g), \quad g \in SL(2,\mathbb{R}), \quad e^{\psi_g(z)} = c(g,z)\right\} ,$$

with group law of $\widetilde{G}$ given by

$$(g,\psi_g)(h,\psi_h) = (gh, \psi_g \circ h + \psi_h) .$$

Notice that the inverse of (g,ψ_g) is

$$(g,\psi_g)^{-1} = (g^{-1},\psi_{g^{-1}}), \quad \psi_{g^{-1}} = -\psi_g \circ g^{-1}$$

Let $\lambda \in \mathbb{R}$. The formula

$$\rho_\lambda(g)F(z) = e^{-\lambda\,\psi_{g^{-1}}(z)} F\big(g^{-1}(z)\big) .$$

defines a representation of $\widetilde{G}$ on $\mathcal{H}(\Pi)$. Notice that for $\lambda = \ell \in \mathbb{N}$, the representation ρ_ℓ lifts down to a representation of G, which coincides with the ρ_ℓ in the Introduction.

For $\lambda > 1$, the representation ρ_λ is unitary (*holomorphic discrete series*). Define the *weighted Bergman space* $\mathcal{H}_\lambda$ as the space of all holomorphic functions F on Π which satisfy

$$\|F\|_\lambda^2 = \int_\Pi |F(x+iy)|^2 y^{\lambda-2} dx dy < +\infty .$$

For $\lambda > 1$, $\mathcal{H}_\lambda$ is not reduced to $\{0\}$ and is a Hilbert space for the norm $\|\ \|_\lambda$. The representation ρ_λ restricted to $\mathcal{H}_\lambda$ can be shown to be unitary and irreducible (see [8] for a presentation of these spaces in the more general context of Hermitian symmetric spaces of tube-type).

There is another realization of the unitary representation ρ_λ, which uses the Laplace transform.

Let f be a smooth function on $\mathbb{R}_+ =]0, +\infty)$. Define its Laplace transform $\mathcal{L}f$ by the formula

$$\mathcal{L}f(z) = \int_0^{+\infty} f(\xi)\, e^{iz\xi}\, d\xi \ .$$

Let $\lambda > 1$ and define

$$L^2_\lambda = \left\{ f : \mathbb{R}_+ \to \mathbb{C}, \quad \int_0^{+\infty} |f(\xi)|^2 \xi^{1-\lambda} d\xi \ < +\infty \right\} .$$

Proposition 5.1 *Let $\lambda > 1$. Then the Laplace transform is an isomorphism from L^2_λ onto $\mathcal{H}_\lambda$. Moreover,*

$$\|\mathcal{L}f\|^2_\lambda = c_\lambda \| f \|^2_\lambda,$$

where $c(\lambda) = 2\pi\, 2^{1-\lambda}\, \Gamma(\lambda - 1)$.

The representation ρ_λ is then transferred on L^2_λ by setting for $g \in G$

$$\widehat{\rho_\lambda}(g) = \mathcal{L}^{-1} \circ \rho_\lambda(g) \circ \mathcal{L} \, .$$

The representation $\widehat{\rho_\lambda}$ can be made explicit for elements in N and A, namely

$$\widehat{\rho_\lambda}\left(\begin{pmatrix} 1 & b \\ 0 & 1 \end{pmatrix}\right) f(\xi) = e^{-b\xi} f(\xi)$$

$$\widehat{\rho_\lambda}\left(\begin{pmatrix} t & 0 \\ 0 & \frac{1}{t} \end{pmatrix}\right) f(\xi) = t^{-2-\lambda,\epsilon} f\left(\frac{\xi}{t^2}\right) .$$

The (completed) tensor product $\rho_\lambda \widehat{\otimes} \rho_\mu$ of two such representations has a similar realization. For $\lambda, \mu > 1$, let $\mathcal{H}_{\lambda,\mu}$ be the space of all holomorphic functions F on $\Pi \times \Pi$ such that

$$\int_{\Pi\times\Pi} |F(x+iy, s+it)|^2 y^{\lambda-2} t^{\mu-2} dx\, dy\; ds\, dt \ < +\infty \, .$$

There is a natural action "'diagonal" of G on this space, which gives a realization of $\rho_\lambda \widehat{\otimes} \rho_\mu$. There is a corresponding realization through the Laplace transform.

Let $\mathbb{R}^2_+ =]0, \infty \times]0, +\infty)$ and set

$$L^2_{\lambda,\mu} = \left\{ f : \mathbb{R}^2_+ \longrightarrow \mathbb{C}, \quad \int_0^{+\infty}\int_0^{+\infty} |f(\xi,\zeta)|^2 \xi^{1-\lambda} \zeta^{1-\mu} d\xi d\zeta \ < +\infty \right\} .$$

The Laplace transform $\mathcal{L}_2$ on $\mathbb{R}^2_+$ is given by

$$\mathcal{L}_2 f(z,w) = \iint_{]0,+\infty)\times]0,+\infty)} f(\xi,\zeta) e^{i(z\xi+w\zeta)} d\xi\, d\zeta\ ,$$

and yields an isomorphism of $L^2_{\lambda,\mu}$ onto $\mathcal{H}_{\lambda,\mu}$, from which follows the realization of the representation $\widehat{\rho_\lambda}\,\widehat{\otimes}\,\widehat{\rho_\mu}$ on $L^2_{\lambda,\mu}$.

Let us first interpret the 0th-order Rankin–Cohen operator (i.e., the restriction map to the diagonal) in its version on the Laplace transform side.

Proposition 5.2 *Let* $\iota :]0,+\infty\times]-1,+1[\longrightarrow \mathbb{R}^2_+$ *be given by*

$$\iota(\rho,\theta) = \left(\rho\,\frac{(1-\theta)}{2}, \rho\,\frac{(1+\theta)}{2}\right) .$$

The map ι *is a diffeomorphism from* $]0,+\infty\times]-1,+1[$ *onto* $\mathbb{R}^2_+$. *The inverse map is given by*

$$\iota^{-1}(\xi,\zeta) = \left(\xi+\zeta\ ,\ \frac{\zeta-\xi}{\xi+\zeta}\right) .$$

Moreover, for $f \in C_c^\infty(\mathbb{R}^2_+)$

$$\int_{]0,+\infty\times]0,+\infty)} f(\xi,\zeta)\, d\xi\, d\zeta \ = \frac{1}{2}\int_{]0,+\infty\times]-1,+1[} (f\circ\iota)(\rho,\theta)\rho\, d\rho\, d\theta\ . \tag{34}$$

For a proof, see [15].

Let $f \in C_c^\infty(\mathbb{R}^2_+)$. Let

$$\mathcal{I}f(\rho) = \frac{\rho}{2}\int_{-1}^{1} f(\iota(\rho,\theta))d\theta\ . \tag{35}$$

Notice that $\mathcal{I}f$ belongs to $C_c^\infty(\mathbb{R}_{>0})$ and that $\mathcal{I} : C_c^\infty(\mathbb{R}^2_+) \longrightarrow C_c^\infty(\mathbb{R}_+)$ is a continuous operator.

Proposition 5.3

$$\mathrm{res}\circ\mathcal{L}_2 = \mathcal{L}\circ\mathcal{I} \tag{36}$$

Proof Let $f \in C_c^\infty(\mathbb{R}^2_+)$. Then

$$(\mathrm{res}\circ\mathcal{L}_2)f\,(z) = \iint_{\mathbb{R}^2_+} e^{iz(\xi+\zeta)} f(\xi,\zeta)\, d\xi\, d\zeta$$

$$= \int_0^\infty e^{iz\rho}\left(\frac{\rho}{2}\int_{-1}^{1} f(\iota(\rho,\theta)\, d\theta\right) d\rho$$

by the change of variables formula (34). □

Lemma 5.4 *Let $\lambda, \mu > 1$. There exists a constant $C > 0$, depending only on (λ, μ), such that for any $f \in C_c^\infty(\mathbb{R}_+^2)$*

$$\|\mathcal{I} f\|_{\lambda+\mu} \leq C\|f\|_{\lambda,\mu} \,.$$

Proof By Cauchy-Schwarz inequality[3]

$$\left|\int_{-1}^{1} f(\iota(\rho,\theta)d\theta\right|^2$$

$$\leq \left(\int_{-1}^{1} |f(\iota(\rho,\theta)|^2\left(\tfrac{1-\theta}{2}\right)^{1-\lambda}\left(\tfrac{1+\theta}{2}\right)^{1-\mu}d\theta\right)\left(\int_{-1}^{1}\left(\tfrac{1-\theta}{2}\right)^{-1+\lambda}\left(\tfrac{1+\theta}{2}\right)^{-1+\mu}d\theta\right)$$

$$\leq C\int_{-1}^{1} |f(\iota(\rho,\theta)|^2\left(\frac{1-\theta}{2}\right)^{1-\lambda}\left(\frac{1+\theta}{2}\right)^{1-\mu}d\theta \,.$$

Hence,

$$\|\mathcal{I} f\|_{\lambda+\mu}^2 = \int_0^{+\infty} |\mathcal{I} f(\rho)\,|^2 \rho^{1-(\lambda+\mu)}\, d\rho$$

$$= \int_0^{+\infty} \frac{\rho^2}{4}\left|\int_{-1}^{1} f(\iota(\rho,\theta)d\theta\right|^2 \rho^{1-(\lambda+\mu)}d\rho$$

$$\leq C\int_0^{+\infty}\int_{-1}^{1} |f(\iota(\rho,\theta)|^2\left(\tfrac{1-\theta}{2}\right)^{1-\lambda}\left(\tfrac{1+\theta}{2}\right)^{1-\mu}\rho^{1-\lambda}\rho^{1-\mu}\rho\, d\rho\, d\theta$$

$$= C\int_{\mathbb{R}_+^2} |f(\xi,\zeta)|^2\xi^{1-\lambda}\zeta^{1-\mu}\, d\xi\, d\zeta = C\|f\|_{\lambda,\mu}^2 \,.$$

□

Hence, the identity (36) can be extended by continuity to the space $L^2_{\lambda,\mu}$. As $\mathcal{L}$ and $\mathcal{L}_2$ are isomorphisms,

$$\operatorname{res} = \mathcal{L}\circ\mathcal{I}\circ\mathcal{L}_2^{-1}\,, \tag{37}$$

as an equality of maps from $\mathcal{H}_{\lambda,\mu}$ into $\mathcal{H}_{\lambda+\mu}$.

An easy extension of Lemma 5.4 yields the following result.

[3] As usual, the letter C in the proof may take different values from one line to the next one

Lemma 5.5 *Let D be a holomorphic differential operator with constant coefficients, homogeneous of degree k on $\mathbb{C}^2$. Then* $\operatorname{res}\circ D$ *extends as a continuous operator from $\mathcal{H}_{\lambda,\mu}$ into $\mathcal{H}_{\lambda+\mu+2k}$.*

For $\lambda, \mu > 1$ and $k \in \mathbb{N}$, *define* the Rankin–Cohen bracket $B^{(k)}_{\lambda,\mu}$ as the bi-differential operator

$$B^{(k)}_{\lambda,\mu} F(z) = b^{(k)}_{\lambda,\mu}\left(\frac{\partial}{\partial z}, \frac{\partial}{\partial w}\right) F\ (z, z)\ .$$

The Rankin–Cohen operators have a counterpart when working with the weighted L^2 spaces on $\mathbb{R}_+$ instead of the weighted Bergman spaces. Define the operator $\widehat{\mathbf{B}}^{(k)}_{\lambda,\mu}$ by the formula

$$\widehat{\mathbf{B}}^{(k)}_{\lambda,\mu} = \mathcal{L}^{-1} \circ \mathbf{B}^{(k)}_{\lambda,\mu} \circ \mathcal{L}_2 = \mathcal{L}^{-1}\ \circ \operatorname{res} \circ\, b^{(k)}_{\lambda,\,\mu}\left(\frac{\partial}{\partial z}, \frac{\partial}{\partial w}\right) \circ \mathcal{L}_2\ , \tag{38}$$

where $b^{(k)}_{\lambda,\mu}$ are defined in Proposition 2.3. Introduce the polynomial $B^{(k)}_{\lambda,\mu}$ on $\mathbb{R}$ defined by

$$B^{(k)}_{\lambda,\mu}(\theta) = b^{(k)}_{\lambda,\mu}\left(\frac{1-\theta}{2}, \frac{1+\theta}{2}\right)\ . \tag{39}$$

so that

$$b^{(k)}_{\lambda,\mu}\iota(\rho,\theta) = \rho^k B^{(k)}_{\lambda,\mu}(\theta) \tag{40}$$

Proposition 5.6 *For $f \in C^\infty_c\left(\mathbb{R}^2_+\right)$*

$$\widehat{\mathbf{B}}^{(k)}_{\lambda,\mu} f(\rho) = \frac{1}{2} i^k \rho^{k+1} \int_{-1}^{1} B^{(k)}_{\lambda,\,\mu}(\theta) f\big(\iota(\rho,\theta)\big) d\theta\ . \tag{41}$$

Proof Recall the elementary formula for the Laplace transform, valid for any holomorphic polynomial p

$$p\left(\frac{\partial}{\partial z}\right)(\mathcal{L}f)(z) = \int_0^\infty p(i\xi) f(\xi) e^{i(z,\xi)}\, d\xi\ .$$

There is an obvious extension of this result for $\mathcal{L}_2$. When applied to the polynomial $b^{(k)}_{\lambda,\mu}$, this yields the following result.

$$b_{\lambda,\,\mu}\left(\frac{\partial}{\partial z}, \frac{\partial}{\partial w}\right) \circ \mathcal{L}_2 = i^k\ \mathcal{L}_2 \circ b_{\lambda,\,\mu}(\xi,\eta)\ . \tag{42}$$

Start from the definition (38) of $\widehat{\mathbf{B}}^{(k)}_{\lambda,\mu}$, use (37) and (42) to obtain

$$\widehat{\mathbf{B}}^{(k)}_{\lambda,\mu} = i^k \left(\mathcal{I} \circ \mathcal{L}_2^{-1}\right) \circ \left(\mathcal{L}_2 \circ c^{(k)}_{\lambda,\mu}(\xi,\eta)\right) = i^k\, \mathcal{J} \circ b^{(k)}_{\lambda,\,\mu}\,.$$

For $f \in C_c^\infty\left(\mathbb{R}^2_+\right)$

$$\widehat{\mathbf{B}}^{(k)}_{\lambda,\mu} f(\rho) = \frac{i^k}{2}\,\rho \int_{-1}^{1} b^{(k)}_{\lambda,\mu}(\iota(\rho,\theta)) f(\iota(\rho,\theta)) d\theta$$

$$= \frac{i^k}{2}\,\rho^{k+1} \int_{-1}^{1} B^{(k)}_{\lambda,\,\mu}(\theta) f\,(\iota(\rho,\theta))\, d\theta\;,$$

and the conclusion follows. □

Needless to say, the operator $\widehat{\mathbf{B}}^{(k)}_{\lambda,\mu}$ extends as a continuous operator from $L^2_{\lambda,\mu}$ into $L^2_{\lambda+\mu+2k}$, as a consequence of Lemma 5.5.

The space $\mathcal{H}_\lambda$ has a *reproducing kernel* equal to

$$K_\lambda(z,w) = \frac{\lambda-1}{4\pi}\left(\frac{z-\overline{w}}{2i}\right)^{-\lambda}.$$

Let

$$\kappa^w_\lambda(z) = K_\lambda(z,w) = \frac{\lambda-1}{4\pi}\left(\frac{z-\overline{w}}{2i}\right)^{-\lambda}, \tag{43}$$

and in particular,

$$\kappa_\lambda(z) = \kappa^i_\lambda(z) = \frac{\lambda-1}{4\pi}\left(\frac{z+i}{2i}\right)^{-\lambda}. \tag{44}$$

Lemma 5.7 *κ_λ is the Laplace transform of the function ϕ_λ given by*

$$\phi_\lambda(\xi) = \frac{1}{4\pi}\frac{2^\lambda}{\Gamma(\lambda-1)}\, e^{-\xi}\, \xi^{\lambda-1}\,.$$

See [8] for a proof in a broader context.

For simplicity, set

$$c(\lambda) = \frac{1}{4\pi}\frac{2^\lambda}{\Gamma(\lambda-1)}\,. \tag{45}$$

Define

$$\kappa_{\lambda,\mu}^{(k)}(z,w) = (z-w)^k\,\kappa_{\lambda+k}(z)\,\kappa_{\mu+k}(w)\,. \tag{46}$$

Lemma 5.8 *The function $\kappa_{\lambda,\mu}^{(k)}$ belongs to $\mathcal{H}_{\lambda,\mu}$.*

Proof Observe that

$$\kappa_{\lambda,\mu}^{(k)}(z,w) = \text{constant}\,\left(\frac{z-w}{(z+i)(w+i)}\right)^k \kappa_\lambda(z)\kappa_\mu(w)\,.$$

Now $\kappa_\lambda(z)\kappa_\mu(w) \in \mathcal{H}_{\lambda,\mu}$, and

$$sup_{z,w\in\Pi}\left|\frac{z-w}{(z+i)(w+i)}\right| < +\infty\,,$$

so that the statement follows. □

Lemma 5.9 *The function $\kappa_{\lambda,\mu}^{(k)}$ is the Laplace transform of the function*

$$\phi_{\lambda,\mu}^{(k)}(\xi,\zeta) = c^{(k)}(\lambda,\mu)\,b_{\lambda,\mu}^{(k)}(\xi,\zeta)\phi_\lambda(\xi)\phi_\mu(\zeta)$$

where

$$c^{(k)}(\lambda,\mu) = \frac{c(\lambda+k)c(\mu+k)}{c(\lambda)c(\mu)}\,i^k \tag{47}$$

Proof By the previous lemma,

$$\kappa_{\lambda+k}(z)\kappa_{\mu+k}(w)$$

$$= c(\lambda+k)c(\mu+k)\iint_{\mathbb{R}_+^2} e^{i(z\xi+w\zeta)}e^{-(\xi+\zeta)}\xi^{\lambda+k-1}\zeta^{\mu+k-1}\,d\xi\,d\zeta\,,$$

and hence by using the classical formula $z\circ\mathcal{L} = i\mathcal{L}\circ\frac{\partial}{\partial\xi}$

$$(z-w)^k\kappa_{\lambda+k}^i(z)\kappa_{\mu+k}^i(w) =$$

$$= c(\lambda+k)c(\mu+k)\,i^k\iint_{\mathbb{R}_+^2} e^{i(z\xi+w\zeta)}\left(\frac{\partial}{\partial\xi}-\frac{\partial}{\partial\zeta}\right)^k\left(e^{-(\xi+\zeta)}\xi^{\lambda+k-1}\zeta^{\mu+k-1}\right)d\xi\,d\zeta$$

$$= c(\lambda+k)c(\mu+k)\,i^k\iint_{\mathbb{R}_+^2} e^{i(z\xi+w\zeta)}e^{-(\xi+\zeta)}\left(\frac{\partial}{\partial\xi}-\frac{\partial}{\partial\zeta}\right)^k\left(\xi^{\lambda+k-1}\zeta^{\mu+k-1}\right)d\xi\,d\zeta$$

$$= c(\lambda+k)c(\mu+k)\, i^k \iint_{\mathbb{R}^2_+} e^{i(z\xi+w\zeta)} e^{-(\xi+\zeta)} b^{(k)}_{\lambda,\mu}(\xi,\zeta)\xi^{\lambda-1}\zeta^{\mu-1}\, d\xi\, d\zeta\ ,$$

$$= \frac{c(\lambda+k)c(\mu+k)}{c(\lambda)c(\mu)}\, i^k \iint_{\mathbb{R}^2_+} e^{i(z\xi+w\zeta)} b^{(k)}_{\lambda,\mu}(\xi,\zeta)\phi_\lambda(\xi)\phi_\mu(\zeta)\, d\xi\, d\zeta\ ,$$

from which the formula follows easily. □

Let $\widehat{B}^{(k)\,\star}_{\lambda,\mu}$ be the adjoint of $\widehat{B}^{(k)}_{\lambda,\mu}$. It is a continuous operator from $L^2_{\lambda+\mu+2k}$ into $L^2_{\lambda,\mu}$.

Lemma 5.10

$$\widehat{\mathbf{B}}^{(k)\,*}_{\lambda,\mu}\, \phi_{\lambda+\mu+2k} = \gamma^{(k)}(\lambda,\mu)\phi^{(k)}_{\lambda,\mu} \tag{48}$$

where

$$\gamma^{(k)}(\lambda,\mu) = (-1)^k\, \frac{c(\lambda+\mu+2k)}{c(\lambda)c(\mu)}\ .$$

Proof For any $\varphi \in L^2_{\lambda,\mu}$, we want to show that

$$\left(\phi_{\lambda+\mu+2k}\, ,\, \widehat{B}^{(k)}_{\lambda,\mu}\varphi\right)_{L^2_{\lambda+\mu+2k}} = \gamma^{(k)}(\lambda,\mu)\left(\phi^{(k)}_{\lambda,\mu}\, ,\, \varphi\right)_{L^2_{\lambda,\mu}}\ . \tag{49}$$

Let compute first the right handside of (49).

$$\left(\phi^{(k)}_{\lambda,\mu}\, ,\, \varphi\right)_{L^2_{\lambda,\mu}} =$$

$$c^{(k)}(\lambda,\mu)\, c(\lambda+k)\, c(\mu+k) \int_{\mathbb{R}^2_+} b_{\lambda,\mu}(\xi,\zeta) e^{-\xi}\xi^{\lambda-1} e^{-\zeta}\zeta^{\mu-1}\overline{\varphi(\xi,\zeta)}\xi^{1-\lambda}\zeta^{1-\mu}\, d\xi\, d\zeta$$

$$= c(\lambda)\, c(\mu)\, i^k \int_0^{+\infty}\!\!\int_0^{\infty} b_{\lambda,\mu}(\xi,\zeta) e^{-(\xi+\zeta)}\overline{\varphi(\xi,\zeta)}\, d\xi\, d\zeta\ .$$

For the left handside, we use (41) to write

$$\left(\phi_{\lambda+\mu+2k}\, ,\, \widehat{B}^{(k)}_{\lambda,\mu}\varphi\right)_{L^2_{\lambda+\mu+2k}} = c(\lambda+\mu+2k)\frac{(-i)^k}{2} \times \cdots$$

$$\int_0^{+\infty} e^{-\rho}\rho^{\lambda+\mu+2k-1}\rho^{k+1}\int_{-1}^{1} B^{(k)}_{\lambda,\mu}(\theta)\overline{\varphi(\iota(\rho,\theta))}d\theta\ \rho^{1-(\lambda+\mu+2k)}d\rho$$

$$= c(\lambda+\mu+2k)\frac{(-i)^k}{2}\int_0^{+\infty}\int_{-1}^{1} b_{\lambda,\mu}^{(k)}(\iota(\rho,\theta))\overline{\varphi(\iota(\rho,\theta)}\, e^{-\rho}\,\rho\, d\rho\, d\theta$$

$$= c(\lambda+\mu+2k)(-i)^k\int_0^{+\infty}\int_0^{\infty} b_{\lambda,\mu}(\xi,\zeta)e^{-(\xi+\zeta)}\overline{\varphi(\xi,\zeta)}\, d\xi\, d\zeta\ ,$$

where we used the change of variable formula (34). The conclusion follows from (45) and (47). □

Corollary 5.11

$$B_{\lambda,\mu}^{(k)\,\star}\kappa_{\lambda+\mu+2k} = \gamma^{(k)}(\lambda,\mu)\kappa_{\lambda,\mu}^{(k)}\ . \tag{50}$$

After these preliminaries, we now come to the proof of the intertwining property of the operators $\mathbf{B}_{\lambda,\mu}^{(k)}$.

As the representations $\rho_\lambda\otimes\rho_\mu$ and $\rho_{\lambda+\mu+2k}$ are unitary, it is equivalent to prove that its adjoint $\mathbf{B}_{\lambda,\mu}^{(k)\ \star}$ satisfy the dual intertwining property, for any $g\in G$

$$\mathbf{B}_{\lambda,\mu}^{(k)\ \star}\circ\rho_{\lambda+\mu+2k}(g) = (\rho_\lambda(g)\otimes\rho_\mu(g))\circ\mathbf{B}_{\lambda,\mu}^{(k)\ \star}\ . \tag{51}$$

As N acts by translations on Π and A^* acts by dilation on Π, the following result is easy to prove.

Lemma 5.12 *The intertwining relation* (51) *is satisfied for* $g\in A^*$ *and for* $g\in N$.

As the functions $\left(\kappa_{\lambda+\mu+2k}^{v}\right)_{v\in\Pi}$ form a total system in $\mathcal{H}_{\lambda+\mu+2k}$, it is enough to verify the intertwining condition on these functions.

Lemma 5.13 *Let* $v\in\Pi$. *Then, for any* $g\in G$

$$\rho_\lambda(g)\kappa_\lambda^{v} = \overline{c(g,v)^{-\lambda}}\ \kappa_\lambda^{g(v)}\ . \tag{52}$$

Proof Let F be an arbitrary function in $\mathcal{H}_\lambda$. Then, by unitarity of the representation ρ_λ

$$\big(F\ ,\ \rho_\lambda(g)\kappa_\lambda^{v}\big) = \big(\rho_\lambda(g^{-1})F\ ,\ \kappa_\lambda^{v}\big) = \big(\rho_\lambda(g^{-1})F\big)(v)$$

$$= c(g,v)^{-\lambda}F(gv) = \big(F\ ,\ \overline{c(g,v)^{-\lambda}}\kappa_\lambda^{g(v)}\big)\ ,$$

which implies the statement. □

For $v\in\Pi$, let $\kappa_{\lambda,\mu}^{(k)\,v}$ be defined by

$$\kappa_{\lambda,\mu}^{(k)\,v}(z,w) = \kappa_{\lambda+k}^{v}(z)(z-w)^k\kappa_{\mu+k}^{v}(w)$$

Proposition 5.14

$$\big(\rho_\lambda(g)\otimes\rho_\mu(g)\big)\kappa^{(k)\,v}_{\lambda,\mu}=\overline{c(g,v)^{-(\lambda+\mu+2k)}}\,\kappa^{(k)\,g(v)}_{\lambda,\mu} \tag{53}$$

Proof

$$\big(\rho_\lambda(g)\otimes\rho_\mu(g)\big)\kappa^{(k)\,v}_{\lambda,\mu}(z,w)$$

$$=c(g^{-1},z)^{-\lambda}c(g^{-1},w)^{-\mu}\kappa^{(k)\,v}_{\lambda,\mu}\big(g^{-1}(z),g^{-1}(w)\big)$$

$$=c(g^{-1},z)^{-\lambda}\kappa^{v}_{\lambda+k}(z)\,\big(g^{-1}(z)-g^{-1}(w)\big)^k\,c(g^{-1},w)^{-\mu}\kappa^{v}_{\mu+k}(w)$$

$$=c(g^{-1},z)^{-\lambda-k}\kappa^{v}_{\lambda+k}(z)(z-w)^k c(g^{-1},w)^{-\mu-k}\kappa^{v}_{\mu+k}(w)$$

$$=\big(\rho_{\lambda+k}(g)\kappa^{v}_{\lambda+\rho}\big)(z)(z-w)^k\big(\rho_{\mu+k}(g)\kappa^{v}_{\mu+k}\big)(w)$$

$$=\overline{c(g,v)^{-\lambda-\mu-2k}}\,\kappa^{(k)\,g(v)}_{\lambda,\mu}(z,w)\,.$$

□

We now verify the intertwining property (51) for the function $\kappa_{\lambda+\mu+2k}$.

Lemma 5.15 *For $v=s+it\in\Pi$, let $p_v=\begin{pmatrix}\sqrt{t} & \frac{s}{\sqrt{t}}\\ 0 & \frac{1}{\sqrt{t}}\end{pmatrix}$ and observe that $p_v(i)=v$ and $p_v\in A^*N$.*

Let $g\in G$ and let $v=g(i)\in\Pi$. Set $\ell=p_v^{-1}g$ so that $\ell(i)=i$. As $g=p_v\ell$, use (52) *for $g=\ell$ and $v=i$*

$$\rho_{\lambda+\mu+2k}(g)\kappa_{\lambda+\mu+2k}=\overline{c(\ell,i)^{-(\lambda+\mu+2k)}}\rho_{\lambda+\mu+2k}(p_v)\kappa_{\lambda+\mu+2k}\,.$$

Hence, by using Lemma (5.12)

$$\big(\mathbf{B}^{(k)\,\star}_{\lambda,\mu}\circ\rho_{\lambda+\mu+2k}(g)\big)\kappa_{\lambda+\mu+2k}$$

$$=\overline{c(\ell,i)^{-(\lambda+\mu+2k)}}\big(\rho_\lambda(p_v)\otimes\rho_\mu(p_v)\big)\mathbf{B}^{(k)\,\star}_{\lambda,\mu}\kappa_{\lambda+\mu+2k}$$

and by (50)

$$=\overline{c(\ell,i)^{-(\lambda+\mu+2k)}}\gamma^{(k)}(\lambda,\mu)\big(\rho_\lambda(p_v)\otimes\rho_\mu(p_v)\big)\kappa^{(k)}_{\lambda,\mu}$$

and by (53)

$$= \overline{c(\ell, i)^{-(\lambda+\mu+2k)}}\gamma^{(k)}(\lambda, \mu)\overline{c(p_v, i)^{-(\lambda+\mu+2k)}}\kappa^{(k)\,v}_{\lambda,\mu}$$

$$= \overline{c(g, i)^{-(\lambda+\mu+2k)}}\gamma^{(k)}(\lambda, \mu)\kappa^{(k)\,g(i)}_{\lambda,\mu}$$

$$= \gamma^{(k)}(\lambda, \mu)\big(\rho_\lambda(g) \otimes \rho_\mu(g)\big)\kappa^{(k)}_{\lambda,\mu}$$

and again by (50)

$$= \big(\rho_\lambda(g) \otimes \rho_\mu(g)\big)B^{(k)\,\star}_{\lambda,\mu}\kappa_{\lambda+\mu+2k}\,.$$

To finish the proof, it remains to prove the intertwining property for $\kappa^v_{\lambda+\mu+2k}$, *where* v *is an arbitrary element in* Π. *Now as*

$$\kappa^v_{\lambda+\mu+2k} = \overline{c(p_v, i)^\lambda}\rho_{\lambda+\mu+2k}(p_v)\kappa_{\lambda+\mu+2k}\,,$$

$$\rho_{\lambda+\mu+2k}(g)\kappa^v_{\lambda+\mu+2k} = \overline{c(p_v, i)^\lambda}\,\rho_{\lambda+\mu+2k}(gp_v)\kappa_{\lambda+\mu+2k}\,.$$

We now may apply the previous result for the element gp_v *to get the required intertwining property on the element* $\kappa^v_{\lambda+\mu+2k}$. *This finishes the proof.*

For a generalization of this construction for the Hermitian symmetric spaces of tube type, see [4].

6 Link with the Jacobi Polynomials

Recall the definition of the Jacobi polynomials (see [9] 8.960)

$$P_k^{\alpha,\beta}(x) = \frac{(-1)^k}{2^k k!}(1-x)^{-\alpha}(1+x)^{-\beta}\frac{d^k}{dx^k}\left((1-x)^{\alpha+k}(1+x)^{\beta+k}\right).$$

Lemma 6.1 *Let* g *be a smooth function on* $]0, +\infty)\times]-1, +1[$ *and let* f *the function on* $\mathbb{R}^2_+$ *such that* $f \circ \iota = g$. *Then*

$$\left(\frac{\partial}{\partial\xi} - \frac{\partial}{\partial\zeta}\right) f = -\frac{2}{\rho}\frac{\partial g}{\partial\rho}\,. \tag{54}$$

Proof For $\xi, \zeta > 0$,

$$f(\xi, \zeta) = g\left(\xi + \zeta, \frac{-\xi + \zeta}{\xi + \zeta}\right).$$

Then

$$\left(\frac{\partial}{\partial \xi} - \frac{\partial}{\partial \zeta}\right) f =$$

$$= \frac{\partial g}{\partial \rho} + \left(\frac{-(\xi+\zeta)-(\xi-\zeta)}{(\xi+\zeta)^2}\right)\frac{\partial g}{\partial \theta} - \frac{\partial g}{\partial \rho} - \left(\frac{(\xi+\zeta)-(\xi-\zeta)}{(\xi+\zeta)^2}\right)\frac{\partial g}{\partial \theta}$$

$$= -\frac{2}{\xi+\zeta}\frac{\partial g}{\partial \theta}\,.$$

□

Proposition 6.2

$$B^{(k)}_{\lambda,\mu}(\theta) = k!\, P_k^{\lambda-1\,,\,\mu-1}(\theta)\,. \tag{55}$$

Proof For $\xi, \zeta > 0$,

$$\xi^{\lambda+k-1}\zeta^{\mu+k-1} = \rho^{\lambda+\mu+2k-2}\left(\frac{1-\theta}{2}\right)^{\lambda+k-1}\left(\frac{1+\theta}{2}\right)^{\mu+k-1}.$$

Apply the previous lemma to obtain

$$\left(\left(\frac{\partial}{\partial \xi} - \frac{\partial}{\partial \zeta}\right)^k \xi^{\lambda+k-1}\zeta^{\mu+k-1}\right)(\iota(\rho,\theta))$$

$$= (-1)^k \frac{2^k}{\rho^k}\,\rho^{\lambda+\mu+2k-2}\left(\frac{d}{d\theta}\right)^k\left(\frac{1-\theta}{2}\right)^{\lambda+k-1}\left(\frac{1+\theta}{2}\right)^{\mu+k-1}.$$

Now by definition,

$$\left(\left(\frac{\partial}{\partial \xi} - \frac{\partial}{\partial \zeta}\right)^k \xi^{\lambda+k-1}\zeta^{\mu+k-1}\right)(\iota(\rho,\theta))$$

$$= \rho^{\lambda+\mu-2}\left(\frac{1-\theta}{2}\right)^{\lambda-1}\left(\frac{1+\theta}{2}\right)^{\mu-1}\rho^k B^{(k)}_{\lambda,\mu}(\theta)$$

so that, by identification

$$B^{(k)}_{\lambda,\mu}(\theta) = \frac{(-1)^k}{2^k}(1-\theta)^{-(\lambda-1)}(1+\theta)^{-(\mu-1)}\left(\frac{d}{d\theta}\right)^k\left((1-\theta)^{(\lambda-1)+k}(1+\theta)^{(\mu-1)+k}\right),$$

and the result follows. □

References

1. Beckmann, R., Clerc, J.-L.: Singular invariant trilinear forms and covariant (bi-) differential operators under the conformal group. J. Funct. Anal. **262**, 4341–4376 (2012)
2. Ben Saïd S., Clerc J-L, Koufany, Kh: Conformally covariant bidifferential operators on a simple Jordan algebra. Int. Math. Res. Notes (2018). https://doi.org/10.193/imrn/rny082
3. Cayley, A.: On linear transformations. Camb. Dublin Math. J. **1**, 104–122 (1846)
4. Clerc, J.-L.: Symmetry breaking differential operators, the source operator and Rodrigues formulæ. Pac. J. Math. **307**, 79–107 (2020)
5. Clerc, J.-L.: Rankin-Cohen brackets on tube-type domains. Tunisian J. Math. **3**, 551–569 (2021)
6. Clerc, J.-L.: Construction à la Ibukiyama of symmetry breaking differential operators. J. Math. Sci. Univ. Tokyo **29**, 51–88 (2022)
7. Cohen, H.: Sums involving the values at negative integers of L-functions of quadratic characters. Math. Ann. **217**, 271–285 (1975)
8. Faraut, J., Korányi, A.: Analysis on Symmetric Cones. Oxford University Press, Oxford (1994)
9. Gradshteyn, I.S., Ryzhik, I.: Table of Integrals, Series and Products, 7th edn. Elsevier/Academic Press, Amsterdam (2007)
10. Hörmander, L.: The Analysis of Linear Partial Differential Operators I. Springer, Berlin (1983)
11. Ibukiyama, T.: On differential operators on automorphic forms and invariant pluriharmonic polynomials. Commun. Math. Univ. Sancti Pauli **48**, 103–117 (1999)
12. Knapp, A.: Representation Theory of Semisimple Groups, an Overview Based on Examples. Princeton University Press, Princeton (1986)
13. Kobayashi, T.: F-method for symmetry breaking operators. Diff. Geom. Appl. **33**, 272–289 (2014)
14. Kobayashi, T., Pevzner, M.: Differential symmetry breaking operators II. Rankin-Cohen operators for symmetric pairs. Selecta Math. **22**, 847–911 (2016)
15. Kobayashi, T., Pevzner, M.: Inversion of Rankin-Cohen operators via holographic transform. Ann. Inst. Fourier (Grenoble) **70**, 2131–2190 (2020)
16. Olver, P.: Classical Invariant Theory. London Mathematical Society Student Texts, vol. 44. Cambridge University Press, Cambridge (1999)
17. Ovsienko, V., Redou, P.: Generalized transvectants, Rankin-Cohen brackets. Lett. Math. Phys. **63**, 19–28 (2003)
18. Peng, L., Zhang, G.: Tensor products of holomorphic representations and bilinear differential operators. J. Funct. Anal. **210**(2004), 171–192 (2003)

Restricting Holomorphic Discrete Series Representations to a Compact Dual Pair

Jan Frahm and Quentin Labriet

Abstract The goal of this chapter is to study the branching problem for a holomorphic discrete series representation of the conformal group of a simple Euclidean Jordan algebra V restricted to the subgroup $\mathrm{PSL}_2(\mathbb{R}) \times \mathrm{Aut}(V)$ where $\mathrm{Aut}(V)$ denotes the compact group of automorphisms of V. We use a realization of the holomorphic discrete series on a space of vector-values L^2-functions as well as the *stratified model* developed by the second author to relate the branching problem to the decomposition of certain representations of the compact group $\mathrm{Aut}(V)$ and to vector-valued orthogonal polynomials.

1 Introduction

The restriction of an irreducible unitary representation of a Lie group G to a non-compact closed subgroup G' does in general not decompose into a direct sum of irreducible representations of G', but rather into a direct integral. However, there are some pairs of groups (G, G') and classes of irreducible unitary representations of G, which do indeed decompose discretely when restricted to G'. A detailed study of such settings was initiated by Kobayashi [5, 6, 7, 8], thus providing a suitable framework for discrete branching problems.

One consequence of Kobayashi's results is that any holomorphic discrete series representation of a real semisimple group G of Hermitian type decomposes discretely when restricted to a subgroup G' of Hermitian type whose corresponding Riemannian symmetric space embeds holomorphically into the one for G. If (G, G') is a symmetric pair, he even provides a formula for the multiplicities occurring in the decomposition. In this chapter, we study a family of subgroups G', which are, in general, not symmetric, but still fall into the framework of discrete decomposability.

J. Frahm (✉) · Q. Labriet
Department of Mathematics, Aarhus University, Aarhus, Denmark
e-mail: frahm@math.au.dk; quentin.labriet@math.au.dk

M. Pevzner, H. Sekiguchi (eds.), *Symmetry in Geometry and Analysis, Volume 2*,
Progress in Mathematics 358, https://doi.org/10.1007/978-981-97-7662-7_4

The subgroups are of the form $G' = \mathrm{PSL}_2(\mathbb{R}) \times H$ for some compact group H, so that $\mathrm{PSL}_2(\mathbb{R})$ and H form a dual pair inside G.

In contrast to Kobayashi's multiplicity formula, which is obtained by algebraic methods, our study is of a more analytic nature. Following Ding–Gross [1], we realize the holomorphic discrete series representations of G on a Hilbert space of vector-valued L^2-functions on a symmetric cone (assuming the group G is of tube type). The key point is to use a particular set of coordinates that is adapted to the subgroup G', yielding the so-called *stratified model* introduced by the second author in [10] following ideas already present in [9]. Loosely speaking, this set of coordinates separates the actions of $\mathrm{PSL}_2(\mathbb{R})$ and H and reduces the branching problem to one for the compact group H. It further relates the decomposition to certain (vector-valued) orthogonal polynomials on a real bounded domain. Using these orthogonal polynomials, we are able to give explicit formulas for the corresponding symmetry breaking resp. holographic operators, i.e., the intertwining operators that project onto the various discrete summands of the representation resp. embed the discrete summands into the holomorphic discrete series representation.

1.1 Statement of the Results

Let us describe the results in more detail. Let S_π be a holomorphic discrete series representation of the conformal group G of a simple Jordan algebra V associated to a representation (π, V_π) of a maximal compact subgroup K of G. Since K and the automorphism group L of the symmetric cone Ω of invertible squares in V have isomorphic complexifications, we can consider π to be a representation of L. Further, let $H = \mathrm{Aut}(V)$ be the group of automorphisms of V, or alternatively the subgroup of L fixing the identity element of the Jordan algebra V. The centralizer of H in G is isomorphic to $\mathrm{PSL}_2(\mathbb{R})$, so the product $G' = \mathrm{PSL}_2(\mathbb{R}) \times H$ is a subgroup of G. We study the restriction of S_π to G'.

Following [1] and [3], we can realize S_π on a space $L^2_\pi(\Omega)$ of V_π-valued functions on Ω which are square-integrable with respect to a certain operator-valued measure on Ω. The second author introduced in [10] a set of coordinates on Ω, which are in this setting given by

$$\iota : \mathbb{R}^+ \times X \to \Omega, \quad \iota(t, v) = \frac{t}{r}(e + v),$$

where e is the identity element of V and X is the real bounded domain in the orthogonal complement $(\mathbb{R} \cdot e)^\perp$ of e in V given by

$$X = \left\{ v \in (\mathbb{R} \cdot e)^\perp \mid e + v \in \Omega \right\}.$$

Using the coordinates ι, we first restrict S_π to $\mathrm{PSL}_2(\mathbb{R})$ and show by explicit computations using the Lie algebra action that it decomposes into a direct sum of holomorphic discrete series representations ρ_λ of parameter $\lambda > 1$, realized on $L^2_\lambda(\mathbb{R}^+) = L^2(\mathbb{R}^+, t^{\lambda-1}\,dt)$, the first coordinate of ι. Moreover, the multiplicity space turns out to be a space of vector valued polynomials on X (see Sect. 3.1). This gives a natural correspondence between symmetry breaking and holographic operators and vector-valued orthogonal polynomials on X. Since $\mathrm{PSL}_2(\mathbb{R})$ and H commute, the latter will act on the multiplicity space, and hence on the vector-valued polynomials on X. This leads to our main result:

Theorem A (See Theorem 3.9) *For every holomorphic discrete series representation S_π, we have the following branching rule:*

$$S_\pi|_{\mathrm{PSL}_2(\mathbb{R})\times H} \simeq \sum_{p\in\mathbb{N}}^{\oplus} \rho_{\alpha+2p} \boxtimes \big(\mathrm{Pol}_p(X) \otimes V_\pi|_H\big).$$

where ρ_λ is the holomorphic discrete series representation of $\mathrm{PSL}_2(\mathbb{R})$ *of parameter λ, α is a constant depending on π, and* $\mathrm{Pol}_p(X)$ *denotes the space of homogeneous polynomials of degree p on X as a representation of H.* □

This result reduces the branching problem to the irreducible decomposition of the tensor product $\mathrm{Pol}_p(X) \otimes V_\pi|_H$, which involves only finite dimensional representations. Assume to have the following decomposition of this tensor product into an orthogonal direct sum of irreducible representations of H (with respect to a certain inner product induced from the inner product on $L^2_\pi(\Omega)$, see Sect. 3.3 for details):

$$\mathrm{Pol}_p(X) \otimes V_\pi|_H = \bigoplus_j F_p^j.$$

Denote by $K_p^j : X \times X \to \mathrm{End}(V_\pi)$ the reproducing kernel of F_p^j, we can then explicitly describe the symmetry breaking and holographic operators as follows.

Theorem B (See Theorem 3.11)

(a) *The operator $\phi_\pi^{p,j} : L^2_\pi(\Omega) \to L^2_{\alpha+2p}(\mathbb{R}^+) \otimes F_p^j$ defined by*

$$\phi_\pi^{p,j} f(t,v) = t^{-p} \int_X K_p^j(u,v) f(\iota(t,u))\Delta(e+u)^{-\frac{n}{r}}\,du,$$

is a symmetry breaking operator between $S_\pi|_{\mathrm{PSL}_2(\mathbb{R})\times H}$ and $\rho_{\alpha+2p} \boxtimes F_p^j$.

(b) *The operator $\Phi^\pi_{p,j} : L^2_{\alpha+2p}(\mathbb{R}^+) \otimes F_p^j \to L^2_\pi(\Omega)$ defined by*

$$\Phi^\pi_{p,j} f(t,v) = t^p f(t,v),$$

is a holographic operator between $\rho_{\alpha+2p} \boxtimes F_p^j$ and $S_\pi|_{\mathrm{PSL}_2(\mathbb{R})\times H}$.

□

We remark that for the case $\mathfrak{g} = \mathfrak{so}(2, n)$, the subgroup $\mathfrak{g}' = \mathfrak{so}(2, 1) \oplus \mathfrak{so}(n-1)$ is actually symmetric, and our results agree with the more general decomposition obtained by Kobayashi in [8].

2 Preliminaries

We recall the basic facts about Euclidean Jordan algebras and their associated groups from [2] and use them to describe L^2-models for holomorphic discrete series representations as in [1].

2.1 *Euclidean Jordan Algebras*

In this subsection, we set up the necessary notation for Jordan algebras and refer to [2] for the precise definitions.

Let V be a simple Euclidean Jordan algebra of dimension n, rank r and with unit element e. Let Ω be its associated symmetric cone, i.e. the interior of the set of squares in V. We denote by $V_{\mathbb{C}}$ the complexification of V and by $T_\Omega = V + i\Omega \subset V_{\mathbb{C}}$ the corresponding tube domain.

Write $\mathrm{tr}(x)$ and $\Delta(x)$ for the Jordan trace and Jordan determinant of $x \in V_{\mathbb{C}}$. Notice that $\mathrm{tr}(e) = r$ and $\Delta(e) = 1$. We denote by $(\cdot|\cdot)$ the trace form on $V_{\mathbb{C}}$ given by $(x|y) = \mathrm{tr}(xy)$ for $x, y \in V_{\mathbb{C}}$.

Further, let $L(x) : V_{\mathbb{C}} \to V_{\mathbb{C}}$ be the multiplication by $x \in V_{\mathbb{C}}$. Define the quadratic representation $P(x)$, and its polarized version $P(x, y)$ for $x, y \in V_{\mathbb{C}}$ by:

$$P(x) = 2L(x)^2 - L(x^2), \qquad P(x, y) = L(x)L(y) + L(y)L(x) - L(xy).$$

Finally, we define the box operator for $x, y \in V_{\mathbb{C}}$ by

$$x \Box y = L(xy) + [L(x), L(y)].$$

Let $G = \mathrm{Aut}(T_\Omega)$ denote the group of biholomorphic automorphisms of T_Ω. It is well-known that G is a simple Lie group with trivial center. It acts transitively on T_Ω and the stabilizer K of ie is a maximal compact subgroup of G. If θ denotes the Cartan involution of G which fixes K, we write

$$\mathfrak{g} = \mathfrak{k} \oplus \mathfrak{p}.$$

for the corresponding Cartan decomposition of the Lie algebra $\mathfrak{g}$ of G.

We consider several subgroups of G. First, consider the subgroup of translations

$$N = \{\tau_u : x \in T_\Omega \mapsto x + u \in T_\Omega \mid u \in V\}.$$

Further, the group

$$L = \{g \in GL(V) \mid g \cdot \Omega \subset \Omega\},$$

acts linearly on T_Ω. Then $H = K \cap L$ is a maximal compact subgroup of L and it equals the automorphism group of the Jordan algebra V, i.e.

$$H = \{g \in \mathrm{GL}(V) : g(x \cdot y) = g(x) \cdot g(y) \text{ for all } x, y \in V\}.$$

The subgroup $P = LN$ is a maximal parabolic subgroup of G with Levi factor L and unipotent radical N. Finally, N and L together with the inversion

$$j : z \in T_\Omega \mapsto -z^{-1} \in T_\Omega,$$

generate the group G. Note that the Cartan involution θ is given by $\theta(g) = jgj^{-1}$ $(g \in G)$,

Finally, we define the group

$$\bar{N} = \left\{\bar{\tau}_u = j\tau_u j^{-1} \mid u \in V\right\}.$$

On the Lie algebra level, we have the Gelfand–Naimark decomposition

$$\mathfrak{g} = \mathfrak{n} \oplus \mathfrak{l} \oplus \bar{\mathfrak{n}}, \tag{1}$$

where $\mathfrak{l} = \mathrm{Lie}(L)$, $\mathfrak{n} = \mathrm{Lie}(N)$ and $\bar{\mathfrak{n}} = \mathrm{Lie}(\bar{N})$. The Lie algebra $\mathfrak{l}$ is generated by the elements $u\Box v$ for $u, v \in V$, and the Lie algebras $\mathfrak{n}$ and $\bar{\mathfrak{n}}$ are spanned by

$$N_u = \frac{d}{dt}\bigg|_{t=0} \tau_{tu} \in \mathfrak{n} \qquad \text{and} \qquad \overline{N}_u = \frac{d}{dt}\bigg|_{t=0} \bar{\tau}_{tu} \in \bar{\mathfrak{n}} \qquad (u \in V),$$

respectively.

Fix a Jordan frame $(c_1, \cdots, c_r)$ of V, and define $h_j = 2L(c_j)$ so that $\sum h_j = 2L(e)$. This leads to a maximal abelian subspace:

$$\mathfrak{a} = \bigoplus_{i=1}^{r} \mathbb{R}h_i$$

of $\mathfrak{l} \cap \mathfrak{p}$, and this is also maximal abelian in $\mathfrak{p}$. The associated root system $\Sigma(\mathfrak{g}, \mathfrak{a})$ is of type C_r, and is given by

$$\left\{\frac{1}{2}(\pm\gamma_j \pm \gamma_k) \mid 1 \leq j < k \leq r\right\} \cup \left\{\pm\gamma_j \mid 1 \leq j \leq r\right\},$$

with $\gamma_i(h_j) = 2\delta_{ij}$, and the subsystem $\Sigma(\mathfrak{l}, \mathfrak{a})$ is

$$\left\{\frac{1}{2}(\gamma_j - \gamma_k) : 1 \leq j \neq k \leq r\right\}.$$

We choose the positive system $\Sigma^+(\mathfrak{g}, \mathfrak{a})$ induced by the ordering $\gamma_r > \cdots > \gamma_1 > 0$ and let $\Sigma^+(\mathfrak{l}, \mathfrak{a}) = \Sigma^+(\mathfrak{g}, \mathfrak{a}) \cap \Sigma(\mathfrak{l}, \mathfrak{a})$.

2.2 L^2-Model for Vector-Valued Holomorphic Discrete Series

In [1] an L^2-model for holomorphic discrete series representations of the group G was constructed. We use a slight variation of this model as introduced in [3].

Choose a finite-dimensional irreducible representation (π, V_π) of L. Its restricted lowest weight is given by

$$-\frac{1}{2}\sum_{i=1}^{r} m_i\gamma_i.$$

We also define

$$\omega(\pi) = m_r \qquad \text{and} \qquad \alpha = \sum_{i=1}^{r} m_i.$$

For an element $x \in \Omega$, we set $\pi(x) = \pi(P(x))$.

Notice that $L(e) = \mathrm{id}_V$ is contained in the center of L, hence by Schur's Lemma one finds:

$$d\pi(L(e)) = -\frac{1}{2}\alpha\,\mathrm{id}_{V_\pi}.$$

As a consequence we get for $t \in \mathbb{R}^+$ and $x \in \Omega$:

$$\pi(tx) = \pi(P(tx)) = \pi(t^2\,\mathrm{id}_V)\pi(x) = t^{-\alpha}\pi(x). \tag{2}$$

For $\omega(\pi) > \frac{2n}{r} - 1$, define the operator Γ_π acting on V_π by the absolutely convergent integral

$$\Gamma_\pi = \int_\Omega e^{-2\,\mathrm{tr}(u)}\pi(u)^{-1}\Delta(u)^{-\frac{2n}{r}}\,du. \tag{3}$$

It is known that if $\pi|_H$ is irreducible then Γ_π is a scalar and is equal to

$$2^{-r}\pi^{r(r-1)\frac{d}{4}}\prod_{j=1}^{r}\Gamma\left(m_j-\frac{n}{r}-(j-1)\frac{d}{2}\right).$$

We introduce the following Hilbert space:

$$L^2_\pi(\Omega):=\left\{f:\Omega\to V_\pi \mid \int_\Omega \langle \Gamma_\pi\pi(u^{\frac{1}{2}})^{-1}f(u)|\pi(u^{\frac{1}{2}})^{-1} \right.$$
$$\left. \times f(u)\rangle\Delta(u)^{-\frac{n}{r}}\,du<\infty\right\}, \tag{4}$$

where $u^{\frac{1}{2}}\in\Omega$ denotes the unique square root of $u\in\Omega$. On $L^2_\pi(\Omega)$ consider the following action of G:

$$S_\pi(\tau_u)f(x)=e^{-i(x|u)}f(x), \qquad (u\in V), \tag{5}$$
$$S_\pi(g)f(x)=\pi(g^*)^{-1}f(g^*x), \qquad (g\in L), \tag{6}$$
$$S_\pi(j)f(x)=\int_\Omega \mathcal{J}_\pi(u,x)f(u)\Delta(u)^{-\frac{n}{r}}\,du. \tag{7}$$

Here $\mathcal{J}_\pi(u,x)$ denotes the operator-valued Bessel function associated to π (see [1, Definition 3.5] and [3, Section 1.6]).

For $\omega(\pi)>\frac{2n}{r}-1$, this representation is equivalent to the holomorphic discrete series representation with highest weight space isomorphic to V_π. If $\dim(V_\pi)=1$, then we call S_π a scalar-valued holomorphic discrete series representation. One recovers the scalar-valued case by choosing $\pi(g)=|\det(g)|^{-\frac{r\lambda}{2n}}$, then $m_1=\cdots=m_r=\lambda$, so that $\alpha=r\lambda$.

On the smooth vectors, the derived representation is given by:

$$\begin{aligned} dS_\pi(N_u)&=-i(x|u),\\ dS_\pi(S)&=\partial_{S^*x}-d\pi(S^*),\\ dS_\pi(\overline{N}_v)&=i(v|\mathcal{B}_\pi).\end{aligned}$$

Here S^* denotes the adjoint of $S\in\mathfrak{l}$ with respect to the trace form and $\mathcal{B}_\pi$ denotes the vector-valued Bessel operator given by

$$(v|\mathcal{B}_\pi)=\sum_{i,j}(v|P(e_i,e_j)x)\frac{\partial^2}{\partial e_i\partial e_j}-2\sum_i d\pi(v\Box e_i)\frac{\partial}{\partial e_i}. \tag{8}$$

The space of K-finite vectors in $L^2_\pi(\Omega)$ is the space of functions of the form

$$x \mapsto p(x)e^{-\operatorname{tr}(x)} \qquad (x \in \Omega),$$

where $p \in \operatorname{Pol}(V, V_\pi)$, a polynomial on V with values in V_π.

We take a closer look at the case case $V = \mathbb{R}$ where $G \simeq \mathrm{PSL}_2(\mathbb{R})$. Here all holomorphic discrete series representations are scalar-valued. The earlier discussion gives a realization ρ_λ $(\lambda > 1)$ on

$$L^2_\lambda(\mathbb{R}^+) = L^2(\mathbb{R}^+, t^{\lambda-1}\,dt)$$

by the following formulas:

$$d\rho_\lambda \begin{pmatrix} 0 & 1 \\ 0 & 0 \end{pmatrix} = -it, \tag{9}$$

$$d\rho_\lambda \begin{pmatrix} 1 & 0 \\ 0 & -1 \end{pmatrix} = 2t\frac{d}{dt} + \lambda, \tag{10}$$

$$d\rho_\lambda \begin{pmatrix} 0 & 0 \\ -1 & 0 \end{pmatrix} = i\mathcal{B}^{\mathbb{R}}_\lambda = i\left(t\frac{d^2}{dt^2} + \lambda\frac{d}{dt}\right). \tag{11}$$

3 Restriction of a Holomorphic Discrete Series

We consider the restriction of a representation S_π of the holomorphic discrete series of G to a non-compact subgroup G' of the form $\mathrm{PSL}_2(\mathbb{R}) \times H$. Here, H is the automorphism group of the cone Ω, and its centralizer in G turns out to be the image of the map

$$\psi : \mathrm{PSL}_2(\mathbb{R}) \to G, \quad \psi\begin{pmatrix} a & b \\ c & d \end{pmatrix} z = \frac{az+b}{cz+d} \qquad (z \in T_\Omega).$$

To do so, we use a specific change of variables introduced in [10] and transfer the L^2-model for the representation into the stratified model.

3.1 Stratification

In [10], a stratification map was introduced to study the restriction of holomorphic discrete series realized on L^2-functions on Ω to the conformal group of a Jordan subalgebra V_1 of V. The stratification map ι is a diffeomorphism from $\Omega_1 \times X$ to

Ω, where Ω_1 is the symmetric cone in V_1 and X is a certain real bounded domain. In this chapter, we focus on the special case where $V_1 = \mathbb{R} \cdot e$.

In this situation, the stratification space is the domain

$$X = \left\{ v \in (\mathbb{R} \cdot e)^{\perp} \mid e + v \in \Omega \right\} \subseteq (\mathbb{R} \cdot e)^{\perp},$$

and the stratification map $\iota : \mathbb{R}^{+} \times X \to \Omega$ becomes

$$\iota(t, v) = \frac{t}{r}(e + v).$$

Choosing an orthonormal basis $\{e_0, \ldots, e_{n-1}\}$ of V such that $e_0 = r^{-\frac{1}{2}} e$ and $(e_1, \cdots, e_{n-1})$ is an orthonormal basis for $(\mathbb{R} \cdot e)^{\perp}$, the inverse of ι can be expressed as

$$\iota^{-1}(x) = (r x_0, x_0^{-1} x'),$$

where $x = x_0 e + \sum_{i>0} x_i e_i$ and $x' = \sum_{i>0} x_i e_i$. If dv denotes Lebesgue measure on X normalized by the trace form $(\cdot|\cdot)$, then the following integral formula holds:

$$\int_{\Omega} f(x)\, dx = r^{\frac{1}{2}-n} \int_{X} \int_{\mathbb{R}^{+}} f(\iota(t, v)) t^{n-1}\, dt\, dv. \tag{12}$$

We first study how the operator-valued gamma function Γ_{π} that is used to define the invariant inner product of $L^2_{\pi}(\Omega)$ behaves with respect to these coordinates.

Lemma 3.1 *The operator Γ_{π} factors as*

$$\Gamma_{\pi} = \Gamma_{\alpha} \Gamma_{\pi, X}, \tag{13}$$

where $\Gamma_{\alpha} = \frac{\Gamma(\alpha - n)}{r^{\alpha - n - \frac{1}{2}} 2^{\alpha - n}}$ and $\Gamma_{\pi, X}$ denotes the operator defined by:

$$\Gamma_{\pi, X} = \int_{X} \pi(e + v)^{-1} \Delta(e + v)^{-\frac{2n}{r}}\, dv. \tag{14}$$

Furthermore, if $\pi|_H$ is irreducible then $\Gamma_{\pi, X}$ is a scalar.

Proof Using (2) and (12), one gets:

$$\begin{aligned}
\Gamma_{\pi} &= \int_{\Omega} e^{-2\,\mathrm{tr}(u)} \pi(u)^{-1} \Delta(u)^{-\frac{2n}{r}}\, du \\
&= \frac{1}{r^{\alpha - n - \frac{1}{2}}} \left(\int_{\mathbb{R}^{+}} e^{-2t} t^{\alpha - n - 1}\, dt \right) \left(\int_{X} \pi(e + v)^{-1} \Delta(e + v)^{-\frac{2n}{r}}\, dv \right).
\end{aligned}$$

The last statement is direct since Γ_{π} already is a scalar in this case. □

This Lemma allows us to identify $L^2_\pi(\Omega)$ with a space of functions on $\mathbb{R}_+ \times X$.

Proposition 3.2 *The pullback ι^* is a scalar multiple of a unitary map between $L^2_\pi(\Omega)$ and $L^2_\alpha(\mathbb{R}^+)\hat{\otimes}L^2_\pi(X)$ where:*

$$L^2_\pi(X) = \left\{ f : X \to V_\pi \,\middle|\, \int_X \langle \Gamma_{\pi,X}\pi((e+v)^{\frac{1}{2}})^{-1}f(v), \pi((e+v)^{\frac{1}{2}})^{-1}f(v)\rangle \right.$$

$$\left. \times\, \Delta(e+v)^{-\frac{n}{r}}\, dv < \infty \right\}.$$

More precisely:

$$\|f\|^2_{L^2_\pi(\Omega)} = \frac{\Gamma_\alpha}{r^{\alpha-\frac{1}{2}}}\|f\circ\iota\|^2_{L^2(\mathbb{R}^+,t^{\alpha-1}\,dt)\hat{\otimes}L^2_\pi(X)}$$

Proof This is a direct computation similar to the one for the previous lemma. □

We now study how the isomorphism between $L^2_\pi(\Omega)$ and $L^2(\mathbb{R}^+, t^{\alpha-1}\,dt)\hat{\otimes} L^2_\pi(X)$ can be used to decompose the restriction of S_π to $\mathrm{PSL}_2(\mathbb{R}) \times H$. Note that H is contained in L, so its action in S_π is given by (6):

- For $k \in H$, we get:

$$S_\pi(k)f(t,v) = \pi(k)f(t,k^{-1}v). \tag{15}$$

On the factor $\mathrm{PSL}_2(\mathbb{R})$, the group action involves the complicated operator-valued Bessel function, so we use the action of the Lie algebra of $\mathrm{PSL}_2(\mathbb{R})$ instead. Note that the Lie algebra of $\mathrm{PSL}_2(\mathbb{R})$ decomposes according to the Gelfand–Naimark decomposition (1) as

$$\mathrm{Lie}(\mathrm{PSL}_2(\mathbb{R})) = \mathfrak{n}_1 \oplus \mathbb{R}L(e) \oplus \bar{\mathfrak{n}}_1,$$

with $\mathfrak{n}_1 \subset \mathfrak{n}$ and $\bar{\mathfrak{n}}_1 \subset \bar{\mathfrak{n}}$ corresponding to the embedding $\mathbb{R}e \subseteq V$. This gives the following actions in the stratified model:

- For the translations of $\mathrm{PSL}_2(\mathbb{R})$:

$$S_\pi(\psi\begin{pmatrix}1 & b\\ 0 & 1\end{pmatrix})f(t,v) = e^{-itb}f(t,v). \tag{16}$$

- The matrix $g = \mathrm{diag}(a, a^{-1}) \in \mathrm{PSL}_2(\mathbb{R})$ acts via scalar multiplication by a^2 on Ω, hence we have $\pi(\mathrm{diag}(a, a^{-1})) = \pi(a^2 \cdot \mathrm{id}) = a^{-\alpha}$, so:

$$S_\pi(\psi\begin{pmatrix}a & 0\\ 0 & a^{-1}\end{pmatrix})f(t,v) = a^\alpha f(a^2t, v). \tag{17}$$

Finally, the $\bar{\mathfrak{n}}_1$ action is given by the following:

Proposition 3.3 *The action of $\bar{\mathfrak{n}}_1 \subseteq \mathfrak{sl}_2(\mathbb{R})$ in the stratified model is given by*

$$dS_\pi(d\psi\begin{pmatrix}0&0\\1&0\end{pmatrix}) = i\left(\mathcal{B}_\alpha^{\mathbb{R}} + t^{-1}D_\pi\right),$$

where $\mathcal{B}_\alpha^{\mathbb{R}}$ denotes the Bessel operator for the one dimensional Jordan algebra $\mathbb{R}$ (see (11)*) and D_π is the following second order differential operator in the variable $v \in X$:*

$$D_\pi = r\sum_{i=1}^{n-1}\frac{\partial^2}{\partial {v_i}^2} + \sum_{1\le i,j\le n-1}\left(r(e_ie_j|v) - v_iv_j\right)\frac{\partial^2}{\partial v_i\partial v_j} - \alpha\sum_{i=1}^{n-1}v_i\frac{\partial}{\partial v_i} - 2r\sum_{i\ge 1}d\pi(L(e_i))\frac{\partial}{\partial v_i}.$$

Proof For a smooth function $f : \Omega \to V_\pi$ we have the formula

$$dS_\pi\begin{pmatrix}0&0\\1&0\end{pmatrix}f(x) = i\left(\sum_{i,j}(e|P(e_i,e_j)x)\frac{\partial^2 f}{\partial e_i\partial e_j} - 2\sum_i d\pi(e\Box e_i)\frac{\partial f}{\partial e_i}\right). \tag{18}$$

In what follows we consider $x = x_0e + \sum x_ie_i$ hence:

$$\frac{\partial f}{\partial e_0} = \frac{1}{r^{\frac{1}{2}}}\frac{\partial f}{\partial x_0} \quad\text{and}\quad \frac{\partial f}{\partial e_i} = \frac{\partial f}{\partial x_i}.$$

The second order part of the Bessel operator, denoted $\mathcal{B}_0$, gives:

$$(e|\mathcal{B}_0f(x)) = (e|P(e_0,e_0)x)\frac{\partial^2 f}{\partial {e_0}^2} + 2\sum_{i=1}^{n-1}(e|P(e_i,e_0)x)\frac{\partial^2 f}{\partial e_0\partial e_i} + \sum_{i,j\ge 1}(e|P(e_i,e_j)x)\frac{\partial^2 f}{\partial e_i\partial e_j}. \tag{19}$$

Hence, using the previous remark and the fact that $(e|e) = r$, we get:

$$(e|\mathcal{B}_0f(x)) = \frac{x_0}{r}\frac{\partial^2 f}{\partial {x_0}^2} + \frac{2}{r}\sum_{i=1}^{n-1}x_i\frac{\partial^2 f}{\partial x_0\partial x_i} + \sum_{i,j\ge 1}(e|P(e_i,e_j)x)\frac{\partial^2 f}{\partial x_i\partial x_j}.$$

Next we compute the derivatives for the variables (t, v) with respect to the variables x_i and we get:

$$\frac{\partial t}{\partial x_0} = r, \qquad \frac{\partial t}{\partial x_i} = 0 \qquad (i \geq 1),$$
$$\frac{\partial v_i}{\partial x_0} = -rt^{-1}v_i, \qquad \frac{\partial v_i}{\partial x_j} = \delta_{ij}rt^{-1} \qquad (i, j \geq 1).$$

Using the chain rule this gives:

$$\begin{aligned}
\frac{\partial f}{\partial x_0} &= r\frac{\partial f}{\partial t} - rt^{-1}\sum_{i=1}^{n-1} v_i \frac{\partial f}{\partial v_i},\\
\frac{\partial f}{\partial x_i} &= rt^{-1}\frac{\partial f}{\partial v_i},\\
\frac{\partial^2 f}{\partial {x_0}^2} &= r^2\left(\frac{\partial^2 f}{\partial t^2} + 2t^{-2}\sum_{i=1}^{n-1} v_i \frac{\partial f}{\partial v_i} - 2t^{-1}\sum_{i=1}^{n-1} v_i \frac{\partial^2 f}{\partial v_i \partial t}\right.\\
&\qquad \left. + t^{-2}\sum_{1 \leq i, j \leq n-1} v_i v_j \frac{\partial^2 f}{\partial v_i \partial v_j}\right),\\
\frac{\partial^2 f}{\partial x_0 \partial x_j} &= r^2\left(t^{-1}\frac{\partial^2 f}{\partial v_j \partial t} - t^{-2}\frac{\partial f}{\partial v_j} - t^{-2}\sum_{i=1}^{n-1} v_i \frac{\partial^2 f}{\partial v_i \partial v_j}\right),\\
\frac{\partial^2 f}{\partial x_i \partial x_j} &= r^2 t^{-2}\frac{\partial^2 f}{\partial v_i \partial v_j}.
\end{aligned}$$

Plugging these derivatives into (19) one gets

$$\begin{aligned}
(e|\mathcal{B}_0 f(x)) = t\frac{\partial^2 f}{\partial t^2} &+ rt^{-1}\sum_{i,j}(P(e_i, e_j)(e + v)|e)\frac{\partial^2 f}{\partial v_i \partial v_j}\\
&- t^{-1}\sum_{1 \leq i, j \leq n-1} v_i v_j \frac{\partial^2 f}{\partial v_i \partial v_j}.
\end{aligned}$$

Finally, we focus our attention on the first-order part of the Bessel operator, and we get in the coordinates (t, v):

$$\begin{aligned}\sum_{i\geq 0} d\pi(e\Box e_i)\frac{\partial f}{\partial e_i} &= d\pi(L(e_0))\frac{\partial f}{\partial e_0} + \sum_{i\geq 1} d\pi(e\Box e_i)\frac{\partial f}{\partial e_i}\\ &= -\frac{\alpha}{2r}\frac{\partial f}{\partial e_0} + \sum_{i\geq 1} d\pi(e\Box e_i)\frac{\partial f}{\partial e_i}\\ &= -\frac{\alpha}{2}\left(\frac{\partial f}{\partial t} - t^{-1}\sum_{i=1}^{n-1} v_i\frac{\partial f}{\partial v_i}\right) + rt^{-1}\sum_{i\geq 1} d\pi(L(e_i))\frac{\partial f}{\partial v_i}.\end{aligned}$$

Adding the contribution of the second order part finishes the proof. □

3.2 *Branching to* $\mathbf{PSL}_2(\mathbb{R})$

Now we consider the branching to the subgroup $\mathrm{PSL}_2(\mathbb{R})$, and we first look at the Casimir operator

$$C = H^2 + 2(XY + YX) = H^2 + 2H + 4YX,$$

where H, X, Y denotes the standard basis of $\mathfrak{sl}_2(\mathbb{R})$ satisfying

$$[H, X] = 2X, \quad [H, Y] = -2Y, \quad [X, Y] = H.$$

Proposition 3.4 *The Casimir operator of $\mathfrak{sl}_2(\mathbb{R})$ acts in the stratified model by*

$$dS_\pi(d\psi(C)) = \alpha(\alpha - 1) - 4D_\pi \tag{20}$$

In particular, this implies that D_π has a self-adjoint extension to $L^2_\pi(X)$.

Proof This is a direct consequence of the formulas for the derived representation and Proposition 3.3. □

The operator D_π can be written as

$$D_\pi = r\Delta_X + \Psi_\pi - E(E + \alpha - 1), \tag{21}$$

where $\Delta_X = \sum_{i=1}^{n-1}\frac{\partial^2}{\partial {v_i}^2}$ denotes the Laplacian on X and $E = \sum_i v_i\frac{\partial}{\partial v_i}$ the Euler operator on $(\mathbb{R}\cdot e)^\perp$, and Ψ_π is the differential operator defined by:

$$\Psi_\pi = \sum_{1\leq i,j\leq n-1} r(e_ie_j|v)\frac{\partial^2}{\partial v_i\partial v_j} - 2r\sum_{i\geq 1} d\pi(L(e_i))\frac{\partial}{\partial v_i}. \tag{22}$$

Let $\mathrm{Pol}(X, V_\pi)$ be the space of polynomials on $(\mathbb{R}\cdot e)^\perp$ with values in V_π restricted to the open subset X.

Proposition 3.5 *The operator D_π acts on the space $\mathrm{Pol}(X, V_\pi)$ and the subspace:*

$$W_p^\pi = \{P \in \mathrm{Pol}(X, V_\pi) |\, \deg(P) \leq p, (P|Q)_{L^2_\pi(X)} = 0 \text{ if } \deg(Q) < p\},$$

is an eigenspace for D_π with eigenvalue $-p(p+\alpha-1)$.

Proof Since D_π is a differential operator with polynomial coefficients, it clearly acts on $\mathrm{Pol}(X, V_\pi)$. It is also immediate from the expression for D_π that $\deg(D_\pi P) \leq \deg(P)$ for all $P \in \mathrm{Pol}(X, V_\pi)$. Since D_π is self-adjoint on $L^2_\pi(X)$, it follows that each W_p^π $(p \geq 0)$ is invariant under D_π. Moreover, as a self-adjoint operator D_π is diagonalizable on W_p^π. Let $P \in W_p^\pi$ be an eigenvector with eigenvalue μ and write $P = P_1 + P_2$ with P_1 homogeneous of degree p and $\deg(P_2) < p$. Then the highest order term in $D_\pi P$ is

$$-E(E-\alpha-1)P_1 = -p(p+\alpha-1)P_1,$$

so $\mu = -p(p-1+\alpha)$. □

Finally, this gives the following branching law, symmetry breaking and holographic operators:

Theorem 3.6 *Let S_π a holomorphic discrete series representation of G.*

(a) *The restriction of S_π to the subgroup $\mathrm{PSL}_2(\mathbb{R})$ decomposes as*

$$S_\pi|_{\mathrm{PSL}_2(\mathbb{R})} \simeq \sum_{p\geq 0}^{\oplus} \left[\binom{n+p-2}{n-2} \dim V_\pi\right] \cdot \rho_{\alpha+2p}.$$

(b) *For any $P \in \mathrm{Pol}(X, V_\pi)$ of degree p, the operator*

$$\phi_\pi^p(P) : L^2_\alpha(\mathbb{R}^+)\hat{\otimes} L^2_\pi(X) \to L^2_{\alpha+2p}(\mathbb{R}^+),$$

$$\phi_\pi^p(P)f(t) = t^{-p} \int_X \langle \Gamma_{\pi,X}\pi((e+v)^{\frac{1}{2}})^{-1} f(t,v), \pi((e+v)^{\frac{1}{2}})^{-1} P(v)\rangle \Delta(e+v)^{-\frac{n}{r}}\, dv.$$

is a symmetry breaking operator between S_π and $\rho_{\alpha+2p}$ if and only if $P \in W_p^\pi$.

(c) *For any $P \in \mathrm{Pol}(X, V_\pi)$ of degree p, the operator*

$$\Phi_p^\pi(P) : L^2_{\alpha+2p}(\mathbb{R}^+) \to L^2_\alpha(\mathbb{R}^+)\hat{\otimes} L^2_\pi(X), \quad \Phi_p^\pi(P)f(t) = t^p f(t)P(v).$$

is a holographic operator between $\rho_{\alpha+2p}$ and S_π if and only if $P \in W_p^\pi$.

□

Proof Since W_p^π is an eigenspace for D_π, the space $L^2_\alpha(\mathbb{R}^+)\otimes W_p^\pi$ is an eigenspace for the action of the Casimir element $dS_\pi(C)$ of $\mathrm{PSL}_2(\mathbb{R})$. Hence, $L^2_\alpha(\mathbb{R}^+)\otimes W_p^\pi$ is a subrepresentation of $S_\pi|_{\mathrm{PSL}_2(\mathbb{R})}$. By (16), (17) and Proposition 3.3, $\mathrm{PSL}_2(\mathbb{R})$ only acts in the variable t on each subrepresentation $L^2_\alpha(\mathbb{R}^+)\otimes W_p^\pi$, so for fixed $P\in W_p^\pi$ we consider the map $\Phi_p^\pi(P) : L^2_{\alpha+2p}(\mathbb{R}^+) \to L^2_\alpha(\mathbb{R}^+)\hat{\otimes} L^2_\pi(X)$ from (c). A short computation using (16), (17) and Proposition 3.3 as well as (9), (10) and (11) shows that this map is intertwining for $\rho_{\alpha+2p}$ and $S_\pi|_{\mathrm{PSL}_2(\mathbb{R})}$ if and only if $P\in W_p^\pi$, so (c) follows. Since $\phi_\pi^p(P)$ is the adjoint of $\Phi_\pi^p(P)$, this also shows (b).

Finally, the K-finite vectors in $L^2_\pi(\Omega)$ are of the form $P(x)e^{-\operatorname{tr}(x)}$ with P a polynomials in $\mathrm{Pol}(\Omega, V_\pi)$. Using the stratification map, the K-finite vectors becomes $Q(t, tv)e^{-t}$ where Q is a polynomial on $\mathbb{R}\times(\mathbb{R}\cdot e)^\perp$. On $L^2_{\alpha+2p}(\mathbb{R}^+)$ the $(K\cap \mathrm{PSL}_2(\mathbb{R}))$-finite vectors are of the form $R(t)e^{-t}$ with R a polynomial on $\mathbb{R}$, so the $(K\cap \mathrm{PSL}_2(\mathbb{R}))$-finite vectors on each $L^2_\alpha(\mathbb{R}^+)\hat{\otimes} W_p^\pi$ are of the form $t^pQ(t,v)e^{-t} = Q(t,tv)e^{-t}$ with Q a polynomial in $\mathbb{C}[t]\otimes W_p^\pi$. This shows that the sum of all the images of $\Phi_p^\pi(P)$, $P\in W_p^\pi$, $p\geq 0$, contains the K-finite vectors of S_π, hence it is dense in $L^2_\pi(\Omega)$. Together with the fact that $\dim(W_p^\pi) = \dim(V_\pi)\cdot\binom{n+p-2}{n-2}$ this shows (a). □

3.3 *Branching to* $\mathrm{PSL}_2(\mathbb{R})\times H$

Recall that the group H acts on $L^2_\pi(X)$ by (15).

Lemma 3.7 *The operator D_π commutes with the action of H on $L^2_\pi(X)$.*

Proof Since H and $\mathrm{PSL}_2(\mathbb{R})$ commute in G, their actions $S_\pi(k)$ $(k\in H)$ and $dS_\pi(d\psi(T))$ $(T\in\mathfrak{sl}_2(\mathbb{R}))$ commute as well. □

It follows that the space W_p^π is a representation of the compact group H. Since W_p^π is difficult to work with, we relate it to the space $\mathrm{Pol}_p(X, V_\pi)$ of homogeneous polynomials of degree p with values in V_π on which H acts by

$$(k\cdot P)(v) = \pi(k)P(k^{-1}v) \qquad (k\in H,\ P\in \mathrm{Pol}_p(X,V_\pi),\ v\in X).$$

Then any polynomial $P\in\mathrm{Pol}(X,V_\pi)$ of degree p can be decomposed into $P=\sum_{i=0}^p P_i$ with $P_i\in\mathrm{Pol}_i(X,V_\pi)$ its homogeneous part of degree i.

Proposition 3.8 *The map $T : W_p^\pi\to\mathrm{Pol}_p(X,V_\pi)$ defined by $T(P)=P_p$ is an H intertwining isomorphism. For $Q\in\mathrm{Pol}_p(X,V_\pi)$, the polynomial $P=T^{-1}(Q)$ is uniquely determined by the following recursion formula:*

$$\Delta_X P_{i+2} + \Psi_\pi P_{i+1} = (i(i+\alpha-1) - p(p+\alpha-1))\,P_i \qquad (0\leq i\leq p-1),$$

where Δ_X is the Laplacian on X and Ψ_π the operator defined in (22).

Proof It is clear that T is an intertwining operator since the action of H is given by the same formula on W_p^π and $\mathrm{Pol}_p(X, V_\pi)$, and this formula preserved the degree of homogeneity. The map T is injective since $T(P) = 0$ implies $\deg(P) < p$, and every $P \in W_p^\pi$ is orthogonal to all polynomials of degree $< p$. Furthermore we have $\dim(W_p^\pi) = \dim(\mathrm{Pol}_p(X, V_\pi))$ so T is a bijection.

From Proposition 3.5 we know that $P = \sum_{i=1}^p P_i \in W_p^\pi$ is equivalent to $D_\pi P = -p(p+\alpha-1)P$. Recalling that $D_\pi = \Delta_X + \Psi_\pi - E(E+\alpha-1)$, this is equivalent to the claimed recursion formula. □

By the previous result, we have the following isomorphism of H-representations:

$$W_p^\pi \simeq \mathrm{Pol}_p(X, V_\pi|_H) \simeq \mathrm{Pol}_p(X) \otimes V_\pi|_H.$$

So to decompose W_p^π into irreducible representations of H one has to decompose $\mathrm{Pol}_p(X)$ and $V_\pi|_H$ into irreducible representations and then decompose the corresponding tensor products. This implies the following branching law:

Theorem 3.9 *For every holomorphic discrete series representation S_π, we have the following branching rule:*

$$S_\pi|_{\mathrm{PSL}_2(\mathbb{R})\times H} \simeq \sum_{p\in\mathbb{N}}^{\oplus} \rho_{\alpha+2p} \boxtimes \left(\mathrm{Pol}_p(X) \otimes V_\pi|_H\right).$$

□

Example 3.10 Since $\mathrm{Pol}_p(X) = S^p(U)$, where $U^* = (\mathbb{R}\cdot e)_{\mathbb{C}}^{\perp}$, the decomposition of $\mathrm{Pol}_p(X)$ is the classical problem of decomposing $S^p(U)$ for the following representation U of H:

1. $V = \mathrm{Sym}(n, \mathbb{R})$. Here $\mathfrak{h}_{\mathbb{C}} = \mathfrak{so}(n, \mathbb{C})$ is acting on $V = \mathrm{Sym}(n, \mathbb{C})$ by conjugation and $U^* = \{X \in \mathrm{Sym}(n, \mathbb{C}) : \mathrm{tr}(X) = 0\}$ which can be identified with $S_0^2(\mathbb{C}^n)$, the trace-free part of the second symmetric power of the standard representation of $\mathfrak{so}(n)$ on $\mathbb{C}^n$. This representation is irreducible and self-dual, so $U = S_0^2(\mathbb{C}^n)$. Its highest weight in terms of the fundamental weights $\omega_1, \dots, \omega_{\lfloor \frac{n}{2} \rfloor}$ is $2\omega_1$.
2. $V = \mathrm{Herm}(n, \mathbb{C})$. Here $\mathfrak{h}_{\mathbb{C}} = \mathfrak{sl}(n, \mathbb{C})$ is acting on $V = M(n\times n, \mathbb{C})$ by conjugation and $U^* = \{X \in M(n \times n, \mathbb{C}) : \mathrm{tr}(X) = 0\}$, which can be identified with $(\mathbb{C}^n \otimes (\mathbb{C}^n)^*)_0$, the trace-free part of the tensor product of the standard representation of $\mathfrak{sl}(n, \mathbb{C})$ on $\mathbb{C}^n$ and its dual. Clearly, $\mathbb{C}^n \otimes (\mathbb{C}^n)^*$ is self-dual, so $U = (\mathbb{C}^n \otimes (\mathbb{C}^n)^*)_0$. This representation is irreducible and its highest weight in terms of the fundamental weights $\omega_1, \dots, \omega_{n-1}$ is $\omega_1 + \omega_{n-1}$.
3. $V = \mathrm{Herm}(n, \mathbb{H})$. Here $\mathfrak{h}_{\mathbb{C}} = \mathfrak{sp}(n, \mathbb{C})$ is acting on $V_{\mathbb{C}} = \{X \in \mathrm{Skew}(2n, \mathbb{C})\}$ by conjugation, so U^* can be identified with $(\mathbb{C}^{2n} \otimes (\mathbb{C}^{2n})^*)_0$, the trace-free part of the tensor product of the standard representation of $\mathfrak{sp}(n, \mathbb{C})$ on $\mathbb{C}^{2n}$ and its dual. Since the standard representation is self-dual, we find $U = S_0^2(\mathbb{C}^{2n})$. This representation is irreducible and its highest weight in terms of the fundamental weights $\omega_1, \dots, \omega_n$ is $2\omega_1$.

4. $V = \mathrm{Herm}(3, \mathbb{O})$. Here $\mathfrak{h}_\mathbb{C} = \mathfrak{f}_4(\mathbb{C})$ is acting on the 26-dimensional space U^*. Since this is the smallest dimension of a nontrivial irreducible representation, and there is precisely one irreducible representation of this dimension, we find that U is the fundamental representation of $\mathfrak{f}_4(\mathbb{C})$ with highest weight ω_1 in terms of the fundamental weights $\omega_1, \omega_2, \omega_3, \omega_4$.
5. $V = \mathbb{R} \times \mathbb{R}^{n-1}$. Here $\mathfrak{h}_\mathbb{C} = \mathfrak{so}(n-1, \mathbb{C})$ is acting on $U^* = \{0\} \times \mathbb{C}^{n-1} = \mathbb{C}^{n-1}$ by the standard representation, which is self-dual, so U^* can be identified with $\mathbb{C}^{n-1}$. This representation is irreducible and its highest weight in terms of the fundamental weights $\omega_1, \ldots, \omega_{\lfloor \frac{n-1}{2} \rfloor}$ is ω_1.

□

To also state formulas for the symmetry breaking and holographic operators, we decompose W_p^π into irreducible representations F_p^j:

$$W_p^\pi = \bigoplus_j F_p^j.$$

We also denote by F_p^j the action of H on the vector space F_p^j, more precisely:

$$F_p^j(k)P(v) = \pi(k)P(k^{-1} \cdot v).$$

Note that this decomposition is not unique since F_p^j might occur with higher multiplicity in W_p^π.

For every $v \in X$ and $\xi \in V_\pi$, the linear form $F_p^j \to \mathbb{C}$, $P \mapsto \langle P(v), \xi \rangle$ is represented by a vector $K_v\xi \in F_p^j$, i.e. $\langle P(v), \xi \rangle = \langle P, K_v\xi \rangle$. We write $K(u, v)\xi = (K_v\xi)(u)$. Since $K(u, v)\xi$ is linear in ξ, we have $K_p^j(u, v) = K(u, v) \in \mathrm{End}(V_\pi)$. This function $K_p^j : X \times X \to \mathrm{End}(V_\pi)$ satisfies

$$K_p^j(k \cdot u, k \cdot v) = \pi(k)K_p^j(u, v)\pi(k)^{-1} \qquad (u, v \in X, k \in H). \tag{23}$$

Theorem 3.11

(a) The operator $\phi_\pi^{p,j} : L_\pi^2(\Omega) \to L_{\alpha+2p}^2(\mathbb{R}^+) \otimes F_p^j$ *defined by*

$$\phi_\pi^{p,j} f(t, v) = t^{-p} \int_X K_p^j(u, v) f(t, u) \Delta(e + u)^{-\frac{n}{r}}\, du,$$

is a symmetry breaking operator between $S_\pi|_{\mathrm{PSL}_2(\mathbb{R}) \times H}$ *and* $\rho_{\alpha+2p} \boxtimes F_p^j$.

(b) The operator $\Phi_{p,j}^\pi : L_{\alpha+2p}^2(\mathbb{R}^+) \otimes F_p^j \to L_\pi^2(\Omega)$ *defined by*

$$\Phi_{p,j}^\pi f(t, v) = t^p f(t, v),$$

is a holographic operator between $\rho_{\alpha+2p} \boxtimes F_p^j$ *and* $S_\pi|_{\mathrm{PSL}_2(\mathbb{R}) \times H}$.

□

Proof The intertwining property for the $\mathrm{PSL}_2(\mathbb{R})$ action is a consequence of Theorem 3.6. The intertwining property for H is a consequence of formula (23). □

3.4 The Case $\mathfrak{g} = \mathfrak{so}(2, n+1)$

We conclude by considering the special case where V is a Euclidean Jordan algebra of rank 2. More precisely, we have $V = \mathbb{R} \times \mathbb{R}^n$, and the Jordan multiplication is given by:

$$(x, u) \cdot (y, v) = (xy + B(u, v), xv + yu),$$

where $B(u, v)$ denotes the usual inner product on $\mathbb{R}^n$. The inner product on the Euclidean Jordan algebra V is given by:

$$(x|y) = \mathrm{tr}(xy) = 2 \sum x_i y_i,$$

for $x = \sum x_i f_i$, $y = \sum y_i f_i$ $(0 \leq i \leq n)$ where $\{f_i\}$ is the canonical basis on $\mathbb{R} \times \mathbb{R}^n$ which is not an orthonormal basis. The symmetric cone is

$$\Omega = \{x \in V \mid Q_{1,n}(x) > 0,\ x_1 > 0\},$$

hence we have $L \simeq \mathrm{SO}_0(1, n)$ and $H = \mathrm{SO}(n)$. This leads to:

$$X = \{(0, v) \mid v \in \mathbb{R}^n,\ \|v\| < 1\} = B^n.$$

In this situation, it is well known that as an $\mathrm{SO}(n)$ representation

$$\mathrm{Pol}_p(X) \simeq \bigoplus_{j=1}^{\lfloor p/2 \rfloor} \mathcal{H}^n_{p-2j},$$

where $\mathcal{H}^n_p$ denotes the irreducible representation of $\mathrm{SO}(n)$ on the space of harmonic polynomials of degree p in n variables. Thus, to find the explicit abstract branching law, one needs to deal with tensor products of the form

$$\mathcal{H}^n_{p-2j} \otimes \pi|_H.$$

This can be decomposed explicitly using, for instance, the results in [4, 11].

Since $f_i \cdot f_j = 0$ for $i, j \geq 1$ and $i \neq j$, and $f_i^2 = e$, the operator D_π is explicitly given by:

$$D_\pi = \sum_{i=1}^{n} \frac{\partial^2}{\partial {x_i}^2} - \sum_{1 \leq i, j \leq n} x_i x_j \frac{\partial^2}{\partial x_i \partial x_j} - \alpha \sum_{i=1}^{n} x_i \frac{\partial}{\partial x_i} - 2 \sum_{i=1}^{n} d\pi(L(f_i)) \frac{\partial}{\partial x_i}.$$

Notice that if we consider the case where π is a one dimensional character, i.e. S_π is a scalar-valued holomorphic discrete series representation, then we recover the operator considered in [10] for the pair $(SO_0(2,n), SO_0(2,n-p))$ for the special case $p = n-1$.

Remark 3.12 In this case, the pair $(\mathfrak{g}, \mathfrak{g}') = (\mathfrak{so}(2, n+1), \mathfrak{so}(2,1) \oplus \mathfrak{so}(n))$ is a symmetric pair with the isomorphism $\mathfrak{so}(2,1) \simeq \mathfrak{sl}(2,\mathbb{R})$. Hence, the abstract branching law is a special case of the more general result of Kobayashi in [8, Lemma 8.8]. His proof relies on the Hua–Kostant–Schmid formula for the space of polynomials $\mathrm{Pol}((e^\perp)^\mathbb{C})$ under the action of $\mathfrak{k} \cap \mathfrak{g}' = \mathfrak{so}(2) \oplus \mathfrak{so}(n)$. In our approach, we took care of the $\mathfrak{so}(2)$-action before studying the action of $\mathfrak{so}(n)$ on $\mathrm{Pol}(e^\perp)$, and this corresponds to the grading of this space by homogeneous polynomials of a fixed degree. □

Acknowledgments Both authors were supported by a research grant from the Villum Foundation (Grant No. 00025373).

References

1. Ding, H., Gross, K.I.: Operator-valued Bessel functions on Jordan algebras. J. Reine Angew. Math. **435**, 157–196 (1993)
2. Faraut, J., Korányi, A.: Analysis on Symmetric Cones. Oxford Mathematical Monographs. The Clarendon Press, Oxford University Press, New York (1994)
3. Frahm, J., Ólafsson, G., Ørsted, B.: The holomorphic discrete series contribution to the generalized Whittaker Plancherel formula. J. Funct. Anal. **286**(6), 110333 (2024)
4. Howe, R., Tan, E.-C., Willenbring, J.F.; Stable branching rules for classical symmetric pairs. Trans. Am. Math. Soc. **357**(4), 1601–1626 (2005)
5. Kobayashi, T.: Discrete decomposability of restriction of $A_\mathfrak{q}(\lambda)$ with respect to reductive subgroups and its applications. Invent. Math. **117**(2), 181–205 (1994)
6. Kobayashi, T.: Discrete decomposability of the restriction of $A_\mathfrak{q}(\lambda)$ with respect to reductive subgroups. III: Restriction of Harish-Chandra modules and associated varieties. Invent. Math. **131**(2), 229–256 (1998)
7. Kobayashi, T.: Discrete decomposability of the restriction of $A_q(\lambda)$ with respect to reductive subgroups. II: Micro-local analysis and asymptotic K-support. Ann. Math. (2) **147**(3), 709–729 (1998)
8. Kobayashi, T.: Multiplicity-Free Theorems of the Restrictions of Unitary Highest Weight Modules with Respect to Reductive Symmetric Pairs. Representation Theory and Automorphic Forms, pp. 45–109. Progress in Mathematics, vol. 255. Birkhäuser, Boston (2008)
9. Kobayashi, T., Pevzner, M.: Inversion of Rankin–Cohen operators via holographic transform. Ann. Inst. Fourier **70**(5), 2131–2190 (2020)
10. Labriet, Q.: A geometrical point of view for branching problems for holomorphic discrete series of conformal Lie groups. Int. J. Math. **33**(10–11), 2250069 (2022)
11. Okada, S.: Pieri rules for classical groups and equinumeration between generalized oscillating tableaux and semistandard tableaux. Electron. J. Comb. **23**(4), P4.43 (2016)

Nets of Standard Subspaces on Non-compactly Causal Symmetric Spaces

Jan Frahm, Karl-Hermann Neeb, and Gestur Òlafsson

Abstract Let G be a connected simple linear Lie group and $H \subset G$ a symmetric subgroup such that the corresponding symmetric space G/H is non-compactly causal. We show that any irreducible unitary representation of G leads naturally to a net of standard subspaces on G/H that is isotone, covariant and has the Reeh–Schlieder and the Bisognano–Wichmann property. We also show that this result extends to the universal covering group of $\mathrm{SL}_2(\mathbb{R})$, which has some interesting application to intersections of standard subspaces associated to representations of such groups. For this, a detailed study of hyperfunction and distribution vectors is needed. In particular, we show that every H-finite hyperfunction vector is in fact a distribution vector.

1 Introduction

This chapter is part of an ongoing project exploring the connections between causal structures on homogeneous spaces, Algebraic Quantum Field Theory (AQFT), modular theory of operator algebras, and unitary representations of Lie groups (cf. [63, 65, 64, 52, 54]).

The main achievement of this chapter is the construction of a net of real subspaces for any irreducible representation of a connected simple linear Lie group G on any corresponding non-compactly causal symmetric space G/H, such that the Reeh–Schlieder and the Bisognano–Wichman condition are satisfied. To explain

J. Frahm
Department of Mathematics, Aarhus University, Aarhus, Denmark
e-mail: frahm@math.au.dk

K.-H. Neeb (✉)
Department of Mathematics, FAU Erlangen-Nürnberg, Erlangen, Germany
e-mail: neeb@math.fau.de; neeb@mi.uni-erlangen.de

G. Òlafsson
Department of Mathematics, Louisiana State University, Baton Rouge, LA, USA
e-mail: olafsson@math.lsu.edu

M. Pevzner, H. Sekiguchi (eds.), *Symmetry in Geometry and Analysis, Volume 2*,
Progress in Mathematics 358, https://doi.org/10.1007/978-981-97-7662-7_5

these concepts, let us call a closed real subspace $\mathtt{V}$ of a complex Hilbert space $\mathcal{H}$ *cyclic* if $\mathtt{V}+i\mathtt{V}$ is dense in $\mathcal{H}$, *separating* if $\mathtt{V}\cap i\mathtt{V}=\{0\}$, and *standard* if it is both. If $\mathtt{V}$ is standard, then the complex conjugation $T_{\mathtt{V}}$ on $\mathtt{V}+i\mathtt{V}$ has a polar decomposition $T_{\mathtt{V}} = J_{\mathtt{V}}\Delta_{\mathtt{V}}^{1/2}$, where $J_{\mathtt{V}}$ is a conjugation (an antilinear involutive isometry) and $\Delta_{\mathtt{V}}$ is a positive selfadjoint operator satisfying $J_{\mathtt{V}}\Delta_{\mathtt{V}}J_{\mathtt{V}} = \Delta_{\mathtt{V}}^{-1}$. The unitary one-parameter group $(\Delta_{\mathtt{V}}^{it})_{t\in\mathbb{R}}$, the *modular group of* $\mathtt{V}$, preserves the subspace $\mathtt{V}$. We refer to Longo [48] and Neeb/Ólafsson [60] for more on standard subspaces.

For a unitary representation $(U, \mathcal{H})$ of a Lie group G and a homogeneous space $M = G/H$, we are interested in families $(\mathsf{H}(O))_{O\subseteq M}$ of closed real subspaces of $\mathcal{H}$, indexed by open subsets $O \subseteq M$; so-called *nets of real subspaces*. For such nets, we consider the following properties:

- (Iso) **Isotony:** $O_1 \subseteq O_2$ implies $\mathsf{H}(O_1) \subseteq \mathsf{H}(O_2)$
- (Cov) **Covariance:** $U_g\mathsf{H}(O) = \mathsf{H}(gO)$ for $g \in G$.
- (RS) **Reeh–Schlieder property:** $\mathsf{H}(O)$ is cyclic if $O \neq \emptyset$.
- (BW) **Bisognano–Wichmann property:** There exists an open subset $W \subseteq M$ (called a *wedge region*), and an element $h \in \mathfrak{g}$ such that $\mathsf{H}(W)$ is standard, and its modular group satisfies

$$\Delta_{\mathsf{H}(W)}^{-it/2\pi} = U(\exp th) \quad \text{for} \quad t \in \mathbb{R}.$$

Currently, we leave locality conditions aside. We refer to Sect. 4.3, where the related difficulties are explained. We plan to address them in subsequent work.

Nets satisfying (Iso) and (Cov) can easily be constructed as follows. The subspace $\mathcal{H}^\infty \subseteq \mathcal{H}$ of vectors $v \in \mathcal{H}$ for which the orbit map

$$U^v \colon G \to \mathcal{H}, \quad g \mapsto U(g)v,$$

is smooth (*smooth vectors*) is dense and carries a natural Fréchet topology for which the action of G on this space is smooth [33, 58]. The space $\mathcal{H}^{-\infty}$ of continuous antilinear functionals $\eta \colon \mathcal{H}^\infty \to \mathbb{C}$ (*distribution vectors*) contains in particular Dirac's kets $\langle\cdot, v\rangle$, $v \in \mathcal{H}$, so that we obtain complex linear embeddings

$$\mathcal{H}^\infty \hookrightarrow \mathcal{H} \hookrightarrow \mathcal{H}^{-\infty},$$

where G acts on all three spaces by representations denoted U^∞, U and $U^{-\infty}$, respectively. Here we follow the convention common in physics to require inner products to be conjugate linear in the first and complex linear in the second argument.

All of the three aforementioned representations can be integrated to the convolution algebra $C_c^\infty(G) := C_c^\infty(G, \mathbb{C})$ of test functions, for instance $U^{-\infty}(\varphi) := \int_G \varphi(g)U^{-\infty}(g)\,dg$, where dg stands for a left Haar measure on G. To any real subspace $\mathtt{E} \subseteq \mathcal{H}^{-\infty}$ and every open subset $O \subseteq G$, we associate the closed real subspace

$$\mathsf{H}_{\mathtt{E}}^G(O) := \overline{\mathrm{span}_{\mathbb{R}} U^{-\infty}(C_c^\infty(O, \mathbb{R}))\mathtt{E}}. \tag{1.1}$$

Note that the operators $U^{-\infty}(\varphi)$ map $\mathcal{H}^{-\infty}$ into $\mathcal{H}$ because they are adjoints of continuous operators $U(\varphi)\colon \mathcal{H} \to \mathcal{H}^{\infty}$ ([26], [63, App. A]). Accordingly, the closure is taken with respect to the topology of $\mathcal{H}$ (cf. in particular Proposition 9). Clearly, this definition does also make sense for real subspaces $\mathsf{E} \subseteq \mathcal{H}$, but the key advantage of working with the larger space $\mathcal{H}^{-\infty}$ of distribution vectors is that it contains finite-dimensional subspaces invariant under ad-diagonalizable elements and non-compact subgroups (cf. Lemma 7). For finite-dimensional subspaces of $\mathcal{H}$, this is excluded by Moore's Theorem [51]. On a homogeneous space $M = G/H$ with the projection map $q\colon G \to M$, we now obtain a "push-forward net"

$$\mathsf{H}^M_{\mathsf{E}}(O) := \mathsf{H}^G_{\mathsf{E}}(q^{-1}(O)). \tag{1.2}$$

The so-obtained net on M thus corresponds to the restriction of the net $\mathsf{H}^G_{\mathsf{E}}$ indexed by open subsets of G to those open subsets $O \subseteq G$ which are H-right invariant in the sense that $O = OH$; these are the inverse images of open subsets of M under q. If, in addition, E is invariant under $U^{-\infty}(H)$, then [63, Lemma 2.11] implies that $\mathsf{H}^G_{\mathsf{E}}(O) = \mathsf{H}^G_{\mathsf{E}}(OH)$ for any open subset $O \subseteq G$, so that $\mathsf{H}^G_{\mathsf{E}}$ can be recovered from the net $\mathsf{H}^M_{\mathsf{E}}$ on M by $\mathsf{H}^G_{\mathsf{E}}(O) = \mathsf{H}^G_{\mathsf{E}}(OH) = \mathsf{H}^M_{\mathsf{E}}(q(O))$. The assignment (1.2) trivially satisfies (Iso), and (Cov) follows from the simple relation

$$U^{-\infty}(g)U^{-\infty}(\varphi) = U^{-\infty}(\delta_g * \varphi),$$

where $(\delta_g * \varphi)(x) = \varphi(g^{-1}x)$ is the left translate of φ. Therefore, a key problem is to specify subspaces E of distribution vectors for which (RS) and (BW) hold as well. According to Morinelli/Neeb [53], the potential generators $h \in \mathfrak{g}$ of the modular groups in (BW) are *Euler elements*, i.e., $\operatorname{ad} h$ defines a 3-grading

$$\mathfrak{g} = \mathfrak{g}_1(h) \oplus \mathfrak{g}_0(h) \oplus \mathfrak{g}_{-1}(h), \qquad \text{where} \qquad \mathfrak{g}_\lambda(h) = \ker(\operatorname{ad} h - \lambda \mathbf{1}).$$

Moreover, if $\ker(U)$ is discrete, then the conjugation $J := J_{\mathsf{H}(W)}$ satisfies

$$JU(g)J = U(\tau_h^G(g)) \quad \text{for} \quad g \in G, \tag{1.3}$$

where $\tau_h^G \in \operatorname{Aut}(G)$ is an involution inducing $\tau_h := e^{\pi i \operatorname{ad} h}|_{\mathfrak{g}}$ on the Lie algebra. So we may fix an Euler element $h \in \mathfrak{g}$ from the start. Throughout, $\mathfrak{g}$ denotes a **simple real Lie algebra** (if not otherwise stated). We choose a Cartan involution θ with $\theta(h) = -h$ and observe that $\tau := \theta e^{\pi i \operatorname{ad} h}$ defines an involution on $\mathfrak{g}$. Writing $\mathfrak{g} = \mathfrak{h} \oplus \mathfrak{q}$ with

$$\mathfrak{h} = \mathfrak{g}^\tau = \{x \in \mathfrak{g}\colon \tau(x) = x\} \quad \text{and} \quad \mathfrak{q} = \mathfrak{g}^{-\tau} = \{x \in \mathfrak{g}\colon \tau(x) = -x\},$$

the subspace $\mathfrak{q}$ contains $e^{\operatorname{ad} \mathfrak{h}}$-invariant convex cones C with $h \in C^\circ$ [54, Thm. 4.21]. We refer to Morinelli/Neeb/Ólafsson [54] for a detailed discussion of the connection between Euler elements and non-compactly causal symmetric spaces.

The *causal homogeneous spaces* we consider in this chapter are symmetric spaces $M = G/H$, where G is a connected Lie group with Lie algebra $\mathfrak{g}$ carrying an involutive automorphism τ^G corresponding to τ, and H is an open subgroup of the group G^τ of τ^G-fixed points, preserving a pointed generating cone $C \subseteq \mathfrak{q} \cong T_{eH}(M)$ with $h \in C^\circ$. Then $V_+(gH) := g.C^\circ \subseteq T_{gH}(M)$ defines a G-invariant field $(V_+(m))_{m\in M}$ of open cones on M; called a *causal structure*. Here

$$G \times T(M) \to T(M), \quad (g, v) \mapsto g.v$$

denotes the canonical lift of the G-action on M to the tangent bundle. This brings us to an additional property of a causal symmetric Lie algebra $(\mathfrak{g}, \tau, C)$. It is called

- *compactly causal (cc)*, if all elements $x \in C^0$ are *elliptic* in the sense that $\operatorname{ad} x$ is semisimple with purely imaginary spectrum. An important Lorentzian example is *anti-de Sitter space*

$$\mathrm{AdS}^d = \mathrm{SO}_{2,d-1}(\mathbb{R})_e / \mathrm{SO}_{1,d-1}(\mathbb{R})_e.$$

- *non-compactly causal (ncc)*, if all elements $x \in C^0$ are *hyperbolic* in the sense that $\operatorname{ad} x$ is diagonalizable over $\mathbb{R}$. Here an important Lorentzian example is *de Sitter space* $\mathrm{dS}^d = \mathrm{SO}_{1,d}(\mathbb{R})_e / \mathrm{SO}_{1,d-1}(\mathbb{R})_e$.

If G is semisimple with finite center, then, for compactly causal spaces, there are closed causal curves, so that no global causal order exists on M, but, for non-compactly causal spaces, there even exists a global order, which is *globally hyperbolic* in the sense that all order intervals are compact [42, Thm. 5.3.5].

In Algebraic Quantum Field Theory in the sense of Haag–Kastler, one considers *nets* of von Neumann algebras $\mathcal{M}(\mathcal{O})$ on a fixed Hilbert space $\mathcal{H}$, associated to regions $\mathcal{O}$ in some space-time manifold M (see Haag [36]). The hermitian elements of the algebra $\mathcal{M}(\mathcal{O})$ represent observables that can be measured in the "laboratory" $\mathcal{O}$. Accordingly, one requires *isotony*, i.e., that

$$\mathcal{O}_1 \subseteq \mathcal{O}_2 \Rightarrow \mathcal{M}(\mathcal{O}_1) \subseteq \mathcal{M}(\mathcal{O}_2).$$

One further assumes the existence of a unitary representation $U : G \to \mathrm{U}(\mathcal{H})$ of a Lie group G, acting as a space-time symmetry group on M, such that

$$U(g)\mathcal{M}(\mathcal{O})U(g)^* = \mathcal{M}(g\mathcal{O}) \quad \text{for } g \in G \quad \text{(covariance)}.$$

In addition, one assumes a $U(G)$-fixed unit vector $\Omega \in \mathcal{H}$, representing typically a vacuum state of a quantum field. The domains $\mathcal{O} \subseteq M$ for which Ω is cyclic and separating for $\mathcal{M}(\mathcal{O})$ are of particular relevance. For these domains $\mathcal{O}$, the Tomita–Takesaki Theorem (Bratelli/Robinson [12, Thm. 2.5.14]) yields for the von

Neumann algebra $\mathcal{M}(O)$ a conjugation (antiunitary involution) J_O and a positive selfadjoint operator Δ_O satisfying

$$J_O\mathcal{M}(O)J_O = \mathcal{M}(O)' \quad \text{and} \quad \Delta_O^{it}\mathcal{M}(O)\Delta_O^{-it} = \mathcal{M}(O) \quad \text{for} \quad t \in \mathbb{R}. \tag{1.4}$$

This defines a standard subspace $\mathtt{V} = \mathrm{Fix}(J_O\Delta_O^{1/2})$ connecting the Tomita–Takesaki theory to the theory of standard subspaces. We also obtain the modular automorphism group defined by

$$\widetilde{\alpha}_t(A) = \Delta_O^{-it/2\pi} A \Delta_O^{it/2\pi}, \quad A \in B(\mathcal{H}).$$

It is now an interesting question when this modular group is "geometric" in the sense that it is implemented by a one-parameter subgroup of G. This is always the case for nets of operator algebras obtained from nets of real subspaces $\mathsf{H}(O)$ satisfying (Iso), (Cov), (RS) and (BW) by some second quantization functor (see Simon [70] for the classical context, and Correa da Silva/Lechner for generalizations [20]). Here the modular group corresponds to the flow defined by $\alpha_t(m) = \exp(th).m$ on $M = G/H$, where h is an Euler element. With a view toward the connection with AQFT, the set of pairs (h, τ), consisting of an Euler element $h \in \mathfrak{g}$ and an involution τ with $\tau(h) = h$, has been studied in Morinelli/Neeb [52]. Although our construction ignores field theoretic interactions, it displays already some crucial features of quantum field theories. This has been explored for the flat case in Neeb/Ólafsson/Ørsted [66], and in Neeb/Ólafsson [63] for the class of compactly causal symmetric spaces and unitary highest weight representations $(U, \mathcal{H})$ of G. In this chapter, we address non-compactly causal spaces and general unitary representations. For non-compactly causal spaces, the group G may not have nontrivial unitary representations in which a non-zero Lie algebra element has semibounded spectrum (e.g. the Lorentz group $\mathrm{SO}_{1,d}(\mathbb{R})_e$ for $d > 2$). Therefore, the methods developed in Neeb/Ólafsson [63, 65] do not apply.

To deal with the Bisognano–Wichmann property (BW), we have to specify suitable wedge regions $W \subseteq M$. As the specific examples in AQFT suggest, the modular flow on W should be timelike future-oriented because the modular flow should correspond to the "flow of time" (see Connes/Rovelli [19] and also Borchers/Buchholz [11], Buchholz/Mund/Summers [16], Borchers [10], and Ciolli/Longo/Ranallo/Ruzzi [18, §3]). In our context this means that the *modular vector field*

$$X_h^M(m) := \frac{d}{dt}\Big|_{t=0} \alpha_t(m) = \frac{d}{dt}\Big|_{t=0} \exp(th).m \tag{1.5}$$

should satisfy

$$X_h^M(m) \in V_+(m) \quad \text{for all} \quad m \in W.$$

Choosing the Euler element h in such a way that $h \in C^\circ$, this condition is satisfied in the base point, so that the connected component $W := W_M^+(h)_{eH}$ of the base point $eH \in M$ in the *positivity region*

$$W_M^+(h) := \{m \in M : X_h^M(m) \in V_+(m)\} \tag{1.6}$$

is the natural candidate for a domain for which (BW) could be satisfied. These "wedge regions" have been studied for compactly in Neeb/Ólafsson [65] and for non-compactly causal symmetric spaces by Neeb/Ólafsson/Morinelli [64, 55].

We start with a unitary representation $(U, \mathcal{H})$ of G that extends to an antiunitary representation of

$$G_{\tau_h} := G \rtimes \{\mathbf{1}, \tau_h^G\} \quad \text{by} \quad U(\tau_h^G) := J,$$

where J is a conjugation with $JU(g)J = U(\tau_h^G(g))$ for $g \in G$ (see (1.3)). We write $\mathtt{V} \subseteq \mathcal{H}$ for the standard subspace specified by

$$\Delta_{\mathtt{V}} = e^{2\pi i \cdot \partial U(h)} \quad \text{and} \quad J_{\mathtt{V}} = J, \tag{1.7}$$

where $\partial U(h)$ is the skew-adjoint infinitesimal generator of the one-parameter group $U(\exp th)$. Denote the open π-strip in $\mathbb{C}$ by $\mathcal{S}_\pi := \{z \in \mathbb{C} : 0 < \operatorname{Im} z < \pi\}$. Then the elements $\xi \in \mathtt{V}$ are characterized among elements of $\mathcal{H}$ by the *KMS like condition* that the orbit map $U_h^\xi(t) := U(\exp th)\xi$ has a continuous extension $U_h^\xi : \overline{\mathcal{S}_\pi} \to \mathcal{H}$ to the closed strip $\overline{\mathcal{S}_\pi}$, which is holomorphic on the interior and satisfies

$$U_h^\xi(\pi i) = J\xi \quad \text{resp.} \quad U_h^\xi(t + \pi i) = JU_h^\xi(t) \quad \text{for} \quad t \in \mathbb{R} \tag{1.8}$$

(cf. [66, Prop. 2.1]).

Denote by $\mathcal{H}^\omega \subseteq \mathcal{H}$ the space of analytic vectors, endowed with its natural locally convex topology (cf. Sect. 2) and by $\mathcal{H}^\infty \subseteq \mathcal{H}$ the space of smooth vectors with the standard topology. Then $\mathcal{H}^{-\omega}$, the space of *hyperfunction vectors*, is its antidual space (the continuous antilinear functionals) and $\mathcal{H}^{-\infty}$, the space of *distribution vectors*, is the the antidual of $\mathcal{H}^\infty$. We note that $\mathcal{H}^{-\omega} \subseteq \mathcal{H}^{-\infty}$ but in general they are not the same.

We now define subspaces

$$\mathcal{H}_{\mathrm{KMS}}^{-\infty} \subseteq \mathcal{H}^{-\infty} \quad \text{and} \quad \mathcal{H}_{\mathrm{KMS}}^{-\omega} \subseteq \mathcal{H}^{-\omega}, \tag{1.9}$$

as the space of hyperfunction, respectively, distribution vectors such that the orbit map extends continuously, with respect to the weak-$*$-topology, to the the closed strip $\overline{\mathcal{S}}_\pi$, weak-$*$ holomorphic in the interior, such that (1.8) is satisfied (cf. Definition 1). For $G = \mathbb{R}$ our main new result on hyperfunction vectors is Theorem 2, asserting that the subspace $\mathcal{H}_{\mathrm{KMS}}^{-\omega}$ of $\mathcal{H}^{-\omega}$ is closed and can be characterized as the annihilator of a concrete real subspace of $\mathcal{H}^\omega$ with respect to the pairing induced by $\mathrm{Im}\langle\cdot,\cdot\rangle$. Further, Theorem 3 provides a characterization of distribution vectors that are limits of analytic vectors in terms of asymptotic growth of the norm on curves $t \mapsto e^{tH}v$, where $v \in \mathcal{H}$, and $H = H^*$ is the selfadjoint generator of the unitary one-parameter group.

We now explain how to find suitable subspaces E of distribution vectors if G is a connected simple real Lie group. Let $K := G^\theta$ be the group of fixed points of a Cartan involution θ, so that $\mathrm{Ad}(K) \subseteq \mathrm{Ad}(G)$ is maximally compact. We consider a finite-dimensional K-invariant subspace $\mathcal{E} \subseteq \mathcal{H}$, which is also J-invariant and set

$$\mathsf{E}_K := \mathcal{E}^J = \{v \in \mathcal{E} \colon Jv = v\}.$$

Note that $\tau_h(K) = K$ implies that J leaves the subspace $\mathcal{H}^{[K]}$ of K-finite vectors invariant. Therefore, J-invariant finite-dimensional K-invariant subspaces exist in abundance. We work with the following two hypotheses:

(H1) $\mathcal{E} \subseteq \bigcap_{x\in\Omega_\mathfrak{p}} \mathcal{D}(e^{i\partial U(x)})$ (Proposition 6), where $\Omega_\mathfrak{p} \subseteq \mathfrak{p}$ consists of all elements for which the spectral radius of $\mathrm{ad}\, x$ is smaller than $\frac{\pi}{2}$. Note that $th \in \Omega_\mathfrak{p}$ if and only if $|t| < \pi/2$.

(H2) In addition, the limits

$$\beta^+(v) := \lim_{t\to\pi/2} e^{-it\partial U(h)}v, \qquad v \in \mathcal{E},$$

which exist in the space $\mathcal{H}^{-\omega}$ of hyperfunction vectors, are actually contained in the space $\mathcal{H}^{-\infty}$ of distribution vectors.

Natural equivariance properties (Proposition 7) then imply that

$$\mathsf{E}_H := \beta^+(\mathsf{E}_K) \subseteq \mathcal{H}^{-\infty} \tag{1.10}$$

is a finite-dimensional H-invariant subspace, and $\mathsf{H}^M_{\mathsf{E}_H}$, defined as in (1.2), specifies a net of real subspaces on $M = G/H$. Our main result (Theorem 8) states that this net also satisfies (RS) and (BW). Here it is of key importance that our construction is such that the real subspace E_H is actually contained in the subspace $\mathcal{H}^{-\infty}_{\mathrm{KMS}}$. The hard part of our argument is to verify that we also have $\mathsf{H}^M_{\mathsf{E}_H}(W) \subseteq \mathcal{H}^{-\infty}_{\mathrm{KMS}}$, which in turn leads to $\mathsf{H}^M_{\mathsf{E}_H}(W) \subseteq \mathsf{V} = \mathcal{H}_{\mathrm{KMS}}$ (cf. Proposition 9).

There are two approaches to verify our hypotheses. For the class of linear simple groups G one can use the Krötz–Stanton Extension Theorem (see Theorem 4 below) to verify Hypothesis (H1). Hypothesis (H2) requires the additional regularity that the subspace E_H, which a priori consists only of hyperfunction vectors, actually consists of distribution vectors. For finite coverings of linear groups, this can be derived from a generalization of the van den Ban–Delorme Automatic Continuity Theorem (Theorem 5). So (H1) and (H2) hold in particular if G is linear, but we expect them to hold in general. In the meantime Tobias Simon verified these hyptheses for all irreducible unitary representations of general connected semisimple Lie groups (see [71]). This problem is discussed in Sect. 6, where we argue that the Casselman Subrepresentation Theorem extends to Harish–Chandra modules for which G does not have to be linear. Therefore, the missing result is a generalization of the Casselman–Wallach Globalization Theorem (see Bernstein/Krötz [7]) to nonlinear groups.

An example that is of particular importance in physics arises for the group $G = \mathrm{SO}_{1,d}(\mathbb{R})_e$ and $H = \mathrm{SO}_{1,d-1}(\mathbb{R})_e$, for which $M = G/H = \mathrm{dS}^d$ is d-dimensional de Sitter space. For $d \geq 3$, the simply connected covering is the spin group $\widetilde{G} \cong \mathrm{Spin}_{1,d}(\mathbb{R}) \subseteq \mathrm{Spin}_{1+d}(\mathbb{C})$. As this group is linear, our results yield in particular for every irreducible unitary representation $(U, \mathcal{H})$ of this group, and any $\mathsf{E}_K \subseteq \mathcal{H}^{[K],J}$ as above, a corresponding net $\mathsf{H}^M_{\mathsf{E}_H}$ on de Sitter space. For spherical representations, i.e., representations where $\mathcal{E}$ can be chosen as $\mathcal{E} = \mathbb{C}v_K$ for a K-fixed vector v_K, the corresponding subspace $\mathsf{E}_H = \mathbb{R}v_H \subseteq \mathcal{H}^{-\infty}$ is spanned by a non-zero H-fixed distribution vector v_H (Proposition 7(c),(d)). Through the natural analytic extension process from functions on G/K to distributions on G/H, these H-fixed distribution vectors now lead to a G-equivariant realization of the representation $(U, \mathcal{H})$ as a Hilbert space of distributions on G/H, on which the scalar product is specified by a distribution kernel $D \in C^{-\infty}(M \times M, \mathbb{C})$. For irreducible unitary representations, these kernels satisfy differential equations (in both variables) coming from the center of the enveloping algebra of $\mathfrak{g}$. In the special situation of de Sitter space dS^d and $G = \mathrm{SO}_{1,d}(\mathbb{R})_e$, one obtains eigendistributions of the corresponding wave operator. These kernels are precisely the perikernels occuring in the work of J. Bros and U. Moschella in the 1990s (cf. [14, 56, 13]). This correspondence also emerges quite naturally in the context of reflection positivity on spheres, where these kernels can be described explicitly in terms of hypergeometric functions (see in particular Neeb/Ólafsson [62] and Sect. 5.1.2 below). For non-spherical representations, it is much harder to match the picture developed in Bros/Moschella [14] with our representation theoretic approach. For some recent progress in this direction, we refer to [25].

Identifying the conformal group $\mathrm{SO}_{2,d}(\mathbb{R})_e$ of Minkowski space with a conformal group acting (locally) on de Sitter space dS^d, our construction yields in particular conformally covariant nets on (domains in) de Sitter space, as they are studied by Guido and Longo in [35]. We expect that many aspects of their work, in particular the "holographic reconstruction" of nets on M from nets on "horizons," have counterparts for "conformally flat" causal symmetric spaces as they are classified by Bertram in [8], but this remains to be explored in more detail. For $d = 2$ this leads to the interesting relation between nets on de Sitter space dS^2, anti-de Sitter space AdS^2 and conformal nets on the circle. We refer to Sect. 5.4, and in particular Remark 18 for more details on this circle of ideas.

In Sect. 5 we show that Hypothesis (H1) always holds if $\dim \mathcal{E} = 1$ ($\mathcal{E} = \mathbb{C}v$, $Jv = v$) for real rank-one groups. We further show that there exists constants $C > 0$ and $N > 0$ such that

$$\|e^{it\partial U(h)}v\| \leq C\Big(\frac{\pi}{2} - t\Big)^{-N} \quad \text{for} \quad 0 \leq t < \frac{\pi}{2}. \tag{1.11}$$

This implies that Hypothesis (H2) also holds. Since all irreducible K-subrepresentations are one-dimensional if $\mathfrak{g} = \mathfrak{sl}_2(\mathbb{R})$, both (H1) and (H2) are satisfied in this case, so that that Theorem 8 also applies to all irreducible unitary representations of any connected Lie group G with Lie algebra $\mathfrak{sl}_2(\mathbb{R})$. As $\mathfrak{sl}_2(\mathbb{R}) \cong \mathfrak{so}_{1,2}(\mathbb{R})$, this covers in particular the situation where M is any covering of the 2-dimensional de Sitter space dS^2. It allows us in particular to solve the SL_2-problem on intersections of standard subspaces, as formulated in Morinelli/Neeb [52, §4], in the affirmative.

The structure of this chapter is as follows. We start in Sect. 2 with a discussion of hyperfunction and distribution vectors for unitary representations $(U, \mathcal{H})$ of a Lie group G and discuss the crucial case $G = \mathbb{R}$ in some detail in Sects. 2.2 and 2.3. In particular Theorem 3 will be used in Sect. 5 to verify Hypotheses (H1) and (H2) for groups with Lie algebra $\mathfrak{sl}_2(\mathbb{R})$. In Sect. 3 we consider for a unitary representation $(U, \mathcal{H})$ of a connected simple Lie group holomorphic extensions of orbits maps $U^v : G \to \mathcal{H}, g \mapsto U(g)v$. Here we introduce Hypotheses (H1) and (H2) and their consequences. In Sect. 4 we turn to the Reeh–Schlieder (RS) and the Bisognano–Wichmann (BW) property of the nets of real subspaces $\mathsf{H}^M_{E_H}$. In particular we prove our main result (Theorem 8), which applies in particular if G is linear, i.e., contained in its complexification. In Sect. 5 this result is extended to covering groups of $\mathrm{SL}_2(\mathbb{R})$. A possible strategy to extend it to general connected simple Lie groups is outlined in Sect. 6. Appendix 1 contains a proof of a version of the van den Ban–Delorme Automatic Continuity Theorem to $\mathfrak{h}$-finite vectors (Theorem 5). Finally, Appendix 2 contains a brief discussion of results on wedge regions in non-compactly causal spaces that have been developed in Morinelli/Neeb/Ólafsson [55].

Notation

- For $r > 0$ we denote the corresponding horizontal strips in $\mathbb{C}$ by

$$\mathcal{S}_r := \{z \in \mathbb{C} : 0 < \operatorname{Im} z < r\} \quad \text{and} \quad \mathcal{S}_{\pm r} := \{z \in \mathbb{C} : |\operatorname{Im} z| < r\}.$$

- For a unitary representation $(U, \mathcal{H})$ of G we write:
  - $\mathcal{H}^{[K]}$ for the space of K-finite vectors, where $K \subseteq G$ is a subgroup.
  - $\partial U(x) = \frac{d}{dt}\big|_{t=0} U(\exp tx)$ for the infinitesimal generator of the unitary one-parameter group $(U(\exp tx))_{t\in\mathbb{R}}$ in the sense of Stone's Theorem.
  - $\mathrm{d}U : \mathcal{U}(\mathfrak{g}_{\mathbb{C}}) \to \mathrm{End}(\mathcal{H}^\infty)$ for the representation of the enveloping algebra $\mathcal{U}(\mathfrak{g}_{\mathbb{C}})$ of $\mathfrak{g}_{\mathbb{C}}$ on the space $\mathcal{H}^\infty$ of smooth vectors. Then $\partial U(x) = \overline{\mathrm{d}U(x)}$ for $x \in \mathfrak{g}$.

2 Analytic and Hyperfunction Vectors

In this section, we discuss preliminaries concerning hyperfunction and distribution vectors. Section 2.1 introduces, for a general Lie group G, the topology on the space $\mathcal{H}^\omega$ of analytic vectors of a unitary representations, its dual space $\mathcal{H}^{-\omega}$ and the G-

representations on these spaces. In Sect. 2.2 we specialize to $G = \mathbb{R}$, for which very explicit descriptions of these spaces can be given in terms of spectral theory. In Sect. 2.3 we further discuss distribution vectors for $G = \mathbb{R}$. We focus on the properties of the spaces $\mathcal{H}^{-\omega}_{\rm KMS}$ and $\mathcal{H}^{-\infty}_{\rm KMS}$ (cf. (1.9)).

For $G = \mathbb{R}$ our main result on hyperfunction vectors is Theorem 2, asserting that the subspace $\mathcal{H}^{-\omega}_{\rm KMS}$ of $\mathcal{H}^{-\omega}$ (cf. 1.9) is closed and can be characterized as the annihilator of a concrete real subspace of $\mathcal{H}^{\omega}$ with respect to the pairing induced by $\mathrm{Im}\langle\cdot,\cdot\rangle$. Further, Theorem 3 provides a characterization of distribution vectors that are limits of analytic vectors in terms of asymptotic growths of the norm on curves $t \mapsto e^{tH}v$, where $v \in \mathcal{H}$ and $H = H^*$ is the self-adjoint generator of the unitary one-parameter group.

2.1 The Space of Analytic Vectors

In this subsection, we briefly discuss the space of analytic vectors of a unitary representation of a Lie group. For analytic vectors of more general representations, we refer to Gimperlein, Krötz and Schlichtkrull [28].

Let $(U, \mathcal{H})$ be a unitary representation of the connected real Lie group G. We write

$$\mathcal{H}^{\omega} = \mathcal{H}^{\omega}(U) \subseteq \mathcal{H}$$

for the space of *analytic vectors*, i.e., those $\xi \in \mathcal{H}$ for which the orbit map

$$U^{\xi}: G \to \mathcal{H}, \quad g \mapsto U(g)\xi,$$

is analytic.

To endow $\mathcal{H}^{\omega}$ with a locally convex topology, we specify subspaces $\mathcal{H}^{\omega}_V$ by open convex 0-neighborhoods $V \subseteq \mathfrak{g}$ as follows. Let $\eta_G: G \to G_{\mathbb{C}}$ denote the universal complexification of G and assume that η_G has discrete kernel (this is always the case if G is semisimple). We assume that V is so small that the map

$$\eta_{G,V}: G_V := G \times V \to G_{\mathbb{C}}, \quad (g, x) \mapsto \eta_G(g)\exp(ix) \tag{2.1}$$

is a covering. Then we endow G_V with the unique complex manifold structure for which $\eta_{G,V}$ is holomorphic.

We now write $\mathcal{H}^{\omega}_V$ for the set of those analytic vectors ξ for which the orbit map $U^{\xi}: G \to \mathcal{H}$ extends to a holomorphic map

$$U^{\xi}_V: G_V \to \mathcal{H}.$$

As any such extension is G-equivariant by uniqueness of analytic continuation, it must have the form

$$U_V^\xi(g, x) = U(g)e^{i\partial U(x)}\xi \quad \text{for} \quad g \in G, x \in V, \tag{2.2}$$

so that $\mathcal{H}_V^\omega \subseteq \bigcap_{x\in V} \mathcal{D}(e^{i\partial U(x)})$. The following lemma shows that we even have equality.

Lemma 1 *If $V \subseteq \mathfrak{g}$ is an open convex 0-neighborhood for which* (2.1) *is a covering, then $\mathcal{H}_V^\omega = \bigcap_{x\in V} \mathcal{D}(e^{i\partial U(x)})$.*

Proof It remains to show that each $\xi \in \bigcap_{x\in V} \mathcal{D}(e^{i\partial U(x)})$ is contained in $\mathcal{H}_V^\omega$. For that, we first observe that the holomorphy of the functions $z \mapsto e^{iz\partial U(x)}v$ on a neighborhood of the closed unit disc in $\mathbb{C}$ implies that the $\mathcal{H}$-valued power series

$$f_\xi(x) := \sum_{n=0}^\infty \frac{i^n}{n!}\partial U(x)^n\xi$$

converges for each $x \in V$. Further, [33, Thm. 1.1] implies that $\xi \in \mathcal{H}^\infty$, so that the functions $x \mapsto \partial U(x)^n\xi = \mathrm{d}U(x)^n\xi$ are homogeneous $\mathcal{H}$-valued polynomials (cf. [9]). Thus [9, Thm. 5.2] shows that the above series defines an analytic function $f_\xi \colon V \to \mathcal{H}$. It follows in particular that ξ is an analytic vector, and the map

$$U_V^\xi \colon G_V \to \mathcal{H}, \quad (g, x) \mapsto U^\xi(g, x) := U(g)e^{i\partial U(x)}\xi$$

is defined. It is clearly equivariant. We claim that it is holomorphic. As it is locally bounded, it suffices to show that, for each $\eta \in \mathcal{H}^\omega$, the function

$$f \colon G_V \to \mathbb{C}, \quad f(g, x) := \langle \eta, U^\xi(g, x)\rangle$$

is holomorphic [57, Cor. A.III.3]. As

$$f(g, x) = \langle U(g)^{-1}\eta, e^{i\partial U(x)}\xi\rangle$$

and the orbit map of η is analytic, f is real analytic. Therefore it suffices to show that it is holomorphic on some 0-neighborhood. This follows from the fact that it is G-equivariant and coincides on some 0-neighborhood with the local holomorphic extension of the orbit map of ξ. Here we use that, for $x, y \in \mathfrak{g}$ sufficiently small, the holomorphic extension U^ξ of the ξ-orbit map satisfies

$$U^\xi(\exp(x * iy)) = U(\exp x)U^\xi(\exp iy) = U(\exp x)f_\xi(y) = U_V^\xi(\exp x, y),$$

where $a * b = a + b + \frac{1}{2}[a, b] + \cdots$ denotes the Baker–Campbell–Hausdorff series. □

We topologize the space $\mathcal{H}_V^\omega$ by identifying it with $\mathcal{O}(G_V, \mathcal{H})^G$, the Fréchet space of G-equivariant holomorphic maps $F: G_V \to \mathcal{H}$, endowed with the Fréchet topology of uniform convergence on compact subsets. Now $\mathcal{H}^\omega = \bigcup_V \mathcal{H}_V^\omega$, and we topologize $\mathcal{H}^\omega$ as the locally convex direct limit of the Fréchet spaces $\mathcal{H}_V^\omega$. If the universal complexification $\eta_G: G \to G_{\mathbb{C}}$ is injective, it is easy to see that we thus obtain the same topology as in [28]. Note that, for any monotone basis $(V_n)_{n\in\mathbb{N}}$ of convex 0-neighborhoods in $\mathfrak{g}$, we then have

$$\mathcal{H}^\omega \cong \varinjlim \mathcal{H}_{V_n}^\omega,$$

so that $\mathcal{H}^\omega$ is a countable locally convex limit of Fréchet spaces. As the evaluation maps

$$\mathcal{O}(G_V, \mathcal{H})^G \to \mathcal{H}, \quad F \mapsto F(e, 0)$$

are continuous, the inclusion $\iota: \mathcal{H}^\omega \to \mathcal{H}$ is continuous.

Remark 1 Alternatively, one could also use the spaces $\mathcal{H}_V^{\omega,b}$ of those vectors for which the extended orbit maps are bounded. They correspond to the spaces $\mathcal{O}(G_V, \mathcal{H})^{b,G}$ of G-equivariant bounded holomorphic maps, which are Banach spaces.

If $V_1 \subseteq V_2$ is relatively compact, then $\mathcal{H}_{V_2}^\omega \subseteq \mathcal{H}_{V_1}^{\omega,b}$, and this implies that also

$$\mathcal{H}^\omega \cong \varinjlim \mathcal{H}_{V_n}^{\omega,b},$$

describing $\mathcal{H}^\omega$ is a locally convex direct limit of Banach spaces.

We write $\mathcal{H}^{-\omega}$ for the space of continuous antilinear functionals $\eta: \mathcal{H}^\omega \to \mathbb{C}$ (called *hyperfunction vectors*) and

$$\langle \cdot, \cdot \rangle : \mathcal{H}^\omega \times \mathcal{H}^{-\omega} \to \mathbb{C}$$

for the natural sesquilinear pairing that is linear in the second argument. We endow $\mathcal{H}^{-\omega}$ with the weak-$*$-topology. We then have natural continuous inclusions

$$\mathcal{H}^\omega \hookrightarrow \mathcal{H} \hookrightarrow \mathcal{H}^{-\omega}.$$

Our specification of the topology on $\mathcal{H}^\omega$ differs from the one in [28] because we do not want to assume that the universal complexification $\eta_G: G \to G_{\mathbb{C}}$ is injective, but both constructions define the same topology. Moreover, the arguments in [28] apply with minor changes to general Lie groups.

Recall that a vector $v \in \mathcal{H}$ is *smooth* if the orbit map $g \mapsto U^v(g) = U(g)v$ is smooth. Denote by $\mathcal{H}^\infty$ the space of smooth vectors. The group G acts smoothly on $\mathcal{H}^\infty$, and we also have a representation of $\mathfrak{g}$ on $\mathcal{H}^\infty$ given by $\mathrm{d}U(x)v = \frac{d}{dt}\big|_{t=0} U(\exp tx)v$. The topology is defined by the seminorms $\|\mathrm{d}U(x)u\|$, $x \in \mathcal{U}(\mathfrak{g})$.

The space of *distribution vectors*, denoted by $\mathcal{H}^{-\infty}$, is the conjugate linear dual of $\mathcal{H}^{\infty}$. We have

$$\mathcal{H}^{\omega} \subseteq \mathcal{H}^{\infty} \subseteq \mathcal{H} \subseteq \mathcal{H}^{-\infty} \subseteq \mathcal{H}^{-\omega},$$

where all inclusions are continuous.

Proposition 1 *The topology on $\mathcal{H}^{\omega}$ has the following properties:*

(a) *The representation $U^{\omega}\colon G \to \mathcal{L}(\mathcal{H}^{\omega})$ of G on $\mathcal{H}^{\omega}$ defines a continuous action.*
(b) *The orbit maps of U^{ω} are analytic.*
(c) *If $B \subseteq G$ is compact, then $U^{\omega}(B) \subseteq \mathcal{L}(\mathcal{H}^{\omega})$ is equicontinuous.*
(d) *$\mathcal{H}^{\omega}$ is complete.*
(e) *For $g \in G$ and a convex 0-neighborhood $V \subseteq \mathfrak{g}$, we have $U(g)\mathcal{H}^{\omega}_V = \mathcal{H}^{\omega}_{\mathrm{Ad}(g)V}$.*
(f) *For $\varphi \in C_c(G, \mathbb{C})$ and $U(\varphi) = \int_G \varphi(g)\,dg$, we have $U(\varphi)\mathcal{H}^{\omega} \subseteq \mathcal{H}^{\omega}$, the operator*

$$U^{\omega}(\varphi) := U(\varphi)|_{\mathcal{H}^{\omega}}\colon \mathcal{H}^{\omega} \to \mathcal{H}^{\omega}$$

is continuous, and it has a weak-$*$-continuous adjoint

$$U^{-\omega}\colon \mathcal{H}^{-\omega} \to \mathcal{H}^{-\omega}, \quad U^{-\omega}(\varphi)\eta := \eta \circ U^{\omega}(\varphi^*)$$

that extends the operator $U^{-\infty}(\varphi) = \int_G \varphi(g) U^{-\infty}(g)\,dg$.

Proof (a), (b) follow from [28, Prop. 3.4] and (c) from its proof. Further (d) follows from [28, Prop. 3.7].
(e) follows easily from the relation

$$U^{U(g)\xi}(h, x) = U(h)e^{i\partial U(x)}U(g)\xi = U(hg)e^{i\partial U(\mathrm{Ad}(g)^{-1}x)}\xi = U^{\xi}(hg, \mathrm{Ad}(g)^{-1}x)$$

(see (2.2) for the notation).
(f) For $v \in \mathcal{H}^{\omega}$, the orbit map $U^{v}\colon G \to \mathcal{H}^{\omega}$ is continuous and $\mathcal{H}^{\omega}$ is complete by (b) and (d). Therefore the weak integral

$$U(\varphi)v := \int_G \varphi(g)U(g)v\,dg$$

exists in $\mathcal{H}^{\omega}$, which means that $U(\varphi)\mathcal{H}^{\omega} \subseteq \mathcal{H}^{\omega}$.

To see that $U^{\omega}(\varphi)$ is continuous, let $B := \mathrm{supp}(\varphi)$. Then $U^{\omega}(B)$ is equicontinuous by (c), i.e., for any 0-neighborhood $W_1 \subseteq \mathcal{H}^{\omega}$ (w.l.o.g. absolutely convex and closed), there exists a 0-neighborhood $W_2 \subseteq \mathcal{H}^{\omega}$ with $U^{\omega}(B)W_2 \subseteq W_1$. As W_1 is closed and absolutely convex, it follows that

$$U^{\omega}(\varphi)W_2 \subseteq \|\varphi\|_1 \cdot W_1$$

and hence, that $U^{\omega}(\varphi)$ is continuous.

This implies in particular the existence of the weak-$*$-continuous operator $U^{-\omega}(\varphi)$. To see that this operator extends the operator

$$U^{-\infty}(\varphi)\colon \mathcal{H}^{-\infty} \to \mathcal{H}^{-\infty}, \quad U^{-\infty}(\varphi)\eta = \int_G \varphi(g)U^{-\infty}(g)\eta\, dg,$$

we evaluate it on some $\eta \in \mathcal{H}^{-\infty} \subseteq \mathcal{H}^{-\omega}$. So let $\xi \in \mathcal{H}^{\omega} \subseteq \mathcal{H}^{\infty}$. Then

$$\begin{aligned}\langle \xi, U^{-\omega}(\varphi)\eta\rangle &= \int_G \varphi(g)\langle \xi, U^{-\omega}(g)\eta\rangle\, dg = \int_G \varphi(g)\langle \xi, U^{-\infty}(g)\eta\rangle\, dg \\ &= \int_G \varphi(g)\langle U(g^{-1})\xi, \eta\rangle\, dg = \langle U(\varphi^*)\xi, \eta\rangle = \langle \xi, U^{-\infty}(\varphi)\eta\rangle.\end{aligned}$$

We conclude that $U^{-\infty}(\varphi)\eta|_{\mathcal{H}^{\omega}} = U^{-\omega}(\varphi)\eta$, i.e., the inclusion $\mathcal{H}^{-\infty} \hookrightarrow \mathcal{H}^{-\omega}$ maps $U^{-\infty}(\varphi)\eta$ to $U^{-\omega}(\varphi)\eta$. Here we use that $\mathcal{H}^{\omega}$ is dense in $\mathcal{H}^{\infty}$ (Theorem 1 below). □

Theorem 1 *The subspace $\mathcal{H}^{\omega}$ is dense in the Fréchet space $\mathcal{H}^{\infty}$.*

The density of $\mathcal{H}^{\omega}$ in $\mathcal{H}$ for finite-dimensional Lie groups follows from Nelson [67, Thm. 4] and from Gårding [27] by convolution with heat kernels (cf. Warner [75, Sect. 4.4.5]). However, the theorem does not follow from this weaker result.

Proof Let $x_1, \ldots, x_n$ be a linear basis of $\mathfrak{g}$ and $\Delta_N := \sum_{j=1}^n x_j^2 \in \mathcal{U}(\mathfrak{g})$ be the *Nelson Laplacian* in the universal enveloping algebra. For the positive selfadjoint operator $A := (\mathbf{1} - \overline{\mathrm{d}U(\Delta_N)})^{1/2}$, we have

$$\mathcal{H}^{\omega}(A) = \mathcal{H}^{\omega} \subseteq \mathcal{H}^{\infty} = \mathcal{H}^{\infty}(A) = \bigcap_{n\in\mathbb{N}} \mathcal{D}(A^n),$$

where $\mathcal{H}^{\omega}(A)$, respectively $\mathcal{H}^{\infty}(A)$, denotes the space of analytic, respectively smooth vectors for the unitary one-parameter group $(e^{itA})_{t\in\mathbb{R}}$. Here the first equality is Goodman's Theorem [33, Thm. 2.1] and the second equality follows from [67, Thm. 3, Cor. 9.3] because A and A^2 have the same set of smooth vectors. So it suffices to show that $\mathcal{H}^{\omega}(A)$ is dense in the Fréchet space $\mathcal{H}^{\infty}(A)$. This reduces the problem to the case of 1-parameter groups.
Case $G = \mathbb{R}$: Let $U_t = e^{itA}$ be a unitary one-parameter group. We consider the normalized Gaussians $\gamma_n(t) := \sqrt{\frac{n}{\pi}}e^{-nt^2}$ and recall that $v_n := U(\gamma_n)v \in \mathcal{H}^{\infty}$ converges to v in the Fréchet space $\mathcal{H}^{\infty}$ (cf. Beltiţă/Neeb [6, Prop. 3.3(ii) and §5]). The relation

$$e^{izA}U(\gamma_n)v = U(\delta_z * \gamma_n)v$$

and the fact that the map $\mathbb{C} \to L^1(\mathbb{R})$, $z \mapsto \delta_z * \gamma_n$ is holomorphic shows that v_n is an analytic vector. Hence, the assertion follows. □

Remark 2 If $(\pi, \mathcal{H})$ is irreducible and G is semisimple, then the subspace $\mathcal{H}^{[K]}$ of K-finite vectors is contained in $\mathcal{H}^\omega$ (Harish–Chandra [37]). The density of $\mathcal{H}^{[K]}$ in $\mathcal{H}^\infty$ with respect to the Fréchet topology on $\mathcal{H}^\infty$ follows from the Peter–Weyl Theorem, applied to the continuous action of K, resp., the compact group $\overline{U(K)\mathbb{T}}$ on $\mathcal{H}^\infty$ (cf. Neeb [58, Cor. 4.5]; see also Harish–Chandra [37]).

2.2 Analytic Vectors for One-Parameter Groups

After the discussion of analytic vectors in the preceding subsection, we now take a closer look at the very special case $G = \mathbb{R}$. Let $U_t = e^{itH}$ be a unitary 1-parameter group on $\mathcal{H}$, where $H = H^*$ is self-adjoint. Let P_H denote the spectral measure of H, so that we obtain for each $v \in \mathcal{H}$ a finite positive Borel measure $\mu_v := \langle v, P_H(\cdot)v\rangle$ on $\mathbb{R}$. For $v \in \mathcal{H}$, we then have

$$\langle v, U_t v\rangle = \int_{\mathbb{R}} e^{it\lambda}\, d\mu_v(\lambda)$$

and v is smooth if and only if the measure μ_v possesses moments of all order. Likewise, v is analytic if and only if there exists an $r > 0$ with

$$\int_{\mathbb{R}} e^{\pm r\lambda}\, d\mu_v(\lambda) < \infty.$$

and then $U^v(x + iy) = e^{(ix-y)H}v$ is defined for $|y| < r/2$ by spectral calculus (Neeb/Ólafsson [61, Lemma A.2.5]). Accordingly, we have

$$\mathcal{H}^\omega = \bigcup_{r>0} \mathcal{H}^\omega(r), \qquad \text{where} \qquad \mathcal{H}^\omega(r) = \{v \in \mathcal{H}\colon v \in \mathcal{D}(e^{\pm rH})\}. \tag{2.3}$$

The space $\mathcal{H}^\omega(r)$ consists of all elements whose orbit map extends to the closed strip $\overline{\mathcal{S}_{\pm r}}$, where

$$\mathcal{S}_{\pm r} = \{z \in \mathbb{C}\colon |\operatorname{Im} z| < r\}$$

[61, Lemma A.2.5], and we topologize $\mathcal{H}^\omega(r)$ by the embedding

$$\mathcal{H}^\omega(r) \hookrightarrow \mathcal{H}^{\oplus 2}, \qquad v \mapsto (e^{rH}v, e^{-rH}v). \tag{2.4}$$

Remark 3 To see that this topology coincides with the one introduced in Sect. 2.1 for general Lie groups, we use the embedding

$$\iota\colon \mathcal{H}^\omega(r) \to \mathcal{O}_\partial(\mathcal{S}_{\pm r}, \mathcal{H})^{\mathbb{R}}, \quad v \mapsto U^v, \quad U^v(z) = e^{izH}v$$

to the space $\mathcal{O}_\partial(\mathcal{S}_{\pm r}, \mathcal{H})^{\mathbb{R}}$ of equivariant continuous functions $\overline{\mathcal{S}_{\pm r}} \to \mathcal{H}$ that are holomorphic on the interior. All these maps are bounded and satisfy

$$\|U^v\|_\infty \leq \max\{\|e^{rH}v\|, \|e^{-rH}v\|\},$$

showing that ι is continuous. Further, ι is bijective [61, Lemma A.2.5] and $\iota^{-1}(F) = F(0)$ is also continuous.

From (2.4) we derive that the dual space $\mathcal{H}^{-\omega}(r)$ of $\mathcal{H}^\omega(r)$ consists of all antilinear functionals of the form

$$\alpha(v) = \langle e^{rH}v, w_1\rangle + \langle e^{-rH}v, w_2\rangle \quad \text{for} \quad w_1, w_2 \in \mathcal{H}. \tag{2.5}$$

Informally, we think of α as a sum $e^{rH}w_1 + e^{-rH}w_2$. Note that

$$\mathcal{H}^{-\omega} = \bigcap_{r>0} \mathcal{H}^{-\omega}(r) \tag{2.6}$$

carries a natural Fréchet space structure. Further (2.5) implies that, for $\eta \in \mathcal{H}^{-\omega}$ and any $t > 0$, we have $e^{-t|H|}\mathcal{H} \subseteq \mathcal{H}^\omega$ because the functions $\lambda \mapsto e^{\pm r\lambda - t|\lambda|}$ are bounded for $|r| < t$, and

$$\eta \circ e^{-t|H|} \in \mathcal{H}. \tag{2.7}$$

Remark 4 Consider $\mathcal{H} = L^2(\mathbb{R}, \mu)$ and $(U_t f)(\lambda) = e^{it\lambda} f(\lambda)$. We claim that

$$\mathcal{H}^{-\omega} = \{f\colon \mathbb{R} \to \mathbb{C}\colon f \text{ measurable}, (\forall t > 0)\ e^{-t|\lambda|} f(\lambda) \in L^2\}.$$

Let $r > 0$. Then every $f \in \mathcal{H}^{-\omega}$ can be written as $\eta = e_r f_1 + e_{-r} f_2$ with $f_1, f_2 \in L^2$ and $e_r(\lambda) = e^{r\lambda}$. For $t > r$ we then have $e^{-t|\lambda|}\eta \in L^2$.

If, conversely, $\eta\colon \mathbb{R} \to \mathbb{C}$ is a measurable function with $e^{-t|\lambda|}\eta \in L^2$ for all $t > 0$, then we write $\eta = \eta_+ + \eta_-$ with $\eta_\pm$ supported on $\pm[0, \infty)$. Then $\eta_+ = e_t(e_{-t}\eta_+)$ and $e_{-t}\eta_+ \in L^2$, and likewise $\eta_- = e_{-t}(e_t\eta_-)$ with $e_t\eta_- \in L^2$.

Lemma 2 *Let $v \in \mathcal{H}^\omega$ and $r > 0$ such that $v \in \mathcal{D}(e^{tH})$ for $0 \leq t < r$. Then the following assertions hold:*

(a) *$\eta := \lim_{t\to r} e^{tH}v$ exists in the space $\mathcal{H}^{-\omega}$ with respect to the weak-$*$-topology.*
(b) *The orbit map of $\eta \in \mathcal{H}^{-\omega}$ extends to a weak-$*$-bounded, weak-$*$-continuous map $U^\eta\colon \overline{\mathcal{S}_r} \to \mathcal{H}^{-\omega}$.*

Proof

(a) Let $w \in \mathcal{H}^\omega = \bigcup_{r>0} \mathcal{H}^\omega(r)$. Then there exists an $\varepsilon \in (0, r)$ with $w \in \mathcal{D}(e^{\varepsilon H})$. Let $w_\varepsilon := e^{\varepsilon H} w$. Then

$$\langle w, e^{tH} v\rangle = \langle e^{-\varepsilon H} w_\varepsilon, e^{tH} v\rangle = \langle w_\varepsilon, e^{(t-\varepsilon)H} v\rangle,$$

and since the orbit map

$$U^v : \{z \in \mathbb{C} : 0 < \operatorname{Re} z < r\} \to \mathcal{H}, \qquad z \mapsto e^{zH} v$$

is holomorphic, we obtain

$$\lim_{t \to r} \langle w, e^{tH} v\rangle = \langle w_\varepsilon, e^{(r-\varepsilon)H} v\rangle.$$

On $\mathcal{H}^\omega(\varepsilon)$ we obtain

$$\eta(w) \le \|w_\varepsilon\| \cdot \|e^{(r-\varepsilon)H} v\|,$$

which implies that $\eta \in \mathcal{H}^{-\omega}(\varepsilon)$. This shows that $\eta \in \mathcal{H}^{-\omega}$ and that $e^{tH} v \to \eta$ in the weak-$*$-topology of $\mathcal{H}^{-\omega}$.

(b) For $z = x + iy$ with $0 \le y \le r$, we put

$$U^\eta(z) := \begin{cases} U_x^{-\omega} \eta & \text{for } y = 0 \\ e^{(r+iz)H} v & \text{for } 0 < y \le r. \end{cases}$$

This map is a holomorphic $\mathcal{H}$-valued on the open strip $\{0 < y < r\}$, weak-$*$-continuous on the semi-closed strip $\{0 \le y < r\}$ and bounded on any strip of the form $\{0 \le y < r'\}$ for $0 < r' < r$. To verify (b), let $w \in \mathcal{D}(e^{\varepsilon H})$, $w_\varepsilon := e^{\varepsilon H} w$ as in (a), and $r' \in (r - \varepsilon, r)$ for $0 < \varepsilon < r$. Then

$$\begin{aligned} \langle w, U_t^{-\omega} \eta\rangle &= \langle U_{-t} w, \eta\rangle = \langle e^{-itH} e^{-\varepsilon H} w_\varepsilon, \eta\rangle = \langle e^{-itH} w_\varepsilon, e^{(r-\varepsilon)H} v\rangle \\ &= \langle w_\varepsilon, e^{(r-\varepsilon+it)H} v\rangle = \lim_{z \to t} \langle w_\varepsilon, e^{(r-\varepsilon+iz)H} v\rangle = \lim_{z \to t} \langle w, e^{(r+iz)H} v\rangle \\ &= \lim_{z \to t} \langle w, U^\eta(z)\rangle \end{aligned}$$

shows that U^η is also continuous in $\mathbb{R}$. The boundedness $z \mapsto \langle w, U^\eta(z)\rangle$ follows from

$$|\langle w, U^\eta(z)\rangle| = |\langle w_\varepsilon, e^{(r-\varepsilon+iz)H} v\rangle| \le \|w_\varepsilon\| \cdot \max\{\|e^{(r-\varepsilon)H} v\|, \|v\|\}$$

for $0 \le \operatorname{Im} z \le r - \varepsilon$.

□

Definition 1 Let $(V_t)_{t\in\mathbb{R}}$ be a strongly continuous one-parameter group of topological isomorphisms of a locally convex space $\mathcal{X}$ and $J\colon \mathcal{X} \to \mathcal{X}$ a continuous antilinear involution. Then we write $\mathcal{X}_{\mathrm{KMS}} \subseteq \mathcal{X}$ for the real linear subspace of those elements $\xi \in \mathcal{X}$ for which the orbit map $V^\xi\colon \mathbb{R} \to \mathcal{X}, t \mapsto V_t\xi$, extends to a continuous map $V^\xi\colon \overline{\mathcal{S}_\pi} \to \mathcal{X}$, holomorphic on $\mathcal{S}_\pi$, that satisfies

$$V^\xi(\pi i) = J\xi, \quad \text{resp.,} \quad V^\xi(\pi i + t) = JV^\xi(t) \quad \text{for} \quad t \in \mathbb{R}.$$

Lemma 3 *If $\xi \in \mathcal{X}_{\mathrm{KMS}}$, then $V^\xi(\pi i/2)$ is fixed by J.*

Proof The two maps $V^\xi\colon \overline{\mathcal{S}_\pi} \to \mathcal{X}$ and $z \mapsto JV^\xi(\pi i + \overline{z})$ are both continuous on $\overline{\mathcal{S}_\pi}$ and holomorphic on $\mathcal{S}_\pi$. As they coincide on $\mathbb{R}$, uniqueness of analytic continuation shows they coincide, i.e.,

$$V^\xi(\pi i + \overline{z}) = JV^\xi(z) \quad \text{for} \quad z \in \overline{\mathcal{S}_\pi}.$$

It follows in particular that J fixes $V^\xi(\pi i/2)$. □

Proposition 2

(a) *If J is a conjugation on $\mathcal{H}$ commuting with $U_\mathbb{R}$, and $v \in \mathcal{H}^J$ is such that the orbit map $U^v\colon \mathbb{R} \to \mathcal{H}$ extends to a holomorphic map $\mathcal{S}_{\pm\pi/2} \to \mathcal{H}$, then*

$$\eta := \lim_{t\to\pi/2} U^v(-it) \in \mathcal{H}^{-\omega}_{\mathrm{KMS}},$$

where $\mathcal{H}^{-\omega}_{\mathrm{KMS}}$ is defined around (1.9).

(b) *If, conversely, $\eta \in \mathcal{H}^{-\omega}_{\mathrm{KMS}}$, then*

$$U^\eta(z) \in \mathcal{H} \quad \textit{for} \quad z \in \mathcal{S}_\pi,$$

and, for $v := U^\eta(\pi i/2) \in \mathcal{H}^J$, we have $\eta = \lim_{t\to\pi/2} U^v(-it)$.

Proof

(a) Lemma 2 shows that the holomorphic map $U^v\colon \mathcal{S}_{\pm\pi/2} \to \mathcal{H} \subseteq \mathcal{H}^{-\omega}$ extends to a weak-$*$-continuous, weak-$*$-bounded map $U^v\colon \overline{\mathcal{S}_{\pm\pi/2}} \to \mathcal{H}^{-\omega}$. For $\eta := U^v(-\pi i/2)$, we then have

$$U^\eta(z) = U^v(z - \pi i/2) \quad \text{for} \quad z \in \overline{\mathcal{S}_\pi}.$$

As J commutes with $U_\mathbb{R}$ and fixes v, we have

$$JU^v(z) = U^v(\overline{z}) \quad \text{and thus} \quad JU^\eta(z) = U^v(\overline{z} + \pi i/2) = U^\eta(\overline{z} + \pi i).$$

(b) For the converse, let $\eta \in \mathcal{H}^{-\omega}$. In view of (2.7), $\eta_t := \eta \circ e^{-t|H|} \in \mathcal{H}$. We claim that η vanishes on all analytic vectors ξ orthogonal to the cyclic subspace

$\mathcal{K}$ generated by these vectors η_t. In fact, there exists a $t > 0$ with $e^{t|H|}\xi \in \mathcal{H}$, so that

$$\eta(\xi) = (\eta \circ e^{-t|H|})(e^{t|H|}\xi) \in \langle \mathcal{K}, \mathcal{K}^{\perp} \rangle = \{0\}.$$

By analytic continuation, the same holds for every $U^{\eta}(z)$, $z \in \mathcal{S}_{\pi}$, hence in particular for $\eta' := U^{\eta}(\pi i/2)$, which is fixed by J (Lemma 3). Then, for each $t > 0$, the functional $\eta' \circ e^{-t|H|} \in \mathcal{H}$ is fixed by J.

We may therefore assume that the representation $(U, \mathcal{H})$ is cyclic with a J-fixed generator. By Bochner's Theorem it is therefore of the form $\mathcal{H} = L^2(\mathbb{R}, \mu)$ with

$$(U_t f)(\lambda) = e^{it\lambda} f(\lambda) \quad \text{and} \quad (Jf)(\lambda) = \overline{f(-\lambda)},$$

where we use that the J-fixed generator corresponds to the constant function 1, so that $J1 = 1$ and $Je_{it} = J(U_t 1) = U_t 1 = e_{it}$ implies the asserted form of J. In particular, the measure μ is symmetric. Now η is represented by a function $\eta \colon \mathbb{R} \to \mathbb{C}$ with the property that $e^{-t|\lambda|}\eta(\lambda)$ is L^2 for every $t > 0$ (Remark 4), and the KMS condition turns into

$$\overline{\eta(-\lambda)} = (J\eta)(\lambda) = e^{-\pi\lambda}\eta(\lambda).$$

We write $\eta = \eta_+ + \eta_-$ with $\eta_{\pm}$ supported in $\pm[0, \infty)$. For $z = x + iy \in \mathcal{S}_{\pi}$, we then have $0 < y < \pi$, so that we have for any $r \in (0, y)$:

$$e_{-y}\eta_+ = e_{-y}e_r e_{-r}\eta_+ = e_{r-y}\underbrace{e_{-r}\eta_+}_{\in L^2} \in L^2.$$

We likewise find for $\lambda \leq 0$:

$$(e_{-y}\eta_-)(\lambda) = e^{-\lambda y}\eta_-(\lambda) = e^{\lambda(\pi - y)}e^{-\pi\lambda}\eta_-(\lambda) = e^{\lambda(\pi-y)}\overline{\eta_+(-\lambda)}.$$

As $\pi - y > 0$ and $\eta_+ \in \mathcal{H}^{-\omega}$, it follows that $e_{-y}\eta_- \in L^2$. This shows that $U^{\eta}(z) = e_{iz}\eta \in L^2$ for $z \in \mathcal{S}_{\pi}$.

For $v := U^{\eta}(\pi i/2)$, we then have $U^{v}(-it) = U^{\eta}((\frac{\pi}{2} - t)i)$, so that the last assertion follows from the continuity of U^{η} on the closed strip $\overline{\mathcal{S}_{\pi}}$.

□

The argument in the proof of the preceding proposition can also be used to obtain:

Corollary 1 *If $\eta \in \mathcal{H}^{-\omega}_{\mathrm{KMS}}$ and the element $J\eta \in \mathcal{H}^{-\omega}$ defined by $(J\eta)(\xi) := \overline{\eta(J\xi)}$ is also contained in $\mathcal{H}^{-\omega}_{\mathrm{KMS}}$, then*

$$\eta \in \mathcal{H} \quad \textit{and} \quad U_t\eta = \eta \quad \textit{for all} \quad t \in \mathbb{R}.$$

Proof As in the proof of Proposition 2, we may assume that

$$(U_t f)(\lambda) = e^{it\lambda} f(\lambda) \quad \text{and} \quad (Jf)(\lambda) = \overline{f(-\lambda)}.$$

For $\eta \in \mathcal{H}^{-\omega}_{\mathrm{KMS}}$, we then have the relation $(J\eta)(\lambda) = e^{-\pi\lambda}\eta(\lambda)$. The same argument, applied to $J\eta$, shows that

$$\eta(\lambda) = J(J\eta)(\lambda) = e^{-\pi\lambda}(J\eta)(\lambda) = e^{-2\pi\lambda}\eta(\lambda).$$

This is only possible if η is supported in $\{0\}$, i.e., if the cyclic representation generated by η is trivial. So $\eta = U_t\eta$ for every $t \in \mathbb{R}$ and $\eta = U^\eta(\pi i/2) \in \mathcal{H}$ (Proposition 2). □

Remark 5 The containment in the complex subspace $\mathcal{H}^{-\omega}_{\mathrm{KMS}} + i\mathcal{H}^{-\omega}_{\mathrm{KMS}}$ need not give any information. In fact, if U is norm-continuous, then $\mathcal{H}^{-\omega} = \mathcal{H}$ and

$$\mathcal{H}^{-\omega}_{\mathrm{KMS}} + i\mathcal{H}^{-\omega}_{\mathrm{KMS}} = \mathcal{H}.$$

Remark 6 We now introduce a construction that will be used in the proof of the theorem below.

(a) On $\mathbb{R}$, we consider the analytic L^1-functions

$$\gamma_a(t) := \frac{\sqrt{a}}{\sqrt{\pi}} e^{-at^2}, \quad a > 0.$$

The (left) translation action

$$(\lambda_x\varphi)(t) := \varphi(t - x)$$

defines a continuous isometric action of $\mathbb{R}$ on $L^1(\mathbb{R})$, and the functions γ_a are entire vectors, i.e., their orbit map extends to a holomorphic map

$$\lambda^{\gamma_a} : \mathbb{C} \to L^1(\mathbb{R}), \quad \lambda^{\gamma_a}(z)(t) = \frac{\sqrt{a}}{\sqrt{\pi}} e^{-a(t-z)^2} = \frac{\sqrt{a}}{\sqrt{\pi}} e^{-at^2 + a2tz - az^2}.$$

(b) Let $(U_t)_{t\in\mathbb{R}}$ be a unitary one-parameter group. Then the continuity of the bilinear map

$$L^1(\mathbb{R}) \times \mathcal{H} \to \mathcal{H}, \quad (\psi, \xi) \mapsto U(\psi)\xi$$

implies that $U(\gamma_a)\mathcal{H} \subseteq \mathcal{H}^{\mathcal{O}} \cong \mathcal{O}(\mathbb{C}, \mathcal{H})^{\mathbb{R}}$ (the space of *entire vectors*), so that we obtain a linear map

$$U(\gamma_a) : \mathcal{H} \to \mathcal{H}^{\mathcal{O}}.$$

We also note that the map

$$\mathbb{C} \times \mathcal{H} \to \mathcal{H}, \quad (z, \xi) \mapsto U(\lambda^{\gamma_a}(z))\xi$$

is holomorphic, so that we obtain a natural map

$$\Upsilon: \mathcal{H} \to \mathcal{O}(\mathbb{C}, \mathcal{H}), \quad \Upsilon(\xi)(z) := U(\lambda^{\gamma_a}(z))\xi.$$

As the orbit map λ^{γ_a} in $L^1(\mathbb{R})$ is locally bounded, Υ is continuous. Identifying $\mathcal{H}^{\mathcal{O}}$ with the closed subspace $\mathcal{O}(\mathbb{C}, \mathcal{H})^{\mathbb{R}} \subseteq \mathcal{O}(\mathbb{C}, \mathcal{H})$ of equivariant maps (by evaluation in 0), it follows that $U(\gamma_a): \mathcal{H} \to \mathcal{H}^{\mathcal{O}}$ is continuous.

(c) For $r > 0$, the natural map $\mathcal{H}^{\mathcal{O}} \to \mathcal{H}^{\omega}(r)$ is continuous (cf. (2.3)). Hence, for each $r > 0$, the linear map

$$U(\gamma_a): \mathcal{H} \to \mathcal{H}^{\omega}(r)$$

is continuous by (b). As a consequence, it induces a continuous linear map

$$U(\gamma_a): \mathcal{H} \to \mathcal{H}^{\omega}.$$

So it has a weak-$*$-continuous adjoint map

$$U^{-\omega}(\gamma_a) = U(\gamma_a)^{\sharp}: \mathcal{H}^{-\omega} \to \mathcal{H}. \tag{2.8}$$

Remark 7 For the following theorem, we recall from the introduction that standard subspaces $\mathtt{V} \subseteq \mathcal{H}$ are specified by pairs $(\Delta_{\mathtt{V}}, J_{\mathtt{V}})$ of a positive selfadjoint operator $\Delta_{\mathtt{V}}$ and a conjugation $J_{\mathtt{V}}$ satisfying $J_{\mathtt{V}}\Delta_{\mathtt{V}}J_{\mathtt{V}} = \Delta_{\mathtt{V}}^{-1}$ via

$$\mathtt{V} = \mathrm{Fix}(J_{\mathtt{V}}\Delta_{\mathtt{V}}^{1/2}).$$

These operators are uniquely determined by $\mathtt{V}$ as the factors in the polar decomposition $T_{\mathtt{V}} = J_{\mathtt{V}}\Delta_{\mathtt{V}}^{1/2}$ of the (closed) *Tomita operator*

$$T_{\mathtt{V}}: \mathtt{V} + i\mathtt{V} \to \mathcal{H}, \quad x + iy \mapsto x - iy \quad \text{for} \quad x, y \in \mathtt{V}.$$

For the dual standard subspace

$$\mathtt{V}' = \{w \in \mathsf{H}: \ \mathrm{Im}\langle w, \mathtt{V}\rangle = \{0\}\}$$

we have $\Delta_{\mathtt{V}'} = \Delta_{\mathtt{V}}^{-1}$ and $J_{\mathtt{V}'} = J_{\mathtt{V}}$.

Theorem 2 *Let J be a conjugation commuting with $U_{\mathbb{R}}$ and $\mathtt{V} \subseteq \mathcal{H}$ be the standard subspace with*

$$J_{\mathtt{V}} := J \quad \textit{and} \quad \Delta_{\mathtt{V}} := e^{-2\pi H}.$$

Then

$$\mathcal{H}^{-\omega}_{\rm KMS} = \{\eta \in \mathcal{H}^{-\omega} : (\forall v \in \mathtt{V}' \cap \mathcal{H}^\omega)\ \eta(v) \in \mathbb{R}\}.$$

This theorem implies in particular that $\mathcal{H}^{-\omega}_{\rm KMS}$ is weak-$*$-closed.

Proof "$\subseteq$": **Step 1:** For each entire vector $w \in \mathcal{H}^{\mathcal{O}} \cong \mathcal{O}(\mathbb{C}, \mathcal{H})^{\mathbb{R}}$ we show that

$$\langle w, U^\eta(z)\rangle = \langle U^w(-\overline{z}), \eta\rangle \quad \text{for} \quad z \in \overline{\mathcal{S}_\pi}. \tag{2.9}$$

First we observe that the orbit map $U^w : \mathbb{C} \to \mathcal{H}$ is holomorphic and $\mathbb{R}$-equivariant. Hence, $U^w(z + z') = U^{U^w(z)}(z')$ shows that U^w is also holomorphic as a map $\mathbb{C} \to \mathcal{H}^{\mathcal{O}}$, hence in particular as a map to $\mathcal{H}^\omega$. Therefore both sides of (2.9) are continuous on $\overline{\mathcal{S}_\pi}$ and holomorphic on the interior (the expression on the left by assumption and the expression on the right by the preceding argument). As they coincide on $\mathbb{R}$, the claim follows.
Step 2: For $\eta \in \mathcal{H}^{-\omega}_{\rm KMS}$, $v \in \mathtt{V}' \cap \mathcal{H}^\omega$ and γ_n as in Remark 6, we have $v_n := U(\gamma_n)v \in \mathcal{H}^{\mathcal{O}} \cap \mathtt{V}'$. Hence Step 1 implies that

$$\langle U^{v_n}(\overline{z}), U^\eta(z)\rangle = \langle U^{v_n}(\overline{z} - \overline{z}), \eta\rangle = \langle v_n, \eta\rangle \quad \text{for} \quad z \in \overline{\mathcal{S}_\pi}.$$

Evaluating this relation in $z = \pi i$, we arrive at

$$\eta(v_n) = \langle U^{v_n}(-\pi i), U^\eta(\pi i)\rangle = \langle J v_n, J\eta\rangle = \overline{\langle v_n, \eta\rangle} = \overline{\eta(v_n)},$$

where we have used that $v_n \in \mathtt{V}'$ to obtain $U^{v_n}(-\pi i) = J v_n$.
Step 3: $\eta(v_n) \to \eta(v)$: We have to show that $v_n \to v$ in $\mathcal{H}^\omega$. So let $r > 0$ with $v \in \mathcal{D}(e^{\pm rH})$. It suffices to show that $v_n \to v$ in $\mathcal{H}^\omega(r)$, but this follows from

$$e^{\pm rH} v_n = e^{\pm rH} U(\gamma_n)v = U(\gamma_n)e^{\pm rH} v \to e^{\pm rH} v.$$

Clearly, Steps 2 and 3 imply $\eta(v) \in \mathbb{R}$, so that "$\subseteq$" holds.
"$\supseteq$": This direction is less straight forward. Let $\eta \in \mathcal{H}^{-\omega}$ be such that $\eta(v) \in \mathbb{R}$ for all $v \in \mathtt{V}' \cap \mathcal{H}^\omega$. We consider the sequence $\eta_n := U^{-\omega}(\gamma_n)\eta$ in $\mathcal{H}$ (cf. (2.8)). Then

$$\eta_n(\mathtt{V}') = \eta(U(\gamma_n)\mathtt{V}') \subseteq \eta(\mathtt{V}' \cap \mathcal{H}^\omega) \subseteq \mathbb{R}$$

implies $\eta_n \in \mathtt{V}$ for every $n \in \mathbb{N}$. So we have in particular $\eta_n \in \mathcal{H}^{-\omega}_{\rm KMS}$.

Further, $\eta_n \to \eta$ in the weak-$*$-topology follows from $v_n = U(\gamma_n)v \to v$ in $\mathcal{H}^\omega$ for every $v \in \mathcal{H}^\omega$ in the topology of $\mathcal{H}^\omega$ (see Step 3 above).

Next we consider the Hilbert space $\mathcal{H}^\omega(r)$, identified with a closed subspace of $\mathcal{H}^{\oplus 2}$ as in (2.4) via $v \mapsto (e^{rH}v, e^{-rH}v)$. The closedness follows from the fact that the range of this map is the graph of the self-adjoint operator e^{-2rH}. On this Hilbert space we obtain by $U^r_t := U_t|_{\mathcal{H}^\omega(r)}$ a unitary one-parameter group for which the above embedding is equivariant. As $\mathcal{H}^{-\omega}$ is a projective limit of the dual spaces $\mathcal{H}^{-\omega}(r)$, it suffices to verify that the restriction $\eta^r := \eta|_{\mathcal{H}^\omega(r)}$ is contained in $\mathcal{H}^{-\omega}(r)_{\rm KMS}$.

As J commutes with $U_{\mathbb{R}}$, it also defines a conjugation on the spaces $\mathcal{H}^\omega(r)$ and $\mathcal{H}^{-\omega}(r)$. Let $\mathtt{V}_r = \mathcal{H}^{-\omega}(r)_{\rm KMS} \subseteq \mathcal{H}^{-\omega}(r)$ be the corresponding standard subspace. We have to show that $\eta \in \mathtt{V}_r$. As $\eta_n \to \eta$ also holds weakly in the Hilbert space $\mathcal{H}^{-\omega}(r)$ and $\eta_n \in \mathtt{V}_r$, we obtain $\eta \in \mathtt{V}_r$. □

Corollary 2 $\mathcal{H} \cap \mathcal{H}^{-\omega}_{\rm KMS} = \mathtt{V}$.

Proof In view of Theorem 2, it suffices to observe that $\mathtt{V}' \cap \mathcal{H}^\omega$ is dense in $\mathtt{V}'$. This follows from the fact that $U(\gamma_n)v \to v$ for $v \in \mathtt{V}'$ and $U(\gamma_n)v \in \mathtt{V}' \cap \mathcal{H}^\omega$. □

2.3 Distribution Vectors for One-Parameter Groups

In this section, we turn to distribution vectors for unitary one-parameter groups. We shall characterize them in terms of approximations by holomorphic extensions of orbits maps in $\mathcal{H}$ to strips in $\mathbb{C}$ and the asymptotics of the norm in boundary points.

For a positive Borel measure ν on $\mathbb{R}$, we consider its Laplace transform

$$\mathcal{L}(\nu)(t) = \int_{\mathbb{R}} e^{-tx}\, d\nu(x).$$

We assume that there exists a $\delta > 0$ with $\mathcal{L}(\nu)(t) < \infty$ for $t \in (0, \delta]$. We are interested in the asymptotics for $t \to 0$. If ν is unbounded, then $1 \notin L^2(\mathbb{R}, \nu)$ and

$$\lim_{t\to 0} \mathcal{L}(\nu)(t) = \infty$$

by the Monotone Convergence Theorem (see Neeb [57, §V.4] for more general arguments of this type).

The following proposition describes different types of asymptotic behavior.

Proposition 3 *We consider the measure μ_s on $[1,\infty)$ with the density x^{-s}, $s \in \mathbb{R}$. This measure is finite if and only if $s > 1$, and, for $s \leq 1$, the asymptotics of $\mathcal{L}(\mu_s)$ for $t \to 0$ is given by*

$$t^{s-1} \ \textit{for}\ s < 1 \quad \textit{and} \quad |\log t| \ \textit{for}\ s = 1.$$

Proof For $s > 1$ we have $\lim_{t\to 0} \mathcal{L}(\mu_s)(t) = \mu_s(\mathbb{R}) > 0$. For $s < 1$ we have

$$\mathcal{L}(\mu_s)(t) = \int_1^\infty e^{-tx}\frac{dx}{x^s} = t^{s-1}\int_t^\infty e^{-x}\frac{dx}{x^s}$$

and therefore

$$\lim_{t\to 0} t^{1-s}\mathcal{L}(\mu_s)(t) = \int_0^\infty e^{-x}\frac{dx}{x^s} \in (0,\infty).$$

For $s = 1$ we obtain

$$\mathcal{L}(\mu_1)(t) = \int_1^\infty e^{-tx}\frac{dx}{x} = \int_t^\infty e^{-x}\frac{dx}{x} = \big|e^{-x}\log(x)\big]_t^\infty - \int_t^\infty (-e^{-x})\log x\,dx$$
$$= -e^{-t}\log t + \int_t^\infty e^{-x}\log x\,dx,$$

so that $\lim_{t\to 0}\frac{\mathcal{L}(\mu_1)(t)}{|\log t|} = 1$ because $\int_0^\infty e^{-x}\log(x)\,dx$ exists. □

Remark 8 Suppose that v is a K-fixed vector in a unitary representation of a semisimple Lie group G with $K \subseteq G$ maximal compact modulo the center and

$$\varphi_v(t) := \langle v, U(\exp th)v\rangle, \quad v \in \mathcal{H}^K$$

for an Euler element $h \in \mathfrak{a} \subseteq \mathfrak{p}$. Writing $\varphi_v(t) = \widehat{\mu}_v(t)$ for a finite positive measure on $\mathbb{R}$ (Bochner's Theorem), we have

$$\varphi_v(z) := \int_{\mathbb{R}} e^{izx}\,d\mu_v(x) \quad \text{for} \quad z \in \mathcal{S}_{\pm\frac{\pi}{2}}.$$

The asymptotics of $\varphi_v(i(\frac{\pi}{2} - \varepsilon))$ for spherical functions φ_v on rank-one groups found in Krötz/Stanton [45, Thm. 5.1] are of the form $|\log\varepsilon|$ and ε^{-s}, $s > 0$.

Proposition 4 *Let μ be a positive Borel measure for which there exists a $\delta > 0$ with*

$$\mathcal{L}(\mu)(t) = \int_{\mathbb{R}} e^{-tx}\,d\mu(x) < \infty \quad \textit{for} \quad t \in (0,\delta].$$

Then μ is tempered, i.e., there exists an $n \in \mathbb{N}$ for which

$$\int_{\mathbb{R}} (1+\lambda^2)^{-n}\,d\mu(\lambda) < \infty,$$

if and only if there exists an $N \in \mathbb{N}$ and $C > 0$ such that

$$\mathcal{L}(\mu)(t) \le Ct^{-N} \quad \textit{for} \quad 0 < t \le \delta. \tag{E$_N$}$$

Proof Our assumption implies that the interval $(-\infty, 1]$ has finite measure and since

$$\lim_{t\to 0}\int_{-\infty}^{1} e^{-tx}\, d\mu(x) = \int_{-\infty}^{1} d\mu(x)$$

by the Monotone Convergence Theorem (applied to the integral over $\mathbb{R}_-$), we may w.l.o.g. assume that μ is supported by $[1, \infty)$.

First we assume that μ is tempered and that $\int_1^\infty \frac{d\mu(x)}{x^N} < \infty$ for an $N \in \mathbb{N}$. As the function $e^{-tx}x^N$ assumes its maximal value on $\mathbb{R}_+$ for $x_0 = \frac{N}{t}$, we have

$$\mathcal{L}(\mu)(t) = \int_1^\infty e^{-tx}\, d\mu(x) = \int_1^\infty e^{-tx}x^N \frac{d\mu(x)}{x^N} \le e^{-N}\frac{N^N}{t^N}\int_1^\infty \frac{d\mu(x)}{x^N}.$$

It follows that $\mathcal{L}(\mu)(t) \le Ct^{-N}$ for some $C > 0$.

Suppose, conversely, that $\mathcal{L}(\mu)(t) \le Ct^{-N}$ for some $C > 0$ and $0 < t \le \delta$. For $M \in \mathbb{N}$, we consider the measure

$$d\mu_M(x) := \frac{d\mu(x)}{x^M} \quad \text{on} \quad [1, \infty).$$

We have to show that one of these measures is finite. Then the Laplace transforms $\mathcal{L}(\mu_M)$ exist on $(0, \delta]$ for some $\delta < 1$ and its M-fold derivative is

$$\mathcal{L}(\mu_M)^{(M)} = (-1)^M \mathcal{L}(\mu).$$

In particular, $\mathcal{L}(\mu_1)' = -\mathcal{L}(\mu)$. First we assume that $N \ge 2$. Then we have

$$\begin{aligned}
\mathcal{L}(\mu_1)(t) &= \mathcal{L}(\mu_1)(\delta) + \int_t^\delta \mathcal{L}(\mu)(x)\, dx \le \mathcal{L}(\mu_1)(\delta) + C\int_t^\delta \frac{dx}{x^N} \\
&= \mathcal{L}(\mu_1)(\delta) + C\Big[\frac{-1}{N-1}\frac{1}{x^{N-1}}\Big|_t^\delta \\
&= \mathcal{L}(\mu_1)(\delta) - \frac{C}{(N-1)\delta^{N-1}} + \frac{C}{(N-1)t^{N-1}} \le C_1\frac{1}{t^{N-1}}
\end{aligned}$$

for some $C_1 > 0$ and every $t \in (0, \delta]$. Iterating this argument, we see that there exists $C_{N-1} > 0$ with

$$\mathcal{L}(\mu_{N-1})(t) \le \frac{C_{N-1}}{t} \quad \text{for} \quad 0 < t \le \delta.$$

This leaves us with the case $N = 1$, where $\mathcal{L}(\mu)(t) \leq \frac{C}{t}$. Then

$$\mathcal{L}(\mu_1)(t) \leq \mathcal{L}(\mu_1)(\delta) + C \int_t^\delta \frac{dx}{x} \leq \mathcal{L}(\mu_1)(\delta) + C(\log \delta - \log t) \leq C'|\log t|$$

for some $C' > 0$ and every $t \in (0, \delta]$. For the measure μ_2, we thus obtain

$$\begin{aligned} \mathcal{L}(\mu_2)(t) &= \mathcal{L}(\mu_2)(t) = \mathcal{L}(\mu_2)(\delta) + \int_t^\delta \mathcal{L}(\mu_1)(x)\,dx \\ &\leq \mathcal{L}(\mu_2)(\delta) + \int_t^\delta C'|\log x|\,dx \leq \mathcal{L}(\mu_2)(\delta) + \int_0^\delta C'|\log x|\,dx. \end{aligned}$$

We conclude that $\mathcal{L}(\mu_2)$ is bounded on $(0, \delta]$, which, by monotone convergence, implies that μ_2 is finite, hence that μ is tempered. □

The preceding proof provides the following quantitative information on condition $(\mathrm{E_N})$ from Proposition 4:

$$\int_1^\infty \frac{d\mu(x)}{x^N} < \infty \quad \Rightarrow \quad (\mathrm{E_N}) \quad \Rightarrow \quad \int_1^\infty \frac{d\mu(x)}{x^{N+1}} < \infty.$$

The preceding proposition has an important consequence:

Theorem 3 *Let $H = H^*$ be a self-adjoint operator on the complex Hilbert space $\mathcal{H}$ and $v \in \mathcal{H}$ such that*

$$v \in \mathcal{D}(e^{tH}) \quad \text{for} \quad t \in [0, b).$$

If there exist $C > 0$ and $N > 0$ such that

$$\|e^{tH}v\| \leq \frac{C}{(b-t)^N} \quad \text{for} \quad t \in [0, b),$$

then $\eta := \lim_{t \to b-} e^{tH}v$ exists in the space $\mathcal{H}^{-\infty}$ of distribution vectors for the unitary one-parameter group $(e^{itH})_{t \in \mathbb{R}}$.

If $v \in \mathcal{D}(e^{bH})$, then we may take $N = 0$ and the limit $e^{bH}v = \lim_{t \to b-} e^{tH}v$ exists in $\mathcal{H}$.

Proof We may assume that v is a cyclic vector, so that $\mathcal{H} \cong L^2(\mathbb{R}, \mu)$ with the multiplication operator $(Hf)(\lambda) = \lambda f(\lambda)$ (Spectral Theorem). To simplify matters, we may assume that $v(\lambda) = e^{-b\lambda}$. Then μ is an unbounded measure whose Laplace transform exists for $t \in [-2b, 0)$. Our assumption then means that

$$\mathcal{L}(\mu)(-2(b-t)) = \|e^{tH}v\|^2 \leq \frac{C^2}{(b-t)^{2N}} \quad \text{for} \quad t \in [0, b).$$

As we may replace N by $\lceil N \rceil \in \mathbb{N}$, Proposition 4 now implies that the measure μ is tempered, and hence, that the constant function $\eta = 1$ is a distribution vector for the unitary one-parameter group $(e^{itH})_{t\in\mathbb{R}}$ (Neeb/Ólafsson [59, Cor. 10.8]).

A vector $f \in L^2(\mathbb{R}, \mu)$ is smooth if and only if $\lambda^n f(\lambda)$ is L^2 for every $n \in \mathbb{N}$. We want to show that, for $e_t(\lambda) := e^{t\lambda}$, we have $\lim_{t\to 0} e_{-t} = 1$ in the space of distribution vectors, i.e., that

$$\lim_{t\to 0} \int_{\mathbb{R}} e^{-t\lambda} f(\lambda)\, d\mu(\lambda) \to \int_{\mathbb{R}} f(\lambda)\, d\mu(\lambda)$$

for every smooth vector f. Choose N such that $(1+\lambda^2)^{-N}$ is μ-integrable. Then

$$\int_{\mathbb{R}} e^{-t\lambda} f(\lambda)\, d\mu(\lambda) = \int_{\mathbb{R}} e^{-t\lambda} f(\lambda)(1+\lambda^2)^N \frac{d\mu(\lambda)}{(1+\lambda^2)^N}.$$

Splitting the integral into the integration over $\mathbb{R}_-$ and $\mathbb{R}_+$, the correct limit behavior over $\mathbb{R}_-$ follows by dominated convergence and the fact that e_{-t} is integrable for some $t > 0$. Over $\mathbb{R}_+$ the corresponding statement follows from the fact that $f(\lambda)(1+\lambda^2)^N$ is L^2 with respect to μ, hence also with respect to the finite measure $\frac{d\mu(\lambda)}{(1+\lambda^2)^N}$, and thus, also L^1 with respect to this measure. Now we can argue with dominated convergence. □

3 Hyperfunction Boundary Values

In this section, we consider for a unitary representation $(U, \mathcal{H})$ of a connected simple Lie group holomorphic extensions of orbits maps $U^v \colon G \to \mathcal{H}, g \mapsto U(g)v$. If $G \subseteq G_{\mathbb{C}}$, i.e., G is linear, then the Krötz–Stanton Extension Theorem implies that U^v extends to the domain $G\exp(i\Omega_{\mathfrak{p}})K_{\mathbb{C}} \subseteq G_{\mathbb{C}}$, where $\Omega_{\mathfrak{p}} \subseteq \mathfrak{p}$ consists of all elements for which the spectral radius of $\operatorname{ad} x$ is smaller than $\frac{\pi}{2}$. This result is best expressed as the extendability of certain vector bundle maps

$$G \times_K \mathcal{E} \to \mathcal{H}$$

to a holomorphic vector bundle $\mathbb{E}$ over the crown Ξ of G/K (Akhiezer/Gindikin [1], Gindikin/Krötz [31]). As this makes also sense for nonlinear groups, we call this extension property Hypothesis (H1) and discuss its consequences. The main conclusion is the existence of limit maps

$$\beta^{\pm} \colon \mathcal{H}^{[K]} \to \mathcal{H}^{-\omega,[H]}_{\mathrm{KMS}}, \quad \beta^{\pm}(v) := \lim_{t\to\pi/2} e^{\mp it\partial U(h)} v$$

from K-finite vectors to H-finite hyperfunction vectors. Combining this with the Automatic Continuity Theorem 17, we obtain finite-dimensional H-invariant subspaces $\mathtt{E}_H \subseteq \mathcal{H}^{-\infty}$ that define nets of real subspaces $(\mathsf{H}^{G/H}_{\mathtt{E}_H}(\mathcal{O}))_{\mathcal{O}\subseteq G/H}$ for which we shall see in the next section that the Reeh–Schlieder and the Bisognano–Wichmann property are both satisfied.

Concretely, we consider the following setup:

- G is a connected semisimple Lie group
- $\eta_G: G \to G_{\mathbb{C}}$ is the universal complexification of G; its kernel is discrete.
- $h \in \mathfrak{g}$ is an Euler element, i.e., $(\operatorname{ad} h)^3 = \operatorname{ad} h$.
- θ is a Cartan involution on G and its Lie algebra $\mathfrak{g}$, and $\mathfrak{g} = \mathfrak{k} \oplus \mathfrak{p}$ is the eigenspace decomposition, and we assume that $\theta(h) = -h$.
- $(\mathfrak{g}, \tau)$ is ncc, where $\tau = \tau_h\theta$, $\tau_h = e^{\pi i \operatorname{ad} h}|_{\mathfrak{g}}$, and the τ-eigenspace decomposition is denoted $\mathfrak{g} = \mathfrak{h} \oplus \mathfrak{q}$ (see Morinelli/Neeb/Ólafsson [54, Thm. 4.21] and the introduction); we also write

$$\mathfrak{h}_{\mathfrak{k}} = \mathfrak{h} \cap \mathfrak{k}, \quad \mathfrak{h}_{\mathfrak{p}} = \mathfrak{h} \cap \mathfrak{p}, \quad \mathfrak{q}_{\mathfrak{k}} = \mathfrak{q} \cap \mathfrak{k}, \quad \mathfrak{q}_{\mathfrak{p}} = \mathfrak{q} \cap \mathfrak{p}.$$

- $C \subseteq \mathfrak{q}$ is the **maximal** $\operatorname{Inn}(\mathfrak{h})$-invariant cone containing h (cf. [54, §3]).
- $H \subseteq G^\tau$ is an open θ-invariant subgroup for which $\operatorname{Ad}(H)C = C$ (which is equivalent to $H_K = H \cap K$ fixing h [54, Cor. 4.6]). If G is given, this means that $H_{\min} \subseteq H \subseteq H_{\max}$, where $H_{\min} = G^\tau_e$ is connected and $H_{\max} = K^{\tau,h}\exp(\mathfrak{h}_{\mathfrak{p}})$.
- $M = G/H$ is the corresponding ncc symmetric space.
- $G_{\tau_h} := G \rtimes \{\mathbf{1}, \tau_h\}$ is the corresponding graded Lie group.
- $(U, \mathcal{H})$ an **irreducible** antiunitary representation of G_{τ_h}, i.e., $J := U(\tau_h)$ is a conjugation on $\mathcal{H}$ and $U(G) \subseteq \mathrm{U}(\mathcal{H})$.

Remark 9 (Existence of Antiunitary Extensions) If $(U, \mathcal{H})$ is an irreducible unitary representation of G, then by Neeb/Ólafsson [60, Thm. 2.11(d)] exactly one of the following cases occurs:

- It extends to an antiunitary representation of G_{τ_h} on $\mathcal{H}$, where the commutant is $\mathbb{R}\mathbf{1}$. We refer to Morinelli/Neeb [52, Thm. 4.24] for a discussion of the case $G = \widetilde{\mathrm{SL}}_2(\mathbb{R})$, asserting that every irreducible unitary representation of this group extends to an antiunitary representation of G_{τ_h}.
- $U \oplus (U^* \circ \tau_h^G)$ extends to an irreducible antiunitary representation V of G_{τ_h} on $\mathcal{H} \oplus \mathcal{H}^*$ by $V(\tau_h)(v, \alpha) = (\Phi^{-1}\alpha, \Phi v)$, where $\Phi: \mathcal{H} \to \mathcal{H}^*$ is given by $\Phi(v)(w) = \langle v, w\rangle$. Its commutant is $\mathbb{C}$ if $U^* \circ \tau_h^G \not\cong U$ and the quaternions $\mathbb{H}$ if $U^* \circ \tau_h^G \cong U$.

Let $(U, \mathcal{H})$ be a unitary representation of the semisimple Lie group G, fix a Cartan involution θ, and write $\mathfrak{g} = \mathfrak{k} \oplus \mathfrak{p}$ for the corresponding Cartan decomposition of the Lie algebra. Then

$$\Omega_{\mathfrak{p}} := \{x \in \mathfrak{p} : \operatorname{Spec}(\operatorname{ad} x) \subseteq (-\pi/2, \pi/2)\}$$

is an open convex subset of $\mathfrak{p}$.

Let $\eta: G \to G_{\mathbb{C}}$ denote the universal complexification of G. Then $G_{\mathbb{C}}$ carries a uniquely determined antiholomorphic involution σ with $\sigma \circ \eta = \eta$, i.e., it induces on the Lie algebra $\mathfrak{g}_{\mathbb{C}}$ the complex conjugation with respect to $\mathfrak{g}$. The holomorphic involution $\theta_{\mathbb{C}}$ on $G_{\mathbb{C}}$ induced by a Cartan involution θ on G via $\theta_{\mathbb{C}} \circ \eta = \eta \circ \theta$, then commutes with σ. We put

$$K_{\mathbb{C}} := (G_{\mathbb{C}}^{\theta_{\mathbb{C}}})_e = \langle \exp \mathfrak{k}_{\mathbb{C}} \rangle.$$

Hence, σ preserves the complex symmetric subspace

$$G_{\mathbb{C}}^{-\theta} = \{g \in G_{\mathbb{C}} : \theta_{\mathbb{C}}(g) = g^{-1}\}$$

of $G_{\mathbb{C}}$ whose identity component is isomorphic to $G_{\mathbb{C}}/G_{\mathbb{C}}^{\theta}$ via the $G_{\mathbb{C}}$-action by $g.h := gh\theta_{\mathbb{C}}(g)^{-1}$.

We consider the crown domain of the Riemannian symmetric space G/K:

$$\Xi := G \times_K i\Omega_{\mathfrak{p}} = (G \times i\Omega_{\mathfrak{p}})/\sim \quad \text{with} \quad (g, ix) \sim (gk, \mathrm{Ad}(k)^{-1}ix), k \in K. \tag{3.1}$$

The complex structure on this domain is determined by the requirement that the map

$$q: \Xi \to G_{\mathbb{C}}^{-\theta}, \quad q([g, ix]) \mapsto g.\mathrm{Exp}(ix) = g\exp(2ix)\theta(g)^{-1},$$

which is a covering of an open subset of $G_{\mathbb{C}}^{-\theta_{\mathbb{C}}}$, is holomorphic.

Remark 10 The construction of Ξ only depends on the Lie algebra $\mathfrak{g}$ of G. It produces the same manifold for the simply connected covering group $\widetilde{G}$ and for the adjoint group $G/Z(G) \cong \mathrm{Ad}(G)$. This is due to the fact that $K = G^{\theta}$ is always connected.

For each finite-dimensional unitary K-representation $(\sigma, \mathcal{E})$, we obtain a vector bundle $\mathbb{E} \to \Xi$ by

$$q_{\mathbb{E}}: \mathbb{E} := (G \times i\Omega_{\mathfrak{p}}) \times_K \mathcal{E} \to \Xi, \quad [g, ix, v] \mapsto [g, ix]. \tag{3.2}$$

Proposition 5 *$\mathbb{E} \to \Xi$ is a holomorphic vector bundle. If, in addition, $J\mathcal{E} = \mathcal{E}$, then the antiholomorphic involution $\overline{\tau}_h$ on Ξ induced by τ_h lifts to an antiholomorphic involution $\overline{\tau}_{\mathbb{E}}$ on $\mathbb{E}$, given by*

$$\overline{\tau}_{\mathbb{E}}([g, ix, v]) := [\tau_h(g), -i\tau_h(x), Jv]. \tag{3.3}$$

Proof In view of Remark 10, we may assume that G is simply connected, which, by polar decomposition, implies that K and its universal complexification $K_{\mathbb{C}}$ are simply connected as well. Then $G_{\mathbb{C}}$ is also simply connected and the universal complexification of K is the natural map $\eta_K \colon K \to \widetilde{K}_{\mathbb{C}}$ whose existence follows from the simple connectedness of K.

Moreover, the fixed point group $G_{\mathbb{C}}^{\theta_{\mathbb{C}}}$ is connected by (Loos [49, Thm. IV.3.4]), hence equal to $K_{\mathbb{C}}$. Thus $(G_{\mathbb{C}}^{-\theta})_e \cong G_{\mathbb{C}}/K_{\mathbb{C}}$, and we have a natural covering map $\Xi \to G.\operatorname{Exp}(i\Omega_{\mathfrak{p}}) \subseteq G_{\mathbb{C}}/K_{\mathbb{C}}$ of an open domain of $G_{\mathbb{C}}/K_{\mathbb{C}}$, that is diffeomorphic to $\eta_G(G) \times_{\eta_G(K)} i\Omega_{\mathfrak{p}}$. Remark 10 implies that this domain is biholomorphic to Ξ, and we may thus consider Ξ as a domain in $G_{\mathbb{C}}/K_{\mathbb{C}}$.

Let $q \colon G_{\mathbb{C}} \to G_{\mathbb{C}}/K_{\mathbb{C}}$ denote the quotient map and

$$\Xi_{G_{\mathbb{C}}} := q^{-1}(\Xi) = q^{-1}(G.\operatorname{Exp}(i\Omega_{\mathfrak{p}})) = G\exp(i\Omega_{\mathfrak{p}})K_{\mathbb{C}}.$$

This is an open subset of $G_{\mathbb{C}}$ that is right $K_{\mathbb{C}}$-invariant, so that $\Xi_{G_{\mathbb{C}}}$ is a $K_{\mathbb{C}}$-principal bundle over Ξ. As Ξ is contractible (it is an affine bundle over the contractible space G/K), the natural homomorphism $\pi_1(K_{\mathbb{C}}) \to \pi_1(\Xi_{G_{\mathbb{C}}})$ is an isomorphism by the long exact homotopy sequence for fiber bundles. We conclude that the simply connected covering $\widetilde{\Xi}_{G_{\mathbb{C}}}$ is a holomorphic $\widetilde{K}_{\mathbb{C}}$-principal bundle over Ξ.

Extending the representation $\sigma^{\mathcal{E}} \colon K \to \mathrm{U}(\mathcal{E})$ to a holomorphic representation $\sigma_{\mathbb{C}}^{\mathcal{E}} \colon \widetilde{K}_{\mathbb{C}} \to \mathrm{GL}(\mathcal{E})$, we obtain a holomorphic vector bundle

$$\mathbb{E}' := \widetilde{\Xi}_{G_{\mathbb{C}}} \times_{\widetilde{K}_{\mathbb{C}}} \mathcal{E} \to \Xi.$$

As G is simply connected, the G-action on $\Xi_{G_{\mathbb{C}}}$ by left multiplications lifts to the simply connected covering $\widetilde{\Xi}_{G_{\mathbb{C}}}$ and the map $\exp \colon i\Omega_{\mathfrak{p}} \to \Xi_{G_{\mathbb{C}}}$ lifts uniquely to a map

$$\widetilde{\exp} \colon i\Omega_{\mathfrak{p}} \to \widetilde{\Xi}_{G_{\mathbb{C}}} \quad \text{with} \quad \widetilde{\exp}(0) = \widetilde{e},$$

where $\widetilde{e}$ is the base point in $\widetilde{\Xi}_{G_{\mathbb{C}}}$ over $e \in G_{\mathbb{C}}$. As lifts are uniquely determined, once the image of the base point is fixed, $\widetilde{\exp} \colon i\Omega_{\mathfrak{p}} \to \widetilde{\Xi}_{G_{\mathbb{C}}}$ is equivariant with respect to the conjugation action by K. We thus obtain a well-defined map

$$\Psi \colon \mathbb{E} \to \mathbb{E}', \quad [g, ix, v] \mapsto [g\widetilde{\exp}(ix), v].$$

This is an equivalence of complex vector bundles over Ξ, so that we obtain a holomorphic vector bundle structure on $\mathbb{E}$.

Now we turn to the antiholomorphic involution $\overline{\tau}_h$ of Ξ. As $\tau_h(\mathfrak{k}) = \mathfrak{k}$ and $\tau_h(\Omega_{\mathfrak{p}}) = \Omega_{\mathfrak{p}}$, the antiholomorphic involution $\overline{\tau}_h(g) = \exp(\pi i h)\sigma^{\mathcal{E}}(g)\exp(-\pi i h)$ of $G_{\mathbb{C}}$ preserves the subgroups $K_{\mathbb{C}}$, G and the subset $\exp(i\Omega_{\mathfrak{p}})$, hence also $\Xi_{G_{\mathbb{C}}}$. Therefore, it induces an antiholomorphic involution on $\widetilde{\Xi}_{G_{\mathbb{C}}}$. If $J\mathcal{E} = \mathcal{E}$, then the relation $JU(k) = U(\tau_h(k))J$ for $k \in K$ implies that $J\sigma_{\mathbb{C}}^{\mathcal{E}}(k) = \sigma_{\mathbb{C}}^{\mathcal{E}}(\overline{\tau}_h(k))J$ for $k \in \widetilde{K}_{\mathbb{C}}$, which leads to an antiholomorphic involution $[m, v] \mapsto [\overline{\tau}_h(m), Jv]$ on $\mathbb{E}'$. Now the assertion follows from the fact that Ψ intertwines this involution with $\overline{\tau}_{\mathbb{E}}$. □

The following hypothesis will turn out to be crucial for the construction of local nets over ncc symmetric spaces.

Hypothesis (H1) *For the finite-dimensional K-invariant subspace $\mathcal{E} \subseteq \mathcal{H}$, the map*

$$G \times_K \mathcal{E} \to \mathcal{H}, \quad [g, v] \mapsto U(g)v$$

extends to a holomorphic map

$$\Psi_{\mathcal{E}} : \mathbb{E} = (G \times i\Omega_{\mathfrak{p}}) \times_K \mathcal{E} \to \mathcal{H}, \quad [g, ix, v] \mapsto U(g)e^{i\partial U(x)}v. \tag{3.4}$$

Theorem 4 (Krötz–Stanton Extension Theorem; [45, Thm. 3.1]) *If U is irreducible, G is simple and $\eta_G : G \to G_{\mathbb{C}}$ is an embedding, then* Hypothesis (H1) *is satisfied.*

The Krötz–Stanton Extension Theorem immediately generalizes to connected linear semisimple groups because their irreducible representations factorize as tensor products of irreducible representations of the simple normal subgroups.

We expect that Hypothesis (H1) also holds if η_G is not injective. In Sect. 5 below, we shall verify this for $G = \widetilde{\mathrm{SL}}_2(\mathbb{R})$, and more generally for $\mathfrak{g} = \mathfrak{so}_{1,d}(\mathbb{R})$, $d \geq 2$.

The following reformulation of Hypothesis (H1) approaches it from a different angle:

Proposition 6 Hypothesis (H1) *is equivalent to $\mathcal{E} \subseteq \bigcap_{x \in \Omega_{\mathfrak{p}}} \mathcal{D}(e^{i\partial U(x)})$.*

Proof Clearly, Hypothesis (H1) implies that any $v \in \mathcal{E}$ is contained in $\mathcal{D}(e^{i\partial U(x)})$ for $x \in \Omega_{\mathfrak{p}}$. Suppose, conversely, that this is the case. Then we obtain a map

$$f : G \times i\Omega_{\mathfrak{p}} \times \mathcal{E} \to \mathcal{H}, \quad f(g, ix, v) := U(g)e^{i\partial U(x)}v.$$

First, we argue as in the proof of Lemma 1 to see that this map is analytic. Then we observe that, for $k \in K$,

$$\begin{aligned} f(gk, \mathrm{Ad}(k)^{-1}ix, U(k)^{-1}v) &= U(gk)e^{i\partial U(\mathrm{Ad}(k)^{-1}x)}U(k)^{-1}v \\ &= U(g)e^{i\partial U(x)}v = f(g, ix, v), \end{aligned}$$

so that f factors through a well-defined real-analytic map

$$F : \mathbb{E} \to \mathcal{H}, \quad [g, ix, v] \mapsto U(g)e^{i\partial U(x)}v.$$

As this map is real-analytic, it suffices to verify its holomorphy in a neighborhood of $[e, 0, v]$. As v is an analytic vector, we have for $x, y \in \mathfrak{p}$ sufficiently small

$$U(\exp x)e^{i\partial U(y)}v = U^v(\exp(x * iy)),$$

where $a * b = a + b + \frac{1}{2}[a, b] + \cdots$ denotes the Baker–Campbell–Hausdorff series. Comparing with the construction of the complex structure on $\mathbb{E}$ in the proof of Proposition 5, it now follows that F is holomorphic. □

Remark 11

(a) Hypothesis (H1) is a condition on the representation U and the K-invariant subspace $\mathcal{E}$. If $\eta\colon G \to G_{\mathbb{C}}$ is injective, then the inclusion $K \hookrightarrow K_{\mathbb{C}}$ is the universal complexification of K and $\Xi_{G_{\mathbb{C}}} = G\exp(i\Omega_{\mathfrak{p}})K_{\mathbb{C}}$. Hence Hypothesis (H1) reduces to the statement that, for every K-finite vector $v \in \mathcal{H}$, the orbit map $U^v\colon G \to \mathcal{H}$ extends to a holomorphic map $\Xi_{G_{\mathbb{C}}} \to \mathcal{H}$.

The preceding discussion shows that Hypothesis (H1) is also satisfied if $\ker(\eta_G) \subseteq \ker(U)$ because it then factors through a representation of a group G that embeds in its complexification.

(b) Theorem 4 has a slightly involved history. In Krötz/Stanton [45] it is based on Conjecture A, asserting that

$$G\exp(i\Omega_{\mathfrak{p}}) \subseteq N_{\mathbb{C}}A_{\mathbb{C}}K_{\mathbb{C}}, \tag{3.5}$$

where $G = NAK$ is an Iwasawa decomposition and $N_{\mathbb{C}}$, $A_{\mathbb{C}}$, $K_{\mathbb{C}}$ are the corresponding complex integral subgroups of $G_{\mathbb{C}}$. The inclusion (3.5) was proved for classical groups in Krötz/Stanton [45]. The general case was obtained by A. Huckleberry in [43, Prop. 2.0.2], using a certain strictly plurisubharmonic function, and with structure theoretic methods by T. Matsuki in [50, Thm.]. It also follows from the Complex Convexity Theorem in Gindikin/Krötz [30] which provides finer information. Another argument for the inclusion (3.5), based on holomorphic extension of eigenfunctions of the Laplace–Beltrami operator, is given by Krötz/Schlichtkrull in [46, Cor. 3.3].

(c) In this context, it is interesting to observe that the Krötz–Stanton Extension theorem is optimal with respect to the domain to which analytic orbit maps may extend. Refining techniques by R. Goodman [33], the following result has been obtained in Beltiţă/Neeb [6]. Consider the 2-dimensional group $\mathrm{Aff}(\mathbb{R})_e \cong \mathbb{R} \rtimes \mathbb{R}_+$ with Lie algebra $\mathbb{R}y + \mathbb{R}h$ where $[h, y] = y$, and a unitary representation $(U, \mathcal{H})$ satisfying the non-degeneracy condition $\ker(\partial U(y)) = \{0\}$. Then any analytic vector v for which $e^{it\partial U(h)}v$ is defined and contained in $\mathcal{H}^\omega$ for $|t| \leq \pi/2$ is zero.

As the Lie algebra $\mathfrak{g}$ of a non-compact semisimple Lie group contains many copies of the non-abelian 2-dimensional Lie algebra, this observation implies that, if $\mathcal{H}^G = \{0\}$, and $v \in \mathcal{H}^\omega$ and $x \in \mathfrak{p}$ are such that $e^{it\partial U(x)}v$ is defined and an analytic vector for $|t| \leq 1$, then all eigenvalues of $\mathrm{ad}\, x$ are $< \pi/2$. Hence, the domain $\Omega_{\mathfrak{p}} \subseteq \mathfrak{p}$ is the maximal domain for which a result as Theorem 4 may hold.

Remark 12 If $J\mathcal{E} = \mathcal{E}$ and Hypothesis (H1) is satisfied, then

$$J \circ \Psi_{\mathcal{E}} = \Psi_{\mathcal{E}} \circ \overline{\tau}_{\mathbb{E}} \tag{3.6}$$

(cf. (3.4)) holds for the antiholomorphic involution $\overline{\tau}_{\mathbb{E}}$ on $\mathbb{E}$, given by

$$\overline{\tau}_{\mathbb{E}}([g, ix, v]) = [\tau_h(g), -i\tau_h(x), Jv] \tag{3.7}$$

as in Proposition 5.

Several parts of the following proposition follow from Gindikin/Krötz/Ólafsson [32, Thm. 2.1.3] if η_G is injective and $G_{\mathbb{C}}$ is simply connected. Our formulation does not require this assumption. The main information is contained in (e).

Proposition 7 *Under* Hypothesis (H1)*, for any K-finite vector $v \in \mathcal{H}^{[K]}$, the limits*

$$\beta^{\pm}(v) := \lim_{t \to \pi/2} e^{\mp it\partial U(h)} v$$

exist in $\mathcal{H}^{-\omega}(U_h)$, where $U_h(t) = \exp(th)$ for $t \in \mathbb{R}$. Moreover, the following assertions hold:

(a) *The maps $\beta^{\pm} : \mathcal{H}^{[K]} \to \mathcal{H}^{-\omega}(U_h) \subseteq \mathcal{H}^{-\omega}$ are injective.*
(b) *The limits $\beta^{\pm}(v)$ exist in the topology of uniform convergence on subsets of the form $U(C)\xi$, where $C \subseteq G$ is compact and $\xi \in \mathcal{H}^{\omega}$.*
(c) *The automorphism $\zeta := e^{-\frac{\pi i}{2} \operatorname{ad} h} \in \operatorname{Aut}(\mathfrak{g}_{\mathbb{C}})$ satisfies*

$$\zeta(\mathfrak{h}_{\mathfrak{k}} + i\mathfrak{q}_{\mathfrak{k}}) = \mathfrak{h}, \tag{3.8}$$

hence in particular $\zeta(\mathfrak{k}_{\mathbb{C}}) = \mathfrak{h}_{\mathbb{C}}$.
(d) *We have the intertwining relation*

$$\beta^{\pm} \circ \mathrm{d}U(x) = \mathrm{d}U^{-\omega}(\zeta^{\pm 1}(x)) \circ \beta^{\pm} : \mathcal{H}^{[K]} \to \mathcal{H}^{-\omega,[H]} \quad \textit{for} \quad x \in \mathfrak{g}_{\mathbb{C}}.$$

(e) *If $\mathcal{E} \subseteq \mathcal{H}^{[K]}$ is finite-dimensional and J-invariant, then the finite-dimensional real subspaces $\beta^{\pm}(\mathcal{E}^J) \subseteq \mathcal{H}^{-\omega}$ are $U^{-\omega}(H)$-invariant. We further have*

$$J\beta^{\pm}(v) = \beta^{\mp}(v) \quad \textit{for} \quad v \in \mathcal{H}^{[K]}.$$

Proof For $|t| < \frac{\pi}{2}$, we have $th \in \Omega_{\mathfrak{p}}$, so that Hypothesis (H1) implies that

$$v \in \mathcal{D}(e^{it\partial U(h)}) \quad \text{for} \quad |t| < \frac{\pi}{2}. \tag{3.9}$$

Hence the existence of the limits $\beta^{\pm}(v)$ in the weak-$*$-topology on $\mathcal{H}^{-\omega}(U_h)$ follows from Lemma 2.

(a) (cf. [32, Thm. 2.1.3]) Suppose that $\beta^{+}(v) = 0$. As $v \in \mathcal{H}^{[K]} \subseteq \mathcal{H}^{\omega}$ (Harish–Chandra [37]) is contained in $\mathcal{D}(e^{ti\partial U(h)})$ for $|t| < \frac{\pi}{2}$, the function

$$f : \mathbb{R} \to \mathbb{C}, \quad f(t) := \langle v, e^{t\partial U(h)} v\rangle$$

extends analytically to the strip $\mathcal{S}_{\pm\pi}$. Our assumption implies that

$$f\big(-\tfrac{\pi i}{2}+t\big) = \beta^+(v)(e^{t\partial U(h)}v) = 0 \quad \text{for} \quad t \in \mathbb{R},$$

so that $f = 0$ by analytic continuation, and thus $0 = f(0) = \|v\|^2$ leads to $v = 0$.

(b) (cf. Gindikin/Krötz/Ólafsson [32, p. 653]) The corollary to Tréves [72, Thm. 3.3.1] (Banach–Steinhaus Theorem) implies that pointwise convergence of a sequence on a Fréchet space implies uniform convergence on compact subsets. To see that the limit defining $\beta^+(v)$ is uniform on subsets of the form $U(C)w$, $w \in \mathcal{H}^\omega$, $C \subseteq G$ compact, we recall from Proposition 1, that the orbit map $U^w : G \to \mathcal{H}^\omega$ is continuous. Actually its proof in Gimperlein/Krötz/Schlichtkrull [28, Prop. 3.4] shows that the restriction of the orbit map to C factors through a continuous map $U^w : C \to \mathcal{H}^\omega_V$, for some V as in Lemma 1. Therefore, the Banach–Steinhaus Theorem applies.

(c) We have $\ker(\operatorname{ad} h) = \mathfrak{h}_{\mathfrak{k}} \oplus \mathfrak{q}_{\mathfrak{p}} = \mathfrak{g}^{\tau_h}$ and $\mathfrak{g}^{-\tau_h} = \mathfrak{h}_{\mathfrak{p}} \oplus \mathfrak{q}_{\mathfrak{k}} = \mathfrak{g}_1(h) \oplus \mathfrak{g}_{-1}(h)$. As $\theta(h) = -h$, we have $\theta(\mathfrak{g}_1(h)) = \mathfrak{g}_{-1}(h)$. So $\mathfrak{q}_{\mathfrak{k}} = \{x + \theta(x) : x \in \mathfrak{g}_1(h)\}$. This shows that

$$\zeta(\mathfrak{q}_{\mathfrak{k}}) = \{i(x - \theta(x)) : x \in \mathfrak{g}_1(h)\} = i\mathfrak{h}_{\mathfrak{p}},$$

and therefore $\zeta(\mathfrak{h}_{\mathfrak{k}} + i\mathfrak{q}_{\mathfrak{k}})) = \mathfrak{h}_{\mathfrak{k}} + \mathfrak{h}_{\mathfrak{p}} = \mathfrak{h}$, which entails $\zeta(\mathfrak{k}_{\mathbb{C}}) = \mathfrak{h}_{\mathbb{C}}$ (cf. [64, Thm. 5.4]).

(d) For a K-finite vector v, we have

$$\begin{aligned}
dU^{-\omega}(\zeta^{\pm 1}(x))\beta^{\pm}(v) &= \lim_{t\to\pi/2} dU(\zeta^{\pm 1}(x))e^{\mp it\partial U(h)}v \\
&= \lim_{t\to\pi/2} e^{\mp it\partial U(h)} dU(e^{\pm it \operatorname{ad} h}\zeta^{\pm 1}(x))v \\
&= \beta^{\pm}\big(dU(e^{\pm i\frac{\pi}{2}\operatorname{ad} h}\zeta^{\pm 1}(x))v\big) = \beta^{\pm}(dU(x)v).
\end{aligned}$$

Here we use that $t \mapsto dU\big(e^{it \operatorname{ad} h}\zeta(x)\big)v$ is a continuous curve in a finite-dimensional subspace.

(e) As $Ji\partial U(h)J = -i\partial U(h)$, we have $J\beta^{\pm}(v) = \beta^{\mp}(Jv)$ for $v \in \mathcal{H}^{[K]}$. The relation

$$J\partial U(z)J = \partial U(\tau_h(\overline{z})) \tag{3.10}$$

shows that, on $\mathcal{H}^{[K]}$, the operator $dU(z)$ for $z \in \mathfrak{h}_{\mathfrak{k}} + i\mathfrak{q}_{\mathfrak{k}}$, commutes with J. By (d), $\beta^{\pm}$ intertwines these operators with $dU^{-\omega}(\mathfrak{h})$. Hence the subspaces $\beta^{\pm}(\mathcal{E}^J)$ are $dU^{-\omega}(\mathfrak{h})$-invariant. The subspace $\mathcal{E}^J \subseteq \mathcal{E}$ is invariant under the subgroup K^{τ_h}. As $H = H_K \exp(\mathfrak{h}_{\mathfrak{p}})$ with $H_K \subseteq K^\tau = K^{\tau_h}$ and $H_K \subseteq K^h$ (Morinelli/Neeb/Ólafsson [55, Lemma 4.11]), the K^h-equivariance of $\beta^\pm$ entails that the $\mathrm{d}U^{-\omega}(\mathfrak{h})$-invariant subspaces $\beta^\pm(\mathcal{E}^J)$ are invariant under $U^{-\omega}(H)$.

□

Remark 13 As a consequence of (3.10), the real subspace $\mathcal{H}^{[K],J}$ of the $\mathfrak{g}$-module $\mathcal{H}^{[K]}$ is invariant under the Lie subalgebra

$$\mathfrak{g}^{\tau_h} + i\mathfrak{g}^{-\tau_h} = \mathfrak{h}_{\mathfrak{k}} + \mathfrak{q}_{\mathfrak{p}} + i\mathfrak{h}_{\mathfrak{p}} + i\mathfrak{q}_{\mathfrak{k}}.$$

Further,

$$\zeta^\pm(\mathfrak{g}^{\tau_h} + i\mathfrak{g}^{-\tau_h}) = \mathfrak{g}^{\tau_h} + \mathfrak{g}^{-\tau_h} = \mathfrak{g},$$

so that the equivariance relation Proposition 7(d) implies that $\beta^\pm(\mathcal{H}^{[K],J})$ are $\mathfrak{g}$-submodules of $\mathcal{H}^{-\omega}$.

Remark 14 The intertwining relation in Proposition 7(d) can be strengthened as follows. We consider the dense subspace

$$\mathcal{D} := \bigcap_{|t|<\pi/2} \mathcal{D}(e^{it\partial U(h)})$$

of $\mathcal{H}$ as a subspace of the locally convex space $\mathcal{H}^{-\omega}$, on which the Lie algebra acts by continuous operators. Hypothesis (H1) asserts that $\mathcal{H}^{[K]} \subseteq \mathcal{D}$, but $\mathcal{D}$ may be substantially larger as we shall see in Sect. 5.2. In particular, it is invariant under $U(G^h)$, where G^h is the centralizer of h in G.

The existence of the limits

$$\beta^\pm : \mathcal{D} \to \mathcal{H}^{-\omega}(U_h) \subseteq \mathcal{H}^{-\omega}$$

in the weak-$*$-topology follows from Lemma 2. The same argument as in the proof of Proposition 7(a) implies that both maps $\beta^\pm$ are injective.

To establish an equivariance relation for the $\mathfrak{g}$-action, let $\mathcal{H}^{-\omega}_{\mathrm{ext}} \subseteq \mathcal{H}^{-\omega}$ be the subspace of those hyperfunction vectors for which the orbit map $U^\eta : \mathbb{R} \to \mathcal{H}^{-\omega}$ extends to a weak-$*$-continuous map on $\overline{\mathcal{S}_\pi}$ that is holomorphic on the interior. For any $\eta \in \mathcal{H}^{-\omega}_{\mathrm{ext}}$ and $x \in \mathfrak{g}$, we have

$$U^\omega(\exp th)\mathrm{d}U^{-\omega}(x)\eta = \mathrm{d}U^{-\omega}(e^{t\,\mathrm{ad}\,h}x)U^\omega(\exp th)\eta. \tag{3.11}$$

Since $\mathrm{d}U^{-\omega}(\mathfrak{g})$ consists of continuous operators on $\mathcal{H}^{-\omega}$ and

$$\mathrm{d}U^{-\omega}(e^{t\,\mathrm{ad}\,h}x) = e^t\mathrm{d}U^{-\omega}(x_1) + \mathrm{d}U^{-\omega}(x_0) + e^{-t}\mathrm{d}U^{-\omega}(x_{-1}),$$

where $x = x_1 + x_0 + x_{-1}$ is the decomposition of x into $\mathrm{ad}\, h$-eigenvectors, the right-hand side of (3.11) extends to a weak-$*$-continuous map on $\overline{\mathcal{S}_\pi}$ that is holomorphic on the interior. It follows that $\eta' := \mathrm{d}U^{-\omega}(x)\eta \in \mathcal{H}^{-\omega}_{\mathrm{ext}}$ with

$$U^{\eta'}(z) = \mathrm{d}U^{-\omega}(e^{z\, \mathrm{ad}\, h}x)U^{\eta}(z) \quad \text{for} \quad z \in \overline{\mathcal{S}_\pi}.$$

For $z = \pi i/2$ and $\eta \in \mathcal{H}$, we obtain in particular

$$\beta^{\pm} \circ \mathrm{d}U^{-\omega}(x) = \mathrm{d}U^{-\omega}(\zeta^{\pm 1}(x)) \circ \beta^{\pm} : \mathcal{D} \to \mathcal{H}^{-\omega}. \tag{3.12}$$

Lemma 4 *$\mathcal{H}^{-\omega}_{\mathrm{KMS}} \subseteq \mathcal{H}^{-\omega}$ is a real $\mathfrak{g}$-invariant subspace.*

Proof Let $\eta \in \mathcal{H}^{-\omega}_{\mathrm{KMS}}$ and $x \in \mathfrak{g}_{\pm 1}(h)$. Then

$$U_h^{-\omega}(t)\mathrm{d}U^{-\omega}(x)\eta = e^{\pm t}\mathrm{d}U^{-\omega}(x)U_h^{-\omega}(t)\eta = e^{\pm t}\mathrm{d}U^{-\omega}(x)U_h^{\eta}(t).$$

As the operator $\mathrm{d}U^{-\omega}(x) : \mathcal{H}^{-\omega} \to \mathcal{H}^{-\omega}$ is continuous and U_h^{η} extends to a continuous map $\overline{\mathcal{S}_\pi} \to \mathcal{H}^{-\omega}$ that is holomorphic on $\mathcal{S}_\pi$, the same holds for the orbit map of $\eta' := \mathrm{d}U^{-\omega}(x)\eta$.

Moreover, this extension satisfies

$$\begin{aligned} JU_h^{\eta'}(t) &= e^{\pm t}\mathrm{d}U^{-\omega}(\tau_h(x))JU_h^{\eta}(t) = e^{\pm t}\mathrm{d}U^{-\omega}(-x)U_h^{\eta}(t + \pi i) \\ &= e^{\pm(t+\pi i)}\mathrm{d}U^{-\omega}(x)U_h^{\eta}(t + \pi i) = U_h^{\eta'}(t + \pi i). \end{aligned}$$

This proves the lemma because $\mathfrak{g} = \mathfrak{g}_1(h) + \mathfrak{g}_0(h) + \mathfrak{g}_{-1}(h)$ and $\mathcal{H}^{-\omega}_{\mathrm{KMS}}$ is obviously invariant under $\mathrm{d}U^{-\omega}(\mathfrak{g}_0(h))$. □

Hypothesis (H2) Hypothesis (H1) *is satisfied for $\mathcal{E}$ and, in addition,*

$$\mathrm{E}_H := \beta^{+}(\mathrm{E}_K) \subseteq \mathcal{H}^{-\infty}.$$

Proposition 7(e) then implies that the subspace E_H is $U^{-\infty}(H)$-invariant.

The following generalization of the van den Ban–Delorme Automatic Continuity Theorem is a key tool to show that Hypothesis (H2) follows from Hypothesis (H2). We indicate a proof of this result in Appendix 1 (Theorem 17). Note that, by Proposition 7(e), the finite-dimensional subspaces $\beta^{\pm}(\mathcal{E}^J)$ of $\mathcal{H}^{-\omega}$ are $\mathfrak{h}$-invariant, hence consist of $\mathfrak{h}$-finite distribution vectors, so that the following theorem implies that $\beta^{\pm}(\mathcal{E}^J) \subseteq \mathcal{H}^{-\infty}$.

Theorem 5 (Automatic Continuity Theorem) *If $\eta_G : G \to \eta_G(G) \subseteq G_{\mathbb{C}}$ is a finite covering and $(U, \mathcal{H})$ is irreducible, then every $\mathfrak{h}$-finite hyperfunction vector is a distribution vector, i.e.,*

$$(\mathcal{H}^{-\omega})^{[\mathfrak{h}]} \subseteq \mathcal{H}^{-\infty}.$$

Proposition 8 *Under* Hypothesis (H2), *the real subspaces* $\beta^{\pm}(\mathcal{H}^{[K],J}) \subseteq \mathcal{H}^{-\infty,[H]}$ *are* $\mathfrak{g}$*-submodules with trivial intersection.*

Proof The real subspace $\beta^{+}(\mathcal{H}^{[K],J}) \subseteq \mathcal{H}^{-\infty,[H]}$ is contained in $\mathcal{H}^{-\omega}(U_h)_{\mathrm{KMS}}$ and $\beta^{-}(\mathcal{H}^{[K],J}) = J\beta^{+}(\mathcal{H}^{[K],J})$, so that Corollary 1 shows that

$$\beta^{+}(\mathcal{H}^{[K],J}) \cap \beta^{-}(\mathcal{H}^{[K],J}) \subseteq \ker(\partial U(h)) \subseteq \mathcal{H}.$$

Moore's Theorem [51] and the unboundedness of $e^{\mathbb{R}\,\mathrm{ad}\,h}$ show that we have $\ker(\partial U(h)) = \{0\}$, so that this intersection is trivial. □

We conclude this section by listing some open research problems.

Problem 1 The map $\beta^{\pm}$ provides a realization of the representation $(U, \mathcal{H})$ from the perspective of H-finite distribution vectors, i.e., the vector bundle

$$\mathbb{E}_M := G \times_H (\mathrm{E}_H)_{\mathbb{C}},$$

constructed from the finite-dimensional H-invariant subspace

$$\mathrm{E}_H = \beta^{+}(\mathrm{E}_K) \subseteq \mathcal{H}^{-\infty},$$

permits a G-equivariant injection

$$\mathcal{H}^{-\infty} \hookrightarrow C^{-\infty}(M, \mathbb{E}_M).^{1}$$

We thus obtain in particular a realization of the unitary representation $(U, \mathcal{H})$ in distributional sections of the vector bundle $\mathbb{E}_M$. It would be interesting to see more concrete expressions of the corresponding bundle-valued distribution kernel specifying this Hilbert space. This is of particular importance to understand the locality properties of the corresponding net $\mathsf{H}^{M}_{\mathrm{E}_H}$ on M.

If $(U, \mathcal{H})$ is irreducible, then $\mathcal{H}^{[K]}$ is an irreducible $\mathfrak{g}$-module (Harish–Chandra [37]). However, Proposition 8 shows that the space $(\mathcal{H}^{-\infty})^{[H]}$ of $\mathfrak{h}$-finite distribution vectors is never an irreducible real $\mathfrak{g}$-module.

Problem 2 Does $\eta \in \mathcal{H}^{-\infty}_{\mathrm{KMS}}$ imply that $U^{\eta}(z) \in \mathcal{H}$ for $z \in \mathcal{S}_{\pi}$? The same question can be asked for $\eta \in \mathcal{H}^{-\omega}_{\mathrm{KMS}}$. This is true if $\eta \in \mathcal{H}^{-\omega}(U_h)_{\mathrm{KMS}}$ (Proposition 2), but it is not clear for every $\eta \in \mathcal{H}^{-\infty}_{\mathrm{KMS}}$.

Problem 3 Is the subspace $\mathcal{H}^{-\omega}(U_h)$ invariant under $\mathrm{d}U^{-\omega}(\mathfrak{g})$? This is clearly the case for the Lie subalgebra $\mathfrak{g}_0(h) \subseteq \mathfrak{g}$. Dualy, one may ask if the subspace $\mathcal{H}^{\omega}(U_h)$ of U_h-analytic vectors in $\mathcal{H}$ is invariant under $\mathfrak{g}$, but this requires that $\mathcal{H}^{\omega}(U_h) \subseteq \mathcal{H}^{\infty}$. Is this always the case for irreducible unitary representations of G?

[1] We plan to discuss these realizations in more detail in the future.

4 Reeh–Schlieder and Bisognano–Wichmann Property

For linear groups $G \subseteq G_{\mathbb{C}}$, we obtain from the Krötz–Stanton Extension Theorem 4, Proposition 7(e) and Theorem 5 that $\beta^{\pm}(\mathcal{H}^{[K]}) \subseteq \mathcal{H}^{-\infty}$. With $\mathtt{E}_H \subseteq \mathcal{H}^{-\infty}$ as in Proposition 7, we thus obtain a *net of real subspaces* $\mathsf{H}^M_{\mathtt{E}_H}$ on $M = G/H$. In this section, we show that this net has the Reeh–Schlieder and the Bisognano–Wichman property.

4.1 The Reeh–Schlieder Property

In this section, we assume Hypothesis (H2), so that the real subspaces $\mathsf{H}^G_{\mathtt{E}_H}(O)$ are defined for every open subset $O \subseteq G$. In this section we show that, whenever $O \neq \emptyset$, this subspace is total.

We shall need the following lemma (Neeb/Ólafsson [63, Lemma 3.7]):

Lemma 5 *Let X be a locally compact space and $f: X \to \mathcal{H}^{-\infty}$ be a weak-$*$-continuous map. Then the following assertions hold:*

(a) $f^\wedge: X \times \mathcal{H}^\infty \to \mathbb{C}$, $f^\wedge(x, \xi) := f(x)(\xi)$, *is continuous.*
(b) *If, in addition, X is a complex manifold and f is antiholomorphic, then $f^\wedge$ is antiholomorphic.*

We shall use the following result from Beltiţă/Neeb [6, Thm. 6.4, Cor. 6.8]:

Theorem 6 *For the real bilinear form $\mathcal{H}^\infty \times \mathcal{H}^{-\infty} \to \mathbb{R}$, $(\xi, \psi) \mapsto \mathrm{Im}\langle \xi, \psi \rangle$, $\mathcal{H}^{-\infty}_{\mathrm{KMS}}$ is the annihilator of $\mathtt{V}' \cap \mathcal{H}^\infty$ and*

$$\mathcal{H}^{-\infty}_{\mathrm{KMS}} \cap \mathcal{H} = \mathtt{V}. \tag{4.1}$$

In our context, this result has the following interesting consequence. Whenever $\eta \in \mathcal{H}^{-\omega}(U_h)_{\mathrm{KMS}}$ is a distribution vector, then the extended orbit map $U^\eta: \overline{\mathcal{S}_\pi} \to \mathcal{H}^{-\infty}$ is automatically weak-$*$-continuous, and holomorphic on the interior, as a map with values in $\mathcal{H}^{-\infty}$.

Corollary 3 $\mathcal{H}^{-\omega}(U_h)_{\mathrm{KMS}} \cap \mathcal{H}^{-\infty} \subseteq \mathcal{H}^{-\infty}_{\mathrm{KMS}}$.

Proof We recall from Theorem 2 that

$$\mathcal{H}^{-\omega}(U_h)_{\mathrm{KMS}} = \{\eta \in \mathcal{H}^{-\omega}(U_h): (\forall v \in \mathtt{V}' \cap \mathcal{H}^\omega(U_h))\ \eta(v) \in \mathbb{R}\}.$$

For $\eta \in \mathcal{H}^{-\omega}(U_h)_{\mathrm{KMS}} \cap \mathcal{H}^{-\infty}$ we therefore have $\eta(v) \in \mathbb{R}$ for $v \in \mathtt{V}' \cap \mathcal{H}^\omega(U_h)$. In view of Theorem 6, we have to show that $\eta(v) \in \mathbb{R}$ holds for any $v \in \mathtt{V}' \cap \mathcal{H}^\infty$.

Now let $v \in \mathtt{V}' \cap \mathcal{H}^\infty$. To see that we also have $\eta(v) \in \mathbb{R}$, we consider the normalized Gaussians $\gamma_n(t) = \sqrt{\frac{n}{\pi}} e^{-nt^2}$ and recall from [6, Prop. 3.5(ii) and §5] that $v_n := U_h(\gamma_n)v \in \mathcal{H}^\infty$ converges to v in the Fréchet space $\mathcal{H}^\infty$. As $\mathtt{V}$ is U_h-

invariant, so ist $\mathtt{V}' = J\mathtt{V}$. Further, U_h commutes with J and γ_n is real, so that $v_n \in \mathtt{V}'$, hence $\eta(v_n) \in \mathbb{R}$ because $v_n \in \mathtt{V}' \cap \mathcal{H}^\omega(U_h)$. Finally the assertion follows from $\eta(v_n) \to \eta(v)$. □

Corollary 4 *$\mathtt{V}^\infty := \mathtt{V} \cap \mathcal{H}^\infty$ is a $\mathfrak{g}$-invariant subspace of $\mathcal{H}^\infty$ and $\mathcal{H}^{-\infty}_{\mathrm{KMS}}$ is a $\mathfrak{g}$-invariant subspace of $\mathcal{H}^{-\infty}$.*

Proof For the first assertion, we observe that $\mathtt{V} = \mathcal{H}_{\mathrm{KMS}}$ (Neeb/Ólafsson/Ørsted [66, Prop. 2.1]) implies

$$\mathtt{V}^\infty = \mathcal{H}_{\mathrm{KMS}} \cap \mathcal{H}^\infty \subseteq \mathcal{H}^{-\omega}_{\mathrm{KMS}} \cap \mathcal{H}^\infty,$$

and the space on the right is $\mathfrak{g}$-invariant by Lemma 4. Therefore, the $\mathfrak{g}$-invariant subspace of $\mathcal{H}^\infty$ generated by $\mathtt{V}^\infty$ is contained in

$$\mathcal{H}^{-\omega}_{\mathrm{KMS}} \cap \mathcal{H}^\infty \subseteq \mathcal{H}^{-\omega}_{\mathrm{KMS}} \cap \mathcal{H} = \mathtt{V}$$

(Corollary 2). Therefore, $\mathtt{V}^\infty$ is $\mathfrak{g}$-invariant.

Replacing h by $-h$, it follows that $\mathtt{V}' \cap \mathcal{H}^\infty$ is also $\mathfrak{g}$-invariant, so that the $\mathfrak{g}$-invariance of $\mathcal{H}^{-\infty}_{\mathrm{KMS}}$, the annihilator of $\mathtt{V}' \cap \mathcal{H}^\infty$, follows from Theorem 6. □

Lemma 6 *Under* Hypothesis (H2), *for any $v \in \mathtt{E}_K = \mathcal{E}^J$, the extended orbit map*

$$U_h^v \colon \overline{\mathcal{S}_{-\frac{\pi}{2},\frac{\pi}{2}}} = \left\{z \in \mathbb{C} \colon |\operatorname{Im} z| \leq \frac{\pi}{2}\right\} \to \mathcal{H}^{-\infty}, \quad z \mapsto e^{-z\partial U(h)}v$$

is continuous and holomorphic on the interior, with respect to the weak-∗-topology on $\mathcal{H}^{-\infty}$.

Proof From Hypothesis (H2), we know that $\mathtt{E}_H \subseteq \mathcal{H}^{-\infty}$. Further,

$$\mathcal{H}^{-\omega}(U_h)_{\mathrm{KMS}} \cap \mathcal{H}^{-\infty} \subseteq \mathcal{H}^{-\infty}_{\mathrm{KMS}}$$

by Corollary 3 and $\mathtt{E}_H \subseteq \mathcal{H}^{-\omega}(U_h)_{\mathrm{KMS}}$ by Lemma 2. For $\eta := \beta^+(v) \in \mathtt{E}_H$, we then have by Proposition 2

$$U_h^v(z) = U_h^\eta\Big(z + \frac{\pi i}{2}\Big),$$

so that the assertion follows from $\eta \in \mathcal{H}^{-\infty}_{\mathrm{KMS}}$. □

Theorem 7 (Reeh–Schlieder Theorem) *Let $(U, \mathcal{H})$ be a unitary representation of G and $\mathtt{F}_K \subseteq \mathcal{H}^{[K]}$ be a real linear subspace. Suppose that*

1. *$\mathtt{F}_K \subseteq \bigcap_{|t|<\pi/2} \mathcal{D}(e^{it\partial U(h)})$, that the limit $\beta^+(v) = \lim_{t\to\pi/2} e^{-it\partial U(h)}v$ exists for every $v \in \mathtt{F}_K$ in the weak-∗-topology of $\mathcal{H}^{-\infty}$, and that*
2. *$U(G)\mathtt{F}_K$ is total in $\mathcal{H}$.*

Then, for every non-empty open subset $O \subseteq G$ and $\mathtt{F}_H := \beta^+(\mathtt{F}_K)$, the subspace $\mathsf{H}^G_{\mathtt{F}_H}(O)$ is total in $\mathcal{H}$.

Proof Let $\xi \in \mathcal{H}$ be orthogonal to $\mathsf{H}^G_{\mathtt{F}_H}(\mathcal{O})$ and $\mathcal{O} \neq \emptyset$. We have to show that $\xi = 0$.
Smooth case: First we assume that ξ is a smooth vector, so that the orbit map

$$U^\xi : G \to \mathcal{H}^\infty, \quad g \mapsto U(g)\xi$$

is smooth, hence in particular continuous with respect to the Fréchet topology on $\mathcal{H}^\infty$ (Neeb [58, Cor. 4.5]). Let $U^v_h(z) = e^{z\partial U(h)}v$ be as above. Then Lemma 5 implies that the map

$$F : \overline{\mathcal{S}_{-\frac{\pi}{2},\frac{\pi}{2}}} \times G \to \mathbb{C}, \quad F(z, g) := \langle U(g)^{-1}\xi, U^v_h(-z)\rangle$$

is continuous for every K-finite vector $v \in \mathtt{F}_K$. We conclude that, for every relatively compact open subset $\mathcal{O}_c \subseteq G$, the map

$$F^\wedge : \overline{\mathcal{S}_{-\frac{\pi}{2},\frac{\pi}{2}}} \to C(\mathcal{O}_c, \mathbb{C}), \quad F^\wedge(z)(g) := F(z, g) = \langle U(g)^{-1}\xi, U^v_h(-z)\rangle$$

is continuous with respect to the topology of uniform convergence on the Banach space $C^b(\mathcal{O}_c, \mathbb{C})$ of bounded continuous functions on $\mathcal{O}_c$, hence in particular locally bounded. Moreover, for every fixed $g \in \mathcal{O}_c$, the map $z \mapsto F^\wedge(z)(g)$ is holomorphic on the interior of the strip, so that [57, Cor. A.III.3] implies that $F^\wedge$ is holomorphic on the interior of the strip.

Suppose that $\overline{\mathcal{O}_c} \subseteq \mathcal{O}$ and choose $\varepsilon > 0$ in such a way that

$$\overline{\mathcal{O}_c \exp([-\varepsilon, \varepsilon]h)} \subseteq \mathcal{O}. \tag{4.2}$$

We claim that $F^\wedge(t + \pi i/2) = 0$ for $t \in \mathbb{R}$. In fact,

$$U^v_h(-t - \pi i/2) = U(\exp(-th))\beta^+(v).$$

For $g \in \mathcal{O}_c$ we therefore have

$$\begin{aligned} F^\wedge\Big(t + \frac{\pi i}{2}\Big)(g) &= \langle U(g)^{-1}\xi, U^{-\infty}(\exp(-th))\beta^+(v)\rangle \\ &= \langle \xi, U^{-\infty}(g\exp(-th))\beta^+(v)\rangle. \end{aligned}$$

This is a continuous function of g, and we now show that it vanishes on $\mathcal{O}_c$ for every t in $[-\varepsilon, \varepsilon]$. For every test function $\varphi \in C^\infty_c(\mathcal{O}_c, \mathbb{R})$ and the translated test function $\psi(g) = \varphi(g\exp(th))$, which is supported in $\mathcal{O}$, we have

$$\begin{aligned} \int_G \varphi(g) F^\wedge\Big(t + \frac{\pi i}{2}\Big)(g)\, dg &= \int_G \varphi(g)\langle \xi, U^{-\infty}(g\exp(-th))\beta^+(v)\rangle\, dg \\ &= \langle \xi, \int_G \varphi(g) U^{-\infty}(g\exp(-th))\, dg\ \beta^+(v)\rangle \\ &= \langle \xi, U^{-\infty}(\psi)\beta^+(v)\rangle \in \langle \xi, \mathsf{H}^G_{\mathtt{F}_H}(\mathcal{O})\rangle = \{0\}. \end{aligned}$$

We conclude that $F^\wedge$ vanishes on the boundary interval $\frac{\pi i}{2} + [-\varepsilon, \varepsilon]$. The Schwarz Reflection Principle now implies that $F^\wedge = 0$. For $z = 0$, we obtain in particular that $\xi \perp U(O_c)\mathtt{F}_K$. Since $\mathtt{F}_K$ consists of analytic vectors, $U(O_c)\mathtt{F}_K$ and $U(G)\mathtt{F}_K$ generate the same closed subspace. We thus obtain $\xi \perp U(G)\mathtt{F}_K$ and hence that $\xi = 0$.
General case: Now we drop the assumption that ξ is a smooth vector. We pick a non-empty, relatively compact subset $O_c \subseteq O$ and observe that there exists a compact symmetric e-neighborhood $O_e \subseteq G$ with $O_e O_c \subseteq O$. Then

$$\langle \xi, U(C_c^\infty(O_e, \mathbb{R}))\mathsf{H}^G_{\mathtt{F}_H}(O_c)\rangle \subseteq \langle \xi, \mathsf{H}^G_{\mathtt{F}_H}(O)\rangle = \{0\},$$

so that, for any test function $\psi \in C_c^\infty(O_e, \mathbb{C})$, the smooth vector $U(\psi)^*\xi$ is orthogonal to $\mathsf{H}^G_{\mathtt{F}_H}(O_c)$, hence zero by the smooth case. As $C_c^\infty(O_e, \mathbb{C})$ contains an approximate identity of G, this implies that $\xi = 0$. □

4.2 *The Bisognano–Wichmann Property*

Let $q\colon G \to G/H$ be the quotient map and

$$W^G := q^{-1}(W) \quad \text{for} \quad W := W_M^+(h)_{eH},$$

where the positivity domain $W_M^+(h)_{eH}$ is defined in (1.6) in the introduction. We want to show that

$$\mathsf{H}^M_{\mathtt{E}_H}(W) = \mathsf{H}^G_{\mathtt{E}_H}(W^G) = \mathtt{V},$$

where $\mathtt{V}$ is the standard subspace specified as in (1.7) (in the introduction) by

$$\Delta_{\mathtt{V}} = e^{2\pi i \partial U(h)} \quad \text{and} \quad J_{\mathtt{V}} = J.$$

4.2.1 A Realization as a Space of Holomorphic Functions

In this subsection, we describe a realization of the representation $(U, \mathcal{H})$ in the space $\mathcal{O}(\mathbb{E}^{\mathrm{op}})$ of holomorphic functions on the holomorphic vector bundle $\mathbb{E}^{\mathrm{op}}$ over Ξ^{op}, obtained by flipping the sign of the complex structure on $\mathbb{E}$ (cf. Proposition 5). We therefore have by Hypothesis (H1) an antiholomorphic map

$$\Psi_{\mathcal{E}}\colon \mathbb{E}^{\mathrm{op}} \to \mathcal{H}, \quad [g, ix, v] \mapsto U(g)e^{i\partial U(x)}v \tag{4.3}$$

that is fiberwise antilinear. This leads to a complex linear map

$$\Phi\colon \mathcal{H} \to \mathcal{O}(\mathbb{E}^{\mathrm{op}}), \quad \Phi(\xi)(v) := \langle \Psi_{\mathcal{E}}(v), \xi\rangle.$$

As $U(G)\mathcal{E}$ spans a dense subspace of $\mathcal{H}$, it maps $\mathcal{H}$ isometrically onto a reproducing kernel Hilbert space $\mathcal{H}_Q \subseteq \mathcal{O}(\mathbb{E}^{\rm op})$ whose kernel $Q\colon \mathbb{E}^{\rm op} \times \mathbb{E}^{\rm op} \to \mathbb{C}$ is sesquiholomorphic. For the evaluation functionals Q_w, $w \in \mathbb{E}$, we find with

$$\langle \Psi_{\mathcal{E}}(w), \xi \rangle = \Phi(\xi)(w) = \langle Q_w, \Phi(\xi)\rangle = \langle \Phi^{-1}(Q_w), \xi \rangle$$

the relation $\Phi^{-1}(Q_w) = \Psi_{\mathcal{E}}(w)$, resp.,

$$Q_w = \Phi(\Psi_{\mathcal{E}}(w)), \quad w \in \mathbb{E}. \tag{4.4}$$

The map Φ is G-equivariant with respect to the natural G-action on $\mathcal{O}(\mathbb{E}^{\rm op})$:

$$\Phi(U(g)\xi)(w) = \langle U(g)^{-1}\Psi_{\mathcal{E}}(w), \xi \rangle = \langle \Psi_{\mathcal{E}}(g^{-1}.w), \xi \rangle = \Phi(\xi)(g^{-1}.w).$$

If $J\mathcal{E} = \mathcal{E}$, then we obtain with Remark 12 and the antiholomorphic involution

$$\tau_{\mathbb{E}}([g, ix, v]) = [\tau_h(g), -i\tau_h(x), Jv]$$

on $\mathbb{E}$ the relation

$$\begin{aligned}\Phi(J\xi)([g, ix, v]) &= \langle \Psi_{\mathcal{E}}([g, ix, v]), J\xi \rangle = \overline{\langle J\Psi_{\mathcal{E}}([g, ix, v]), \xi \rangle} \\ &= \overline{\langle \Psi_{\mathcal{E}}([\tau_h(g), -i\tau_h(x), Jv]), \xi \rangle} = \overline{\Phi(\xi)([\tau_h(g), -i\tau_h(x), Jv])},\end{aligned}$$

so that

$$\Phi(J\xi) = \overline{\Phi(\xi) \circ \tau_{\mathbb{E}}}. \tag{4.5}$$

We now obtain

$$JQ_w = J\Phi(\Psi_{\mathcal{E}}(w)) = \Phi(J\Psi_{\mathcal{E}}(w)) = \Phi(\Psi_{\mathcal{E}}(\tau_{\mathbb{E}}(w))) = Q_{\tau_{\mathbb{E}}(w)}.$$

We conclude that $w \in \mathbb{E}^{\tau_{\mathbb{E}}}$ implies $Q_w \in \mathcal{H}_Q^J$. For $w = [g, ix, v]$, the relation $\tau_{\mathbb{E}}(w) = w$ is equivalent to

$$[g, ix, v] = [\tau_h(g), -i\tau_h(x), Jv],$$

which means that $q_{\mathbb{E}}(w) \in \Xi^{\overline{\tau}_h}$ and that $Jv = v$ holds in the fiber $\mathbb{E}_{[g,ix]}$ over the $\overline{\tau}_h$-fixed point $[g, ix] \in \Xi$.

For $m \in \Xi^{\overline{\tau}_h}$ the modular flow $\alpha_{it}(m) = \exp(ith).m$ is defined in Ξ for $|t| < \frac{\pi}{2}$ (cf. Morinelli/Neeb/Ólafsson [55, §8]) and the same holds for its lift to the bundle $\mathbb{E}$, given by

$$\alpha_t([g, ix, v]) = [\exp(th)g, ix, v].$$

So we obtain for each $w \in \mathbb{E}^{\tau_\mathbb{E}}$ a holomorphic extension

$$\alpha^w : \mathcal{S}_{\pm\pi/2} \to \mathbb{E}^{\rm op}.$$

For each $w \in \mathbb{E}^{\tau_\mathbb{E}}$, we thus obtain by Hypothesis (H1) with Lemma 2 a limit

$$\beta^{\pm}(\Psi_{\mathcal{E}}(w)) = \lim_{t \to \mp\pi/2} e^{it\partial U(h)} \Psi_{\mathcal{E}}(w) \in \mathcal{H}^{-\omega}(U_h)_{\rm KMS}, \tag{4.6}$$

where $\mathcal{H}^{-\omega}(U_h)_{\rm KMS}$ is defined as in Definition 1 with respect to the weak-$*$-topology on $\mathcal{H}^{-\omega}(U_h)$ (see also (1.9) in the introduction).

4.2.2 Verification of the Bisognano-Wichmann Property

In this section, we show that the Bisognano-Wichmann property introduced in the introduction holds for the net $\mathsf{H}^M_{\mathsf{E}_H}(\mathcal{O})$. The main results are collected in Theorem 8.

Lemma 7 *Under* Hypothesis (H2), *the map*

$$\Upsilon := \beta^+ \circ \Psi_{\mathcal{E}} : \mathbb{E}^{\tau_\mathbb{E}} \to \mathcal{H}^{-\omega}(U_h)_{\rm KMS}$$

has the following properties:

(a) *It is weak-$*$-analytic.*
(b) *It is G^h_e-equivariant.*
(c) *The automorphism $\zeta = e^{-\frac{\pi i}{2}\,\mathrm{ad}\,h}$ of $\mathfrak{g}_\mathbb{C}$ maps $i\Omega_{\mathfrak{h}_\mathfrak{p}} := i\Omega_\mathfrak{p} \cap \mathfrak{h}$ to*

$$\Omega_{\mathfrak{q}_\mathfrak{k}} := \{x \in \mathfrak{q}_\mathfrak{k} : \mathrm{Spec}(\mathrm{ad}\,x) \subseteq (-\pi/2, \pi/2)i\},$$

and for $x \in \Omega_\mathfrak{p}$ and $v \in \mathsf{E}_K$, we have

$$U^{-\omega}(\exp\zeta(ix))\beta^+(v) = \beta^+(e^{i\partial U(x)}v). \tag{4.7}$$

(d) $U^{-\omega}(W^G)\beta^+(\mathsf{E}_K) \subseteq \mathcal{H}^{-\omega}(U_h)_{\rm KMS}$.

Proof

(a) Let $\xi \in \mathcal{H}^\omega(U_h)$ and $w \in \mathbb{E}^{\tau_\mathbb{E}}$. Then there exists an $\varepsilon \in (0, \pi/2)$ with $\xi \in \mathcal{D}(e^{-i\varepsilon\partial U(h)})$. Let $\xi_\varepsilon := e^{-i\varepsilon\partial U(h)}\xi$. Then

$$\begin{aligned}\langle \xi, \Upsilon(w)\rangle &= \langle e^{i\varepsilon\partial U(h)}\xi_\varepsilon, \Upsilon(w)\rangle = \langle \xi_\varepsilon, e^{i\varepsilon\partial U(h)}\Upsilon(w)\rangle \\ &= \langle \xi_\varepsilon, \Upsilon(\alpha_{i\varepsilon}w)\rangle = \langle \xi_\varepsilon, \Psi_{\mathcal{E}}(\alpha_{i(\varepsilon-\pi/2)}w)\rangle,\end{aligned}$$

so that the assertion follows from the analyticity of $\Psi_{\mathcal{E}}$ on $\mathbb{E}$ and the analyticity of the map

$$\alpha_{i(\varepsilon-\pi/2)} \colon \mathbb{E}^{\tau_{\mathbb{E}}} \to \mathbb{E}.$$

Finally we observe that, by Theorem 2, the subspace $\mathcal{H}^{-\omega}(U_h)_{\mathrm{KMS}}$ is closed in $\mathcal{H}^{-\omega}(U_h)$ and since Υ takes values in this subspace, it is real analytic as an $\mathcal{H}^{-\omega}(U_h)_{\mathrm{KMS}}$-valued map.

(b) follows from the fact that the action of the subgroup $G_e^h = G_e^{\tau_h}$ on $\mathbb{E}$ preserves the fixed point set $\mathbb{E}^{\tau_{\mathbb{E}}}$ and commutes with $U_h(\mathbb{R})$.

(c) First we recall from Proposition 1(b) that the orbit maps of U^ω are analytic, hence that the orbit maps of $U^{-\omega}$ are weak-$*$-analytic. Hence, for a fixed $v \in \mathcal{E}^J$, the map

$$\Omega_{\mathfrak{p}} \to \mathcal{H}^{-\omega}(U), \quad x \mapsto U^{-\omega}(\exp \zeta(ix))\beta^+(v) = U^{-\omega}(\exp \zeta(ix))\Upsilon([e, 0, v]) \tag{4.8}$$

is weak-$*$-analytic. Further, (a) and the weak-$*$-continuity of the canonical inclusion $\mathcal{H}^{-\omega}(U_h) \to \mathcal{H}^{-\omega}(U)$ imply that the map

$$\Omega_{\mathfrak{p}} \to \mathcal{H}^{-\omega}(U), \qquad x \mapsto \Upsilon([e, ix, v]) \tag{4.9}$$

is analytic. The n-th order terms in the Taylor expansion of (4.8) in 0 are given by

$$\frac{1}{n!}\mathrm{d}U^{-\omega}(\zeta(ix))^n\beta^+(v)$$

and for (4.9) by

$$\frac{1}{n!}\beta^+(\mathrm{d}U(ix)^n v) = \frac{1}{n!}\mathrm{d}U^{-\omega}(i\zeta(x))^n\beta^+(v),$$

where we have used Proposition 7(d) for the last equality. We conclude that both maps have the same Taylor expansion in $[e, 0, v]$, hence that they coincide because their domain is connected. This proves (4.7).

(d) As the right-hand side of (4.7) is contained in the subspace $\mathcal{H}^{-\omega}(U_h)_{\mathrm{KMS}}$ by (4.6), (4.7) implies that

$$U^{-\omega}(\exp(\Omega_{\mathfrak{q}_{\mathfrak{k}}}))\beta^+(\mathrm{E}_K) \subseteq \mathcal{H}^{-\omega}(U_h)_{\mathrm{KMS}}. \tag{4.10}$$

Now $W^G = G_e^h \exp(\Omega_{\mathfrak{q}_{\mathfrak{k}}})H$ by (8.2) in Appendix 2. implies

$$\begin{aligned} U^{-\omega}(W^G)\beta^+(\mathtt{E}_K) &= U^{-\omega}(G_e^h)U^{-\omega}(\exp(\Omega_{\mathfrak{q}_{\mathfrak{k}}}))U^{-\omega}(H)\beta^+(\mathtt{E}_K) \\ &\overset{\text{Prop. 7(e)}}{=} U^{-\omega}(G_e^h)U^{-\omega}(\exp(\Omega_{\mathfrak{q}_{\mathfrak{k}}}))\beta^+(\mathtt{E}_K) \\ &\subseteq U^{-\omega}(G_e^h)\mathcal{H}^{-\omega}(U_h)_{\mathrm{KMS}} \subseteq \mathcal{H}^{-\omega}(U_h)_{\mathrm{KMS}}. \end{aligned}$$

□

The following proposition leads to a characterization of those subspaces $\mathtt{E}$ with $\mathsf{H}_{\mathtt{E}}^M(W) \subseteq \mathtt{V}$ in Corollary 3.

Proposition 9 *For an open subset $O \subseteq G$ and a real subspace $\mathtt{E} \subseteq \mathcal{H}^{-\infty}$, the following are equivalent:*

(a) $\mathsf{H}_{\mathtt{E}}^G(O) \subseteq \mathtt{V}$.
(b) *For all $\varphi \in C_c^\infty(O, \mathbb{R})$ we have $U^{-\infty}(\varphi)\mathtt{E} \subseteq \mathtt{V}$.*
(c) *For all $\varphi \in C_c^\infty(O, \mathbb{R})$ we have $U^{-\infty}(\varphi)\mathtt{E} \subseteq \mathcal{H}_{\mathrm{KMS}}^{-\infty}$.*
(d) *$U^{-\infty}(g)\mathtt{E} \subseteq \mathcal{H}_{\mathrm{KMS}}^{-\infty}$ for every $g \in O$.*

Proof It is clear that (a) implies (b) by the definition of $\mathsf{H}_{\mathtt{E}}^G(O)$. (b) implies (c) because $U^{-\infty}(\varphi)\mathtt{E} \subseteq \mathcal{H}$ and $\mathtt{V} = \mathcal{H} \cap \mathcal{H}_{\mathrm{KMS}}^{-\infty}$ (Theorem 6).

For the implication (c) $\Rightarrow$ (d), let $(\delta_n)_{n\in\mathbb{N}}$ be a δ-sequence in $C_c^\infty(G, \mathbb{R})$. Then $U(\delta_n)\xi \to \xi$ in $\mathcal{H}^\infty$ and hence also in $\mathcal{H}^{-\infty}$. It follows in particular that

$$U^{-\infty}(\delta_n * \delta_g)\eta = U^{-\infty}(\delta_n)U^{-\infty}(g)\eta \to U^{-\infty}(g)\eta \quad \text{for} \quad \eta \in \mathcal{H}^{-\infty}.$$

Hence $\overline{\mathtt{V}} \subseteq \mathcal{H}_{\mathrm{KMS}}^{-\infty}$, resp., the closedness of $\mathcal{H}_{\mathrm{KMS}}^{-\infty}$, shows that (c) implies (d). Here we use that $\delta_n * \delta_g \in C_c^\infty(O, G)$ for $g \in O$ if n is sufficiently large.

As the G-orbit maps in $\mathcal{H}^{-\infty}$ are continuous and $\mathcal{H}_{\mathrm{KMS}}^{-\infty}$ is closed, hence stable under integrals over compact subsets and $U^{-\infty}(C_c^\infty(O, \mathbb{R}))\mathcal{H}^{-\infty} \subset \mathcal{H}^\infty$, we see that (d) implies (a). □

Proposition 10 *Under* Hypothesis (H2), *the space $\mathtt{E}_H = \beta^+(\mathtt{E}_K)$ consists of distribution vectors and $\mathsf{H}_{\mathtt{E}_H}^G(W^G) \subseteq \mathtt{V}$.*

Proof By Lemma 7(d), we have $U^{-\omega}(W^G)\mathtt{E}_H \subseteq \mathcal{H}^{-\omega}(U_h)_{\mathrm{KMS}}$. If, in addition, $\mathtt{E}_H \subseteq \mathcal{H}^{-\infty}$, then Corollary 3 yields

$$U^{-\infty}(W^G)\mathtt{E}_H = U^{-\omega}(W^G)\mathtt{E}_H \subseteq \mathcal{H}^{-\omega}(U_h)_{\mathrm{KMS}} \cap \mathcal{H}^{-\infty} \subseteq \mathcal{H}_{\mathrm{KMS}}^{-\omega} \cap \mathcal{H}^{-\infty} = \mathcal{H}_{\mathrm{KMS}}^{-\infty},$$

and therefore Proposition 9(a),(d) imply the assertion. □

We are now ready to state our main theorem for the case where the simple Lie group G is linear, i.e., $\eta : G \to G_{\mathbb{C}}$ is injective.

Theorem 8 *Let* $(U, \mathcal{H})$ *be an irreducible antiunitary representation of*

$$G_{\tau_h} := G \rtimes \{\mathbf{1}, \tau_h\},$$

let $\mathcal{E}$ *be a finite-dimensional subspace invariant under* K *and* J, *and* $\mathtt{E}_K := \mathcal{E}^J$. *If* G *is linear, then* Hypothesis (H2) *is satisfied, so that* $\mathtt{E}_H = \beta^+(\mathtt{E}_K) \subseteq \mathcal{H}^{-\infty}$. *Then the net* $\mathsf{H}^M_{\mathtt{E}_H}$ *on the non-compactly causal symmetric space satisfies* (Iso), (Cov), (RS) *and* (BW), *where* $W = W_M^+(h)_{eH}$ *is the connected component of the positivity domain of* h *on* M, *containing the base point.*

Proof (Iso) and (Cov) are trivially satisfied and (RS) follows from Theorem 7. It remains to verify (BW).

As W is invariant under the modular flow on M, the subspace

$$\mathsf{H} := \mathsf{H}^M_{\mathtt{E}_H}(W) = \mathsf{H}^G_{\mathtt{E}_H}(W^G)$$

is invariant under $U(\exp th) = \Delta_{\mathtt{V}}^{-it/2\pi}$ for $t \in \mathbb{R}$. Further, $\mathsf{H} \subseteq \mathtt{V}$ by Proposition 10, so that H is separating, hence standard because it is also cyclic by the Reeh–Schlieder Theorem 7. Then Neeb/Ólafsson [63, Lemma 3.4] implies that $\mathsf{H} = \mathtt{V}$ because H is invariant under the modular group $U(\exp \mathbb{R}h)$ of $\mathtt{V}$. □

Remark 15 ("Independence" of the Net from H**)** In the context of Theorem 8, the real subspace $\mathtt{E}_H \subseteq \mathcal{H}^{-\infty}$ is invariant under $U^{-\infty}(H)$. For any open subset $\mathcal{O}_G \subseteq G$ we therefore have

$$\mathsf{H}^G_{\mathtt{E}_H}(\mathcal{O}) = \mathsf{H}^G_{\mathtt{E}_H}(\mathcal{O}H)$$

by [63, Lemma 2.11]. Hence the inclusions $H_{\min} \subseteq H \subseteq H_{\max}$ from the introduction to Sect. 3 imply that

$$\mathsf{H}^G_{\mathtt{E}_H}(\mathcal{O}) = \mathsf{H}^G_{\mathtt{E}_H}(\mathcal{O}H) = \mathsf{H}^G_{\mathtt{E}_H}(\mathcal{O}H_{\max}).$$

Here we use that the real subspace $\mathtt{E} \subseteq \mathcal{H}^{-\infty}$ is invariant under $H_{\max}$ because Proposition 7(e) also applies to $H_{\max}$, which is the maximal choice for H. For the covering

$$q_m : G/H \to M_{\min} := G/H_{\max}$$

it therefore follows that the net $\mathsf{H}^M_{\mathtt{E}_H}$ on M can be recovered from its pushforward $\mathsf{H}^{M_{\min}}_{\mathtt{E}_H}$ to $M_{\min}$ because

$$\mathsf{H}^M_{\mathtt{E}_H}(\mathcal{O}) = \mathsf{H}^M_{\mathtt{E}_H}(q_m^{-1}(q_m(\mathcal{O}))) = \mathsf{H}^{M_{\min}}_{\mathtt{E}_H}(q_m(\mathcal{O}))$$

for any open subset $\mathcal{O} \subseteq M$.

4.3 Locality

In the context of quantum field theories on space-time manifolds, one considers for nets $\mathcal{O} \mapsto \mathsf{H}(\mathcal{O})$ of real subspaces the *locality condition*

$$\mathcal{O}_1 \subseteq \mathcal{O}_2' \quad \Rightarrow \quad \mathsf{H}(\mathcal{O}_1) \subseteq \mathsf{H}(\mathcal{O}_2)',$$

where $\mathsf{H}(\mathcal{O}_2)' := \mathsf{H}(\mathcal{O}_2)^{\perp_\omega}$ is the symplectic orthogonal space with respect to $\omega = \mathrm{Im}\langle\cdot,\cdot\rangle$ and $\mathcal{O}_2'$ is the *causal complement of* $\mathcal{O}_2$, i.e., the largest open subset that cannot be connected to $\mathcal{O}_2$ with causal curves.

Presently, we do not know if this strong locality condition is satisfied for our nets $\mathsf{H}^M_{\mathsf{E}_H}$ on non-compactly causal symmetric spaces $M = G/H$. However, there is some information that can be formulated as follows. Let $W := W_M^+(h)_{eH}$ be the wedge region in M associated to the Euler element h. To see natural candidates for its "causal complement" W', we note that our construction implicitly uses an embedding of M into the "boundary" of the crown domain Ξ. We actually have two natural embeddings, that can be described by

$$M_\pm := \alpha_{\mp\pi i/2}(G/K) \subseteq \partial\,\Xi$$

whenever Ξ is embedded into a complex homogeneous space $G_{\mathbb{C}}/K_{\mathbb{C}}$ and

$$\alpha_z(m) = \exp(zh).m.$$

Note that these embeddings are highly non-unique and have different geometric properties. In particular, the boundary G-orbits $M_\pm$ in $\partial\,\Xi$ may coincide or not. Such embeddings are also studied by Gindikin and Krötz in [31] and in Neeb/Ólafsson [64, Thm. 5.4], where Ξ is identified with a natural tube domain of $M = G/H$.

Identifying M with M_+, a natural candidate for the "causal complement" W' is the domain

$$W' = \alpha_{\pi i}(W) \subseteq M_-.$$

For locality issues, one now has to distinguish between the two cases $M_+ = M_-$ and $M_+ \neq M_-$. A simple example with $M_+ \neq M_-$ arises for

$$M_+ = \mathbb{R}_+, \qquad \text{where} \qquad \Xi = \mathbb{C}_+ = \{z \in \mathbb{C} \colon \mathrm{Im}\, z > 0\}$$

is the complex upper half plane (the crown of the Riemannian space $i\mathbb{R}_+$), $\alpha_t(z) = e^t z$, and $M_- = \mathbb{R}_- = (-\infty, 0)$. If $M = \mathbb{R}^{1,n}$ is $(n+1)$-dimensional Minkowski space, $G = \mathbb{R}^{1,n} \rtimes \mathrm{SO}_{1,n}(\mathbb{R})_e$ is the connected Poincaré group,

$$W = \{(x_0, \mathbf{x}) \colon x_1 > |x_0|\}$$

is the standard right wedge and α_t is the corresponding group of boosts, then $\alpha_{\pi i}(W) = -W = W'$, $\Xi = \mathbb{R}^{1,n} + iV_+$ ($V_+ \subseteq \mathbb{R}^{1,n}$ is the open future light cone), and $M_+ = M_-$. Although these examples do not come from simple groups, they represent the two different situations in an elementary way.

Our construction produces nets of real subspaces $\mathsf{H}^{M_+}_{\mathsf{E}_H}$ on M_+ and $\mathsf{H}^{M_-}_{J\mathsf{E}_H}$ on M_-. For the wedge regions we then derive from the (BW) property that

$$\mathsf{H}^{M_+}_{\mathsf{E}_H}(W) = \mathsf{V} \quad \text{and} \quad \mathsf{H}^{M_-}_{J\mathsf{E}_H}(W') = J\mathsf{V} = \mathsf{V}'.$$

Let

$$\mathcal{W}(M_+) := \{g.W : g \in G\}$$

denote the *wedge space of* M_+ and, likewise

$$\mathcal{W}(M_-) := \{g.W' : g \in G\}$$

the wedge space of M_- (cf. Morinelli/Neeb/Ólafsson [52, 55]). Putting $(g.W)' := g.W'$ for $g \in G$, we obtain by covariance and isotony the following property:

(L_W) Wedge-Locality: If there exists a wedge domain W_1 such that $\mathcal{O}_1 \subseteq W_1$ and $\mathcal{O}_2 \subseteq W_1'$, then $\mathsf{H}^M_{\mathsf{E}}(\mathcal{O}_1) \subseteq \mathsf{H}^{M_-}_{\mathsf{E}}(\mathcal{O}_2)'$.

In fact, $\mathcal{O}_1 \subseteq W_1 = g.W$ and $\mathcal{O}_2 \subseteq W_1' = g.W'$ lead to

$$\mathsf{H}^M_{\mathsf{E}}(\mathcal{O}_1) \subseteq U(g)\mathsf{V} \quad \text{and} \quad \mathsf{H}^{M_-}_{\mathsf{E}}(\mathcal{O}_2) \subseteq U(g)\mathsf{V}' = (U(g)\mathsf{V})'.$$

Of course, this is most interesting if $M_+ = M_-$, so that duality relates domains in the same homogeneous space. We plan to explore the locality condition in subsequent work.

5 Covering Groups of $\mathrm{SL}_2(\mathbb{R})$

In this section, we show by different methods that Hypothesis (H2) is also satisfied for all connected Lie groups G with Lie algebra $\mathfrak{g} = \mathfrak{sl}_2(\mathbb{R})$. Our argument is based on the observation that, for any simple Lie group G of real rank one and any K-eigenvector v in an irreducible G-representation $(U, \mathcal{H})$, we have

$$v \in \bigcap_{x \in \Omega_{\mathfrak{p}}} \mathcal{D}(e^{i\partial U(x)}),$$

which by Proposition 6 is Hypothesis (H1) for $\mathcal{E} = \mathbb{C}v$. We further show that there exists a constant $C > 0$ such that

$$\|e^{it\partial U(h)}v\| \leq C\Big(\frac{\pi}{2} - t\Big)^{-N} \quad \text{for} \quad 0 \leq t < \frac{\pi}{2}. \tag{5.1}$$

This implies that the limit $\beta^+(v)$ is contained in $\mathcal{H}^{-\infty}(U_h) \subseteq \mathcal{H}^{-\infty}$ (Theorem 3). The inclusion $U(W^G)\beta^+(v) \subseteq \mathcal{H}^{-\infty}_{\rm KMS}$ is obtained by verifying that actually

$$e^{i\partial U(\Omega_{\mathfrak{p}})}v \subseteq \bigcap_{|t|<\pi/2} \mathcal{D}(e^{it\partial U(h)})$$

and that (5.1) extends to any $e^{i\partial U(x)}v$ for $x \in \Omega_{\mathfrak{p}}$ (Sect. 5.2). Then Theorem 8 applies and shows that the net defined by $\mathtt{E}_H := \mathbb{R}\beta^+(v)$ satisfies (Iso), (Cov), (RS), and (BW).

We address these issues as follows. Let G be a connected simple Lie group of real rank one, $\mathfrak{g} = \mathfrak{k} + \mathfrak{p}$ be a Cartan decomposition and $\mathfrak{a} \subseteq \mathfrak{p}$ a maximal abelian subspace, which is 1-dimensional by assumption. Accordingly, the restricted root system is either $\Sigma(\mathfrak{g}, \mathfrak{a}) = \{\pm\alpha\}$ or $\Sigma(\mathfrak{g}, \mathfrak{a}) = \{\pm\alpha, \pm\alpha/2\}$. We choose a basis element $h \in \mathfrak{a}$ normalized by $\alpha(h) = 1$. We now consider an irreducible unitary representation $(U, \mathcal{H})$ of G and assume that $v \in \mathcal{H}$ is a normalized K-eigenvector. We write

$$\chi\colon K \to \mathbb{T}$$

for the corresponding character. Then

$$\varphi\colon G \to \mathbb{C}, \quad \varphi(g) := \langle v, U(g)v\rangle$$

is called a χ*-spherical function* and a *spherical function* if $\chi = 1$. It satisfies

$$\varphi(k_1gk_2) = \chi(k_1)\varphi(g)\chi(k_2) \quad \text{for} \quad g \in G, k_1, k_2 \in K.$$

As $G = KAK = K\exp(\mathbb{R}h)K$, it is determined by its values on the one-parameter group defined by $a_t := \exp(th)$. We shall see further that the values $\varphi(a_t)$ are given by hypergeometric functions and that (5.1) will follow from estimates for

$$\|e^{it\partial U(h)}v\|^2 = \varphi(a_{2it}) \quad \text{for} \quad |t| < \pi/2. \tag{5.2}$$

Clearly, nontrivial characters χ occur only if the Lie algebra $\mathfrak{k}$ has non-trivial center, i.e., if $\mathfrak{g}$ is hermitian (G/K is a bounded complex symmetric domain). Then the rank-one condition implies $\mathfrak{g} \cong \mathfrak{su}_{1,n}(\mathbb{C})$ for some $n \in \mathbb{N}$. For $n = 1$, we have $\mathfrak{su}_{1,1}(\mathbb{C}) \cong \mathfrak{sl}_2(\mathbb{R}) \cong \mathfrak{so}_{1,2}(\mathbb{R})$. The only rank-one Lie algebras containing Euler elements are those for which $\alpha/2$ is not a root, i.e., $\mathfrak{g} = \mathfrak{so}_{1,n}(\mathbb{R})$. Note that only $\mathfrak{so}_{1,2}(\mathbb{R})$ is hermitian, and, for $n > 2$, the character χ is trivial.

As the spherical functions can be written as Gauss hypergeometric functions ${}_2F_1$, we start our discussion by a short overview of those (Sect. 5.1). We then apply this to the spherical functions. For χ-spherical functions we use results of Shimeno [69]. For $\chi = 0$ the estimates for the spherical functions $\varphi(a_{it})$ for $t \to \pm\pi i$ also follow from Krötz/Opdam [44, Thm. 7.2].

We actually expect that the results that we use in this section for rank-one spaces hold more generally:

Conjecture 1 Let $(U, \mathcal{H})$ be an irreducible unitary representation of the connected real simple Lie group G, $h \in \mathfrak{g}$ an Euler element, and $v \in \mathcal{H}^{[K]}$ be a K-finite vector. Then there exists a constant $C > 0$ and $N > 0$ such that

$$\|e^{it\partial U(h)}v\| \leq C\Big(\frac{\pi}{2} - t\Big)^{-N} \quad \text{for} \quad 0 \leq t < \frac{\pi}{2}. \tag{5.3}$$

5.1 *Growth Estimates and Hypergeometric Functions*

5.1.1 The Hypergeometric Functions

Our standard references for this section are Erdélyi/Magnus/Oberhettinger/Tricomi [23, Ch. II] and Lebedev/Silverman [47, Ch. 9].

The Gauß hypergeometric functions, or simply the *hypergeometric function*, ${}_2F_1(\alpha, \beta; \gamma; z)$ is given by the series

$${}_2F_1(\beta, \alpha; \gamma; z) = {}_2F_1(\alpha, \beta; \gamma; z) = \sum_{k=0}^{\infty} \frac{(\alpha)_k(\beta)_k}{k!(\gamma)_k} z^k, \quad \gamma \neq 0, -1, -2, \ldots \tag{5.4}$$

where

$$(a)_k = a(a+1)\cdots(a+k-1), \quad k = 0, 1, 2, \ldots.$$

We note the following, see [23, Ch. II] and [47, Ch. 9]

Lemma 8

(a) *If* α *or* β *is contained in* $-\mathbb{N}_0$, *then the series in* (5.4) *is finite and* ${}_2F_1$ *is a polynomial function.*
(b) *The series in* (5.4) *converges absolutely for* $|z| < 1$.
(c) *The hypergeometric function* ${}_2F_1(\alpha, \beta; \gamma; z)$ *is the unique solution to the differential equation*

$$z(1-z)u'' + (\gamma - (\alpha + \beta + 1)z)u' - \alpha\beta u = 0$$

which is regular at $z = 0$ *and takes the value* 1 *at that point.*

The following can be found in [47, p. 241, 246, 245]. For the second theorem, we use in our formulation that $\arg(1 - z) < \pi$ is equivalent to $z \in \mathbb{C} \setminus [1, \infty)$.

Theorem 9 *The hypergeometric function* ${}_2F_1(\alpha, \beta; \gamma; z)$ *extends holomorphically to the slit plane* $\mathbb{C} \setminus [1, \infty)$*, and for fixed* $z \in \mathbb{C} \setminus [1, \infty)$*, the function*

$$(\alpha, \beta, \gamma) \mapsto \frac{{}_2F_1(\alpha, \beta; \gamma; z)}{\Gamma(\gamma)}$$

is an entire function of α*,* β *and* γ*.*

In particular, the functions $\gamma \mapsto {}_2F_1(\alpha, \beta; \gamma; z)$ are meromorphic with simple poles at $\gamma \in -\mathbb{N}_0$. Some more facts that we will use are collected in the following theorem.

Theorem 10 *The following assertions hold:*

(i) *If* $\mathrm{Re}(\gamma - \alpha - \beta) > 0$ *and* $\gamma \notin -\mathbb{N}_0$*, then*

$$\lim_{t \to 1-} {}_2F_1(\alpha, \beta; \gamma; t) = \frac{\Gamma(\gamma)\Gamma(\gamma - \alpha - \beta)}{\Gamma(\gamma - \alpha)\Gamma(\gamma - \beta)}.$$

(ii) *For* $z \in \mathbb{C} \setminus [1, \infty)$*, we have*

$${}_2F_1(\alpha, \beta; \gamma; z) = (1 - z)^{\gamma - \alpha - \beta} {}_2F_1(\gamma - \alpha, \gamma - \beta; \gamma; z).$$

(iii) *If* $\gamma = \alpha + \beta \notin -\mathbb{N}_0$*, then*

$$\lim_{t \to 1-} \frac{{}_2F_1(\alpha, \beta; \gamma; t)}{-\log(1 - t)} = \frac{\Gamma(\alpha + \beta)}{\Gamma(\alpha)\Gamma(\beta)} = \frac{\Gamma(\gamma)}{\Gamma(\alpha)\Gamma(\beta)}.$$

Proof

(i) is [47, p. 244 (9.3.4)].
(ii) is [47, p. 248]. Here we use that $\arg(1 - z) < \pi$ if and only if $z \in \mathbb{C} \setminus [1, \infty)$ (see also [21, (15.4.20,15.4.22]).
(iii) We use [21, (15.4.21)] with $\gamma = \alpha + \beta$.

□

If $\mathrm{Re}(\gamma - \alpha - \beta) < 0$, then we can use the hypergeometric identity (ii) to evaluate limits as in (i) because

$$\gamma - (\gamma - \alpha) - (\gamma - \beta) = -\gamma + \alpha + \beta$$

has positive real part.

5.1.2 Spherical Functions

We start with a discussion of spherical functions of real rank-one groups (see the introduction to this section for notation). Following the notation in Ólafsson/Pasquale [68, p. 1158] we fix a positive root α such that α and possibly $\alpha/2$ are the only positive roots and normalize $h \in \mathfrak{a}$ by $\alpha(h) = 1$ and identify $\mathfrak{a}_{\mathbb{C}}$ and $\mathbb{C}$ using the isomorphism $x \mapsto \alpha(x)$ with inverse $z \mapsto zh$.

Let

$$m_\alpha := \dim \mathfrak{g}_\alpha, \quad m_{\alpha/2} := \dim \mathfrak{g}_{\alpha/2} \quad \text{and} \quad \rho := \frac{1}{2}\left(m_\alpha + \frac{1}{2}m_{\alpha/2}\right).$$

We also let $a_t = \exp th$ and consider a spherical function

$$\varphi(g) = \langle v, U(g)v\rangle$$

for an irreducible unitary representation and a K-fixed unit vector v. According to [68, p. 1158] and Helgason [41, Ch. IV, Ex. B.8], there exists a $\lambda \in \mathbb{C}$ such that

$$\varphi(a_t) = \varphi_\lambda(a_t) := {}_2F_1\Big(\rho + \lambda, \rho - \lambda; \frac{m_{\alpha/2} + m_\alpha + 1}{2}; -\sinh^2(t/2)\Big). \tag{5.5}$$

By (5.4), this function is constant 1 if $\lambda = \pm\rho$.

The result of Kostant stated as follows (see Flensted-Jensen/Koornwinder [24, Thm. 1] for a rather direct proof) characterizes the values of λ that occur in our context, i.e., those for which the corresponding spherical function is positive definite.

Theorem 11 (Kostant's Characterization Theorem) *Suppose that $\Sigma = \{\alpha\}$ or $\{\alpha, \alpha/2\}$. Let*

$$s_0 := \begin{cases} \rho = \frac{1}{2}\left(m_\alpha + \frac{1}{2}m_{\alpha/2}\right) & \text{if } m_{\alpha/2} = 0 \\ \frac{1}{2}(1 + \frac{m_{\alpha/2}}{2}) \leq \rho & \text{if } m_{\alpha/2} > 0. \end{cases}$$

Then the spherical function φ_λ is positive definite if and only if

$$\lambda \in i\mathbb{R} \cup ([-s_0, s_0] \cup \{\pm\rho\}).$$

Note that $\varphi_{\pm\rho} \equiv 1$ by (5.5), and this function is trivially positive definite.

As they appear in (5.5), we recall in the following table the root multiplicities for the real rank-one simple Lie algebras (cf. Helgason [40]):

$\mathfrak{g}$	$\mathfrak{so}_{1,n}(\mathbb{R})$	$\mathfrak{su}_{1,n}(\mathbb{C})$	$\mathfrak{u}_{1,n}(\mathbb{H})$	$\mathfrak{f}_{4(-20)}$
m_α	$n-1$	1	3	7
$m_{\alpha/2}$	0	$2(n-1)$	$4(n-1)$	8

It follows in particular that $m_\alpha = 1$ occurs only for $\mathfrak{g} = \mathfrak{su}_{1,n}(\mathbb{C})$. Furthermore, the only real rank-one Lie algebras containing an Euler element are the Lie algebras

$\mathfrak{so}_{1,n}(\mathbb{R})$ (see Table 3 in Morinelli/Neeb/Ólafsson [54, §4]). Note that $\mathfrak{su}_{1,1}(\mathbb{C}) \cong \mathfrak{so}_{1,2}(\mathbb{R})$.

Theorem 12 *The following assertions hold for $\lambda \neq \pm\rho$:*

(1) *If $m_\alpha > 1$, then*

$$\lim_{t \to \pi-} \cos(t/2)^{m_\alpha - 1} \cdot \varphi_\lambda(a_{it}) = 2^{\frac{m_\alpha - 1}{2}} \frac{\Gamma\left(\frac{m_{\alpha/2}+m_\alpha+1}{2}\right)\Gamma\left(\frac{m_\alpha-1}{2}\right)}{\Gamma(\rho-\lambda)\Gamma(\rho+\lambda)}.$$

(2) *If $m_\alpha = 1$, then*

$$\lim_{t \to \pi-} \frac{\varphi_\lambda(a_{it})}{-\log(\pi - t)} = \frac{2 \cdot \Gamma\left(1 + \frac{m_{\alpha/2}}{2}\right)}{\Gamma(\rho-\lambda)\Gamma(\rho+\lambda)}.$$

Proof We put

$$a := \frac{1}{4}(2 + m_{\alpha/2}), \quad b := \frac{1}{2}(m_\alpha - 1) \quad \text{and} \quad c := \frac{1}{2}(m_{\alpha/2} + m_\alpha + 1).$$

Using that

$$c - (\rho + \lambda) - (\rho - \lambda) = c - 2\rho = \frac{1 - m_\alpha}{2} = -b, \quad \text{and}$$

$$c - (\rho \pm \lambda) = \frac{m_{\alpha/2} + m_\alpha + 1}{2} - (\rho \pm \lambda) = \frac{2 + m_{\alpha/2}}{4} \mp \lambda = a \mp \lambda,$$

we get with (5.5) and Theorem 10(ii):

$$\begin{aligned}\varphi_\lambda(a_{it}) &= {}_2F_1\Big(\rho + \lambda, \rho - \lambda; c; \sin^2(t/2)\Big) \\ &= 2^b \cos^{-2b}(t/2) \cdot {}_2F_1\big(a - \lambda, a + \lambda; c; \sin^2(t/2)\big).\end{aligned} \tag{5.6}$$

Now (1) follows from (5.6) and Theorem 10(i) because

$$c - a = \frac{1}{2}(m_{\alpha/2} + m_\alpha + 1) - \frac{1}{4}(2 + m_{\alpha/2}) = \frac{m_{\alpha/2} + 2m_\alpha}{4}$$

and

$$c - (a - \lambda) - (a + \lambda) = c - 2a = \frac{m_\alpha - 1}{2}.$$

Note that, by Theorem 11, we have for $\lambda \in \mathbb{R}$ that

$$|\lambda| \leq \rho = \frac{1}{4}(2m_\alpha + m_{\alpha/2})$$

with equality only for $\lambda = \pm\rho$. So the Γ-functions in the denominator are singular only for $\lambda = \pm\rho$, and in this case $\varphi_{\pm\rho} \equiv 1$.
For (2), we observe that $m_\alpha = 1$ implies that $\rho = \frac{1}{2} + \frac{m_{\alpha/2}}{4}$ and $c = 2\rho$. So we have

$$\lim_{t\to\pi-} \frac{\varphi_\lambda(a_{it})}{-\log(\pi - t)} = \lim_{t\to\pi-} \frac{{}_2F_1\big(\rho+\lambda, \rho-\lambda; c; \sin^2(t/2)\big)}{-\log(\pi - t)}.$$

Now (2) follows from Theorem 10(iii) and the fact that

$$\lim_{t\to\pi-} \frac{-\log(1-\sin^2(t/2))}{-\log(\pi - t)} = \lim_{t\to\pi-} \frac{\log(\cos^2(t/2))}{\log(\pi - t)}$$
$$= \lim_{t\to\pi-} \frac{\frac{1}{\cos^2(t/2)}(-\cos(t/2)\sin(t/2))}{\frac{-1}{\pi - t}}$$
$$= \lim_{t\to\pi-} \frac{\pi - t}{\cos(t/2)} = \lim_{t\to\pi-} \frac{-1}{-\sin(t/2)\frac{1}{2}} = 2.$$

□

5.1.3 χ-Spherical Functions

Now we turn to χ-spherical functions where χ is non-trivial, so that $\mathfrak{g} \cong \mathfrak{su}_{1,n}(\mathbb{C})$. This case was treated by Shimeno [69]. The general case for $\chi \neq 1$ and G/K a bounded symmetric domain was discussed in Heckman/Schlichtkrull [39]. We use the normalization from [69, pp. 383] and set

$$h := \frac{1}{2}\begin{pmatrix} 0 & 0 & 1 \\ 0 & \mathbf{0}_{n-1} & 0 \\ 1 & 0 & 0 \end{pmatrix} \quad \text{and} \quad \tilde{z} = i\begin{pmatrix} 1 & 0 \\ 0 & -\frac{1}{n}\mathbf{1}_n \end{pmatrix}.$$

Then $\mathfrak{a} = \mathbb{R}h$ is maximal abelian in $\mathfrak{p}$ and $\mathfrak{z}(\mathfrak{k}) = \mathbb{R}\tilde{z}$ is the center of $\mathfrak{k}$. Here the normalization of α is that $\alpha(h) = 1$ and the positive roots are α and, if $n \geq 2$, also $\alpha/2$. The unitary characters of K are parametrized by

$$\chi_\ell(\exp t\tilde{z}) = e^{i\ell t}, \quad \ell \in \mathbb{R}.$$

In view of [69, p. 384], the χ_ℓ-spherical functions take on $a_t = \exp(th)$ the form by

$$\varphi_{\ell,\lambda}(a_t) = (\cosh t/2)^{-\ell} \cdot {}_2F_1\Big(\frac{n-\ell+\lambda}{2}, \frac{n-\ell-\lambda}{2}; n; -\sinh^2(t/2)\Big)$$

and hence

$$(\cos t/2)^{\ell}\varphi_{\ell,\lambda}(a_{it}) = {}_2F_1\Big(\frac{n-\ell+\lambda}{2}, \frac{n-\ell-\lambda}{2}; n; \sin^2(t/2)\Big)$$

Theorem 13 *Assume that $\ell \neq 0$. Then we have*

$$\lim_{t\to\pi-} \cos^{|\ell|}(t/2)\varphi_{\ell,\lambda}(\exp ith) = \frac{\Gamma(|\ell|)(n-1)!}{\Gamma(\frac{1}{2}(n+|\ell|-\lambda))\Gamma(\frac{1}{2}(n+|\ell|+\lambda))}.$$

Proof Assume first that $\ell > 0$. Then $\mathrm{Re}(\gamma-\alpha-\beta) = \ell > 0$ and the claim follows from Theorem 10(i). For $\ell < 0$ we use Theorem 10(ii) to write

$$\begin{aligned}
&{}_2F_1\left(\frac{1}{2}(n-\ell+\lambda), \frac{1}{2}(n-\ell-\lambda); n; \sin^2(t/2)\right)\\
&\quad = \cos^{2\ell}(t/2)\cdot {}_2F_1\left(\frac{1}{2}(n+\ell-\lambda), \frac{1}{2}(n+\ell+\lambda); n; \sin^2(t/2)\right)\\
&\quad = \cos^{-2|\ell|}(t/2)\cdot {}_2F_1\left(\frac{1}{2}(n-|\ell|-\lambda), \frac{1}{2}(n-|\ell|+\lambda); n; \sin^2(t/2)\right)
\end{aligned}$$

and the claim follows as above. □

Corollary 5 *Let $(U,\mathcal{H})$ be an irreducible unitary representation of the rank-one group G. Any K-eigenvector is contained in $\bigcap_{x\in\Omega_{\mathfrak{p}}} \mathcal{D}(e^{i\partial U(x)})$ and* Hypothesis (H2) *is satisfied.*

Proof The first assertion follows from

$$\mathcal{D}(e^{i\partial U(\mathrm{Ad}(k)x)}) = U(k)\mathcal{D}(e^{i\partial U(x)}) \quad \text{for} \quad k\in K, x\in\mathfrak{p}$$

and $\Omega_{\mathfrak{p}} = \mathrm{Ad}(K)(-\pi/2,\pi/2)h$ for any Euler element $h\in\mathfrak{p}$. It follows in particular that, for each K-eigenvector v, the hyperfunction vector $\beta^+(v)\in\mathcal{H}^{-\omega}$ is defined. Further, the preceding discussion implies the existence of an $N>0$ such that

$$\sup_{|t|<\pi/2}\Big(\frac{\pi}{2}-t\Big)^N \|e^{it\partial U(h)}v\| = \sup_{|t|<\pi/2}\Big(\frac{\pi}{2}-t\Big)^N \varphi(a_{2it})^{1/2} < \infty.$$

This in turn implies with Theorem 3 that $\beta^+(v)\in\mathcal{H}^{-\infty}(U_h)\subseteq\mathcal{H}^{-\infty}$, i.e., that Hypothesis (H2) is also satisfied. □

5.2 More General Asymptotics

We keep our assumption that $\mathfrak{g}$ is of real rank-one. Let $v \in \mathcal{H}$ be a K-eigenvector fixed by J for which

$$q_N(v) := \sup_{|t|<\pi/2} \Big(\frac{\pi}{2} - t\Big)^N \|e^{it\partial U(h)}v\| < \infty \tag{5.7}$$

(Corollary 5). We then consider $\mathcal{E} = \mathbb{C}v$ and $\mathrm{E}_K = \mathbb{R}v$.

Lemma 9 *The functions*

$$\ell([g, ix]) := \|U(g)e^{i\partial U(x)}v\| = \|e^{i\partial U(x)}v\| \quad \textit{on} \quad \Xi \cong G \times_K i\Omega_{\mathfrak{p}}$$

and

$$\delta \colon \Xi \to \Big[0, \frac{\pi}{2}\Big), \quad \delta([g, ix]) = \|\operatorname{ad} x\| \tag{5.8}$$

satisfy

$$\ell(z) \leq q_N(v)\Big(\frac{\pi}{2} - \delta(z)\Big)^{-N} \quad \textit{for} \quad z \in \Xi. \tag{5.9}$$

Proof Clearly, both functions ℓ and δ on Ξ are G-invariant, hence defines an $\mathrm{Ad}(K)$-invariant function on $i\Omega_{\mathfrak{p}}$ and thus are determined by their values on elements of the form $[e, ith]$, $0 \leq t < \frac{\pi}{2}$. By (5.7), we have on multiples of h the estimate

$$\ell([e, ith]) \leq q_N(v)\Big(\frac{\pi}{2} - t\Big)^{-N} = q_N(v)\Big(\frac{\pi}{2} - \delta([e, ith]\Big)^{-N} \quad \text{for} \quad 0 \leq t < \frac{\pi}{2}, \tag{5.10}$$

so that the lemma follows by G-invariance. □

For $\mathfrak{g} = \mathfrak{so}_{1,n}(\mathbb{R})$, we consider the Euler elements defined by

$$h(x_0, \ldots, x_n) = (x_1, x_0, 0, \ldots, 0) \quad \text{and} \quad h_n(x_0, \ldots, x_n) = (x_n, 0, \ldots, 0, x_0),$$

so that the corresponding involution acts on $\mathbb{C}^{1+n}$ by

$$\tau_h(z_0, z_1, \ldots, z_n) = (-z_0, -z_1, z_2, \ldots, z_n).$$

Its action on $\mathfrak{g}$ satisfies $\tau_h(h) = h$ and $\tau_h(h_n) = -h_n$, so that $h \in \mathfrak{q}_{\mathfrak{p}}$ and $h_n \in \mathfrak{h}_{\mathfrak{p}}$.

On $\Xi = G \times_K i\Omega_{\mathfrak{p}}$, the antiholomorphic involution $\overline{\tau}_h([g, x]) = [\tau_h(g), -\tau_h(x)]$ has the fixed point set

$$\Xi^{\overline{\tau}_h} = G^h_e.\mathrm{Exp}(i\Omega_{\mathfrak{p}}^{-\tau_h}) = G^h_e.\mathrm{Exp}(i\Omega_{\mathfrak{h}_{\mathfrak{p}}})$$

(see Morinelli/Neeb/Ólafsson [55, §8] and the context of Neeb/Ólafsson [64, Thm. 6.1]). We have

$$\Omega_{\mathfrak{h}_{\mathfrak{p}}} = e^{\operatorname{ad} \mathfrak{h}_{\mathfrak{k}}} \Big\{ s h_n \colon |s| < \frac{\pi}{2} \Big\}.$$

Assume that $|s| < \frac{\pi}{2}$ and consider the curve

$$\gamma \colon \Big(-\frac{\pi}{2}, \frac{\pi}{2} \Big) \to \Xi, \quad \gamma(t) := \alpha_{it} \operatorname{Exp}(ish_n) \in \Xi.$$

To relate the functions ℓ and δ on the curve γ, we use a concrete model for $G/K \cong \mathbb{H}^n$, $G/H \cong \mathrm{dS}^n$ and $G_{\mathbb{C}}/K_{\mathbb{C}} \cong \mathbb{S}^n_{\mathbb{C}} \subseteq \mathbb{C}^{n+1}$. For $G := \mathrm{SO}_{1,n}(\mathbb{R})_e$, we have

$$G/K \cong \mathbb{H}^n := G.i\mathbf{e}_0 \subseteq i\mathbb{R}^{1+n}$$

and

$$G/H \cong \mathrm{dS}^n := G.\mathbf{e}_n = \{(x_0, \mathbf{x}) \colon x_0^2 - \mathbf{x}^2 = -1\} \subseteq \mathbb{R}^{1,n}.$$

We then have

$$\Xi = \mathbb{S}^n_{\mathbb{C}} \cap (\mathbb{R}^{1,n} + iV_+) \quad \text{for} \quad V_+ = \{(x_0, \mathbf{x}) \colon x_0 > 0, x_0^2 > \mathbf{x}^2\}. \tag{5.11}$$

Note that

$$\overline{\tau}_h(z_0, z_1, \ldots, z_n) = (-\overline{z}_0, -\overline{z}_1, \overline{z}_2, \ldots, \overline{z}_n)$$

is an antiholomorphic involution fixing $i\mathbf{e}_0 \in \mathbb{H}^n$ and $\mathbf{e}_n \in \mathrm{dS}^n$, but it maps $\mathbf{e}_1$ to $-\mathbf{e}_1$. We have

$$(\mathbb{C}^{1+n})^{\overline{\tau}_h} = i\mathbb{R}\mathbf{e}_0 + i\mathbb{R}\mathbf{e}_1 + \mathbb{R}\mathbf{e}_2 + \cdots + \mathbb{R}\mathbf{e}_n.$$

Now

$$\alpha_{it}(z_0, \ldots, z_n) = (\cos t \cdot z_0 + i \sin t \cdot z_1, i \sin t \cdot z_0 + \cos t \cdot z_1, z_2, \ldots, z_n).$$

For $x \in \mathbb{R}^{1+n}$, we therefore get

$$\operatorname{Im}((\alpha_{it} x)_0) = \sin t \cdot x_1 > 0 \quad \text{for} \quad 0 < t < \pi, \quad x \in W = \{x \in \mathbb{R}^{1,n} \colon x_1 > |x_0|\},$$

so that $\alpha_{it} x \in \Xi$. We also see that

$$\alpha_{\frac{\pi i}{2}} x = (ix_1, ix_0, x_2, \ldots, x_n) \in \Xi^{\overline{\tau}_h},$$

where

$$\Xi^{\overline{\tau}_h} = \{(ix_0, ix_1, x_2, \dots, x_n) \colon x_0 > |x_1|, -x_0^2 + x_1^2 - x_2^2 - \dots - x_n^2 = -1\}$$
$$= \{(ix_0, ix_1, x_2, \dots, x_n) \colon x_0 > |x_1|, \underbrace{x_0^2 - x_1^2}_{>0} + x_2^2 + \dots + x_n^2 = 1\}.$$

It follows in particular that

$$x_0^2 - x_1^2 \in (0, 1].$$

Lemma 10 *The G-invariant function δ on $\Xi = \mathbb{S}^n_{\mathbb{C}} \cap (\mathbb{R}^{1,n} + iV_+)$ is given by*

$$\delta(z) = \arccos(\sqrt{\beta(\operatorname{Im} z)}) = \arccos(\sqrt{1 + \beta(\operatorname{Re} z)}), \tag{5.12}$$

where $\beta(x) := x_0^2 - x_1^2 - \dots - x_n^2$.

Proof Since G-orbits in V_+ (cf. (5.11)) intersect $\mathbb{R}_+\mathbf{e}_0$, it suffices to evaluate on elements $z \in \Xi$ with $\operatorname{Im} z \in \mathbb{R}_+\mathbf{e}_0$. With the stabilizer $\mathrm{SO}_n(\mathbb{R})$ of $\mathbf{e}_0$ we move $\operatorname{Re} z$ into $\mathbb{R}\mathbf{e}_0 + [0, \infty]\mathbf{e}_n$, so that

$$z = (z_0, 0, \cdots, 0, z_n) \quad \text{with} \quad \operatorname{Im} z_0 \in i\mathbb{R}_+\mathbf{e}_0, \quad z_n \geq 0, \quad -z_0^2 + z_n^2 = 1.$$

It follows that $z_0^2 = z_n^2 - 1 \in \mathbb{R}$, so that $z_0 \in i\mathbb{R}$. Now $z = (ai, 0, \dots, 0, b)$ with $a, b \in \mathbb{R}$ and $a^2 + b^2 = 1$. For the Euler element h_n, we have

$$\exp(sih_n).i\mathbf{e}_0 = \begin{pmatrix} \cos s & 0 & i \sin s \\ 0 & \mathbf{1} & 0 \\ i \sin s & 0 & \cos s \end{pmatrix} i\mathbf{e}_0 = i \cos s \cdot \mathbf{e}_0 - \sin s \cdot \mathbf{e}_n. \tag{5.13}$$

For $a = \cos s$ and $b = -\sin s$, we thus obtain for $z = \exp(ish_n).i\mathbf{e}_0$, which corresponds to $[e, ish_n] \in \Xi$, the relation

$$\delta(z) = |s| = \arccos(\operatorname{Im} z_0) = \arccos\left(\sqrt{1 - z_n^2}\right). \tag{5.14}$$

This leads to the identity (5.12). □

Starting with a $\overline{\tau}_h$-fixed element $z = (ix_0, ix_1, x_2, \dots, x_n)$ in Ξ, this leads to

$$\alpha_{it}(ix_0, ix_1, x_2, \dots, x_n) = (\cos t \cdot ix_0 - \sin t \cdot x_1, -\sin t \cdot x_0 + \cos t \cdot ix_1, x_2, \dots, x_n)$$

with imaginary part

$$(x_0 \cos t, x_1 \cos t, 0, \dots, 0).$$

We thus obtain for $0 \leq t < \frac{\pi}{2}$:

$$\delta(\alpha_{it}z) = \arccos\Big(\cos t \cdot \sqrt{x_0^2 - x_1^2}\Big) \in \Big[\arccos\Big(\sqrt{x_0^2 - x_1^2}\Big), \frac{\pi}{2}\Big). \tag{5.15}$$

We need the asymptotics of this function for $t \to \frac{\pi}{2}$. So let

$$\lambda := \sqrt{x_0^2 - x_1^2} \in (0, 1]$$

and consider the function

$$g(t) := \delta(\alpha_{it}z) = \arccos(\lambda \cos t) \quad \text{for} \quad 0 \leq t < \frac{\pi}{2}.$$

Then g extends smoothly to $(0, \pi)$ with

$$g'\Big(\frac{\pi}{2}\Big) = \arccos'(0)\lambda \cos'\Big(\frac{\pi}{2}\Big) = \frac{1}{\cos'(\frac{\pi}{2})}\lambda \cos'\Big(\frac{\pi}{2}\Big) = \lambda.$$

We conclude that, for $t \to \frac{\pi}{2}$, we have

$$\lambda\Big|\frac{\pi}{2} - t\Big| \sim \Big|\frac{\pi}{2} - g(t)\Big| = \frac{\pi}{2} - \delta(\alpha_{it}z). \tag{5.16}$$

For $z = \exp(sih_n).i\mathbf{e}_0$ as above, (5.13) shows that $\lambda = \cos(s)$.

Lemma 11 *$q_N(e^{i\partial U(x)}v) < \infty$ for $x \in \Omega_{\mathfrak{h}_\mathfrak{p}}$.*

Proof Since q_N is invariant under $U(H_K)$ and $\mathfrak{q}_\mathfrak{p} = e^{\mathrm{ad}\,\mathfrak{h}_\mathfrak{k}}(\mathbb{R}h_n)$, it suffices to show that

$$q_N(e^{is\partial U(h_n)}v) < \infty \quad \text{for} \quad |s| < \pi/2,$$

which means that

$$t \mapsto \Big(\frac{\pi}{2} - t\Big)^N \|e^{it\partial U(h)}e^{is\partial U(h_n)}v\| = \Big(\frac{\pi}{2} - t\Big)^N \ell(\alpha_{it}\,\mathrm{Exp}(ish_n)) \tag{5.17}$$

is bounded on the interval $|t| < \frac{\pi}{2}$. In view of (5.9), it suffices to show for $z_s := \mathrm{Exp}(ish_n)$ that

$$\Big(\frac{\pi}{2} - t\Big)^N \Big(\frac{\pi}{2} - \delta(\alpha_{it}z_s)\Big)^{-N}$$

is bounded, which follows immediately from (5.16). □

In view of Theorem 3, Lemma 11 implies for $\mathtt{E}_K = \mathbb{R}v$ that

$$\beta^+(e^{i\partial U(\Omega_{\mathfrak{h}_\mathfrak{p}})}\mathtt{E}_K) \subseteq \mathcal{H}^{-\infty}(U_h)_{\mathrm{KMS}}.$$

In particular, Hypothesis (H2) is satisfied. Next we observe that $U^{-\omega}(G_e^h)$ commutes with J and U_h, hence leaves $\mathcal{H}^{-\infty}(U_h)_{\mathrm{KMS}}$ invariant. We thus obtain with $W^G = G_e^h \exp(\Omega_{\mathfrak{q}_\mathfrak{k}})H$ (see (8.2) in Appendix 2) and Lemma 7(c):

$$\begin{aligned} U^{-\infty}(W^G)\mathtt{E}_H &= U^{-\infty}(G_e^h)e^{\partial U(\Omega_{\mathfrak{q}_\mathfrak{k}})}\mathtt{E}_H = U^{-\infty}(G_e^h)\beta^+(e^{\partial U(i\Omega_{\mathfrak{h}_\mathfrak{p}})}\mathtt{E}_K) \\ &\subseteq U^{-\infty}(G_e^h)\mathcal{H}^{-\infty}(U_h)_{\mathrm{KMS}} = \mathcal{H}^{-\infty}(U_h)_{\mathrm{KMS}} \subseteq \mathcal{H}^{-\infty}_{\mathrm{KMS}}. \end{aligned}$$

Theorem 14 *Let G be a connected Lie group with Lie algebra $\mathfrak{g} = \mathfrak{sl}_2(\mathbb{R})$ and $(U, \mathcal{H})$ be an irreducible unitary representation of G. Then U extends to an antiunitary representation of $G_{\tau_h} = G \rtimes \{\mathbf{1}, \tau_h\}$. Let $v \in \mathcal{H}$ be a J-fixed K-eigenvector, $\mathcal{E} := \mathbb{C}v$ and $\mathtt{E}_K := \mathbb{R}v$. Then* Hypothesis (H2) *is satisfied, so that $\mathtt{E}_H = \beta^+(\mathtt{E}_K) \subseteq \mathcal{H}^{-\infty}$. Further, the net $\mathsf{H}^M_{\mathtt{E}_H}$ on the non-compactly causal symmetric space $M = G/H$ satisfies* (Iso), (Cov), (RS) *and* (BW)*, where $W = W_M^+(h)_{eH}$ is the connected component of the positivity domain of h on M, containing the base point.*

Proof First, we derive the existence of the antiunitary extension of U from Morinelli/Neeb [52, Thm. 4.24]. As all K-types in $\mathcal{H}$ are 1-dimensional, they consist of eigenvectors. Let $\mathcal{H}_\chi = \mathbb{C}v_\chi$ be the eigenspace corresponding to the character $\chi \in \widehat{K}$. As $\tau_h(k) = k^{-1}$ for $k \in K$, all K-eigenspaces are J-invariant, hence they all contain J-fixed vectors. The preceding discussion shows that Hypothesis (H2) holds for $\mathcal{E} = \mathbb{C}v$, so that the theorem follows from Theorem 8. □

Remark 16 (The Possibilities for H) For $m \in \mathbb{N} \cup \{\infty\}$, let G_m be a connected Lie group with Lie algebra $\mathfrak{g} = \mathfrak{sl}_2(\mathbb{R})$ and $|Z(G_m)| = m$. For $m \in \mathbb{N}$ this means that $Z(G_m) \cong \mathbb{Z}/m\mathbb{Z}$ and G_m is an m-fold covering of $\mathrm{Ad}(G_m) \cong \mathrm{PSL}_2(\mathbb{R}) \cong G_1$. Note that $G_2 \cong \mathrm{SL}_2(\mathbb{R})$. Further $G_\infty \cong \widetilde{\mathrm{SL}}_2(\mathbb{R})$ is simply connected with $Z(G_\infty) \cong \mathbb{Z}$.

We consider the Cartan involution $\theta(x) = -x^\top$, the Euler element

$$h = \frac{1}{2}\begin{pmatrix} 1 & 0 \\ 0 & -1 \end{pmatrix} \quad \text{and} \quad z = \begin{pmatrix} 0 & -\pi \\ \pi & 0 \end{pmatrix} \in \mathfrak{k} = \mathfrak{so}_2(\mathbb{R}),$$

which satisfies $e^z = -\mathbf{1}$. Then

$$K = \exp(\mathbb{R}z), \quad Z(G_m) = \exp(\mathbb{Z}z), \quad \text{and} \quad \tau_h(\exp tz) = \tau(\exp tz) = \exp(-tz)$$

because $\tau = \theta\tau_h$. We conclude that

$$K^\tau = \{e\} \quad \text{if} \quad m = \infty \quad \text{and} \quad K^\tau = \{e, \exp(mx/2)\} \quad \text{otherwise.}$$

For $m = \infty$, $H = G_m^\tau$ is connected. For $m \in \mathbb{N}$, the group $G_m^\tau = K^\tau \exp(\mathfrak{h})$ has two connected components, but if m is odd, then K^τ does not fix the Euler element $h \in C^\circ$. Therefore only $H := \exp(\mathfrak{h})$ leads to a causal symmetric space G_m/H. If m is even, then H can be either $(G_m)_e^\tau$ or G_m^τ.

In $G_1 \cong \mathrm{PSL}_2(\mathbb{R})$, the subgroup H corresponds to $\mathrm{SO}_{1,1}(\mathbb{R})_e$ and the noncompactly causal symmetric space $G_1/H \cong \mathrm{dS}^2$ is the 2-dimensional de Sitter space.

The universal covering $\widetilde{\mathrm{dS}}^2$ is obtained for $m = \infty$, $G_\infty = \widetilde{\mathrm{SL}}_2(\mathbb{R})$ and then $H = \exp(\mathfrak{h})$ is connected. All other coverings of dS^2 are obtained as G_m/H for $H = \exp(\mathfrak{h})$.

Remark 17 If $\mathfrak{g} = \mathfrak{sl}_2(\mathbb{R})$, then $G_{\mathrm{ad}} = \mathrm{PSL}_2(\mathbb{R}) \cong \mathrm{SO}_{1,2}(\mathbb{R})_e$ (the Möbius group), and $H_{\mathrm{ad}} = \exp(\mathbb{R}h)$, then $G_{\mathrm{ad}}/H_{\mathrm{ad}} \cong \mathrm{dS}^2$ and it follows easily that $W_{M_{\mathrm{ad}}}^+(h)$ is connected; see Morinelli/Neeb/Ólafsson [55, Thm. 7.1] for a generalization of this observation to non-compactly causal spaces. Therefore $W = W_{M_{\mathrm{ad}}}^+(h)$ in the preceding theorem. If $Z(G)$ is nontrivial, then the connected components of $W_M^+(h)$ can be labeled by the elements of $Z(G)$ because this subgroup acts non-trivially on $M = G/H$, leaving the positivity region $W_M^+(h)$ invariant. In any irreducible representation $(U, \mathcal{H})$ we have $U(Z(G)) \subseteq \mathbb{T}$, but this subgroup preserves the standard subspace $\mathtt{V}$ if and only if it is contained in $\{\pm 1\}$.

5.3 *An Application to Intersections of Standard Subspaces*

In $\mathfrak{sl}_2(\mathbb{R})$, we consider the two Euler elements

$$h := \frac{1}{2}\begin{pmatrix} 1 & 0 \\ 0 & -1 \end{pmatrix} \quad \text{and} \quad h_1 := \frac{1}{2}\begin{pmatrix} 0 & 1 \\ 1 & 0 \end{pmatrix} \in \mathfrak{h} = \mathfrak{so}_{1,1}(\mathbb{R}). \tag{5.18}$$

Let $(U, \mathcal{H})$ be a unitary representation of group $G := \widetilde{\mathrm{SL}}_2(\mathbb{R})$. By Theorem 14, U extends to an (anti-)unitary representation of $\widetilde{\mathrm{GL}}_2(\mathbb{R}) := \widetilde{\mathrm{SL}}_2(\mathbb{R}) \rtimes \{\mathbf{1}, \tau_h\}$, and we put

$$J := U(\tau_h), \quad \Delta := e^{2\pi i \cdot \partial U(h)} \quad \text{and} \quad \mathtt{V} := \mathrm{Fix}(J\Delta^{1/2}).$$

The following proposition solves the SL_2-problem from Morinelli/Neeb [52, §4] in the affirmative. From Guido/Longo [34, Thm. 1.1, Cor. 1.3(c)], one can deduce this result for principal series representations and lowest and highest weight representations, but it was not known to hold for the complementary series.

Theorem 15 *Let $(U, \mathcal{H})$ be an irreducible unitary representation of $\widetilde{\mathrm{SL}}_2(\mathbb{R})$. For every $t \in \mathbb{R}$, the intersection $U(\exp th_1)\mathtt{V} \cap \mathtt{V}$ is a standard subspace.*

Proof We pick a J-fixed K-eigenvector $v \in \mathcal{H}$ and define $\mathtt{E}_H := \beta^+(v)$ as above. By Theorem 14 we then obtain a net $\mathsf{H}_{\mathtt{E}_H}^M(\mathcal{O})$ of real subspace indexed by open subsets $\mathcal{O} \subseteq M = G/H$. Moreover,

$$\mathtt{V} = \mathsf{H}^M_{\mathrm{E}_H}(W)$$

holds for the basic connected component $W = W^+_M(h)_{eH}$ of the positivity domain of the modular vector field on M. For each $t \in \mathbb{R}$, the intersection $W_t := W \cap \exp(th_1)W$ is non-empty because $\exp(th_1) \in H$ fixes the base point in G/H. Therefore,

$$\begin{aligned} U(\exp th_1)\mathtt{V} \cap \mathtt{V} &= U(\exp th_1)\mathsf{H}^M_{\mathrm{E}_H}(W) \cap \mathsf{H}^M_{\mathrm{E}_H}(W) \\ &= \mathsf{H}^M_{\mathrm{E}_H}(\exp(th_1)W) \cap \mathsf{H}^M_{\mathrm{E}_H}(W) \supseteq \mathsf{H}^M_{\mathrm{E}_H}(W_t). \end{aligned}$$

As $\mathsf{H}^M_{\mathrm{E}_H}(W_t)$ is cyclic by the Reeh–Schlieder property (Theorem 14), it follows that $U(\exp th_1)\mathtt{V} \cap \mathtt{V}$ is cyclic, hence standard because it is contained in the standard subspace $\mathtt{V}$. □

5.4 *Positive Energy Representations of* $\mathrm{PSL}_2(\mathbb{R})$

The nontrivial positive energy representations $(U_s, \mathcal{H}_s)$, $s = 2, 4, 6, \ldots$ of the Möbius group $G = \mathrm{PSL}_2(\mathbb{R})$ can be realized on the reproducing kernel Hilbert spaces $\mathcal{H}_s \subseteq \mathcal{O}(\mathbb{C}_+)$ on the upper half plane with the reproducing kernel

$$Q(z, w) = \Big(\frac{z - \overline{w}}{2i}\Big)^{-s}, \tag{5.19}$$

for which positive definiteness follows from the fact that the function $\big(\frac{z}{i}\big)^{-s}$ is the Laplace transform of the Riesz measure

$$d\mu_s(p) = \Gamma(s)^{-1} p^{s-1}\, dp \quad \text{on} \quad (0, \infty),$$

see Neeb/Ólafsson/Ørsted [66, §6]. Point evaluations on $\mathcal{H}_s$ take the form

$$f(w) = \langle Q_w, f\rangle \quad \text{with} \quad Q_w(z) = Q(z, w) = \Big(\frac{z - \overline{w}}{2i}\Big)^{-s} \quad \text{for} \quad z, w \in \mathbb{C}_+.$$

Note that the Fourier transform defines a unitary isomorphism

$$\mathcal{F}\colon L^2(\mathbb{R}_+, \mu_s) \to \mathcal{H}_s \subseteq \mathcal{O}(\mathbb{C}_+), \quad \mathcal{F}(f)(z) := 2^s \int_0^\infty e^{izp} f(p)\, d\mu_s(p)$$

[66, Lemma 3.10] which extends to an isomorphism

$$\mathcal{F}\colon L^2(\mathbb{R}_+, \mu_s)^{-\infty} \to \mathcal{H}_s^{-\infty} \subseteq \mathcal{O}(\mathbb{C}_+)$$

and this exhibits the image of $\mathcal{H}_s^{-\infty}$ under the boundary values on $\mathbb{R}$ as a space of tempered distributions on $\mathbb{R}$ (cf. [66, Rem. 3.7]). The representation U_s (lifted to $\mathrm{SL}_2(\mathbb{R})$) is given by

$$(U_s(g)f)(z) := (a - cz)^{-s} f(g^{-1}.z) = (a - cz)^{-s} f\Big(\frac{dz - b}{a - cz}\Big)$$

for

$$g = \begin{pmatrix} a & b \\ c & d \end{pmatrix}, \quad g.z = \frac{az + b}{cz + d}.$$

Identifying $g \in G$ with the corresponding Möbius transformation, the derivative of g is $g'(z) = (cz + d)^{-2}$, so that we obtain

$$(U_s(g^{-1})f)(z) := g'(z)^{s/2} f(g.z) \quad \text{for} \quad g \in G, z \in \mathbb{C}_+$$

and in particular

$$U_s(g)Q_z = \overline{g'(z)^{s/2}} Q_{g.z} \quad \text{for} \quad g \in G, z \in \mathbb{C}_+. \tag{5.20}$$

Since the $(Q_z)_{z \in \mathbb{C}_+}$ are analytic vectors, we obtain an embedding

$$\iota\colon \mathcal{H}_s^{-\omega} \hookrightarrow \mathcal{O}(\mathbb{C}_+), \quad \iota(\alpha)(z) := \alpha(Q_z),$$

extending the inclusion $\mathcal{H}_s \hookrightarrow \mathcal{O}(\mathbb{C}_+)$ in a G-equivariant way. As we shall see further, the holomorphic functions $(Q_x)_{x \in \mathbb{R}}$ defined by $Q_x(z) = Q(z, x) = \big(\frac{z-x}{2i}\big)^{-s}$ are distribution vectors, and the action of G on these distribution vectors satisfies

$$U_s^{-\omega}(g)Q_x = g'(x)^{s/2} Q_{g.x} \quad \text{for} \quad g \in G, x, g.x \in \mathbb{R}. \tag{5.21}$$

It follows in particular that Q_x is an eigenvector for the action of the parabolic subgroup

$$G^x := \{g \in G\colon g.x = x\}.$$

We consider the Euler elements h and h_1 from (5.18) with $H = \exp(\mathbb{R}h_1)$, and extend U_s to an antiunitary representation of G_{τ_h} by setting $U_s(\tau_h) := J$, where J is the conjugation

$$(JF)(z) := e^{\pi is/2}\overline{F(-\overline{z})} = (-1)^{s/2}\overline{F(-\overline{z})}, \tag{5.22}$$

which satisfies $JU_s(g)J = U_s(\tau_h(g))$ for $g \in G$.

The subgroup $K = \exp(\mathfrak{so}_2(\mathbb{R})) = \mathrm{PSO}_2(\mathbb{R})$ fixes the point $i \in \mathbb{C}_+$ and by (5.20)

$$Q_i(z) = \Big(\frac{z+i}{2i}\Big)^{-s}$$

is a K-eigenfunction, so that we may put $\mathcal{E} := \mathbb{C}Q_i$. As $JQ_i = e^{\pi is/2}Q_i$, we obtain the fixed point space

$$\mathsf{E}_K = \mathcal{E}^J = \mathbb{R}e^{\pi is/4}Q_i.$$

To determine $\mathsf{E}_H \subseteq \mathcal{H}^{-\infty}$, we have to determine the limit of $e^{it\partial U_s(h)}Q_i$ for $t \to \pm\pi/2$. From $\exp(th).z = e^t h$, we get for $t \in \mathbb{R}$ with (5.20) that

$$U_s(\exp th)Q_i = e^{st/2}Q_{e^t i}.$$

Analytic continuation leads for $|t| < \pi/2$ to

$$e^{it\partial U(h)}Q_i = e^{ist/2}Q_{e^{it}i},$$

and for $t \to -\pi/2$ we get the limit $e^{-s\pi i/4}Q_1 = i^{-s/2}Q_1$. In particular, Corollary 5 implies that $Q_1 \in \mathcal{H}_s^{-\infty}$ and we find

$$\mathsf{E}_H = \mathbb{R}Q_1.$$

Since the orbit of Q_1 under the group of translations is $(Q_x)_{x\in\mathbb{R}}$, all these functions are distribution vectors.

We also note that, for $x > 0$, and $\Delta = e^{2\pi i\partial U(h)}$, we obtain in $\mathcal{H}_s^{-\infty}$ the relation

$$\Delta^{1/2}Q_x = (-1)^{s/2}Q_{-x} = JQ_x,$$

so that

$$Q_x \in \mathcal{H}_{\mathrm{KMS}}^{-\infty} \quad \text{for} \quad x > 0. \tag{5.23}$$

As the family $(Q_x)_{x>0}$ is the orbit of Q_1 under the modular group $\exp(\mathbb{R}h)$, acting with test functions on $\mathbb{R}_+$ generates $\mathtt{V}$ (Longo [48, Prop. 3.10]). Likewise $(Q_x)_{x<0}$ generates $\mathtt{V}' = J\mathtt{V}$ (here we use that $s \in 2\mathbb{N}$). It follows that

$$\begin{aligned}\mathtt{V}^\infty = \mathtt{V} \cap \mathcal{H}_s^\infty &= \{f \in \mathcal{H}_s^\infty \colon \langle f, \mathtt{V}' \cap \mathcal{H}_s^\infty \rangle \subseteq \mathbb{R}\} \\ &= \{f \in \mathcal{H}_s^\infty \colon (\forall x < 0)\ \langle Q_x, f \rangle = f(x) \in \mathbb{R}\}.\end{aligned}$$

We conclude that

$$\mathtt{V} = \{f \in \mathcal{H}_s \colon f(\mathbb{R}_-) \subseteq \mathbb{R}\}$$

in the sense of boundary values on $\mathbb{R}$ (cf. (3.42) in Neeb/Ólafsson/Ørsted [66]).

As Q_1 is an eigenvector of the parabolic subgroup $P_+ := G^1 \supseteq H$ (the stabilizer of $1 \in \mathbb{R}$), we obtain $\mathsf{H}^G_{\mathrm{E}_H}(O) = \mathsf{H}^G_{\mathrm{E}_H}(OP_+)$ for every open subset $O \subseteq G$ with Neeb/Ólafsson [63, Lemma 2.11], so that the net $\mathsf{H}^G_{\mathrm{E}_H}$ on G actually descends to a net on $G/P_+ \cong \mathbb{S}^1$.

Likewise, negative energy representations lead to nets on $\mathbb{S}^1 \cong G/P_-$ for $P_- := G^{-1}$ (the stabizer of $-1 \in \mathbb{R}$). As $P_+ \cap P_- = H$, we obtain an equivariant embedding

$$\mathrm{dS}^2 \cong G/H \hookrightarrow G/P_+ \times G/P_- \cong \mathbb{T}^2.$$

The aforementioned construction thus leads to nets on dS^2 that can be reconstructed from their pushforwards under the projections $p_\pm \colon \mathrm{dS}^2 \to \mathbb{S}^1$.

Remark 18 We find a similar situation whenever $\mathfrak{g}$ is a simple hermitian Lie algebra, i.e., unitary highest weight representations (positive energy representations in the physics terminology) exist. Then the existence of an Euler element in $\mathfrak{g}$ implies that it is of tube type (Morinelli/Neeb [52, Prop. 3.11(b)]) and G/H is a causal symmetric space of Cayley type that embeds into a product

$$p = (p_+, p_-) \colon G/H \to G/P_+ \times G/P_-,$$

where $G/P_\pm$ are causal homogeneous spaces (compactifications of euclidean Jordan algebras), for which G is the conformal group (or some covering). As $G/P_+ \cong G/P_-$ carries two G-invariant causal structures, the above embedding leads to causal embeddings of G/H, endowed with a compactly causal structure, and also with a non-compactly causal structure (cf. Hilgert/Ólafsson [42]). For $\mathfrak{g} = \mathfrak{sl}_2(\mathbb{R})$, this corresponds to the embedding of 2-dimensional de Sitter space dS^2 (which is non-compactly causal) and also of the dual anti-de Sitter space AdS^2 into the product $G/P_+ \times G/P_-$. In Neeb/Ólafsson [63, §5.2] covariant nets on $G/P_\pm$ corresponding to positive energy representations are constructed along the lines indicated above for $\mathrm{PSL}_2(\mathbb{R})$.

For the conformal group $G = \mathrm{SO}_{2,d}(\mathbb{R})_e$ of compactified Minkowski space, this leads to nets of standard subspaces which yield by second quantization conformally covariant nets of operator algebras, as they are discussed in Guido/Longo [35]. It is an interesting question when, and to which extent, a net $\mathsf{H}^M_{\mathrm{E}_H}$ on M can be reconstructed from its pushforward nets $\mathsf{H}^{G/P_\pm}_{\mathrm{E}_H}$. On $G/P_\pm$, the results of [35] suggest "nice situations" for positive/negative energy representations of G and these properties can probably be characterized in terms of isotony requirements on the nets. In particular, we expect Morinelli/Neeb/Ólafsson [55] to provide the necessary geometric background to analyse nets in terms of KMS properties of geodesic observers in wedge regions.

6 Outlook: Beyond Linear Simple Lie Groups

Our main result Theorem 8 is formulated under the assumption of Hypothesis (H2), which can be derived for linear simple Lie groups from the Krötz–Stanton Extension Theorem (Theorem 4) and the Automatic Continuity Theorem (Theorem 17). Both are limited to linear groups. In this section, we outline one possible strategy to extend these results to the non-linear case, hence in particular to connected simple Lie groups with infinite center. We shall see in Sect. 6.1 that the Casselman-Subrepresentation Theorem extends to Harish–Chandra modules for which G does not have to be linear, but the Casselman–Wallach Globalization Theorem (Bernstein/Krötz [7]), respectively its presently available proofs, does not extend in any obvious fashion; see Sect. 6.3 for a discussion.

Let $(U, \mathcal{H})$ be a unitary representation of a connected semisimple Lie group G. Note that the center $Z(G)$ may be infinite and G may not be linear. We fix a Cartan decomposition $\mathfrak{g} = \mathfrak{k} \oplus \mathfrak{p}$ and the corresponding integral subgroup $K := \exp \mathfrak{k}$. Then $\mathrm{Ad}(K)$ is compact, but K need not be compact. However, we have a polar diffeomorphism $K \times \mathfrak{p} \to G$, $(k, x) \mapsto k \exp x$. In particular $\pi_1(K) \cong \pi_1(G)$.

6.1 Casselman's Theorem for Non-linear Groups

The notion of an *(admissible)* $(\mathfrak{g}, K)$*-module* generalizes to our context in the obvious fashion. As all irreducible unitary K-representations are finite-dimensional, this causes no problem if admissibility is defined as the finiteness of all K-multiplicities.

In [37, Thm. 5] Harish-Chandra shows that, if U is *quasi-simple*, i.e., $U(Z(G)) \subseteq \mathbb{T}$ and it has an infinitesimal character in the sense that

$$\mathrm{d}U(Z(\mathcal{U}(\mathfrak{g}))) \subseteq \mathbb{C}\mathbf{1},$$

then all K-finite vectors are analytic and the multiplicity of each K-type is finite [37, Lemma 33]. This assumption is in particular satisfied for any irreducible unitary representation, so that the space $\mathcal{H}^{[K]}$ of K-finite vectors is an admissible $(\mathfrak{g}, K)$-module.

Let $\mathfrak{p}_{\min} = \mathfrak{m} + \mathfrak{a} + \mathfrak{n} \subseteq \mathfrak{g}$ be a minimal parabolic, where $\mathfrak{a} \subseteq \mathfrak{p}$ is maximal abelian, $\mathfrak{m} = \mathfrak{z}_{\mathfrak{k}}(\mathfrak{a})$, and $\mathfrak{n}$ is the sum of all positive root spaces for a positive system $\Sigma^+(\mathfrak{g}, \mathfrak{a})$ of restricted roots. Osborne's Lemma (Wallach [73, Prop. 3.7.1]) asserts the existence of a finite-dimensional subspace $F \subseteq \mathcal{U}(\mathfrak{g}_{\mathbb{C}})$ such that

$$\mathcal{U}(\mathfrak{g}_{\mathbb{C}}) = \mathcal{U}(\mathfrak{n}_{\mathbb{C}})FZ(\mathcal{U}(\mathfrak{g}_{\mathbb{C}}))\mathcal{U}(\mathfrak{k}_{\mathbb{C}}).$$

This implies that every finitely generated admissible $\mathfrak{g}$-module V with an infinitesimal character is finitely generated as an $\mathcal{U}(\mathfrak{n}_{\mathbb{C}})$-module, hence in particular that

$$\dim(V/\mathfrak{n}V) < \infty.$$

In [5] Beilinson and Bernstein show that $V \neq \mathfrak{n}V$ for non-zero admissible $(\mathfrak{g}, K)$-modules, where K does not have to be compact. Their argument makes heavy use of $\mathcal{D}$-modules. Another proof of this fact is given in [73, Thm. 3.8.3]. Wallach states it for real reductive groups, i.e., finite coverings of linear groups, but all arguments in its proof work for general connected reductive groups, where Harish–Chandra modules are replaced by admissible $(\mathfrak{g}, K)$-modules, where K is connected but not necessarily compact.

As $\mathfrak{n} \trianglelefteq \mathfrak{p}_{\min}$ is an ideal, the finite-dimensional space $V/\mathfrak{n}V$, associated to a finitely generated admissible $(\mathfrak{g}, K)$-module, carries a natural structure of a $(\mathfrak{p}_{\min}, M)$-module. We thus obtain an irreducible M-subrepresentation (σ, W_σ) of $V/\mathfrak{n}V$ and a $\lambda \in \mathfrak{a}^*_{\mathbb{C}}$ such that $P_{\min} := MAN$ acts on W_σ by

$$\sigma_\lambda(man) := a^{\lambda+\rho}\sigma(m).$$

Write $W_{\sigma,\lambda}$ for the corresponding representation of $P_{\min}$, so that we have a $(\mathfrak{p}_{\min}, M)$-morphism

$$\psi \colon V \to W_{\sigma,\lambda}$$

and thus, by Frobenius reciprocity, a $(\mathfrak{g}, K)$-morphism into a principal series

$$\Psi \colon V \to V^{[K]}_{\sigma,\lambda},$$

where

$$V^\infty_{\sigma,\lambda} := \{f \in C^\infty(G, W_{\sigma,\lambda}) \colon$$
$$(\forall g \in G, m \in M, a \in A, n \in N)\ f(mang) = a^\lambda \sigma(m) f(g)\},$$

endowed with the right translation action of G. Note that evaluation in the identity is a $P_{\min}$-equivariant map

$$\mathrm{ev}_e\colon V^{\infty}_{\sigma,\lambda} \to W_{\sigma,\lambda}, \quad f \mapsto f(e).$$

Therefore, the Casselman Subrepresentation Theorem (cf. [73, §3.8.3]) extends to connected semisimple Lie groups G:

Theorem 16 *If G is a connected semisimple Lie group and $(U, \mathcal{H})$ an irreducible unitary representation, then the corresponding $(\mathfrak{g}, K)$-module $\mathcal{H}^{[K]}$ embeds in a $(\mathfrak{g}, K)$-equivariant way into a principal series representation $V^{[K]}_{\sigma,\lambda}$.*

In [38], Harish–Chandra shows that every irreducible admissible $\mathfrak{g}$-module appears as a subquotient of a principal series representation (cf. also [73, Thm. 3.5.6]). His arguments do not use linearity of the group.

Problem 4 Show that the $(\mathfrak{g}, K)$-morphism $\mathcal{H}^{[K]} \to V^{\infty}_{\sigma,\lambda}$ from Theorem 16 extends to a continuous linear map on $\mathcal{H}^{\infty}$. This is contained in Casselman/Osborne [17, Cor. 5.2] for the linear case and in Wallach [74, Cor. 11.5.4] for connected real reductive groups, which include connected semisimple groups with finite center.

6.2 Growth Control

Let G be a connected Lie group. A *scale on G* is submultiplicative function $s\colon G \to \mathbb{R}_+$ for which s and s^{-1} are locally bounded. The set of scales on G carries an order defined by

$$s \prec s' \quad \text{if} \quad (\exists N \in \mathbb{N}) \ \sup_{g\in G} \frac{s(g)}{s'(g)^N} < \infty.$$

Two scales s and s' are said to be *equivalent* if $s \prec s'$ and $s' \prec s$. The equivalence class $[s]$ of s is called a *scale structure*. Any left invariant Riemannian metric d on G defines by $s_{\max}(g) := e^{d(g,e)}$ a maximal scale structure on G (cf. Bernstein/Krötz [7, §2.1]). Simple scales are functions of the form $s(k \exp x) = e^{\|x\|}$ for $k \in K$, $x \in \mathfrak{p}$, where $\|\cdot\|$ is an $\mathrm{Ad}(K)$-invariant norm on $\mathfrak{p}$.

Remark 19 On $G = \mathbb{R}$, the maximal scale structure is defined by $s_1(x) := e^{|x|}$, but there are others, such as the scale structure defined by

$$s_2(x) := \left\| \begin{pmatrix} 1 & x \\ 0 & 1 \end{pmatrix} \right\| \leq 1 + |x| \left\| \begin{pmatrix} 0 & 1 \\ 0 & 0 \end{pmatrix} \right\|,$$

which grows only linearly.

Any continuous representation π of G on a seminormed space (E, p) specifies a scale by $s_\pi(g) := \|\pi(g)\|$ and (π, E) is called *s-bounded* if $s_\pi \prec s$.

Let $(G, [s])$ be a Lie group with scale structure. An *F-representation* of $(G, [s])$ is a continuous representation (π, E) of G on a Fréchet space E whose topology is defined by a countable family of s-bounded seminorms $(p_n)_{n\in\mathbb{N}}$ which are *G-continuous* in the sense that the representation of G on the seminormed space (E, p_n) is continuous for every $n \in \mathbb{N}$. Not every continuous representation on a Fréchet spaces is an F-representation (cf. [7, Rem. 2.8]).

Further, we assume that G is **connected semisimple** and that $[s]$ is maximal.

An F-representation (π, E) is called *smooth* if $E = E^\infty$ as topological vector space. Smooth F-representations are called *SF-representations*.

An element $f \in L^1(G)$ is called *rapidly decreasing* if all functions $g \mapsto s(g)^n f(g)$, are L^1. The subspace $\mathcal{R}(G) \subseteq L^1(G)$ of rapidly decreasing functions then carries an F-representation

$$((L \times R)(g_1, g_2)f)(g) := f(g_1^{-1} g g_2) \tag{6.1}$$

of $G \times G$ and it is a Fréchet algebra under convolution. Further, any F-representation (π, E) of $(G, [s])$ integrates to a continuous algebra representation

$$\mathcal{R}(G) \times E \to E, \quad (f, v) \mapsto \pi(g)v \quad \text{for} \quad \pi(f) = \int_G f(g)\pi(g)\, dg.$$

For $u \in \mathcal{U}(\mathfrak{g})$, we write L_u and R_u for the corresponding operators on $C^\infty(G)$ defined by the derived representation of $G \times G$ and define the *Schwartz space of G* by

$$\mathcal{S}(G) := \mathcal{R}(G)^\infty \subseteq C^\infty(G)$$

for the $G \times G$-representation (6.1) on $\mathcal{R}(G)$. Note that the smoothness of the elements of $\mathcal{R}(G)^\infty \subseteq L^1(G)^\infty$ follows from the local Sobolev Lemma.

On any SF-representation of $(G, [s])$ we then have representations of the three convolution algebras

$$C_c^\infty(G) \subseteq \mathcal{S}(G) \subseteq \mathcal{R}(G)$$

and the Dixmier–Malliavin Theorem [22] implies that $\pi(C_c^\infty(G))E = E^\infty$, which in turn implies

$$E^\infty = \pi(\mathcal{S}(G))E.$$

6.3 Localization in Central Unitary Characters

If G is a connected simple Lie group and $\eta_G\colon G \to G_{\mathbb{C}}$ its universal complexification, then $\eta_G(G) \subseteq G_{\mathbb{C}}$ is a linear simple group and $\eta_G\colon G \to \eta_G(G) \cong G/\ker(\eta_G)$ is a covering with discrete central kernel. Therefore, extending results from linear simple groups to general connected simple Lie groups requires a localization in central characters. We outline the main ideas in this section.

Let G be a locally compact group, $Z \subseteq Z(G)$ a closed subgroup and $\chi : Z \to \mathbb{T}$ a unitary character. We write $q_{G/Z}\colon G \to G/Z$ for the canonical projection and

$$\begin{aligned} C_c(G)_\chi := \{&f \in C(G, \mathbb{C})\colon \\ &(\forall g \in G, z \in Z)\ f(gz) = \chi(z)^{-1} f(g), q_{G/Z}(\mathrm{supp}(f))\ \text{compact}\}. \end{aligned}$$

If $\mathbb{L}_\chi := G \times_Z \mathbb{C}$ is the line bundle over G/Z obtained by factorization of the Z-action $z.(g, w) = (gz, \chi(z)^{-1} w)$, then

$$C_c(G)_\chi \cong C_c(G/Z, \mathbb{L}_\chi)$$

corresponds to the space of compactly supported sections of $\mathbb{L}_\chi$.

On $C_c(G)_\chi$, we define a convolution product by

$$(f_1 * f_2)(g) := \int_{G/Z} f_1(x) f_2(x^{-1}g)\, dx,$$

where we use that, as a function of x, the integrand factors through a function on G/Z. Therefore, the integral exists by compactness of $q_{G/Z}(\mathrm{supp}(f_1))$. The subspace $C_c(G)_\chi$ is also invariant under the involution defined by

$$f^*(g) := \overline{f(g^{-1})} \Delta_G(g)^{-1}.$$

We thus obtain a $*$-algebra $(C_c(G)_\chi, *)$. Completing with respect to the norm

$$\|f\|_1 := \int_{G/Z} |f(gZ)|\, d(gZ)$$

leads to the Banach $*$-algebra $L^1(G)_\chi$.

A unitary representation $(\pi, \mathcal{H})$ is called a *χ-representation* if

$$\pi(z) = \chi(z)\mathbf{1} \quad \text{for all} \quad z \in Z.$$

Then

$$\pi(f) := \int_{G/Z} f(g)\pi(g)\, d(gZ)$$

is well-defined because the integrand factors through a function on G/Z. The arguments used for $Z = \{e\}$ carry over to this context and show that we thus obtain a contractive $*$-representation of the Banach $*$-algebra $L^1(G)_\chi$ on $\mathcal{H}$.

Remark 20 All concepts from Sect. 6.2 carry over to the χ-local situation if we work with scales on the quotient group G/Z. An important special case is $Z = \ker(\eta_G)$, where $\eta\colon G \to G_{\mathbb{C}}$ is the universal complexification of a connected semisimple Lie group. This leads in particular to the Fréchet convolution algebras

$$\mathcal{S}(G)_\chi \subseteq \mathcal{R}(G)_\chi \subseteq L^1(G)_\chi.$$

Remark 21 Suppose that (π, E) is a χ-representation, $f \in C_c(G)$ and

$$f_\chi(g) := \int_Z f(gz)\chi(z)\, dz.$$

For $v \in \mathcal{H}$, we then have

$$\begin{aligned}\pi(f)v &= \int_G f(g)\pi(g)v\, dg = \int_{G/Z}\int_Z f(gz)\pi(gz)v\, dz\, d(gZ) \\ &= \int_{G/Z}\Big(\int_Z f(gz)\chi(z)\, dz\Big)\pi(g)v\, d(gZ) \qquad (6.2)\\ &= \int_{G/Z} f_\chi(g)\pi(g)v\, d(gZ) = \pi(f_\chi)v\,. \qquad (6.3)\end{aligned}$$

As the map $C_c^\infty(G) \to C_c^\infty(G)_\chi, f \mapsto f_\chi$, is surjective, for any SF-χ-representation, we derive from the Dixmier–Malliavin Theorem

$$E^\infty = \pi(C_c^\infty(G))E \overset{(6.2)}{=} \pi(C_c^\infty(G)_\chi)E = \pi(\mathcal{S}(G)_\chi)E.$$

Problem 5 Show the existence of a unique minimal SF-globalization of χ-HC-modules. This would generalize Bernstein/Krötz [7] that deals with the case of linear groups.

Inspection of the arguments in [7] shows that their Theorem 4.5 and Corollary 5.6 remain true in the χ-context. The key point is a generalization of their Proposition 11.2.

From [7, Cor. 2.16] we obtain:

Proposition 11 *If $(\pi, \mathcal{H})$ is an irreducible unitary χ-representations, then $(\pi^\infty, \mathcal{H}^\infty)$ is an SF globalization of $\mathcal{H}^{[K]}$.*

Problem 6 Suppose that $(\pi_{\sigma,\lambda}, V_{\sigma,\lambda})$ is a χ-principal series, i.e., the representation $\sigma\colon M = Z_K(A) \to \mathrm{U}(W_\sigma)$ satisfies

$$\sigma|_{Z(G)} = \chi\mathbf{1}.$$

Are the closures E^∞ of irreducible $(\mathfrak{g}, K)$-submodules $E \subseteq V_{\sigma,\lambda}^{[K]}$ in $V_{\sigma,\lambda}^\infty$ obtained with Theorem 16 initial SF globalizations? This would lead to a G-morphism

$$E^\infty \to \mathcal{H}^\infty \subseteq \mathcal{H}$$

which is precisely what we need to transfer the existence of the holomorphic extension map $\Phi \colon \mathbb{E} \to E^\infty \subseteq V_{\sigma,\lambda}^\infty$ that we should get from the complex Iwasawa decomposition (following Krötz/Stanton [45]) to a holomorphic map $\mathbb{E} \to \mathcal{H}$.

Given a unitary representation $(U, \mathcal{H})$, one may alternatively work with the minimal globalization technique developed of Gimperlein/Krötz/Kuit/Schlichtkrull [29] to show that we have a continuous intertwiner $E^\omega \hookrightarrow \mathcal{H}^\omega$.

For connected real reductive groups (including connected semisimple groups with finite center), the solution of Problems 5 and 6 is well-known and stated as the two parts of Theorem 11.6.7 in Wallach [74].

Let $V := \mathcal{H}^{[K]}$ be the irreducible $(\mathfrak{g}, K)$-module underlying our representation and $v \in V$ a cyclic vector. Then

$$\mathcal{S}(G)_{\chi,v} := \{f \in \mathcal{S}(G)_\chi : \pi(f)v = 0\}$$

is a closed left ideal in $\mathcal{S}(G)_\chi$ and

$$V_{\min}^\infty := \mathcal{S}(G)_\chi / \mathcal{S}(G)_{\chi,v}$$

is an SF-representation of G with $V_{\min}^{\infty,[K]} = V$. It clearly satisfies

$$\pi(\mathcal{S}(G)_\chi) V_{\min} = V_{\min}$$

(cf. Bernstein/Krötz [7, Rem. 2.19]).

The proof of [7, Prop. 11.2] carries over to our context and implies that, for every smooth $\mathcal{S}(G)_\chi$-representation (π_E, E), the restriction map

$$\mathrm{Hom}_G(V_{\min}^\infty, E) \to \mathrm{Hom}_{(\mathfrak{g},K)}(V, E^{[K]}) \tag{6.4}$$

is a linear isomorphism. Therefore, the remaining problem is to identify $V_{\min}^\infty$, for an irreducible unitary representation with $\mathcal{H}^\infty$ and for a subrepresentation of a minimal principal series with the corresponding space of smooth vectors.

7 Appendix 1: An Automatic Continuity Theorem

Definition 2 A reductive Lie group G with finitely many connected components is said to be *of Harish–Chandra class* if $\mathrm{Ad}(G) \subseteq \mathrm{Aut}(\mathfrak{g}_\mathbb{C})_e$ and the semisimple commutator group (G, G) has finite center.

In Wallach [73, §2.1], a real Lie group G is called *real reductive* if it is a finite covering of an open subgroup of a reductive real algebraic subgroup $G_{\mathbb{R}} \subseteq \mathrm{GL}_n(\mathbb{R})$.

If G is semisimple and connected, it is real reductive if and only if its center $Z(G)$ is finite, but it always is of Harish–Chandra class; (as required in van den Ban [2]). So the Casselman–Wallach Globalization Theorem (Wallach [74, Thm. 11.6.7]) holds for G if $Z(G)$ is finite. Let τ be an involution on G and H an open subgroup of its group G^τ of fixed points. In this appendix, we show the following automatic continuity result:

Theorem 17 *Let G be a real reductive group of Harish–Chandra class, τ an involutive automorphisms of $\mathfrak{g}$ and $\mathfrak{h} := \mathfrak{g}^\tau$. Let $(U, \mathcal{H})$ be an irreducible unitary representation of G, $\mathcal{H}^{-\omega}$ the space of hyperfunction vectors and $(\mathcal{H}^{-\omega})^{[\mathfrak{h}]}$ the subspace of $\mathfrak{h}$-finite hyperfunction vectors. Then*

$$(\mathcal{H}^{-\omega})^{[\mathfrak{h}]} \subseteq \mathcal{H}^{-\infty}.$$

Remark 22 A special case of the preceding theorem is the automatic continuity of $\mathfrak{h}$-fixed hyperfunction vectors (in contrast to $\mathfrak{h}$-finite vectors):

$$(\mathcal{H}^{-\omega})^{\mathfrak{h}} \subseteq \mathcal{H}^{-\infty}. \tag{AC}$$

This statement also follows from the Automatic Continuity Theorem of van den Ban–Brylinski–Delorme. In fact, their result is more general than (AC), as they are dealing with linear functionals on $\mathcal{H}^{[K]}$. In Brylinski/Delorme [15, Thm.] it is shown that every linear functional on an irreducible $(\mathfrak{g}, K)$-module V, that is H-fixed, extends continuously to its Casselman–Wallach globalization V^∞. The proof uses in a crucial way the main result of van den Ban/Delorme [4] which constructs the $\mathfrak{g}$-equivariant embedding Φ of V into $C^\infty(G/H)$ from the $\mathfrak{h}$-fixed functional on V. This construction can also be generalized to the case of $\mathfrak{h}$-finite vectors. However, since we already start with $\mathfrak{h}$-finite hyperfunction vectors, we can use them to embed V into the space of smooth sections of $G \times_H W$ in terms of matrix coefficients and thus avoid a lengthy discussion of the extension of [4] to $\mathfrak{h}$-finite vectors. We therefore discuss automatic continuity only in the generality we need in this chapter, i.e., starting with the $\mathfrak{h}$-finite hyperfunction vecrors instead of $\mathfrak{h}$-finite linear functionals on the Harish–Chandra module $\mathcal{H}^{[K]}$.

For the subspace $(\mathcal{H}^{-\omega})^{\mathfrak{h}} \subseteq (\mathcal{H}^{-\omega})^{[\mathfrak{h}]}$ of $\mathfrak{h}$-fixed hyperfunction vectors, the inclusion $(\mathcal{H}^{-\omega})^{\mathfrak{h}} \subseteq \mathcal{H}^{-\infty}$ follows from the Automatic Continuity Theorem of Brylinski–Delorme [15, Thm. 1]. The proof we present follows closely their argument, in particular we use van den Ban's results [3] about the asymptotic expansion of matrix coefficients.

We first observe that the Automatic Continuity Theorem 17 is equivalent to the following statement: every $\mathfrak{h}$-equivariant map $\varphi : \mathcal{H}^\omega \to W$ to a finite-dimensional representation (π_W, W) of $\mathfrak{h}$ extends continuously to $\mathcal{H}^\infty \to W$. For its proof, we consider the space

$$C^\infty(G \times_H W) = \{f \in C^\infty(G, W) : f(gh) = \pi_W(h)^{-1} f(g) \text{ for all } g \in G, h \in H\},$$

whose elements represent smooth sections of the vector bundle $G \times_H W \to G/H$ associated with W. We now have to extend the corresponding matrix coefficient map

$$\Phi : \mathcal{H}^\omega \to C^\infty(G \times_H W), \quad \Phi(v)(g) = \varphi(U(g^{-1})v)$$

continuously to $\Phi : \mathcal{H}^\infty \to C^\infty(G \times_H W)$. Then

$$\mathrm{ev}_e \circ \Phi : \mathcal{H}^\infty \to W, \quad v \mapsto \Phi(v)(e)$$

is the desired extension of φ.

Since $\mathcal{H}^\infty$ is a smooth admissible Fréchet representation of moderate growth (SAF representation in the sense of Bernstein/Krötz [7]), the extension of

$$\Phi : \mathcal{H}^\omega \to C^\infty(G \times_H W)$$

to $\mathcal{H}^\infty$ will follow from the functoriality of the Casselman–Wallach Globalization Functor once we have shown that Φ maps $\mathcal{H}^\omega$ into a certain SAF representation. For this matter, we introduce a scale of smooth Fréchet subrepresentations of $C^\infty(G \times_H W)$ of moderate growth.

Let θ be a Cartan involution on G that commutes with τ and fix an embedding $\iota : G \hookrightarrow \mathrm{GL}_n(\mathbb{R})$ under which $\iota(\theta(g)) = (g^\top)^{-1}$. Then

$$\|g\| = \|\iota(g) \oplus \iota(\theta(g))\| \qquad (g \in G),$$

with the operator norm $\|\cdot\|$ on $M_{2n}(\mathbb{R})$, defines a norm on G with the following properties:

$$\|g^{-1}\| = \|g\| = \|\tau(g)\| \qquad \text{for all } g \in G,$$

$$\|xy\| \leq \|x\|\|y\| \qquad \text{for all } x, y \in G,$$

$$\|k_1 \exp(tX) k_2\| = \|\exp X\|^t \qquad \text{for all } k_1, k_2 \in K, X \in \mathfrak{p}, t \geq 0.$$

Fixing a norm $\|\cdot\|_W$ on W, we define for $N \in \mathbb{N}$, $D \in \mathcal{U}(\mathfrak{g})$ and $f \in C^\infty(G \times_H W)$:

$$p_{N,D}(f) = \sup_{g \in G} \|g\|^{-N} \|(L_D f)(g)\|_W \in [0, \infty],$$

where L_D is the (right invariant) differential operator on $C^\infty(G, W)$ specified by the left multiplication action of G on itself. For $N \in \mathbb{N}$ we put

$$\mathcal{A}_N(G \times_H W) = \{f \in C^\infty(G \times_H W) : p_{N,D}(f) < \infty \text{ for all } D \in \mathcal{U}(\mathfrak{g})\}.$$

Endowed with the seminorms $(p_{N,D})_{D \in \mathcal{U}(\mathfrak{g})}$, the space $\mathcal{A}_N(G \times_H W)$ becomes a Fréchet space. The following result is proven along the same lines as Brylinski/Delorme [15, Lemme 1]:

Lemma 12 *The left regular representation of G on $\mathcal{A}_N(G \times_H W)$ is a smooth Fréchet representation of moderate growth.*

In order to extend $\Phi : \mathcal{H}^\omega \to C^\infty(G \times_H W)$ to $\mathcal{H}^\infty$, we show that Φ maps the subspace $\mathcal{H}^{[K]}$ of K-finite vectors into one of the moderate growth representations $\mathcal{A}_N(G \times_H W)$. This is done using van den Ban's results about the asymptotic behavior of matrix coefficients in van den Ban [2].

Proposition 12 *There exists $N \in \mathbb{N}$ such that $\Phi(\mathcal{H}^{[K]}) \subseteq \mathcal{A}_N(G \times_H W)$.*

Proof Let $\mathfrak{a} \subseteq \mathfrak{p} \cap \mathfrak{q}$ be a maximal abelian subspace and $A = \exp(\mathfrak{a})$. For any choice $\Delta^+ \subseteq \Delta(\mathfrak{g}^{\tau\theta}, \mathfrak{a})$ of positive roots where $\mathfrak{g}^{\tau\theta} = \mathfrak{k} \cap \mathfrak{h} + \mathfrak{p} \cap \mathfrak{q}$ we consider the negative Weyl chamber

$$A^- = A^-(\Delta^+) = \{a \in A : a^\alpha < 1 \text{ for all } \alpha \in \Delta^+\}.$$

Then the Cartan decomposition $G = KAH$ holds (see e.g. [2, Cor. 1.4]). Now let $v \in \mathcal{H}^{[K]}$, then v is contained in a finite-dimensional K-invariant subspace $E \subseteq \mathcal{H}$. Consider the function

$$F : G \to E^* \otimes W \simeq \operatorname{Hom}_{\mathbb{C}}(E, W), \quad F(g) = \varphi \circ U(g)^{-1}|_E.$$

We note that $F(g)(v) = \Phi(v)(g)$, so it suffices to estimate $F(g)$. The function F is (μ_1, μ_2)-spherical in the sense of [2, page 230] with $\mu_1(k) = (U(k^{-1})|_E)^*$ and $\mu_2 = \pi_W$:

$$F(kgh) = \pi_W(h)^{-1} F(g) U(k^{-1})|_E \quad \text{for} \quad k \in K, g \in G, h \in H.$$

Moreover, since F is a matrix coefficient of the irreducible representation $(U, \mathcal{H})$, it is annihilated by a cofinite ideal I of the center of $\mathcal{U}(\mathfrak{g})$ (see [2, pp. 230–231]). We may therefore apply the results of [2].

Applying [2, Thm. 6.1] yields a character ω of A and $C > 0$ with the property that

$$\|F(a)\| \leq C\omega(a) \qquad \text{for all } a \in \overline{A^-}.$$

Here $\|\cdot\|$ denotes the tensor product norm on $E^* \otimes W$ induced by a K-invariant norm on E and the norm $\|\cdot\|_W$ on W. We note that the character ω only depends on the cofinite ideal I annihilating the representation U and not on the vector $v \in \mathcal{H}^{[K]}$. Bounding the character ω by a power of the norm $\|\cdot\|$ on A (see Wallach [73, Lemma 2.A.2.3]) and applying this argument to all of the finitely many choices of positive systems $\Delta^+ \subseteq \Delta(\mathfrak{g}^{\tau\theta}, \mathfrak{a})$ and hence to all Weyl chambers in A, shows that there exist $N \in \mathbb{N}$ and $C > 0$ such that

$$\|F(a)\| \leq C\|a\|^N \qquad \text{for all } a \in A.$$

To extend this estimate to an estimate on all of G we note that the finite-dimensional representation (μ_2, W) of H is of moderate growth and hence there exist $C' > 0$ and $M \in \mathbb{N}$ such that

$$\|\mu_2(h)w\|_W \leq C'\|h\|^M\|w\|_W \quad \text{for} \quad h \in H, w \in W.$$

Together with the unitarity of μ_1 this implies

$$\|F(kah)\| = \|\pi_W(h)^{-1}F(a)U(k^{-1})\| \leq C'\|h\|^M\|F(a)\| \leq CC'\|h\|^M\|a\|^N.$$

But the properties of the norm imply that for $a \in A$ and $h \in H$:

$$\|ah\| = \|\tau(ah)\| = \|a^{-1}h\|$$

and hence

$$\|a\|^2 = \|a^2\| = \|ahh^{-1}a\| \leq \|ah\|\|h^{-1}a\| = \|ah\|\|a^{-1}h\| = \|ah\|^2.$$

This also implies

$$\|h\| = \|a^{-1}ah\| \leq \|a^{-1}\|\|ah\| = \|a\|\|ah\| \leq \|ah\|^2,$$

so we obtain

$$\|F(kah)\| \leq CC'\|ah\|^{N+2M} = CC'\|kah\|^{N+2M} \qquad (k \in K, a \in A, h \in H).$$

Since $F(g)(v) = \Phi(v)(g)$, this shows that $\Phi(\mathcal{H}^{[K]}) \subseteq \mathcal{A}_{N+2M}(G \times_H W)$. □

Proof of Theorem 17 Since $\Phi : \mathcal{H}^{[K]} \to \mathcal{A}_N(G \times_H W)$ and $\mathcal{A}_N(G \times_H W)$ is a smooth Fréchet representation of moderate growth, the closure $\overline{\Phi(\mathcal{H}^{[K]})} \subseteq \mathcal{A}_N(G \times_H W)$ is a smooth admissible Fréchet representation of moderate growth. By the Casselman–Wallach Globalization Theorem (Wallach [74, Thm. 11.6.7]), the map $\Phi : \mathcal{H}^{[K]} \to \overline{\Phi(\mathcal{H}^{[K]})}$ extends uniquely to a continuous intertwining operator $\Phi : \mathcal{H}^\infty \to \overline{\Phi(\mathcal{H}^{[K]})} \subseteq \mathcal{A}_N(G \times_H W)$. Now the map

$$\mathrm{ev}_e \circ \Phi : \mathcal{H}^\infty \to W, \quad v \mapsto \Phi(v)(e)$$

is the desired extension of $\mathcal{H}^\omega \to W$. □

Remark 23 We note that in Sections 1–5 of van den Ban [2] the assumption on G to be of Harish-Chandra's class is not used, so the results about the asymptotic expansion of matrix coefficients also hold for the universal covering G of a Hermitian Lie group. Since the center $Z(G)$ of G acts by a unitary character χ in an irreducible unitary representation $(U, \mathcal{H})$, the arguments in the proof of Theorem 17 can be adapted to show that $\mathcal{H}^\omega$ embeds for some $N \in \mathbb{N}$ into

$$\mathcal{A}_{\chi,N}(G \times_H W)$$
$$= \left\{ f \in C^\infty(G \times_H W) : \begin{array}{l} f(cg) = \chi(c) f(g) \text{ for all } g \in G, c \in Z(G), \\ p_{N,D}(f) < \infty \text{ for all } D \in \mathcal{U}(\mathfrak{g}) \end{array} \right\}.$$

Here, $p_{N,D}$ is defined using a norm function on a linear quotient of G. If a version of the Casselman–Wallach Globalization Theorem holds for the group G, the embedding $\mathcal{H}^\omega \hookrightarrow \mathcal{A}_{\chi,N}(G \times_H W)$ extends to $\mathcal{H}^\infty \hookrightarrow \mathcal{A}_{\chi,N}(G \times_H W)$.

8 Appendix 2: Wedge Regions in Non-compactly Causal Spaces

In this appendix, we put some of the results from Morinelli/Neeb/Ólafsson [54, 55] into the context in which they are used in the present paper.

As aforementioned, G denotes a connected simple Lie group, $h \in \mathfrak{g}$ is an Euler element, $\tau = \theta\tau_h$ for a Cartan involution θ satisfying $\theta(h) = -h$ and $M = G/H$ is a corresponding non-compactly causal symmetric space, where the causal structure is specified by a maximal $\mathrm{Ad}(H)$-invariant closed convex cone $C \subseteq \mathfrak{q}$ satisfying $h \in C^\circ$ (cf. [54, Thm. 4.21]).

First, we consider the "minimal" space associated to the triple $(\mathfrak{g}, \tau, C)$. It is obtained as

$$M_{\mathrm{min}} := G_{\mathrm{ad}}/H_{\mathrm{ad}},$$

where

$$G_{\mathrm{ad}} := \mathrm{Ad}(G) = \mathrm{Inn}(\mathfrak{g}) \quad \text{and} \quad H_{\mathrm{ad}} := K_{\mathrm{ad}}^h \exp(\mathfrak{h}_{\mathfrak{p}}) \subseteq G_{\mathrm{ad}}^\tau$$

(see [54, Rem. 4.20(b)] for more details). In this space the positivity domain $W_{M_{\mathrm{min}}}^+(h)$ is connected by [55, Thm. 7.1]. Further, [55, Thm. 8.2, Prop. 8.3] imply that

$$W_{M_{\mathrm{min}}}^+(h) = G_e^h.\mathrm{Exp}(\Omega_{\mathfrak{q}_{\mathfrak{p}}}).$$

By [54, Rem. 4.20(a)], we have $H = H_K \exp(\mathfrak{h}_\mathfrak{p})$ with $H_K \subseteq K^h$, so that $\mathrm{Ad}(H) \subseteq H_{\mathrm{ad}}$. Therefore

$$q_M : M \to M_{\min}, \quad gH \mapsto \mathrm{Ad}(g)H_{\mathrm{ad}} \in M_{\min}$$

defines a covering of causal symmetric spaces. The stabilizer in G of the base point in $M_{\min}$ is the subgroup

$$H^\sharp := \mathrm{Ad}^{-1}(H_{\mathrm{ad}}) = K^h \exp(\mathfrak{h}_\mathfrak{p})$$

because $Z(G) = \ker(\mathrm{Ad}) \subseteq K^h$. Note that $H^\sharp$ need not be contained in G^τ because τ may act nontrivially on K^h. A typical example if $G = \widetilde{\mathrm{SL}}_2(\mathbb{R})$ with $K \cong \mathbb{R}$ and $\tau(k) = k^{-1}$ for $k \in K$. So we may consider $M_{\min}$ as the homogeneous G-space

$$M_{\min} \cong G/H^\sharp.$$

As q_M is a G-equivariant covering of causal manifolds,

$$W_M^+(h) = q_M^{-1}(W_{M_{\min}}^+(h)) = q_M^{-1}(G_e^h.\mathrm{Exp}(\Omega_{\mathfrak{q}_\mathfrak{p}}))$$

and the inverse image under the map $q : G \to G/H = M$ is

$$\begin{aligned} q^{-1}(W_M^+(h)) &= G_e^h \exp(\Omega_{\mathfrak{q}_\mathfrak{p}})H^\sharp = G_e^h \exp(\Omega_{\mathfrak{q}_\mathfrak{p}})K^h \exp(\mathfrak{h}_\mathfrak{p}) \\ &= G_e^h K^h \exp(\Omega_{\mathfrak{q}_\mathfrak{p}}) \exp(\mathfrak{h}_\mathfrak{p}) = G^h \exp(\Omega_{\mathfrak{q}_\mathfrak{p}}) \exp(\mathfrak{h}_\mathfrak{p}). \end{aligned}$$

Next we recall from [55, Prop. 8.3] that the map

$$G_e^h \times_{K_e^h} \Omega_{\mathfrak{q}_\mathfrak{k}} \to W_{M_{\mathrm{ad}}}^+(h), \quad [g, x] \mapsto g \exp(x) H_{\mathrm{ad}} \tag{8.1}$$

is a diffeomorphism. Therefore $W_{M_{\mathrm{ad}}}^+(h)$ is an affine bundle over the Riemannian symmetric space G_e^h/K_e^h, hence contractible and therefore simply connected. So its inverse image $W_M^+(h)$ in M is a union of open connected components, all of which are mapped diffeomorphically onto $W_{M_{\mathrm{ad}}}^+(h)$ by q_M. It follows in particular that the diffeomorphism (8.1) lifts to a diffeomorphism

$$G_e^h \times_{K_e^h} \Omega_{\mathfrak{q}_\mathfrak{k}} \to W = W_M^+(h)_{eH}, \quad [g, x] \mapsto g \exp(x) H. \tag{8.2}$$

Acknowledgments We are particularly indebted to Job Kuit for several discussions concerning the Automatic Continuity Theorem outlined in Appendix 1. We also thank Erik van den Ban and Patrick Delorme for helpful comments on this matter. We further thank Jacques Faraut for communicating a proof of Proposition 4. We are also indepted to Tobias Simon for abundant comments on a first draft of this chapter and also for comments by Joachim Hilgert.

Last, but not least, we wish to thank both referees for very inspiring and most useful reports. Their suggestions helped us to improve the exposition significantly.

The first named author was supported by a research grant from the Villum Foundation (Grant No. 00025373). The second named author was supported by DFG-grant NE 413/10-2. The third named author was supported by Simons grant 586106.

References

1. Akhiezer, D.N., Gindikin, S.G.: On Stein extensions of real symmetric spaces. Math. Ann. **286**, 1–12 (1990)
2. van den Ban, E.: Asymptotic behaviour of matrix coefficients related to reductive symmetric spaces. Nederl. Akad. Wetensch. Indag. Math. **49**(3), 225–249 (1987)
3. van den Ban, E.: The principal series for a reductive symmetric space. I. H-fixed distribution vectors. Ann. Sci. École Norm. Sup. (4) **21**(3), 359–412 (1988)
4. van den Ban, E., Delorme, P.: Quelques propriétés des représentations sphériques pour les espaces symétriques réductifs. J. Funct. Anal. **80**, 284–307 (1988)
5. Beilinson, A., Bernstein, J.A.: A generalization of Casselman's submodule theorem. In: Representation Theory of Reductive Groups (Park City, Utah, 1982). Progress in Mathematics, vol. 40, pp. 35–52. Birkhäuser Boston, Boston (1983)
6. Beltiţă, D., Neeb, K.-H.: Holomorphic extension of one-parameter operator groups. Pure Appl. Funct. Anal. arXiv:2304.09597
7. Bernstein, J., Krötz, B.: Smooth Fréchet globalizations of Harish–Chandra modules. Israel J. Math. **199**, 45–111 (2014)
8. Bertram, W.: On some causal and conformal groups. J. Lie Theory **6**, 215–247 (1996)
9. Bochnak, J., Siciak, J.: Analytic functions in topological vector spaces. Stud. Math. **39**, 77–112 (1971)
10. Borchers, H.-J.: On the net of von Neumann algebras assiciated with a wedge and wedge-causal manifold (2009). http://www.theorie.physik.uni-goettingen.de/forschung/qft/publications/2009
11. Borchers, H.-J., Buchholz, D.: Global properties of vacuum states in de Sitter space. Ann. Poincare Phys. Theor. **A70**, 23–40 (1999)
12. Bratteli, O., Robinson, D.W.: Operator Algebras and Quantum Statistical Mechanics I, 2nd edn. Texts and Monographs in Physics. Springer, Berlin (1987)
13. Bros, J., Epstein, H., Moschella, U.: Analyticity properties and thermal effects for general quantum field theory on de Sitter space-time. Commun. Math. Phys. **196**(3), 535–570 (1998)
14. Bros, J., Moschella, U.: Two-point functions and quantum fields in de Sitter universe. Rev. Math. Phys. **8**(3), 327–391 (1996)
15. Brylinski, J.-K., Delorme, P.: Vecteurs distributions H-invariants pour les séries principales généralisées d'espaces symétriques réductifs et prolongement méromorphe d'intégrales d'Eisenstein. Invent. Math. **109**(3), 619–664 (1992)
16. Buchholz, D., Mund, J., Summers, S.J.: Transplantation of local nets and geometric modular action on Robertson-Walker Space-Times. Fields Inst. Commun. **30**, 65–81 (2001)
17. Casselman, W., Osborne, M.S.: The restriction of admissible representations to $\mathfrak{n}$. Math. Ann. **233**, 193–198 (1978)
18. Ciolli, F., Longo, R., Ranallo, A., Ruzzi, G.: Relative entropy and curved spacetimes. J. Geom. Phys. **172**, 104416 (2022)
19. Connes, A., Rovelli, C.: von Neumann algebra automorphisms and time-thermodynamics relation in generally covariant quantum theories. Classical Quantum Gravity **11**(12), 2899–2917 (1994)
20. Correa da Silva, R., Lechner, G.: Modular Structure and Inclusions of Twisted Araki-Woods Algebras. Commun. Math. Phys. **402**(3), 2339–2386 (2023). arXiv:2212.02298
21. Digital Library of Mathematical Functions. https://dlmf.nist.gov

22. Dixmier, J., Malliavin, P.: Factorisations de fonctions et de vecteurs indéfiniment différentiables. Bull. Soc. Math. 2e série **102**, 305–330 (1978)
23. Erdélyi, A., Magnus, W., Oberhettinger, F., Tricomi, F.G.: Higher Transcendental Functions, vol. I. McGraw-Hill, New York-Toronto-London (1953)
24. Flensted-Jensen, M., Koornwinder, T.H.: Positive definite spherical functions on a noncompact, rank one symmetric space. In: Analyse Harmonique sur les Groupes de Lie (Sém., Nancy-Strasbourg 1976–1978), II. Lecture Notes in Mathematics, vol. 739, pp. 249–282. Springer, Berlin (1979)
25. Frahm, J., Neeb, K.-H., Ólafsson, G.: Realization of unitary representations of the Lorentz group on de Sitter space. Preprint, arXiv:2401.17140
26. Gårding, L.: Note on continuous representations of Lie groups. Proc. Nat. Acad. Sci. U.S.A. **33**, 331–332 (1947)
27. Gårding, L.: Vecteurs analytiques dans les représentations des groupes de Lie. Bull. Soc. Math. France **88**, 73–93 (1960)
28. Gimplerlein, H., Krötz, B., Schlichtkrull, H.: Analytic representation theory of Lie groups: general theory and analytic globalization of Harish–Chandra modules. Comp. Math. **147**(5), 1581–1607 (2011); corrigendum ibid. **153**(1), 214–217 (2017); arXiv:1002.4345v2
29. Gimperlein, H., Krötz, B., Kuit, J., Schlichtkrull, H.: A Paley-Wiener theorem for Harish-Chandra modules. Camb. J. Math. **10**(3), 689–742 (2022)
30. Gindikin, S., Krötz, B.: Invariant Stein domains in Stein symmetric spaces and a nonlinear complex convexity theorem. Int. Math. Res. Not. **18**, 959–971 (2002)
31. Gindikin, S., Krötz, B.: Complex crowns of Riemannian symmetric spaces and non-compactly causal symmetric spaces. Trans. Am. Math. Soc. **354**(8), 3299–3327 (2002)
32. Gindikin, S., Krötz, B., Ólafsson, G.: Holomorphic H-spherical distribution vectors in principal series representations. Invent. Math. **158**(3), 643–682 (2004)
33. Goodman, R.: Analytic and entire vectors for representations of Lie groups. Trans. Am. Math. Soc. **143**, 55–76 (1969)
34. Guido, D., Longo, R.: An algebraic spin and statistics theorem. Commun. Math. Phys. **172**(3), 517–533 (1995)
35. Guido, D., Longo, R.: A converse Hawking–Unruh effect and dS2/CFT correspondence. Ann. Henri Poincaré **4**, 1169–1218 (2003)
36. Haag, R.: Local Quantum Physics. Fields, Particles, Algebras, 2nd edn. Texts and Monographs in Physics. Springer, Berlin (1996)
37. Harish-Chandra: Representations of semisimple Lie groups on a Banch space. I. Trans. Am. Math. Soc. **75**, 185–243 (1953)
38. Harish-Chandra: Representations of semisimple Lie groups. II. Trans. Am. Math. Soc. **76**, 26–65 (1954)
39. Heckman, G., Schlichtkrull, H.: Harmonic Analysis and Special Functions on Symmetric Spaces. Perspectives in Mathematics, vol. 16. Academic Press, New York (1994)
40. Helgason, S.: Differential Geometry, Lie Groups. and Symmetric Spaces. Acadamy Press, London (1978)
41. Helgason, S.: Groups and Geometric Analysis. Academic Press, London (1984)
42. Hilgert, J., Ólafsson, G.: Causal Symmetric Spaces, Geometry and Harmonic Analysis. Perspectives in Mathematics, vol. 18. Academic Press, London (1997)
43. Huckleberry, A.: On certain domains in cycle spaces of flag manifolds. Math. Ann. **323**(4), 797–810 (2002)
44. Krötz, B., Opdam, E.: Analysis on the crown domain. Geom. Funct. Anal. **18**(4), 1326001421 (2008)
45. Krötz, B., Stanton, R.J.: Holomorphic extensions of representations. I. Automorphic functions. Ann. Math. **159**, 641–724 (2004)
46. Krötz, B., Schlichtkrull, H.: Holomorphic extension of eigenfunctions. Math. Ann. **345**(4), 835–841 (2009)
47. Lebedev, N.N., Silverman, R.A.: Special Functions and their Application. Prentice-Hall (1965)

48. Longo, R.: Real Hilbert subspaces, modular theory, SL(2, $\mathbb{R}$) and CFT. In: Von Neumann Algebras in Sibiu. Theta Ser. Adv. Math., vol. 10, pp. 33–91. Theta, Bucharest (2008)
49. Loos, O.: Symmetric Spaces I: General Theory. W. A. Benjamin, New York, Amsterdam (1969)
50. Matsuki, T.: Stein extensions of Riemann symmetric spaces and some generalizations. J. Lie Theory **13**, 563–570 (2003)
51. Moore, C.C.: The Mautner phenomenon for general unitary representations. Pac. J. Math. **86**(1), 155–169 (1980)
52. Morinelli, V., Neeb, K.-H.: Covariant homogeneous nets of standard subspaces. Commun. Math. Phys. **386**, 305–358 (2021). arXiv:math-ph.2010.07128
53. Morinelli, V., Neeb, K.-H.: From local nets to Euler elements. Adv. Math. **458**, 109960 (2024). arXiv:2312.12182
54. Morinelli, V., Neeb, K.-H., Ólafsson, G.: From Euler elements and 3-gradings to non-compactly causal symmetric spaces. J. Lie theory **33**, 377–432 (2023). arXiv:2207.14034
55. Morinelli, V., Neeb, K.-H., Ólafsson, G.: Modular geodesics and wedge domains in general non-compactly causal symmetric spaces. Ann. Global Anal. Geometry **65**(9), (2024). https://doi.org/10.1007/s10455-023-09937-6. arXiv:2307.00798
56. Moschella, U.: New results on de Sitter quantum field theory. Ann. Inst. H. Poincaré **63**(4), 411–426 (1995)
57. Neeb, K.-H.: Holomorphy and Convexity in Lie Theory. Expositions in Mathematics, vol. 28. de Gruyter, Berlin (2000)
58. Neeb, K.-H.: On differentiable vectors for representations of infinite dimensional Lie groups. J. Funct. Anal. **259**, 2814–2855 (2010)
59. Neeb, K.-H., Ólafsson, G.: Reflection positive one-parameter groups and dilations. Complex Anal. Oper. Theory **9**(3), 653–721 (2015)
60. Neeb, K.-H., Ólafsson, G.: Antiunitary representations and modular theory. In : Grabowska, K., et al. (eds.) 50th Sophus Lie Seminar Banach Center Publications, vol. 113, pp. 291–362 (2017). arXiv:math-RT:1704.01336
61. Neeb, K.-H., Ólafsson, G.: Reflection Positivity. A Representation Theoretic Perspective. Springer Briefs in Mathematical Physics, vol. 32 (2018)
62. Neeb, K.-H., Ólafsson, G.: Reflection positivity on spheres. Anal. Math. Phys. **10**(1), 9 (2020)
63. Neeb, K.-H., Ólafsson, G.: Nets of standard subspaces on Lie groups. Adva. Math. **384**, 107715 (2021). arXiv:2006.09832
64. Neeb, K.-H., Ólafsson, G.: Wedge domains in non-compactly causal symmetric spaces. Geometriae Dedicata **217**(2), 30 (2023). arXiv:2205.07685
65. Neeb, K.-H., Ólafsson, G.: Wedge domains in compactly causal symmetric spaces. Int. Math. Res. Notices **2023**(12), 10209–10312 (2023). arXiv:math-RT:2107.13288
66. Neeb, K.-H., Ólafsson, G., Ørsted, B.: Standard subspaces of Hilbert spaces of holomorphic functions on tube domains. Commun. Math. Phys. **386**, 1437–1487 (2021). arXiv:2007.14797
67. Nelson, E.: Analytic vectors. Ann. Math. **70**(3), 572–615 (1959)
68. Ólafsson, G., Pasquale, A.: Ramanujan's master theorem for hypergeometric fourier transform assoiciated with root systems. J. Fourier Anal. Appl. **19**, 1150–1183 (2013)
69. Shimeno, N.: The Plancherel Formula for spherical functions with a one-dimensional K-type on a simply connected Lie group of Hermitian type. J. Funct. Anal. **121**, 330–388 (1994)
70. Simon, B.: The $P(\Phi)_2$ Euclidean (Quantum) Field Theory. Princeton University Press (1974)
71. Simon, T.: Asymptotic behaviour of holomorphic extensions of matrix coefficients at the boundary of the complex crown domain. Preprint, arXiv:2403.13572
72. Trèves, F.: Topological Vector Spaces, Distributions, and Kernels. Academic Press, New York (1967)
73. Wallach, N.: Real Reductive Groups I. Academic Press, Boston (1988)
74. Wallach, N.: Real Reductive Groups II. Academic Press, Boston (1988)
75. Warner, G.: Harmonic Analysis on Semisimple Lie Groups I. Springer, Berlin, Heidelberg, New York (1972)

Heisenberg Parabolically Induced Representations of Hermitian Lie Groups, Part II: Next-to-Minimal Representations and Branching Rules

Jan Frahm, Clemens Weiske, and Genkai Zhang

Abstract Every simple Hermitian Lie group has a unique family of spherical representations induced from a maximal parabolic subgroup whose unipotent radical is a Heisenberg group. For most Hermitian groups, this family contains a complementary series, and at its endpoint sits a proper unitarizable subrepresentation. We show that this subrepresentation is next-to-minimal in the sense that its associated variety is a next-to-minimal nilpotent coadjoint orbit. Moreover, for the Hermitian groups $SO_0(2, n)$ and $E_{6(-14)}$ we study some branching problems of these next-to-minimal representations.

1 Introduction

Minimal representations of simple Lie groups are well studied and have several equivalent descriptions. The most natural one is by their relation to minimal nilpotent coadjoint orbits via the orbit philosophy. They are often unique and show up naturally in relation to the theta correspondence, unipotent representations, and the quantization of nilpotent coadjoint orbits. Moreover, they occur as the local Archimedean components of certain automorphic representations of reductive groups over global fields. Minimal representations of real groups have been studied extensively from various different perspectives such as classical harmonic analysis, partial differential equations, complex analysis, and conformal geometry (see, e.g., [1, 9, 14, 19, 23, 24, 26]). From the representation theoretic point of view, one particularly important question in this context is how minimal representations decompose when restricted to certain subgroups. If the subgroup arises from a dual

J. Frahm (✉)
Department of Mathematics, Aarhus University, Aarhus C, Denmark
e-mail: frahm@math.au.dk

C. Weiske · G. Zhang
Mathematical Sciences, Chalmers University of Technology and Mathematical Sciences, Göteborg University, Göteborg, Sweden
e-mail: weiske@chalmers.se; genkai@chalmers.se

M. Pevzner, H. Sekiguchi (eds.), *Symmetry in Geometry and Analysis, Volume 2*, Progress in Mathematics 358, https://doi.org/10.1007/978-981-97-7662-7_6

pair, this question falls into the framework of the celebrated theta correspondence (see, e.g., [16, 18, 25, 27]). On the other hand, the restriction to symmetric subgroups has also turned out to reveal interesting new results (see, e.g., [22, 28]).

Much less studied are *next-to-minimal representations*; these ought to correspond to nilpotent coadjoint orbits whose closure is the union of the orbit itself and the trivial and minimal orbits; see the precise definition below. For some groups it has been shown that certain next-to-minimal representations also occur as local Archimedean components of global automorphic representations (see [13]), and their automorphic realizations seem to be of growing interest, in particular for exceptional groups (see [3, 11, 12, 29]). It is therefore desirable to gain a better understanding of next-to-minimal representations, both globally and locally. The purpose of this paper is to study some branching laws for next-to-minimal representations of Hermitian groups.

While minimal representations of Hermitian Lie groups turn out to be unitary highest or lowest weight representations, there do exist next-to-minimal representations which are neither highest nor lowest weight representations. This makes them more difficult to construct and understand. In our previous work [8], where we studied Heisenberg parabolically induced representations of Hermitian groups, we exhibited some interesting unitary representations showing up at the end of the complementary series. We proved that the representations can be realized on Hilbert spaces of distributions on the Heisenberg group whose Weyl transforms have rank one as operators on Fock spaces. Thus, they have similar properties as the last point in the Wallach set for scalar unitary highest weight representations described using the Euclidean Fourier transform [6, 15, 17].

In this paper, we show that they are in fact next-to-minimal representations. For the Hermitian groups $SO_0(2, n)$ and $E_{6(-14)}$, we further study corresponding branching problems when restricting these next-to-minimal representations to certain symmetric subgroups. Let us describe our results in more detail.

1.1 Next-to-Minimal Representations

Let $\mathfrak{g}$ be a real form of a complex simple Lie algebra $\mathfrak{g}^{\mathbb{C}}$ and $(\mathfrak{g}^{\mathbb{C}})^*$ the dual space of $\mathfrak{g}^{\mathbb{C}}$. The nilpotent cone in $(\mathfrak{g}^{\mathbb{C}})^*$ decomposes into finitely many nilpotent coadjoint orbits, and they can be partially ordered by the closure relation: $\mathcal{O}_1 \leq \mathcal{O}_2$ if $\mathcal{O}_1 \subseteq \overline{\mathcal{O}_2}$. Among all nontrivial nilpotent orbits, there is a unique minimum $\mathcal{O}_{\min}$ called the *minimal* nilpotent orbit. It is characterized by the property that its closure equals $\mathcal{O}_{\min} \sqcup \{0\}$, or, equivalently, that $\dim \mathcal{O}_{\min}$ is minimal among all nontrivial nilpotent coadjoint orbits. Following [12, 13], we define next-to-minimal nilpotent coadjoint orbits to be orbits larger than the minimal orbit but not larger than any other orbit.

Definition A nilpotent coadjoint orbit $\mathcal{O} \subseteq (\mathfrak{g}^{\mathbb{C}})^*$ is called *next-to-minimal* if its closure is equal to $\mathcal{O} \sqcup \mathcal{O}_{\min} \sqcup \{0\}$. An irreducible unitary representation π of a

Lie group G with Lie algebra $\mathfrak{g}$ is called *next-to-minimal* if its associated variety in $(\mathfrak{g}^{\mathbb{C}})^*$ is the closure of a next-to-minimal nilpotent coadjoint orbit. □

Note that in contrast to the minimal nilpotent coadjoint orbit, next-to-minimal orbits are in general not unique and there might be next-to-minimal orbits of different dimensions. One could also define next-to-minimal orbits as the orbits of next-to-minimal dimension, possibly excluding some orbits of higher dimension that are next-to-minimal in the closure relation. However, in view of the applications in [3, 11, 12, 29], it is only the closure relation of next-to-minimal orbits that is relevant for studying Fourier coefficients of automorphic forms attached to next-to-minimal representations. This justifies the above definition.

Now let G be a simple Hermitian Lie group with $\mathfrak{g} \not\simeq \mathfrak{sp}(n, \mathbb{R})$, i.e., $\mathfrak{g}$ is one of the following Lie algebras:

$$\mathfrak{su}(p,q), \quad \mathfrak{so}(2,n), \quad \mathfrak{so}^*(2n), \quad \mathfrak{e}_{6(-14)}, \quad \mathfrak{e}_{7(-25)}.$$

Up to conjugation, G has a unique maximal parabolic subgroup $P = MAN$ whose unipotent radical N is a Heisenberg group. We consider the degenerate principal series representations:

$$\pi_\nu = \mathrm{Ind}_P^G(1 \otimes e^\nu \otimes 1) \qquad (\nu \in (\mathfrak{a}^{\mathbb{C}})^*),$$

where $\mathfrak{a}$ denotes the Lie algebra of A. Here, π_ν is normalized such that π_ν contains the trivial representation as a quotient for $\nu = \rho$ and as a subrepresentation for $\nu = -\rho$, ρ being the half sum of positive roots with respect to $\mathfrak{a}$, and π_ν is unitary for $\nu \in i\mathfrak{a}^*$. Excluding the case $\mathfrak{g} \simeq \mathfrak{su}(p,q)$ with $p - q$ odd, there exists by Frahm et al. [8, Theorem 4.1] an interval $(-\nu_0, \nu_0) \subseteq \mathfrak{a}^*$ such that π_ν, $\nu \in \mathfrak{a}^*$, is irreducible and unitarizable if and only if $\nu \in (-\nu_0, \nu_0)$. Let $A_\nu : \pi_\nu \to \pi_{-\nu}$ denote the Knapp–Stein standard intertwining operators.

Theorem A (see Sect. 3) $\pi_{\mathrm{ntm}} = \mathrm{Ker}\, A_{-\nu_0} = \mathrm{Im}\, A_{\nu_0} \subsetneq \pi_{-\nu_0}$ *is a proper irreducible and unitarizable subrepresentation which is spherical and next-to-minimal. Its K-type decomposition is given in Theorem 2.1.* □

For $G = \mathrm{O}(p,q)$ with $\min(p,q) \geq 4$, a next-to-minimal representation was constructed in [36], but their construction does not extend to the case $\min(p,q) = 2$. The reason is that $\mathrm{O}(p,q)$ has two different next-to-minimal nilpotent coadjoint orbits. For $\min(p,q) = 2$, the next-to-minimal orbit considered in [36] does not have real points, so there cannot exist a unitary representation whose associated variety is equal to the closure of this orbit. Our next-to-minimal representation has the other next-to-minimal orbit as associated variety.

We further remark that whenever the rank of G is at least 3, the analytic continuation of the scalar type holomorphic discrete series of G contains a next-to-minimal representation, namely, the one corresponding to the next-to-minimal discrete point in the Wallach set. Our representation is different from this one as it is neither highest nor lowest weight module.

1.2 Branching $G \searrow \mathrm{SL}(2,\mathbb{R}) \times M$

If $P = MAN$ is a Langlands decomposition of the Heisenberg parabolic subgroup P of G, then the centralizer of M in G is a subgroup locally isomorphic to $\mathrm{SL}(2,\mathbb{R})$. In fact, the two subgroups M and $\mathrm{SL}(2,\mathbb{R})$ form a dual pair inside G. We study the restriction of π_{ntm} to $\mathrm{SL}(2,\mathbb{R}) \times M$ for the two cases $G = \mathrm{SO}_0(2,n)$ and $G = E_{6(-14)}$.

Let us first consider $G = \mathrm{SO}_0(2,n)$, then $M = \mathrm{SL}(2,\mathbb{R}) \times \mathrm{SO}(n-2)$. In fact, the subgroup $\mathrm{SL}(2,\mathbb{R})$ of M is conjugate to the centralizer of M, and we simply write $\mathrm{SL}(2,\mathbb{R}) \times M$ as $\mathrm{SL}(2,\mathbb{R}) \times \mathrm{SL}(2,\mathbb{R}) \times \mathrm{SO}(n-2)$ without distinguishing between the two copies of $\mathrm{SL}(2,\mathbb{R})$. To state the decomposition of π_{ntm}, let $\eta_k^{\mathrm{SO}(n-2)}$ denote the irreducible representation of $\mathrm{SO}(n-2)$ on the space $\mathcal{H}^k(\mathbb{R}^{n-2})$ of harmonic homogeneous polynomials on $\mathbb{R}^{n-2}$ of degree k. Moreover, for $\mu \in i\mathbb{R}$ and $\varepsilon \in \mathbb{Z}/2\mathbb{Z}$, let $\tau_{\mu,\varepsilon}^{\mathrm{SL}(2,\mathbb{R})}$ be the unitary principal series of $\mathrm{SL}(2,\mathbb{R})$, spherical for $\varepsilon = 0$ and nonspherical for $\varepsilon = 1$, and for $\ell \in \mathbb{Z}$, $|\ell| \geq 2$, let $\tau_\ell^{\mathrm{SL}(2,\mathbb{R})}$ be the discrete series of $\mathrm{SL}(2,\mathbb{R})$ of parameter ℓ. We write $\mathrm{par}(k) \in \mathbb{Z}/2\mathbb{Z}$ for the parity of $k \in \mathbb{Z}$.

Theorem B (see Theorem 4.6) *The restriction of the next-to-minimal representation π_{ntm} of $G = \mathrm{SO}_0(2,n)$ to $\mathrm{SL}(2,\mathbb{R}) \times \mathrm{SL}(2,\mathbb{R}) \times \mathrm{SO}(n-2)$ is given by*

$$\pi_{\mathrm{ntm}}|_{\mathrm{SL}(2,\mathbb{R})\times\mathrm{SL}(2,\mathbb{R})\times\mathrm{SO}(n-2)} \simeq \bigoplus_{k=0}^{\infty} \Bigg(\int_{i\mathbb{R}_{\geq 0}}^{\oplus} \tau_{\mu,\mathrm{par}(k)}^{\mathrm{SL}(2,\mathbb{R})} \boxtimes \tau_{\mu,\mathrm{par}(k)}^{\mathrm{SL}(2,\mathbb{R})} \, d\mu \oplus \bigoplus_{\substack{2\leq|\ell|\leq k \\ \ell\equiv k \mod 2}} \tau_\ell^{\mathrm{SL}(2,\mathbb{R})} \boxtimes \tau_\ell^{\mathrm{SL}(2,\mathbb{R})} \Bigg) \boxtimes \eta_k^{\mathrm{SO}(n-2)}.$$

Now let $G = E_{6(-14)}$, then $M = \mathrm{SU}(5,1)$. Let $P_M = M_M A_M N_M$ be a minimal parabolic subgroup of M. Then, M_M is a double cover of $\mathrm{U}(4)$ whose irreducible representations are parameterized by tuples $(\nu_1, \nu_2, \nu_3, \nu_4)$ with $\nu_j \in \frac{1}{2}\mathbb{Z}$ and $\nu_i - \nu_j \in \mathbb{N}$ for all $1 \leq i < j \leq 4$. Moreover, the irreducible unitary characters of A_M are parameterized by $\mu \in i\mathbb{R}$. We denote the corresponding parabolically induced representation of $\mathrm{SU}(5,1)$ by $\tau_{(\nu_1,\nu_2,\nu_3,\nu_4),\mu}^{\mathrm{SU}(5,1)}$. Keeping the notation for representations of $\mathrm{SL}(2,\mathbb{R})$ from the previous discussion, we can identify part of the spectrum for the decomposition of the restriction of π_{ntm} to $\mathrm{SL}(2,\mathbb{R}) \times \mathrm{SU}(5,1)$:

Theorem C (see Theorem 4.9) *The restriction of the next-to-minimal representation π_{ntm} of $E_{6(-14)}$ to $\mathrm{SL}(2,\mathbb{R}) \times \mathrm{SU}(5,1)$ contains for every $k > 0$ and every $0 < m \leq k$ a representation of the form:*

$$\int_{i\mathbb{R}_{\geq 0}}^{\oplus} \tau_{-l}^{\mathrm{SL}(2,\mathbb{R})} \boxtimes \tau_{(\frac{k}{2},\frac{k}{2}-m,-\frac{k}{2},-\frac{k}{2}),\mu} \, d\mu \oplus \int_{i\mathbb{R}_{\geq 0}}^{\oplus} \tau_{l}^{\mathrm{SL}(2,\mathbb{R})} \boxtimes \tau_{(\frac{k}{2},\frac{k}{2},m-\frac{k}{2},-\frac{k}{2}),\mu} \, d\mu$$

for some $l \geq 1$ as a direct summand. □

The proofs of both Theorem B and C make use of the realization of π_{ntm} in terms of the Heisenberg group Fourier transform. Indeed, since the restriction problem is mostly of analytic nature, the Fourier transform is an effective tool. It relates the decomposition of $\pi_{\mathrm{ntm}}|_M$ to the decomposition of the tensor product of some lowest/highest weight representations of M that occur in the restriction to M of a metaplectic representation. In both cases this tensor product can be decomposed completely. Finally, the contribution of the $\mathrm{SL}(2,\mathbb{R})$-factor is obtained by conjugating $\mathrm{SL}(2,\mathbb{R})$ inside G to a subgroup of M and comparing with the decomposition when restricting π_{ntm} to M. For $G = \mathrm{SO}_0(2,n)$, we can identify the action of the $\mathrm{SL}(2,\mathbb{R})$-factor explicitly, while for $G = E_{6(-14)}$, we are only able to describe its restriction to a parabolic subgroup of $\mathrm{SL}(2,\mathbb{R})$.

We remark that, unlike for minimal representations where branching to dual pairs usually yields a one-to-one correspondence between certain representations of the members of the dual pair, the restriction of π_{ntm} to $\mathrm{SL}(2,\mathbb{R}) \times M$ does not seem to yield such a correspondence. Let us explain this in the case of Theorem B. While a representation of M of the form $\tau \boxtimes \eta_k^{\mathrm{SO}(n-2)}$ corresponds only to the representation τ of $\mathrm{SL}(2,\mathbb{R})$, a representation τ of $\mathrm{SL}(2,\mathbb{R})$ corresponds to representations $\tau \boxtimes \eta_k^{\mathrm{SO}(n-2)}$ of M for infinitely many k. Thus, we obtain a map $\tau \boxtimes \eta_k^{\mathrm{SO}(n-2)} \mapsto \tau$ from certain representations of M to representations of $\mathrm{SL}(2,\mathbb{R})$ which is not injective and therefore does not yield a one-to-one correspondence.

Structure of This Article

We first recall the main results of our previous work [8] in Sect. 2. In Sect. 3 we compute the Gelfand–Kirillov dimension of π_{ntm} using its K-type decomposition and use this information to show that the associated variety of π_{ntm} is indeed a next-to-minimal nilpotent coadjoint orbit in $(\mathfrak{g}^{\mathbb{C}})^*$. Finally, Sect. 4 is concerned with the restriction of the next-to-minimal representation π_{ntm} of G to the subgroup $\mathrm{SL}(2,\mathbb{R}) \times M$. While Sect. 4.1 contains some general observation on how to use the Heisenberg Fourier transform for this problem, we treat the case $G = \mathrm{SO}_0(2,n)$ in Sect. 4.2, and in Sect. 4.3 we discuss the case $G = E_{6(-14)}$.

Notation

For a unitarizable Casselman–Wallach representation π, we abuse notation and denote its unitary closure also by π, while usually suppressing the notation of the underlying Fréchet resp. Hilbert spaces. In this sense, if not stated otherwise, all direct sums and tensor products of representations are to be considered in the category of Hilbert spaces.

2 Preliminaries

We recall the results about Heisenberg parabolically induced representations of Hermitian Lie groups from [35] as well as their realization using the Heisenberg group Fourier transform obtained in [8].

2.1 Heisenberg Parabolically Induced Representations

Let G be a simple Hermitian Lie group with Lie algebra $\mathfrak{g}$ not isomorphic to $\mathfrak{sl}(2, \mathbb{R})$. Then, G has a unique conjugacy class of parabolic subgroups whose unipotent radical is a Heisenberg group. Let $P = MAN$ be the Langlands decomposition of one of them and write $\mathfrak{m}$, $\mathfrak{a}$, $\mathfrak{n}$ for the Lie algebras of M, A, N.

The one-dimensional subalgebra $\mathfrak{a}$ is spanned by an element $H \in \mathfrak{a}$ such that

$$\mathfrak{g} = \bar{\mathfrak{n}}_2 \oplus \bar{\mathfrak{n}}_1 \oplus (\mathfrak{m} \oplus \mathfrak{a}) \oplus \mathfrak{n}_1 \oplus \mathfrak{n}_2$$

is a decomposition into eigenspaces of $\mathrm{ad}(H)$ with eigenvalues $-2, -1, 0, 1, 2$, and $\mathfrak{n} = \mathfrak{n}_1 \oplus \mathfrak{n}_2$. We further write $\bar{\mathfrak{n}} = \bar{\mathfrak{n}}_1 \oplus \bar{\mathfrak{n}}_2$ for the Lie algebra of the opposite unipotent radical $\bar{N}$. Note that $\mathfrak{n}_2$ and $\bar{\mathfrak{n}}_2$ equal the center of $\mathfrak{n}$ and $\bar{\mathfrak{n}}$, respectively, and are therefore one-dimensional and we choose $E \in \mathfrak{n}_2$ and $F \in \bar{\mathfrak{n}}_2$ such that $[E, F] = H$. Hence $\{E, H, F\}$ form an $\mathfrak{sl}(2)$-triple. We identify $(\mathfrak{a}^{\mathbb{C}})^*$ with $\mathbb{C}$ by $\nu \mapsto \nu(H)$; then, the half sum of positive roots is given by $\rho = d_1 + 1$, where $d_1 = \frac{1}{2} \dim_{\mathbb{R}} \mathfrak{n}_1 \in \mathbb{N}$ (see Table 1). Consider the degenerate principal series representations (smooth normalized parabolic induction):

$$\pi_\nu = \mathrm{Ind}_P^G(1 \otimes e^\nu \otimes 1) \qquad (\nu \in (\mathfrak{a}^{\mathbb{C}})^* \simeq \mathbb{C}),$$

on the space

$$I(\nu) = \{f \in C^\infty(G);\ f(gman) = a^{-\nu-\rho} f(g)\, \forall man \in MAN\}.$$

To describe the K-type structure of π_ν, let θ be a Cartan involution on G which leaves MA invariant and maps E to $-F$ and denote by $K \subseteq G$ the corresponding maximal compact subgroup. Its Lie algebra $\mathfrak{k}$ decomposes into $\mathfrak{k} = \mathfrak{z}(\mathfrak{k}) \oplus \mathfrak{k}'$ with $\mathfrak{k}' = [\mathfrak{k}, \mathfrak{k}]$ the semisimple part and $\mathfrak{z}(\mathfrak{k})$ the center.

The Lie algebra $\mathfrak{k}'$ has a Hermitian symmetric pair $(\mathfrak{k}', \mathfrak{l}')$ of rank 2. Let $2\alpha_1 > 2\alpha_2$ be the Harish–Chandra strongly orthogonal roots and note that the restricted root system for the pair $(\mathfrak{k}', \mathfrak{l}')$ is of type C_2 or BC_2:

$$\{\pm 2\alpha_1, \pm 2\alpha_2\} \cup \{\pm\alpha_1 \pm \alpha_2\} \quad \Big[\cup \{\pm\alpha_1, \pm\alpha_2\}\Big].$$

The multiplicity of $\pm 2\alpha_1$ and $\pm 2\alpha_2$ is 1, and we write a_1 resp. $2b_1$ for the root multiplicities of $\pm\alpha_1 \pm \alpha_2$ resp. $\pm\alpha_1$ and $\pm\alpha_2$. The values of a_1 and b_1 for the different Hermitian groups are listed in Table 1.

We further let α_0 be a linear functional on the center $\mathfrak{z}(\mathfrak{k}^{\mathbb{C}})$ normalized such that it has value 1 on the central element with eigenvalues ± 1 on $\mathfrak{p}^{\mathbb{C}}$. Writing $W_{\mu_1,\mu_2,\ell}$

Table 1 Hermitian Lie algebras $\mathfrak{g}$ with subalgebras $\mathfrak{k}$ and $\mathfrak{m}$ and structure constants a_1, b_1, and d_1

$\mathfrak{g}$	$\mathfrak{k}$	$\mathfrak{m}$	(a_1, b_1)	d_1
$\mathfrak{su}(p,q)$	$\mathfrak{s}(\mathfrak{u}(p)+\mathfrak{u}(q))$	$\mathfrak{u}(p-1,q-1)$	$(0,p-2),(0,q-2)$	$p+q-2$
$\mathfrak{so}^*(2n)$	$\mathfrak{u}(n)$	$\mathfrak{so}^*(2n-4)+\mathfrak{su}(2)$	$(2,n-4)$	$2n-4$
$\mathfrak{sp}(n,\mathbb{R})$	$\mathfrak{u}(n)$	$\mathfrak{sp}(n-1,\mathbb{R})$	$(0,n-2)$	$n-1$
$\mathfrak{so}(n,2)$	$\mathfrak{so}(n)+\mathfrak{so}(2)$	$\mathfrak{sl}(2,\mathbb{R})+\mathfrak{so}(n-2)$	$(n-4,0)$	$n-2$
$\mathfrak{e}_{6(-14)}$	$\mathfrak{spin}(10)+\mathfrak{so}(2)$	$\mathfrak{su}(5,1)$	$(4,2)$	10
$\mathfrak{e}_{7(-25)}$	$\mathfrak{e}_6+\mathfrak{so}(2)$	$\mathfrak{so}(10,2)$	$(6,4)$	16

for the irreducible representation of K with highest weight $\mu_1\alpha_1+\mu_2\alpha_2+\ell\alpha_0$, the K-type decomposition of π_ν can be written as follows:

$$\pi_\nu|_K \simeq \bigoplus_{\substack{\mu_1\geq\mu_2\geq|\ell| \\ \mu_1\equiv\mu_2\equiv\ell \mod 2}} W_{\mu_1,\mu_2,\ell} \qquad \text{if } \mathfrak{g}\not\simeq\mathfrak{sp}(n,\mathbb{R}),\mathfrak{su}(p,q),$$

$$\pi_\nu|_K \simeq \bigoplus_{\substack{\mu_1,\mu_2\geq|\ell|, \\ \mu_1\equiv\mu_2\equiv\ell \mod 2}} W_{\mu_1,\mu_2,\ell} \qquad \text{if } \mathfrak{g}\simeq\mathfrak{su}(p,q).$$

2.2 Intertwining Operators

Let

$$w_0 = \exp\left(\frac{\pi}{2}(E-F)\right) \in K;$$

then, $\mathrm{Ad}(w_0)P = \bar{P} = MA\bar{N}$ is the parabolic subgroup opposite to P. Let $A_\nu : I(\nu) \to I(-\nu)$ be the standard intertwining operator, which is for $\mathrm{Re}(\nu) > \rho$, $f \in I(\nu)$ and $g \in G$ given by the convergent integral:

$$A_\nu f(g) = \int_{\bar{N}} f(gw_0\bar{n})\, d\bar{n}$$

and extended meromorphically to all $\nu \in \mathbb{C}$. For $\nu \in \mathbb{R}$, the corresponding Hermitian form on $I(\nu)$ given by

$$\langle f_1, f_2\rangle_\nu = \langle A_\nu f_1, f_2\rangle_{L^2(\bar{N})} = \int_{\bar{N}} A_\nu f_1(\bar{n})\overline{f_2(\bar{n})}\, d\bar{n} \tag{1}$$

is G-invariant. In [8] we studied unitarizability and the composition series of the representations π_ν using the operators A_ν. We recall the main results.

Theorem 2.1 (see [8, Theorem 4.1 and Theorem 4.3]) *Assume* $\mathfrak{g} \not\simeq \mathfrak{sp}(n, \mathbb{R})$ *and let*

$$\nu_0 = \begin{cases} 1 & \text{if } \mathfrak{g} \simeq \mathfrak{su}(p,q) \text{ and } p-q \text{ is even,} \\ a_1 + 1 & \text{if } \mathfrak{g} \not\simeq \mathfrak{su}(p,q). \end{cases}$$

(i) For $\nu \in \mathbb{R}$*, the representation* π_ν *belongs to the complementary series, i.e., it is irreducible and unitarizable, if and only if* $|\nu| < \nu_0$.
(ii) At the endpoint $-\nu_0$ *of the complementary series, there is a proper unitarizable irreducible subrepresentation* $\pi_{\mathrm{ntm}} := \operatorname{Ker} A_{-\nu_0} = \operatorname{Im} A_{\nu_0}$ *with* K*-type decomposition:*

$$\pi_{\mathrm{ntm}}|_K \simeq \bigoplus_{\substack{\mu \geq |\ell| \\ \mu \equiv \ell \mod 2}} W_{\mu,\mu,\ell} \qquad \text{if } \mathfrak{g} \not\simeq \mathfrak{su}(p,q),$$

$$\pi_{\mathrm{ntm}}|_K \simeq \bigoplus_{\substack{\mu_1,\mu_2 \geq |\ell|, \\ \mu_1 - \mu_2 = q-p, \\ \mu_1 \equiv \mu_2 \equiv \ell \mod 2}} W_{\mu_1,\mu_2,\ell} \qquad \text{if } \mathfrak{g} \simeq \mathfrak{su}(p,q),\ p-q \text{ even.}$$

2.3 The Heisenberg Group Fourier Transform

Since $\bar{N}MAN$ is open and dense in G, the restriction from G to $\bar{N}$ defines an embedding of $I(\nu)$ into $C^\infty(\bar{N})$, the so-called non-compact picture of π_ν.

Following [8] we further use the notation $V_1 = \bar{\mathfrak{n}}_1$ and note that V_1 has a canonical complex structure given by the complex structure for the Hermitian symmetric subpair $(\mathfrak{m}, \mathfrak{m} \cap \mathfrak{k})$ of $(\mathfrak{g}, \mathfrak{k})$. We can identify $\bar{\mathfrak{n}} = \bar{\mathfrak{n}}_1 \oplus \bar{\mathfrak{n}}_2 = V_1 \oplus \mathbb{R}F$ with $V_1 \times \mathbb{R}$ by

$$(v, t) \mapsto v + tF \qquad (v \in V_1, t \in \mathbb{R}).$$

The infinite-dimensional irreducible unitary representations of $\bar{N}$ are parameterized by their central characters in $i\bar{\mathfrak{n}}_2^*$. We identify $\bar{\mathfrak{n}}_2^*$ with $\mathbb{R}^\times$ by $\lambda \mapsto \lambda(F)$. Write σ_λ for the representation with central character $-i\lambda$ which we realize on a Fock space $\mathcal{F}_\lambda(V_1)$ of (anti-)holomorphic (depending on the sign of λ) functions on V_1 which are square-integrable with respect to a Gaussian measure.

The Heisenberg group Fourier transform is the unitary isomorphism:

$$\mathcal{F}\colon L^2(\bar{N}) \to \int_{\mathbb{R}^\times}^{\oplus} \mathcal{F}_\lambda(V_1) \otimes \mathcal{F}_\lambda(V_1)^* \, d\lambda$$

given by

$$\mathcal{F}u(\lambda) = \sigma_\lambda(u) = \int_{\bar{N}} u(\bar{n})\sigma_\lambda(\bar{n})\, d\bar{n} \qquad (u \in L^1(\bar{N}) \cap L^2(\bar{N})),$$

where we identify the Hilbert space tensor product $\mathcal{F}_\lambda(V_1) \otimes \mathcal{F}_\lambda(V_1)^*$ with the Hilbert space $\mathrm{HS}(\mathcal{F}_\lambda(V_1))$ of Hilbert–Schmidt operators on $\mathcal{F}_\lambda(V_1)$. More precisely, using the Hilbert–Schmidt norm $\|T\|_{\mathrm{HS}} = \mathrm{tr}(TT^*)^{\frac{1}{2}}$, we have:

$$\|u\|^2_{L^2(\bar{N})} = \mathrm{const} \times \int_{\mathbb{R}^\times} \|\sigma_\lambda(u)\|^2_{\mathrm{HS}} |\lambda|^{d_1}\, d\lambda \qquad (u \in L^2(\bar{N})), \tag{2}$$

the constant only depending on the normalization of measures and the dimension $d_1 = \dim_{\mathbb{C}} V_1$.

The Fourier transform can be extended to tempered distributions on $\overline{N}$. For this, we identify all spaces $\mathcal{F}_\lambda(V_1)$ $(\lambda \in \mathbb{R}^\times)$ with a fixed Hilbert space $\mathcal{H}$, so that the Fourier transform can be viewed as a unitary isomorphism:

$$\mathcal{F}\colon L^2(\overline{N}) \to L^2(\mathbb{R}^\times; |\lambda|^{d_1}\, d\lambda) \hat{\otimes}\, \mathrm{HS}(\mathcal{H}),$$

the tensor product being the tensor product of Hilbert spaces. This can for instance be done by using the Schrödinger model of σ_λ on $\mathcal{H} = L^2(\Lambda)$ for a Lagrangian real subspace $\Lambda \subseteq V_1$. The identification should be chosen in a way so that the Fréchet space $\mathcal{H}^\infty$ of smooth vectors is independent of λ (for $\mathcal{H} = L^2(\Lambda)$, we have $\mathcal{H}^\infty = \mathcal{S}(\Lambda)$, the Schwartz space). Write $\mathcal{H}^{-\infty}$ for the dual space of $\mathcal{H}^\infty$ (for $\mathcal{H} = L^2(\Lambda)$, this is the space $\mathcal{H}^{-\infty} = \mathcal{S}'(\Lambda)$ of tempered distributions). We endow the space $\mathrm{Hom}(\mathcal{H}^\infty, \mathcal{H}^{-\infty})$ of continuous linear operators from $\mathcal{H}^\infty$ to $\mathcal{H}^{-\infty}$ with the topology of bounded convergence (for $\mathcal{H} = L^2(\Lambda)$, we have $\mathrm{Hom}(\mathcal{H}^\infty, \mathcal{H}^{-\infty}) = \mathrm{Hom}(\mathcal{S}(\Lambda), \mathcal{S}'(\Lambda)) \simeq \mathcal{S}'(\Lambda \times \Lambda)$ by Trèves [31, Corollary 51.6]). It is shown in [7, Corollary 3.5.3] that the Fourier transform extended to a continuous linear map:

$$\mathcal{F}\colon \mathcal{S}'(\bar{N}) \to \mathcal{D}'(\mathbb{R}^\times) \hat{\otimes} \mathrm{Hom}(\mathcal{H}^\infty, \mathcal{H}^{-\infty}),$$

where we embed $L^2(\mathbb{R}^\times; |\lambda|^{d_1}\, d\lambda)$ into $\mathcal{D}'(\mathbb{R}^\times)$ simply by using the Lebesgue measure on $\mathbb{R}^\times \subseteq \mathbb{R}$. Since $\mathcal{D}'(\mathbb{R}^\times)$ is nuclear (see [31, Corollary 51.5]), it does not matter which topological tensor product is used above (see [31, Theorem 50.1]). By Frahm [7, Corollary 3.5.3] the extension of the Fourier transform to distributions is injective on $I(\nu)$ for $\mathrm{Re}\, \nu > -\rho$.

2.4 The Metaplectic Representation

The real space V_1 is naturally a symplectic vector space with symplectic form ω given by

$$[v, w] = \omega(v, w)F \qquad (v, w \in V_1 = \bar{\mathfrak{n}}_1).$$

Let $\mathrm{Sp}(V_1, \omega)$ denote its symplectic group with Lie algebra $\mathfrak{sp}(V_1, \omega)$. For $\lambda \in \mathbb{R}^\times$ let $\omega_{\mathrm{met},\lambda}$ be the unique projective representation of $\mathrm{Sp}(V_1, \omega)$ on $\mathcal{F}_\lambda(V_1)$, such that

$$\sigma_\lambda(gv, t) = \omega_{\mathrm{met},\lambda}(g) \circ \sigma_\lambda(v, t) \circ \omega_{\mathrm{met},\lambda}(g)^{-1}.$$

This representation is called *metaplectic representation*, and its equivalence class only depends on the sign of λ. Let $d\omega_{\mathrm{met},\lambda}$ be the derived $\mathfrak{sp}(V_1, \omega)$-representation. In particular, $d\omega_{\mathrm{met},\lambda}$ is given by skew symmetric holomorphic resp. anti-holomorphic differential operators on V_1 for $\lambda > 0$ resp. $\lambda < 0$. The underlying Harish–Chandra module is the space $\mathcal{P}(V_1)$ of holomorphic resp. anti-holomorphic polynomials on V_1. We recall the main result of [8], which gives the explicit decomposition of $d\omega_{\mathrm{met},\lambda}$ restricted to $\mathfrak{m}$, the Lie algebra of M. Let therefore $L := K \cap M$ with Lie algebra $\mathfrak{l}$ and let $\mathfrak{t}_{\mathfrak{l}}$ be a Cartan subalgebra of $\mathfrak{l}^{\mathbb{C}}$.

Theorem 2.2 (see [8, Theorem 2.2]) *Let $\lambda > 0$ and assume $\mathfrak{g} \not\simeq \mathfrak{su}(p, q), \mathfrak{sp}(n, \mathbb{R})$. Then, the restriction $d\omega_{\mathrm{met},\lambda}|_{\mathfrak{m}}$ of the metaplectic representation of $\mathfrak{sp}(V_1, \omega)$ to $\mathfrak{m}$ decomposes as*

$$d\omega_{\mathrm{met},\lambda}|_{\mathfrak{m}} = \bigoplus_{k=0}^{\infty} \tau_{-k\delta_0 - \frac{1}{2}\zeta_0},$$

where τ_μ denotes the unitary highest weight representation of $\mathfrak{m}$ with highest weight $\mu \in \mathfrak{t}_{\mathfrak{l}}^*$, δ_0 is the lowest root of V_1 and ζ_0 is the central character of $\mathfrak{l}$ obtained by restriction of the trace of the defining complex linear action of $\mathfrak{u}(V_1) \subseteq \mathfrak{sp}(V_1, \omega)$ on V_1 to $\mathfrak{t}_{\mathfrak{l}}$.

To ease notation, we denote the k-th part in the decomposition of $d\omega_{\mathrm{met},\lambda}|_{\mathfrak{m}}$ resp. $\omega_{\mathrm{met},\lambda}|_M$, in the sense of the theorem above, by $d\omega_{\mathrm{met},\lambda,k}$ resp. $\omega_{\mathrm{met},\lambda,k}$.

2.5 The Fourier Transform of Intertwining Operators

From Theorem 2.2 it follows that the Fock space $\mathcal{F}_\lambda(V_1)$ decomposes into the direct sum of representation spaces for the unitary highest weight representations $\tau_{-k\delta_0 - \frac{1}{2}\zeta_0}$:

$$\mathcal{F}_\lambda(V_1) = \bigoplus_{k \geq 0} \mathcal{F}_{\lambda,k}(V_1). \tag{3}$$

We denote by $P_k : \mathcal{F}_\lambda(V_1) \to \mathcal{F}_{\lambda,k}(V_1)$ the orthogonal projections. Since the Fourier transform is injective on $I(\nu)$ and $I(-\nu)$ for $\mathrm{Re}(\nu) \in (-\rho, \rho)$, we have that the Fourier transform $\widehat{A}_\nu$ of the intertwining operator A_ν, which is defined by $\widehat{A}_\nu \sigma_\lambda(u) := \sigma_\lambda(A_\nu u)$, is a diagonal operator with respect to the decomposition (3).

Theorem 2.3 (see [8, Theorem 3.1]) *For* $\mathfrak{g} \not\simeq \mathfrak{su}(p, q), \mathfrak{sp}(n, \mathbb{R})$ *the operator* $\widehat{A}_\nu$ *is given by*

$$\widehat{A}_\nu = \mathrm{const} \times |\lambda|^{-\nu} \sum_{k \geq 0} a_k(\nu) \cdot P_k,$$

with positive constant only depending on the structure constants $\rho = d_1 + 1$ *and* a_1 *(see Table 1) and*

$$a_k(\nu) = \frac{2^\nu \left(\frac{-\nu+a_1+1}{2}\right)_k \Gamma(\frac{\nu-\rho+a_1+2}{2})\Gamma(\frac{\nu}{2})}{\Gamma(\frac{-\nu+\rho}{2})\Gamma(\frac{\nu+a_1+1}{2}+k)}.$$

Combined with (2), Theorem 2.3 shows the following identity for the invariant Hermitian form $\langle \cdot, \cdot \rangle_\nu$ from (1):

$$\langle u, u \rangle_\nu = \mathrm{const} \times \sum_{k \geq 0} a_k(\nu) \int_{\mathbb{R}^\times} \| P_k \circ \sigma_\lambda(u) \|_{\mathrm{HS}}^2 |\lambda|^{d_1 - \nu} \, d\lambda \qquad (u \in I(\nu)). \tag{4}$$

In particular, at the endpoint of the complementary series $\nu = \nu_0 = a_1 + 1$, we have: $a_k(\nu_0) = 0$ for all $k \neq 0$, so

$$\langle u, u \rangle_{\nu_0} = \mathrm{const} \times a_0(\nu_0) \int_{\mathbb{R}^\times} \| P_0 \circ \sigma_\lambda(u) \|_{\mathrm{HS}}^2 |\lambda|^{d_1 - \nu_0} \, d\lambda \qquad (u \in I(\nu)). \tag{5}$$

Remark 2.4 For $\mathfrak{g} = \mathfrak{su}(p, q)$ a similar decomposition as in Theorem 2.2 is obtained in [8, Theorem 2.3] and is also given by a multiplicity free direct sum of unitary highest weight representations. The parametrization is slightly different, so we omit the statement. The eigenvalues of the intertwining operator in this case are found in [8, Theorem 3.1].

In the symplectic case $\mathfrak{g} = \mathfrak{sp}(n, \mathbb{R})$, we have $\mathfrak{m} = \mathfrak{sp}(V_1, \omega)$, and the Fock space decomposes into two nonequivalent irreducible subrepresentations, the even and the odd part of the metaplectic representations. The eigenvalues for the standard intertwining operator are given in [8, Theorem 3.8].

However, the results for $\mathfrak{su}(p, q)$ and $\mathfrak{sp}(n, \mathbb{R})$ will not be used in the present paper. □

3 Associated Varieties

We compute the associated variety of the representation π_{ntm} that occurs as proper subrepresentation at the endpoint of the complementary series. Recall that the associated variety $\mathcal{V}(\pi)$ of an irreducible unitary representation π of G with annihilator $\mathrm{Ann}(\pi) \subseteq \mathcal{U}(\mathfrak{g}^{\mathbb{C}})$ is the subvariety of $(\mathfrak{g}^{\mathbb{C}})^*$ corresponding to the graded ideal:

$$\mathrm{gr}\,\mathrm{Ann}(\pi) \subseteq \mathrm{gr}\,\mathcal{U}(\mathfrak{g}^{\mathbb{C}}) \simeq S(\mathfrak{g}^{\mathbb{C}}) \simeq \mathbb{C}[(\mathfrak{g}^{\mathbb{C}})^*].$$

By Vogan [33, Corollary 4.7], $\mathcal{V}(\pi)$ is the closure of a single nilpotent coadjoint orbit. Further, by Borho and Kraft [2, Satz 3.2 (b) and Korollar 5.4], its dimension equals the Gelfand–Kirillov dimension of $\mathcal{U}(\mathfrak{g}^{\mathbb{C}})/\mathrm{Ann}(\pi)$ which in turn equals twice the Gelfand–Kirillov dimension of π by Vogan [32, Corollary 4.7].

We briefly recall the definition of the Gelfand–Kirillov dimension of a finitely generated $\mathcal{U}(\mathfrak{g}^{\mathbb{C}})$-module V (see, e.g., the discussion in [32, Section 2]). Let $\{\mathcal{U}_n(\mathfrak{g}^{\mathbb{C}})\}_{n\geq 0}$ denote the natural filtration of the universal enveloping algebra $\mathcal{U}(\mathfrak{g}^{\mathbb{C}})$ of $\mathfrak{g}^{\mathbb{C}}$. For a finite-dimensional subspace $V_0 \subseteq V$ that generates V, we let $V_n = \mathcal{U}_n(\mathfrak{g}^{\mathbb{C}})V_0$ $(n \in \mathbb{N})$. Then, the function $n \mapsto \dim V_n$ agrees for large n with a polynomial whose degree d is independent of the choice of V_0. The integer $\mathrm{GKDIM}(V) = d$ is called *Gelfand–Kirillov dimension of* V.

Our strategy is to compute the Gelfand–Kirillov dimension of π_{ntm} by studying the growth of the dimensions of K-types, and to compare it with the dimensions of nilpotent coadjoint orbits. It turns out that the associated variety always is the closure of a next-to-minimal nilpotent coadjoint orbit $\mathcal{O}_{\mathrm{ntm}} \subseteq (\mathfrak{g}^{\mathbb{C}})^*$. The computation of the Gelfand–Kirillov dimension of π_{ntm} is done uniformly for all cases, but for the comparison with coadjoint orbits, we carry out a case-by-case analysis.

3.1 *The Gelfand–Kirillov Dimension*

Assume that $\mathfrak{g} \not\simeq \mathfrak{sp}(n,\mathbb{R}), \mathfrak{su}(p,q)$, the case $\mathfrak{su}(p,q)$ is treated separately in Sect. 3.6. By Theorem 2.1 the K-types of π_{ntm} are $W_{\mu,\mu,\ell}$ with $\mu \geq |\ell|$, $\mu \equiv \ell$ mod 2. Note that $\dim W_{\mu,\mu,\ell}$ is independent of ℓ since ℓ only parameterizes the action of the center of $\mathfrak{k}$ which is by a scalar. By the Weyl dimension formula, the function $\mu \mapsto \dim W_{\mu,\mu,\ell}$ is a polynomial. We first compute its degree.

Lemma 3.1 *The degree of the polynomial* $\mu \mapsto \dim W_{\mu,\mu,\ell}$ *is* $a_1 + 4b_1 + 2$. □

Proof Recall the Weyl dimension formula:

$$\dim W_{\mu,\mu,\ell} = \prod_{\alpha\in\Delta^+(\mathfrak{k}^{\mathbb{C}},\mathfrak{t}^{\mathbb{C}})} \frac{\langle \lambda+\rho,\alpha\rangle}{\langle \rho,\alpha\rangle},$$

where $\mathfrak{t} \subseteq \mathfrak{k}$ is a Cartan subalgebra and $\lambda \in (\mathfrak{t}^{\mathbb{C}})^*$ denotes the highest weight of $W_{\mu,\mu,\ell}$ with respect a system $\Delta^+(\mathfrak{k}^{\mathbb{C}}, \mathfrak{t}^{\mathbb{C}})$ of positive roots. It follows that $\dim W_{\mu,\mu,\ell}$ is a polynomial in μ whose degree is the number of positive roots which are *not* orthogonal to λ. Passing to the restricted root system

$$\{\pm 2\alpha_1, \pm 2\alpha_2\} \cup \{\pm\alpha_1 \pm \alpha_2\} \quad \Big[\cup \{\pm\alpha_1, \pm\alpha_2\}\Big]$$

with multiplicities 1, a_1, and $2b_1$, we can express λ as $\mu(\alpha_1+\alpha_2)+\ell\alpha_0$. The positive restricted roots not orthogonal to $\alpha_1 + \alpha_2$ are $2\alpha_1$ and $2\alpha_2$, each with multiplicity 1, $\alpha_1 + \alpha_2$ with multiplicity a_1, and possibly α_1 and α_2, each with multiplicity $2b_1$. Adding up the multiplicities shows the claim. □

Proposition 3.2 *The Gelfand–Kirillov dimension of* π_{ntm} *is* $a_1 + 4b_1 + 4$. □

Proof Let $V_0 = W_{0,0,0}$ and $V_n = \mathcal{U}_n(\mathfrak{g}^{\mathbb{C}})V_0$. From [35, Theorem 3.1], it follows that

$$V_n = \bigoplus_{\substack{|\ell|\leq\mu\leq n \\ \mu\equiv\ell \mod 2}} W_{\mu,\mu,\ell},$$

and hence

$$\dim V_n = \sum_{\mu=0}^{n} \sum_{\substack{\ell=-\mu \\ \ell\equiv\mu \mod 2}}^{\mu} \dim W_{\mu,\mu,\ell} \sim \sum_{\mu=0}^{n} \sum_{\substack{\ell=-\mu \\ \ell\equiv\mu \mod 2}}^{\mu} \mu^{a_1+4b_1+2}$$

$$= \sum_{\mu=0}^{n} (\mu+1)\mu^{a_1+4b_1+2} \sim n^{a_1+4b_1+4}.$$

Hence, $\dim V_n$ is of degree $a_1 + 4b_1 + 4$. □

Now we know that the associated variety of π_{ntm} is the closure of a finite union of nilpotent coadjoint orbits of dimension $2\cdot\mathrm{GKDIM}(\pi_{\mathrm{ntm}}) = 2(a_1+4b_1+4)$. In each of the cases, we show that there is a unique nilpotent orbit $\mathcal{O}_{\mathrm{ntm}}$ whose dimension is minimal among all next-to-minimal orbits and equal to $2(a_1 + 4b_1 + 4)$. It follows that $\mathcal{O}_{\mathrm{ntm}}$ is the unique nilpotent coadjoint orbit of dimension $2 \cdot \mathrm{GKDIM}(\pi_{\mathrm{ntm}})$, so the associated variety of π_{ntm} equals $\overline{\mathcal{O}_{\mathrm{ntm}}}$.

3.2 *The Case* $\mathfrak{g} = \mathfrak{so}(2, n)$

We have $a_1 = n - 4$ and $b_1 = 0$, so

$$\mathrm{GKDIM}(\pi_{\mathrm{ntm}}) = a_1 + 4b_1 + 4 = n.$$

By Collingwood and McGovern [4, Theorem 6.2.5], there are two next-to-minimal nilpotent coadjoint orbits in $\mathfrak{g}^{\mathbb{C}} \simeq \mathfrak{so}(n+2, \mathbb{C})$, and they are associated with the partitions $(2^4, 1^{n-6})$ and $(3^1, 1^{n-1})$. Using [4, Corollary 6.1.4], we find that the dimension of the orbit associated with $(2^4, 1^{n-6})$ is $2(2n-6)$, while for $(3^1, 1^{n-1})$, it equals $2n$. It follows that the associated variety of π_{ntm} is the next-to-minimal nilpotent coadjoint orbit associated with the partition $(3^1, 1^{n-1})$.

Remark 3.3 In [36], a next-to-minimal representation of $\mathrm{O}(p, q)$ is constructed under the assumption that $\min(p, q) \geq 4$. The associated variety of this representation corresponds to the partition $(2^4, 1^{p+q-8})$. The assumption $\min(p, q) \geq 4$ is necessary for the existence of such a representation, because it is equivalent to the nilpotent orbit associated with $(2^4, 1^{p+q-8})$ having real points. This implies that for $\min(p, q) = 2$, there is no irreducible unitary representation of $G = \mathrm{O}(p, q)$ with associated variety equal to the next-to-minimal orbit associated with the partition $(2^4, 1^{n-6})$. In this sense, our representation π_{ntm} provides a replacement of the next-to-minimal representation of [36] for the case $\min(p, q) = 2$. □

3.3 *The Case* $\mathfrak{g} = \mathfrak{so}^*(2n)$

Here $a_1 = 2$ and $b_1 = n - 4$, so

$$\mathrm{GKDIM}(\pi_{\mathrm{ntm}}) = a_1 + 4b_1 + 4 = 4n - 10.$$

Since $\mathfrak{g}^{\mathbb{C}} = \mathfrak{so}(2n, \mathbb{C})$, the discussion in Sect. 3.2 shows that there are precisely two next-to-minimal nilpotent coadjoint orbits, and their dimensions are $2(4n-10)$ and $2(2n-2)$, respectively. The first one belongs to the partition $(2^4, 1^{2n-8})$, and its closure is the associated variety of π_{ntm}.

3.4 *The Case* $\mathfrak{g} = \mathfrak{e}_{6(-14)}$

In this case $a_1 = 4$ and $b_1 = 2$, so

$$\mathrm{GKDIM}(\pi_{\mathrm{ntm}}) = a_1 + 4b_1 + 4 = 16.$$

By Collingwood and McGovern [4, table on p. 129], there is a unique next-to-minimal nilpotent coadjoint orbit in $\mathfrak{g}^{\mathbb{C}} \simeq \mathfrak{e}_6(\mathbb{C})$, and it has dimension $32 = 2 \cdot 16$ and Bala–Carter label $2A_1$. This shows that the associated variety of π_{ntm} is equal to the closure of this orbit.

3.5 The Case $\mathfrak{g} = \mathfrak{e}_{7(-25)}$

Here $a_1 = 6$ and $b_1 = 4$, so

$$\mathrm{GKDIM}(\pi_{\mathrm{ntm}}) = a_1 + 4b_1 + 4 = 26.$$

By Collingwood and McGovern [4, table on p. 130], there is a unique next-to-minimal nilpotent coadjoint orbit in $\mathfrak{g}^{\mathbb{C}} \simeq \mathfrak{e}_7(\mathbb{C})$, and it has dimension $52 = 2 \cdot 26$ and Bala–Carter label $2A_1$. This shows that the associated variety of π_{ntm} is equal to the closure of this orbit.

3.6 The Case $\mathfrak{g} = \mathfrak{su}(p, q)$

This case differs slightly from the other cases in the sense that the dependence of $W_{\mu_1,\mu_2,\ell}$ on ℓ is not just by the central character. The root system $\Delta(\mathfrak{k}^{\mathbb{C}}, \mathfrak{t}^{\mathbb{C}})$ is of type $A_{p-1} \times A_{q-1}$ and we write:

$$\Delta(\mathfrak{k}^{\mathbb{C}}, \mathfrak{t}^{\mathbb{C}}) = \{\pm(e_i - e_j) : 1 \leq i < j \leq p\} \cup \{\pm(f_i - f_j) : 1 \leq i < j \leq q\}.$$

With respect to the positive system

$$\Delta^+(\mathfrak{k}^{\mathbb{C}}, \mathfrak{t}^{\mathbb{C}}) = \{e_i - e_j : 1 \leq i < j \leq p\} \cup \{f_i - f_j : 1 \leq i < j \leq q\}$$

the highest weight of $W_{\mu_1,\mu_2,\ell}$ is

$$\left(\frac{\mu_1 + \ell}{2} e_1 - \frac{\mu_1 - \ell}{2} e_p\right) + \left(\frac{\mu_2 + \ell}{2} f_1 - \frac{\mu_2 - \ell}{2} f_q\right)$$

and its dimension equals

$$\frac{(\mu_1 + p - 1)(\mu_2 + q - 1)}{(p-1)(q-1)} \binom{\frac{\mu_1+\ell}{2} + p - 2}{p-2} \binom{\frac{\mu_1-\ell}{2} + p - 2}{p-2} \binom{\frac{\mu_2+\ell}{2} + q - 2}{q-2}$$
$$\times \binom{\frac{\mu_2-\ell}{2} + q - 2}{q-2} \sim \mu_1\mu_2(\mu_1^2 - \ell^2)^{p-2}(\mu_2^2 - \ell^2)^{q-2}.$$

The K-types of π_{ntm} are given by $W_{\mu_1,\mu_2,\ell}$ with $\mu_1 - \mu_2 = q - p$ and $|\ell| \leq \mu_1, \mu_2$, $\mu_1 \equiv \mu_2 \equiv \ell \mod 2$. Summing the dimensions of $W_{\mu,\mu,\ell}$ over $|\ell| \leq \mu_1, \mu_2 \leq n$ with $\mu_1 - \mu_2 = q - p$ in a similar way as in the previous cases shows that

$$\mathrm{GKDIM}(\pi_{\mathrm{ntm}}) = 2p + 2q - 4.$$

By Collingwood and McGovern [4, Theorem 6.2.5], the unique next-to-minimal nilpotent coadjoint orbit in $\mathfrak{g}^{\mathbb{C}} \simeq \mathfrak{sl}(p+q,\mathbb{C})$ is associated with the partition $(2^2, 1^{p+q-4})$, and by Collingwood and McGovern [4, Corollary 6.1.4] it has dimension $4(p+q)-8 = 2(2p+2q-4)$. This implies that the associated variety of π_{ntm} is equal to the closure of this orbit.

4 Restriction from G to $\mathrm{SL}(2,\mathbb{R}) \times M$

Recall that the elements $E \in \mathfrak{n}_2$, $F \in \bar{\mathfrak{n}}_2$ and $H \in \mathfrak{a}$ form an $\mathfrak{sl}(2)$-triple that commutes with $\mathfrak{m}$. Writing $\mathfrak{sl}(2,\mathbb{R})_A = \mathrm{span}\{E, F, H\}$, we obtain a subalgebra $\mathfrak{sl}(2,\mathbb{R})_A \oplus \mathfrak{m}$ and a corresponding subgroup $\mathrm{SL}(2,\mathbb{R})_A \times M$. The goal of this section is to understand the restriction of π_{ntm} to $\mathrm{SL}(2,\mathbb{R})_A \times M$. We first make some general observations before specializing to the cases $\mathfrak{g} = \mathfrak{so}(2,n)$ and $\mathfrak{g} = \mathfrak{e}_{6(-14)}$.

4.1 The Fourier Transformed Model of the Next-to-Minimal Representation

To study the restriction of π_{ntm} to $\mathrm{SL}(2,\mathbb{R})_A \times M$, we use the Heisenberg Fourier transform:

$$\mathcal{F} \colon \mathcal{S}'(\bar{N}) \to \mathcal{D}'(\mathbb{R}^\times)\hat{\otimes}\mathrm{Hom}(\mathcal{H}^\infty, \mathcal{H}^{-\infty}),$$

where $\mathcal{F}_\lambda(V_1) \simeq \mathcal{H}$ for all $\lambda \in \mathbb{R}^\times$ (see Sect. 2.3 for details). We rewrite the decomposition (3) of $\mathcal{F}_\lambda(V_1)$ as

$$\mathcal{H} = \bigoplus_k \mathcal{H}_k,$$

assuming that the isomorphism $\mathcal{F}_\lambda(V_1) \simeq \mathcal{H}$ maps each $\mathcal{F}_{\lambda,k}(V_1)$ onto $\mathcal{H}_k$.

In what follows we assume that $\mathfrak{g} \not\simeq \mathfrak{sp}(n,\mathbb{R}), \mathfrak{su}(p,q)$.

Lemma 4.1 *Viewing π_{ntm} as a quotient of $I(\nu_0)$, the Heisenberg group Fourier transform $\mathcal{F}$ induces a unitary (up to a scalar) isomorphism:*

$$\mathcal{F} \colon \pi_{\mathrm{ntm}} \to L^2(\mathbb{R}^\times; |\lambda|^{2b_1+1}\, d\lambda)\hat{\otimes}\, \mathrm{HS}(\mathcal{H}_0, \mathcal{H}), \quad \mathcal{F}u(\lambda) = \sigma_\lambda(u),$$

where $\mathrm{HS}(\mathcal{H}_0, \mathcal{H})$ is the Hilbert space of Hilbert–Schmidt operators from $\mathcal{H}_0$ to $\mathcal{H}$. □

Proof Following [8, Section 4], the representation π_{ntm} is a quotient of the principal series $(\pi_{\nu_0}, I(\nu_0))$ for $\nu_0 = a_1 + 1$. Explicitly it is given as the quotient by the kernel of the intertwining operator $A_{\nu_0} : I(\nu_0) \to I(-\nu_0)$. The inner product on the quotient $I(\nu_0)/\operatorname{Ker} A_{\nu_0} = \pi_{\mathrm{ntm}}$ is by (5) explicitly given for $f_1, f_2 \in \pi_{\mathrm{ntm}}$ as

$$\langle f_1, f_2\rangle_{\nu_0} = \mathrm{const} \times \int_{\mathbb{R}^\times} \mathrm{tr}\left(\sigma_\lambda(f_1) \circ P_0 \circ \sigma_\lambda(f_2)^*\right) |\lambda|^{d_1-\nu_0}\, d\lambda,$$

where $P_0 : \mathcal{F}_\lambda(V_1) \to \mathcal{F}_{\lambda,0}(V_1)$ is the orthogonal projection. Since $d_1 = a_1+2b_1+2$ and $\nu_0 = a_1 + 1$, the claim follows. □

To employ the Heisenberg group Fourier transform for the decomposition of the restriction of π_{ntm} to $\mathrm{SL}(2, \mathbb{R})_A \times M$, we need to understand how it behaves with respect to the action of $\mathrm{SL}(2, \mathbb{R})_A$ and M. The action of a general element of $\mathrm{SL}(2, \mathbb{R})_A$ turns out to be quite complicated, but the action of the parabolic subgroup

$$B := \exp(\mathbb{R}H)\exp(\mathbb{R}F) \subseteq \mathrm{SL}(2, \mathbb{R})_A \tag{6}$$

of $\mathrm{SL}(2, \mathbb{R})_A$ is rather simple.

Lemma 4.2 (see [7, Proposition 3.5.5]) *Let $u \in I(\nu)$.*

1. *(B-action) For $t \in \mathbb{R}$, we have:*

$$\sigma_\lambda(\pi_\nu(e^{tH})u) = e^{(\nu-\rho)t}\delta_{e^t} \circ \sigma_{e^{-2t}\lambda}(u) \circ \delta_{e^{-t}},$$

$$\sigma_\lambda(\pi_\nu(e^{tF})u) = e^{-i\lambda t}\sigma_\lambda(u),$$

where $\delta_s\zeta(z) = \zeta(sz)$.

2. *(M-action) For $m \in M$ we have:*

$$\sigma_\lambda(\pi_\nu(m)u) = \omega_{\mathrm{met},\lambda}(m) \circ \sigma_\lambda(u) \circ \omega_{\mathrm{met},\lambda}(m)^{-1}.$$

By the last formula in Lemma 4.2, the decomposition of $\pi_{\mathrm{ntm}}|_M$ is related to the decomposition of the tensor product representation $\omega^*_{\mathrm{met},\lambda,0} \otimes \omega_{\mathrm{met},\lambda}|_M$. For this, we specialize to $\mathfrak{g} = \mathfrak{so}(2, n)$ or $\mathfrak{g} = \mathfrak{e}_{6(-14)}$. Note that $\omega^*_{\mathrm{met},\lambda,0} \simeq \omega_{\mathrm{met},-\lambda,0}$ since $\omega_{\mathrm{met},\lambda}$ and $\omega_{\mathrm{met},-\lambda}$ are contragredient to each other.

4.2 The Case $G = \mathrm{SO}_0(2, n)$

Let $G = \mathrm{SO}_0(2, n)$, $n > 4$, then $\mathfrak{m} = \mathfrak{sl}(2, \mathbb{R}) \oplus \mathfrak{so}(n-2)$. To distinguish the copy of $\mathfrak{sl}(2, \mathbb{R})$ in M from $\mathfrak{sl}(2, \mathbb{R})_A$, we denote it by $\mathfrak{sl}(2, \mathbb{R})_M$ and similarly on the group level.

As abstract M-representation, the metaplectic representation $\omega_{\mathrm{met},\lambda}$ and its subrepresentations $\omega_{\mathrm{met},\lambda,k}$ do not depend on the parameter λ, but only on its sign.

Hence, we might suppress the parameter in the following whenever convenient and write:

$$\omega_{\mathrm{met}}^{\mathrm{sgn}(\lambda)} \cong \omega_{\mathrm{met},\lambda} \qquad \text{and} \qquad \omega_{\mathrm{met},k}^{\mathrm{sgn}(\lambda)} \cong \omega_{\mathrm{met},\lambda,k}.$$

The decomposition of $\omega_{\mathrm{met}}^{+}|_M$ is made explicit in [8, Proposition B.2]. Denote by $\eta_k^{\mathrm{SO}(m)}$ the irreducible representation of $\mathrm{SO}(m)$ on the space $\mathcal{H}^k(\mathbb{R}^m)$ of homogeneous harmonic polynomials on $\mathbb{R}^m$ of degree k. Moreover, let $\tau_{\mu,\varepsilon}^{\mathrm{SL}(2,\mathbb{R})}$, $\mu \in i\mathbb{R}$, $\varepsilon \in \mathbb{Z}/2\mathbb{Z}$, be the unitary principal series of $\mathrm{SL}(2,\mathbb{R})$, spherical for $\varepsilon = 0$ and nonspherical for $\varepsilon = 1$. We further write $\tau_k^{\mathrm{SL}(2,\mathbb{R})}$ ($k \in \mathbb{Z} \setminus \{0\}$) for the (limit of) discrete series of $\mathrm{SL}(2,\mathbb{R})$ of parameter k, where the notation is such that $\tau_k^{\mathrm{SL}(2,\mathbb{R})}$ is a holomorphic resp. anti-holomorphic discrete series for $k \leq -2$ resp. $k \geq 2$. Then

$$\omega_{\mathrm{met}}^{+}|_M = \bigoplus_{k=0}^{\infty} \omega_{\mathrm{met},\lambda,k} = \bigoplus_{k=0}^{\infty} \tau_{-k-(n-2)}^{\mathrm{SL}(2,\mathbb{R})} \boxtimes \eta_k^{\mathrm{SO}(n-2)}. \tag{7}$$

Since $\omega_{\mathrm{met}}^{+}$ is contragredient to $\omega_{\mathrm{met}}^{-}$, this also implies the corresponding decomposition of $\omega_{\mathrm{met}}^{-}$.

Write $\mathrm{par}(k) \in \mathbb{Z}/2\mathbb{Z}$ for the parity of $k \in \mathbb{Z}$.

Lemma 4.3 *The following tensor product decomposition for representations of* $M = \mathrm{SL}(2,\mathbb{R})_M \times \mathrm{SO}(n-2)$ *holds:*

$$\omega_{\mathrm{met},0}^{\mp} \otimes \omega_{\mathrm{met}}^{\pm}|_M \simeq \bigoplus_{k\geq 0} \int_{i\mathbb{R}_{\geq 0}}^{\oplus} \tau_{\mu,\mathrm{par}(k)}^{\mathrm{SL}(2,\mathbb{R})} \boxtimes \eta_k^{\mathrm{SO}(n-2)}\, d\mu \oplus$$

$$\bigoplus_{l\geq 2} \tau_{\mp l}^{\mathrm{SL}(2,\mathbb{R})} \boxtimes \left(\bigoplus_{\substack{k\geq l \\ l\equiv k \mod 2}} \eta_k^{\mathrm{SO}(n-2)} \right).$$

Proof This follows from the classical formulas for tensor products of holomorphic and anti-holomorphic discrete series of $\mathrm{SL}(2,\mathbb{R})$ (see, e.g., [30]) and the decomposition (7). □

To also obtain information about the action of the other factor $\mathrm{SL}(2,\mathbb{R})_A$ on the isotypic components, we make the following observation.

Proposition 4.4 *The two copies* $\mathrm{SL}(2,\mathbb{R})_A$ *and* $\mathrm{SL}(2,\mathbb{R})_M$ *of* $\mathrm{SL}(2,\mathbb{R})$ *in* G *are conjugate via an element of* $\mathrm{SO}(2) \times \mathrm{S}(\mathrm{O}(2) \times \mathrm{O}(n-2)) \subseteq \mathrm{SO}(2) \times \mathrm{SO}(n) \subseteq G$. □

Proof Note that it suffices to show that the two copies of $\mathfrak{so}(2)$ in $\mathfrak{sl}(2,\mathbb{R})$ are conjugate. Choose, for example, explicitly $H = E_{1,n+1} + E_{2,n+2} + E_{n+1,1} + E_{n+2,2}$. Then, $\mathfrak{k} \cap (\mathfrak{sl}(2,\mathbb{R})_A \oplus \mathfrak{m}) = \mathfrak{so}(2) \oplus \mathfrak{so}(2) \oplus \mathfrak{so}(n-2)$ is realized in diagonal blocks in $\mathfrak{g}$ in the following way. One copy of $\mathfrak{so}(2)$ is spanned by $\mathrm{diag}(X, X, 0_{n-2})$ and the other one by $\mathrm{diag}(X, -X, 0_{n-2})$, where

$$X = \begin{pmatrix} 0 & 1 \\ -1 & 0 \end{pmatrix}.$$

Such two elements are conjugate via the matrix $g = \mathrm{diag}(1_2, y, 1_{n-2}) \in \mathrm{O}(2, n)$, where

$$y = \begin{pmatrix} 0 & 1 \\ 1 & 0 \end{pmatrix},$$

and conjugation with g is trivial on $\mathrm{SO}(n-2)$. Since $n > 2$, the product of g with some element in $\mathrm{O}(n-2)$ of determinant -1 does the job. (The explicit matrix realizations of all subgroups can also be found in [7, Appendix B.2].) □

Recall the subgroup $B \subseteq \mathrm{SL}(2, \mathbb{R})$ defined in (6). By the classification of the irreducible unitary representations of B (see, e.g., [10]), there exist exactly two nonequivalent infinite dimensional irreducible unitary representations of B. They can be realized on $L^2(\mathbb{R}_\pm, |\lambda|^{\mathrm{Re}\,\alpha-1}\, d\lambda)$ for any $\alpha \in \mathbb{C}$ by the action

$$(e^{tH}.\varphi)(\lambda) = e^{(\alpha-2)t}\varphi(e^{-2t}\lambda),$$
$$(e^{tF}.\varphi)(\lambda) = e^{-i\lambda t}\varphi(\lambda),$$

where $\varphi \in L^2(\mathbb{R}_\pm, |\lambda|^{\mathrm{Re}\,\alpha-1}\, d\lambda)$ and $t \in \mathbb{R}$. We denote the equivalence classes by σ_B^+ and σ_B^-.

Lemma 4.5 *The restriction of a unitary principal series representation* $\tau_{\mu,\varepsilon}^{\mathrm{SL}(2,\mathbb{R})}$ *(*$\mu \in i\mathbb{R}$, $\varepsilon \in \mathbb{Z}/2\mathbb{Z}$*) or a lowest/highest weight representation* $\tau_{\pm k}^{\mathrm{SL}(2,\mathbb{R})}$ *(*$k > 0$*) of* $\mathrm{SL}(2, \mathbb{R})$ *to* B *is given by*

$$\tau_{\mu,\varepsilon}^{\mathrm{SL}(2,\mathbb{R})}|_B \cong \sigma_B^+ \oplus \sigma_B^-, \qquad \tau_{\pm k}^{\mathrm{SL}(2,\mathbb{R})}|_B \cong \sigma_B^\mp.$$

Proof The unitary principal series of $\mathrm{SL}(2, \mathbb{R})$ can be realized on $L^2(\mathbb{R})$, where the action on a function $f \in L^2(\mathbb{R})$ is given by (see, e.g., [21, Chapter II, §5])

$$\begin{pmatrix} a & 0 \\ c & a^{-1} \end{pmatrix}.f(x) = \mathrm{sgn}(a)^\varepsilon |a|^{1+\mu} f(a^2x - ac) \qquad (a \in \mathbb{R}^\times, c \in \mathbb{R}).$$

Applying the Euclidean Fourier transform on the real line turns this action into the one of $\sigma_B^+ \oplus \sigma_B^-$ on $L^2(\mathbb{R}) = L^2(\mathbb{R}_+) \oplus L^2(\mathbb{R}_-)$ described above.

Let $\Pi = \{z = x+iy \in \mathbb{C} : y > 0\} \subseteq \mathbb{C}$ be the upper half plane. The holomorphic discrete series $\tau_{-k}^{\mathrm{SL}(2,\mathbb{R})}$, $k \geq 2$ can be realized on the weighted Bergman space $H_k^2(\Pi)$ on Π, given by

$$H_k^2(\Pi) = \mathcal{O}(\Pi) \cap L^2(\Pi, y^{k-2}\, dx\, dy).$$

Explicitly $\mathrm{SL}(2, \mathbb{R})$ acts on a function $f \in H_k^2(\Pi)$ by

$$(\pi_{-k}^{\mathrm{SL}(2,\mathbb{R})}(g)f)(z) = (-bz + d)^{-k} f\left(\frac{az - c}{-bz + d}\right),$$

where

$$g = \begin{pmatrix} a & b \\ c & d \end{pmatrix} \in \mathrm{SL}(2, \mathbb{R}).$$

For any $k > 1$, consider the Laplace transform $\mathcal{L}$ on $L^2(\mathbb{R}_+, |\lambda|^{k-1}\, d\lambda)$, given by

$$\mathcal{L}g(z) = \int_0^\infty g(\lambda)e^{i\lambda z}\, d\lambda.$$

By Faraut and Korányi [6, Theorem XIII.1.1] the Laplace transform gives a surjective isometry onto $H_k^2(\Pi)$, and it is easily checked that it intertwines the action of B given by σ_B^+. For the anti-holomorphic discrete series, the statement follows by taking complex conjugates. For the limits of discrete series, we have that $\pi_{-1}^{\mathrm{SL}(2,\mathbb{R})} \oplus \pi_1^{\mathrm{SL}(2,\mathbb{R})}$ is a unitary principal series which restricts to B as $\sigma_B^+ \oplus \sigma_B^-$, and the statement follows by a small modification of the argument. □

We can finally combine all the gathered information to obtain the full decomposition of π_{ntm} restricted to $\mathrm{SL}(2, \mathbb{R})_A \times M = \mathrm{SL}(2, \mathbb{R})_A \times \mathrm{SL}(2, \mathbb{R})_M \times \mathrm{SO}(n-2)$:

Theorem 4.6 *The restriction of the next-to-minimal representation of* $\mathrm{SO}_0(2, n)$ *to* $\mathrm{SL}(2, \mathbb{R})_A \times \mathrm{SL}(2, \mathbb{R})_M \times \mathrm{SO}(n-2)$ *is given by*

$$\pi_{\mathrm{ntm}}|_{\mathrm{SL}(2,\mathbb{R})\times\mathrm{SL}(2,\mathbb{R})\times\mathrm{SO}(n-2)} \simeq \bigoplus_{k=0}^{\infty}\left(\int_{i\mathbb{R}_{\geq 0}}^{\oplus} \tau_{\mu,\mathrm{par}(k)}^{\mathrm{SL}(2,\mathbb{R})} \boxtimes \tau_{\mu,\mathrm{par}(k)}^{\mathrm{SL}(2,\mathbb{R})}\, d\mu \right.$$
$$\left. \oplus \bigoplus_{\substack{2\leq|\ell|\leq k \\ \ell\equiv k \bmod 2}} \tau_\ell^{\mathrm{SL}(2,\mathbb{R})} \boxtimes \tau_\ell^{\mathrm{SL}(2,\mathbb{R})}\right) \boxtimes \eta_k^{\mathrm{SO}(n-2)}.$$

Proof By Lemma 4.1, we have:

$$\pi_{\mathrm{ntm}} \simeq L^2(\mathbb{R}^\times, \mathcal{F}_{-\lambda,0}(V_1) \otimes \mathcal{F}_\lambda(V_1), |\lambda|\, d\lambda),$$

so in view of Lemma 4.2 (2), the restriction of π_{ntm} to M can be written as

$$\pi_{\mathrm{ntm}}|_M \simeq \left(L^2(\mathbb{R}_+, |\lambda|\,d\lambda) \boxtimes (\omega^-_{\mathrm{met},0} \otimes \omega^+_{\mathrm{met}}|_M)\right)$$
$$\oplus \left(L^2(\mathbb{R}_-, |\lambda|\,d\lambda) \boxtimes (\omega^+_{\mathrm{met},0} \otimes \omega^-_{\mathrm{met}}|_M)\right),$$

by dividing $\mathbb{R}^\times$ into the positive and negative axes. Here M is acting trivially on $L^2(\mathbb{R}_\pm, |\lambda|\,d\lambda)$. It follows that the action of B preserves this decomposition and acts on each M-isotypic component of $\omega^\mp_{\mathrm{met},0} \otimes \omega^\pm_{\mathrm{met}}|_M$ by unitary automorphisms. The action on $\varphi \in L^2(\mathbb{R}^\times, |\lambda|\,d\lambda)$ is explicitly given in Lemma 4.2 (1) for $\nu = \nu_0 = n - 3 = \rho - 2$:

$$(e^{tH} \cdot \varphi)(\lambda) = e^{-2t}\varphi(e^{-2t}\lambda), \qquad (e^{tF} \cdot \varphi)(\lambda) = e^{-i\lambda t}\varphi(\lambda),$$

so that

$$\pi_{\mathrm{ntm}}|_{B\times M} \simeq \left(\sigma_B^+ \boxtimes (\omega^-_{\mathrm{met},0} \otimes \omega^+_{\mathrm{met}}|_M)\right) \oplus \left(\sigma_B^- \boxtimes (\omega^+_{\mathrm{met},0} \otimes \omega^-_{\mathrm{met}}|_M)\right). \tag{8}$$

Since the subgroups B of the two copies of $\mathrm{SL}(2,\mathbb{R})$ in $\mathrm{SL}(2,\mathbb{R}) \times M$ are conjugate by Proposition 4.4, we have by Lemma 4.5

$$\pi_{\mathrm{ntm}}|_{B\times B\times \mathrm{SO}(n-2)} \simeq \bigoplus_{k\geq 0} \int_{i\mathbb{R}_{\geq 0}}^{\oplus} \left((\sigma_B^+ \oplus \sigma_B^-) \boxtimes (\sigma_B^+ \oplus \sigma_B^-) \boxtimes \eta_k^{\mathrm{SO}(n-2)}\right) d\mu$$
$$\oplus \bigoplus_{|k|\geq 2} \sigma_B^{-\,\mathrm{sgn}(k)} \boxtimes \sigma_B^{-\,\mathrm{sgn}(k)} \boxtimes \left(\bigoplus_{\substack{l\geq |k|\\ l\equiv k \mod 2}} \eta_l^{\mathrm{SO}(n-2)}\right).$$

Since even the whole two copies of $\mathrm{SL}(2,\mathbb{R})$ are conjugate, and since we know the action of the left copy by (8) and Lemma 4.3, this implies the theorem by Lemma 4.5. □

4.3 *The Case $G = E_{6(-14)}$*

We now study the same branching problem for $G = E_{6(-14)}$. For this, we first fix some notation regarding the exceptional Lie algebra $\mathfrak{e}_{6(-14)}$.

4.3.1 The Subalgebra $\mathfrak{su}(5,1) \subseteq \mathfrak{e}_{6(-14)}$

Recall again our convention from [8, 35] that the Heisenberg parabolic subalgebra is constructed using the lowest Harish–Chandra root γ_1. More precisely—in the present case—let $\gamma_1 < \gamma_2$ be Harish–Chandra strongly orthogonal roots for the symmetric pair $(\mathfrak{g}, \mathfrak{k}) = (\mathfrak{e}_{6(-14)}, \mathfrak{so}(2) + \mathfrak{spin}(10))$ and $e_{\pm 1}$ the root vectors for $\pm\gamma_1$ with $[e_1, e_{-1}]$ being the corresponding co-root. Our H in Sect. 2 is then $H = e_1 + e_{-1}$. The Levi subalgebra is $\mathfrak{m} = \mathfrak{su}(5, 1)$ and its maximal compact subalgebra is $\mathfrak{l} = \mathfrak{u}(5) = \mathfrak{k} \cap \mathfrak{m}$. To specify the relevant roots, we use the Dynkin diagrams of of $\mathfrak{e}_6$ and $\mathfrak{m}^{\mathbb{C}}$. The diagram is

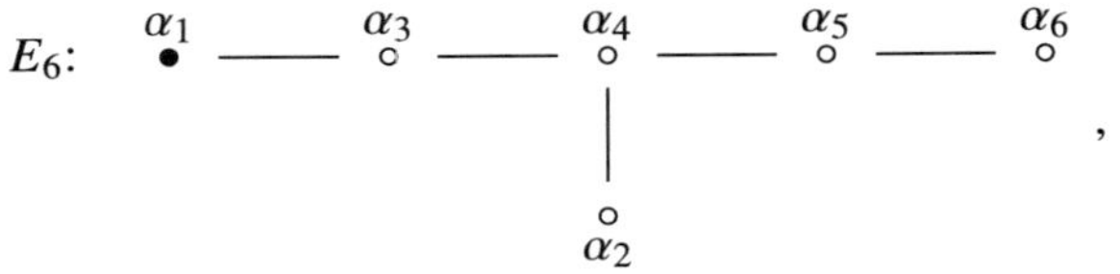

$$E_6: \quad \overset{\alpha_1}{\bullet} \text{ --- } \overset{\alpha_3}{\circ} \text{ --- } \overset{\alpha_4}{\circ} \text{ --- } \overset{\alpha_5}{\circ} \text{ --- } \overset{\alpha_6}{\circ}, \qquad \alpha_2 \ (\circ) \text{ attached to } \alpha_4$$

where the circled roots $\{\alpha_2, \ldots, \alpha_6\}$ are the compact roots and the black root $\gamma_1 = \alpha_1$ is the non-compact lowest root. The compact and non-compact positive roots are (see, e.g., [5])

$$\Delta_c^+ = \{\varepsilon_i \pm \varepsilon_j; 5 \geq i > j \geq 1\},$$

$$\Delta_n^+ = \left\{\frac{1}{2}\Big(\sum_{i=1}^{5}(-1)^{\nu_i}\varepsilon_i - \varepsilon_6 - \varepsilon_7 + \varepsilon_8\Big); \sum_{i=1}^{5} \nu_i \text{ is even}\right\},$$

with

$$\gamma_1 = \alpha_1 = \frac{1}{2}(\varepsilon_1 - \varepsilon_2 - \varepsilon_3 - \varepsilon_4 - \varepsilon_5 - \varepsilon_6 - \varepsilon_7 + \varepsilon_8), \qquad \alpha_2 = \varepsilon_1 + \varepsilon_2,$$

$$\alpha_j = \varepsilon_{j-1} - \varepsilon_{j-2}, \ (3 \leq j \leq 6).$$

The Harish–Chandra strongly orthogonal roots are $\gamma_1 < \gamma_2$, where

$$\gamma_2 = \frac{1}{2}(-\varepsilon_1 + \varepsilon_2 + \varepsilon_3 + \varepsilon_4 - \varepsilon_5 - \varepsilon_6 - \varepsilon_7 + \varepsilon_8).$$

We shall need the opposite Harish–Chandra roots starting from the highest one. This is the pair $\tilde{\gamma}_2 > \tilde{\gamma}_1$, with

$$\tilde{\gamma}_2 = \frac{1}{2}(\varepsilon_1 + \varepsilon_2 + \varepsilon_3 + \varepsilon_4 + \varepsilon_5 - \varepsilon_6 - \varepsilon_7 + \varepsilon_8),$$

$$\tilde{\gamma}_1 = \frac{1}{2}(-\varepsilon_1 - \varepsilon_2 - \varepsilon_3 - \varepsilon_4 + \varepsilon_5 - \varepsilon_6 - \varepsilon_7 + \varepsilon_8).$$

These formulas can easily be checked since there are five non-compact positive roots orthogonal to the highest root $\tilde{\gamma}_2$, and $\tilde{\gamma}_1$ is the highest one among them with our given ordering.

The positive roots of $\mathfrak{m}^{\mathbb{C}} = \mathfrak{sl}(6, \mathbb{C})$ are

$$\Delta_c^+(\mathfrak{m}^{\mathbb{C}}) = \{\varepsilon_j - \varepsilon_i; 5 \geq j > i \geq 2\} \cup \{\varepsilon_i + \varepsilon_1; 5 \geq i \geq 1\},$$

$$\Delta_n^+(\mathfrak{m}^{\mathbb{C}}) = \Big\{\frac{1}{2}(\sum_{i=1}^{5}(-1)^{\nu_i}\varepsilon_i - \varepsilon_6 - \varepsilon_7 + \varepsilon_8) \in \Delta_n^+; (-1)^{\nu_1} - \sum_{i=2}^{5}(-1)^{\nu_i} + 3 = 0\Big\}.$$

The roots $\beta_1 = \alpha_2, \beta_2 = \alpha_4, \beta_3 = \alpha_5, \beta_4 = \alpha_6$ form a system of simple compact roots, and together with $\beta_5 = \gamma_2$, we get a system of simple roots for $\mathfrak{m}^{\mathbb{C}}$, all orthogonal to γ_1. The Dynkin diagram for $\mathfrak{m}^{\mathbb{C}}$ is now

$$\mathfrak{m}^{\mathbb{C}}:\quad \overset{\beta_1}{\circ} \text{——} \overset{\beta_2}{\circ} \text{——} \overset{\beta_3}{\circ} \text{——} \overset{\beta_4}{\circ} \text{——} \overset{\beta_5}{\bullet}\,.$$

Let λ_2 and λ_5 be the corresponding fundamental weights dual to β_2 and β_5 as representations of $\mathfrak{l}^{\mathbb{C}} = \mathfrak{gl}(5, \mathbb{C})$. Then

$$\lambda_2 = \frac{1}{3}(2\beta_1 + 4\beta_2 + 3\beta_3 + 2\beta_4 + \beta_5), \qquad \lambda_5 = \frac{1}{6}(\beta_1 + 2\beta_2 + 3\beta_3 + 4\beta_4 + 5\beta_5).$$

Considered as a character of $\mathfrak{l}$ and $\mathfrak{l}^{\mathbb{C}}$, λ_5 is the fundamental central character of $\mathfrak{l}^{\mathbb{C}}$.

The Harish–Chandra strongly orthogonal root for $\mathfrak{m} = \mathfrak{su}(5, 1)$ is then $\beta_3 = \alpha_5$, and the highest non-compact root is $\tilde{\gamma}_2$; here $\mathfrak{m}$ is not of tube type so $\beta_5 = \gamma_2$ is the lowest non-compact root orthogonal to γ_1, and there are three non-compact roots between $\beta_5 = \gamma_2$ and $\tilde{\gamma}_2$. They are

$$\beta_5 + \beta_4, \quad \beta_2 + (\beta_4 + \beta_3), \quad \beta_2 + (\beta_4 + \beta_3 + \beta_2).$$

The Harish–Chandra decomposition for $\mathfrak{m} = \mathfrak{su}(5, 1)$ is

$$\mathfrak{m}^{\mathbb{C}} = \mathfrak{sl}(6, \mathbb{C}) = \bar{\mathbb{C}}^5 + \mathfrak{gl}(5, \mathbb{C}) + \mathbb{C}^5.$$

The space V_1 has lowest weight:

$$\delta_0 = \frac{1}{2}(-\varepsilon_1 + \varepsilon_2 - \varepsilon_3 - \varepsilon_4 - \varepsilon_5 - \varepsilon_6 - \varepsilon_7 + \varepsilon_8),$$

and the central character $\operatorname{tr}\operatorname{ad}_{V_1}$ of $\mathfrak{gl}(5, \mathbb{C})$ is easily found to be

$$\frac{1}{2}\operatorname{tr}\operatorname{ad}_{V_1} = 3\lambda_5 = \frac{1}{2}(5\beta_5 + 4\beta_4 + 3\beta_3 + 2\beta_2 + \beta_1).$$

4.3.2 Tensor Products of Highest and Lowest Weight Representations of $\mathfrak{su}(5,1)$

The branching of the metaplectic representation ω_{met}^{+} of $\text{Sp}(V_1, \omega)$ restricted to $M = \text{SU}(5,1)$ is explicitly obtained in [8], and it is a sum of the holomorphic representations $\omega_{\text{met},k}^{+} = \pi_{-k\delta_0 - 3\lambda_5}$ of $\mathfrak{su}(5,1)$ with highest weight $-k\delta_0 - 3\lambda_5$, $k \geq 0$. As in the previous section, we first find the decomposition of the tensor product representation $\omega_{\text{met},0}^{-} \otimes \omega_{\text{met},k}^{+}|_M \simeq \pi_{-3\lambda_5}^{*} \otimes \pi_{-k\delta_0 - 3\lambda_5}$.

Proposition 4.7

1. *The highest weight representations $(\pi_{-k\delta_0-3\lambda_5}, \mathfrak{su}(5,1))$ do not belong to the discrete series. They are contained in the continuous range of the analytic continuation of the discrete series, i.e., they are not reduction points.*
2. *The tensor product $\pi_{-3\lambda_5}^{*} \otimes \pi_{-k\delta_0-3\lambda_5}$ is unitarily equivalent to the induced representation:*

$$L^2(\text{SU}(5,1)/\text{U}(5), -k\delta_0) := \text{Ind}_{\text{U}(5)}^{\text{SU}(5,1)}(-k\delta_0).$$

Proof

1. By computing the inner product with simple roots, we find that the highest weight $-\delta_0$ of the dual representation V_1' is $-\delta_0 = -\lambda_5 + \lambda_2$. (Namely it is the representation $\det^{-1} \boxtimes \wedge^2 \mathbb{C}^5$ of $\mathfrak{u}(5) = \mathfrak{u}(1) + \mathfrak{su}(5)$, each factor acting on the corresponding factor in the tensor product.) The discrete series condition can be easily checked; see, e.g., [5].
2. We realize $\pi_{-k\delta_0-3\lambda_5}$ on holomorphic sections of the homogeneous vector bundle over $\text{SU}(5,1)/\text{U}(5)$ induced by the representation $W_{-k\delta_0}$ of $\text{U}(5)$ of highest weight $-k\delta_0$. Similarly, we realize $\pi_{-3\lambda_5}^{*}$ on anti-holomorphic functions on $\text{SU}(5,1)/\text{U}(5)$. Consider the restriction operator:

$$R : \pi_{-k\delta_0-3\lambda_5} \otimes \pi_{-3\lambda_5}^{*} \to C^{\infty}(\text{SU}(5,1)/\text{U}(5); -k\delta_0),$$
$$RF(z) = (1 - |z|^2)^3 F(z, z),$$

where $C^{\infty}(\text{SU}(5,1)/\text{U}(5); -k\delta_0)$ is the space of smooth sections of the vector bundle induced by the representation $W_{-k\delta_0}$. A direct computation shows that R is $\text{SU}(5,1)$-intertwining. Consider then the corresponding L^2-space $L^2(\text{SU}(5,1)/\text{U}(5), -k\delta_0)$. Since $-k\delta_0 = k(-4\lambda_5 + \lambda_2) + 3k\lambda_5$, this is the space of $\odot^k(\wedge^3 T_0^{(1,0)}(B_5)') \otimes \mathbb{C}_{3k\lambda_5}^{\iota}$-valued functions on the bounded symmetric domain $B_5 = \text{SU}(5,1)/\text{U}(5)$ with the square norm:

$$\begin{aligned} \|f\|^2 &= \int_{B_5} \langle (1 - |z|^2)^{-3k} \otimes^k (\wedge^3 B(z, z)^t) f(z), f(z) \rangle \, d\iota(z) \\ &= \int_{B_5} \langle \otimes^k (\wedge^3 (I - \bar{z} z^t)) f(z), f(z) \rangle \, d\iota(z), \end{aligned} \tag{9}$$

where $B(z,z)^t = (1-|z|^2)(I-\bar{z}z^t)$ is the metric on $T_z^{(1,0)}(B_5)'$ dual to the Bergman metric on the holomorphic tangent space $T_z^{(1,0)}(B_5)$ with $T_0^{(1,0)}(B_5)$ being viewed as a representation of $U(5)$, and

$$d\iota(z) = \frac{dm(z)}{(1-|z|^2)^6}$$

is the invariant measure on B_5 with $dm(z)$ the Lebesgue measure; see, e.g., [20]. Now the space $\pi_{-k\delta_0-3\lambda_5}$ contains all $W_{-k\delta_0}$-valued holomorphic polynomials f_1, and $\pi_{-3\lambda_5}$ all scalar holomorphic polynomials f_2, since $\pi_{-k\delta_0-3\lambda_5}$ are not reduction points in the analytic continuation of the holomorphic discrete series. We prove that $F = R(f_1 \otimes \overline{f_2})$ is in the space $L^2(G/K, -k\delta_0)$. Indeed we have:

$$\langle F, F\rangle = \int_{B_5} \langle \otimes^k(\wedge^3(I-zz^*))(1-|z|^2)^3 f_1(z)\overline{f_2}(z), (1-|z|^2)^3 f_1(z)\overline{f_2}(z)\rangle \, d\iota(z)$$

$$= \int_{B_5} \langle \otimes^k(\wedge^3(I-zz^*)) f_1(z)\overline{f_2}(z), f_1(z)\overline{f_2}(z)\rangle \, dm(z).$$

Now the polynomials f_1, f_2 are bounded on B_5 as well as the matrix $\otimes^k(\wedge^3(I-zz^*))$, and B_5 is of finite Lebesgue measure. Thus $\langle F, F\rangle < \infty$ and $F \in L^2(B_5, -k\delta_0)$. The rest of the argument is done by abstract argument by using the polar decomposition of the (unbounded) densely defined closed operator R with dense image; see, e.g., [34] for details about this technique.

□

Next, we determine the decomposition of $L^2(\mathrm{SU}(5,1)/\mathrm{U}(5), -k\delta_0)$ into irreducible representations of $\mathrm{SU}(5,1)$. The representations that occur are unitary principal series of $\mathrm{SU}(5,1)$, so we fix a (minimal) parabolic subgroup $P_M = M_M A_M N_M$ of $M = \mathrm{SU}(5,1)$. Then M_M is a double cover of $\mathrm{U}(4)$ whose irreducible representations are parameterized by tuples $(\nu_1, \nu_2, \nu_3, \nu_4)$ with $\nu_j \in \frac{1}{2}\mathbb{Z}$ and $\nu_i - \nu_j \in \mathbb{Z}_{\geq 0}$ for all $1 \leq i < j \leq 4$. Moreover, the irreducible unitary characters of A_M are parameterized by $\mu \in i\mathbb{R}$. We denote the corresponding parabolically induced representation of $\mathrm{SU}(5,1)$ by $\tau^{\mathrm{SU}(5,1)}_{(\nu_1,\nu_2,\nu_3,\nu_4),\mu}$.

Proposition 4.8 *For every $k \geq 0$, the representation $L^2(\mathrm{SU}(5,1)/\mathrm{U}(5), -k\delta_0)$ of $\mathrm{SU}(5,1)$ and its dual $L^2(\mathrm{SU}(5,1)/\mathrm{U}(5), -k\delta_0)^*$ decompose into irreducible representations as follows:*

$$L^2(\mathrm{SU}(5,1)/\mathrm{U}(5), -k\delta_0) \simeq \bigoplus_{m=0}^{k} \int_{i\mathbb{R}_{\geq 0}}^{\oplus} \tau^{\mathrm{SU}(5,1)}_{(\frac{k}{2}, \frac{k}{2}-m, -\frac{k}{2}, -\frac{k}{2}),\mu} \, d\mu,$$

$$L^2(\mathrm{SU}(5,1)/\mathrm{U}(5), -k\delta_0)^* \simeq \bigoplus_{m=0}^{k} \int_{i\mathbb{R}_{\geq 0}}^{\oplus} \tau^{\mathrm{SU}(5,1)}_{(\frac{k}{2}, \frac{k}{2}, \frac{k}{2}-m, -\frac{k}{2}),\mu} \, d\mu.$$

Proof First note that the second decomposition follows from the first one by passing to the contragredient representation and using the fact that

$$(\tau^{\mathrm{SU}(5,1)}_{(\nu_1,\nu_2,\nu_3,\nu_4),\mu})^* \simeq \tau_{(-\nu_4,-\nu_3,-\nu_2,-\nu_1),-\mu} \simeq \tau_{(-\nu_4,-\nu_3,-\nu_2,-\nu_1),\mu}.$$

To show the first decomposition, we apply the Plancherel formula for the real rank one group $\mathrm{SU}(5,1)$ (see, e.g., [21, Theorem 13.5]): the representation of $\mathrm{SU}(5,1) \times \mathrm{SU}(5,1)$ on $L^2(\mathrm{SU}(5,1))$ by left and right translation decomposes as

$$L^2(\mathrm{SU}(5,1)) \simeq \bigoplus_{\nu_1,\nu_2,\nu_3,\nu_4} \int^{\oplus}_{i\mathbb{R}_{\geq 0}} \tau^{\mathrm{SU}(5,1)}_{(\nu_1,\nu_2,\nu_3,\nu_4),\mu} \boxtimes (\tau^{\mathrm{SU}(5,1)}_{(\nu_1,\nu_2,\nu_3,\nu_4),\mu})^* \, d\mu \oplus \bigoplus_{\tau \text{d.s.}} \tau \boxtimes \tau^*,$$

where the first direct sum is over all $\nu_1, \nu_2, \nu_3, \nu_4 \in \frac{1}{2}\mathbb{Z}$ with $\nu_i - \nu_j \in \mathbb{Z}_{\geq 0}$ for all $1 \leq i < j \leq 4$ and the second direct sum is over all equivalence classes of discrete series representations τ of $\mathrm{SU}(5,1)$. The space $L^2(\mathrm{SU}(5,1)/\mathrm{U}(5), -k\delta_0)$ can be viewed as the subspace $(L^2(\mathrm{SU}(5,1)) \otimes W_{-k\delta_0})^{\mathrm{U}(5)}$ of $L^2(\mathrm{SU}(5,1)) \otimes W_{-k\delta_0}$ consisting of functions that are invariant under the diagonal action of $\mathrm{U}(5)$, acting on $L^2(\mathrm{SU}(5,1))$ from the right and on $W_{-k\delta_0}$ by the representation with highest weight $-k\delta_0$. Since the right action of $\mathrm{U}(5)$ is on the second tensor factor in the Plancherel formula and $(\tau^* \otimes W_{-k\delta_0})^{\mathrm{U}(5)} \simeq \mathrm{Hom}_{\mathrm{U}(5)}(\tau|_{\mathrm{U}(5)}, W_{-k\delta_0})$ for an irreducible unitary representation τ of $\mathrm{SU}(5,1)$, this implies that $L^2(\mathrm{SU}(5,1)/\mathrm{U}(5), -k\delta_0)$ decomposes as

$$\bigoplus_{\nu_1,\nu_2,\nu_3,\nu_4} \int^{\oplus}_{i\mathbb{R}_{\geq 0}} \tau^{\mathrm{SU}(5,1)}_{(\nu_1,\nu_2,\nu_3,\nu_4),\mu} \otimes \mathrm{Hom}_{\mathrm{U}(5)}(\tau^{\mathrm{SU}(5,1)}_{(\nu_1,\nu_2,\nu_3,\nu_4),\mu}|_{\mathrm{U}(5)}, W_{-k\delta_0}) \, d\mu$$

$$\oplus \bigoplus_{\tau \text{d.s.}} \tau \otimes \mathrm{Hom}_{\mathrm{U}(5)}(\tau|_{\mathrm{U}(5)}, W_{-k\delta_0}).$$

Note that the second tensor factor is a multiplicity space without group action of $\mathrm{SU}(5,1)$. Its dimension equals the multiplicity of the K-type $W_{-k\delta_0}$ in the principal series $\tau^{\mathrm{SU}(5,1)}_{(\nu_1,\nu_2,\nu_3,\nu_4),\mu}$ resp. the discrete series τ.

We first show that the highest weight $-k\delta_0$ is not the lowest K-type of any discrete series. In the standard notation for $\mathfrak{sl}(6,\mathbb{C})$, the $\mathfrak{gl}(5,\mathbb{C})$-highest weight $-k\delta_0$ is

$$-k\delta_0 = k(-\lambda_5 + \lambda_2) = \frac{k}{2}(1, 1, -1, -1, -1, 1). \tag{10}$$

With the existing ordering of the roots of $\mathfrak{sl}(6,\mathbb{C})$, we have that a discrete series of $\mathrm{SU}(5,1)$ is determined by a non-singular highest weight $\mu := (\mu_1, \mu_2, \mu_3, \mu_4, \mu_5, \mu_6)$ with $\mu_1 > \cdots > \mu_5 > \mu_6$ with lowest K-type:

$$\mu + \rho_{\mathfrak{sl}(6,\mathbb{C})} - 2\rho_{\mathfrak{gl}(5,\mathbb{C})} = \frac{1}{2}(2\mu_1 - 3, 2\mu_2 - 1, 2\mu_3 + 1, 2\mu_4 + 1, 2\mu_5 + 3, 2\mu_6 - 5).$$

(This statement can for instance be found in [21, Theorem 9.20 and 12.21].) Here $\mu_1 > \cdots > \mu_5 > \mu_6$ are all half-integers and $\sum_j \mu_j = 0$. That the space $L^2(\mathrm{SU}(5,1)/\mathrm{U}(5), -k\delta_0)$ has a discrete series representation is equivalent to that a permutation $s(-k\delta_0)$ of the weight $-k\delta_0$ by a Weyl group element $s \in S_6$ produces the above lowest K-type:

$$s(-k\lambda_0) = \frac{1}{2}(2\mu_1 - 3, 2\mu_2 - 1, 2\mu_3 + 1, 2\mu_4 + 1, 2\mu_5 + 3, 2\mu_6 - 5).$$

The only possible choice is then

$$2\mu_1 - 3 = k, \qquad 2\mu_2 - 1 = k, \qquad 2\mu_3 + 1 = k,$$
$$2\mu_4 + 1 = -k, \qquad 2\mu_5 + 3 = -k, \qquad 2\mu_6 - 5 = -k,$$

i.e.,

$$\mu_1 = \frac{1}{2}(k+3), \qquad \mu_2 = \frac{1}{2}(k+1), \qquad \mu_3 = \frac{1}{2}(k-1),$$
$$\mu_4 = \frac{1}{2}(-k-3), \qquad \mu_5 = \frac{1}{2}(-k-5), \qquad \mu_6 = \frac{1}{2}(-k+5).$$

But then μ is not dominant; thus, $L^2(\mathrm{SU}(5,1)/\mathrm{U}(5), -k\delta_0)$ has no discrete spectrum.

It remains to show that the K-type $W_{-k\delta_0}$ occurs in the unitary principal series $\tau^{\mathrm{SU}(5,1)}_{(\nu_1,\nu_2,\nu_3,\nu_4),\mu}$ if and only if $(\nu_1, \nu_2, \nu_3, \nu_4)$ is of the form $(\frac{k}{2}, \frac{k}{2}-m, -\frac{k}{2}, -\frac{k}{2})$, and in that case it occurs with multiplicity one. By Frobenius reciprocity, the multiplicity of $W_{-k\delta_0}$ in $\tau^{\mathrm{SU}(5,1)}_{(\nu_1,\nu_2,\nu_3,\nu_4),\mu}$ equals the multiplicity of the M_M-representation with highest weight $(\nu_1, \nu_2, \nu_3, \nu_4)$ in $W_{-k\delta_0}$. In view of (10) and the standard branching rules for the pair $(\mathrm{U}(5), \mathrm{U}(4))$, this implies the claim. □

This finally allows us to prove the following result about the restriction of the next-to-minimal representation π_{ntm} of $G = E_{6(-14)}$ to $\mathrm{SL}(2,\mathbb{R}) \times \mathrm{SU}(5,1)$.

Theorem 4.9 *The restriction of the next-to-minimal representation π_{ntm} of $E_{6(-14)}$ to $\mathrm{SL}(2,\mathbb{R}) \times \mathrm{SU}(5,1)$ contains for every $k > 0$ and every $0 < m \leq k$ a representation of the form:*

$$\int^{\oplus}_{i\mathbb{R}_{\geq 0}} \tau^{\mathrm{SL}(2,\mathbb{R})}_{-l} \boxtimes \tau_{(\frac{k}{2},\frac{k}{2}-m,-\frac{k}{2},-\frac{k}{2}),\mu}\, d\mu \oplus \int^{\oplus}_{i\mathbb{R}_{\geq 0}} \tau^{\mathrm{SL}(2,\mathbb{R})}_{l} \boxtimes \tau_{(\frac{k}{2},\frac{k}{2},m-\frac{k}{2},-\frac{k}{2}),\mu}\, d\mu$$

for some $l \geq 1$ as a direct summand. □

Proof Recall the subgroup $B \subseteq \mathrm{SL}(2,\mathbb{R})_A$ and its equivalence classes of infinite dimensional unitary irreducible representations $\sigma_B^{\pm}$. Following the same line of argumentation as in Sect. 4.2, we have that

$$\pi_{\mathrm{ntm}}|_{B\times M} \simeq \left(\sigma_B^+ \boxtimes (\omega_{\mathrm{met},0}^- \otimes \omega_{\mathrm{met}}^+|_M)\right) \oplus \left(\sigma_B^- \boxtimes (\omega_{\mathrm{met},0}^+ \otimes \omega_{\mathrm{met}}^-|_M)\right)$$

which is isomorphic to

$$\bigoplus_{k\geq 0} \left(\sigma_B^+ \boxtimes L^2(\mathrm{SU}(5,1)/\mathrm{U}(5), -k\delta_0)\right) \oplus \left(\sigma_B^- \boxtimes L^2(\mathrm{SU}(5,1)/\mathrm{U}(5), -k\delta_0)^*\right)$$

by Proposition 4.7. Decomposing $L^2(\mathrm{SU}(5,1)/\mathrm{U}(5), -k\delta_0)$ and its dual using Proposition 4.8 shows that this is in turn isomorphic to

$$\bigoplus_{k\geq 0}\bigoplus_{m=0}^{k}\left(\int_{i\mathbb{R}_{\geq 0}}^{\oplus} \sigma_B^+ \boxtimes \tau^{\mathrm{SU}(5,1)}_{(\frac{k}{2},\frac{k}{2}-m,-\frac{k}{2},-\frac{k}{2}),\mu}\, d\mu \oplus \int_{i\mathbb{R}_{\geq 0}}^{\oplus} \sigma_B^- \boxtimes \tau^{\mathrm{SU}(5,1)}_{(\frac{k}{2},\frac{k}{2},m-\frac{k}{2},-\frac{k}{2}),\mu}\, d\mu\right).$$

Now, each representation $\tau_{(\frac{k}{2},\frac{k}{2}-m,-\frac{k}{2},-\frac{k}{2}),\mu}$ resp. $\tau_{(\frac{k}{2},\frac{k}{2},m-\frac{k}{2},-\frac{k}{2}),\mu}$ of $\mathrm{SU}(5,1)$ with $k > 0$ and $0 \leq m \leq k$ occurs precisely once in this decomposition. Since $\mathrm{SL}(2,\mathbb{R})_A$ commutes with $\mathrm{SU}(5,1)$, it acts on the corresponding isotypic component $\sigma_B^+ \boxtimes \tau_{(\frac{k}{2},\frac{k}{2}-m,-\frac{k}{2},-\frac{k}{2}),\mu}$ resp. $\sigma_B^- \boxtimes \tau_{(\frac{k}{2},\frac{k}{2},m-\frac{k}{2},-\frac{k}{2}),\mu}$. The only unitary representations of $\mathrm{SL}(2,\mathbb{R})$ restricting to $\sigma_B^{\pm}$ are $\tau_{\mp l}$ $(l > 0)$ by Lemma 4.5. In the unitary dual of $\mathrm{SL}(2,\mathbb{R})$, these representations are separated, so l has to be constant in every single direct integral. This shows the claim. □

Remark 4.10 The previous statement intentionally excludes the case $m = 0$, because in this case, the representations $\tau_{(\frac{k}{2},\frac{k}{2}-m,-\frac{k}{2},-\frac{k}{2}),\mu}$ and $\tau_{(\frac{k}{2},\frac{k}{2},m-\frac{k}{2},-\frac{k}{2}),\mu}$ are equal, so

$$\left(\sigma_B^+ \boxtimes \tau_{(\frac{k}{2},\frac{k}{2},-\frac{k}{2},-\frac{k}{2}),\mu}\right) \oplus \left(\sigma_B^- \boxtimes \tau_{(\frac{k}{2},\frac{k}{2},-\frac{k}{2},-\frac{k}{2}),\mu}\right) \simeq \left(\sigma_B^+ \oplus \sigma_B^-\right) \boxtimes \tau_{(\frac{k}{2},\frac{k}{2},-\frac{k}{2},-\frac{k}{2}),\mu}$$

and it is not clear which representation of $\mathrm{SL}(2,\mathbb{R})_A$ that restricts to $\sigma_B^+ \oplus \sigma_B^-$ acts on the first factor. □

Remark 4.11 At this stage, it is not clear how to determine the parameter l for every $k > 0$ and $0 < m \leq k$. One idea would be to use a subgroup $\mathrm{SL}(2,\mathbb{R})_M$ of $M = \mathrm{SU}(5,1)$ isomorphic to $\mathrm{SU}(1,1) \simeq \mathrm{SL}(2,\mathbb{R})$ and conjugate to $\mathrm{SL}(2,\mathbb{R})_A$ inside G. The restriction of $\tau_{(\frac{k}{2},\frac{k}{2}-m,-\frac{k}{2},-\frac{k}{2}),\mu}$ to this subgroup $\mathrm{SL}(2,\mathbb{R})_M$ might contain some discrete series representations of $\mathrm{SL}(2,\mathbb{R})$ which might be linked to the parameter l by conjugating the subgroup to $\mathrm{SL}(2,\mathbb{R})_A$. However, since the restriction of $\tau_{(\frac{k}{2},\frac{k}{2}-m,-\frac{k}{2},-\frac{k}{2}),\mu}$ to $\mathrm{SL}(2,\mathbb{R})_M$ will contain several different discrete series representations, it is not clear to us how to use this information. □

Acknowledgments We thank the anonymous referees for their suggestions and comments that helped to improve the paper. The first named author was supported by a research grant from the Villum Foundation (Grant No. 00025373), the second named author was supported by a research grant from the Knut and Alice Wallenberg foundation (KAW 2020.0275), and the third named author was partially supported by the Swedish Research Council (VR, Grants 2018-03402, 2022-02861).

References

1. Ben Saïd, S., Kobayashi, T., Ørsted, B.: Laguerre semigroup and Dunkl operators. Compos. Math. **148**(4), 1265–1336 (2012)
2. Borho, W., Kraft, H.: Über die Gelfand-Kirillov-Dimension. Math. Ann. **220**(1), 1–24 (1976)
3. Bossard, G., Pioline, B.: Exact $\nabla^4\mathcal{R}^4$ couplings and helicity supertraces. J. High Energy Phys. **1**, 050 (2017). front matter+40
4. Collingwood, D.H., McGovern, W.M.: Nilpotent orbits in semisimple Lie algebras. In: Van Nostrand Reinhold Mathematics Series. Van Nostrand Reinhold Co., New York (1993)
5. Enright, T., Howe, R., Wallach, N.: A classification of unitary highest weight modules. In: Representation Theory of Reductive Groups (Park City, Utah, 1982). Progress of Mathematics, vol. 40, pp. 97–143. Birkhäuser Boston, Boston, MA (1983)
6. Faraut, J., Korányi, A.: Analysis on symmetric cones. In: Oxford Mathematical Monographs. The Clarendon Press, Oxford University Press, New York (1994). Oxford Science Publications
7. Frahm, J.: Conformally invariant differential operators on Heisenberg groups and minimal representations. Mém. Soc. Math. Fr. (N.S.) **180** (2024)
8. Frahm, J., Weiske, C., Zhang, G.: Heisenberg parabolically induced representations of Hermitian Lie groups Part I: Intertwining operators and Weyl transform. Adv. Math. **422**, Paper No. 109001 (2023)
9. Frenkel, I., Libine, M.: Quaternionic analysis and the Schrödinger model for the minimal representation of $O(3, 3)$. Int. Math. Res. Not. IMRN **21**, 4904–4923 (2012)
10. Gelfand, I.M., Naimark, M.A.: Unitary representations of the group of linear transformations of the straight line. C. R. (Doklady) Acad. Sci URSS (N.S.) **55**, 567–570 (1943)
11. Gourevitch, D., Gustafsson, H.P.A., Kleinschmidt, A., Persson, D., Sahi, S.: Eulerianity of Fourier coefficients of automorphic forms. Represent. Theory **25**, 481–507 (2021)
12. Gourevitch, D., Gustafsson, H.P.A., Kleinschmidt, A., Persson, D., Sahi, S.: Fourier coefficients of minimal and next-to-minimal automorphic representations of simply-laced groups. Canad. J. Math. **74**(1), 122–169 (2022)
13. Green, M.B., Miller, S.D., Vanhove, P., Small representations, string instantons, and Fourier modes of Eisenstein series. J. Number Theory **146**, 187–309 (2015)
14. Hilgert, J., Kobayashi, T., Möllers, J., Ørsted, B.: Fock model and Segal-Bargmann transform for minimal representations of Hermitian Lie groups. J. Funct. Anal. **263**(11), 3492–3563 (2012)
15. Howe, R.: On a notion of rank for unitary representations of the classical groups. In: Harmonic Analysis and Group Representations, Liguori, Naples, pp. 223–331 (1982)
16. Howe, R.: Transcending classical invariant theory. J. Amer. Math. Soc. **2**(3), 535–552 (1989)
17. Howe, R.: Some recent applications of induced representations. In: Group Representations, Ergodic Theory, and Mathematical Physics: A Tribute to George W. Mackey, Contemporary Mathematics, vol. 449, American Mathematical Society, Providence, RI, pp. 173–191 (2008)
18. Huang, J.-S., Pandžić, P., Savin, G.: New dual pair correspondences. Duke Math. J. **82**(2), 447–471 (1996)
19. Hunziker, M., Sepanski, M.R., Stanke, R.J.: The minimal representation of the conformal group and classical solutions to the wave equation. J. Lie Theory **22**(2), 301–360 (2012)

20. Hwang, S., Liu, Y., Zhang, G.: Hilbert spaces of tensor-valued holomorphic functions on the unit ball of C-n. Pacific J. Math. **214**(2), 303–322 (2004)
21. Knapp, A.W.: Representation theory of semisimple groups. In: Princeton Mathematical Series, vol. 36. Princeton University Press, Princeton, NJ (1986). An overview based on examples
22. Kobayashi, T.: Multiplicity in restricting minimal representations. In: Lie Theory and Its Applications in Physics, Springer Proceedings of the Mathematical Statistics, vol. 396. Springer, Singapore, pp. 3–20 (2022).
23. Kobayashi, T., Mano, G.: The Schrödinger model for the minimal representation of the indefinite orthogonal group $O(p, q)$. Mem. Am. Math. Soc. **213**, 1000 (2011)
24. Kobayashi, T., Ørsted, B.: Analysis on the minimal representation of $O(p, q)$. I. Realization via conformal geometry. Adv. Math. **180**(2), 486–512 (2003)
25. Kobayashi, T., Ørsted, B.: Analysis on the minimal representation of $O(p, q)$. II. Branching laws. Adv. Math. **180**(2), 513–550 (2003)
26. Kobayashi, T., Ørsted, B.: Analysis on the minimal representation of $O(p, q)$. III. Ultrahyperbolic equations on $\mathbb{R}^{p-1,q-1}$. Adv. Math. **180**(2), 551–595 (2003)
27. Li, J.-S.: The correspondences of infinitesimal characters for reductive dual pairs in simple Lie groups. Duke Math. J. **97**(2), 347–377 (1999)
28. Möllers, J., Oshima, Y.: Discrete branching laws for minimal holomorphic representations. J. Lie Theory **25**(4), 949–983 (2015)
29. Pollack, A.: Modular forms on indefinite orthogonal groups of rank three. J. Number Theory **238**, 611–675 (2022). With appendix "Next to minimal representation" by Gordan Savin
30. Repka, J.: Tensor products of unitary representations of $SL_2(R)$. Am. J. Math. **100**(4), 747–774 (1978)
31. Trèves, F.: Topological Vector Spaces, Distributions and Kernels. Dover Publications, Inc., Mineola, NY (2006)
32. Vogan, D.A., Jr.: Gelfand-Kirillov dimension for Harish-Chandra modules. Invent. Math. **48**(1), 75–98 (1978)
33. Vogan, D.A., Jr.: Associated varieties and unipotent representations. In: Harmonic Analysis on Reductive Groups (Brunswick, ME, 1989), Progress in Mathematics, vol. 101, pp. 315–388. Birkhäuser Boston, Boston, MA (1991)
34. Zhang, G.: Berezin transform on real bounded symmetric domains. Trans. Am. Math. Soc. **353**, 3769–3787 (2001)
35. Zhang, G.: Principal series of Hermitian Lie groups induced from Heisenberg parabolic subgroups. J. Funct. Anal. **282**(8), Paper No. 109399, 39 (2022)
36. Zhu, C.-B., Huang, J.-S.: On certain small representations of indefinite orthogonal groups. Represent. Theory **1**, 190–206 (1997)

Quantum-Classical Correspondences for Locally Symmetric Spaces

Joachim Hilgert

Abstract We review the status of a program, outlined and motivated in the introduction, for the study of correspondences between spectral invariants of partially hyperbolic flows on locally symmetric spaces and their quantizations. Further, we briefly sketch an extension of the program to graphs and quotients of higher rank affine buildings. Finally, we formulate a number of concrete problems which may be viewed as possible further steps to be taken in order to complete the program.

1 Introduction

There are very different mathematical models for the dynamics of physical systems depending on whether one is looking at classical or quantum mechanics. Nevertheless, relations between mathematical objects on both sides can be found. In particular, if the quantum model contains the Planck constant $\hbar$ explicitly, one studies the behavior of the model for $\hbar$ tending to zero in order to understand to which extent the classical model is a limit of the quantum model. There is no such canonical procedure to describe a quantum version of a classical physical system—if there were such a canonical quantization, then quantum mechanics would be unnecessary in the first place.

A paradigmatic class of examples comes from Riemannian manifolds $\mathcal{X}$. Their cotangent bundles $T^*\mathcal{X}$ carry canonical symplectic forms so that smooth functions f on $T^*\mathcal{X}$ define Hamiltonian vector fields X_f which can be integrated to give flows. Interpreting f as the energy observable on the state space $T^*\mathcal{X}$, classical Hamiltonian mechanics says that the flow of X_f is the time evolution of the system. On the other hand, one has linear differential operators D with smooth coefficients on $\mathcal{X}$ which define unbounded operators on the space $L^2(\mathcal{X}, \text{vol})$ of functions square integrable with respect to the volume form coming with the Riemannian metric. If

J. Hilgert (✉)
Institut für Mathematik, Universität Paderborn, Paderborn, Germany
e-mail: hilgert@upb.de; hilgert@math.uni-paderborn.de

M. Pevzner, H. Sekiguchi (eds.), *Symmetry in Geometry and Analysis, Volume 2*,
Progress in Mathematics 358, https://doi.org/10.1007/978-981-97-7662-7_7

such a D is (essentially) self-adjoint, it can be interpreted as the energy operator of a quantum system whose time evolution is given by the unitary one-parameter group e^{itD} with infinitesimal generator iD. Associated with D is its symbol σ_D which is a function on $T^*\mathcal{X}$.

It is a legitimate question to ask whether one can read off properties of a quantum system given by D from properties of the classical system given by σ_D. This question received particular attention in the context of classically chaotic dynamical systems when it was observed in computer experiments that quantizations of chaotic systems have different spectral properties than quantizations of completely integrable classical systems. Mathematically this is difficult to pin down as quantum systems are linear by definition and linear systems do not feature chaotic behavior in the established sense. Trying to give rigorous explanations for the observed phenomena is the subject of the field of *quantum chaos*.

A class of examples studied very closely in quantum chaos are hyperbolic surfaces together with their Laplace-Beltrami operators and the geodesic flows generated by their symbols. These surfaces have negative curvature which tends to produce chaotic behavior. Moreover, they occur as moduli spaces of complex tori which are of great relevance in number theory, whence they have been studied very intensely. So there is a lot of information available. In particular, it is a well-established fact that hyperbolic surfaces are a meeting point of many mathematical fields such as dynamical systems, complex and harmonic analysis, number theory, and algebraic and differential geometry. The wealth of documented knowledge and available mathematical tools made hyperbolic surfaces an excellent toy model to study quantum chaos.

Studying examples like hyperbolic surfaces, one realizes that the presence of symmetry restricts the possibilities to quantize a classical system. Actually, one can take the success of the orbit method in representation theory as an argument supporting this view. We take it as a guideline to look for correspondences between dynamical invariants of quantum systems with many symmetries and their classical counterparts. It turns out that this leads to very interesting results even far beyond hyperbolic surfaces.

The major part of this article will deal with such quantum-classical correspondences for systems built from locally symmetric spaces. The symmetry groups in question are (mostly semisimple) real Lie groups. It turns out that one can find non-Archimedean counterparts given in terms of affine buildings which may serve as toy models because of their combinatorial nature. First steps in establishing these toy models will also be described.

1.1 Some Motivating Examples

The first examples of quantum-classical correspondences were typically not designed as such, but need to be reinterpreted to actually show this nature.

Example 1.1 (Poisson Summation Formula) The classical Poisson summation formula

$$\sum_{n\in\mathbb{Z}^\ell} f(x+n) = \sum_{m\in\mathbb{Z}^\ell} \hat{f}(m)e^{2\pi imx},$$

say for Schwartz functions f on $\mathbb{R}^\ell$ and their Fourier transforms $\hat{f}$, can be interpreted as follows: The $n \in \mathbb{Z}^\ell$ parameterize the closed geodesics in the ℓ-torus $\mathbb{T}^\ell := \mathbb{R}^\ell/\mathbb{Z}^\ell$ passing through the origin via $n \mapsto \mathbb{R}n + \mathbb{Z}^\ell$. On the other hand, the $m \in \mathbb{Z}^\ell$ parameterize the eigenvectors $x \mapsto e^{2\pi imx}$ of the Laplace operator $\Delta = \sum_{j=1}^{\ell} \frac{\partial^2}{\partial x_j^2}$ in $L^2(\mathbb{T}^\ell)$. Thus, the summation formula relates the spectrum of the quantum object Δ with the geometric "spectrum" of oriented closed geodesics (up to translation). The flexibility of this correspondence lies in the fact that one can choose the test functions f freely.

Also the next example has its origin in harmonic analysis and is of importance in particular in number theory.

Example 1.2 (Selberg Trace Formula) In his paper [Se56], Selberg established a formula for compact locally symmetric spaces (with noncompact semisimple symmetry group) which, similar to the torus case, has a geometric side that can be interpreted as a sum over classes of closed geodesics and a spectral side that involves the spectrum of invariant operators. That paper started a spectacular development which has been described in a number of excellent survey articles and books. As trace formulas are not in the focus of this article, we refrain from giving any details but refer to [Wa76, Gu77] as good starting points to read up and [DKV79] for the interpretation in terms of geodesics.

We come to an example, which so far has not been interpreted as a quantum-classical correspondence, but does motivate such a correspondence (see Example 1.4).

Example 1.3 (Modular Cusp Forms and Period Polynomials) We refer to [MR20, Chap. 2] for details on the material in this example. In particular we follow the notation and conventions laid out in the "Introductory Roadmap" of [MR20] here as well as in the following Example 1.4. A *modular cusp form of weight* $k \in 2\mathbb{N}$ is a holomorphic function F on the upper half-plane $\mathbb{H} := \{z \in \mathbb{C} \mid \mathrm{Im} z > 0\}$ such that

$$\forall a, b, c, d \in \mathbb{Z}, ad - bc = 1, z \in \mathbb{H}: \quad F\Big(\frac{az+b}{cz+d}\Big) = (cz+d)^k F(z),$$

$F(z) \longrightarrow 0$ for z tending to $i\infty$ in a strip of bounded width, and F admits a Fourier expansion of the form:

$$F(z) = \sum_{n=1}^{\infty} a(n)e^{2\pi inz}.$$

We denote the space of modular cusp forms of weight k by S_k.

The *period polynomial* P of a modular cusp form F of weight k is given by the integral:

$$P(X) := \int_0^{i\infty} (z - X)^{k-2} F(z)\, dz.$$

with $X \in \mathbb{C}$. It satisfies the identities:

$$P + (P|_{2-k}T) = 0, \tag{1}$$

$$P + (P|_{2-k}TS) + (P|_{2-k}(TS)^2) = 0, \tag{2}$$

where

$$T := \begin{pmatrix} 0 & -1 \\ 1 & 0 \end{pmatrix} \quad \text{and} \quad S := \begin{pmatrix} 1 & 1 \\ 0 & 1 \end{pmatrix}$$

and

$$\left(f\Big|_m \begin{pmatrix} a & b \\ c & d \end{pmatrix}\right)(z) := (cz+d)^m f\left(\frac{az+b}{cz+d}\right)$$

for $\begin{pmatrix} a & b \\ c & d \end{pmatrix} \in \mathrm{SL}_2(\mathbb{Z})$, $m \in \mathbb{Z}$ and f a function on $\mathbb{H}$. The set of polynomials occurring as period polynomials of modular cusp forms of weight k is the space of polynomials of degree at most $k-2$ satisfying the equations (1) and (2). We denote it by $\mathbb{P}_{k-2}$.

As a corollary to the Eichler cohomology theorem, one finds that

$$S_k \oplus S_k \cong \mathbb{P}_{k-2}/\mathbb{C}[X^{k-2} - 1].$$

More precisely, [MR20, Cor. 2.53] gives an isomorphism between $S_k \oplus S_k$ and the *parabolic cohomology* $H^1_{\mathrm{par}}(\mathrm{SL}_2(\mathbb{Z}), \mathbb{C}[X]_{k-2})$, where $\mathbb{C}[X]_{k-2}$ denotes the space of polynomials of degree less or equal to $k-2$. The cocycles defining *Eichler cohomology classes* in $H^1(\mathrm{SL}_2(\mathbb{Z}), \mathbb{C}[X]_{k-2})$ are $\mathbb{C}[X]_{k-2}$-valued functions $M \mapsto P_M$ on $\mathrm{SL}_2(\mathbb{Z})$ satisfying the *cocycle identity*:

$$\forall M, N \in \mathrm{SL}_2(\mathbb{Z}): \quad P_{MN} = (P_M\big|_{2-k}N) + P_N.$$

The *coboundaries* are the functions P_M satisfying

$$\exists Q \in \mathbb{C}[X]_{k-2}\ \forall M \in \mathrm{SL}_2(\mathbb{Z}): \quad P_M = (Q\big|_{2-k}M) - Q.$$

A cocycle is called *parabolic* if

$$\exists Q \in \mathbb{C}[X]_{k-2}: \quad P_S = (Q\big|_{2-k}S) - Q.$$

The parabolic Eichler cohomology $H^1_{\mathrm{par}}(\mathrm{SL}_2(\mathbb{Z}), \mathbb{C}[X]_{k-2})$ consists of the cohomology classes represented by parabolic cocycles. Then, one has a short exact sequence:

$$0 \longrightarrow \mathbb{C}[X^{k-2} - 1] \longrightarrow \mathbb{P}_{k-2} \underset{P \mapsto (P_M)}{\longrightarrow} H^1_{\mathrm{par}}(\mathrm{SL}_2(\mathbb{Z}), \mathbb{C}[X]_{k-2}) \longrightarrow 0,$$

where the second arrow is determined by $P_T = P$ and $P_S = 0$.

Example 1.4 (Lewis-Zagier Correspondence) We keep the notation from Example 1.3 and refer to [MR20, Chaps. 3–5] for details on the material in this example. A *Maass cusp form* is an $\mathrm{SL}_2(\mathbb{Z})$-invariant function F on the upper half-plane $\mathbb{H}$ such that there exists a $\lambda \in \mathbb{C}$ with $\Delta F = \lambda F$, where $\Delta := -y^2\left(\frac{\partial^2}{\partial x^2} + \frac{\partial^2}{\partial y^2}\right)$ is the *Laplace-Beltrami operator* for the upper half-plane, and $F(z) \longrightarrow 0$ for z tending to $i\infty$ in a strip of bounded width. Each Maass cusp form is real analytic and square integrable as a function on the *modular surface* $\mathrm{SL}_2(\mathbb{Z})\backslash\mathbb{H}$. The Δ-eigenvalues are of the form $\lambda = s(1-s)$ with $\mathrm{Re}\, s = \frac{1}{2}$. The number $s \in \mathbb{C}$ is sometimes called the *spectral parameter* of corresponding eigenfunction. Lewis and Zagier proved in [LZ01] that there is a linear bijection between the space of Maass cusp forms with spectral parameter s and the space of real analytic solutions of the *three-term functional equation*:

$$\psi(x) = \psi(x+1) + (x+1)^{-2s}\psi\left(\frac{x}{x+1}\right)$$

on $\mathbb{R}_{>0}$ which satisfy the growth conditions

$$\psi(x) = \begin{cases} o(x^{-1}) & x \longrightarrow 0, \\ o(1) & x \longrightarrow \infty. \end{cases}$$

The functions ψ were called *period functions* by Lewis and Zagier as they have various properties analogous to the properties of period polynomials (see [LZ01, § 2]). For example, the functional equation can be rewritten as

$$\psi = \psi\big|_{-2s}S + \psi\big|_{-2s}STS$$

and for an even Maass cusp form F according to [LZ01, (2.2)], the associated period function has the integral representation:

$$\psi(x) = \int_0^\infty \frac{xy^s}{(x^2+y^2)^{s+1}} F(iy)\, \mathrm{d}t.$$

In fact, in the follow-up paper [BLZ15] to [LZ01] Bruggeman, Lewis and Zagier also established an interpretation of period functions in terms of parabolic cohomology.

While we do not try at this point to relate parabolic cohomology to the geodesic flow on the modular surface (that such a relation exists is suggested for instance by the work of Juhl [Ju01] and Bunke-Olbrich [BO99]), we note that [LZ01, § 3] contains an interpretation of period functions as eigenfunctions of transfer operators which are built from a symbolic dynamics associated with the geodesic flow on the modular surface. This gives the Lewis-Zagier correspondence an interpretation as a quantum-classical correspondence; see also the survey article [PZ20].

In view of this interpretation of the Lewis-Zagier correspondence, a natural question to ask is whether one can give a reasonable quantum system having modular cusp forms in the sense of Example 1.3 as states and such that it quantizes a classical system to which one can naturally associate period polynomials.

The Lewis-Zagier correspondence from Example 1.4 has been set up specifically for the modular surface, but meanwhile period functions have been generalized in various ways (see [MR20] for a fairly up-to-date compilation). Also the transfer operator approach has been extended substantially, mostly through the work of Pohl and her collaborators (see, e.g., [Po12]).

Example 1.5 (Distributions Invariant Under the Horocycle Flow) Let $\mathcal{X}$ be a hyperbolic surface of finite volume and $S\mathcal{X}$ its tangent unit sphere bundle. In [FF03] Flaminio and Forni studied the space $\mathcal{I}(S\mathcal{X})$ of compactly supported distributions on $S\mathcal{X}$ which are invariant under the *horocycle flow* generated by the matrix:

$$U := \begin{pmatrix} 0 & 1 \\ 0 & 0 \end{pmatrix} \in \mathfrak{sl}_2(\mathbb{R}),$$

where $S\mathcal{X}$ is realized as a quotient $\Gamma\backslash S\mathbb{H} \cong \Gamma\backslash \mathrm{PSL}_2(\mathbb{R})$ for a discrete subgroup Γ of $\mathrm{PSL}_2(\mathbb{R})$ acting without fixed points on $\mathbb{H}$. They showed that $\mathcal{I}(S\mathcal{X})$ decomposes as

$$\mathcal{I}(S\mathcal{X}) = \bigoplus_{\mu \in \mathrm{spec}_{\mathrm{pp}}(\Delta_{\mathcal{X}})} \mathcal{I}_\mu \oplus \bigoplus_{n \in \mathbb{N}} \mathcal{I}_n^d \oplus \bigoplus_{c \in C} \mathcal{I}_c, \tag{3}$$

where $\mathrm{spec}_{\mathrm{pp}}(\Delta_{\mathcal{X}})$ is the pure point spectrum of the Laplace-Beltrami operator $\Delta_{\mathcal{X}}$ of $\mathcal{X}$ and C is the (finite) set of cusps of $\mathcal{X}$. The spaces $\mathcal{I}_n^d$ are closely related to the spaces of holomorphic sections of powers of the canonical line bundle over $\mathcal{X}$. This decomposition is derived from the Plancherel decomposition of $L^2(S\mathcal{X}) \cong L^2(\Gamma\backslash \mathrm{PSL}_2(\mathbb{R}))$.

For $0 < \mu < \frac{1}{4}$, the spaces $\mathcal{I}_\mu$ decompose further as $\mathcal{I}_\mu = \mathcal{I}_\mu^+ \oplus \mathcal{I}_\mu^-$ with $\dim(\mathcal{I}_\mu^\pm)$ equal to the multiplicity of the eigenvalue μ. For $\mu \geq \frac{1}{4}$, the multiplicity of the eigenvalue μ is equal to $\dim(\mathcal{I}_\mu)$. Finally, the space $\mathcal{I}_0$ is spanned by the $\mathrm{PSL}_2(\mathbb{R})$-invariant volume.

In the case of compact surfaces, $\mathrm{spec}_{\mathrm{pp}}(\Delta_{\mathcal{X}})$ is the entire spectrum of $\Delta_{\mathcal{X}}$ and C is empty. In the noncompact case, $\mathrm{spec}(\Delta_{\mathcal{X}})$ has an absolutely continuous part consisting of $[\frac{1}{4}, \infty[$ with multiplicity equal to the number of cusps. Thus in both cases one can extract quantum information, namely, the spectral data of the Laplace-Beltrami operator, from $\mathcal{J}(S\mathcal{X})$.

Flaminio and Forni also observe that the splitting (3) is invariant under the geodesic flow. More precisely, they show that it is essentially the spectral decomposition of the geodesic flow (see [FF03, Thm. 1.4] for details). Thus, $\mathcal{J}(S\mathcal{X})$ yields a spectral quantum-classical correspondence for finite area hyperbolic surfaces.

Example 1.6 (Dyatlov-Faure-Guillarmou Correspondence) Let $\mathcal{X}$ be a compact hyperbolic surface and let X be the vector field generating the geodesic flow φ_t on the unit tangent bundle $S\mathcal{X}$ of $\mathcal{X}$. The linear operator

$$\mathcal{L}_t : \begin{cases} C_c^\infty(S\mathcal{X}) \to C_c^\infty(S\mathcal{X}) \\ \qquad f \quad \mapsto \ f \circ \varphi_{-t} \end{cases}$$

is called the *transfer operator* of the geodesic flow, and the vector field $-X$ is its generator.

For $f_1, f_2 \in C_c^\infty(S\mathcal{X})$, one defines the *correlation functions*:

$$C_X(t; f_1, f_2) := \int_{S\mathcal{X}} (\mathcal{L}_t f_1) \cdot f_2 \,\mathrm{d}\mu_L$$

where μ_L is the Liouville measure (invariant by φ_t). By [BL07, FS11, DZ16] the Laplace transform

$$R_X(\lambda; f_1, f_2) := -\int_0^\infty e^{-\lambda t} C_X(t; f_1, f_2) dt$$

extends meromorphically from $\mathrm{Re}(\lambda) > 0$ to $\mathbb{C}$. For $\mathrm{Re}(\lambda) > 0$, we have $R_X(\lambda; f_1, f_2) = \langle(-X-\lambda)^{-1} f_1, f_2\rangle$, and $R_X(\lambda)$ gives a meromorphic extension of the Schwartz kernel of the resolvent of $-X$. The poles are called *Ruelle resonances* and the *residue operator* $\Pi^X_{\lambda_0} : C_c^\infty(S\mathcal{X}) \to \mathcal{D}'(S\mathcal{X})$ defined by

$$\forall f_1, f_2 \in C_c^\infty(S\mathcal{X}) : \quad \langle \Pi^X_{\lambda_0} f_1, f_2\rangle := \mathrm{Res}_{\lambda_0} R_X(\lambda; f_1, f_2)$$

has finite rank, commutes with X, and $(-X - \lambda)$ is nilpotent on its range. The elements in the range of $\Pi^X_{\lambda_0}$ are called *generalized Ruelle resonant states*. By results in [BL07, FS11, DZ16], the poles can be identified with the discrete spectrum of $-X$ in certain Hilbert spaces and the generalized resonant states with generalized eigenfunctions.

In [DFG15] Dyatlov, Faure, and Guillarmou show that if $\mathrm{spec}(\Delta_{\mathcal{X}}) = \{s_j(1 - s_j)\}$ for a sequence $\{s_j \in [0,1] \cup (\frac{1}{2} + i\mathbb{R}) \mid j \in \mathbb{N}_0\}$, then the Ruelle resonances of $S\mathcal{X}$ in $\mathbb{C} \setminus (-1 - \frac{1}{2}\mathbb{N}_0)$ are given by

$$\{\lambda_{j,m} := -m - 1 + s_j \mid j, m \in \mathbb{N}_0\}.$$

The sets $\{\lambda_{j,m} \mid j \in \mathbb{N}_0\}$ are called the *bands* of the Ruelle resonance spectrum. One refers to $\{\lambda_{j,0} \mid j \in \mathbb{N}_0\}$ as the *first band*. It is characterized by the fact that the corresponding resonant states are annihilated by the horocycle vector field generated by $U \in \mathfrak{sl}_2(\mathbb{R})$; see Example 1.5. Moreover, the points $\lambda \in -1 - \frac{1}{2}\mathbb{N}_0$ excluded from this correspondence between the spectrum of the Laplacian and the Ruelle resonance spectrum of the geodesic flow are called *exceptional points*.

In fact, [DFG15] contains generalizations of the above results to compact real hyperbolic spaces of arbitrary dimension.

1.2 Outline of the Program and its Purpose

Let $\mathcal{X} = \Gamma\backslash G/K$ be a locally symmetric space and $\mathcal{A}$ the algebra of G-invariant smooth functions on the cotangent bundle $T^*(G/K)$. The elements of $\mathcal{A}$ may be considered as Poisson commuting symbols of G-invariant pseudo-differential operators on G/K but, when Γ-invariant, also as functions on $T^*(\mathcal{X})$. The goal is now to study commutative algebras Ψ of pseudo-differential operators on $\mathcal{X}$ quantizing $\mathcal{A}$ and to relate dynamical invariants of $\mathcal{A}$ (classical side) and Ψ (quantum side).

On the classical side, the primary dynamical systems to be considered are the Hamiltonian flows associated with the elements of $\mathcal{A}$. Derived from those one has restrictions of flow-invariant submanifolds such as Γ-quotients of G-orbits or subbundles given, e.g., by hyperbolic properties of the Hamiltonian flows. This may lead to restrictions in the flow variables as well (e.g., "positive time" only). For these derived dynamical systems, alternative descriptions may be available, for instance via symbolic dynamics. Given any of these dynamical systems, one may consider dynamical invariants such as compact orbits, dynamical zeta functions and their divisors, transfer (or Koopman) operators, and their spectral theory or classical scattering. As in the higher rank situation one typically has multiparameter dynamical actions, one wants to establish multiparameter versions of these invariants rather than simply using the invariants coming with one-parameter restrictions of the dynamical systems.

On the quantum side, one expects the pseudo-differential operators to generate multiparameter flows of linear operators, whose spectral theory needs to be studied. G-invariance allows to decompose the dynamical system and leads to questions of representation theory. Here spectral invariants can involve parameters of representations such as infinitesimal characters, dimensions of associated spaces (e.g., consisting of Γ-invariant distributions), or spectral invariants of the flow operators themselves. The latter could be ordinary spectra, resonances, or derived objects such as divisors of meromorphically continued scattering operators.

The general strategy to relate the quantum and the classical side is to use the G-invariance properties to connect the spectral invariants with G-representations and then use representation theoretic knowledge about intertwining of representations to relate two such invariants. A particularly successful example in this direction has been to relate the classical dynamics with principal series representations and then use Poisson transform to connect them to eigenspace representations which in turn are intimately related to the quantum dynamics.

Apart from supporting the philosophy that the presence of symmetry narrows down the possibilities to quantize given classical systems, the main motivation for establishing quantum-classical correspondences is to create bridges which allow to solve problems on one side using the tool boxes from the other side. An example of this kind of application is the proof in [HWW21] of uniform spectral gaps of Weyl chamber flows in higher rank.

Finally, one should note that given specific results for the rigid symmetric situations, one can by deformation sometimes guess or even prove results for situations where the symmetry is broken. This is the case for instance in [KW20], where Betti numbers are given as the dimensions of Ruelle resonant states. Another example are the cases of quantum-classical correspondences for finite graphs in [BHW23, AFH23a] which take a similar form as the ones for compact rank one locally symmetric spaces but need no homogeneity.

2 Locally Symmetric Spaces

Locally symmetric spaces are quotients of symmetric spaces by discrete subgroups of the group of isometries. In general they have an orbifold structure, but we will always make assumptions on the discrete subgroups which guarantee that the quotient is actually a manifold. A natural way to study locally symmetric spaces is to lift objects to the symmetric space, where one can use group invariance as a tool and check results for invariance under the discrete subgroup allowing to interpret them as results for the locally symmetric space. Thus, a detailed description of the symmetric spaces and (geometric) objects derived from them will be useful.

2.1 Symmetric Spaces

Consider a real semisimple Lie group G, connected and of noncompact type, and let $G = KAN$ be an Iwasawa decomposition with A abelian, K a compact maximal subgroup and N nilpotent. Then $A \cong \mathbb{R}^\ell$ and ℓ is called the *real rank* of G. Let $\mathfrak{a}$ be the Lie algebra of A and consider the adjoint action of $\mathfrak{a}$ on $\mathfrak{g}$ which leads to the definition of a finite set $\Sigma \subset \mathfrak{a}^*$ of *restricted roots*. More precisely, for $0 \neq \alpha \in \mathfrak{a}^*$ set

$$\mathfrak{g}_\alpha := \{X \in \mathfrak{g} \mid \forall D \in \mathfrak{a} : \ [D, X] = \alpha(D)X\}.$$

If $\mathfrak{g}_\alpha \neq \{0\}$, then α is a *restricted root* of $(\mathfrak{a}, \mathfrak{g})$. An element X of $\mathfrak{a}$ is *regular* if $\alpha(X) \neq 0$ for all $\alpha \in \Sigma$. For each regular element X_o, one can define a set of *positive roots* via

$$\Sigma^+ := \{\alpha \in \Sigma \mid \alpha(X_o) > 0\}.$$

Then Σ is the disjoint union of Σ^+ and $\Sigma^- := -\Sigma^+$, the set of *negative roots*. We fix some choice of positive roots and define the corresponding *positive Weyl chamber* by

$$\mathfrak{a}_+ := \{H \in \mathfrak{a} \mid \forall \alpha \in \Sigma_+ : \alpha(H) > 0\}.$$

Note that $\mathfrak{a}$ is a *Cartan subspace*, i.e., a maximal abelian subspace of $\mathfrak{p} := \mathfrak{k}^\perp \subseteq \mathfrak{g}$, where the orthogonal complement is taken with respect to the Killing form. Further let $M' := N_K(\mathfrak{a})$ be the normalizer of $\mathfrak{a}$ in K and $M := Z_K(\mathfrak{a})$ the centralizer of $\mathfrak{a}$ in K. Then $Z_K(\mathfrak{a})$ is normal in $N_K(\mathfrak{a})$. The quotient group $W := W(\mathfrak{a}, K) := N_K(\mathfrak{a})/Z_K(\mathfrak{a})$ is the *Weyl group*.

For later use we set:

$$\mathfrak{k}_\alpha := \mathfrak{k} \cap (\mathfrak{g}_\alpha + \mathfrak{g}_{-\alpha}), \qquad \mathfrak{p}_\alpha := \mathfrak{p} \cap (\mathfrak{g}_\alpha + \mathfrak{g}_{-\alpha})$$

for $\alpha \in \mathfrak{a}^*$.

Remark 2.1 (Quotients by Discrete Torsion-free Subgroups) If one now considers a torsion-free discrete subgroup $\Gamma < G$, one can define the locally symmetric space $\mathcal{X} := \Gamma\backslash G/K$ and the manifold $\mathcal{M} := \Gamma\backslash G/M$. If G is of real rank one, then $\mathcal{M}$ can be identified with the sphere bundle $S\mathcal{X}$. In higher rank the situation is more complicated as we will explain below.

2.1.1 Equivariant Differential Geometry on G/K

We review some preparatory material on differential analysis of the natural G-action on G/K which can found in [Hi05]. The action of G on G/K induces natural actions of G on the tensor bundles of G/K. Viewing $g \in G$ as a diffeomorphism on G/K, the action on the tangent bundle $T(G/K)$ is given by the derivative $g' \colon T(G/K) \to T(G/K)$. Then the action on the cotangent bundle $T^*(G/K)$ is defined via the canonical pairing:

$$\begin{aligned}\forall \xi \in T_x^*(G/K), v \in T_x(G/K) : \ &\langle g \cdot \xi, v\rangle_{G/K} = \langle \xi, g^{-1} \cdot v\rangle_{G/K} \\ &= \langle \xi, (g^{-1})'(x)(v)\rangle_{G/K}.\end{aligned}$$

Note that in this way all the canonical projections are G-equivariant and G acts on all bundles by bundle maps covering the original action, so we obtain induced actions on the spaces of sections. For example, G acts on the space $\mathcal{V}(G/K)$ of smooth vector fields via

$$(g \cdot X)(x) = g \cdot (X(g^{-1} \cdot x)). \tag{4}$$

It is now clear what is meant by a G-invariant vector field, metric, symplectic form, and so on.

Remark 2.2 (Homogeneous Vector Bundles and Invariant Vector Fields) Let H be a closed subgroup of G. Then $T(G/H) \cong G \times_H (\mathfrak{g}/\mathfrak{h})$, where $G \times_H (\mathfrak{g}/\mathfrak{h})$ is the homogeneous vector bundle over G/H with respect to the H-action on $\mathfrak{g}/\mathfrak{h}$. More precisely, $G \times_H (\mathfrak{g}/\mathfrak{h})$ is the set of H-orbits on $G \times (\mathfrak{g}/\mathfrak{h})$ under the right H-action:

$$(g, X + \mathfrak{h}) \cdot h = (gh, h^{-1} \cdot X + \mathfrak{h}),$$

where $h \cdot X$ denotes the adjoint action. We denote the equivalence class of $(g, X+\mathfrak{h})$ by $[g, X + \mathfrak{h}]$. Then $g_1 \cdot [g_2, X + \mathfrak{h}] := [g_1 g_2, X + \mathfrak{h}]$ defines a left action of G on $G \times_H (\mathfrak{g}/\mathfrak{h})$. The derivative of the left translation

$$\lambda_g \colon G/H \to G/H, \quad g_1 H \mapsto g g_1 H$$

is an isomorphism denoted by $\lambda'_g(x) \colon T_x(G/H) \to T_{g\cdot x}(G/H)$, and we identify $G \times_H (\mathfrak{g}/\mathfrak{h})$ with $T(G/H)$ via the equivariant correspondence:

$$[g, X + \mathfrak{h}] \longleftrightarrow \lambda'_g(o)(X + \mathfrak{h}), \tag{5}$$

where $o = H \in G/H$ is the canonical base point of G/H.

The dimensions in the stratification of $T(G/H)$ by G-orbits are

$$\dim(G \cdot [g, X + \mathfrak{h}]) = \dim(G/H) + \dim(K \cdot (X + \mathfrak{h}))$$

Moreover, $[g, X + \mathfrak{h}]$ and $[\tilde{g}, \tilde{X} + \mathfrak{h}]$ belong to the same G-orbit in $G \times_H (\mathfrak{g}/\mathfrak{h})$ if and only if $X + \mathfrak{h}$ and $\tilde{X} + \mathfrak{h}$ belong to the same H-orbit in $\mathfrak{g}/\mathfrak{h}$.

Let $\mathrm{pr}_H : G \to G/H$ be the canonical projection. Its derivative $\mathrm{pr}'_H : TG \to T(G/H)$ is surjective and given by $\mathrm{pr}'_H(g, X) = [g, X + \mathfrak{h}]$ in the notation of homogeneous vector bundles. Suppose that we have an H-invariant decomposition $\mathfrak{g} = \mathfrak{h} + \mathfrak{q}$. Then we can identify $T_o(G/H) \cong \mathfrak{g}/\mathfrak{h}$ with $\mathfrak{q}$. This gives an identification $T(G/H) \cong G \times_H \mathfrak{q}$ via

$$[g, X] \longleftrightarrow \lambda'_g(o)X.$$

The vector fields on G/H correspond to smooth functions $\overline{X}: G \to \mathfrak{g}/\mathfrak{h}$ satisfying

$$\forall g \in G, h \in H: \quad \overline{X}(gh) = h^{-1} \cdot \overline{X}(g).$$

The vector field is then given by $gH \mapsto [g, \overline{X}(g)]$. A vector field $gH \mapsto [g, \overline{X}(g)]$ is G-invariant if and only if $\overline{X}$ is a constant map:

$$[x, \overline{X}(x)] = \left(g \cdot [\bullet, \overline{X}(\bullet)]\right)(x) \overset{(4)}{=} g \cdot [g^{-1}x, \overline{X}(g^{-1}x)] = [x, \overline{X}(g^{-1}x)].$$

In that case the value of $\overline{X}$ has to be an H-fixed point in $\mathfrak{g}/\mathfrak{h}$. Therefore the map

$$\mathcal{V}(G/H)^G \to (\mathfrak{g}/\mathfrak{h})^H, \quad \overline{X} \mapsto \overline{X}(o),$$

where $\mathcal{V}(G/H)^G$ denotes the G-invariant vector fields and $(\mathfrak{g}/\mathfrak{h})^H$ the H-fixed points, is a linear isomorphism. If $\overline{X}(o) = X + \mathfrak{h}$ with $X \in \mathfrak{g}$, then the flow of $\overline{X}$ is given by

$$\varphi_t^X(g \cdot o) = g(\exp tX) \cdot o. \tag{6}$$

In particular, the invariant vector fields are complete.

Using the description of $T(G/K)$ given in Remark 2.2, we can show the following proposition.

Proposition 2.3 (cf. [Hi05, Lemmas 8.4 & 8.6]) *Let G act on the left of $G/M \times \mathfrak{a}$ via left translation on the first factor. Then the following holds.*

(i) *The map*

$$\Phi: G/M \times \mathfrak{a} \to T(G/K) \cong G \times_K \mathfrak{p}, \quad (gM, X) \mapsto \lambda_g'(o)X = [g, X]$$

is G-equivariant, surjective, and the following two statements are equivalent:

(1) $\Phi(gM, X) = \Phi(\tilde{g}M, \tilde{X})$.
(2) *There exists $k \in K$ with $gk = \tilde{g}$ and* $\mathrm{Ad}(k)X = \tilde{X}$.

In particular, $\tilde{X}$ is contained in the Weyl group orbit of X, and we have a bijection of orbit spaces:

$$G\backslash T(G/K) \longleftrightarrow W(\mathfrak{a}, K)\backslash\mathfrak{a}.$$

(ii) *Φ is smooth and surjective. Nevertheless, Φ is not a covering. The derivative $\Phi'(gM, X)$ is bijective if and only if X is regular. More precisely,*

$$\ker \Phi'(e, X) = \sum_{X \in \ker \alpha} \mathfrak{k}_\alpha \quad \textit{and} \quad \operatorname{im} \Phi'(e, X) = \mathfrak{p} \times (\mathfrak{a} + \sum_{X \notin \ker \alpha} \mathfrak{p}_\alpha).$$

Set $P := \exp \mathfrak{p}$ and note that the Cartan decomposition $G = PK$ yields a retraction:

$$j : G/K \to P, \quad pK \mapsto p$$

for the canonical quotient map pr_K. We identify G/K with $P \subseteq G$, so that $T(G/K)$ can be viewed as a subset of $TG \cong G \times \mathfrak{g}$. Then the derivative $\exp' : T\mathfrak{p} \cong \mathfrak{p} \times \mathfrak{p} \to TP \subseteq TG \cong G \times \mathfrak{g}$ is given by

$$\exp'(X, Y) \overset{(5)}{\cong} \Big(\exp X, \sum_{k=1}^{\infty} \tfrac{(-\operatorname{ad} X)^{k-1}}{k!} Y\Big) =: (\exp X, \Psi(X)Y).$$

and yields a coordinate system for $T(G/K)$.

We recall descriptions the G-actions on G/K and $T(G/K)$ in our various identifications. We consider the projection:

$$\mathrm{pr}_P : G = PK \to P, \quad pk \mapsto p$$

and note that $p^2 = (pk)\theta(pk)^{-1}$, where $\theta \colon G \to G$ is the Cartan involution corresponding to the Cartan decomposition $G = PK$. The G-action on G/K by left translation becomes $g \cdot p = \mathrm{pr}_P(gp)$ under the above identification of G/K with $P = \exp \mathfrak{p} \cong \mathfrak{p}$. Thus

$$(g \exp X)\theta(g \exp X)^{-1} = g(\exp X)\theta(\exp X)^{-1}\theta(g)^{-1} = g(\exp 2X)\theta(g)^{-1}$$

for $g \in G$ and $X \in \mathfrak{p}$ shows that the induced action on $\mathfrak{p}$ is given by

$$g \cdot X = \tfrac{1}{2} \log(g(\exp 2X)\theta(g)^{-1}). \tag{7}$$

Lemma 2.4 ([Hi05, Lemma 8.2]) *The action of G on $T(G/K) \cong \mathfrak{p} \times \mathfrak{p}$ is given by*

$$k \cdot (X, Y) = (\mathrm{Ad}(k)X, \mathrm{Ad}(k)Y)$$

$$p \cdot (X, Y) = \Big(\tfrac{1}{2} \log(p(\exp 2X)p), \big(\mathrm{Ad}(p) \circ \Psi(\log(p(\exp 2X)p))\big)^{-1} \Psi(2X)Y\Big)$$

for $k \in K$, $p \in P$ and $X, Y \in \mathfrak{p}$.

The Riemannian metric on G/K allows us to identify $T(G/K)$ and the cotangent bundle $T^*(G/K)$. The cotangent bundle carries a natural symplectic structure which we will have to study in some detail. In this context the iterated tangent bundle $T(T(G/K))$ will be of importance. [Pa99, § 1.3] contains a detailed geometric description. Here we give a description in group theoretic terms which allows explicit calculations for Weyl chamber flows. To this end we consider the derivative:

$$f' \colon T(T\mathfrak{p}) = (\mathfrak{p} \times \mathfrak{p}) \times (\mathfrak{p} \times \mathfrak{p}) \to T(TG) \cong (G \times \mathfrak{g}) \times (\mathfrak{g} \times \mathfrak{g})$$

of the map $f := \exp'$. It is given by

$$f'(X, Y)(A, B) \cong (\exp X, \Psi(X)Y, \Psi(X)A, (\Psi'(X)A)Y + \Psi(X)B).$$

For $X = 0$, the formula for f' simplifies to

$$f'(0, Y)(A, B) = (\mathbf{1}, Y, A, B - \tfrac{1}{2}[A, Y]).$$

Under the identification of the tangent bundle $T(TG)$ with $(G \times \mathfrak{g}) \times (\mathfrak{g} \times \mathfrak{g})$ and the corresponding identification of $T(T(G/K))$ as a subset of $(G \times \mathfrak{g}) \times (\mathfrak{g} \times \mathfrak{g})$ for $Y = (0, Y) \in T_o(G/K) \cong \mathfrak{p}$, we have:

$$T_Y(T(G/K)) = \{(\mathbf{1}, Y, A, -\tfrac{1}{2}[A, Y] + B) \mid A, B \in \mathfrak{p}\}. \tag{8}$$

The other tangent spaces are then determined via G-invariance.

We consider the canonical projection $\mathrm{pr}_{T^*} : T^*(G/K) \to G/K$ and its derivative $\mathrm{pr}'_{T^*} : T(T^*(G/K)) \to T(G/K)$. Then the set

$$\{\xi \in T(T^*(G/K)) \mid \mathrm{pr}'_{T^*}(\xi) = 0\}$$

is called the *vertical subbundle* of $T(T^*(G/K))$. Analogously one has a vertical subbundle of $T(T(G/K))$ using the canonical projection $\mathrm{pr}_T : T(G/K) \to G/K$. It is clear that the identification $T(G/K) \cong T^*(G/K)$ preserves the vertical bundles. Since $T_v(T(G/K))$ is canonically identified with $T_{\mathrm{pr}_T(v)}(G/K) \times T_{\mathrm{pr}_T(v)}(G/K)$, the Riemannian metric on G/K induces a Riemannian metric on $T(T(G/K))$. Therefore it makes sense to speak about the *horizontal subbundle*, i.e., the bundle of elements orthogonal to the elements of the vertical subbundle.

Using the above identifications, the vertical and horizontal bundles can be described explicitly in group theoretic terms.

Lemma 2.5 ([Hi05, Lemma 8.1]) *Viewing $T_Y(T(G/K))$ as a subset of $T(TG) = (G \times \mathfrak{g}) \times (\mathfrak{g} \times \mathfrak{g})$, we have:*

(i) *The space of vertical vectors is* $\{(\mathbf{1}, Y, 0, B) \mid B \in \mathfrak{p}\}$.
(ii) *The space of horizontal vectors is* $\{(\mathbf{1}, Y, A, -\frac{1}{2}[A, Y]) \mid A \in \mathfrak{p}\}$.

Lemma 2.6 ([Hi05, Lemma 8.3]) *Under the identification*

$$T_{(0,Y)}(\mathfrak{p} \times \mathfrak{p}) = \mathfrak{p} \times \mathfrak{p} \to \mathfrak{g} \times \mathfrak{g} \cong T_{(\mathbf{1},Y)}(G \times \mathfrak{g}), \;\; (A, B) \mapsto (A, B - \tfrac{1}{2}[A, Y])$$

we have:

(i) *The vector fields* $\tilde{Z} \in \mathcal{V}(T(G/K))$ *induced by the action of* $\exp \mathbb{R}Z$ *are given by*

$$\forall Z \in \mathfrak{k}, Y \in \mathfrak{p}: \ \tilde{Z}(\mathbf{1}, Y) = (\mathbf{1}, Y, 0, [Z, Y])$$
$$\forall Z \in \mathfrak{p}, Y \in \mathfrak{p}: \ \tilde{Z}(\mathbf{1}, Y) = (\mathbf{1}, Y, Z, -\tfrac{1}{2}[Z, Y]).$$

In particular, $\tilde{Z}$ *is a section of the vertical bundle for* $Z \in \mathfrak{k}$ *and a section of the horizontal bundle for* $Z \in \mathfrak{p}$.

(ii) *The tangent space* $T_{(\mathbf{1},Y)}(G \cdot (\mathbf{1}, Y))$ *is given by*

$$\{(\mathbf{1}, Y, A, [B, Y] - \tfrac{1}{2}[A, Y]) \mid A \in \mathfrak{p}, B \in \mathfrak{k}\} \cong \mathfrak{p} \times \mathrm{ad}(Y)\mathfrak{k}.$$

It contains all the horizontal vectors in $T_{(\mathbf{1},Y)}(T(G/K))$.

Remark 2.7 (The Right A-action on $\mathcal{M}$) As A commutes with M, the space $\mathcal{M}$ from Remark 2.1 carries a right A-action. The tangent bundle $T(G/M)$ can be canonically identified with the homogeneous bundle $G \times_M (\mathfrak{g}/\mathfrak{m})$. With respect to this identification, the derived A-action on $T(G/M)$ is given by $[g, X] \cdot a = [ga, \mathrm{Ad}(a^{-1})X]$. To describe the properties of this action in more detail, we consider the Lie algebra $\mathfrak{n} = \sum_{\alpha \in \Sigma^+} \mathfrak{g}_\alpha$ of N and its "opposite" $\overline{\mathfrak{n}} := \sum_{-\alpha \in \Sigma^+} \mathfrak{g}_\alpha$. The decomposition $\mathfrak{g} = \overline{\mathfrak{n}} + \mathfrak{m} + \mathfrak{a} + \mathfrak{n}$ allows us to identify $\mathfrak{g}/\mathfrak{m}$ with $\overline{\mathfrak{n}} + \mathfrak{a} + \mathfrak{n}$. Note that each of these spaces is M-invariant, so $T(G/M)$ is decomposed as the Whitney sum:

$$T(G/M) = (G \times_M \overline{\mathfrak{n}}) \oplus (G \times_M \mathfrak{a}) \oplus (G \times_M \mathfrak{n}).$$

As the left Γ-actions do not interfere with the right A-action, this sum descends to $\mathcal{M}$:

$$T\mathcal{M} = (\Gamma\backslash G \times_M \overline{\mathfrak{n}}) \oplus (\Gamma\backslash G \times_M \mathfrak{a}) \oplus (\Gamma\backslash G \times_M \mathfrak{n}). \tag{9}$$

This decomposition will show that the A-action is an Anosov flow in the sense of Sect. 3.1.1. Note that $T_e(G/M)$, where $e := M \in G/M$ is the canonical base point, can be identified with $\mathfrak{g}/\mathfrak{m}$. This in turn can also be identified with $\sum_{\alpha \in \Sigma^+} \mathfrak{k}_\alpha$.

2.1.2 Invariant Differential Operators

Let $\mathbb{D}(G/K)$ be the algebra of *G-invariant differential operators* on G/K, i.e., differential operators commuting with the natural G-action on $C^\infty(G/K)$. Then we have an algebra isomorphism $\mathrm{HC}\colon \mathbb{D}(G/K) \to I(\mathfrak{a}_\mathbb{C}^*)$ from $\mathbb{D}(G/K)$ to the W-invariant complex polynomials on $\mathfrak{a}_\mathbb{C}^*$ which is called *Harish-Chandra* *homomorphism* (see [He84, Ch. II, Theorem 5.18]). For $\lambda \in \mathfrak{a}_{\mathbb{C}}^*$ let χ_λ be the character of $\mathbb{D}(G/K)$ defined by $\chi_\lambda(D) := \mathrm{HC}(D)(\lambda)$. Obviously, $\chi_\lambda = \chi_{w\lambda}$ for $w \in W$. Furthermore, the χ_λ exhaust all characters of $\mathbb{D}(G/K)$ (see [He84, Ch. III, Lemma 3.11]). We define the space of joint eigenfunctions:

$$E_\lambda := \{f \in C^\infty(G/K) \mid \forall D \in \mathbb{D}(G/K) : \ Df = \chi_\lambda(D)f\}. \tag{10}$$

Note that E_λ is G-invariant.

Remark 2.8 (Harish-Chandra Isomorphism) Combining the Harish-Chandra isomorphism with a result of Chevalley, we find isomorphisms:

$$\mathbb{D}(G/K) \cong \mathcal{E}'_K(G/K) \cong U(\mathfrak{a})^W \cong I(\mathfrak{a}_{\mathbb{C}}^*) \cong \mathbb{C}[X_1, \ldots, X_\ell],$$

where $\ell = \dim_{\mathbb{R}} \mathfrak{a}$ is the rank of G/K, $\mathcal{E}'_K(G/K)$ denotes the space of distributions supported in the base point, and $U(\mathfrak{a})$ is the universal enveloping algebra of $\mathfrak{a}_{\mathbb{C}}$. In fact, the generators of $I(\mathfrak{a}_{\mathbb{C}}^*)$ can be chosen real, i.e., as real polynomials on $\mathfrak{a}^*$, which then come from differential operators with real-valued coefficients.

2.2 Representation Theory and Harmonic Analysis

In the context of quantum-classical correspondences for locally symmetric spaces, three aspects of representation theory and harmonic analysis play key roles. According to the Plancherel formula for symmetric spaces, the *spherical principal series representations* are the building blocks of $L^2(G/K)$, so they may be considered as quantum objects. At the same time, they are realized on the boundary of the symmetric space. In rank one the set of geodesics can be parameterized by two boundary points, namely, their limits for time t tending to $\pm\infty$. This ties the boundary to the underlying classical geometry of the sphere bundle. Moreover, relating geodesics to pairs of boundary points allows to interpret *standard intertwining operators* (often called *Knapp-Stein operators*) as classical scattering operators. It is possible to move functions on the boundary to functions on the symmetric space preserving the symmetries and hence relating associated spectral invariants. The objects doing this are the *Poisson transforms*. They are instrumental in the quantum-classical correspondences we will describe.

The intuitive picture just drawn for rank one symmetric spaces does have higher rank generalizations which, however, are a little less intuitive. Moreover, G-equivariance properties of the maps involved will allow to move from symmetric spaces to locally symmetric spaces.

2.2.1 Principal Series Representations for Minimal Parabolics

The concept of a principal series representation is an important tool in representation theory of semisimple Lie groups. It can be described using different pictures. We start with the *induced picture*: Pick $\lambda \in \mathfrak{a}_{\mathbb{C}}^*$ and an irreducible unitary representation (σ, V_σ) of M. We define:

$$V_{\sigma,\lambda} := \left\{ f\colon G \overset{\text{cont.}}{\to} V_\sigma \;\middle|\; \begin{array}{c} \forall g \in G, m \in M, a \in A, n \in N: \\ f(gman) = e^{-(\lambda+\rho)\log a}\sigma(m)^{-1} f(g) \end{array} \right\}$$

endowed with the norm $\|f\|^2 := \int_K \|f(k)\|_\sigma^2 \mathrm{d}k$, where $\mathrm{d}k$ is the normalized Haar measure on K, 2ρ is the sum of all positive restricted roots weighted by multiplicity, and $\|\cdot\|_\sigma$ is the norm on V_σ. The group G acts on $V_{\sigma,\lambda}$ by the left regular representation. The completion $H_{\sigma,\lambda}$ of $V_{\sigma,\lambda}$ with respect to the norm is called *induced picture of the (not necessarily unitary) principal series representation* with respect to (σ, λ). We also write $\pi_{\sigma,\lambda}$ for this representation. If σ is the trivial representation, then we write H_λ and π_λ and call it the *spherical principal series* with respect to λ. Note that for equivalent irreducible unitary representations σ_1, σ_2 of M, the corresponding principal series representations are equivalent as representations as well. In particular, the Weyl group W acts on the unitary dual of M by $m_w\sigma(m) = \sigma(m_w^{-1} m m_w)$, where $w \in W$ is given by a representative $m_w \in M'$ and therefore $H_{\lambda, w\sigma}$ is well-defined up to equivalence.

The *compact picture* is given by restricting the function $f\colon G \to V_\sigma$ to K, i.e., a dense subspace is given by

$$\{f\colon K \to V_\sigma \text{ cont.} \mid \forall k \in K, m \in M:\; f(km) = \sigma(m)^{-1} f(k)\}$$

with the same norm as above. In this picture the G-action is given by

$$\pi_{\sigma,\lambda}(g) f(k) = e^{-(\lambda+\rho)H(g^{-1}k)} f(k_{KAN}(g^{-1}k)), \quad g \in G, k \in K,$$

where k_{KAN} is the K-component in the Iwasawa decomposition $G = KAN$.

Recall the associated homogeneous vector bundle $\mathcal{V}_\sigma := G \times_M V_\sigma$ over G/M built in analogy to $G \times_K \mathfrak{p}$ considered in Remark 2.2. We can identify the principal series representation $H_{\sigma,\lambda}$ with L^2-sections of an associated bundle. If $\mathcal{V}_\sigma^{K/M}$ denotes the restriction of the vector bundle $\mathcal{V}_\sigma$ over G/M to $K/M \subseteq G/M$, we obtain the principal series representation with parameters (σ, λ) as the Hilbert space of L^2-sections of the bundle $\mathcal{V}_\sigma^{K/M}$ over the Furstenberg boundary K/M of G/K with the action

$$\overline{\pi_{\sigma,\lambda}(g) f}(k) = e^{-(\lambda+\rho)H(g^{-1}k)} \overline{f}(k_{KAN}(g^{-1}k)).$$

Remark 2.9 (Extension of Continuous G-representations) For a continuous G-representation on some locally convex complete Hausdorff topological vector space

V, one considers the spaces V^∞ of *smooth vectors* consisting of the $v \in V$ for which the map $g \mapsto gv$ is smooth. If one equips V^∞ with the topology induced by the embedding $V^\infty \hookrightarrow C^\infty(G, V)$, then V^∞ is a G-invariant Fréchet space and the G-representation on V^∞ is continuous (see, e.g., [Wa72, § 4.4.1]). Recall the *contragredient representation* $\check{V}$ of G from [Wa72, § 4.1.2], where

$$\check{V} := \{\nu \in V' \mid G \to V'_{\mathrm{b}},\ g \mapsto \nu \circ g^{-1} \text{ cont.}\}$$

and V'_{b} is the topological dual of V equipped with the topology of bounded convergence. Then the strong dual of $\check{V}^\infty$ is denoted by $V^{-\infty}$ and called the space of *distribution vectors* of V.

Similarly one defines the space V^ω of *analytic vectors* in V and the space of *hyperfunction vectors* $V^{-\omega}$ of V (see [Wa72, § 4.4.5]). Together one finds the following chain of continuous inclusions (cf. [Ju01, p. 220]):

$$V^\omega \hookrightarrow V^\infty \hookrightarrow V \hookrightarrow V^{-\infty} \hookrightarrow V^{-\omega}.$$

Remark 2.10 (Poisson Transforms) The representation of G on the eigenspace E_λ from (10) can be described via the *Poisson transform*: The Poisson transform $\mathcal{P}_\lambda$ is a G-equivariant map $H_{-\lambda}^{-\omega} \to E_\lambda$. It is given by

$$\mathcal{P}_\lambda f(xK) := \int_K f(k) e^{-(\lambda+\rho)H(x^{-1}k)}\, \mathrm{d}k$$

if f is a sufficiently regular function in the compact picture of the principal series. If f is given in the induced picture, then $\mathcal{P}_\lambda f(xK)$ simply is $\int_K f(xk)dk$.

It is important to know for which values of $\lambda \in \mathfrak{a}_{\mathbb{C}}^*$ the Poisson transform is a bijection. By [K+78] we have that $\mathcal{P}_\lambda \colon H_{-\lambda}^{-\omega} \to E_\lambda$ is a bijection if

$$-\frac{2\langle\lambda, \alpha\rangle}{\langle\alpha, \alpha\rangle} \notin \mathbb{N}_{>0} \quad \text{for all} \quad \alpha \in \Sigma^+. \tag{11}$$

Here $\langle\cdot, \cdot\rangle$ is the inner product on $\mathfrak{a}^*$ induced from the Killing form. So, in particular, $\mathcal{P}_\lambda$ is a bijection if $\mathrm{Re}\lambda$ is in the closure $\overline{\mathfrak{a}_+^*}$ of $\mathfrak{a}_+^* = \{\lambda \in \mathfrak{a}^* \mid \forall \alpha \in \Sigma^+ : \langle\lambda, \alpha\rangle > 0\}$.

2.2.2 Intertwining Operators

For $\lambda \in i\mathfrak{a}^*$ and $w \in W$, we know that π_λ and $\pi_{w\lambda}$ are equivalent. An *intertwining operator* $\mathsf{A}(\lambda, w)$ can be defined as follows. For $w \in W$, we set $\bar{N}_w := \bar{N} \cap m_w^{-1} N m_w$, where as before $m_w \in M'$ is a representative of w, and $\bar{N} = \theta(N)$.

Recall here that θ is the Cartan involution on G. We normalize the Haar measure on $\bar{N}_w$ by $\int_{\bar{N}_w} e^{-2\rho(H(\bar{n}_w))}\,\mathrm{d}\bar{n}_w = 1$. For $\lambda \in \mathfrak{a}_{\mathbb{C}}^*$ and

$$f \in \mathcal{D}_\lambda(G) := \left\{ f \in \mathrm{C}^\infty(G) \;\middle|\; \begin{matrix} \forall g \in G, m \in M, a \in A, n \in N : \\ f(gman) = e^{(\lambda-\rho)\log a} f(g) \end{matrix} \right\}, \tag{12}$$

we define the intertwining integrals:

$$\mathsf{A}(\lambda, w) f(g) := \int_{\bar{N}_w} f(g m_w \bar{n}_w)\,\mathrm{d}\bar{n}_w.$$

This integral is absolutely convergent in the region:

$$C(w) := \{\lambda \in \mathfrak{a}_{\mathbb{C}}^* \mid \forall\, \alpha \in \Sigma(w) :\ \mathrm{Re}\langle \lambda, \alpha\rangle > 0\,\},$$

where $\Sigma(w) := \{\alpha \in \Sigma^+ \mid w \cdot \alpha < 0\}$.

With the topology induced from that of $\mathrm{C}^\infty(G)$, $\mathcal{D}_\lambda(G)$ is Fréchet, and is isomorphic to $\mathrm{C}^\infty(G/MAN)$. Using the decomposition $G = KAN$, we see that this linear isomorphism is just the restriction map $\phi \mapsto \phi|_K$ with the inverse mapping given by $\psi \longmapsto \phi(kan) := e^{(\lambda-\rho)\log a}\psi(k)$, $(k, a, n) \in K \times A \times N$. To $\phi \in C_c^\infty(G)$, we can associate $\phi_\lambda \in \mathrm{C}^\infty(G/MAN)$ which is defined by

$$\phi_\lambda(kM) := \int_{MAN} f(kman) e^{(\lambda+\rho)\log a}\,\mathrm{d}m\,\mathrm{d}a\,\mathrm{d}n.$$

Observe that the mapping $\phi \mapsto \phi_\lambda$ is surjective (see [Bou04, Chapter 7, §2, Proposition 2]). This then leads to the corresponding identification at the level of distributions as follows.

We denote by $\mathcal{D}'(G/MAN)$ the space of distributions on $G/MAN \cong K/M$. Given $T \in \mathcal{D}'(G/MAN)$, we define $T_\lambda \in \mathcal{D}'(G)$ by $T_\lambda(\phi) = T(\phi_\lambda)$, where $\phi_\lambda \in \mathrm{C}^\infty(G/MAN)$ is as above. Then

$$T_\lambda \in \mathcal{D}'_\lambda(G) := \{T \in \mathcal{D}'(G) \mid \forall m \in M, a \in A, n \in N :\ R_{man} T = e^{(\rho-\lambda)\log a} T\},$$

the space of distributions on G that are equivariant under the right regular action R_{man} of MAN in the specified way, and $T \longmapsto T_\lambda$ is an isomorphism between $\mathcal{D}'(G/MAN)$ and $\mathcal{D}'_\lambda(G)$. Thus, we can and we will move freely between the two spaces.

It can be shown (see, e.g., [Kn86, §§VII.3–7] and [STS76, Cor. to Prop. 5]) that we have the following properties.

Proposition 2.11 (Standard Intertwining Operators) *For $\lambda \in C(w)$:*

(i) $\mathsf{A}(\lambda, w) : \mathcal{D}'_\lambda(G) \longrightarrow \mathcal{D}'_{w\cdot\lambda}(G)$ *is a continuous G-homomorphism.*
(ii) $\mathsf{A}(\lambda, w_1 w_2) = \mathsf{A}(w_2 \cdot \lambda, w_1)\mathsf{A}(\lambda, w_2)$ *for a minimal decomposition $w = w_1 w_2$.*

(iii) $\mathsf{A}(\lambda, w) \circ \pi_\lambda = \pi_{w\cdot\lambda} \circ \mathsf{A}(\lambda, w)$.
(iv) *By considering the principal series representations in the compact realization on $\mathcal{D}'(G/P)$, so that the space is independent of the parameter λ, the operators $\mathsf{A}(\lambda, w)$ extend meromorphically in λ to all of $\mathfrak{a}^*_\mathbb{C}$, and the properties (i), (ii), and (iii) still hold.*

For each $w \in W$, we can define an intertwining operator $\mathsf{A}(w)$ on $L^2(i\mathfrak{a}^*; L^2(K/M))$ by

$$(\mathsf{A}(w)f)(\lambda) := \mathsf{A}(\lambda, w)f(\lambda),$$

where we interpret $\mathsf{A}(\lambda, w)$ as an operator on $L^2(K/M)$.

2.2.3 Γ-Cohomology

It is obvious that analysis on a locally symmetric space $\mathcal{X} = \Gamma\backslash G/K$ leads to the question how to find Γ-invariant objects in spaces defined from G/K in a G-equivariant way. If the constructions are functorial but not exact, cohomological obstructions to the existence of Γ-invariant objects may occur. Thus it is not surprising that Γ-cohomology plays a role in the context of locally symmetric spaces.

Recall the definition of *Γ-cohomology $H^\bullet(\Gamma, V)$* with coefficients in a Γ-module V, e.g., from [Ju01, p. 218]. For $p \geq 0$ the space $C^p(\Gamma, V)$ of *homogeneous p-cochains* consists of all functions $c : \Gamma^{p+1} \to V$. The *coboundary operator* $\partial : C^p(\Gamma, V) \to C^{p+1}(\Gamma, V)$ is given by

$$(\partial c)(\gamma_0, \ldots, \gamma_{p+1}) := \sum_{j=0}^{p+1} (-1)^j c(\gamma_0, \ldots, \widehat{\gamma_j}, \ldots, \gamma_{p+1}),$$

where $\widehat{\gamma_j}$ means that this entry is omitted. Then $\partial \circ \partial = 0$ and Γ acts on $C^p(\Gamma, V)$ via

$$(\gamma \cdot c)(\gamma_0, \ldots, \gamma_p) := \gamma(c(\gamma^{-1}\gamma_0, \ldots, \gamma^{-1}\gamma_p)).$$

The spaces of Γ-invariants form a complex:

$$0 \longrightarrow C^0(\Gamma, V)^\Gamma \xrightarrow{\partial} C^1(\Gamma, V)^\Gamma \xrightarrow{\partial} C^2(\Gamma, V)^\Gamma \xrightarrow{\partial} \ldots$$

whose cohomology is $H^\bullet(\Gamma, V)$. In particular, $H^0(\Gamma, V)$ coincides with the space V^Γ of Γ-invariants in V.

Remark 2.12 (Patterson Conjecture) In [Ju01] Γ-cohomologies are mostly taken with coefficients related to principal series representations with the goal to explain Patterson's conjectures on cohomological descriptions of the divisor of dynamical zeta functions and the proofs of these conjectures by Bunke and Olbrich (see in particular [Ju01, § 3.4 and Chap. 8]). While the conjecture relates two dynamical objects belonging to the classical domain, the proof given by Bunke and Olbrich involves Poisson transforms and thus eigenspaces of Laplacians which belong to the quantum world.

Remark 2.13 (Lewis-Zagier Correspondences) In the paper [BLZ15] Bruggeman, Lewis, and Zagier generalized results from [LZ01] to various discrete subgroups Γ of $G := \mathrm{PSL}_2(\mathbb{R})$. They concentrated on proving linear isomorphisms between spaces of Maass cusp forms and Γ-cohomology spaces with coefficients in principal series representations. In the case of hyperbolic surfaces with cusps, they showed that as for the modular surface, *parabolic* Γ-cohomology (see [BLZ15, § 11.3]) is a suitable cohomology theory to use for this purpose.

Instead of recalling the definition of parabolic Γ-cohomology used in [BLZ15], we make a little detour which allows us to also introduce *cuspidal* Γ*-cohomology* which played a decisive role in [DH05], a first attempt to extend the Lewis-Zagier correspondence to higher rank situations.

As we want to consider arithmetic groups, we now assume that G is algebraic. Let Γ be an arithmetic subgroup of G and assume that Γ is torsion-free. Then every Γ-module V induces a local system or locally constant sheaf C_V on $\mathcal{X} = \Gamma\backslash G/K$. In the étale picture, the sheaf C_V equals $C_V = \Gamma\backslash((G/K) \times V)$ (diagonal action). Let $\overline{\mathcal{X}}^{\mathrm{BS}}$ denote the Borel-Serre compactification [BS73] of $\mathcal{X}$, then Γ also is the fundamental group of $\overline{\mathcal{X}}^{\mathrm{BS}}$ and V induces a sheaf also denoted by C_V on $\overline{\mathcal{X}}^{\mathrm{BS}}$. This notation is consistent as the sheaf on $\mathcal{X}$ is indeed the restriction of the one on $\overline{\mathcal{X}}^{\mathrm{BS}}$. Let $\partial^{\mathrm{BS}}\mathcal{X}$ denote the boundary of the Borel-Serre compactification. There are natural identifications:

$$H^j(\Gamma, V) \cong H^j(\mathcal{X}, C_V) \cong H^j(\overline{\mathcal{X}}^{\mathrm{BS}}, C_V).$$

Now *parabolic cohomology* of a Γ-module V can be defined as the kernel of the restriction to the boundary:

$$H^j_{\mathrm{par}}(\Gamma, V) := \ker\Big(H^j(\overline{\mathcal{X}}^{\mathrm{BS}}, C_V) \longrightarrow H^j(\partial^{\mathrm{BS}}\mathcal{X}, C_V)\Big).$$

The long exact sequence of the pair $(\overline{\mathcal{X}}^{\mathrm{BS}}, \partial^{\mathrm{BS}}\mathcal{X})$ gives rise to

$$\begin{aligned} \ldots \longrightarrow H^j_c(\mathcal{X}, C_V) &\longrightarrow H^j(\mathcal{X}, C_V) \\ &= H^j(\overline{\mathcal{X}}^{\mathrm{BS}}, C_V) \longrightarrow H^j(\partial^{\mathrm{BS}}\mathcal{X}, C_V) \longrightarrow \ldots, \end{aligned}$$

where H^j_c denotes cohomology with compact supports. The image of the cohomology with compact supports under the natural map is called the *interior cohomology* of $\Gamma\backslash X$ and is denoted by $H_!^j(X, C_V)$. The exactness of the above sequence shows that

$$H_{\mathrm{par}}^j(\Gamma, V) \cong H_!^j(X, C_V).$$

It is shown in [DH05, § 2] that $H^\bullet(\Gamma, V)$ is computed by a standard complex of $(\mathfrak{g}, K)$-cohomology (see [BW00, § I.5]) if V is a *smooth G-representation*, i.e., if $V = V^\infty$. More precisely, one then has

$$H^\bullet(\Gamma, V) \cong H^\bullet(\mathfrak{g}, K, C^\infty(\Gamma\backslash G)\hat{\otimes}V),$$

where $C^\infty(\Gamma\backslash G)\hat{\otimes}V$ is the completed projective tensor product isomorphic to $C^\infty(\Gamma\backslash G, V)$ (see [Gr55, p. 80/81]).

Recall the space of automorphic cusp forms $L^2_{\mathrm{cusp}}(\Gamma\backslash G)$ (see, e.g., [Ha68, § I.2]), which is a closed G-invariant subspace of $L^2(\Gamma\backslash G)$. Then, for a smooth G-representation V, the *cuspidal Γ-cohomology* is defined by

$$H^\bullet_{\mathrm{cusp}}(\Gamma, V) := H^\bullet(\mathfrak{g}, K, L^2_{\mathrm{cusp}}(\Gamma\backslash G)^\infty\hat{\otimes}V).$$

Note that there is a natural map $H^\bullet_{\mathrm{cusp}}(\Gamma, V) \to H^\bullet(\Gamma, V)$ which is not necessarily injective (see [DH05, Cor. 5.2]). Its image is called the *reduced cuspidal Γ-cohomology* and denoted by $\tilde{H}^\bullet_{\mathrm{cusp}}(\Gamma, V)$. By [DH05, Prop. 2.2], we have:

$$\tilde{H}^\bullet_{\mathrm{cusp}}(\Gamma, V) \subseteq H^\bullet_{\mathrm{par}}(\Gamma, V).$$

The representation $L^2_{\mathrm{cusp}}(\Gamma\backslash G)$ decomposes discretely into irreducibles with finite multiplicity (see, e.g., [Ar79, Part I]). We denote the multiplicity of an irreducible G-representation π in $L^2_{\mathrm{cusp}}(\Gamma\backslash G)$ by $N_\Gamma(\pi)$.

Theorem 2.14 ([DH05, Theorems 0.2 & 4.3]) *Let Γ be a torsion-free arithmetic subgroup of a split semisimple Lie group G and let $\pi \in \hat{G}$ be an irreducible unitary principal series representation. Then*

$$N_\Gamma(\pi) = \dim H^{d-\ell}_{\mathrm{cusp}}(\Gamma, \pi^\omega) = \dim H^\ell_{\mathrm{cusp}}(\Gamma, \pi^{-\omega}),$$

where $d = \dim G/K$ and ℓ is the real rank of G.

For G non-split the assertion of Theorem 2.14 remains true for a generic set of representations π.

3 Dynamical Systems

In this section we discuss some mathematical specifics of dynamical systems entering the classical and quantum evolutions of our class of examples.

3.1 *Classical Dynamics*

The mathematics of modeling classical mechanics revolves around vector fields and their integration. We start by collecting some background material before we describe our key examples, the geodesic and the Weyl chamber flows.

3.1.1 Anosov Flows

Let $\mathcal{M}$ be a compact manifold, $A \simeq \mathbb{R}^\ell$ an abelian group and let $\tau : A \to \mathrm{Diffeo}(\mathcal{M})$ be a smooth locally free group action. If $\mathfrak{a} := \mathrm{Lie}(A) \cong \mathbb{R}^\ell$, we can define a *generating map*:

$$X : \begin{cases} \mathfrak{a} & \to C^\infty(\mathcal{M}; T\mathcal{M}) \\ H & \mapsto X_H := \frac{d}{dt}_{|t=0}\tau(\exp(tH)), \end{cases}$$

so that for each basis $H_1, \ldots, H_\ell$ of $\mathfrak{a}$, $[X_{H_j}, X_{H_k}] = 0$ for all j, k. For $H \in \mathfrak{a}$ we denote by $\varphi_t^{X_H}$ the flow of the vector field X_H. Notice that, as a differential operator, we can view X as a map:

$$X : C^\infty(\mathcal{M}) \to C^\infty(\mathcal{M}; \mathfrak{a}^*), \quad (Xu)(H) := X_H u.$$

It is customary to call the action *Anosov* if there is an $H \in \mathfrak{a}$ such that there is a continuous $d\varphi_t^{X_H}$-invariant splitting:

$$T\mathcal{M} = E_0 \oplus E_{\mathrm{u}} \oplus E_{\mathrm{s}}, \tag{13}$$

where $E_0 := \mathrm{span}(X_{H_1}, \ldots, X_{H_\ell})$, and there exists a $C > 0, \nu > 0$ such that for each $x \in \mathcal{M}$

$$\forall w \in E_{\mathrm{s}}(x), \forall t \geq 0 : \quad \|d\varphi_t^{X_H}(x)w\| \leq Ce^{-\nu t}\|w\|,$$
$$\forall w \in E_{\mathrm{u}}(x), \forall t \leq 0 : \quad \|d\varphi_t^{X_H}(x)w\| \leq Ce^{-\nu|t|}\|w\|.$$

Here the norm on $T\mathcal{M}$ is fixed by choosing any smooth Riemannian metric on $\mathcal{M}$. We say that such an H is *transversely hyperbolic*. The letters s and u in E_{s} and E_{u} stand for *stable* and *unstable*. The splitting is invariant by the entire action. This does not mean that all $H \in \mathfrak{a}^*$ have this transversely hyperbolic behavior. In fact, there is a maximal open convex cone $\mathfrak{a}_+^\tau \subset \mathfrak{a}$ containing H such that for all $H' \in \mathfrak{a}_+^\tau$, $X_{H'}$ is also transversely hyperbolic with the same splitting as H ([GGHW21, Lemma 2.2]); $\mathfrak{a}_+^\tau$ is called a *positive Weyl chamber*. This name is motivated by the classical examples of such Anosov actions that are the Weyl chamber flows for locally symmetric spaces of rank ℓ as described in Sect. 3.1.3. For other classes of examples, see, e.g., [KS94, SV19].

3.1.2 Symplectic Geometry and Hamiltonian Actions

Let $(\mathcal{N}, \omega)$ be a symplectic manifold with tangent bundle $T\mathcal{N}$ and cotangent bundle $T^*\mathcal{N}$. Then for each smooth function $f \in C^\infty(\mathcal{N})$, one defines its *symplectic gradient* or *Hamiltonian vector field* X_f via

$$-\omega(X_f, X) = \langle df, X \rangle_{\mathcal{N}},$$

where $X \in \mathcal{V}(\mathcal{N})$ is a vector field on $\mathcal{N}$ and $\langle \cdot, \cdot \rangle_{\mathcal{N}}$ denotes the canonical pairing between $T^*\mathcal{N}$ and $T\mathcal{N}$ (for the basic facts of symplectic geometry, see, e.g., [AM87]). The Hamiltonian vector fields satisfy:

$$X_{f_1 f_2} = f_2 X_{f_1} + f_1 X_{f_2}. \tag{14}$$

For $f_1, f_2 \in C^\infty(\mathcal{N})$, one has the *Poisson bracket* $\{f_1, f_2\} \in C^\infty(\mathcal{N})$ defined by

$$\{f_1, f_2\} := \omega(X_{f_1}, X_{f_2}).$$

It defines a Lie algebra structure on $C^\infty(\mathcal{N})$, and the equality $\{f_1, f_2\} = df_2(X_{f_1})$ shows that for two Poisson commuting functions, the one function (and hence its level sets) is invariant under the flow of the other's Hamiltonian vector field.

Suppose that a group G acts on $\mathcal{N}$ by diffeomorphisms such that ω is G-invariant. If $f \in C^\infty(\mathcal{N})$ is G-invariant as well, then the Hamiltonian vector field $X_f \in \mathcal{V}(\mathcal{N})$ is again G-invariant. As a consequence we see that the flow $\varphi_t^{X_f}(x)$ of X_f is G-equivariant, i.e.,

$$\varphi_t^{X_f}(g \cdot x) = g \cdot \varphi_t^{X_f}(x) \quad \text{for } (x, t) \in \mathcal{N} \times \mathbb{R},$$

where (x, t) is restricted to the domain of definition for the flow.

Let $\mathcal{A} \subseteq C^\infty(\mathcal{N})$ be a finitely generated associative subalgebra consisting of Poisson commuting functions. Let $f_1, \ldots, f_k$ be a set of generators and define $F\colon \mathcal{N} \to \mathbb{R}^k$ by $F := (f_1, \ldots, f_k)$. Then an arbitrary element of $\mathcal{A}$ can be written as $f = \sum_{\alpha \in \mathbb{N}_0^n} c_\alpha F^\alpha$, where we use the usual multi-index notation. For any $v \in \mathbb{R}^k$, we consider the closed subset $\mathcal{N}_{(v)} := F^{-1}(v)$ of $\mathcal{N}$. Since the f_j Poisson commute, we see that all the $\mathcal{N}_{(v)}$ are stable under all the flows of the X_{f_j}'s.

If now $\mathcal{N}_{(v)}$ is a submanifold of $\mathcal{N}$, then the Hamiltonian vector fields X_{f_j} are all tangent to $\mathcal{N}_{(v)}$, i.e., restrict to vector fields on $\mathcal{N}_{(v)}$. Note that all elements of $\mathcal{A}$ are constant on $\mathcal{N}_{(v)}$. But then the identity (14) implies that the restriction of any X_f with $f \in \mathcal{A}$ is a linear combination of the restrictions of the X_{f_j}.

In general, the $\mathcal{N}_{(v)}$ will not be manifolds. It is, however, always possible to find submanifolds of $\mathcal{N}$ contained in $\mathcal{N}_{(v)}$ such that all the vector fields X_{f_k} restrict to vector fields of these submanifolds. To this end we recall the notion of a $\mathcal{D}$-orbit for a family $\mathcal{D}$ of vector fields from [Su73]. We set:

$$\mathcal{D} := \{X_f \mid f \in \mathcal{A}\} \tag{15}$$

and note that according to (14) the linear spans of $\{X_f(x) \mid f \in \mathcal{A}\}$ and $\{X_{f_1}(x), \ldots, X_{f_k}(x)\}$ in $T_x\mathcal{N}$ agree for all $x \in \mathcal{N}$. We denote this linear span by $\Delta(x)$. Then $x \mapsto \Delta(x)$ is a smooth distribution in the sense of [Su73]. It is involutive since $\mathcal{D}$ is a commutative family of vector fields. Moreover, this shows that Δ satisfies the condition (e) from [Su73, Thm. 4.2]. Therefore the $\mathcal{D}$-orbits $\mathcal{D} \cdot x$ are maximal integral manifolds of Δ. They satisfy:

$$\dim(\mathcal{D} \cdot x) = \dim \Delta(x). \tag{16}$$

From the construction of the $\mathcal{D}$-orbits, it is clear that they are invariant under the flows of the $X_f \in \mathcal{D}$. Since these flows preserve the $\mathcal{N}_{(v)}$, we see also that the $\mathcal{D}$-orbits are contained in single $\mathcal{N}_{(v)}$'s. This, finally, shows that the restriction of X_f with $f \in \mathcal{A}$ to any $\mathcal{D}$-orbit is a linear combination (with coefficients depending on v) of the restrictions of the $X_{f_1}, \ldots, X_{f_k}$.

Suppose that y is a regular value of F, i.e., the derivative $F'(x)\colon T_x\mathcal{N} \to \mathbb{R}^k$ is surjective for all $x \in F^{-1}(y)$. Then $\mathcal{N}_{(F(x))}$ is a closed submanifold of $\mathcal{N}$ with $\dim \mathcal{N}_{(F(x))} = \dim \mathcal{N} - k$. On the other hand, k is equal to

$$\dim \operatorname{span}\{f_1'(x), \ldots, f_k'(x)\} = \dim \operatorname{span}\{X_{f_1}(x), \ldots, X_{f_k}(x)\} = \dim(\mathcal{D} \cdot x)$$

so that

$$\dim(\mathcal{D} \cdot x) + \dim \mathcal{N}_{(F(x))} = \dim \mathcal{N}. \tag{17}$$

The commuting vector fields $X_{f_1}, \ldots, X_{f_k}$ define a (local) action of $\mathbb{R}^k$ on $\mathcal{N}$ which leaves the $\mathcal{D}$-orbits invariant.

Remark 3.1 (Hamiltonian Actions on Cotangent Bundles) Let $\mathcal{N}$ be any manifold. Recall the *canonical symplectic form* $\omega = d\Theta$ on $T^*\mathcal{N}$, where the 1-form Θ is given by the formula

$$\forall v \in T_\xi(T^*\mathcal{N}), \xi \in T^*\mathcal{N}: \quad \langle \Theta(\xi), v\rangle = \langle \xi, \mathrm{pr}'_{T^*}(v)\rangle.$$

with the canonical projection $\mathrm{pr}_{T^*} : T^*\mathcal{N} \to \mathcal{N}$. For a diffeomorphism $h\colon \mathcal{N}_1 \to \mathcal{N}_2$, the induced map $h^*\colon T^*\mathcal{N}_2 \to T^*\mathcal{N}_1$ is a *canonical transformation*, i.e., it preserves the symplectic structures. Thus the induced map

$$C^\infty(T^*\mathcal{N}_1) \to C^\infty(T^*\mathcal{N}_2), \quad f \mapsto h_* f := f \circ h^*$$

is a Poisson isomorphism satisfying $h_* X_f = X_{h_* f}$. Therefore the push-forward of vector fields $h_*\colon \mathcal{V}(T^*\mathcal{N}_1) \to \mathcal{V}(T^*\mathcal{N}_2)$ maps Hamiltonian vector fields to Hamiltonian vector fields.

If now G acts smoothly on $\mathcal{N}$, these considerations show that the Hamiltonian vector fields of G-invariant functions on $T^*\mathcal{N}$ are G-invariant Hamiltonian vector fields on $T^*\mathcal{N}$. Moreover, the action of G on $T^*\mathcal{N}$ is Hamiltonian with moment map $J\colon T^*\mathcal{N} \to \mathfrak{g}^*$ given by

$$\langle J(\eta), X\rangle := \langle \eta, \tilde{X} \circ \mathrm{pr}_{T^*}(\eta)\rangle_{\mathcal{N}},$$

where $X \in \mathfrak{g}$ and $\tilde{X} \in \mathcal{V}(\mathcal{N})$ is the vector field on $\mathcal{N}$ induced by the derived action of G on $\mathcal{N}$ via

$$\tilde{X}(x) := \frac{d}{dt}|_{t=0}(\exp tX) \cdot x.$$

If $T^*\mathcal{N}$ and $T\mathcal{N}$ are identified via a Riemannian metric g, the formula for Θ reads:

$$\langle \Theta(\xi), v\rangle = \langle \xi, \mathrm{pr}'_{T^*}(v)\rangle = g\big(\xi, \mathrm{pr}_{T'}(v)\big)$$

for $\xi \in T^*\mathcal{N} \cong T\mathcal{N}$, $v \in T_\xi(T^*\mathcal{N}) \cong T_\xi(T\mathcal{N})$ and the canonical projection $\mathrm{pr}_T : T\mathcal{N} \to \mathcal{N}$.

In the case of $\mathcal{N} = G/K$, using $T^*(G/K) \cong G \times_K \mathfrak{p}^*$ and identifying $\mathfrak{p}$ and $\mathfrak{p}^*$ via the Killing form gives:

$$T^*(G/K) \cong G \times_K \mathfrak{p} \cong T(G/K).$$

Under these identifications the moment map is given by $J([g, X]) = \mathrm{Ad}(g)X$.

3.1.3 Geodesic and Weyl Chamber Flow

Recall the notation for locally symmetric spaces and the right A-action on $T\mathcal{M}$ from Sect. 2. All elements of the positive Weyl chamber $\mathfrak{a}_+$ are transversely hyperbolic elements sharing the same stable and unstable distributions given by the associated homogeneous vector bundles:

$$E_0 = \Gamma\backslash G \times_M \mathfrak{a}, \quad E_{\mathrm{s}} = \Gamma\backslash G \times_M \mathfrak{n}, \quad E_{\mathrm{u}} = \Gamma\backslash G \times_M \overline{\mathfrak{n}}.$$

Thus, if $\mathcal{X} = \Gamma\backslash G/K$ is compact, the A-action is Anosov. Below, we will recall the symplectic interpretation of the A-action from [Hi05]. We will not discuss the origin of the name Weyl chamber flow for the A-action in detail. Let it suffice to say that the objects involved can be reinterpreted in such a way that A acts on the set of Weyl chambers.

We give a symplectic interpretation of the right A-action on $\mathcal{M} = \Gamma\backslash G/M$. The natural symplectic form on $T(T(G/K))$ is given by

$$\omega_{(\mathbf{1},Y)}\big((A_1, B_1 - \tfrac{1}{2}[A_1, Y]), (A_2, B_2 - \tfrac{1}{2}[A_2, Y])\big) = B(A_1, B_2) - B(A_2, B_1), \tag{18}$$

where B denotes the Killing form.

Let $f \in C^\infty(T(G/K))$ be G-invariant and $h \in C^\infty(\mathfrak{p})$ the restriction of f to $T_o(G/K) = \mathfrak{p}$ (which determines f uniquely). Then the Hamiltonian vector field X_f is determined by

$$X_f(X) = \left(\operatorname{grad} h(X), -\tfrac{1}{2}[\operatorname{grad} h(X), X]\right), \tag{19}$$

where the gradient is taken with respect to the Killing form on $\mathfrak{p}$. In particular, X_f is horizontal, so by Lemma 2.6 the flow of X_f preserves G-orbits. Writing such an orbit as G/H, according to (6), we obtain a flow $\varphi^{X_f} : \mathbb{R} \times G/H \to G/H$ of X_f on $G/H \subseteq T(G/K)$ of the form:

$$\varphi_t^{X_f}([g, X]) = [g \exp(t \operatorname{grad} h(X)), X] \tag{20}$$

and preserves G-orbits. Therefore the flow can be studied on these orbits separately. Note further that (18) and (19) show that any two G-invariant functions $f_1, f_2 \in C^\infty(T(G/K))$ commute under the Poisson bracket, i.e.,

$$\forall f_1, f_2 \in C^\infty(T(G/K))^G : \quad \{f_1, f_2\} = 0. \tag{21}$$

Theorem 3.2 ([Hi05, Thm. 8.1]) *Consider the algebra $\mathcal{A}_{\mathrm{pf}}$ of G-invariant functions in $C^\infty(T^*(G/K))$ which restrict to polynomials on the fiber $\mathfrak{p}$. Then $\mathcal{A}_{\mathrm{pf}}$ is finitely generated, commutative under the Poisson bracket, and the joint level sets of these functions are precisely the G-orbits in $T^*(G/K)$.*

Let $\mathcal{A}$ be the algebra of G-invariant functions in $C^\infty(T^*(G/K))$. In view of Dadok's smooth Chevalley theorem, [Da82], it is easy to show that restriction yields an isomorphism between $\mathcal{A}$ and $C^\infty(\mathfrak{a}^*)^W$.

By Remark 2.8 the associative subalgebra of $C^\infty(T(G/K)) \cong C^\infty(T^*(G/K))$ generated by the principal symbols of elements of $\mathbb{D}(G/K)$ coincides with the complexification of the algebra $\mathcal{A}_{\mathrm{pf}}$ introduced in Theorem 3.2. Note that for a K-invariant smooth function $h\colon \mathfrak{p} \to \mathbb{R}$, we have $\operatorname{grad} h(X) \in \mathfrak{a}$ for all $X \in \mathfrak{a}$, so that

$\operatorname{grad} h(X) = \operatorname{grad} h|_{\mathfrak{a}}(X)$ (see, e.g., [Hi05, Lemma 8.5]). Together with (20), this shows that we have a map:

$$\mathcal{F}\colon \mathcal{A} \times \mathfrak{a} \to \mathfrak{a}, \quad (f, X) \mapsto \operatorname{grad} f|_{\mathfrak{a}}(X)$$

such that the flow $\varphi^{X_f}: \mathbb{R} \times T(G/K) \to T(G/K)$ of X_f on $T(G/K)$ is given by

$$\varphi_t^{X_f}([g, X]) = [g \exp(t\,\mathcal{F}(f, X)), X]. \tag{22}$$

In particular, $\mathcal{A}$ is Poisson-commutative as well. Extending the right A-action on G/M trivially to $G/M \times \mathfrak{a}$ via

$$(gM, X) \cdot a := \rho(a)(gM, X) := (gaM, X)$$

we obtain the following equivariance property of $\Phi\colon G/M \times \mathfrak{a} \to T(G/K)$ under φ^{X_f}:

$$\Phi \circ \rho\left(e^{t \operatorname{grad} f(X)}\right) = \varphi_t^{X_f} \circ \Phi. \tag{23}$$

Recall, e.g., from [Hi05, Ex. 8.1], that the geodesic flow φ_t on the tangent bundle $T(G/K) \cong G \times_K \mathfrak{p}$ can be written:

$$\varphi_t([g, X]) = [g \exp tX, X].$$

This implies in particular that each geodesic, viewed as a curve in $T(G/K)$, i.e., as an orbit of the geodesic flow, is completely contained in a G-orbit.

If G/K has rank 1, i.e., if $\dim \mathfrak{a} = 1$, the K-invariant polynomials on $\mathfrak{p}$ are generated by $X \mapsto \|X\|^2$, so that the G-orbits in $T(G/K)$ are simply the sphere bundles (and the zero section). As this function essentially represents the kinetic energy, the G-orbits can be viewed as energy shells being preserved by the flow. The relevant $T^*(G/K)_{(v)}$ for nonzero v is (up to scaling) the sphere bundle $S(G/K)$ of G/K in $T(G/K)$. Left translation gives an identification of $S(G/K)$ with G/M. The geodesic flow restricted to the sphere bundle is then given by the right multiplication of A. For $v = 0$ the bundle is the zero section, and the geodesic flow restricted to this section is trivial.

In higher rank we have not only the kinetic energy function which is invariant by the A-action. All the elements of $\mathcal{A}_{\mathrm{pf}}$ are invariant under the A-flow, and the joint level sets may rightfully be called *generalized energy shells*. By Theorem 3.2 they agree with the G-orbits. Studying the multiparameter flow on the G-orbits separately allows to give more explicit descriptions. To this end we fix $X \in \mathfrak{a}$ and recall from [Hi05] the description of the sets $\{\operatorname{grad} h(X) \mid h \in \mathbb{R}[\mathfrak{a}]^W\}$, where $\mathbb{R}[\mathfrak{a}]^W$ denotes the W-invariant real polynomials on $\mathfrak{a}$.

Note that the centralizer $\mathfrak{k}_X$ of X in $\mathfrak{k}$ is given by

$$\mathfrak{z}_{\mathfrak{k}}(X) := \{Y \in \mathfrak{k} \mid [X, Y] = 0\}.$$

Analogously, given $X \in \mathfrak{a}$, consider the centralizer $\mathfrak{g}_X := \mathfrak{z}_{\mathfrak{g}}(X) := \{Y \in \mathfrak{g} \mid [Y, X] = 0\}$. If $\Sigma_X = \{\alpha \in \Sigma \mid \alpha(X) = 0\}$, then $\mathfrak{g}_X = \mathfrak{m}+\mathfrak{a}+\sum_{\alpha\in\Sigma_X} \mathfrak{g}_\alpha$. Consider the center $\mathfrak{z}(\mathfrak{g}_X)$ of the reductive θ-invariant algebra $\mathfrak{g}_X$ and $\mathfrak{a}(X) := [\mathfrak{g}_X, \mathfrak{g}_X] \cap \mathfrak{a}$. Then $\mathfrak{a}(X)$ is a Cartan subspace for $[\mathfrak{g}_X, \mathfrak{g}_X]$ and $\mathfrak{a} = \mathfrak{a}(X) \oplus \mathfrak{a}_X$, where

$$\mathfrak{a}_X := \mathfrak{p} \cap \mathfrak{z}(\mathfrak{g}_X). \tag{24}$$

Proposition 3.3 ([Hi05, Lemma 8.9]) *Let $f_1, \ldots, f_\ell$ be a set of algebraically independent generators of $\mathcal{A}_{\mathrm{pf}}$ and $X \in \mathfrak{a}$. Then*

$$\mathfrak{a}_X = \sum_{j=1}^{\ell} \mathbb{R} \operatorname{grad} f_j(X).$$

Theorem 3.4 ([Hi05, Thm. 8.2]) *The smooth distribution Δ associated with the family of vector fields $\mathcal{D} = \{X_f \mid f \in \mathcal{A}_{\mathrm{pf}}\}$ introduced in (15) is given by*

$$\Delta([\mathbf{1}, X]) = \mathfrak{a}_X \times \{0\} \in \mathfrak{p} \times \mathfrak{p} \cong T_{[\mathbf{1},X]}(G \times_K \mathfrak{p}) \cong T_{[\mathbf{1},X]}(T(G/K))$$

for $X \in \mathfrak{a}$.

Remark 3.5 (Singular Flows) Recall that each G-orbit in $T^*(G/K) = G \times_K \mathfrak{p}^*$ is of the form $G \cdot [\mathbf{1}, \xi]$ with $\xi \in \mathfrak{a}^*$. The stabilizer of $[\mathbf{1}, \xi]$ in G coincides with the stabilizer K_ξ of ξ with respect to the coadjoint action. Fix some $\xi \in \mathfrak{a}^*$ and identify $\mathfrak{g}$ with $\mathfrak{g}^*$. Under this identification $\mathfrak{a}$ gets identified with $\mathfrak{a}^*$, so we can set $A_\xi := \exp \mathfrak{a}_\xi$. Then A_ξ commutes with K_ξ so that we have a right A_ξ-action on $G \cdot [\mathbf{1}, \xi] \cong G/K_\xi$ via

$$(gK_\xi, a) \mapsto gaK_\xi =: (gK_\xi) \cdot a.$$

If ξ is regular, then $\mathfrak{a}_\xi = \mathfrak{a}$ and $K_\xi = M$. Thus we recover the Weyl chamber flow. If $\xi = 0$, then $K_\xi = K$ and $A_\xi = \{\mathbf{1}\}$, so that we find the trivial action on G/K.

3.2 Quantum Dynamics

While classical dynamics is based on flows of Hamiltonian vector fields, quantum dynamics is based on unitary linear one-parameter groups of operators generated by skew-adjoint densely defined operators via Stone's theorem; see, e.g., [La02, § 35].

Multiplication by i allows to shift attention from skew-adjoint to self-adjoint operators. As quantization procedures often start with differential operators, an important question is to decide which of the differential operators one considers can actually be completed to self-adjoint operators. Such operators are called *essentially self-adjoint*. There are theorems relating essential self-adjointness of a quantized Hamiltonian to the geodesic completeness of the underlying classical Hamiltonian vector field. This is why some people call essential self-adjointness *quantum completeness*. A very nice discussion of this subject can be found in Tao's blog [Ta11].

In the situation we are interested in, namely, viewing invariant differential operators as generating the quantum evolution, we may refer to [vdB87], where van den Ban shows that formally self-adjoint invariant differential operators are essentially self-adjoint, i.e., quantum complete. This should be viewed as a counterpart of the fact that the Hamiltonian vector fields associated with functions from the algebra $\mathcal{A}_{\mathrm{pf}}$ introduced in Sect. 3.1.3 are complete.

4 Quantum-Classical Correspondences

In this section we will reach the quantum-classical correspondences for locally symmetric spaces which are in the focus of this paper. Before we can describe these correspondences, we have to introduce the dynamical invariants on the quantum and the classical side which will be connected.

Recall that both kinds of dynamical systems we consider are built from linear operators. In the quantum case, this is built into the mathematical formulation of quantum mechanics. In the classical case, we concentrate on Hamiltonian mechanics which is based on vector fields, i.e., first-order differential operators. Thus considering spectral properties of these operators is a reasonable starting point.

For the Hamiltonian operators generating the quantum dynamics on Hilbert spaces, there is a canonical spectral theory to consider. But already the example of the Dyatlov-Faure-Guillarmou correspondence for compact hyperbolic surfaces (see Example 1.6) shows that setting up the right kind of spectral invariants for the first-order differential operators is subtle. The situation becomes even more complicated if we consider noncompact surfaces. In [GHW18] it was shown that it is possible to extend the theory of Ruelle resonances to noncompact surfaces without cusps, but in order to obtain a correspondence, one has to replace the no longer discrete ordinary L^2-spectrum of the hyperbolic Laplacian by the discrete *resonance spectrum* of this operator.

The dynamical invariants we will discuss here are all of a spectral nature. Resonances will play a distinguished role, but we will also consider decompositions of representations as in Example 1.5. Finally, we will also briefly look at zeta functions.

4.1 Quantum Invariants

The kind of dynamical invariants that comes to mind first in the context of quantum mechanics is the spectrum of the (often unbounded) linear operators involved. Things get more complicated when one also looks at commutants of Hamiltonians and noncommutative harmonic analysis enters because of their symmetries.

4.1.1 Joint Spectrum of the Algebra of Invariant Differential Operators in the Cocompact Case

In the rank one case, the quantization of the geodesic flow is given by the Laplacian on G/K. In the higher rank case, we have to consider the algebra of G-invariant differential operators on G/K which we denote by $\mathbb{D}(G/K)$. As an abstract algebra, this is a polynomial algebra with $\ell = \operatorname{rank}(G/K)$ algebraically independent operators, among them the Laplace operator of Remark 2.8. These operators descend to $\mathcal{X} = \Gamma\backslash G/K$, and in analogy to (10), we can define the joint eigenspace:

$$^{\Gamma}E_{\lambda} := \{f \in C^{\infty}(\Gamma\backslash G/K) \mid \forall D \in \mathbb{D}(G/K) : \ Df = \chi_{\lambda}(D)f\}, \tag{25}$$

where χ_λ is a character of $\mathbb{D}(G/K)$ parametrized by $\lambda \in \mathfrak{a}^*_{\mathbb{C}}/W$ with the Weyl group W. Here χ_ρ is the trivial character. Let σ_{Q} denote the corresponding *quantum spectrum* $\{\lambda \in \mathfrak{a}^*_{\mathbb{C}} \mid {}^{\Gamma}E_\lambda \neq \{0\}\}$.

4.1.2 Laplace Resonances for Noncompact Locally Symmetric Spaces

Resonance spectra for the Laplace-Beltrami operators $\Delta_{\mathcal{X}}$ play a role as soon as our locally symmetric space $\mathcal{X}$ is no longer compact. This applies in particular to the global symmetric space G/K, i.e., the case of trivial Γ.

One way to introduce *resonances* of the Laplace-Beltrami operator $\Delta_{\mathcal{X}}$ for a noncompact Riemannian manifold $\mathcal{X}$ is to consider the Schwartz kernel $R(z)$ of the resolvent $(\Delta_{\mathcal{X}} - z^2 - \|\rho\|^2)^{-1}$. Under good circumstances it is possible to extend $z \mapsto R(z)$ to a meromorphic family of Schwartz kernels whose poles are then called the resonances of $\Delta_{\mathcal{X}}$.

In the case of (global) Riemannian symmetric spaces, noncommutative harmonic analysis allows for an explicit description of the resolvent as an integral operator. For real ranks 1 and 2, the task of doing the meromorphic continuation turns out to be managable.

Theorem 4.1 ([HP09, Thm. 3.8]) *Let $\mathcal{X} = G/K$ be a Riemannian symmetric space of rank* 1*:*

(i) *If $\Sigma^+ = \{\alpha\}$ with m_α even, then $R(z)$ extends to an entire function (no poles).*

(ii) *In the other cases, let α be the simple root in Σ^+. Then $R(z)$ extends to a meromorphic function with simple poles at the points $z_k = \lambda_k\|\alpha\|$ for $k \in \mathbb{N}_0$ with $\lambda_k = -i(\rho + j_X k)$, where $\rho = \frac{1}{2}m_\alpha + m_{2\alpha}$ is the weighted half-sum of positive roots and $j_X \in \{1, 2\}$ depends on X.*

(iii) *The residue operator $R_k := \mathrm{res}_{z=z_k} R(z)$ is of finite rank and can be expressed explicitly as a convolution with the spherical function associated with λ_k.*

The paper [HP09] was written following a suggestion of M. Zworski, who had done the calculation for the upper half-plane. It later turned out that Miatello and Will had already done (in [MW00]) the case of Damek-Ricci spaces, which include the rank one Riemannian symmetric spaces. The methods of [HP09], however, are applicable in higher rank as well—at least in principle. The technical effort to do the meromorphic continuation via contour shifts in higher dimension turns out to be formidable. In fact, it took a series of papers just to deal with rank 2 (see [HPP16, HPP17a, HPP17b]).

Among the noncompact locally symmetric spaces, the class of spaces for which one has good control over the resonances obtained as resolvent poles is even more restricted. The following result was established in the context of asymptotically hyperbolic manifolds (see [MM87]). For hyperbolic surfaces this means one needs to exclude cusps, i.e., restrict attention to the so-called convex cocompact surfaces.

Theorem 4.2 (cf. [GHW18, Thm. 4.2]) *Let $X = \Gamma\backslash\mathbb{H}^2$ be a convex cocompact hyperbolic surface:*

(i) *The nonnegative Laplacian Δ_X on X has a resolvent $(\Delta_X - s(1-s))^{-1}$ that admits a meromorphic extension from $\{s \in \mathbb{C} \mid \mathrm{Re}(s) > 1/2\}$ to $\mathbb{C}$ as a family of bounded operators $R(z) : C_c^\infty(X) \to C^\infty(X)$.*

(ii) *For a pole $z_0 \neq \frac{1}{2}$, the polar part of the resolvent is of the form:*

$$\sum_{j=1}^{J(z_0)} \frac{(\Delta_X - z_0(1-z_0))^{j-1}(1-2z_0)}{(z(1-z) - z_0(1-z_0))^j} \mathrm{res}_{z=z_0} R(z)$$

for some $J(z_0) \in \mathbb{N}$.

(iii) *If $z_0 = \frac{1}{2}$ is a pole, the polar part of the resolvent at z_0 is of the form:*

$$\frac{1}{z(1-z) - z_0(1-z_0)} \mathrm{res}_{z=z_0} R(z).$$

Note that in this result, the poles are not determined explicitly, but one can say more about the location of the resonances and the order of the poles (see [GHW18, Prop. 4.3]).

Theorem 4.2 is used in the quantum-classical correspondence for convex cocompact hyperbolic surfaces established in [GHW18]. Motivated by that correspondence, Hadfield in the process of proving a quantum-classical correspondence for convex cocompact real hyperbolic spaces obtained similar but more complicated results for Bochner Laplacians acting on symmetric tensors (see [Ha17, Ha20]).

Coming from a different motivation, also Bunke and Olbrich have a result on the meromorphic continuation of resolvents. They consider Casimir operators of G acting on homogeneous vector bundles over a rank one locally symmetric space $\mathcal{X} = \Gamma\backslash G/K$ with convex cocompact Γ; see [BO12, § 6]. Their result is that the resolvent of the Casimir C meromorphically extends to a branched cover of $\mathbb{C}$ with branching points given by the values of C on certain principal series representations.

4.1.3 Laplace Resonances vs. Scattering Poles

In Sect. 4.1.2 we introduced resonances of Laplacians as poles of meromorphic continuations of the resolvent. Physicists typically define resonances of Laplacians as poles of the scattering matrix of an associated scattering problem. They tend to take these as synonymous concepts which, strictly speaking, is not the case. However, under certain circumstances one can show in a mathematically rigorous way that resolvent poles and scattering poles are closely related.

The reason such connections are of interest in our context is the desire to generalize quantum resonances to higher rank. The case of compact locally symmetric spaces (see Sect. 4.3.1) shows that instead of considering just the Laplace-Beltrami operator, one should rather consider the commutative algebra of all invariant differential operators. While it is unclear what the resolvent of this commutative (associative) algebra should be, there are candidates for scattering operators derived from a suitable multitemporal wave equation. For the symmetric spaces, these are the Knapp-Stein intertwining operators (see Sect. 2.2.2), which are well-known to depend meromorphically on as many parameters as the invariant differential operators (via the Harish-Chandra isomorphism; see [Wa92, Chap. 10]).

Scattering operators for surfaces were studied in the context of automorphic form by Lax and Phillips in [LP76] as an application of their general scattering theory. Guillopé and Zworski proved analytic extension of families of scattering operators for very general hyperbolic surfaces in [GZ95, GZ97]. Patterson and Perry treated the case of higher-dimensional real hyperbolic spaces in [PP99]. In [JS00] Joshi and Sá Barreto provided a generalization to all conformally compact spaces. On the other hand, for classical rank one convex cocompact locally symmetric spaces, Bunke and Olbrich constructed scattering operators which also have meromorphic continuation to all of $\mathbb{C}$; see [BO00, Thm. 5.10].

Suppose that $\mathcal{X}$ is a rank one Riemannian symmetric space. Then $\mathfrak{a}_{\mathbb{C}}^*$ can be identified with $\mathbb{C}$, and the spherical principal series representations π_λ and $\pi_{-\lambda}$ for $\lambda \in \mathfrak{a}_{\mathbb{C}}^*$ are intertwined by the Knapp-Stein intertwiner $\mathsf{A}(\lambda, w)$. Since we are in rank one, the only nontrivial Weyl group element is $w = -1$. Then the *scattering operator* can be defined by

$$S(z) := \mathbf{c}(\lambda)^{-1}\mathsf{A}(\lambda, -1), \quad \lambda = iz, \tag{26}$$

where $\mathbf{c}$ is the Harish-Chandra c-function. The scattering operator depends meromorphically on z. We have the following result on *scattering poles*, that is, the possible poles of the $S(z)f$, $f \in \mathbb{C}^\infty(K/M)$.

Theorem 4.3 ([HHP19, Thm. 7.1]) *Let $\mathcal{X}$ be a rank one Riemannian symmetric space. The scattering poles are at most simple and located on the imaginary axis minus the origin. The scattering poles in* $\mathrm{Im} z > 0$ *are precisely the resonances. At a resonance z_0, $\beta_{\rho+iz_0}$ maps the range of* $\mathrm{res}_{z=z_0} R(z)$ *isomorphically onto the range of* $\mathrm{res}_{z=z_0} S(z)$*; in particular, the residues of the resolvent and of the scattering matrix have the same finite rank. The non-resonance scattering poles are the poles of the standard intertwiner* $\mathsf{A}(iz, -1)$*; they are contained in* $(2i)^{-1}\mathbb{N}$. *At* $\mathrm{Im} z < 0$*, the residue* $\mathrm{res}_{z=z_0} S(z)$ *is a nonzero multiple of the residue* $\mathrm{res}_{z=z_0} \mathsf{A}(iz, -1)$.

As for resolvent poles, one also has results on the relation between resolvent poles and scattering poles for asymptotically hyperbolic manifolds (see [BP02, Gu05]).

4.1.4 Microlocal Lifts, Wigner and Patterson-Sullivan Distributions

For compact hyperbolic surfaces Anantharaman and Zelditch in [AZ07] considered dynamical zeta functions as meromorphic families of distributions by integrating over the primitive closed geodesics. The resulting residues are what they call Patterson-Sullivan distributions. Their study was motivated by quantum ergodicity as the Patterson-Sullivan distributions asymptotically agree with the Wigner distributions, which can be seen by giving an alternative construction in terms of boundary values and Poisson transforms. This alternative construction was generalized in [Sch10] to compact locally symmetric spaces of rank one and extended to higher rank in [HHS12].

For $\mathcal{X} = \Gamma\backslash G/K$, we consider sequences, $(\varphi_h)_h \subset L^2(\mathcal{X})$, of normalized joint eigenfunctions which belong to the principal spectrum of the algebra of invariant differential operators. Using a h-pseudo-differential calculus Op_h on $\mathcal{X}$, in [HHS12] we define and study *lifted quantum limits* or *microlocal lifts* as weak$*$-limit points of *Wigner distributions*:

$$W_h : a \mapsto \big(\mathrm{Op}_h(a)\varphi_h \mid \varphi_h\big)_{L^2(\mathcal{X})}.$$

Here, h^{-1} is the norm of a spectral parameter associated with φ_h, and $h \downarrow 0$ through a strictly decreasing null sequence. Lifted quantum limits are positive Radon measures supported in the cosphere bundle. The problem of quantum ergodicity asks for a description of the lifted quantum limits. Using the boundary values of the φ_h on $G/MAN = K/M$, we construct *Patterson-Sullivan distributions* on $\mathcal{M} = \Gamma\backslash G/M$. In the context of quantum ergodicity, Patterson-Sullivan distributions are relevant because they are asymptotically equivalent to lifted quantum limits and satisfy invariance properties. In fact, for compact hyperbolic surfaces, $\mathcal{X} = \Gamma\backslash\mathbb{H}$, the asymptotic equivalence of lifted Wigner distributions and Patterson–Sullivan distributions was observed by Anantharaman and Zelditch [AZ07]. While it was known from earlier work that lifted quantum limits on compact hyperbolic surfaces are invariant under geodesic flows, it turned out that Patterson-Sullivan distributions are themselves invariant under the geodesic flow. Moreover, in [AZ07] it is shown that they have an interpretation in terms of dynamical zeta functions which can be defined completely in terms of the geodesic flow.

We assume that Γ is cocompact and torsion-free. Under the diagonal action, there is a unique open G-orbit $(G/MAN)^{(2)} \cong G/MA$ in $G/MAN \times G/MAN$. For rank one spaces, $(G/MAN)^{(2)}$ is the set of pairs of distinct boundary points. In this case each geodesic of $\mathcal{X}$ has a unique forward limit point and a unique backward limit point in G/MAN. In particular, one can identify $(G/MAN)^{(2)}$ with the space of geodesics. In higher rank the geometric interpretation is more complicated. It involves the Weyl chamber flow rather than the geodesic flow.

Joint eigenfunctions come with a spectral parameter $\lambda \in \mathfrak{a}_{\mathbb{C}}^*$. The principal part of the spectrum comes from the purely imaginary spectral parameters. We assume that the spectral parameter of φ_h is $i\nu_h/h \in i\mathfrak{a}^*$, $|\nu_h| = 1$. The Patterson-Sullivan distribution $PS_h^\Gamma \in \mathcal{D}'(\Gamma\backslash G/M)$ associated with φ_h is constructed as follows. The Poisson transform (see Remark 2.10) allows us to write $\varphi_h(x) = \mathcal{P}_{i\frac{\nu_h}{h}}(T_h)$, where $T_h \in \mathcal{D}'(G/MAN)$ is the boundary value of φ_h. Consider the *weighted Radon transform* $\mathcal{R}_h : C_c^\infty(G/M) \to C_c^\infty(G/MA)$ defined by

$$(\mathcal{R}_h f)(gMA) = \int_A d_h(gaM, \nu_h) f(gaM)\, da$$

with a weight function d_h given explicitly in terms of the Iwasawa decomposition, [HHS12, Def. 4.1]. Denote by $\mathcal{R}_h' : \mathcal{D}'(G/MAN \times G/MAN) \to \mathcal{D}'(G/M)$ the dual of $\mathcal{R}_h$. The Patterson-Sullivan distribution $PS_h^\Gamma \in \mathcal{D}'(\Gamma\backslash G/M)$ is defined as the Γ-average of $\mathcal{R}_h'(T_h \otimes \tilde{T}_h)$, where $\tilde{T}_h$ is given by the complex conjugate $\overline{\varphi_h} = \mathcal{P}_{i\frac{w_0\cdot\nu_h}{h}}(\tilde{T}_h)$ of φ_h. Here w_0 is again the longest element of W.

Let $W_0 := \lim_h W_h \in \mathcal{D}'(T^*\mathcal{X})$ be a lifted quantum limit which, after passing to a subsequence if necessary, has a regular limit direction $\nu_0 = \lim_h \nu_h$. In addition, assume:

$$\nu_h = \nu_0 + O(h) \quad \text{as } h \downarrow 0.$$

To link W_0 to the sequence $(PS_h^\Gamma)_h$ of Patterson-Sullivan distributions, we make use of the natural G-equivariant map $\Phi\colon G/M \times \mathfrak{a}^* \to T^*X$ from Proposition 2.3. For regular $\nu_0 \in \mathfrak{a}^*$, this induces a push-forward of distributions:

$$\Phi(\cdot, \nu_0)_* : \mathcal{D}'(\Gamma\backslash G/M) \to \mathcal{D}'(T^*\mathcal{X}).$$

Then a simplified version of the main result on Patterson-Sullivan distributions in [HHS12] can be stated as follows:

$$W_0 = \kappa(w_0 \cdot \nu_0) \lim_{h \downarrow 0} (2\pi h)^{\dim N/2} \Phi(\cdot, \nu_0)_* PS_h^\Gamma \quad \text{in } \mathcal{D}'(T^*\mathcal{X}). \tag{27}$$

Here κ is a normalizing function defined in terms of structural data of G/K. We point out that [HHS12, Thm 7.4] is more general as it also describes the situation arising from off-diagonal Wigner distributions $\big(\mathrm{Op}_{\Gamma,h}(a)\varphi_h \mid \varphi_h'\big)_{L^2(\mathcal{X})}$.

In the course of proving (27) the following theorem, which is of relevance in the context of *quantum unique ergodicity* (see [SV07] and in particular [Si15, § 5]) was established.

Theorem 4.4 ([HHS12, Thm. 6.7], [SV07, Theorem 1.6(3)], [AS13, Theorem 1.3]) *Assume that $(\varphi_h)_h$ has the lifted quantum limit W_0. Then* $\mathrm{supp}(W_0) \subset S^*\mathcal{X}$, *and W_0 is invariant under the geodesic flow. Moreover,* $\mathrm{supp}\, W_0$ *is contained in a joint level set of $\mathcal{A}$, i.e., in a G-orbit in $S^*\mathcal{X}$. Moreover, for every $f \in \mathcal{A}$, W_0 is invariant under the Hamiltonian flow generated by f. If the direction $\nu_0 \in \mathfrak{a}^*$ of W_0 is regular, then W_0 is A-invariant.*

4.2 Classical Invariants

In contrast to the situation in quantum mechanics, it is clear from the very beginning that classical dynamics offers a variety of different invariants studied in the theory of dynamical system. Examples are the lengths of closed geodesics, invariant measures, dynamical zeta functions, and so on. But also spectral invariants of vector fields, viewed as differential operators, come in. As for the quantum case, extra room for investigation is added when one looks at commutants, which here enter as flow invariant functions, and symmetry properties of Hamiltonians.

4.2.1 Ruelle-Taylor Resonances for Compact Riemannian Manifolds

Let $\tau : A \to \mathrm{Diffeo}(\mathcal{M})$ be a smooth abelian Anosov action on a compact Riemannian manifold $\mathcal{M}$ with positive Weyl chamber $\mathfrak{a}_+^\tau$ as in Sect. 3.1.1. Associated with a basis $H_1, \ldots, H_\ell$ for $\mathfrak{a}$, we have commuting vector fields $X_{H_1}, \ldots, X_{H_\ell}$ which we view as first-order differential operators. It is natural to consider a joint spectrum for this family. We may choose the H_j's to be transversely hyperbolic with the same splitting. Guided by the case of a single Anosov flow (done in [BL07, FS11, DZ16]), we define $E_\mathrm{u}^* \subset T^*\mathcal{M}$ to be the subbundle such that[1] $E_\mathrm{u}^*(E_\mathrm{u} \oplus E_0) = 0$. We shall

[1] This may look strange as E_u^* is not the dual space of E_u, but it should be read rather as $(E^*)_\mathrm{u}$, the unstable part of E^*.

say that $\lambda = (\lambda_1 \ldots, \lambda_\ell) \in \mathbb{C}^\kappa$ is a *joint Ruelle resonance* for the Anosov action if there is a nonzero distribution $u \in \mathcal{D}'(\mathcal{M})$ with wavefront set $\mathrm{WF}(u) \subset E_\mathrm{u}^*$ such that

$$\forall j = 1, \ldots, \ell : \quad (X_{H_j} + \lambda_j)u = 0. \tag{28}$$

The distribution u is called a *joint Ruelle resonant state*. We will denote the space of distributions u with $\mathrm{WF}(u) \subset E_\mathrm{u}^*$ by $\mathcal{D}'_{E_\mathrm{u}^*}(\mathcal{M})$.

A basis-free way to define joint Ruelle resonances is to call an element $\lambda \in \mathfrak{a}_\mathbb{C}^*$ joint Ruelle resonance of τ if there is a nonzero $u \in \mathcal{D}'_{E_\mathrm{u}^*}(\mathcal{M})$ with

$$\forall H \in \mathfrak{a} : \quad (X_H + \lambda(H))u = 0.$$

We denote the Ruelle resonance spectrum of τ by $\sigma_\mathrm{R}(\tau)$. It is a priori not clear that the set of joint Ruelle resonances is discrete—or non-empty for that matter—nor that the dimension of joint resonant states is finite. In the case of compact Riemannian manifolds $\mathcal{M}$, this, however, turns out to be the case. In that case one has normalized volume (total volume 1) which allows to identify $C^\infty(\mathcal{M})$ with the space of smooth densities on $\mathcal{M}$ and hence $\mathcal{D}'(\mathcal{M})$ with the space $C^{-\infty}(\mathcal{M})$ of generalized functions. Note that $C^\infty(\mathcal{M}) \subseteq C^{-\infty}(\mathcal{M})$.

Theorem 4.5 ([GGHW21, Thm. 1]) *Let $\tau : A \to \mathrm{Diffeo}(\mathcal{M})$ be a smooth abelian Anosov action on a compact Riemannian manifold $\mathcal{M}$ with positive Weyl chamber $\mathfrak{a}_+^\tau$. Then the set $\sigma_\mathrm{R}(\tau)$ of joint Ruelle resonances $\lambda \in \mathfrak{a}_\mathbb{C}^*$ is a discrete set contained in*

$$\bigcap_{H \in \mathfrak{a}_+^\tau} \{\lambda \in \mathfrak{a}_\mathbb{C}^* \mid \mathrm{Re}(\lambda(H)) \leq 0\}. \tag{29}$$

Moreover, for each joint Ruelle resonance $\lambda \in \mathfrak{a}_\mathbb{C}^*$, the space of joint Ruelle resonant states is finite dimensional.

The Ruelle resonance spectrum $\sigma_\mathrm{R}(\tau)$ always contains $\lambda = 0$ (with $u = 1$ being the joint eigenfunction). When $\mathcal{M} = \Gamma\backslash G/M$ for some compact locally symmetric space $\Gamma\backslash G/K$ and τ is the Weyl chamber action, it contains infinitely many joint Ruelle resonances (see [HWW21, Theorem 1.1]).

This theorem is by no means a straightforward extension of the case of a single Anosov flow. It relies on a deeper result based on the theory of joint spectrum and joint functional calculus developed by Taylor [Ta70b, Ta70a]. This theory allows to set up a good Fredholm problem on certain functional spaces by using Koszul complexes. In fact, one defines $X + \lambda$, for $\lambda \in \mathfrak{a}_\mathbb{C}^*$, as an operator:

$$X + \lambda : C^\infty(\mathcal{M}) \to C^\infty(\mathcal{M}; \mathfrak{a}_\mathbb{C}^*), \quad ((X + \lambda)u)(A) := (X_A + \lambda(A))u.$$

For each $\lambda \in \mathfrak{a}_{\mathbb{C}}^*$, this yields differential operators:

$$d_{(X+\lambda)} : C^\infty(\mathcal{M}; \Lambda^j \mathfrak{a}_{\mathbb{C}}^*) \to C^\infty(\mathcal{M}; \Lambda^{j+1} \mathfrak{a}_{\mathbb{C}}^*)$$

via $d_{(X+\lambda)}(u \otimes \omega) := ((X + \lambda)u) \wedge \omega$ for $u \in C^\infty(\mathcal{M})$ and $\omega \in \Lambda^j \mathfrak{a}_{\mathbb{C}}^*$. Due to the commutativity of the family of vector fields X_H for $H \in \mathfrak{a}$, it can be checked that $d_{(X+\lambda)} \circ d_{(X+\lambda)} = 0$. Moreover, as a differential operator, it extends to a continuous map:

$$d_{(X+\lambda)} : C^{-\infty}_{E_u^*}(\mathcal{M}; \Lambda^j \mathfrak{a}_{\mathbb{C}}^*) \to C^{-\infty}_{E_u^*}(\mathcal{M}; \Lambda^{j+1} \mathfrak{a}_{\mathbb{C}}^*)$$

and defines an *associated Koszul complex*

$$0 \longrightarrow C^{-\infty}_{E_u^*}(\mathcal{M}) \overset{d_{(X+\lambda)}}{\longrightarrow} C^{-\infty}_{E_u^*} \otimes \Lambda^1 \mathfrak{a}_{\mathbb{C}}^* \overset{d_{(X+\lambda)}}{\longrightarrow} \dots \overset{d_{(X+\lambda)}}{\longrightarrow} C^{-\infty}_{E_u^*}(\mathcal{M}) \otimes \Lambda^\ell \mathfrak{a}_{\mathbb{C}}^* \longrightarrow 0. \tag{30}$$

In this setting one has the following theorem.

Theorem 4.6 ([GGHW21, Thm. 2]) *Let τ be a smooth abelian Anosov action on a closed manifold $\mathcal{M}$ with generating map X. Then for each $\lambda \in \mathfrak{a}_{\mathbb{C}}^*$ and $j = 0, \dots, \ell$, the cohomology*

$$\Big(\ker d_{(X+\lambda)}|_{C^{-\infty}_{E_u^*}(\mathcal{M}) \otimes \Lambda^j \mathfrak{a}_{\mathbb{C}}^*}\Big) \Big/ \Big(\operatorname{ran} d_{(X+\lambda)}|_{C^{-\infty}_{E_u^*}(\mathcal{M}) \otimes \Lambda^{j-1} \mathfrak{a}_{\mathbb{C}}^*}\Big)$$

is finite dimensional. It is nontrivial only at a discrete subset of $\{\lambda \in \mathfrak{a}_{\mathbb{C}}^* \mid \forall H \in \mathfrak{a}_+^\tau : \operatorname{Re}(\lambda(H)) \leq 0\}$.

The statement about the cohomologies in Theorem 4.6 is not only a stronger statement than Theorem 4.5, but the cohomological setting is in fact a fundamental ingredient in proving the discreteness of the resonance spectrum and its finite multiplicity. The proof in [GGHW21] relies on the theory of joint *Taylor spectrum*, defined using such Koszul complexes carrying a suitable notion of Fredholmness and furthermore provides a good framework for a parametrix construction via microlocal methods. More precisely, the parametrix construction is not done on the topological vector spaces $C^{-\infty}_{E_u^*}(\mathcal{M})$ but on a scale of Hilbert spaces $\mathcal{H}_{NG}$, depending on the choice of an escape function $G \in C^\infty(T^*\mathcal{M})$ and a parameter $N \in \mathbb{R}_{\geq 0}$, by which one can in some sense approximate $C^{-\infty}_{E_u^*}(\mathcal{M})$. The spaces $\mathcal{H}_{NG}$ are *anisotropic Sobolev spaces* which roughly speaking allow $H^N(\mathcal{M})$ Sobolev regularity in all directions except in E_u^* where one allows for $H^{-N}(\mathcal{M})$ Sobolev regularity. They can be rigorously defined using microlocal analysis, following the techniques of Faure-Sjöstrand [FS11]. By further use of pseudo-differential and Fourier integral operator theory, one can then construct a parametrix $Q(\lambda)$, which is a family of bounded operators on $\mathcal{H}_{NG} \otimes \Lambda\mathfrak{a}^*_{\mathbb{C}}$ depending holomorphically on $\lambda \in \mathfrak{a}^*_{\mathbb{C}}$ and satisfying

$$d_{(X+\lambda)}Q(\lambda) + Q(\lambda)d_{(X+\lambda)} = \mathrm{Id} + K(\lambda). \tag{31}$$

Here $K(\lambda)$ is a holomorphic family of compact operators on $\mathcal{H}_{NG} \otimes \Lambda\mathfrak{a}^*_{\mathbb{C}}$ for λ in a suitable domain of $\mathfrak{a}^*_{\mathbb{C}}$ that can be made arbitrarily large letting $N \to \infty$. Even after having this parametrix construction, the fact that the joint spectrum is discrete and intrinsic (i.e., independent of the precise construction of the Sobolev spaces) is more difficult than for an Anosov flow (the rank 1 case): this is because holomorphic functions in $\mathbb{C}^\ell$ do not have discrete zeros when $\ell \geq 2$ and we are lacking a good notion of resolvent, while for one operator the resolvent is an important tool. Due to the link with the theory of the Taylor spectrum, we call $\lambda \in \mathfrak{a}^*_{\mathbb{C}}$ a *Ruelle-Taylor resonance* for the Anosov action if for some $j = 0, \ldots, \kappa$ the j-th cohomology is nontrivial

$$\Big(\ker d_{(X+\lambda)}|_{C^{-\infty}_{E^*_u}(\mathcal{M})\otimes\Lambda^j\mathfrak{a}^*_{\mathbb{C}}}\Big)\Big/\Big(\mathrm{ran}\ d_{(X+\lambda)}|_{C^{-\infty}_{E^*_u}\otimes\Lambda^{j-1}\mathfrak{a}^*_{\mathbb{C}}}\Big) \neq 0,$$

and we call the nontrivial cohomology classes *Ruelle-Taylor resonant states*. Note that the definition of joint Ruelle resonances precisely means that the 0-th cohomology is nontrivial. Thus, any joint Ruelle resonance is a Ruelle-Taylor resonance. The converse statement is not obvious but turns out to be true; see [GGHW21, Prop. 4.15]: if the cohomology of degree $j > 0$ is not 0, then the cohomology of degree 0 is not trivial.

We conclude this subsection with some applications of Ruelle-Taylor resonances to dynamical systems. In view of (29), such a resonance is called a *leading resonance* when its real part vanishes. The leading resonances carry important information about the dynamics as they are related to a special type of invariant measures as well as to mixing properties of these measures.

Let v_g denote the Riemannian measure associated with the given Riemannian metric g on $\mathcal{M}$. A τ-invariant probability measure μ on $\mathcal{M}$ is called a *physical measure* if there is $v \in C^\infty(\mathcal{M})$ nonnegative such that for any continuous function f and any open cone $C \subset \mathfrak{a}^\tau_+$,

$$\mu(f) = \lim_{T\to\infty} \frac{1}{\mathrm{Vol}(C_T)} \int_{A\in C_T} \int_{\mathcal{M}} f(\varphi_1^{-X_A}(x))v(x)\, dv_g(x)\, dA \tag{32}$$

where $C_T := \{A \in C \,|\, |A| \leq T\}$, and here $|\cdot|$ denotes a fixed Euclidean norm on $\mathfrak{a}$. In other words, μ is the weak Cesaro limit of a Lebesgue type measure under the dynamics. As an application of the methods developed for the proof of Theorem 4.6, one can prove the following result.

Theorem 4.7 ([GGHW21, Thm. 3]) *Let* $\tau : A \to \mathrm{Diffeo}(\mathcal{M})$ *be a smooth abelian Anosov action on a compact Riemannian manifold* $\mathcal{M}$ *with generating map* X *and positive Weyl chamber* $\mathfrak{a}^\tau_+$*:*

(i) *The linear span over* $\mathbb{C}$ *of the physical measures is isomorphic (as a* $\mathbb{C}$*-vector space) to* $\ker d_X|_{C^{-\infty}_{E_u^*}(\mathcal{M})}$, *the space of joint Ruelle resonant states at* $\lambda = 0 \in \mathfrak{a}^*_{\mathbb{C}}$. *In particular, it is finite dimensional. The dimension can be expressed in dynamical terms; see [GBGW21, Theorem 3].*

(ii) *A probability measure* μ *is a physical measure if and only if it is* τ*-invariant and* μ *has wavefront set* $\mathrm{WF}(\mu) \subset E_s^*$, *where* $E_s^* \subset T^*\mathcal{M}$ *is defined by* $E_s^*(E_s \oplus E_0) = 0$.

(iii) *Assume that there is a unique physical measure* μ *(or by (i) equivalently that the space of joint resonant states at 0 is one dimensional). Then the following are equivalent:*

(1) *The only Ruelle-Taylor resonance on* $i\mathfrak{a}^*$ *is zero.*
(2) *There exists* $H \in \mathfrak{a}$ *such that* $\varphi_t^{X_H}$ is weakly *mixing with respect to* μ.
(3) *For any* $H \in \mathfrak{a}_+^\tau$, $\varphi_t^{X_H}$ *is* strongly *mixing with respect to* μ.

(iv) $\lambda \in i\mathfrak{a}^*$ *is a joint Ruelle resonance if and only if there is a complex measure* μ_λ *with* $\mathrm{WF}(\mu_\lambda) \subset E_s^*$ *satisfying for all* $H \in \mathfrak{a}_+^\tau, t \in \mathbb{R}$ *the following equivariance under push-forwards of the action:* $(\varphi_t^{X_H})_*\mu_\lambda = e^{-\lambda(H)t}\mu_\lambda$. *Moreover, such measures are absolutely continuous with respect to the physical measure obtained by taking* $v = 1$ *in* (32).

(v) *If* $\mathcal{M}$ *is connected and if there exists a smooth invariant measure* μ *with* $\mathrm{supp}(\mu) = \mathcal{M}$, *we have for any* $j = 0, \ldots, \ell$

$$\dim\left(\left(\ker d_X|_{C^{-\infty}_{E_u^*}(\mathcal{M})\otimes\Lambda^j\mathfrak{a}^*_{\mathbb{C}}}\right)\Big/\left(\mathrm{ran}\ d_X|_{C^{-\infty}_{E_u^*}\otimes\Lambda^{j-1}\mathfrak{a}^*_{\mathbb{C}}}\right)\right) = \binom{\ell}{j}.$$

The isomorphism stated in (i) and the existence of the complex measures in (iv) can be constructed explicitly in terms of spectral projectors built from the parametrix (31).

In the case of a single Anosov flow, physical measures are known to coincide with SRB measures (Sinai-Ruelle-Bowen; see, e.g., [Yo02] and references therein). The latter are usually defined as invariant measures that can locally be disintegrated along the stable or unstable foliation of the flow with absolutely continuous conditional densities.

In [GBGW21] it is proved that the microlocal characterization Theorem 4.7(ii) of physical measures via their wavefront sets implies that the physical measures of an Anosov action are exactly those invariant measures that allow an absolutely continuous disintegration along the stable manifolds. Moreover, [GBGW21, Theorem 3] says that for each physical/SRB measure, there is a basin $B \subset \mathcal{M}$ of positive Lebesgue measure such that for all $f \in C^0(\mathcal{M})$, all proper open subcones $C \subset \mathfrak{a}_+^\tau$ and all $x \in B$, one has the convergence:

$$\mu(f) = \lim_{T\to\infty} \frac{1}{\mathrm{Vol}(C_T)} \int_{A\in C_T} f(\varphi_1^{-X_A}(x))\, \mathrm{d}A. \tag{33}$$

Moreover, [GBGW21, Theorem 2] says that the measure μ can be written as an infinite weighted sum of the Dirac measures on the periodic tori of the action, showing an equidistribution of periodic tori in the support of μ. This measure has full support in $\mathcal{M}$ if the action is positively transitive in the sense that there is a dense orbit $\cup_{H\in\mathfrak{a}_+^\tau}\varphi_1^{X_H}(x)$ for some $x \in \mathcal{M}$. The existence of such a measure is in fact considered as an important step toward the resolution of the Katok-Spatzier rigidity conjecture, [KS94].

Remark 4.8 (Cocompact Locally Symmetric Spaces and First Band Resonant States) The theory of Ruelle-Taylor resonances sketched in this subsection applies to compact locally symmetric spaces $\mathcal{X} = \Gamma\backslash G/K$ with the Weyl chamber flow, i.e., A-action from Remark 2.7 on $\mathcal{M} = \Gamma\backslash G/M$.

We denote the space $\{u \in \mathcal{D}'_{E_u^*}(\mathcal{M}) \mid \forall H \in \mathfrak{a} : (X_H + \lambda(H))u = 0\}$ of resonant states by $\mathrm{Res}_X(\lambda)$. Then the space $\mathrm{Res}_X^0(\lambda)$ of *first band resonant states* is defined as those resonant states that are in addition horocyclically invariant:

$$\begin{aligned}\mathrm{Res}_X^0(\lambda) := \{u \in \mathcal{D}'_{E_u^*}(\mathcal{M}) \mid \forall H \in \mathfrak{a}, U \in C^\infty(\mathcal{M}; E_u) :\\ (X_H + \lambda(H))u = 0, Uu = 0\}\end{aligned}$$

We call a Ruelle-Taylor resonance a *first band resonance* iff $\mathrm{Res}_X^0(\lambda) \neq 0$.

Working with horocycle operators and vector valued Ruelle-Taylor resonances, it is possible to show that all resonances with real part in a certain neighborhood of zero in $\mathfrak{a}^*$ are always first band resonances (see [HWW21, Prop. 3.7]). In view of the symplectic interpretation of the Weyl chamber flow, we consider the set σ_{RT} of Ruelle-Taylor resonances for that flow as a *classical spectrum*.

4.2.2 Ruelle Resonances for Noncompact Locally Hyperbolic Spaces

We state a special case of a result of Dyatlov and Guillarmou from [DG16] which yields Ruelle resonances and resonant states for convex cocompact hyperbolic surfaces.

Theorem 4.9 (cf. [GHW18, Thm. 4.1]) *If $\mathcal{X} = \Gamma\backslash\mathbb{H}^2$ is a convex cocompact hyperbolic surface, then the generator X of the geodesic flow on $\mathcal{M} = S\mathcal{X}$ has a resolvent $R_X(\lambda) := (-X - \lambda)^{-1}$ that admits a meromorphic extension from $\{\lambda \in \mathbb{C} \mid \mathrm{Re}(\lambda) > 0\}$ to $\mathbb{C}$ as a family of bounded operators $C_c^\infty(S\mathcal{X}) \to \mathcal{D}'(S\mathcal{X})$. The resolvent $R_X(\lambda)$ has finite rank polar part at each pole λ_0 and the polar part is of the form:*

$$-\sum_{j=1}^{J(\lambda_0)} \frac{(-X-\lambda_0)^{j-1}\Pi_{\lambda_0}^X}{(\lambda-\lambda_0)^j}, \qquad J(\lambda_0) \in \mathbb{N}$$

for some finite rank projector $\Pi^X_{\lambda_0}$ commuting with X. Moreover, $u \in \mathcal{D}'(S\mathcal{X})$ is in the range of $\Pi^X_{\lambda_0}$ if and only if $(X+\lambda_0)^{J(\lambda_0)}u = 0$ with u supported in the outgoing tail Λ_+ of the geodesic flow and $\mathrm{WF}(u) \subset E^*_{\mathrm{u}}$.

Now one can define *Ruelle resonance*, *generalized Ruelle resonant state*, and *Ruelle resonant state* as, respectively, a pole λ_0 of $R(\lambda)$, an element in $\mathrm{Im}(\Pi^X_{\lambda_0})$, and an element in $\mathrm{Im}(\Pi^X_{\lambda_0}) \cap \ker(-X-\lambda_0)$. Define the spaces:

$$\mathrm{Res}^j_X(\lambda) := \{u \in \mathcal{D}'(S\mathcal{X}) \mid \mathrm{supp}(u) \subset \Lambda_+,\ \mathrm{WF}(u) \subset E^*_{\mathrm{u}},\ (-X-\lambda)^j u = 0\}, \tag{34}$$

$$V^j_m(\lambda) := \{u \in \mathrm{Res}^j_X(\lambda) \mid U_-^{m+1} u = 0\} \tag{35}$$

with $U_- := \begin{pmatrix} 0 & 0 \\ 1 & 0 \end{pmatrix}$. As in the compact case, the operator $(-X-\lambda_0)$ is nilpotent on the finite-dimensional space $\mathrm{Res}_X(\lambda_0) := \cup_{j\geq 1}\mathrm{Res}^j_X(\lambda_0)$ and λ_0 is a Ruelle resonance if and only if $\mathrm{Res}^1_X(\lambda_0) \neq 0$. The presence of Jordan blocks for λ_0 is equivalent to having $\mathrm{Res}^k_X(\lambda_0) \neq \mathrm{Res}^1_X(\lambda_0)$ for some $k > 1$.

Remark 4.10 (Asymptotically Hyperbolic Spaces) In [Ha17] Hadfield constructed Ruelle resonances for negatively curved manifolds which asymptotically at infinity behave as real hyperbolic spaces. This includes the class of convex cocompact quotients of hyperbolic space.

Remark 4.11 (Hyperbolic Manifolds with Cusps) Bonthonneau and Weich in [BW22] constructed Ruelle resonances for Riemannian manifolds with finite volume, negative curvature and the unbounded part consisting of finitely many real hyperbolic cusps.

4.2.3 Γ-Invariant Distributions for Principal Series Representations

Consider compact locally symmetric spaces and recall the first band resonant states from Remark 4.8. It turns out that these are closely related to Γ-invariant distribution vectors in principal series representations, Remark 2.9. In rank one the following proposition was established in [GHW21].

Proposition 4.12 ([GHW21, Prop. 3.8]) *There is an isomorphism of finite-dimensional vector spaces*

$$\mathrm{Res}^0_X(\lambda) \cong {}^{\Gamma}(H^{-\infty}_{-(\lambda+\rho)}),$$

where ${}^{\Gamma}(H^{-\infty}_{-(\lambda+\rho)})$ denotes the spaces of Γ-invariant distributional vectors in the spherical principal series with spectral parameter $\mu = -(\lambda+\rho)$.

In [HWW21] it was shown that this result extends to higher rank. In fact, using lifts to vector bundles (as they were discussed in [GGHW21]), one even gets a version for nonspherical principal series representations; see [HWW21, Lemma 3.10] for a precise formulation.

4.2.4 Dynamical Zeta Functions

Dynamical zeta functions are generating functions counting dynamical objects such as closed orbits. The dynamical zeta functions associated with the geodesic flow count closed geodesics using different weight functions:

(i) *Ruelle zeta function*:

$$Z_{\mathrm{R}}(\lambda) = \prod_c (1 - e^{-\lambda \ell(c)})^{\pm 1} = \exp\Big(\pm \sum_c \sum_{k\geq 1} \frac{e^{-k\lambda\ell(c)}}{k} \Big),$$

where the first summation is over the primitive closed geodesics c and $\ell(c)$ is the length of c.

(ii) *Selberg zeta function*:

$$\begin{aligned} Z_{\mathrm{S}}(\lambda) &= \prod_c \prod_{k\geq 0} \det\Big(1 - S^k(P_c^+) e^{-(\lambda+\rho)\ell(c)}\Big)^{\pm 1} \\ &= \exp\Big(- \sum_c \sum_{j\geq 1} \frac{e^{-j(\lambda+\rho)\ell(c)}}{j \det(1-(P_c^+)^j)} \Big), \end{aligned}$$

where $P_c = P_c^+ \oplus \mathrm{id} \oplus P_c^- \in \mathrm{End}(T_v(\Gamma\backslash G/M))$ is the monodromy (i.e., Poincaré) map and S^k denotes the k-th symmetric power.

(iii) *Dynamical zeta function*:

$$Z_{\mathrm{dyn}}(\lambda) = \exp\Big(- \sum_c \sum_{j\geq 1} \frac{e^{-j\lambda\ell(c)}}{j \left|\det(1-(P_c^+)^j)\right|} \Big).$$

There are more versions with additional twists and weights (see, e.g., [BO95, Chap. 3] and [AZ07]). For the relation between dynamical and Selberg zeta functions, we refer to [FT17]. A key feature of such zeta functions is that they often admit meromorphic continuation (see, e.g., [GLP13, DZ16] and the literature cited there). The resulting divisors may serve as a classical spectral invariant of the underlying flows.

4.2.5 Transfer Operators

Transfer operators were introduced in statistical mechanics as a means to find equilibrium distributions. In our context they were used primarily to study symbolic dynamical systems derived from the geodesic flow on hyperbolic surfaces. As is observed in [PZ20], for suitable symbolic systems, they do resemble weighted Laplacians on graphs which gives them a double meaning, one classical, one quantum mechanical. For the modular surface, this eventually leads to the first dynamical interpretation of Maass cusp forms in [LZ01]. We refer to the introduction in [Po14] for the extended history of this development.

In general, for a discrete dynamical system $(\mathcal{N}, f)$ with $f : \mathcal{N} \to \mathcal{N}$ any map, one associates a *transfer operator* $\mathcal{L}_\psi$ on suitable spaces function spaces on $\mathcal{N}$ to a *weight map* $\psi : \mathcal{N} \to \mathbb{C}$ by

$$\mathcal{L}_\psi \phi(x) = \sum_{y \in f^{-1}(x)} e^{\psi(y)} \phi(y). \tag{36}$$

Here "suitable" depends on the convergence of the sum in (36).

For hyperbolic surfaces the construction of the transfer operator typically is a two-step process. First one constructs a discrete dynamical system with invertible f whose suspension flow (see [KH95, § 0.3]) is basically conjugate to the geodesic flow. Then one applies a reduction leading to a discrete dynamical system for which f is no longer invertible. In [CM99, § 2.2] the process for the modular surface goes as follows. They quote the symbolic dynamics for the geodesic flow from earlier work and start with

$$\begin{aligned} \tilde{f} :]0,1[\times]0,1[\times\{\pm 1\} &\to\ ]0,1[\times]0,1[\times\{\pm 1\} \\ (a,b,\epsilon) &\mapsto \big(a^{-1} \bmod 1, (b + \lfloor a^{-1} \rfloor)^{-1}, -\epsilon\big). \end{aligned}$$

and then work with the reduced dynamical system given by

$$f :]0,1[\to]0,1[, \quad x \mapsto x^{-1} \bmod 1.$$

As weight functions they use $x \mapsto x^{2\beta}$ with $\mathrm{Re}\beta > \frac{1}{2}$. The constructions in [Po14] are much more general, but also more involved. While it is quite clear that the reduction in [CM99] means that the authors concentrate on the expanding part of the dynamics, it is less transparent how to relate the reduction in [Po14, § 9] to the hyperbolic nature of the geodesic flow.

In the transfer operator approach to quantum-classical correspondences, it is the spectral theory of the transfer operators which take the role of Ruelle resonances and resonant states. It should be noted that while the scope of locally symmetric spaces that can be dealt with using the transfer operator approach is almost exclusively restricted to surfaces (see, however, [Po20] for some higher rank examples), it works well in many noncompact situations.

The analogs of geodesic and Weyl chamber flows for graphs and higher rank non-Archimedean counterparts of locally symmetric spaces, i.e., quotients of affine buildings, are of a combinatorial nature. So it is not surprising that transfer operators seem to be suitable tools to establish quantum-classical correspondences in those contexts. In the long run, if one wants to deal with adelic situations, microlocal and transfer methods may have to be used simultaneously in a coordinated way.

For yet another approach to quantum-classical correspondences via transfer operators, we refer to the book [FT15] by Faure and Tsujii.

4.3 Correspondences

In this section we finally come back to the quantum-classical correspondences for locally symmetric spaces that were alluded to in the motivating examples. It turns out that there are partial results of varying generality depending on the specifications of the locally symmetric spaces considered. The most general results are available for compact locally symmetric spaces. We start with that case and then describe what can be done at the moment for noncompact locally symmetric spaces.

4.3.1 The Compact Case for Generic Spectral Parameters

Recall the right A-action on $T\mathcal{M}$ from Remark 2.7 and assume that Γ is cocompact. Then the A-action is Anosov as we remarked in Sect. 3.1.1. For $H \in \mathfrak{a}$ let X_H again be the vector field on $\mathcal{M}$ defined by the right A-action. As was explained in Sect. 4.2.1, the *Ruelle-Taylor resonances* of this Anosov action are then given by

$$\sigma_{\mathrm{RT}} = \{\lambda \in \mathfrak{a}_{\mathbb{C}}^* \mid \exists u \in \mathcal{D}'_{E_{\mathrm{u}}^*}(\mathcal{M}) \setminus \{0\}\, \forall H \in \mathfrak{a} : (X_H + \lambda(H))u = 0\},$$

where $\mathcal{D}'_{E_{\mathrm{u}}^*}(\mathcal{M})$ is the set of distributions with wavefront set contained in the annihilator $E_{\mathrm{u}}^* \subseteq T^*\mathcal{M}$ of $E_0 \oplus E_{\mathrm{u}}$. The distributions $u \in \mathcal{D}'_{E_{\mathrm{u}}^*}(\mathcal{M})$ satisfying $(X_H + \lambda(H))u = 0$ for all $H \in \mathfrak{a}$ are the resonant states of λ, and the dimension of the space of all such distributions is called the *multiplicity* $m(\lambda)$ of the resonance λ. According to Theorem 4.6, $\sigma_{\mathrm{RT}} \subset \mathfrak{a}_{\mathbb{C}}^*$ is discrete and all resonances have finite multiplicity. Moreover, the real parts of the resonances are located in the negative dual cone $-\mathfrak{a}_+^\star \subset \mathfrak{a}^*$ of the positive Weyl chamber $\mathfrak{a}_+$.

We have the following correspondence between the space of classical first band resonant states and the joint quantum eigenspace (see Sect. 4.1.1):

Theorem 4.13 ([HWW21, Thm. 1.3]) *Let $\lambda \in \mathfrak{a}_{\mathbb{C}}^*$ be outside the* exceptional set $\mathcal{E} := \{\lambda \in \mathfrak{a}_{\mathbb{C}}^* \mid \exists \alpha \in \Sigma^+ : \frac{2\langle \lambda+\rho,\alpha\rangle}{\langle \alpha,\alpha\rangle} \in -\mathbb{N}_{>0}\}$. *Then there is a bijection between the finite-dimensional vector spaces:*

$$\pi_* : \mathrm{Res}_X^0(\lambda) \to {}^\Gamma E_{-\lambda-\rho},$$

where π_* is the push-forward of distributions along the canonical projection $\pi : \Gamma\backslash G/M \to \Gamma\backslash G/K$.

Using the 1:1-correspondence in Theorem 4.13 and results about the quantum spectrum, one can obtain obstructions and existence results on the Ruelle-Taylor resonances. Notably, in [HWW21] we used results of Duistermaat-Kolk-Varadarajan [DKV79] on the spectrum σ_Q, but we also use refined information on the quantum spectrum such as L^p-bounds for spherical functions obtained from asymptotic expansions [vdBS87] and L^p-bounds for matrix coefficients based on work by Cowling and Oh [Co79, Oh02]. Theorems 4.14 and 4.15 as stated below give only a rough version of the information on the Ruelle-Taylor resonances that we can actually obtain. As the full results require some further notation, we refrain from stating them here, but rather refer to [HWW21, Thm. 5.1].

The first application says that for any Weyl chamber flow, there exist infinitely many Ruelle-Taylor resonances by providing a Weyl lower bound on an appropriate counting function.

Theorem 4.14 ([HWW21, Thm. 1.1]) *Let ρ be the half-sum of the positive restricted roots, W the Weyl group, and for $t > 0$ let*

$$N(t) := \sum_{\lambda\in\sigma_{\mathrm{RT}},\mathrm{Re}(\lambda)=-\rho,\|\mathrm{Im}(\lambda)\|\le t} m(\lambda).$$

Then for $d := \dim(G/K)$

$$N(t) \ge |W|\mathrm{Vol}(\Gamma\backslash G/K)\left(2\sqrt{\pi}\right)^{-d}\frac{1}{\Gamma(d/2+1)}t^d + O(t^{d-1}).$$

More generally, let $\Omega \subseteq \mathfrak{a}^*$ be open and bounded such that $\partial\Omega$ has finite $(n-1)$-dimensional Hausdorff measure. Then

$$\sum_{\lambda\in\sigma_{\mathrm{RT}},\mathrm{Re}(\lambda)=-\rho,\mathrm{Im}(\lambda)\in t\Omega} m(\lambda) \ge |W|\mathrm{Vol}(\Gamma\backslash G/K)(2\pi)^{-d}\,\mathrm{Vol}(\mathrm{Ad}(K)\Omega)t^d + O(t^{d-1}).$$

The second application is the existence of a uniform spectral gap.

Theorem 4.15 ([HWW21, Thm. 1.2]) *Let G be a real semisimple Lie group with finite center; then for any cocompact torsion-free discrete subgroup $\Gamma \subset G$, there is a neighborhood $\mathcal{G} \subset \mathfrak{a}^*$ of 0 such that*

$$\sigma_{\mathrm{RT}} \cap (\mathcal{G} \times i\mathfrak{a}^*) = \{0\}.$$

If G furthermore has Kazhdan's property (T) (e.g., if G is simple of higher rank), then the spectral gap $\mathcal{G}$ can be taken uniformly in Γ and only depends on the group G.

This theorem has an extension to vector bundles based on the lifted version of Ruelle-Taylor resonances discussed in [GGHW21]; see [HWW21, Thm. 5.1]. In the following remark, we comment on the reasons why such extensions are relevant.

Remark 4.16 (Homogeneous Vector Bundles) One of the key features setting hyperbolic surfaces apart from higher dimensional, let alone higher rank, locally symmetric spaces is that the sphere bundle $\mathrm{SL}_2(\mathbb{R})/\{\pm\mathbf{1}\}$ of its simply connected covering $\mathbb{H} = \mathrm{SL}_2(\mathbb{R})/\mathrm{SO}_2$ is actually a group. Thus compared to the general rank one case, where the sphere bundle is G/M, it has extra symmetries, which simplify the analysis in many ways. In particular, the unipotent Iwasawa group N acts from the right on G/M giving the horocyclic flow which in general does not live on the sphere bundle of the locally symmetric space but only on $\Gamma\backslash G$.

The tensor bundles relevant for the differential analysis on G/M are G-homogeneous bundles associated with M-representations. Their sections can be described as functions on G with M-equivariance properties. This suggests to view G as a principal bundle over G/M with fiber M and analyze the right G-action in terms of its effects on the tensor algebra. This is what was done in [DFG15], where the authors determined the band structure of the Ruelle resonances for compact hyperbolic spaces and in [Em14], where Emonds worked out a dynamical interpretation of Patterson-Sullivan distributions generalizing observations from [AZ07] for hyperbolic surfaces. Similarly, it was used in the study [AH23] of quantum-classical correspondences for exceptional spectral parameters. It is also implicit in the work of Bunke and Olbrich.

For compact locally symmetric spaces of rank one, Küster and Weich in [KW21] undertook a systematic study of Ruelle resonant states consisting of distributions with values in M-representations leading up to a quantum-classical correspondence with eigenspaces of the Bochner Laplacian on G-homogeneous vector bundles over K associated with K-representations intertwining nontrivially with the M-representation used for the Ruelle resonances. In particular, they define horocyclic operators generalizing the ones from [DFG15], thus opening a path to clarify the precise band structure of Ruelle resonances for general compact locally symmetric spaces of rank one.

In the companion paper [KW20], Küster and Weich apply special cases of their results to write the first Betti number of negatively curved compact Riemannian manifolds obtained by deformation from hyperbolic ones in terms of Ruelle resonances.

4.3.2 Results for Noncompact Locally Symmetric Spaces

The first quantum-classical correspondences involving quantum resonant states were the ones on convex cocompact hyperbolic spaces alluded to in Sect. 4.1.2. We state explicitly the correspondence established in [GHW18] for hyperbolic surfaces and refer to [Ha17, Ha20] for extensions to higher dimension.

Let the manifold $\mathcal{X} = \Gamma\backslash\mathbb{H}^2$ be a noncompact complete smooth hyperbolic surface with infinite volume, but finitely many topological ends. $\mathcal{X}$ can be compactified to the smooth manifold $\overline{\mathcal{X}} = \Gamma\backslash(\mathbb{H}^2 \cup \Omega_\Gamma)$, where $\Omega_\Gamma \subset \mathbb{S}^1$ is the *set of discontinuity* of Γ, the complement of the limit set $\Lambda_\Gamma \subset \mathbb{S}^1$ of Γ. Note that $\mathcal{X}$ is *conformally compact* in the sense of Mazzeo-Melrose [MM87]: there is a smooth boundary defining function r such that $\bar{g} := r^2 g$ extends as a smooth metric on $\overline{\mathcal{X}}$. Here g denotes the Riemannian metric on $\mathcal{X}$.

In this case the group Γ is a subgroup of $\mathrm{PSL}_2(\mathbb{C})$ and acts on the Riemann sphere $\overline{\mathbb{C}} := \mathbb{C} \cup \{\infty\}$ as conformal transformations, it preserves the unit disk $\mathbb{H}^2$ and its complement $\overline{\mathbb{C}} \setminus \mathbb{H}^2$. Equivalently, by conjugating by $(z - i)/(z + i)$, Γ acts by conformal transformations on $\overline{\mathbb{C}}$ as a subgroup of $\mathrm{PSL}_2(\mathbb{R}) \subset \mathrm{PSL}_2(\mathbb{C})$ and it preserves the half-planes $\mathbb{H}^2_\pm := \{z \in \mathbb{C} \mid \pm\mathrm{Im}(z) > 0\}$. The half-planes are conformally equivalent through $z \mapsto \bar{z}$ if we put opposite orientations on $\mathbb{H}^2_+$ and $\mathbb{H}^2_-$. In this model the boundary is the compactified real line $\partial\mathbb{H}_\pm = \overline{\mathbb{R}} := \mathbb{R} \cup \{\infty\}$ and the limit set is a closed subset Λ_Γ of $\overline{\mathbb{R}}$. Its complement in $\overline{\mathbb{R}}$ is still denoted by Ω_Γ. Since $\overline{\gamma(\bar{z})} = \gamma(z)$ for each $\gamma \in \Gamma$, the quotients $\mathcal{X}_\pm := \Gamma\backslash(\mathbb{H}^2_\pm \cup \Omega_\Gamma)$ are smooth surfaces with boundaries, equipped with a natural conformal structures and $\mathcal{X}_+$ is conformally equivalent to $\mathcal{X}_-$. The surface $\Gamma\backslash(\overline{\mathbb{C}} \setminus \Lambda_\Gamma)$ is a compact surface diffeomorphic to the gluing $\mathcal{X}_2 := \mathcal{X}_+ \cup \mathcal{X}_-$ of $\mathcal{X}_+$ and $\mathcal{X}_-$ along their boundaries; moreover, it is equipped with a smooth conformal structure which restricts to that of $\mathcal{X}_\pm$.

We denote by $\mathcal{I} : \mathcal{X}_2 \to \mathcal{X}_2$ the involution fixing $\partial\overline{\mathcal{X}}$ and derived from $z \mapsto \bar{z}$ when viewing Γ as acting in $\overline{\mathbb{C}} \setminus \Lambda_\Gamma$. The interiors of $\mathcal{X}_+$ and $\mathcal{X}_-$ are isometric if we put the hyperbolic metric $|dz|^2/(\mathrm{Im}(z))^2$ on $\mathbb{H}^2_\pm$, and they are isometric to the hyperbolic surface $\mathcal{X}$. The conformal class of $\mathcal{X}_\pm$ corresponds to the conformal class of $\bar{g}$ on $\overline{\mathcal{X}}$ as defined above. We identify $\mathcal{X}_+$ with $\overline{\mathcal{X}}$ and define $H_{\pm n}(\mathcal{X})$ as the following finite-dimensional real vector spaces:

$$\begin{aligned} H_n(\mathcal{X}) &:= \{f|_{\mathcal{X}_+} \mid f \in C^\infty(\mathcal{X}_2; \mathcal{K}^n),\ \bar{\partial} f = 0,\ \mathcal{I}^* f = \bar{f}\}, \\ H_{-n}(\mathcal{X}) &:= \{f|_{\mathcal{X}_+} \mid f \in C^\infty(\mathcal{X}_2; \mathcal{K}^{-n}),\ \partial f = 0,\ \mathcal{I}^* f = \bar{f}\}, \end{aligned} \tag{37}$$

where $\mathcal{K}$ is the canonical line bundle; see [GH78, § 1.1].

With the notation established in Sect. 4.2.2, we have the following quantum-classical correspondence.

Theorem 4.17 (cf. [GHW18, Thm. 4.2]) *Let $\mathcal{X} = \Gamma\backslash\mathbb{H}^2$ be a smooth oriented convex cocompact hyperbolic surface and let $\mathcal{M} = S\mathcal{X}$ be its unit tangent bundle:*

(i) *For each* $\lambda \in \mathbb{C} \setminus (-\frac{1}{2} - \frac{1}{2}\mathbb{N}_0)$ *the pushforward map* $\pi_* : \mathcal{D}'(S\mathcal{X}) \to \mathcal{D}'(\mathcal{X})$ *for the projection* $\pi : \mathcal{M} \to \mathcal{X}$ *restricts to a linear isomorphism of complex vector spaces for each* $j \geq 1$

$$\pi_{0*} : V_0^j(\lambda) \to \mathrm{Res}^j_{\Delta_\mathcal{X}}(\lambda_0 + 1), \tag{38}$$

where $\Delta_\mathcal{X}$ *is the Laplacian on* $\mathcal{X}$ *acting on functions.*

(ii) *For each* $\lambda = -\frac{1}{2} - k$ *with* $k \in \mathbb{N}$, $V_0^j(\lambda) = 0$ *and* $\mathrm{Res}^j_{\Delta_\mathcal{X}}(\lambda + 1) = 0$ *for all* $j \in \mathbb{N}$.

(iii) *For* $\lambda = -\frac{1}{2}$, *there are no Jordan blocks, i.e.* $V_0^j(-\frac{1}{2}) = 0$ *for* $j > 1$, *and the map*

$$\pi_{0*} : V_0^1(-\tfrac{1}{2}) \to \mathrm{Res}^1_{\Delta_\mathcal{X}}(1/2) \tag{39}$$

is a linear isomorphism of complex vector spaces.

(iv) *For* $\lambda = -n \in -\mathbb{N}$, *if* Γ *is nonelementary (i.e., there are no invariant sets of cardinality* 1 *or* 2*), there are no Jordan blocks, i.e.,* $V_0^j(-n) = 0$ *if* $j > 1$, *and the following map is an isomorphism of real vector spaces:*

$$i^{n+1}\pi_{n*} : V_0^1(-n) \to H_n(\mathcal{X}) \tag{40}$$

where $H_n(\mathcal{X})$ *is defined by* (37).

Remark 4.18 (QCC via Transfer Operators) The Lewis-Zagier correspondence from Example 1.4 is an example of a quantum-classical correspondence established via transfer operator methods. It has been generalized in various directions. As these methods are not in the focus of this article, I do not present further details here, but refer to the introduction of [Po14] for a brief discussion of further developments as well as references.

Remark 4.19 (Cuspidal Γ-cohomology) Theorem 2.14 may be viewed as a weak quantum-classical correspondence since it proves for irreducible principal series representations π the multiplicity space $\mathrm{Hom}_G(\pi, L^2_{\mathrm{cusp}}(\Gamma\backslash G))$ is isomorphic to $H^r_{\mathrm{cusp}}(\Gamma, \pi^{-\omega})$. However, the isomorphism is only guaranteed by the equality of dimensions and not described in any natural way.

4.3.3 Exceptional Spectral Parameters

In the quantum-classical correspondences described in Sects. 4.3.1 and 4.3.2, one always had to make restrictions on the spectral parameters describing classical resonances. The reason is that the method of proof in all these examples depends on (scalar) Poisson transforms which for generic parameters are bijective. For the exceptional parameters leading to Poisson transforms which are not bijective, the proofs break down and one has to find a replacement for the scalar Poisson transforms.

The strategy for an extension of the quantum-classical correspondence to exceptional spectral parameters that has been successfully applied to compact rank one locally symmetric spaces in [AH23] is as follows. As in the generic case (see [GHW21, § 3.2]), we start by lifting the first band Ruelle resonances to Γ-invariant distributions on the global symmetric space. The lifted spaces can be interpreted in terms of spherical principal series (that part works for all spectral parameters; see [GHW21, Prop. 3.8]), and the first band resonant states $\mathrm{Res}_X^0(-\lambda-\rho)$ correspond to the space ${}^\Gamma H_\lambda^{-\infty}$ of Γ-invariant distribution vectors of the corresponding principal series. For an exceptional spectral parameter λ, the corresponding principal series H_λ is no longer irreducible. But it has a manageable composition series, and it turns out that the Γ-invariant distribution vectors are all contained in the *socle* (i.e., the sum of all irreducible subrepresentations) of the representation. In each of the rank one cases except $\mathrm{SO}_0(2,1)$ (the case of surfaces; see [GHW18]), the socle turns out to be irreducible with a unique minimal K-type τ_λ and one can show that the vector-valued Poisson transform associated with this K-type (sum of K-types in the case of surfaces) is injective. The image consists of spaces of Γ-invariant sections of vector bundles over G/K, and we have a quantum-classical correspondence as soon as we have characterized the image of this Poisson transform.

Theorem 4.20 ([AH23, Thm. 6.1]) *Consider the case of a cocompact rank one locally symmetric space with* $G = \mathrm{SO}_0(n,1)$, $n \geq 3$. *Given an exceptional parameter* $\lambda = -(\rho + \ell\alpha)$, $\ell \in \mathbb{N}_0$, *the socle* $\mathrm{soc}(H_\lambda)$ *of* H_λ *is irreducible and unitary, and its* K*-types are given by the spaces* Y_k *of harmonic homogeneous polynomials of degree* k *for* $k \geq \ell + 1$. *The minimal* K*-type is* $Y_{\ell+1}$ *and the corresponding Poisson transform induces an isomorphism:*

$$P_\lambda^{Y_{\ell+1}} : {}^\Gamma(\mathrm{soc}(H_\lambda))^{-\infty} \xrightarrow{\cong} {}^\Gamma\{f \in C^\infty(G \times_K Y_{\ell+1}) \mid \mathrm{d}_- f = 0,\ \mathrm{D} f = 0\},$$

with differential operators d_- *and* D *which can be described explicitly.*

From the study of the zeros of dynamical zeta functions, one expects the multiplicities of exceptional resonances to carry topological information. In [GHW18] this expectation was confirmed for hyperbolic surfaces defined by cocompact and convex cocompact discrete subgroups of $\mathrm{PSL}_2(\mathbb{R})$.

Example 4.21 (Topological Interpretation [GHW18]) Let $\Gamma\backslash G/K$ be a compact hyperbolic surface. Then

$$\dim \mathrm{Res}_X^0(-n) = \begin{cases} |\chi(\Gamma\backslash G/K)| + 2 & \text{if } n = 1, \\ (2n-1)|\chi(\Gamma\backslash G/K)| & \text{if } 1 < n \in \mathbb{N}, \end{cases}$$

where $\chi(\Gamma\backslash G/K)$ is the Euler characteristic of $\Gamma\backslash G/K$.

In higher dimension we do not have a topological interpretation of the resonance multiplicities except in very special examples boiling down to Hodge theory.

Example 4.22 (Topological Interpretation [AH23]) In the situation of Theorem 4.20 for the first exceptional parameter $\lambda = -\rho$, we get ($Y_1 \cong \mathfrak{p}^*$)

$$P^{Y_1}_{-\rho} : {}^{\Gamma}(\mathrm{soc}(H_{-\rho}))^{-\infty} \xrightarrow{\cong} \{f \in C^{\infty}(\Lambda^1(\Gamma\backslash G/K)) \mid \delta f = 0,\ df = 0\},$$

where $\Lambda^1(\Gamma\backslash G/K)$ denotes the bundle of one forms and (δ resp.) d is the (co)-differential. The dimension is given by the first Betti number $b_1(\Gamma\backslash G/K)$ in this case.

The quantum-classical correspondences for exceptional spectral parameters get more complicated for the other compact rank one locally symmetric spaces. This is due to the fact that the system of differential equations characterizing the image of the vector valued Poisson transform is more involved.

Remark 4.23 (Cocompact Locally Symmetric Spaces of Rank One) Results analogous to Theorem 4.20 are available for all the remaining cocompact rank one locally symmetric spaces, i.e., for the cases:

(i) $G = \mathrm{SU}(n, 1)$, $n \geq 2$, see [AH23, Theorems 6.4 & 6.6],
(ii) $G = \mathrm{Sp}(n, 1)$, $n \geq 2$, see [AH23, Thm. 6.8],
(iii) $G = \mathrm{F}_{4(-20)}$, see [AH23, Thm. 6.10].

In the symplectic and the exceptional cases, however, the description of the image of the vector valued Poisson transformation is a little more complicated.

The representations showing up as socles of spherical principal series representation for exceptional parameters can be determined explicitly in terms of their Langlands parameters. In fact, they have a uniform description as was observed by Jan Frahm when he saw the list of Langlands parameters.

Theorem 4.24 ([AH23, Thm. 4.7]) *There is a one-to-one correspondence between the representations* $\mathrm{soc}(H_{\lambda_\ell})$ *with* λ_ℓ *exceptional and the relative discrete series of the associated pseudo-Riemannian symmetric spaces* G/H. *More precisely, each of the representations corresponds to a minimal closed invariant subspace of* $L^2(G/H)$ *with* $H = \mathrm{SO}_0(n-1, 1)$, $\mathrm{S}(\mathrm{U}(1) \times \mathrm{U}(n-1, 1)) \cong \mathrm{U}(n-1, 1)$, $\mathrm{Sp}(1) \times \mathrm{Sp}(n-1, 1)$, *or* $\mathrm{Spin}(1, 8)$.

4.3.4 Applications

Most applications of quantum-classical correspondences as described in this section that are on record so far make use of the possibility to compute "classical" multiplicities from "quantum"' multiplicities. First examples of such applications were already given in Sect. 4.3.1 (see [HWW21, Theorems 1.1 & 1.2]).

In [DGRS20], which gives a proof of the Fried conjecture (in dimension 3) relating analytic torsions to twisted dynamical zeta values, the authors need to exclude 0 as a Ruelle resonance for the geodesic flow in compact real hyperbolic manifolds of dimension 3. This allows them to perturb the generator of the geodesic flow and thus prove Fried's conjecture for some 3-manifolds with variable curvature. The proof given in [DGRS20, Prop. 7.7] for the fact that 0 is not a Ruelle resonance uses the Selberg trace formula, but the authors point out in [DGRS20, Rem. 10] that they could also use [DFG15, KW21].

Extending ideas from [KW20] on the deformation of geodesic flows (see Sect. 4.3.1), Cekić et al. in [CDDP22] showed that the order of vanishing at zero of the Ruelle zeta function is not stable under generic deformations and thus provided counterexamples to Fried's conjecture for compact hyperbolic 3-manifolds with nonvanishing first Betti number. In this work the quantum-classical correspondence is used to calculate the vanishing order (which can be expressed in terms of multiplicities of Ruelle resonances) for the geodesic flow which is being deformed.

Finally, we mention an application due to Schütte, Weich, and Barkhofen who in [SWB23] introduce weighted zeta functions for flows which are hyperbolic on their trapped set and prove that they continue meromorphically to all of $\mathbb{C}$. They show that poles of the zeta functions are Ruelle resonances of the flows and describe the residues of the zeta functions in terms of the residue operators of the Ruelle resonance (cf. Example 1.6) associated with the resolvent of the generating vector fields. Combining these residue formulas with [GHW21, Cor. 6.1] which expresses the *flat trace* of residue operators in terms of Patterson-Sullivan distributions, they are able to extend the scope of a result of Anantharaman-Zelditch [AZ07] from compact hyperbolic surfaces to all compact Riemannian locally symmetric spaces of rank one (see [SWB23, Thm. 4.1]). Note that [GHW21, Cor. 6.1] is a corollary of the pairing formula [GHW21, Thm. 6.1] relating the product of a resonant and a coresonant state of a Ruelle resonance to the pairing of their push-forwards along the projection to the base space (cf. Theorem 4.13).

Actually, versions of the aforementioned results of from [AZ07] and [SWB23] were formulated by physicists long ago (see [EFMW92]). Arguing with semiclassical trace formulas, they predicted residues of weighted zeta functions to be given by quantum phase space distributions in a semiclassical limit. In the companion paper [BSW22] to [SWB23], the same authors explain how their [SWB23, Thm. 4.1] is a mathematically rigorous version of this prediction.

5 Quotients of Trees and Affine Buildings

Trees and affine buildings come up in the context of p-adic Lie groups where they play a role analogous to symmetric spaces for real Lie groups (see, e.g., [BT72, Ca73, FN91, Se80, Pa06]). Homogeneous trees correspond to rank one situations; the affine buildings may be viewed as suitable higher rank versions of homogeneous trees. The quotients of trees and buildings then correspond to the locally symmetric spaces which are in the focus of this article.

It should be noted that quotients of trees amount to connected graphs which are used very often as approximations or toy models for manifolds. This is in particular the case in dynamical systems (see, e.g., [LP16, An17]) but also in microlocal analysis (see, e.g., [AL15, LM14]). It is thus reasonable to expect that meaningful results about buildings hold, which are analogous to the quantum-classical correspondences we have discussed in Sect. 4.3. At the same time, one can expect that the necessary analysis is less involved compared to the real case. One can hope for clues how to proceed in the real case in places where progress has been impeded by technical difficulties. Moreover, recent results (see [BHW22, BHW23, AFH23a]) support the expectation that one can weaken the symmetry conditions and still obtain nontrivial correspondences.

5.1 Graphs of Bounded Degree

A natural starting point for research in this direction are homogeneous trees. For these, key tools like the Poisson transform have been available for a long time; see, e.g., [FN91]. It turns out that in the case of trees, one can weaken the symmetry assumptions quite a bit and still get Poisson transforms as well as quantum-classical correspondences.

We consider graphs $\mathfrak{G} = (\mathfrak{X}, \mathfrak{E})$ consisting of a set $\mathfrak{X}$ of vertices and a set $\mathfrak{E} \subseteq \mathfrak{X}^2$ of directed edges. We assume that the graph is *symmetric*, i.e., the set $\mathfrak{E} \subseteq \mathfrak{X}^2$ is invariant under the switch of coordinates. Further we assume that the graph contains *no loops*, i.e., $(x, x) \notin \mathfrak{E}$ for all $x \in \mathfrak{X}$, and has *no dead ends*, which means that each vertex has at least two neighbors.

For each directed edge $\mathbf{e} = (a, b)$, we call $a = \iota(\mathbf{e})$ the initial and $b = \tau(\mathbf{e})$ the terminal point of $\mathbf{e}$. A *path* in $\mathfrak{G}$ is a (finite or infinite) sequence $(\mathbf{e}_j)$ of edges such that $\tau(\mathbf{e}_j) = \iota(\mathbf{e}_{j+1})$, but $\iota(\mathbf{e}_j) \neq \tau(\mathbf{e}_{j+1})$ (no backtracking!). We assume that any two vertices in $\mathfrak{X}$ can be connected by a finite path of edges.

As an analog of geodesic rays in Riemannian manifolds, we introduce the space:

$$\mathfrak{P}_+ := \{(\mathbf{e}_j)_{j\in\mathbb{N}}\}$$

of infinite paths starting at some vertex. We refer to such paths as *geodesic rays* in $\mathfrak{G}$. The assumption that $\mathfrak{G}$ has no dead ends together with the symmetry implies that any finite path can be extended to a geodesic ray. This gives a kind of geodesic completeness of the graph.

The shift $(\mathbf{e}_1, \mathbf{e}_2, \ldots) \mapsto (\mathbf{e}_2, \mathbf{e}_3, \ldots)$ of geodesic rays is one possible analog of the geodesic flow in the graph. To obtain spectral invariants for this geodesic flow, one can consider the transfer operator $\mathcal{L} : \mathbb{C}^{\mathfrak{P}_+} \to \mathbb{C}^{\mathfrak{P}_+}$ defined by

$$\mathcal{L}f(\mathbf{e}_1, \mathbf{e}_2, \ldots) := \sum_{\tau(\mathbf{e}_0)=\iota(\mathbf{e}_1),\iota(\mathbf{e}_0)\neq\tau(\mathbf{e}_1)} f(\mathbf{e}_0, \mathbf{e}_1, \mathbf{e}_2, \ldots),$$

where $\mathbb{C}^{\mathfrak{P}_+}$ is the space of all complex valued functions on $\mathfrak{P}_+$. The space $\mathfrak{P}_+$ carries a natural metric topology (see, e.g., [BHW23]), from which one can derive a number of natural function spaces. One of them is the space $C^{\mathrm{lc}}(\mathfrak{P}_+)$ of locally constant functions, which can be viewed as an analog of the space of smooth functions on a smooth manifold.

The analog of the Laplace-Beltrami operator for the graph $\mathfrak{G}$ is the Laplacian $\Delta_{\mathfrak{X}} : \mathbb{C}^{\mathfrak{X}} \to \mathbb{C}^{\mathfrak{X}}$ defined by

$$\Delta f(x) := \frac{1}{1+q_x} \sum_{d(x,y)=1} f(y),$$

where q_x for a vertex $x \in \mathfrak{X}$ is the number of neighboring vertices in the graph and $d(x, y)$ is the minimal numbers of edges needed to connect the vertices x and y by a path.

In order to describe a graph version of a quantum-classical correspondence, for any spectral parameter $0 \neq z \in \mathbb{C}$, we introduce the *multiplier functions* $\delta_z : \mathbb{C}^{\mathfrak{X}} \to \mathbb{C}^{\mathfrak{X}}$ given by

$$\delta_z f(x) := \frac{z + z^{-1} q_x}{1+q_x} f(x).$$

Then we have the following theorem.

Theorem 5.1 ([BHW23, Thm. 11.5, Cor. 9.4], [AFH23a, Thm. 4.4]) *Let $\mathfrak{G}$ be a finite graph satisfying the hypotheses explained above. Then the restriction $\mathcal{L}_{\mathrm{lc}}$ of $\mathcal{L}$ to $C^{\mathrm{lc}}(\mathfrak{P}_+)$ satisfies:*

$$\forall z \notin \{0, \pm 1\} : \quad \{\mathcal{L}_{\mathrm{lc}} = z\} \cong \{\Delta_{\mathfrak{X}} = \delta_z\},$$

where $\cong$ abbreviates an explicitly known isomorphism of vector spaces.

Note that while the algebraic equalizer $\{\mathcal{L}_{\mathrm{lc}} = z\}$ of $\mathcal{L}_{\mathrm{lc}}$ and $z \cdot \mathrm{id}$ on $C^{\mathrm{lc}}(\mathfrak{P}_+)$ is simply the eigenspace of $\mathcal{L}_{\mathrm{lc}}$ for the eigenvalue z, the algebraic equalizer $\{\Delta = \delta_z\}$ of Δ and the multiplication operator $f \mapsto \delta_z f$ is an eigenspace only if δ_z is a constant function which is the case if the graph is *homogeneous*, i.e., if the function $\mathfrak{X} \to \mathbb{N},\ x \mapsto q_x$ is constant.

While the isomorphism in Theorem 5.1 can be obtained directly by combinatorial considerations (see [BHW23]), it can also be obtained in a fashion parallel to the case of compact locally symmetric spaces (see [AFH23a]). The strategy is then to consider the tree $\widetilde{\mathfrak{G}} = (\widetilde{\mathfrak{X}}, \widetilde{\mathfrak{E}})$ which is obtained as the "universal covering" of $\mathfrak{G} = (\mathfrak{X}, \mathfrak{E})$ and the (scalar) Poisson transforms defined for that tree.

To make this precise, note first that symmetric graphs of the type we consider here are in one-to-one correspondence with undirected graphs having corresponding properties (connected, no loops, no dead ends). We simply replace pairs of directed edges $(x, y), (y, x) \in \mathfrak{X}^2$ by the corresponding two-element subset $\{x, y\} \subseteq \mathfrak{X}$ defining an undirected edge. In this way we can use concepts defined for undirected graphs also in our setting. For example, a connected undirected graph is called a *tree* if it does not contain any cycles. Given an undirected graph, we can construct a true universal covering which is a tree. Then we replace the undirected edges by pairs of opposite directed edges and thus arrive at $\widetilde{\mathfrak{G}} = (\widetilde{\mathfrak{X}}, \widetilde{\mathfrak{E}})$.

Recall that one can define the *boundary* of the tree as the set of equivalence classes of geodesic rays, where we call two geodesic rays equivalent if they have infinitely many edges in common. We call the boundary point defined from a geodesic ray its *limit* and denote the boundary of $\widetilde{\mathfrak{G}}$ by $\widetilde{\Omega}$. Then there exists a *horocycle bracket*:

$$\langle\cdot,\cdot\rangle : \widetilde{\mathfrak{X}} \times \widetilde{\Omega} \to \mathbb{Z}, \ (x, \omega) \mapsto d(o, y) - d(x, y)$$

where $o, x, y \in \widetilde{\mathfrak{X}}$ are such that the uniquely determined geodesic rays $[o, \omega[, [x, \omega[, [y, \omega[$ representing $\omega \in \widetilde{\Omega}$ and starting respectively at o, x, and y satisfy $[o, \omega[\cap[x, \omega[= [y, \omega[$. Then for any $z \in \mathbb{C} \setminus \{0\}$, we obtain a *Poisson kernel*:

$$\tilde{p}_z : \widetilde{\mathfrak{X}} \times \widetilde{\Omega} \to \mathbb{C}, \quad (x, \omega) \mapsto z^{\langle x,\omega\rangle}.$$

The boundary $\widetilde{\Omega}$ inherits a natural topology from the space of geodesic rays in $\widetilde{\mathfrak{G}}$, so it makes sense to talk about the space $C^{\mathrm{lc}}(\widetilde{\Omega})$ of locally constant functions on $\widetilde{\Omega}$. Following traditions in harmonic analysis on p-adic groups, we denote the *algebraic* dual space of $C^{\mathrm{lc}}(\widetilde{\Omega})$ by $\mathcal{D}'(\widetilde{\Omega})$. By [BHW22, Prop. 3.9] the space $\mathcal{D}'(\widetilde{\Omega})$ is naturally isomorphic to the space of finitely additive measures on $\widetilde{\Omega}$. Locally constant functions can be integrated against such measures, so that we obtain the *scalar Poisson transforms*:

$$\widetilde{\mathcal{P}}_z : \mathcal{D}'(\widetilde{\Omega}) \to \mathbb{C}^{\widetilde{\mathfrak{X}}}, \quad \mu \mapsto \int_{\widetilde{\Omega}} \tilde{p}_z(\cdot, \omega)\, \mathrm{d}\mu(\omega),$$

which take their values in the respective algebraic equalizer $\{\Delta_{\widetilde{\mathfrak{X}}} = \tilde{\delta}_z\}$. Here $\tilde{\delta}_z : \widetilde{\mathfrak{X}} \to \mathbb{C}$ is the analog of δ_z for the tree $\widetilde{\mathfrak{G}}$. In analogy to the symmetric space case, the Poisson transforms have natural factorizations of the form $\widetilde{\mathcal{P}}_z = \tilde{\pi}_* \circ \tilde{p}_\mu \circ \widetilde{B}^*$, where $\tilde{\pi}_* : \mathcal{D}'(\widetilde{\mathfrak{P}}_+) \to \mathrm{Maps}(\widetilde{\mathfrak{X}})'$ is the push-forward of the canonical projection:

$$\tilde{\pi} : \widetilde{\mathfrak{P}}_+ \to \widetilde{\mathfrak{X}}, \quad (\mathbf{e}_j)_{j\in\mathbb{N}} \mapsto \iota(\mathbf{e}_1),$$

$\tilde{p}_\mu$ is viewed as a multiplier, and $\widetilde{B}^*$ is the pullback by the *endpoint map* $\widetilde{B} : \widetilde{\mathfrak{P}}_+ \to \widetilde{\Omega}$ mapping a geodesic ray to its limit in the boundary. Then the strategy to prove the quantum-classical correspondence is as follows:

- Work with Γ-invariant lifts, where Γ is the group of deck transformations for the covering $\widetilde{\mathfrak{G}} \to \mathfrak{G}$.
- Apply the factorization for bijective $\mathcal{P}_z$, i.e., for $z \notin \{0, \pm 1\}$.
- Show that $\tilde{\pi}_*$, when restricted to Γ-invariant objects, defines the linear isomorphisms of the quantum-classical correspondence.

Apart from being parallel to the case of compact locally symmetric spaces, this strategy of proving a quantum-classical correspondence for finite graphs has the advantage of generalizing also to *exceptional spectral parameters*, i.e., to $z = \pm 1$, for which the scalar Poisson transform fails to be bijective. In fact, for $\mathbf{e} \in \widetilde{\Omega}$, we consider the set:

$$\partial_+\mathbf{e} := \{\omega \in \widetilde{\Omega} \mid \omega = \widetilde{B}(\mathbf{e}, \mathbf{e}_2, \ldots)\}$$

of all boundary points which are limits of geodesic rays starting with $\mathbf{e}$. Then for $z \in \mathbb{C} \setminus \{0\}$, we define the *edge Poisson kernel*:

$$\tilde{p}_z^{\mathrm{e}}(\mathbf{e}, \omega) := z^{\langle \iota(\mathbf{e}), \omega\rangle} I_{\partial_+\mathbf{e}}(\omega),$$

where $I_{\partial_+\mathbf{e}}$ is the indicator function of the set $\partial_+\mathbf{e}$. Integration against the edge Poisson kernel gives the *edge Poisson transform*:

$$\widetilde{\mathcal{P}}_z^{\mathrm{e}} : \mathcal{D}'(\widetilde{\Omega}) \to \mathbb{C}^{\widetilde{\mathfrak{E}}}, \quad \mu \mapsto \int_{\widetilde{\Omega}} \tilde{p}_z^{\mathrm{e}}(\cdot, \omega)\, \mathrm{d}\mu(\omega).$$

The images of the edge Poisson transforms are eigenspaces of the *edge Laplacian* $\Delta_{\widetilde{\mathfrak{E}}} : \mathbb{C}^{\widetilde{\mathfrak{E}}} \to \mathbb{C}^{\widetilde{\mathfrak{E}}}$ defined by

$$\Delta_{\widetilde{\mathfrak{E}}} f(\mathbf{e}) := \sum_{\iota(\mathbf{e}')=\tau(\mathbf{e}),\, \tau(\mathbf{e}')\neq\iota(\mathbf{e})} f(\mathbf{e}').$$

Lemma 5.2 ([AFH23a, Prop. 2.11 & 2.18]) *For $0 \neq z \in \mathbb{C}$, we have:*

(i) $\widetilde{\mathcal{P}}_z^{\mathrm{e}} = \tilde{\pi}_*^{\widetilde{\mathfrak{E}}} \circ \tilde{p}_z \circ \widetilde{B}^*$ *for* $\tilde{\pi}^{\widetilde{\mathfrak{E}}} : \widetilde{\mathfrak{P}}_+ \to \widetilde{\mathfrak{E}},\ (\mathbf{e}_j)_{j\in\mathbb{N}} \mapsto \mathbf{e}_1$.
(ii) $\widetilde{\mathcal{P}}_z^{\mathrm{e}} : \mathcal{D}'(\widetilde{\Omega}) \to \{\Delta_{\widetilde{\mathfrak{E}}} = z\}$ *is a linear isomorphism.*

Replacing $\mathcal{P}_z$ for $z \notin \{0, \pm 1\}$ in the above strategy by $\widetilde{\mathcal{P}}^{\mathrm{e}}_z$ for $z \neq 0$ then yields a quantum-classical correspondence for finite graphs which is valid also for exceptional parameters.

Theorem 5.3 ([AFH23a, Thm. 4.5]) *Let $\mathfrak{G}$ be a finite graph satisfying the hypotheses explained above. Then*

$$\forall z \neq 0: \quad \{\mathcal{L}_{\mathrm{lc}} = z\} \cong \{\Delta_{\mathfrak{E}} = z\},$$

where $\Delta_{\mathfrak{E}}$ is the edge Laplacian for $\mathfrak{G}$ and $\cong$ abbreviates an explicitly known isomorphism of vector spaces.

In the case of finite graphs, we can substantiate the expectation that eigenspaces for exceptional spectral values carry topological information as suggested by examples in the case of locally symmetric spaces (see Sect. 4.3.3).

Let $c(\mathfrak{G})$ the *cyclomatic number* of the finite graph $\mathfrak{G}$, i.e., the minimal number of edges which need to be removed from the graph in order to break all cycles.

Theorem 5.4 ([AFH23a, Theorems C & D]) *Let $\mathfrak{G}$ be a finite graph satisfying the hypotheses explained above:*

(i) $\dim\{\Delta_{\mathfrak{E}} = 1\} = \begin{cases} c(\mathfrak{G}) & \text{if } c(\mathfrak{G}) \neq 1, \\ c(\mathfrak{G}) + 1 & \text{if } c(\mathfrak{G}) = 1. \end{cases}$

(ii) *If $c(\mathfrak{G}) \neq 1$ or $\mathfrak{G}$ is not bipartite, then*

$$\dim\{\Delta_{\mathfrak{E}} = -1\} = \begin{cases} c(\mathfrak{G}) & \text{if } \mathfrak{G} \text{ is bipartite,} \\ c(\mathfrak{G}) - 1 & \text{if } \mathfrak{G} \text{ is not bipartite.} \end{cases}$$

(iii) *If $c(\mathfrak{G}) = 1$ and $\mathfrak{G}$ is bipartite, then* $\dim\{\Delta_{\mathfrak{E}} = -1\} = 2$.

5.2 *Affine Buildings*

Affine buildings of higher rank occur for instance as Bruhat-Tits buildings of p-adic algebraic groups beyond $\mathrm{SL}_2(\mathbb{K})$. In contrast to the case of trees, there are few affine buildings of higher rank which are *not* Bruhat-Tits buildings. Thus it makes sense to make use of group theory and harmonic analysis to study quantum and classical analogs of geodesic and Weyl chamber flows for higher rank affine buildings.

The quantum side is well established with Hecke algebras taking the place of invariant differential operators; see, e.g., [Ma71]. Geometric versions trying to avoid group theory appeared later; see, e.g., [Pa06]. They add new points of view but do not extend the scope of the theory a lot.

While there is some literature on higher rank Cartan actions (see, e.g., [Mo95]), analogs of Weyl chamber flows suitable to build quantum-classical correspondences in the spirit of [HWW21] do not seem to be available as of now. What is available are analogs of the representation theoretic tools used in the context of locally symmetric spaces. For instance S. Kato's papers [Ka81, Ka82] contain scalar Poisson transforms from unramified principal series representations to Hecke eigenspaces and a characterization of the spectral parameters for which these Poisson transforms are bijective. Geometric versions of Poisson transforms were studied by Mantero and Zappa (see [MZ17] and the references given there), mostly for rank two buildings.

Thus, in order to establish quantum-classical correspondences in the context of higher rank affine buildings, one first has to do some more foundational work on the underlying dynamical systems.

It should be noted, however, that there is related material that may become relevant in our context at some point, such as for instance the multivariate geometric zeta functions for higher rank p-adic groups Deitmar and Kang introduced in [DK14].

6 Open Problems

The existing results on quantum-classical correspondences for locally symmetric spaces were obtained with a variety of different methods and have very different degrees of generality. The most complete set of results is available for hyperbolic surfaces. They range from trivial and thin to cocompact fundamental groups. The methods employed range from symbolic dynamics and classical Fourier analysis to noncommutative harmonic analysis and microlocal analysis. The known extensions to higher dimension and higher rank work for different settings which depend on the methods employed. Trying to further extend the scope, one meets a number of difficulties. Some are more of a technical nature, such as explicitly determining composition series of principal series representations; some are truly conceptual such as finding a suitable replacement for resolvents when going from a single operator to a set of commuting operators.

In this section we formulate a number of concrete challenges that could be seen as possible next steps in realizing the program outlined in Sect. 1.2. The focus will be on problems in establishing quantum-classical correspondences via microlocal methods. Given the results described in the previous sections, there will be a number of problems asking for various types of generalizations such as:

- globally vs. locally symmetric spaces
- higher rank vs. rank one
- higher dimension vs. surfaces
- Γ general vs. Γ thin or Γ cocompact
- Orbits or parameters general vs. orbits or parameters generic (regular)

There are also some problems asking for clarification of the relation between the methods and results laid out here with the ones contained in previous work.

6.1 Quantum Resonances in Higher Rank

The Laplacian of a noncompact Riemannian symmetric space G/K can be diagonalized by the non-Euclidean Fourier transform. Thus its resolvent can be written as an explicit integral operator with the integrand depending on a complex parameter. For spaces of low rank (actually 1 and 2), this representation can be used to construct meromorphic continuations via suitable deformations of contours, [HP09, HPP16, HPP17a, HPP17b]. Already for rank 2 spaces, the construction of suitably deformed contours turns out to be quite tricky with no obvious pattern of generalization.

Problem 6.1 (Resolvent Poles for Laplacians) Extend the method used in [HP09, HPP16, HPP17a, HPP17b] to determine quantum resonances for the Laplacian of symmetric spaces of arbitrary rank (cf. Sect. 4.1.2):

(i) Find a systematic way to do the contour deformations working in arbitrary rank.[2]
(ii) Describe the poles of the resulting meromorphic continuation, i.e., the quantum resonances of the Laplacian.
(iii) Determine the G-representations obtained from the residue operators at the quantum resonances.

In view of the Plancherel decomposition of the regular representation of G on $L^2(G/K)$, it seems plausible that one should aim for quantum resonances which are spectral invariants not just for the Laplacian, but rather for the entire commutative algebra $\mathbb{D}(G/K)$ of invariant differential operators. In rank one this requires no essential change since then $\mathbb{D}(G/K)$ is generated by the Laplacian. If the rank is ℓ, the algebra $\mathbb{D}(G/K)$ is isomorphic a polynomial algebra in ℓ variables due to results of Harish-Chandra and Chevalley; see Remark 2.8. In fact, one can still use the non-Euclidean Fourier transform to diagonalize $\mathbb{D}(G/K)$ simultaneously. The resulting eigenvalues are provided by the Harish-Chandra isomorphism, i.e., the spectral parameters are elements of $\mathfrak{a}^*_{\mathbb{C}}$.

Problem 6.2 (Resolvent Cohomology Classes for Invariant Differential Operators) Modify the definition of the Taylor spectrum so that it applies to the commuting family $\mathbb{D}(G/K)$ of differential operators. A good starting point might be $G = \mathrm{SL}_3(\mathbb{R})$, since in that case $\mathbb{D}(G/K)$ can be described very explicitly; see [BCH21]:

(i) Use the Harish-Chandra isomorphism to express the Taylor spectrum of $\mathbb{D}(G/K)$ in a way independent of a choice of generators.

[2] Note that Mazzeo and Vasy in [MV05] studied analytic continuations of the resolvent of the Laplacian in arbitrary rank. They do, however, not solve the problem formulated here, since they do not go around possible singularities but introduce cuts and branched coverings wherever a pole might occur.

(ii) Follow [Sa03, § 2] to construct a resolvent cohomology class for $\mathbb{D}(G/K)$ depending on a parameter in $\mathfrak{a}_{\mathbb{C}}^*$.
(iii) Work out the G-invariance and equivariance properties of the spaces involved in the construction of the resolvent class.
(iv) Build a framework in which one can talk about meromorphic continuation of the resolvent class and show that the resolvent class has meromorphic continuation to all of $\mathfrak{a}_{\mathbb{C}}^*$.
(v) Determine the divisor of the meromorphic continuation of the resolvent class.
(vi) Follow the methods of [GH78, Chap. 5] to calculate suitable residues of the meromorphically continued resolvent classes.
(vii) Give the residues from (vi) a representation theoretic interpretation.

There are other conceivable ways to describe resolvents of (commuting) families of operators and their meromorphic continuations.

Problem 6.3 (Multivariate Resolvents via Laplace Transform) Take the description of the resolvent of a single operator A via the Laplace transform of the one-parameter semigroup $S(t)$ generated by A (cf. Example 1.6 and [La02, § 34.1]) as a model and construct higher rank resolvents as multivariate Laplace transforms. Then study meromorphic continuations and residues.

The view that the "correct" approach to describe quantum resonances for symmetric spaces is to study the spectral theory not only of the Laplacian but the simultaneous spectral theory of $\mathbb{D}(G/K)$ is supported by the relation between resonances and scattering poles predicted by physics. In fact, following [STS76] one may view the *standard intertwining operators* for spherical principal series representations as scattering matrices for the multitemporal wave equation:

$$\forall D \in \mathbb{D}(G/K): \quad D_x u(tx) = \Gamma(D)_{\partial t} u(t, x)$$

on the symmetric space coming from the Harish-Chandra isomorphism HC (see Sect. 2.1.2) by letting $\Gamma(D)_{\partial t}$ be the constant coefficient differential operator with symbol $HC(D)$. The standard intertwining operators depend on parameters in $\mathfrak{a}_{\mathbb{C}}^*$, and it is well-known (see, e.g., [Wa92, Chap. 10]) that they have meromorphic continuation to all of $\mathfrak{a}_{\mathbb{C}}^*$.

Problem 6.4 (Resolvent Poles vs. Scattering Poles for Higher Rank Symmetric Spaces) Extend the results of [HHP19] (see Theorem 4.3) to higher rank using standard intertwining operators as scattering matrices and the resolvent poles of Problems 6.2 or 6.3 as quantum resonances.

In [BO12, § 6] Bunke and Olbrich study meromorphic continuations of resolvent kernels for the Laplacian on convex cocompact locally symmetric spaces. In the process they compare the resolvent kernel of the locally symmetric space with the resolvent kernel of the corresponding symmetric space. Moreover, they make use of the scattering matrix for convex cocompact locally symmetric spaces they had defined in [BO99, BO00].

Problem 6.5 (Resolvent Poles vs. Scattering Poles for Convex Cocompact Locally Symmetric Spaces) Use the techniques of [BO99, BO00, BO12] to extend the results of [HHP19] to convex cocompact locally symmetric spaces of rank 1.

Convex cocompact locally symmetric spaces are rare in higher rank (basically there are only products of rank 1 spaces). As a replacement various authors (see, e.g., [GW12, KLP18]) have studied *Anosov representations* of discrete groups.

Problem 6.6 (Bunke-Olbrich for Higher Rank) Generalize the techniques of [BO99, BO00, BO12] to locally symmetric spaces whose fundamental group is the image of an Anosov representation.

A detailed scattering theory for locally symmetric spaces generalizing [STS76] would be of interest in the study of automorphic forms; see, e.g., [LP76] for the case of hyperbolic surfaces. In [JZ01] Ji and Zworski take up the subject for locally symmetric spaces of $\mathbb{Q}$-rank 1.

Problem 6.7 (Automorphic Scattering in Higher Rank) Extend the scattering theory of Semenov-Tian-Shansky laid out in [STS76] for symmetric spaces to locally symmetric spaces (see [LP76] for the case of hyperbolic surfaces).

6.2 *Ruelle-Taylor Resonances*

The obvious problems in this context are:

- Construct Ruelle-Taylor resonances for the A-action from Remark 2.7 for any locally symmetric space $\mathcal{X} = \Gamma\backslash G/K$.
- Show that these Ruelle-Taylor resonances have a natural band structure.
- Establish a quantum-classical correspondence for the corresponding resonant states.

Attacking these problems head-on is way too ambitious for the moment. We formulate some more modest problems.

Problem 6.8 (Band Structure of Ruelle-Taylor Resonances) Give a precise description of the band structure of Ruelle-Taylor resonances in the spirit of [DFG15] (cf. the comments in Remark 4.16).

Problem 6.9 (Quantum-Classical Correspondences for Real Hyperbolic Manifolds) As mentioned in Remarks 4.10 and 4.11, Bonthonneau and Weich constructed Ruelle resonances for Riemannian manifolds with finite volume, negative curvature and the unbounded part consisting of finitely many real hyperbolic cusps, whereas Hadfield constructed Ruelle resonances for convex cocompact real hyperbolic manifolds; see [BW22, Ha17]. Moreover, in [Ha20] Hadfield established quantum-classical correspondences for the latter manifolds:

(i) Construct Ruelle resonances for general real hyperbolic manifolds.
(ii) Extend Hadfield's quantum-classical correspondences to general real hyperbolic manifolds.

Problem 6.10 (Ruelle Resonances for Convex Cocompact Rank One Locally Symmetric Spaces) Construct Ruelle resonances for general convex cocompact rank one locally symmetric spaces.

Problem 6.11 (Ruelle-Taylor Resonances for Thin Γ in Higher Rank) Construct Ruelle-Taylor resonances for locally symmetric spaces with fundamental group in the image of an Anosov representation.

6.3 *Dynamical Zeta Functions, Γ-Cohomology, and Transfer Operators*

Following pioneering work of Patterson [Pa89] and Juhl [Ju93, Ju01], in the late 1990s Bunke and Olbrich as well as Patterson and Perry established links between the dynamical zeta functions and Γ-cohomology with coefficients in G-representations (see Sects. 2.2.3 and 4.2.4 and [BO99, PP99, BO00, BO12]).

The focus of Juhl's books [Ju93, Ju01] is on dynamical and Selberg zeta functions for the geodesic flow on compact hyperbolic manifolds. In [Ju01, Chap. 3] Juhl establishes a Lefschetz-type formula, where the geometric side is expressed in terms of the contractive part of the Poincaré map of closed geodesics and the spectral side is expressed in terms of the $\mathfrak{n}$-cohomologies of the irreducible components of $L^2(\Gamma\backslash G)$. As an application he gets formulas for the divisors of (twisted) Selberg zeta functions in terms of $\mathfrak{n}$-cohomologies, which he then rephrases in terms of Γ-cohomologies with coefficients in principal series representations. To that end he uses de Rham-type complexes to compute such Γ-cohomologies. The resulting characterization of the zeta divisors in terms of the spectral decomposition of $L^2(\Gamma\backslash G)$ obscures the dynamical meaning, so Juhl sets out in [Ju01, Chap. 4] to develop a kind of Hodge theory for the sphere bundle $S(\Gamma\backslash G) = \Gamma\backslash G/M = \mathcal{M}$ using the Anosov property of the geodesic flow. It is in the detailed description of the complexes and operators on $\mathcal{M}$ given in [Ju01, Chap. 5] that Poisson transforms make their appearance.

The published work of Bunke and Olbrich exclusively deals with locally symmetric spaces of rank one. It in particular describes zeros of dynamical zeta functions in terms of Γ-cohomology with coefficients in spaces of distributions on the boundary (see, e.g., [BO99, Thm. 1.3]), which again might be viewed as a result purely on the classical side. But even more than Juhl, Bunke and Olbrich make extensive use not only of the Poisson transform, but also of the scattering matrix in establishing such results.

So far the wealth of information Juhl, Bunke, and Olbrich have generated in this context has not found its way into the efforts of the more PDE and dynamical systems oriented communities in spectral geometry.[3] One reason for this may be that they use heavy group and representation theoretic machinery as well as results originating in the algebraic analysis in the spirit of [SKK73] which to this day remains to be presented in a form accessible to a wider mathematical audience.

Problem 6.12 (Relation with the Works of Juhl and Bunke-Olbrich) Work out the precise relation between the cohomological methods of Juhl and Bunke-Olbrich on the one side and the dynamical respectively microlocal methods leading to resonances on the other side. A starting point could be to compare the descriptions of the zeta divisors given in [BO99, § 5] and [GHW18, § 5] (the latter is only for surfaces).

Considering the Weyl chamber action as the suitable analog of the geodesic flow, the dynamical zeta functions should count compact A-orbits (see [Mo70, Mo73] for results on the frequency of such orbits). One may follow various approaches:

1. Count according to volume.
   This leads to one-variable zeta functions (see [De00, DKV79] for background).
2. Count according to the geometry.
   A compact A-orbit $c = \xi \cdot A$ is diffeomorphic to a torus and the stabilizer of a point is a lattice in A. Let $\ell(c) = \big(\ell_1(c), \ldots, \ell_r(c)\big)$ be successive length minima of the lattice (use the Killing form to fix a metric on A). Then one might define:

$$Z_{\mathrm{R}}(\lambda) = \prod_c \big(1 - e^{-\lambda \ell(c)}\big)^{\pm 1}$$

   for $\lambda \in \mathbb{C}^r$. In order to generalize the other zeta functions from Sect. 4.2.4, one needs to have a reasonable Poincaré map. Given a lattice $L \subseteq A$, one might try to take the following steps:

   (a) Lift the action to $T(\Gamma\backslash G/M)$.
   (b) Define a Poincaré map for each element of L.
   (c) For $a \in L$ let $\xi \cdot a = \xi \in \Gamma g M$. Then there exist $\gamma \in \Gamma$ and $m \in M$ such that $gam = \gamma g$ and right multiplication ρ_{am} by am yields the identification of ρ_{am} with $\mathrm{Ad}(am)$. Thus the monodromy on $T(\Gamma\backslash G/M) = \Gamma\backslash G \times_M (\mathfrak{n} + \mathfrak{a} + \overline{\mathfrak{n}})$ is given by $\mathrm{Ad}(am)$. Then follow [BO95, De00].

3. Write the zeta function as a Fredholm determinant of a suitable (transfer) operator.
   This works fine for subshift dynamics and requires symbolic dynamics in the case of geodesic flows (known for specific hyperbolic surfaces; see [Po14, Po16]).

[3] Neither has it been taken up a lot in the literature on noncommutative harmonic analysis and representation theory.

Note that the possibility to write zeta functions as (Fredholm) determinants is the reason why symbolic dynamics is used in the effort to construct meromorphic extensions of dynamical zeta functions.

The following three problems take up aspects of the approach to dynamical zeta functions for Weyl chamber flows just outlined.

Problem 6.13 (Geometric Invariants for A-orbits) Find other geometric invariants of compact A-orbits according to which one can count. Is it possible to use the restriction of the Sasaki metric on $\Gamma\backslash G/M \subseteq T(\Gamma\backslash G/K)$ to the orbit to construct suitable invariants?

Problem 6.14 (Multivariate Dynamical Zeta Functions) Work out a basic theory of multivariate dynamical zeta functions:

(a) for flows and torus orbits
(b) for higher rank subshifts of finite type

Problem 6.15 (Symbolic Dynamics for Weyl Chamber Flows) Construct symbolic dynamics for Weyl chamber flows in rank 2 examples. The product of Schottky surfaces is dealt with in [Po20]. One could start with $\mathrm{SL}_3(\mathbb{Z})\backslash\mathrm{SL}_3(\mathbb{R})/\mathrm{SO}_3$.

The treatment of the transfer operator for the modular surface by Chang and Mayer in [CM99] shows a clear connection with the hyperbolic nature of the geodesic flow. This is a lot less transparent for the transfer operators derived from the symbolic dynamics other discrete subgroups of $\mathrm{PSL}_2(\mathbb{R})$.

Problem 6.16 (Hyperbolic Nature of Pohl's Symbolic Dynamics) Make the connection between Pohl's reduced symbolic dynamics for the geodesic flow and the hyperbolic nature of the geodesic flow explicit (cf. the discussion in Sect. 4.2.5).

The next problem is suggested by the similarities between Maass and modular forms.

Problem 6.17 (Quantum Interpretation of Modular Cusp Form) Give a reasonable quantum system having modular cusp forms in the sense of Example 1.3 as states and such that it quantizes a classical system to which one can naturally associate period polynomials.

The last problem in this section describes a strategy to extend the Lewis-Zagier correspondence to higher rank with Remark 4.19 as point of departure.

Problem 6.18 (Cuspidal Cohomology)

(i) Use the multiplicity space from Remark 4.19 and K-invariant vectors of spherical principal series representations to relate the cuspidal Γ-cohomology spaces for spherical principal series representations to Maass cusp forms.
(ii) Follow the isomorphisms proving Theorem 2.14 through to construct a natural isomorphism of vector spaces between Maass cusp forms and cuspidal Γ-cohomology spaces.

(iii) Clarify the relation between the various parabolic Γ-cohomologies from [BLZ15] and cuspidal Γ-cohomology in the case of $G = \mathrm{PSL}_2(\mathbb{R})$.
(iv) Give conditions under which cuspidal Γ-cohomology and parabolic Γ-cohomology agree.

6.4 Horocycle Flow

Let $\mathcal{X} = \Gamma\backslash G/K$ be a locally symmetric space of finite volume. The *horocycle flow* of N on $\Gamma\backslash G$ is given by right multiplication. In contrast to the Weyl chamber flow, it does not descend to $\Gamma\backslash G/M$. Following [FF03] (see Example 1.5), one can define the space $\mathcal{J}(\Gamma\backslash G)$ of compactly supported distributions on $\Gamma\backslash G$ which are invariant under the horocycle flow. In addition we consider the subspace $\mathcal{J}(\Gamma\backslash G)^M$ of M-invariant distributions in $\mathcal{J}(\Gamma\backslash G)$. This amounts to taking MN-invariant distributions on $\Gamma\backslash G$.

Problem 6.19 Check the following claims:

(i) $\mathcal{J}(\Gamma\backslash G)$ is invariant under the natural MA-action on the distributions on $\Gamma\backslash G$ coming from the right MA-action on $\Gamma\backslash G$.
(ii) $\mathcal{J}(\Gamma\backslash G)$ and $\mathcal{J}(\Gamma\backslash G)^M$ have compatible direct integral decompositions coming from the right regular representation of G on $C_c^\infty(\Gamma\backslash G)$.
(iii) The summands in the decompositions from (ii) are invariant under the MA-action.

Problem 6.20

(i) Determine the decompositions from Problem 6.19(ii) in terms of conical distributions, [He94, § II.4], and describe the MA-actions on summands.
(ii) Derive a quantum-classical correspondence for $\mathcal{X}$ from (i) and compare the result in the cocompact case with [HWW21, Thm. 5.1].

It may very well turn out that instead of just considering conical distributions in Problem 6.20(ii), one will have to consider M-finite vectors in the space of N-invariant distributions as in [Em14]. Also one should be looking for a connection with the $\mathfrak{n}$-cohomology techniques applied by Juhl and Bunke-Olbrich (cf. the comments in Sect. 6.3).

6.5 Patterson-Sullivan Distributions

In [GHW21, Thm. 5.2] it is shown that the Patterson-Sullivan distributions on compact rank one locally symmetric spaces can be easily obtained from the intertwiners between Laplace eigenfunctions and the generalized Ruelle resonant states from the first band. Of course one would like to have a higher rank generalization.

Problem 6.21 Use the first band Ruelle resonant states in compact locally symmetric spaces to generalize [GHW21, Thm. 5.2] to higher rank.

To generalize the realizations of Patterson-Sullivan distributions as residues of their zeta functions from hyperbolic surfaces to rank one locally symmetric spaces turned out to be a nontrivial task as neither the symbolic dynamics nor the weighted trace formula used in [AZ07] have immediate generalizations in higher dimension. For compact hyperbolic manifolds, steps in this direction have been taken in [Em14]. As was mentioned in Sect. 4.3.4, the approach taken in [SWB23] using [GHW21] and [DZ16] was actually more successful.

Problem 6.22 Extend the results of [SWB23] to general Riemannian locally symmetric spaces.

Even though the microlocal approach turned out to be more efficient in the problem at hand, there remain open questions about the harmonic analysis approach taken in [AZ07] and [Em14].

Anantharaman and Zelditch [AZ07] make use of a pseudo-differential calculus due to Zelditch (see [Ze84]) on the basis of the non-Euclidean Fourier-Helgason transform. Although lifted quantum limits do not depend on the specific pseudo-differential calculus chosen for their definition, it is useful, for establishing invariance properties, to have an equivariant calculus. For hyperbolic surfaces, based on the non-Euclidean Fourier analysis and closely following the Euclidean model, such a calculus was provided by Zelditch [Ze84]. In [Sch10] this calculus was extended to rank one symmetric spaces. Using this calculus the construction of the Patterson-Sullivan distributions and the proof of the asymptotic equivalence from [AZ07] can be generalized. However, due to singularities arising from Weyl group invariance, it is difficult to construct an equivariant non-Euclidean pseudo-differential calculus in higher rank. Silberman and Venkatesh [SV07], generalizing work of Zelditch and Wolpert for surfaces to compact locally symmetric spaces, introduced a representation theoretic lift as a replacement for a microlocal lift. They sketch, in [SV07, Remark 1.7(4) and §5.4], a proof that the representation theoretic lift asymptotically gives the same result as a microlocal lift using pseudo-differential operators.

If one had a formula intertwining Patterson-Sullivan distributions PS_h^Γ into lifted Wigner distributions W_h, one might be able to deduce (27) as a corollary. Presumably, an intertwining formula holds only for special pseudo-differential calculi.

Problem 6.23 Establish a non-Euclidean pseudo-differential calculus for general Riemannian symmetric spaces of noncompact type. First steps were done in [Sch10], but the problems encountered there lead to the use of a different geometric calculus in [HHS12].

Problem 6.24 Silberman and Venkatesh introduced a *representation theoretic lift* as an alternative to the microlocal lift in proving invariance properties of limit measures; see [SV07, Si15] and [BO06]. As pointed out in [Si15, § 5], the representation theoretic lift works for finite volume and does not need compactness of the locally symmetric space:

(i) Work out the argument for the equivalence of microlocal and representation theoretic lifts sketched in [Si15, § 5.2] for compact locally symmetric spaces.
(ii) Extend the theory of Patterson-Sullivan distributions to locally symmetric spaces of finite volume.

6.6 Singular Situations

We have discussed several singular situations in this article. The first was singular G-orbits $G \cdot [\mathbf{1}, \xi] \cong G/K_\xi$ in $T^*(G/K)$ which results in the right A_ξ-action on G/K_ξ, see Remark 3.5 and [Si15]. This action descends to $\mathcal{M}_\xi := \Gamma\backslash G/K_\xi$. Theorem 3.4 provides this action with a symplectic, i.e. classical, interpretation.

Problem 6.25 (Singular Generalized Energy Shells) Extend the various results presented in this article for the regular generalized energy shells $\mathcal{M}$ to singular generalized energy shells $\mathcal{M}_\xi$.

The second kind of singular situation we have encountered are exceptional spectral parameters, for which Poisson transforms fail to be bijective. These exceptional parameters required separate treatment in the search for quantum-classical correspondences; see Sect. 4.3.3. Some examples considered there suggest that the multiplicities of exceptional Ruelle resonances have topological interpretations.

Problem 6.26 (Topological Interpretation of Exceptional Multiplicities) Give a topological interpretation of $\dim({}^\Gamma(\mathrm{soc}(H_\mu))^{-\infty})$ for the exceptional spectral parameters μ (see Theorem 4.20 and Remark 4.23).

In view of the cohomological constructions done in [TW89], the observation from Theorem 4.24 that the representations associated with the exceptional spectral parameters are discrete series representation of associated non-Riemannian symmetric spaces, may be helpful in attacking this problem.

All the results described in Sect. 4.3.3 are restricted to rank one spaces. One reason for this was that in [AH23], we needed explicit information on the spherical principal series representations involved. So the first problem to solve when trying to extend the results from [AH23] to higher rank is purely representation theoretic.

Problem 6.27 (Spherical Principal Series for Exceptional Parameters) Let π_λ be a spherical principal series representation of G with exceptional spectral parameter λ:

(i) It is known that π_λ is not irreducible. Determine the composition series and the socle of π_λ.
(ii) Determine the kernel of the scalar Poisson transform $\mathcal{P}_\lambda$.
(iii) Determine the minimal K-types of π_λ.
(iv) Show that the vector valued Poisson transform $\mathcal{P}_\lambda^{\mathrm{v}}$ associated with λ and the sum of all minimal K-types is injective on the kernel of $\mathcal{P}_\lambda$.
(v) Show that $\mathcal{P}_\lambda^{\mathrm{v}}$ is injective on the socle of π_λ.
(vi) Determine the image of the socle of π_λ under $\mathcal{P}_\lambda^{\mathrm{v}}$.

In Problem 6.27 the discrete subgroup Γ of G does not appear. In the next problem, it does and we do not specify which further assumptions one should make. Probably the simplest case is to choose Γ cocompact.

Problem 6.28 (Γ-Invariant Distribution Vectors in Spherical Principal Series Representations) Let π_λ be a spherical principal series representation of G with exceptional spectral parameter λ:

(i) Show that all Γ-invariant distribution vectors for π_λ are in the socle of π_λ.
(ii) Show that the vector valued Poisson transform $\mathcal{P}_\lambda^{\mathrm{v}}$ associated with λ and the sum of all minimal K-types is injective on the socle of π_λ.
(iii) Determine the image of the space of Γ-invariant distribution vectors for π_λ under $\mathcal{P}_\lambda^{\mathrm{v}}$.

When we discussed Patterson-Sullivan distributions in Sect. 4.1.4, we saw that the asymptotic equivalence of lifted quantum limits and Patterson-Sullivan distributions depended on the lifted quantum limit being regular.

Problem 6.29 (Singular Lifted Quantum Limits) Extend [HHS12, Thm. 7.4] to singular lifted quantum limits.

The final problem we want to state here showed up implicitly in many places in this article. It is related to the boundary values of functions on a locally symmetric space which occur for instance in the inverse of regular Poisson transforms. The most general construction of these boundary value maps has been given in [K+78], which depends massively on [SKK73]. While the main result of [K+78], the solution of the so-called Helgason conjecture characterizing the bijectivity of the Poisson transform $\mathcal{P}_\lambda$ in terms of λ, has been reproved several times in ways independent of [SKK73], except for the rank one case (done already by Helgason; see also [HHP19]), there is still no elementary construction of the boundary value map (see [HHP18] for a discussion and a PDE construction of boundary value maps for generic parameters).

Problem 6.30 (Boundary Value Maps) Give a PDE-construction of the boundary value maps from [K+78].

6.7 Graphs and Buildings

So far the quantum-classical correspondences established for graphs (see [BHW23, AFH23a]) deal only with finite graphs. One reason is that resonances for the graph Laplacians have not been studied so far.

Problem 6.31 (Quantum Resonances for Infinite Graphs) Establish meromorphic continuation of the resolvent of graph Laplacians for infinite graphs and determine its poles (resonances) and residue operators.

The geodesic flow acting on geodesic rays is only one possible analog of the geodesic flow on Riemannian manifolds. Another possibility would be to consider paths $(\mathbf{e}_j)_{j\in\mathbb{Z}}$ as *geodesic lines*. In that case the geodesic flow is generated by the two-sided shift. The analogy to the classical side of the quantum-classical correpondence for locally symmetric spaces suggests that one has to find a way to formulate hyperbolic properties of the shift dynamics on the space of geodesic lines. Note here that the use of geodesic rays and the transfer operator rather mimic the procedure given in [CM99] for the modular surface, which throws out the stable part of the dynamics altogether and only keeps the unstable part.

Problem 6.32 (Hyperbolicity of Geodesic Flows on Graphs) Formulate hyperbolicity properties of the two-sided shift on the space of geodesic lines in a graph.

A possible approach to this question is to build stable and unstable foliations using horocycles of the covering tree. Using these one might try to construct anisotropic Sobolev spaces analogous to the ones used for locally symmetric spaces (see Sect. 4.2.1).

Further goals could be to introduce resolvents of the shift operator and prove meromorphic continuation. The residues could then be used to introduce graph analogs of *invariant Ruelle densities*, which in the case of compact rank one locally symmetric spaces occur as flat traces of the residue operators of Ruelle resonances (cf. [GHW21, § 2]).

The space of geodesic lines just considered came up already in [Ah88], where one also finds a tree version of the Radon transform. Recall from Sect. 6.5 that for compact rank one locally symmetric spaces, Ruelle resonant states could be used to describe Patterson-Sullivan distributions. In that case they appear as products of distributions on the space of geodesic lines. All this suggests that it is possible to construct Patterson-Sullivan distributions for finite graphs.

Problem 6.33 (Patterson-Sullivan Distributions for Finite Graphs) Build a theory of Patterson-Sullivan distributions for finite graphs.

In particular, it should be clarified how the Patterson-Sullivan distributions are related to eigenfunctions of the transfer operator and invariant Ruelle densities (cf. Problem 6.32). Moreover, one would want an analog of the residue formula [SWB23, Thm. 4.1] connecting Patterson-Sullivan distributions with dynamical zeta functions. Note here that a graph analog of the pairing formula [GHW21,

Thm. 6.1] mentioned in Sect. 4.3.4 has already been established in [AFH23b] for homogeneous graphs.

As was mentioned already in Sect. 5.2, we are lacking an analog of Weyl chamber flows for affine buildings:

Problem 6.34 (Weyl Chamber Flows for Affine Buildings) Establish a theory of Weyl chamber flows for affine buildings.

(i) Use these Weyl chamber flows to prove quantum-classical correspondences for compact quotients of affine buildings in the spirit of [HWW21].
(ii) Define multivariate dynamical zeta functions for these Weyl chamber flows and, for Bruhat-Tits buildings of p-adic groups of $\mathbb{Q}$-rank 1, relate them to the geometric multivariate zeta functions of Deitmar and Kang, [DK14].

Of course, at this point one may repeat all the questions and problems listed for the real case also for the p-adic situation. In fact, one could even reformulate many of them in an adelic context. We refrain from doing this but emphasize the point that the reduced analytic complexity of the non-Archimedean context may allow us to find solutions for the conceptual problems around higher rank resolvents that can then be transferred to the Archimedean context.

Acknowledgments While this article does not explicitly refer to the work of Toshiyuki Kobayashi, it still owes a lot to him. His broad way of looking at representation theory with an open mind enabling him to merge algebraic, analytic, and geometric methods such as derived functor modules, microlocal analysis, and complex geometry has always been a shining example for me.

As for specific mathematical insights pertaining to the subject matter of this article, I thank my coauthors for patiently sharing their knowledge with me and enabling me to attack problems with tools I could not have handled by myself.

This work was partially funded by the Deutsche Forschungsgemeinschaft (DFG, German Research Foundation)–Project-ID 491392403—TRR 358.

References

[AM87] Abraham, R., Marsden, J.E.. Foundations of Mechanics, 2nd edn. Addison Wesley, Redwood City (1987)

[Ah88] Ahumada Bustamante, G.: Analyse harmonique sur l'espace des chemins d'un arbre. Dissertation 1988. Université de Paris-Sud, Orsay (1988)

[An17] Anantharaman, N.: Some Relations between the Spectra of Simple and Non-backtracking Random Walks. arXiv:1703.03852 (2017)

[AL15] Anantharaman, N., Le Masson, E.: Quantum ergodicity on large regular graphs. Duke Math. J. **164**, 723–765 (2015)

[AS13] Anantharaman, N., Silberman, L.: A Haar Component for Quantum Limits in Locally Symmetric Spaces. Israel Math. J. **195**, 393–447 (2013)

[AZ07] Anantharaman, N., Zelditch, S.: Patterson-Sullivan Distributions and Quantum Ergodicity. Ann. Henri Poincaré **8**, 361–426 (2007)

[AFH23a] Arends, C., Frahm, J., Hilgert, J.: Edge Laplacians and vector valued Poisson transforms for graphs. arxiv:2312.09101 (2023)

[AFH23b] Arends, C., Frahm, J., Hilgert, J.: A pairing formula for resonant states on finite regular graphs. arxiv:2312.10509 (2023)
[AH23] Arends, C., Hilgert, J.: Spectral correspondences for rank 1 locally symmetric spaces—the case of exceptional parameters. J. de l'École polytechnique—Mathématiques **10**, 335–403 (2023)
[Ar79] Arthur, J.: Eisenstein series and the trace formula. In: Automorphic Forms, Representations, and L-Functions". Proceedings of Symposia in Pure Mathematics XXXIII , pp. 253–274 (1979)
[BSW22] Barkhofen, S., Schütte, P., Weich, T.: Semiclassical formulae for Wigner distributions. J. Phys. A: Math. Theor. **55**, 244007 (2022)
[BW22] Bonthonneau, Y., Weich, T.: Ruelle-Pollicott resonances for manifolds with cusps. J. EMS **24**, 851–923 (2022)
[BS73] Borel, A., Serre, J.P.: Corners and Arithmetic Groups. Comment. Math. Helv. **48**, 436–491 (1973)
[BW00] Borel, A., Wallach, N.: Continuous Cohomology, Discrete Groups, and Representations of Reductive Groups, 2nd edn. American Mathematical Society, New York (2000)
[BP02] Borthwick, D., Perry, P.: Scattering poles for asymptotically hyperbolic manifolds. Trans. Am. Math. Soc. **354**, 1215–1231 (2002)
[Bou04] Bourbaki, N.: Integration II. Chapters 7–9. Springer, Berlin (2004)
[BCH21] Brennecken, D., Ciardo, L., Hilgert, J.: Algebraically Independent Generators for the Algebra of Invariant Differential operators on $\mathrm{SL}_n(\mathbb{R})/\mathrm{SO}_n(\mathbb{R})$. J. Lie Theory **31**, 459–468 (2021)
[BLZ15] Bruggeman, R., Lewis, J., Zagier, D.: Period functions for Maass wave forms and cohomology. Mem. Am. Math. Soc. **237**, xii + 132 pp. (2015)
[BT72] Bruhat, F., Tits, J.: Groupes réductifs sur un corps local. Inst. Hautes Études Sci. Publ. Math. **41**, 5–251 (1972)
[BO95] Bunke, U., Olbrich, M.: Selberg Zeta and Theta Functions. Akademie Verlag, Berlin (1995)
[BO99] Bunke, U., Olbrich, M.: Group cohomology and the singularities of the Selberg zeta function associated to a Kleinian group. Annals Math. **149**, 627–689 (1999)
[BO00] Bunke, U., Olbrich, M.: The spectrum of Kleinian manifolds. J. Funct. Anal. **172**, 76–164 (2000)
[BO06] Bunke, U., Olbrich, M.: On Quantum Ergodicity for Vector Bundles. Acta Appl. Math. **90**, 19–41 (2006)
[BO12] Bunke, U., Olbrich, M.: Towards the trace formula for convex-cocompact groups. In: Blomer, V., Mihăilescu, P. (eds.) Contributions in Analytic and Algebraic Number Theory. Springer Proceedings of the Mathematical, vol. 9, pp. 97–148. Springer, New York (2012)
[BL07] Butterley, O., Liverani, C.: Smooth Anosov flows: correlation spectra and stability. J. Mod. Dyn. **1**, 301–322 (2007)
[BHW22] Bux, K.-U., Hilgert, J., Weich, T.: Poisson transforms for trees of bounded degree. J. Spectr. Theory **12**, 659–681 (2022)
[BHW23] Bux, K.-U., Hilgert, J., Weich, T.: Spectral correspondences for finite graphs without dead ends. Indagationes Math. (in press)
[Ca73] Cartier, P.: Géométrie et analyse sur les arbres. In: Séminaires Bourbaki. LNM, vol. 407, pp. 123–140 (1973)
[CDDP22] Cekić, M., Delarue, B., Dyatlov, S., Paternain, G.: The Ruelle zeta function at zero for nearly hyperbolic 3-manifolds. Invent. math. **229**, 303–394 (2022)
[CM99] Chang, C.H., Mayer, D.: The transfer operator approach to Selberg's zeta function and modular and Maass wave forms for $\mathrm{PSL}(2, \mathbb{Z})$. In: Hejhal, D., et al. (eds.) Emerging Applications of Number Theory, pp. 73–141. Springer, New York (1999)

[Co79] Cowling, M.: Sur les coefficients des representations unitaires des groupes de Lie simples. In: Eymard, P. et al. (eds.) Analyse Harmonique sur les Groupes de Lie II, pp. 132–178. Springer, Berlin (1979)
[Da82] Dadok, J.: On the C^{∞} Chevalley's Theorem. Advances Math. **44**, 121–131 (1982)
[DGRS20] Dang, N.V., Guillarmou, C., Rivière, G., Shen, S.: The Fried coonjecture in small dimensions. Invent. math. **220**, 525–579 (2020)
[De00] Deitmar, A.: Geometric zeta functions of locally symmetric spaces. Am. J. Math. **122**, 887–926 (2000)
[DH05] Deitmar, A., Hilgert, J.: Cohomology of arithmetic groups with infinite dimensional coefficient spaces. Documenta Math. **10**, 199–216 (2005) and **11**, 241 (2006) (erratum)
[DK14] Deitmar, A., Kang, M.-H.: Geometric zeta functions for higher rank p-adic groups. Illinois J. Math. **58**, 719–738 (2014)
[DKV79] Duistermaat, J., Kolk, J., Varadarajan, V.: Spectra of compact locally symmetric manifolds of negative curvature. Invent. Math. **52**, 27–93 (1979)
[DFG15] Dyatlov, S., Faure, F., Guillarmou, C.: Power spectrum of the geodesic flow on hyperbolic manifolds. Anal. PDE **8**, 923–1000 (2015)
[DG16] Dyatlov, S., Guillarmou, C.: Pollicott-Ruelle resonances for open systems. Annales Henri Poincaré **17**, 3089–3146 (2016)
[DZ16] Dyatlov, S., Zworski, M.: Dynamical zeta functions for Anosov flows via microlocal analysis. Ann. Sci. Ec. Norm. Supér. **49**, 543–577 (2016)
[EFMW92] Eckhardt, B., Fishman, S., Müller, K., Wintgen, D.: Semiclassical matrix elements from periodic orbits. Phys. Rev. A **45**, 3531–3539 (1992)
[Em14] Emonds, J.: A dynamical interpretation of Patterson-Sullivan distributions. Dissertation 2014, Universität Paderborn & Université Lorraine-Metz. arxiv:1407.5450
[FS11] Faure, F., Sjöstrand, J.: Upper bound on the density of Ruelle resonances for Anosov flows. Comm. Math. Phys. **308**, 325–364 (2011)
[FT15] Faure, F., Tsujii, M.: Prequantum transfer operator for symplectic Anosov diffeomorphism. Astérisque **375**, ix + 222 pp. (2015)
[FT17] Faure, F., Tsujii, M.: The semiclassical zeta function for geodesic flows on negatively curved manifolds. Invent. Math. **208**, 851–998 (2017)
[FN91] Figa-Talamanca, A., Nebbia, C.: Harmonic Analysis and Representation Theory for Groups Acting on Homogeneous Trees. Cambridge University Press, Cambridge (1991)
[FF03] Flaminio, L., Forni, G.: Invariant distributions and the time averages for horocycle flows. Duke Math. J. **119**, 465–526 (2003)
[FS23] Frahm, J., Spilioti, P.: Resonances and residue operators for pseudo-Riemannian hyperbolic spaces. J. Math. Pures Appl. **177**, 178–197 (2023)
[GLP13] Giuletti, P., Liverani, C., Pollicott, M.: Anosov flows and dynamical zeta functions. Annals Math. **178**, 687–773 (2013)
[GH78] Griffith, P., Harris, J.: Principles of Algebraic Geometry. Wiley-Interscience, New York (1978)
[Gr55] Grothendieck, A.: Produits tensoriels topologiques et espaces nucléaires. Mem. Am. Math. Soc. **16**, 336 (1955)
[GGHW21] Bonthonneau, G.Y., Guillarmou, C., Hilgert, J., Weich, T.: Ruelle-Taylor resonances of Anosov actions. J. EMS, to appear
[GBGW21] Bonthonneau, G.Y., Guillarmou, C., Weich, T.: SRB Measures for Anosov Actions. J. Differential Geom. **128**, 959–1026 (2024)
[GW12] Guichard, O., Wienhard, A.: Anosov representations: domains of discontinuity and applications. Invent. Math. **190**, 357–438 (2012)
[Gu05] Guillarmou, C.: Resonances and scattering poles on asymptotically hyperbolic manifolds. Math. Res. Lett. **12**, 103–119 (2005)
[GHW18] Guillarmou, C., Hilgert, J., Weich, T.: Classical and quantum resonances for hyperbolic surfaces. Math. Annalen **370**, 1231–1275 (2018)

[GHW21] Guillarmou, C., Hilgert, J., Weich, T.: High frequency limits for invariant Ruelle densities. Annales Henri Lebesgue **4**, 81–119 (2021)

[Gu77] Guillemin, V.: Lectures on spectral theory of elliptic operators. Duke Math. J. **44**, 485–517 (1977)

[GZ95] Guillopé, L., Zworski, M.: Upper bounds on the number of resonances for non-compact Riemann surfaces. J. Funct. Anal. **129**, 364–389 (1995)

[GZ97] Guillopé, L., Zworski, M.: Scattering asymptotics for Riemann surfaces. Ann. of Math. **145**, 597–660 (1997)

[Ha17] Hadfield, C.: Resonances for symmetric tensors an asymptotically hyperbolic spaces. Analysis & PDE **10**, 1877–1922 (2017)

[Ha20] Hadfield, C.: Ruelle and quantum resonances for open hyperbolic manifolds. Int. Math. Res. Not. IRMN **2020**, 1445–1480 (2020)

[HHP18] Hansen, S., Hilgert, J., Parthasarathy, A.: Boundary Values of Eigenfunctions on Riemannian Symmetric Spaces. arxiv:1807.07131 (2018)

[HHP19] Hansen, S., Hilgert, J., Parthasarathy, A.: Resonances and Scattering Poles in Symmetric Spaces of Rank One. Int. Math. Res. Not. IRMN **2019**, 6362–6389 (2019)

[HHS12] Hansen, S., Hilgert, J., Schröder, M.: Patterson–Sullivan distributions in higher rank. Math. Z. **272**, 607–643 (2012)

[Ha68] Harish-Chandra. Automorphic Forms on Semisimple Lie Groups. Lecture Notes in Mathematics, vol. 62. Springer, Berlin (1968)

[He84] Helgason, S.: Groups and geometric analysis: integral geometry, invariant differential operators, and spherical functions. Academic Press, Orlando (1984)

[He94] Helgason, S.: Geometric analysis on symmetric spaces. In: Mathematical Surveys and Monographs, vol. 39. American Mathematical Society, Providence, RI (1994)

[Hi05] Hilgert, J.: An Ergodic Arnold–Liouville Theorem for locally symmetric spaces. In: Ali, S.T., et al. (eds.) Twenty Years of Bialowieza: A Mathematical Anthology. World Scientific Monograph Series in Mathematics, vol. 8, pp. 163–184. World Scientific Publishing, Hackensack, NJ (2005)

[HP09] Hilgert, J., Pasquale, A.: Resonances and residue operators for symmetric spaces of rank one. J. Math. Pures Appl. **91**, 495–507 (2009)

[HPP16] Hilgert, J., Pasquale, A., Przebinda, T.: Resonances for the Laplacian: the cases BC_2 and C_2 (except $\mathrm{SO}_0(p, 2)$ with $p > 2$ odd). In: Kielanowski, P., et al. (eds.) Geometric Methods in Physics, pp. 159–182. Birkhäuser/Springer, Cham (2016)

[HPP17a] Hilgert, J., Pasquale, A., Przebinda, T.: Resonances for the Laplacian on Riemannian symmetric spaces: the case of $\mathrm{SL}(3, \mathbb{R})/\mathrm{SO}(3)$. Represent. Theory **21**, 416–457 (2017)

[HPP17b] Hilgert, J., Pasquale, A., Przebinda, T.: Resonances for the Laplacian on products of two rank one Riemannian symmetric spaces. J. Funct. Anal. **272**, 1477–1523 (2017)

[HWW21] Hilgert, J., Weich, T., Wolf, L.: Higher rank quantum-classical correspondence. Anal. PDE **16**, 2241–2265 (2023)

[JZ01] Ji, L., Zworski, M.: Scattering matrices and scattering geodesics of locally symmetric spaces. Ann. Sci. Ec. Norm. Supér. **34**, 441–469 (2001)

[JS00] Joshi, M., Sá Barreto, A.: Inverse scattering on asymptotically hyperbolic manifolds. Acta Math. **184**, 41–86 (2000)

[Ju93] Juhl, A.: Zeta-Funktionen, Index-Theorie und hyperbolische Dynamik. Habilitationsschrift HU Berlin, (1993)

[Ju01] Juhl, A.: Cohomological theory of dynamical zeta functions. Birkhäuser, Basel (2001)

[KLP18] Kapovich, M., Leeb, B., Porti, J.: Dynamics on flag manifolds: domains of proper discontinuity and cocompactness. Geom. Topol. **22**, 157–234 (2018)

[K+78] Kashiwara, M., Kowata, A., Minemura, K., Okamoto, K., Oshima, T., Tanaka, M.: Eigenfunctions of invariant differential operators on a symmetric space. Annals Math. **107**, 1–39 (1978)

[Ka81] Kato, S.: On eigenspaces of the Hecke algebra with respect to a good maximal compact subgroup of a p-adic reductive group. Math. Ann. **257**, 1–7 (1981)

[Ka82] Kato, S.: Irreducibility of principal series representations for Hecke algebras of affine type. J. Fac. Sci. Univ. Tokyo Sect. Math. **28**, 929–943 (1982)
[KH95] Katok, A., Hasselblatt, B.: Modern Theory of Dynamical Systems. Cambridge University Press, Cambridge (1995)
[KS94] Katok, A., Spatzier, R.: First cohomology of Anosov actions of higher rank abelian groups and applications to rigidity. Publ. Math. Inst. Hautes Études Sci., **79**, 131–156 (1994)
[Kn86] Knapp, A.W.: Representation theory of semisimple groups—an overview based on examples. Princeton University Press, New Jersey (1986)
[KW20] Küster, B., Weich, T.: Pollicott-Ruelle resonant states and Betti numbers. Comm. Math. Phys. **378**, 917–941 (2020)
[KW21] Küster, B., Weich, T.: Quantum-classical correspondence on associated vector bundles over locally symmetric spaces. Int. Math. Res. Not. IRMN **2021**, 8225–8296 (2021)
[La02] Lax, P.: Functional analysis. Wiley-Interscience, New York (2002)
[LP76] Lax, P., Phillips, R.: Scattering Theory for Automorphic Forms. Princeton University Press, Princeton (1976)
[LM14] Le Masson, E.: Pseudo-differential calculus on homogeneous trees. Annales Henri Poincaré **15**, 1697–1732 (2014)
[LZ01] Lewis, J., Zagier, D.: Period functions for Maass wave forms I. Annals Math. **153**, 191–258 (2001)
[LP16] Lubetzky, E., Peres, Y.: Cutoff on all Ramanujan graphs. Geom. Funct. Anal. **26**, 1190–1216 (2016)
[Ma71] Macdonald, I.G.: Spherical functions on a group of p-adic type. Publication Ramanujan Institute, Madras (1971)
[MZ17] Mantero, A., Zappa, A.: Boundary behavior of generalized Poisson integrals on buildings of type $\tilde{A}_2$. Boll. Unione Mat. Ital. **10**, 681–724 (2017)
[MM87] Mazzeo, R., Melrose, R.: Meromorphic extension of the resolvent on complete spaces with asymptotically constant negative curvature. J. Funct. Anal. **75**, 260–310 (1987)
[MV05] Mazzeo, R., Vasy, A.: Analytic continuation of the resolvent of the Laplacian on symmetric spaces of noncompact type. J. Funct. Anal.**228**, 311–368 (2005)
[MW00] Miatello, R., Will, C.: The residues of the resolvent on Damek-Ricci spaces. Proc. Am. Math. Soc. **128**, 1221–1229 (2000)
[Mo70] Mostow, G.D.: Intersections of discrete subgroups with Cartan subgroups. J. Indian Math. Soc. **34**, 203–214 (1970)
[Mo73] Mostow, G.D.: Strong rigidity of locally symmetric spaces. Princeton University Press, Princeton (1973)
[Mo95] Mozes, S.: Actions of Cartan subgroups. Israel J. Math. **90**, 253–294 (1995)
[MR20] Mühlenbruch, T., Raji, W.: On the Theory of Maass Wave Forms. Springer, Cham (2020)
[Oh02] Oh, H.: Uniform pointwise bounds for matrix coefficients of unitary representations and applications to Kazhdan constants. Duke Math. J. **113**, 133–192 (2002)
[Pa06] Parkinson, J.: Spherical harmonic analysis on affine buildings. Math. Zeitschrift **253**, 571–606 (2006)
[Pa99] Paternain, G.: Geodesic Flows. Birkhäuser, Boston (1999)
[Pa89] Patterson, S.J.: The Selberg zeta-function of a Kleinian group. In: Aubert, K.E., Bombieri, E., Goldfeld, D. (eds.) Number Theory, Trace Formulas and Discrete Groups, pp. 409–441. Academic Press, Boston (1987)
[PP99] Patterson, S.J., Perry, P.: The divisor of Selberg's zeta function for Kleinian groups. Duke Math. J. **106**, 321–390 (2001)
[Po12] Pohl, A.: A dynamical approach to Maass cusp forms. J. Mod. Dyn. **6**, 563–596 (2012)
[Po14] Pohl, A.: Symbolic dynamics for the geodesic ow on two-dimensional hyperbolic good orbifolds. Discrete Contin. Dyn. Syst. **34**, 2173–2241 (2014)

[Po16] Pohl, A.: Symbolic dynamics, automorphic functions, and Selberg zeta functions with unitary representations. In: Dynamics and Numbers. Contemporary Mathematics, vol. 669, pp. 205–236. American Mathematical Society, Providence, RI (2016)

[Po20] Pohl, A.: Symbolic dynamics and transfer operators for Weyl chamber flows: a class of examples. arXiv:2011.14098 (2020)

[PZ20] Pohl, A., Zagier, D.: Dynamics of geodesics, and Maass cusp forms. L'Enseignement Math. **66**, 305–340 (2020)

[Ro22] Roby, S.: Resonances of the Laplace operator on homogeneous vector bundles on symmetric spaces of real rank-one. Advances Math. **408**, 108555 (2022)

[Sa03] Sandberg, S.: On non-holomorphic functional calculus for commuting operators. Math. Scand. **93**, 109–135 (2003)

[SKK73] Sato, M., Kawai, T., Kashiwara, M.: Micro-functions and pseudo-differential equations. In: Proceedings of the Conference at Katata, 1971. Lecture Notes in Mathematics, vol. 287, pp. 265–529 (1973)

[Sch10] Schröder, M.: Patterson-Sullivan distributions for symmetric spaces of the noncompact type. Dissertation 2010, Universität Paderborn & Université Lorraine-Metz. arxiv:1012.1113

[SWB23] P. Schütte, Weich, T., Barkhofen, S.: Meromorphic continuation of weighted zeta functions on open hyperbolic systems. Commun. Math. Phys. **398**, 655–678 (2023)

[Se56] Selberg, A.: Harmonic analysis and discontinuous groups in weakly symmetric Riemannian spaces with applications to Dirichlet series. J. Indian Math. Soc. B **20**, 47–87 (1956)

[STS76] Semenov-Tjan-Šanskiĭ, M.: Harmonic analysis on Riemannian symmetric spaces of negative curvature, and scattering theory. Izv. Akad. Nauk SSSR Ser. Mat. **40**, 562–592 (1976)

[Se80] Serre, J.-P.: Trees. Springer, Berlin (1980)

[Si15] Silberman, L.: Quantum Unique Ergodicity on Locally Symmetric Spaces: the Degenerate Lift. Canad. Math. Bull. **58**, 632–650 (2015)

[SV07] Silberman, L., Venkatesh, A.: Quantum unique ergodicity on locally symmetric spaces. Geom. Funct. Anal. **17**, 960–998 (2007)

[SV19] Spatzier, R., Vinhage, K.: Cartan actions of higher rank abelian groups and their classification. arXiv:1901.06559 (2019)

[Su73] Sussmann, H.: Orbits of families of vector fields and integrability of distributions. Trans. Am. Math. Soc. **180**, 171–188 (1973)

[Ta11] Tao, T.: The spectral theorem and its converses for unbounded symmetric operators. https://terrytao.wordpress.com/2011/12/20/the-spectral-theorem-and-its-converses-for-unbounded-symmetric-operators/

[Ta70a] Taylor, J.L.: The analytic functional calculus for several commuting operators. Acta Math. **125**, 1–38 (1970)

[Ta70b] Taylor, J.L.: A joint spectrum for several commuting operators. J. Funct. Anal. **6**, 172–191 (1970)

[TW89] Tong, Y.L., Wang, S.P.: Geometric realization of discrete seiries for semisimple symmetric spaces. Invent. math. **96**, 425–458 (1989)

[vdB87] van den Ban, E.P.: Invariant differential operators on a semisimple symmetric space and finite multiplicities in a Plancherel formula. Ark. Mat. **25**, 175–187 (1987)

[vdBS87] van den Ban, E.P., Schlichtkrull, H.: Asymptotic expansions and boundary values of eigenfunctions on Riemannian symmetric spaces. J. reine angewandte Math. **380**, 108–165 (1987)

[Wa76] Wallach, N.R.: On the Selberg trace formula in the case of a compact quotient. Bull. Am. Math. Soc. **82**, 171–195 (1976)

[Wa88] Wallach, N.R.: Real reductive groups I. In: Pure and Applied Mathematics, vol. 132, Academic Press, Inc., Boston (1988)

[Wa92] Wallach, N.R.: Real reductive groups II. In: Pure and Applied Mathematics, vol. 132. Academic Press, Inc., Boston (1992)

[Wa72] Warner, G.: Harmonic analysis on semi-simple Lie groups I. Springer, New York (1972)
[Yo02] Young, L.S.: What are SRB measures, and which dynamical systems have them? J. Stat. Phys. **108**, 733–754 (2002)
[Ze84] Zelditch, S.: Pseudo-differential analysis on hyperbolic surfaces. J. Funct. Anal. **68**, 72–105 (1986)

Classification of K-type Formulas for the Heisenberg Ultrahyperbolic Operator $\square_s$ for $\widetilde{SL}(3, \mathbb{R})$ and Tridiagonal Determinants for Local Heun Functions

Toshihisa Kubo and Bent Ørsted

Dedicated to Professor Toshiyuki Kobayashi, a "Fanstasista" in the field of representation theory and Lie theory

Abstract The K-type formulas of the space of K-finite solutions to the Heisenberg ultrahyperbolic equation $\square_s f = 0$ for the nonlinear group $\widetilde{SL}(3, \mathbb{R})$ are classified. This completes a previous study of Kable for the linear group $SL(m, \mathbb{R})$ in the case of $m = 3$, as well as generalizes our earlier results on a certain second-order differential operator. As a by-product we also show "functional equations" of certain sequences $\{P_k(x; y)\}_{k=0}^{\infty}$ and $\{Q_k(x; y)\}_{k=0}^{\infty}$ of tridiagonal determinants, whose generating functions are given by local Heun functions.

Keywords Intertwining differential operator · Heisenberg ultrahyperbolic operator · Hypergeometric differential equation · Heun's differential equation · Tridiagonal determinant · Sylvester determinant · Cayley continuant · Krawtchouk polynomial

MSC (2020) Primary 22E46; Secondary 17B10, 05B20, 33C05, 33E30

T. Kubo (✉)
Faculty of Economics, Ryukoku University, Kyoto, Japan
e-mail: toskubo@econ.ryukoku.ac.jp

B. Ørsted
Department of Mathematics, Aarhus University, Aarhus, Denmark
e-mail: orsted@math.au.dk

M. Pevzner, H. Sekiguchi (eds.), *Symmetry in Geometry and Analysis, Volume 2*, Progress in Mathematics 358, https://doi.org/10.1007/978-981-97-7662-7_8

1 Introduction

Let G be a real reductive group and P a parabolic subgroup of G. The aim of this paper is to study the representations realized on the solution space (kernel) of a certain family $\square_s$ of intertwining differential operators between the spaces of smooth sections for line bundles over G/P.

The space of solutions to an intertwining differential operator has been attracting people's attention, especially for the aim of constructing minimal representations (see, for instance, [25, Introduction] and the references therein). For example, in the series [19, 20, 21] of papers, Kobayashi and the second author realized the minimal representation of $O(p, q)$ on the solution space of the Yamabe operator and ultrahyperbolic Laplace operator. The authors of the cited paper studied the minimal representation in great detail from various points of views such as conformal geometry, branching laws, harmonic analysis, and algebraic analysis.

In this paper, we consider a one-parameter family $\square_s$ of natural invariant differential operators and explicit spaces of solutions. It turns out that these will be related to classical function theory, namely, hypergeometric functions and local Heun functions, and some of the relevant identities analogous to classical tridiagonal determinants of Sylvester type and Cayley type. Further, the representations obtained from the solution space to the differential equation $\square_s f = 0$ include ones, which are to be thought of as minimal representations (in some sense). We provide a new aspect on a connection between the representation theory of reductive groups, ordinary differential equations in the complex domains, and sequences of polynomials.

The aim of this paper is twofold. One of them is the classification of K-type formulas of the space of K-finite solutions to the Heisenberg ultrahyperbolic equation $\square_s f = 0$ for $\widetilde{SL}(3, \mathbb{R})$. The other is a study of certain sequences $\{P_k(x; y)\}_{k=0}^{\infty}$ and $\{Q_k(x; y)\}_{k=0}^{\infty}$ of tridiagonal determinants arising from the study of the K-finite solutions to $\square_s f = 0$. We describe these two topics in detail now.

1.1 Heisenberg Ultrahyperbolic Operator $\square_s$

For a moment, let $\mathfrak{g} = \mathfrak{sl}(m, \mathbb{C})$ with $m \geq 3$ and take a real form $\mathfrak{g}_0$ of $\mathfrak{g}$ such that there exists a parabolic subalgebra $\mathfrak{p}_0 = \mathfrak{m}_0 \oplus \mathfrak{a}_0 \oplus \mathfrak{n}_0$ with Heisenberg nilpotent radical $\mathfrak{n}_0$, namely, $\dim_{\mathbb{R}}[\mathfrak{n}_0, \mathfrak{n}_0] = 1$. The Heisenberg condition on $\mathfrak{n}_0$ forces $\mathfrak{g}_0$ to be either $\mathfrak{sl}(m, \mathbb{R})$ or $\mathfrak{su}(p, q)$ with $p + q = m$. We write $\mathfrak{p} = \mathfrak{m} \oplus \mathfrak{a} \oplus \mathfrak{n}$ for the complexification of $\mathfrak{p}_0 = \mathfrak{m}_0 \oplus \mathfrak{a}_0 \oplus \mathfrak{n}_0$. We write $\mathcal{U}(\mathfrak{g})$ for the universal enveloping algebra of $\mathfrak{g}$.

In [14], under the framework of $\mathfrak{g}$ and $\mathfrak{p}$ as above, Kable introduced a one-parameter family $\square_s$ of differential operators with $s \in \mathbb{C}$ as an example of conformally invariant systems [2, 3]. The operator $\square_s$ is referred to as the

Heisenberg ultrahyperbolic operator [14]. For instance, for $m = 3$, it is defined as

$$\square_s = R((XY + YX) + s[X, Y]), \tag{1.1}$$

where R denotes the infinitesimal right translation and X, Y are certain nilpotent elements in $\mathfrak{sl}(3, \mathbb{R})$ (see (3.1)). The differential operator $\square_s$ for $m = 3$ in (1.1) in particular recovers the second-order differential operator studied in [25] as the case of $s = 0$. Some algebraic and analytic properties of the Heisenberg ultrahyperbolic operator $\square_s$ as well as its generalizations are investigated in [13, 14, 15, 16] for a linear group $SL(m, \mathbb{R})$.

From a viewpoint of intertwining operators, the Heisenberg ultrahyperbolic operator $\square_s$ is an intertwining differential operator between certain parabolically induced representations for $G \supset P$, where G and P are Lie groups with Lie algebras $\mathfrak{g}_0$ and $\mathfrak{p}_0$, respectively. Thus the space $\mathcal{S}ol(\square_s)$ of smooth solutions to the equation $\square_s f = 0$ in the induced representation is a subrepresentation of G, and also the space $\mathcal{S}ol(\square_s)_K$ of K-finite solutions is a $(\mathfrak{g}, K)$-module. We then consider the following problem.

Problem 1.1 Classify the K-type formulas for $\mathcal{S}ol(\square_s)_K$.

In [14], an attempt for this direction was made for $G = SL(m, \mathbb{R})$. In the paper, Kable introduced a notion of *$\mathcal{H}$-modules* and developed an algebraic theory for them. Although the theory is powerful to compute the dimension of the $\mathfrak{k} \cap \mathfrak{m}$-invariant subspace of the space of K-finite solutions for each K-type in $\mathcal{S}ol(\square_s)_K$, the determination of the explicit K-type formulas was not achieved.

In this paper, we consider $G = \widetilde{SL}(3, \mathbb{R})$, the universal covering group of $SL(3, \mathbb{R})$. By making use of a technique from our earlier paper [25], we successfully classified the K-type formulas of $\mathcal{S}ol(\square_s)_K$ for $\widetilde{SL}(3, \mathbb{R})$. In order to describe our results in more detail, we next introduce some notation.

1.2 K-Type Formulas

For the rest of this introduction, let $G = \widetilde{SL}(3, \mathbb{R})$ with Lie algebra $\mathfrak{g}_0$. Fix a minimal parabolic subgroup B of G with Langlands decomposition $B = MAN$. Here the subgroup M is isomorphic to the quaternion group Q_8 of order 8. Let K be a maximal compact subgroup of G so that $G = KAN$ is an Iwasawa decomposition of G. We have $K \simeq SU(2) \simeq Spin(3)$.

Let $\mathrm{Irr}(M)$ and $\mathrm{Irr}(K)$ denote the set of equivalence classes of irreducible representations of M and K, respectively. As $M \simeq Q_8$, the set $\mathrm{Irr}(M)$ may be given as

$$\mathrm{Irr}(M) = \{(+,+), (+,-), (-,+), (-,-), \mathbb{H}\},$$

where $(\pm, \pm)$ are some characters (see Sect. 3.5 for the definition) and $\mathbb{H}$ stands for the unique two-dimensional genuine representation of M. The character $(+, +)$ is, for instance, the trivial character. Let $(n/2)$ denote the irreducible finite-dimensional representation of $K \simeq Spin(3)$ with dimension $n + 1$. Then we have:

$$\mathrm{Irr}(K) = \left\{(n/2) : n \in \mathbb{Z}_{\geq 0}\right\}.$$

For $\sigma \in \mathrm{Irr}(M)$ and a character λ of A, we write:

$$I(\sigma, \lambda) = \mathrm{Ind}_B^G(\sigma \otimes (\lambda + \rho) \otimes \mathbb{1})$$

for the representation of G induced from the representation $\sigma \otimes (\lambda + \rho) \otimes \mathbb{1}$ of $B = MAN$, where ρ is half the sum of the positive roots corresponding to B and $\mathbb{1}$ denotes the trivial representation of N. We realize the induced representation $I(\sigma, \lambda)$ on the space of smooth sections for a G-equivariant homogeneous vector bundle over G/B. Let $\mathrm{Diff}_G(I(\sigma_1, \lambda_1), I(\sigma_2, \lambda_2))$ denote the space of intertwining differential operators from $I(\sigma_1, \lambda_1)$ to $I(\sigma_2, \lambda_2)$.

Let $\mathfrak{g}$ denote the complexified Lie algebra of $\mathfrak{g}_0 = \mathfrak{sl}(3, \mathbb{R})$. Since $\mathfrak{g} = \mathfrak{sl}(3, \mathbb{C})$, the Heisenberg ultrahyperbolic operator $\square_s$ is given as in (1.1). We then set

$$D_s := (XY + YX) + s[X, Y] \in \mathcal{U}(\mathfrak{g}), \tag{1.2}$$

so that $\square_s = R(D_s)$. It will be shown in Proposition 4.1 that we have

$$R(D_s) \in \mathrm{Diff}_G(I((+, +), -\widetilde{\rho}(s)), I((-, -), \widetilde{\rho}(-s))),$$

where $\widetilde{\rho}(s)$ is a certain weight determined by $\widetilde{\rho} := \rho/2$ (see (4.1)). It is further shown that, for $\sigma \in \mathrm{Irr}(M)$,

$$R(D_s) \otimes \mathrm{id}_\sigma \in \mathrm{Diff}_G(I(\sigma, -\widetilde{\rho}(s)), I((-, -) \otimes \sigma, \widetilde{\rho}(-s))),$$

where id_σ denotes the identity map on σ.

For notational convenience we consider $R(D_{\bar{s}})$ in place of $R(D_s)$, where $\bar{s}$ denotes the complex conjugate of $s \in \mathbb{C}$ (see Sect. 4.1 for the details). We define:

$$\begin{aligned} \mathcal{S}ol(s; \sigma) &:= \text{the space of smooth solutions to } (R(D_{\bar{s}}) \otimes \mathrm{id}_\sigma) f = 0, \\ \mathcal{S}ol(s; \sigma)_K &:= \text{the space of } K\text{-finite solutions to } (R(D_{\bar{s}}) \otimes \mathrm{id}_\sigma) f = 0. \end{aligned} \tag{1.3}$$

It follows from a Peter–Weyl theorem for the solution space (Theorem 2.5) that the $(\mathfrak{g}, K)$-module $\mathcal{S}ol(s; \sigma)_K$ decomposes as

$$\mathcal{S}ol(s; \sigma)_K \simeq \bigoplus_{n \in \mathbb{Z}_{\geq 0}} (n/2) \otimes \mathrm{Hom}_M (\mathrm{Sol}(s; n), \sigma), \tag{1.4}$$

where $\mathrm{Sol}(s; n)$ is the space of K-type solutions to D_s (without complex conjugation on s) on the K-type $(n/2) = (\delta, V_{(n/2)}) \in \mathrm{Irr}(K)$, that is,

$$\mathrm{Sol}(s; n) = \{v \in V_{(n/2)} : d\delta(D_s^{\flat})v = 0\} \tag{1.5}$$

(see (4.6) and (4.5)). Here $d\delta$ is the differential of δ and $D_s^{\flat}$ denotes the compact model of D_s (see Definition 2.2). Then the K-type decomposition $\mathcal{S}ol(s; \sigma)_K$ for $(s, \sigma) \in \mathbb{C} \times \mathrm{Irr}(M)$ is explicitly given as follows.

Theorem 1.2 *The following conditions on $(\sigma, s) \in \mathrm{Irr}(M) \times \mathbb{C}$ are equivalent:*

(i) $\mathcal{S}ol(s; \sigma) \neq \{0\}$.
(ii) *One of the following conditions holds:*

- $\sigma = (+, +) : s \in \mathbb{C}$.
- $\sigma = (-, -) : s \in \mathbb{C}$.
- $\sigma = (-, +) : s \in 1 + 4\mathbb{Z}$.
- $\sigma = (+, -) : s \in 3 + 4\mathbb{Z}$.
- $\sigma = \mathbb{H} : \quad s \in 2\mathbb{Z}$.

Further, the K-type formulas for $\mathcal{S}ol(s; \sigma)_K$ are given as follows:

(1) $\sigma = (+, +)$:

$$\mathcal{S}ol(s; (+, +))_K \simeq \bigoplus_{n \in \mathbb{Z}_{\geq 0}} (2n) \qquad \text{for all } s \in \mathbb{C}.$$

(2) $\sigma = (-, -)$:

$$\mathcal{S}ol(s; (-, -))_K \simeq \bigoplus_{n \in \mathbb{Z}_{\geq 0}} (1 + 2n) \quad \text{for all } s \in \mathbb{C}.$$

(3) $\sigma = (-, +)$:

$$\mathcal{S}ol(s; (-, +))_K \simeq \bigoplus_{n \in \mathbb{Z}_{\geq 0}} ((|s| + 1)/2 + 2n) \quad \text{for } s \in 1 + 4\mathbb{Z}.$$

(4) $\sigma = (+, -)$:

$$\mathcal{S}ol(s; (+, -))_K \simeq \bigoplus_{n \in \mathbb{Z}_{\geq 0}} ((|s| + 1)/2 + 2n) \quad \text{for } s \in 3 + 4\mathbb{Z}.$$

(5) $\sigma = \mathbb{H}$:

$$\mathcal{S}ol(s; \mathbb{H})_K \simeq \bigoplus_{n \in \mathbb{Z}_{\geq 0}} ((|s| + 1)/2 + 2n) \quad \text{for } s \in 2\mathbb{Z}.$$

Table 1 Comparison between [14] and this paper for $\mathfrak{g}_0 = \mathfrak{sl}(3, \mathbb{R})$

$\mathfrak{g}_0 = \mathfrak{sl}(3, \mathbb{R})$	$\dim_{\mathbb{C}} \mathrm{Sol}(s; n)$	K-type decomposition of $\mathcal{S}ol(s; \sigma)_K$
[14]	Done for $SL(3, \mathbb{R})$	Not obtained
this paper	Done for $\widetilde{SL}(3, \mathbb{R})$	Done for $\widetilde{SL}(3, \mathbb{R})$

We shall deduce Theorem 1.2 from Theorem 5.16 at the end of Sect. 5. The proof is not case-by-case analysis on $\sigma \in \mathrm{Irr}(M)$; each M-representation σ is treated uniformly via a recipe for the K-type decomposition of $\mathcal{S}ol(s; \sigma)_K$. See Sect. 4.4 for the details of the recipe. In regard to Theorem 1.2, we shall also classify the space $\mathrm{Sol}(s; n)$ of K-type solutions for all $n \in \mathbb{Z}_{\geq 0}$. This is done in Theorems 5.4 and 6.4. Recall from (1.3) that $\mathcal{S}ol(s; \sigma)_K$ concerns $R(D_{\bar{s}})$. Theorem 1.2 shows that the K-type decompositions are in fact independent of taking the complex conjugate of the parameter $s \in \mathbb{C}$.

There are several remarks on the space $\mathrm{Sol}(s; n)$ of K-type solutions and the K-type decompositions of $\mathcal{S}ol(s; \sigma)_K$. First, as mentioned above, the dimensions of $\mathfrak{k} \cap \mathfrak{m}$-invariant subspaces of the spaces of K-type solutions are determined in [14] for $SL(m, \mathbb{R})$ with arbitrary rank $m \geq 3$. In particular, as $\mathfrak{k} \cap \mathfrak{m} = \{0\}$ for $m = 3$, the dimensions $\dim_{\mathbb{C}} \mathrm{Sol}(s; n)$ for $n \in 2\mathbb{Z}_{\geq 0}$ are obtained in the cited paper [14, Thm. 5.13]. We note that our normalization on $s \in \mathbb{C}$ is different from one for $z \in \mathbb{C}$ in [14] by $s = -2z$. In the paper, factorization formulas of certain tridiagonal determinants are essential to compute $\dim_{\mathbb{C}} \mathrm{Sol}(s; n)$. For further details on the factorization formulas, see the remark after (1.16) below. By making use of differential equations, we determine the dimensions $\dim_{\mathbb{C}} \mathrm{Sol}(s; n)$ for all $n \in \mathbb{Z}_{\geq 0}$ independently to the results of [14].

Table 1 summarizes the results of [14] and this paper, concerning the K-type decompositions of $\mathcal{S}ol(s; \sigma)_K$ for $\mathfrak{g}_0 = \mathfrak{sl}(3, \mathbb{R})$.

Secondly, the K-type formula for $\mathcal{S}ol(s; \mathbb{H})_K$ is recently determined also by Tamori, as part of his thorough study on minimal representations [36, Thm. 5.3 (2)]. We remark that, both in [36] and in this paper, the K-type formula for $\mathcal{S}ol(s; \mathbb{H})_K$ is obtained by realizing the K-type solutions in $\mathrm{Sol}(s; n)$ as hypergeometric polynomials; nevertheless, the methods are rather different. For instance, some combinatorial computations are carried out in [36], whereas we simply solve the hypergeometric equation so that we treat all irreducible M-representations $\sigma = (\pm, \pm), \mathbb{H}$, uniformly. Further, we also give the K-type solutions by Heun polynomials. We illustrate our methods in detail in Sect. 1.3.

Lastly, the K-type formulas $\mathcal{S}ol(0; \sigma)_K$ for the case of $s = 0$ are previously classified by ourselves in [25, Thm. 1.6]. In this case only three M-representations $\sigma = (+,+), (-,-), \mathbb{H}$ contribute to $\mathcal{S}ol(0; \sigma) \neq \{0\}$ with explicit K-type formulas:

$$\mathcal{S}ol(0; (+,+))_K \simeq \bigoplus_{n \in \mathbb{Z}_{\geq 0}} (2n),$$

$$\mathcal{S}ol(0; (-,-))_K \simeq \bigoplus_{n \in \mathbb{Z}_{\geq 0}} (1+2n)\,,$$

$$\mathcal{S}ol(0; \mathbb{H})_K \simeq \bigoplus_{n \in \mathbb{Z}_{\geq 0}} ((1/2)+2n)\,.$$

It was quite mysterious that only the series of $(3/2) + 2\mathbb{Z}_{\geq 0}$ does not appear in the K-type formulas. Theorem 1.2 now gives an answer for this question.

The representations realized on $\mathcal{S}ol(0; \sigma)$ for $\sigma = (+,+), (-,-), \mathbb{H}$ are known to be unitarizable, and the resulting representations are the ones attached to the minimal nilpotent orbit [31]. For instance, the representation realized on $\mathcal{S}ol(0; \mathbb{H})$ is the genuine representation so-called Torasso's representation [38]. Here are two remarks on a recent progress on the unitarity of the representations on $\mathcal{S}ol(0; \sigma)$. First, Dahl constructed in [7] the unitary structures for the three unitarizable representations on $\mathcal{S}ol(0; \sigma)$ by using the Knapp–Stein intertwining operator and the Fourier transform on the Heisenberg group. Furthermore, Frahm recently gives the L^2-models of these unitary representations in [9], as part of his intensive study on the L^2-realizations of the minimal representations realized on the solution space of conformally invariant systems constructed in [2].

1.3 Hypergeometric and Heun's Differential Equations

The main idea for accomplishing Theorem 1.2 is use of the decomposition formula (1.4) and a refinement of the method applied to the case $s = 0$ in [25]. The central problem is classifying the space $\mathrm{Sol}(s; n)$ of K-type solutions. In order to do so, one needs to proceed with the following two steps:

Step 1: Classify $(s, n) \in \mathbb{C} \times \mathbb{Z}_{\geq 0}$ so that $\mathrm{Sol}(s; n) \neq \{0\}$.
Step 2: Classify the M-representations on $\mathrm{Sol}(s; n) \neq \{0\}$.

In [25] (the case $s = 0$), via the polynomial realization

$$\mathrm{Irr}(K) \simeq \{(\pi_n, \mathrm{Pol}_n[t]) : n \in \mathbb{Z}_{\geq 0}\} \tag{1.6}$$

of $\mathrm{Irr}(K)$ (see (3.18)), we carried out the two steps by realizing $\mathrm{Sol}(0; n)$ in (1.5) as

$$\mathrm{Sol}(0; n) = \{p(t) \in \mathrm{Pol}_n[t] : d\pi_n^{\mathrm{I}}(D_0^\flat)p(t) = 0\},$$

the space of polynomial solutions to an ordinary differential equation $d\pi_n^{\mathrm{I}}(D_0^\flat)$ $p(t) = 0$ via a certain identification $\Omega^{\mathrm{I}} \colon \mathfrak{k} \stackrel{\sim}{\to} \mathfrak{sl}(2, \mathbb{C})$ (see Sects. 3.2 and 4.2). It turned out that the differential equation $d\pi_n^{\mathrm{I}}(D_0^\flat)p(t) = 0$ is a hypergeometric differential equation. Steps 1 and 2 were then easily carried out by observing the Gauss hypergeometric functions arising from $d\pi_n^{\mathrm{I}}(D_0^\flat)p(t) = 0$.

In this paper we apply the same idea for the general case $s \in \mathbb{C}$. Nonetheless, in the general case, Steps 1 and 2 do not become as simple as the case of $s = 0$, as the differential equation $d\pi_n^{\mathrm{I}}(D_s^\flat)p(t) = 0$ turns out to be Heun's differential equation:

$$\mathcal{D}_H(-1, -\frac{ns}{4}; -\frac{n}{2}, -\frac{n-1}{2}, \frac{1}{2}, \frac{1-n-s}{2}; t^2)p(t) = 0 \tag{1.7}$$

with the P-symbol

$$P \begin{Bmatrix} 0 & 1 & -1 & \infty & & \\ 0 & 0 & 0 & -\frac{n}{2} & t^2 & -\frac{ns}{4} \\ \frac{1}{2} & \frac{1+n+s}{2} & \frac{1+n-s}{2} & -\frac{n-1}{2} & & \end{Bmatrix}.$$

(See Sect. 8.1 for the definition of $\mathcal{D}_H(a, q; \alpha, \beta, \gamma, \delta; z)$.) As such, one needs to deal with local Heun functions at 0 [32]; these are not as easy to handle as hypergeometric functions for our purpose. For instance, in Step 1, one needs to classify $(s, n) \in \mathbb{C} \times \mathbb{Z}_{\geq 0}$ for which the local Heun functions in consideration are polynomials; however, as opposed to hypergeometric functions, classifying such parameters is not an easy problem at all, since only known are necessary conditions in which local Heun functions are reduced to polynomials. To resolve this problem, inspired by a work [36] of Tamori, we use a different identification $\Omega^{\mathrm{II}} : \mathfrak{k} \xrightarrow{\sim} \mathfrak{sl}(2, \mathbb{C})$ so that the differential equation $d\pi_n^{\mathrm{II}}(D_s^\flat)p(t) = 0$ becomes again a hypergeometric equation, namely,

$$\mathcal{D}_F(-\frac{n}{2}, -\frac{n+s-1}{4}, \frac{3-n+s}{4}; t^2)p(t) = 0, \tag{1.8}$$

where $\mathcal{D}_F(a, b, c; z)$ denotes the differential operator

$$\mathcal{D}_F(a, b, c; z) = z(1-z)\frac{d^2}{dz^2} + (c - (a+b+1)z)\frac{d}{dz} - ab \tag{1.9}$$

such that

$$\mathcal{D}_F(a, b, c; z)f(z) = 0$$

is the hypergeometric differential equation. For the details, see Sect. 3.2 and 4.3. We accomplished the K-type formulas of $\mathcal{S}ol(s; \sigma)_K$ by using the hypergeometric model $d\pi_n^{\mathrm{II}}(D_s^\flat)p(t) = 0$ (see Theorem 5.16).

The Heun model $d\pi_n^{\mathrm{I}}(D_s^\flat)p(t) = 0$ and the hypergeometric model $d\pi_n^{\mathrm{II}}(D_s^\flat)p(t) = 0$ are related by a Cayley transform $\pi_n(k_0)$ given by an element k_0 of K (3.12). Namely, we have:

$$d\pi_n^{\mathrm{II}}(D_s^\flat)p(t) = 0 \iff d\pi_n^{\mathrm{I}}(D_s^\flat)\pi_n(k_0)p(t) = 0$$

(see Proposition 4.6). For instance, put:

$$u_{[s;n]}(t) := Hl(-1, -\frac{ns}{4}; -\frac{n}{2}, -\frac{n-1}{2}, \frac{1}{2}, \frac{1-n-s}{2}; t^2) \tag{1.10}$$

and

$$a_{[s;n]}(t) := F(-\frac{n}{2}, -\frac{n+s-1}{4}, \frac{3-n+s}{4}; t^2),$$

where $Hl(a, q; \alpha, \beta, \gamma, \delta; z)$ denotes the local Heun function at $z = 0$ and $F(a, b, c; z) \equiv {}_2F_1(a, b, c; z)$ is the Gauss hypergeometric function. Then, for

$$\mathcal{D}_H^{[s;n]} := \mathcal{D}_H(-1, -\frac{ns}{4}; -\frac{n}{2}, -\frac{n-1}{2}, \frac{1}{2}, \frac{1-n-s}{2}; t^2)$$

in (1.7) and

$$\mathcal{D}_F^{[s;n]} := \mathcal{D}_F(-\frac{n}{2}, -\frac{n+s-1}{4}, \frac{3-n+s}{4}; t^2)$$

in (1.8), we have

$$\mathcal{D}_H^{[s;n]}u_{[s;n]}(t) = 0 \quad \text{and} \quad \mathcal{D}_F^{[s;n]}a_{[s;n]}(t) = 0. \tag{1.11}$$

It will be shown in Lemma 6.1 that, for appropriate $(s, n) \in \mathbb{C} \times \mathbb{Z}_{\geq 0}$ at which $u_{[s;n]}(t)$ and $a_{[s;n]}(t)$ are both polynomials, the two functions $u_{[s;n]}(t)$ and $a_{[s;n]}(t)$ are related as

$$\pi_n(k_0)a_{[s;n]}(t) \in \mathbb{C}u_{[s;n]}(t), \tag{1.12}$$

equivalently,

$$(1-\sqrt{-1}t)^n a_{[s;n]}\left(\frac{1+\sqrt{-1}t}{1-\sqrt{-1}t}\cdot\sqrt{-1}\right) \in \mathbb{C}u_{[s;n]}(t).$$

Via the transformation $\pi_n(k_0)$, we also give in detail the space $\mathrm{Sol}(s; n)$ of K-type solutions for the Heun model $d\pi_n^{\mathrm{I}}(D_s^\flat)p(t) = 0$ (Sect. 6).

1.4 Sequences $\{P_k(x; y)\}_{k=0}^{\infty}$ and $\{Q_k(x; y)\}_{k=0}^{\infty}$ of Tridiagonal Determinants

Recall from (1.11) that the local Heun function $u_{[s;n]}(t)$ in (1.10) is a local solution at $t = 0$ to the Heun differential equation (1.7). Let $v_{[s;n]}(t)$ be the second solution (see (4.24)). It will be shown in Proposition 7.3 that $u_{[s;n]}(t)$ and $v_{[s;n]}(t)$ have power series representations:

$$u_{[s;n]}(t) = \sum_{k=0}^{\infty} P_k(s; n)\frac{t^{2k}}{(2k)!} \qquad \text{and} \qquad v_{[s;n]}(t) = \sum_{k=0}^{\infty} Q_k(s; n)\frac{t^{2k+1}}{(2k+1)!},$$

where $P_k(x; y)$ and $Q_k(x; y)$ are certain $k \times k$ tridiagonal determinants (see Sect. 7.1). For instance, the $\frac{n}{2} \times \frac{n}{2}$ tridiagonal determinant $P_{\frac{n}{2}}(x; n)$ for $y = n \in 2\mathbb{Z}_{\geq 0}$ is given as $P_0(x; 0) = 1$, $P_1(x; 2) = 2x$, and for $n \in 4 + 2\mathbb{Z}_{\geq 0}$,

$$P_{\frac{n}{2}}(x; n) = \begin{vmatrix} nx & 1\cdot 2 & & & & \\ -(n-1)n & (n-4)x & 3\cdot 4 & & & \\ & -(n-3)(n-2) & (n-8)x & 5\cdot 6 & & \\ & & \cdots & \cdots & \cdots & \\ & & & -5\cdot 6 & -(n-8)x & (n-3)(n-2) \\ & & & & -3\cdot 4 & -(n-4)x \end{vmatrix}. \tag{1.13}$$

Similarly, the $\frac{n-2}{2} \times \frac{n-2}{2}$ tridiagonal determinant $Q_{\frac{n-2}{2}}(x; n)$ for $y = n \in 2(1+\mathbb{Z}_{\geq 0})$ is given as $Q_0(x; 2) = 1$, $Q_1(x; 4) = 2x$, and for $n \in 6 + 2\mathbb{Z}_{\geq 0}$,

$$Q_{\frac{n-2}{2}}(x; n) = \begin{vmatrix} (n-2)x & 2\cdot 3 & & & & \\ -(n-2)(n-1) & (n-6)x & 4\cdot 5 & & & \\ & -(n-4)(n-3) & (n-10)x & 6\cdot 7 & & \\ & & \cdots & \cdots & \cdots & \\ & & & -6\cdot 7 & -(n-10)x & (n-4)(n-3) \\ & & & & -4\cdot 5 & -(n-6)x \end{vmatrix}. \tag{1.14}$$

In 1854, Sylvester observed that an $(n+1) \times (n+1)$ centrosymmetric tridiagonal determinant

$$\mathrm{Sylv}(x;n) := \begin{vmatrix} x & 1 & & & & & \\ n & x & 2 & & & & \\ & n-1 & x & 3 & & & \\ & & & \cdots\cdots\cdots & & & \\ & & & 3 & x & n-1 & \\ & & & & 2 & x & n \\ & & & & & 1 & x \end{vmatrix} \tag{1.15}$$

satisfies the following factorization formula (cf. [1, 8, 10, 17, 28, 34, 37]):

$$\mathrm{Sylv}(x;n) = \begin{cases} (x^2-1^2)(x^2-3^2)\cdots(x^2-(n-2)^2)(x^2-n^2) & \text{if } n \text{ is odd,} \\ x(x^2-2^2)(x^2-4^2)\cdots(x^2-(n-2)^2)(x^2-n^2) & \text{if } n \text{ is even.} \end{cases} \tag{1.16}$$

By utilizing certain results for the Heun model $d\pi_n^{\mathrm{I}}(D_0^\flat)p(t) = 0$, we show that $P_{[(n+2)/2]}(x;n)$ and $Q_{[(n+1)/2]}(x;n)$ enjoy similar but more involved factorization formulas. See Theorems 7.5 and 7.7 for the details. Based on the factorization formula, we also express $P_{(n+1)/2}(x;n)$ for n odd in terms of the Sylvester determinant $\mathrm{Sylv}(x;n)$ (Corollary 7.8).

Here are some remarks on the factorization formulas in order. First, the factorization formulas $P_{(n+2)/2}(x;n)$ and $Q_{n/2}(x;n)$ for $n \in 2\mathbb{Z}$ also follow from a more general formula in [14, Prop. 5.11], which is obtained via some clever trick based on a change of variables and some general formula on tridiagonal determinants (see, for instance, [5, p. 52] and [14, Lem. 5.10]). As mentioned in the remark after Theorem 1.2, the factorization formulas for $P_{(n+2)/2}(x;n)$ and $Q_{n/2}(x;n)$ play an essential role in [14] to determine $\dim_{\mathbb{C}} \mathrm{Sol}(s;n)$. In this paper we somewhat do this process backward including the case of n odd. Schematically, the difference is described as follows:

- Kable [14]: "factorization $\to \dim_{\mathbb{C}} \mathrm{Sol}(s;n)$" $(n \in 2\mathbb{Z}_{\geq 0})$
- this paper: "$\mathrm{Sol}(s;n) \to$ Heun $\to$ factorization" $(n \in \mathbb{Z}_{\geq 0})$

(It is recalled that, as the linear group $SL(m,\mathbb{R})$ is considered in [14], only even $n \in 2\mathbb{Z}_{\geq 0}$ appear for the space $\mathrm{Sol}(s;n)$ of K-type solutions for $SO(3) \subset SL(3,\mathbb{R})$ in the case of $m = 3$, whereas we handle all $n \in \mathbb{Z}_{\geq 0}$ in this paper, as $Spin(3) \subset \widetilde{SL}(3,\mathbb{R})$ is in consideration.)

The techniques used in [14] can also be applied for the case of n odd. Nevertheless, it requires some involved computations; for instance, the trick used in the cited paper does not make the situation as simple as for the case of n even, and further, the general formula ([5, p. 52] and [14, Lem. 5.10]) cannot be applied either. By simply applying the results for the Heun model $d\pi_n^{\mathrm{I}}(D_s^\flat)p(t) = 0$, we successfully avoid such computations.

We shall further discuss "functional equations" of $\{P_k(x; y)\}_{k=0}^\infty$ and $\{Q_k(x; y)\}_{k=0}^\infty$ in Sect. 7. The functional equations of sequences of polynomials discussed in this paper seem new to the literature. We then give two more examples of such functional equations, that is, we show that Cayley continuants $\{\mathrm{Cay}_k(x; y)\}_{k=0}^\infty$ and Krawtchouk polynomials $\{\mathcal{K}_k(x; y)\}_{k=0}^\infty$ also admit the functional equations. We now briefly summarize them as follows.

Given $\ell \in \mathbb{Z}_{\geq 0}$, define a map:

$$\tau_\ell \colon \{0, 1, \ldots, \ell\} \to \{0, 1, \ldots, \ell\}, \quad k \mapsto \ell - k.$$

For $k \in \mathbb{Z}_{\geq 0}$, we let

$$\widetilde{P}_{[s;n]}(k) = \frac{P_k(s; n)}{(2k)!}, \qquad \widetilde{Q}_{[s;n]}(k) = \frac{Q_k(s; n)}{(2k+1)!},$$

$$\widetilde{\mathrm{Cay}}_{[s;n]}(k) = \frac{\mathrm{Cay}_k(s; n)}{k!}, \quad \widetilde{\mathcal{K}}_{[s;n]}(k) = \mathcal{K}_k(s; n).$$

The details of the following theorems will be discussed in Sects. 7.4 and 9.

Theorem 1.3 (Theorem 7.9) *Let $n \in 2\mathbb{Z}_{\geq 0}$ and $s \in \mathcal{S}\mathrm{ol}_{\frac{n+2}{2}}(P; n)$. For $k \leq \frac{n}{2}$, we have:*

$$\widetilde{P}_{[s;n]}(\tau_{\frac{n}{2}}(k)) = \theta_{(P;n)}(s)\widetilde{P}_{[s;n]}(k) \quad \text{with } \theta_{(P;n)}(s) \in \{\pm 1\}.$$

Theorem 1.4 (Theorem 7.11) *Let $n \in 2(1 + \mathbb{Z}_{\geq 0})$ and $s \in \mathcal{S}\mathrm{ol}_{\frac{n}{2}}(Q; n)$. For $k \leq \frac{n-2}{2}$, we have:*

$$\widetilde{Q}_{[s;n]}(\tau_{\frac{n-2}{2}}(k)) = \theta_{(Q;n)}(s)\widetilde{Q}_{[s;n]}(k) \quad \text{with } \theta_{(Q;n)}(s) \in \{\pm 1\}.$$

Theorem 1.5 (Theorem 7.13) *Let $n \in 1 + 2\mathbb{Z}_{\geq 0}$ and $s \in \mathcal{S}\mathrm{ol}_{\frac{n+1}{2}}(P; n)$. For $k \leq \frac{n-1}{2}$, we have:*

$$\begin{pmatrix} \widetilde{P}_{[s;n]}(\tau_{\frac{n-1}{2}}(k)) \\ \widetilde{Q}_{[s;n]}(\tau_{\frac{n-1}{2}}(k)) \end{pmatrix} = \begin{pmatrix} 0 & \widetilde{P}_{[s;n]}(\frac{n-1}{2}) \\ \widetilde{Q}_{[s;n]}(\frac{n-1}{2}) & 0 \end{pmatrix} \begin{pmatrix} \widetilde{P}_{[s;n]}(k) \\ \widetilde{Q}_{[s;n]}(k) \end{pmatrix}.$$

Theorem 1.6 (Theorem 9.2) *Let $n \in \mathbb{Z}_{\geq 0}$ and $s \in \mathcal{S}\mathrm{ol}_{n+1}(\mathrm{Cay}; n)$. For $k \leq n$, we have:*

$$\widetilde{\mathrm{Cay}}_{[s;n]}(\tau_n(k)) = (-1)^{\frac{n-s}{2}}\widetilde{\mathrm{Cay}}_{[s;n]}(k).$$

Theorem 1.7 (Theorem 9.6) *Let $n \in \mathbb{Z}_{\geq 0}$ and $s \in \mathrm{Sol}_{n+1}(\mathcal{K}; n)$. For $k \leq n$, we have:*

$$\widetilde{\mathcal{K}}_{[s;n]}(\tau_n(k)) = (-1)^s \widetilde{\mathcal{K}}_{[s;n]}(k).$$

1.5 Organization

We now outline the rest of this paper. This paper consists of nine sections including this introduction. In Sect. 2, we overview a general framework established in [25] for the Peter–Weyl theorem (1.4) for the space of K-finite solutions to intertwining differential operators. In Sect. 3, we collect necessary notation and normalizations for $\widetilde{SL}(3, \mathbb{R})$. Two identifications $\Omega^J : \mathfrak{k} \xrightarrow{\sim} \mathfrak{sl}(2, \mathbb{C})$ for $J = \mathrm{I}, \mathrm{II}$ are discussed in this section.

The purpose of Sect. 4 is to recall from [14] the Heisenberg ultrahyperebolic operator $\square_s = R(D_s)$ for $\widetilde{SL}(3, \mathbb{R})$ and to study the associated differential equations $d\pi_n^J(D_s^\flat)p(t) = 0$. In this section we identify $d\pi_n^J(D_s^\flat)p(t) = 0$ with Heun's differential equation ($J = \mathrm{I}$) and the hypergeometric differential equation ($J = \mathrm{II}$) via the identifications $\Omega^J : \mathfrak{k} \xrightarrow{\sim} \mathfrak{sl}(2, \mathbb{C})$. At the end of this section, we give a recipe to classify the space $\mathrm{Sol}(s; n)$ of K-type solutions to $\square_s = R(D_s)$.

In accordance with the recipe given in Sect. 4, we show the K-type decompositions of $\mathcal{S}ol(s; \sigma)_K$ in the hypergeometric model $d\pi_n^{\mathrm{II}}(D_s^\flat)p(t) = 0$ in Sect. 5. This is accomplished in Theorem 5.16. We then convert results in the hypergeometric model $d\pi_n^{\mathrm{II}}(D_s^\flat)p(t) = 0$ to the Heun model $d\pi_n^{\mathrm{I}}(D_s^\flat)p(t) = 0$ by a Cayley transform $\pi_n(k_0)$ in Sect. 6.

In Sect. 7, by utilizing the results in the Heun model $d\pi_n^{\mathrm{I}}(D_s^\flat)p(t) = 0$, we investigate two sequences $\{P_k(x; n)\}_{k=0}^\infty$ and $\{Q_k(x; n)\}_{k=0}^\infty$ of tridiagonal determinants. In particular we give an expression of $P_{\frac{n+1}{2}}(x; n)$ for n odd in terms of the Sylvester determinant $\mathrm{Sylv}(x; n)$ (Corollary 7.8).

The last two sections are appendices. In order to study $\{P_k(x; n)\}_{k=0}^\infty$ and $\{Q_k(x; n)\}_{k=0}^\infty$, we use some general facts on the coefficients of a power series expression of a local Heun function. We collect those facts in Sect. 8. In Sect. 9, we briefly discuss the functional equations of Cayley continuants $\{\mathrm{Cay}_k(x; y)\}_{k=0}^\infty$ and Krawtchouk polynomials $\{\mathcal{K}_k(x; y)\}_{k=0}^\infty$.

2 Peter–Weyl Theorem for the Space of K-finite Solutions

The aim of this section is to recall a general framework established in [25]. In particular, we give a Peter–Weyl theorem for the space of K-finite solutions to intertwining differential operators between parabolically induced representations. This is done in Theorem 2.5.

2.1 General Framework

Let G be a reductive Lie group with Lie algebra $\mathfrak{g}_0$. Choose a Cartan involution $\theta : \mathfrak{g}_0 \to \mathfrak{g}_0$ and write $\mathfrak{g}_0 = \mathfrak{k}_0 \oplus \mathfrak{s}_0$ for the Cartan decomposition of $\mathfrak{g}_0$ with respect to θ. Here $\mathfrak{k}_0$ and $\mathfrak{s}_0$ stand for the $+1$ and -1 eigenspaces of θ, respectively. We take maximal abelian subspaces $\mathfrak{a}_0^{\min} \subset \mathfrak{s}_0$ and $\mathfrak{t}_0^{\min} \subset \mathfrak{m}_0^{\min} := Z_{\mathfrak{k}_0}(\mathfrak{a}_0^{\min})$ such that $\mathfrak{h}_0 := \mathfrak{a}_0^{\min} \oplus \mathfrak{t}_0^{\min}$ is a Cartan subalgebra of $\mathfrak{g}_0$.

For a real Lie algebra $\mathfrak{y}_0$, we denote by $\mathfrak{y}$ the complexification of $\mathfrak{y}_0$. For instance, the complexifications of $\mathfrak{g}_0$, $\mathfrak{h}_0$, $\mathfrak{a}_0^{\min}$, and $\mathfrak{m}_0^{\min}$ are denoted by $\mathfrak{g}$, $\mathfrak{h}$, $\mathfrak{a}^{\min}$, and $\mathfrak{m}^{\min}$, respectively. We write $\mathcal{U}(\mathfrak{y})$ for the universal enveloping algebra of a Lie algebra $\mathfrak{y}$.

Let $\Delta \equiv \Delta(\mathfrak{g}, \mathfrak{h})$ be the set of roots with respect to the Cartan subalgebra $\mathfrak{h}$ and $\Sigma \equiv \Sigma(\mathfrak{g}_0, \mathfrak{a}_0^{\min})$ denote the set of restricted roots with respect to $\mathfrak{a}_0^{\min}$. We choose a positive system Δ^+ and Σ^+ in such a way that Δ^+ and Σ^+ are compatible.

Let $\mathfrak{n}_0^{\min}$ be the nilpotent subalgebra of $\mathfrak{g}_0$ corresponding to Σ^+, so that $\mathfrak{p}_0^{\min} := \mathfrak{m}_0^{\min} \oplus \mathfrak{a}_0^{\min} \oplus \mathfrak{n}_0^{\min}$ is a Langlands decomposition of a minimal parabolic subalgebra $\mathfrak{p}_0^{\min}$ of $\mathfrak{g}_0$. Fix a standard parabolic subalgebra $\mathfrak{p}_0 \supset \mathfrak{p}_0^{\min}$ with Langlands decomposition $\mathfrak{p}_0 = \mathfrak{m}_0 \oplus \mathfrak{a}_0 \oplus \mathfrak{n}_0$. Let P be a parabolic subgroup of G with Lie algebra $\mathfrak{p}_0$. We write $P = MAN$ for the Langlands decomposition of P corresponding to $\mathfrak{p}_0 = \mathfrak{m}_0 \oplus \mathfrak{a}_0 \oplus \mathfrak{n}_0$.

For $\mu \in \mathfrak{a}^* \simeq \mathrm{Hom}_{\mathbb{R}}(\mathfrak{a}_0, \mathbb{C})$, we define a one-dimensional representation $\mathbb{C}_\mu$ of A as $a \mapsto e^{\mu}(a) := e^{\mu(\log a)}$ for $a \in A$. Then, for a finite-dimensional representation $W_\sigma = (\sigma, W)$ of M and weight $\lambda \in \mathfrak{a}^*$, we define an MA-representation $W_{\sigma,\lambda}$ as

$$W_{\sigma,\lambda} := W_\sigma \otimes \mathbb{C}_\lambda.$$

(We note that the definition of $W_{\sigma,\lambda}$ was slightly different from the one in [25].) By letting N act on $W_{\sigma,\lambda}$ trivially, we regard $W_{\sigma,\lambda}$ as a representation of P. We identify the Fréchet space $C^\infty\left(G/P, \mathcal{W}_{\sigma,\lambda}\right)$ of smooth sections for the G-equivariant homogeneous vector bundle $\mathcal{W}_{\sigma,\lambda} := G \times_P W_{\sigma,\lambda} \to G/P$ as $C^\infty\left(G/P, \mathcal{W}_{\sigma,\lambda}\right) \simeq (C^\infty(G) \otimes W_{\sigma,\lambda})^P$, that is,

$$\begin{aligned} &C^\infty\left(G/P, \mathcal{W}_{\sigma,\lambda}\right) \\ &\simeq \left\{ f \in C^\infty(G) \otimes W_{\sigma,\lambda} : f(gman) = \sigma(m)^{-1} e^{-\lambda}(a) f(g) \text{ for all } man \in MAN \right\}, \end{aligned}$$

where G acts by left translation. Let $\rho = (1/2)\mathrm{Trace}(\mathrm{ad}|_{\mathfrak{n}}) \in \mathfrak{a}^*$. Then we realize a parabolically induced representation:

$$I_P(\sigma, \lambda) := \mathrm{Ind}_P^G(\sigma \otimes (\lambda + \rho) \otimes \mathbb{1}) \tag{2.1}$$

of G on $C^\infty\left(G/P, \mathcal{W}_{\sigma,\lambda+\rho}\right)$.

For (σ, λ), (η, ν) with finite-dimensional representations σ, η of M and weights $\lambda, \nu \in \mathfrak{a}^*$, we write $\mathrm{Hom}_G(I_P(\sigma, \lambda), I_P(\eta, \nu))$ for the space of intertwining operators from $I_P(\sigma, \lambda)$ to $I_P(\eta, \nu)$. Then we set:

$$\mathrm{Diff}_G(I_P(\sigma, \lambda), I_P(\eta, \nu)) := \mathrm{Diff}(I_P(\sigma, \lambda), I_P(\eta, \nu)) \cap \mathrm{Hom}_G(I_P(\sigma, \lambda), I_P(\eta, \nu)),$$

where $\mathrm{Diff}(I_P(\sigma, \lambda), I_P(\eta, \nu))$ is the space of differential operators from $I_P(\sigma, \lambda)$ to $I_P(\eta, \nu)$.

Let $\mathrm{Irr}(M)_{\mathrm{fin}}$ be the set of equivalence classes of irreducible finite-dimensional representations of M. As M is not connected in general, we write M_0 for the identity component of M. Let $\mathrm{Irr}(M/M_0)$ denote the set of irreducible representations of the component group M/M_0. Via the surjection $M \twoheadrightarrow M/M_0$, we regard $\mathrm{Irr}(M/M_0)$ as a subset of $\mathrm{Irr}(M)_{\mathrm{fin}}$. Let id_ξ denote the identity map on $\xi \in \mathrm{Irr}(M/M_0)$. Lemma 2.1 shows that the tensored operator $\mathcal{D} \otimes \mathrm{id}_\xi$ to $\mathcal{D} \in \mathrm{Diff}_G(I_P(\sigma, \lambda), I_P(\eta, \nu))$ is also an intertwining differential operator.

Lemma 2.1 ([25, Lems. 2.17, 2.21]) *Let* $\mathcal{D} \in \mathrm{Diff}_G(I_P(\sigma, \lambda), I_P(\eta, \nu))$. *For any* $\xi \in \mathrm{Irr}(M/M_0)$, *we have:*

$$\mathcal{D} \otimes \mathrm{id}_\xi \in \mathrm{Diff}_G(I_P(\sigma \otimes \xi, \lambda), I_P(\eta \otimes \xi, \nu)).$$

If the M-representation σ is the trivial character $\sigma = \chi_{\mathrm{triv}}$, then Lemma 2.1 in particular shows that differential operator $\mathcal{D} \in \mathrm{Diff}_G(I_P(\chi_{\mathrm{triv}}, \lambda), I_P(\eta, \nu))$ yields:

$$\mathcal{D} \otimes \mathrm{id}_\xi \in \mathrm{Diff}_G(I_P(\xi, \lambda), I_P(\eta \otimes \xi, \nu)) \quad \text{for } \xi \in \mathrm{Irr}(M/M_0). \tag{2.2}$$

2.2 *Peter–Weyl Theorem for* $\mathcal{S}ol_{(u;\lambda)}(\xi)_K$

For a later purpose, we specialize the M-representations σ, η to be $\sigma = \chi_{\mathrm{triv}}$ and $\eta = \chi$, where χ is some character of M. In this situation, it follows from (2.2) that, for $\mathcal{D} \in \mathrm{Diff}_G(I_P(\chi_{\mathrm{triv}}, \lambda), I_P(\chi, \nu))$ and $\xi \in \mathrm{Irr}(M/M_0)$, we have:

$$\mathcal{D} \otimes \mathrm{id}_\xi \in \mathrm{Diff}_G(I_P(\xi, \lambda), I_P(\chi \otimes \xi, \nu)).$$

By the duality theorem between intertwining differential operators and homomorphisms between generalized Verma modules (see, for instance, [6, 18, 22]), any intertwining differential operator $\mathcal{D} \in \mathrm{Diff}_G(I_P(\chi_{\mathrm{triv}}, \lambda), I_P(\chi, \nu))$ is of the form $\mathcal{D} = R(u)$ for some $u \in \mathcal{U}(\bar{\mathfrak{n}})$, where R denotes the infinitesimal right translation of $\mathcal{U}(\mathfrak{g})$ and $\bar{\mathfrak{n}}$ is the opposite nilpotent radical to $\mathfrak{n}$, namely,

$$\mathrm{Diff}_G(I_P(\chi_{\mathrm{triv}}, \lambda), I_P(\chi, \nu)) \subset \{R(u) \text{ for } u \in \mathcal{U}(\bar{\mathfrak{n}})\}.$$

In particular, intertwining differential operators $\mathcal{D}$ are determined by the complex Lie algebra $\mathfrak{g}$ and independent of its real forms.

It follows from [11, Lem. 2.1] and [25, Lem. 2.24] that there exists a $(\mathcal{U}(\mathfrak{k} \cap \mathfrak{m}), K \cap M)$-isomorphism:

$$\mathcal{U}(\mathfrak{k}) \otimes_{\mathcal{U}(\mathfrak{k}\cap\mathfrak{m})} \mathbb{C}_{\chi_{\mathrm{triv}}} \xrightarrow{\sim} \mathcal{U}(\mathfrak{g}) \otimes_{\mathcal{U}(\mathfrak{p})} \mathbb{C}_{\chi_{\mathrm{triv}}, -(\lambda+\rho)},$$
$$u \otimes \mathbb{1}_{\chi_{\mathrm{triv}}} \mapsto u \otimes (\mathbb{1}_{\chi_{\mathrm{triv}}} \otimes \mathbb{1}_{-(\lambda+\rho)}), \tag{2.3}$$

where $K \cap M$ acts on $\mathcal{U}(\mathfrak{k}) \otimes_{\mathcal{U}(\mathfrak{k}\cap\mathfrak{m})} \mathbb{C}_{\chi_{\mathrm{triv}}}$ and $\mathcal{U}(\mathfrak{g}) \otimes_{\mathcal{U}(\mathfrak{p})} \mathbb{C}_{\chi_{\mathrm{triv}}, -(\lambda+\rho)}$ diagonally. Via the isomorphism (2.3), we define a *compact model* $u^\flat \in \mathcal{U}(\mathfrak{k})$ of $u \in \mathcal{U}(\bar{\mathfrak{n}})$ as follows.

Definition 2.2 For $u \in \mathcal{U}(\mathfrak{g})$, we denote by $u^\flat \in \mathcal{U}(\mathfrak{k})$ an element of $\mathcal{U}(\mathfrak{k})$ such that the identity

$$u^\flat \otimes (\mathbb{1}_{\chi_{\mathrm{triv}}} \otimes \mathbb{1}_{-(\lambda+\rho)}) = u \otimes (\mathbb{1}_{\chi_{\mathrm{triv}}} \otimes \mathbb{1}_{-(\lambda+\rho)}) \tag{2.4}$$

holds in $\mathcal{U}(\mathfrak{g}) \otimes_{\mathcal{U}(\mathfrak{p})} \mathbb{C}_{\chi_{\mathrm{triv}}, -(\lambda+\rho)}$.

Remark 2.3 We remark that a compact model $u^\flat$ of u is not unique; any choice of $u^\flat$ satisfying the identity (2.4) is acceptable.

Let $\mathrm{Irr}(K)$ be the set of equivalence classes of irreducible representations of the maximal compact subgroup K with Lie algebra $\mathfrak{k}_0$. For $V_\delta := (\delta, V) \in \mathrm{Irr}(K)$ and $u \in \mathcal{U}(\bar{\mathfrak{n}})$, we define a subspace $\mathrm{Sol}_{(u)}(\delta)$ of V_δ by

$$\mathrm{Sol}_{(u)}(\delta) := \{v \in V_\delta : d\delta(\tau(u^\flat))v = 0\}. \tag{2.5}$$

Here $d\delta$ denotes the differential of δ and τ denotes the conjugation $\tau : \mathfrak{g} \to \mathfrak{g}$ with respect to the real form $\mathfrak{g}_0$, that is, $\tau(X_1 + \sqrt{-1}X_2) = X_1 - \sqrt{-1}X_2$ for $X_1, X_2 \in \mathfrak{g}_0$.

Lemma 2.4 ([25, Lem. 2.41]) *Let $u \in \mathcal{U}(\bar{\mathfrak{n}})$ with $R(u) \in \mathrm{Diff}_G(I_P(\chi_{\mathrm{triv}}, \lambda), I_P(\chi, \nu))$. Then the space $\mathrm{Sol}_{(u)}(\delta)$ is a $K \cap M$-representation.*

We set:

$$\mathrm{Sol}_{(u)}^{\mathfrak{k}\cap\mathfrak{m}}(\delta) := \mathrm{Sol}_{(u)}(\delta) \cap V_\delta^{\mathfrak{k}\cap\mathfrak{m}},$$

where $V_\delta^{\mathfrak{k}\cap\mathfrak{m}}$ is the subspace of $\mathfrak{k} \cap \mathfrak{m}$-invariant vectors of V_δ. Clearly $\mathrm{Sol}_{(u)}^{\mathfrak{k}\cap\mathfrak{m}}(\delta)$ is a $K \cap M$-subrepresentation of $\mathrm{Sol}_{(u)}(\delta)$. Further, since $\mathfrak{k} \cap \mathfrak{m}$ acts on $\xi \in \mathrm{Irr}(M/M_0)$ trivially, we have:

$$\mathrm{Hom}_{K\cap M}\left(\mathrm{Sol}_{(u)}(\delta), \xi\right) \neq \{0\} \quad \text{if and only if} \quad \mathrm{Hom}_{K\cap M}\left(\mathrm{Sol}_{(u)}^{\mathfrak{k}\cap\mathfrak{m}}(\delta), \xi\right) \neq \{0\} \tag{2.6}$$

via the composition of maps $K \cap M \hookrightarrow M \twoheadrightarrow M/M_0$.

Let $I_P(\chi_{\text{triv}}, \lambda)_K$ denote the $(\mathfrak{g}, K)$-module consisting of the K-finite vectors of $I_P(\chi_{\text{triv}}, \lambda)$. Then, given $R(u) \in \text{Diff}_G(I_P(\chi_{\text{triv}}, \lambda), I_P(\chi, \nu))$ and $\xi \in \text{Irr}(M/M_0)$, we set:

$$\mathcal{S}ol_{(u;\lambda)}(\xi)_K := \{f \in I_P(\xi, \lambda)_K : (R(u) \otimes \text{id}_\xi)f = 0\}.$$

For the sake of future convenience, we now give a slight modification of a Peter–Weyl theorem [25, Thm. 1.2] for $\mathcal{S}ol_{(u;\lambda)}(\xi)_K$, although such a modification is not necessary for the main objective of this paper. We remark that this modified version is a more direct generalization of the argument given after the proof of [14, Thm. 2.6] than [25, Thm. 1.2].

Theorem 2.5 *Let $R(u) \in \text{Diff}_G(I_P(\chi_{\text{triv}}, \lambda), I_P(\chi, \nu))$ and $\xi \in \text{Irr}(M/M_0)$. Then the $(\mathfrak{g}, K)$-module $\mathcal{S}ol_{(u;\lambda)}(\xi)_K$ can be decomposed as a K-representation as*

$$\mathcal{S}ol_{(u;\lambda)}(\xi)_K \simeq \bigoplus_{\delta \in \text{Irr}(K)} V_\delta \otimes \text{Hom}_{K\cap M}\left(\text{Sol}_{(u)}^{\mathfrak{k}\cap\mathfrak{m}}(\delta), \xi\right). \tag{2.7}$$

Proof It follows from [25, Thm. 1.2] that the space $\mathcal{S}ol_{(u;\lambda)}(\xi)_K$ can be decomposed as

$$\mathcal{S}ol_{(u;\lambda)}(\xi)_K \simeq \bigoplus_{\delta \in \text{Irr}(K)} V_\delta \otimes \text{Hom}_{K\cap M}\left(\text{Sol}_{(u)}(\delta), \xi\right).$$

Now the equivalence (2.6) concludes the theorem.

If G is a split real group and $P = MAN$ is a minimal parabolic subgroup of G, we have $K \cap M = M$, $M/M_0 = M$, and $\mathfrak{k} \cap \mathfrak{m} = \{0\}$. Thus in this case Theorem 2.5 is simplified as follows.

Corollary 2.6 Suppose that G is split real and $P = MAN$ is a minimal parabolic subgroup of G. Then, for $R(u) \in \text{Diff}_G(I_P(\chi_{\text{triv}}, \lambda), I_P(\chi, \nu))$ and $\sigma \in \text{Irr}(M)$, the $(\mathfrak{g}, K)$-module $\mathcal{S}ol_{(u;\lambda)}(\sigma)_K$ can be decomposed as

$$\mathcal{S}ol_{(u;\lambda)}(\sigma)_K \simeq \bigoplus_{\delta \in \text{Irr}(K)} V_\delta \otimes \text{Hom}_{M}\left(\text{Sol}_{(u)}(\delta), \sigma\right). \tag{2.8}$$

3 Specialization to $(\widetilde{SL}(3, \mathbb{R}), B)$

The purpose of this section is to specialize the general theory discussed in Sect. 2 to a pair $(\widetilde{SL}(3, \mathbb{R}), B)$, where B is a minimal parabolic subgroup of $\widetilde{SL}(3, \mathbb{R})$. In this section we in particular give a recipe for computing K-type formulas of the space $\mathcal{S}ol_{(u;\lambda)}(\sigma)_K$ of K-finite solutions to $R(u) \in \text{Diff}_G(I_B(\chi_{\text{triv}}, \lambda), I_B(\chi, \nu))$. It is described in Sect. 3.7.

3.1 Notation and Normalizations

We begin by recalling from [25, Sect. 4.1] the notation and normalizations for $\widetilde{SL}(3, \mathbb{R})$. Let $G = \widetilde{SL}(3, \mathbb{R})$ with Lie algebra $\mathfrak{g}_0 = \mathfrak{sl}(3, \mathbb{R})$. Fix a Cartan involution $\theta\colon \mathfrak{g}_0 \to \mathfrak{g}_0$ such that $\theta(U) = -U^t$. We write $\mathfrak{k}_0$ and $\mathfrak{s}_0$ for the $+1$ and -1 eigenspaces of θ, respectively, so that $\mathfrak{g}_0 = \mathfrak{k}_0 \oplus \mathfrak{s}_0$ is a Cartan decomposition of $\mathfrak{g}_0$. We put $\mathfrak{a}_0 := \mathrm{span}_{\mathbb{R}}\{E_{ii} - E_{i+1,i+1} : i = 1, 2\}$ and $\mathfrak{n}_0 := \mathrm{span}_{\mathbb{R}}\{E_{12}, E_{23}, E_{13}\}$, where E_{ij} denote matrix units. Then $\mathfrak{b}_0 := \mathfrak{a}_0 \oplus \mathfrak{n}_0$ is a minimal parabolic subalgebra of $\mathfrak{g}_0$.

Let K, A, and N be the analytic subgroups of G with Lie algebras $\mathfrak{k}_0$, $\mathfrak{a}_0$, and $\mathfrak{n}_0$, respectively. Then $G = KAN$ is an Iwasawa decomposition of G. We write $M = Z_K(\mathfrak{a}_0)$, so that $B := MAN$ is a minimal parabolic subgroup of G with Lie algebra $\mathfrak{b}_0$.

We denote by $\mathfrak{g}$ the complexification of the Lie algebra $\mathfrak{g}_0$ of G. A similar convention is employed also for subgroups of G; for instance, $\mathfrak{b} = \mathfrak{a} \oplus \mathfrak{n}$ is a Borel subalgebra of $\mathfrak{g} = \mathfrak{sl}(3, \mathbb{C})$. We write $\bar{\mathfrak{n}}$ for the nilpotent radical opposite to $\mathfrak{n}$.

Let $\Delta \equiv \Delta(\mathfrak{g}, \mathfrak{a})$ denote the set of roots of $\mathfrak{g}$ with respect to $\mathfrak{a}$. We denote by Δ^+ and Π the positive system corresponding to $\mathfrak{b}$ and the set of simple roots of Δ^+, respectively. Then we have $\Pi = \{\varepsilon_1 - \varepsilon_2, \varepsilon_2 - \varepsilon_3\}$, where ε_j are the dual basis of E_{jj} for $j = 1, 2, 3$. The root spaces $\mathfrak{g}_{\varepsilon_1-\varepsilon_2}$ and $\mathfrak{g}_{\varepsilon_2-\varepsilon_3}$ are then given as $\mathfrak{g}_{\varepsilon_1-\varepsilon_2} = \mathbb{C}E_{12}$ and $\mathfrak{g}_{\varepsilon_2-\varepsilon_3} = \mathbb{C}E_{23}$. We write ρ for half the sum of the positive roots, namely, $\rho = \varepsilon_1 - \varepsilon_3$.

We define $X, Y \in \mathfrak{g}$ as

$$X = \begin{pmatrix} 0 & 0 & 0 \\ 1 & 0 & 0 \\ 0 & 0 & 0 \end{pmatrix} \quad \text{and} \quad Y = \begin{pmatrix} 0 & 0 & 0 \\ 0 & 0 & 0 \\ 0 & 1 & 0 \end{pmatrix}. \tag{3.1}$$

Then X and Y are root vectors for $-(\varepsilon_1 - \varepsilon_2)$ and $-(\varepsilon_2 - \varepsilon_3)$, respectively. The opposite nilpotent radical $\bar{\mathfrak{n}}$ is thus given as $\bar{\mathfrak{n}} = \mathrm{span}\{X, Y, [X, Y]\}$.

3.2 Two Identifications for $\mathfrak{so}(3, \mathbb{C}) \simeq \mathfrak{sl}(2, \mathbb{C})$

As $\mathfrak{k}_0 = \mathfrak{so}(3) \simeq \mathfrak{su}(2)$, we have $\mathfrak{k} = \mathfrak{so}(3, \mathbb{C}) \simeq \mathfrak{sl}(2, \mathbb{C})$. For later applications we consider two identifications of $\mathfrak{so}(3, \mathbb{C})$ with $\mathfrak{sl}(2, \mathbb{C})$:

$$\Omega^J : \mathfrak{so}(3, \mathbb{C}) \xrightarrow{\sim} \mathfrak{sl}(2, \mathbb{C}) \quad \text{for } J = \mathrm{I}, \mathrm{II}.$$

Let E_+, E_-, and E_0 be the elements of $\mathfrak{sl}(2, \mathbb{C})$ defined as

$$E_+ := \begin{pmatrix} 0 & 1 \\ 0 & 0 \end{pmatrix}, \quad E_- := \begin{pmatrix} 0 & 0 \\ 1 & 0 \end{pmatrix}, \quad E_0 := \begin{pmatrix} 1 & 0 \\ 0 & -1 \end{pmatrix}. \tag{3.2}$$

We now describe the identifications Ω^{I} and Ω^{II} in detail separately.

3.2.1 Identification $\mathfrak{k} \simeq \mathfrak{sl}(2, \mathbb{C})$ via Ω^{I}

We start with the identification $\Omega^{\mathrm{I}} \colon \mathfrak{k} \xrightarrow{\sim} \mathfrak{sl}(2, \mathbb{C})$. First observe that $\mathfrak{k}_0 = \mathfrak{so}(3)$ is spanned by the three matrices:

$$B_1 := \begin{pmatrix} 0 & 0 & -1 \\ 0 & 0 & 0 \\ 1 & 0 & 0 \end{pmatrix}, \quad B_2 := \begin{pmatrix} 0 & 0 & 0 \\ 0 & 0 & -1 \\ 0 & 1 & 0 \end{pmatrix}, \quad B_3 := \begin{pmatrix} 0 & -1 & 0 \\ 1 & 0 & 0 \\ 0 & 0 & 0 \end{pmatrix}$$

with commutation relations

$$[B_1, B_2] = B_3, \quad [B_1, B_3] = -B_2, \quad \text{and} \quad [B_2, B_3] = B_1. \tag{3.3}$$

On the other hand, the Lie algebra $\mathfrak{su}(2)$ is spanned by

$$A_1 := \begin{pmatrix} \sqrt{-1} & 0 \\ 0 & -\sqrt{-1} \end{pmatrix}, \quad A_2 := \begin{pmatrix} 0 & 1 \\ -1 & 0 \end{pmatrix}, \quad A_3 := \begin{pmatrix} 0 & \sqrt{-1} \\ \sqrt{-1} & 0 \end{pmatrix}$$

with commutation relations

$$[A_1, A_2] = 2A_3, \quad [A_1, A_3] = -2A_2, \quad \text{and} \quad [A_2, A_3] = 2A_1.$$

Then one may identify $\mathfrak{k}_0$ with $\mathfrak{su}(2)$ via the linear map:

$$\Omega_0^{\mathrm{I}} \colon \mathfrak{k}_0 \xrightarrow{\sim} \mathfrak{su}(2), \quad B_j \mapsto \tfrac{1}{2} A_j \quad \text{for } j = 1, 2, 3. \tag{3.4}$$

Let Z_+, Z_-, Z_0 be the elements of $\mathfrak{k} = \mathfrak{so}(3, \mathbb{C})$ defined as

$$Z_+ := B_2 - \sqrt{-1} B_3, \quad Z_- := -(B_2 + \sqrt{-1} B_3), \quad Z_0 := [Z_+, Z_-] = -2\sqrt{-1} B_1. \tag{3.5}$$

Then we have:

$$\mathfrak{k} = \operatorname{span}_{\mathbb{C}}\{Z_+, Z_-, Z_0\}. \tag{3.6}$$

As $A_2 - \sqrt{-1} A_3 = 2E_+$ and $-(A_2 + \sqrt{-1} A_3) = 2E_-$, the isomorphism (3.4) yields a Lie algebra isomorphism:

$$\Omega^{\mathrm{I}} \colon \mathfrak{k} \xrightarrow{\sim} \mathfrak{sl}(2, \mathbb{C}), \quad Z_j \mapsto E_j \quad \text{for } j = +, -, 0. \tag{3.7}$$

3.2.2 Identification $\mathfrak{k} \simeq \mathfrak{sl}(2, \mathbb{C})$ via Ω^{II}

We next discuss the identification $\Omega^{\mathrm{II}} \colon \mathfrak{k} \xrightarrow{\sim} \mathfrak{sl}(2, \mathbb{C})$. As for Ω^{I}_0, one may identify $\mathfrak{k}_0$ with $\mathfrak{su}(2)$ via the linear map

$$\Omega^{\mathrm{II}}_0 \colon \mathfrak{k}_0 \xrightarrow{\sim} \mathfrak{su}(2), \quad B_1 \mapsto \tfrac{1}{2}A_2, \quad B_2 \mapsto \tfrac{1}{2}A_1, \quad B_3 \mapsto -\tfrac{1}{2}A_3. \tag{3.8}$$

Let W_+, W_-, and W_0 be the elements of $\mathfrak{so}(3, \mathbb{C})$ defined by

$$W_+ := B_1 + \sqrt{-1}B_3, \quad W_- := -B_1 + \sqrt{-1}B_3, \quad W_0 := [W_+, W_-] = -2\sqrt{-1}B_2. \tag{3.9}$$

Then we have:

$$\mathfrak{k} = \mathrm{span}_{\mathbb{C}}\{W_+, W_-, W_0\}. \tag{3.10}$$

The isomorphism (3.8) yields a Lie algebra isomorphism:

$$\Omega^{\mathrm{II}} \colon \mathfrak{k} \xrightarrow{\sim} \mathfrak{sl}(2, \mathbb{C}), \quad W_j \mapsto E_j \quad \text{for } j = +, -, 0. \tag{3.11}$$

Let

$$k_0 := \frac{1}{\sqrt{2}} \begin{pmatrix} \sqrt{-1} & 1 \\ -1 & -\sqrt{-1} \end{pmatrix} \in SU(2). \tag{3.12}$$

We have:

$$\mathrm{Ad}(k_0)A_1 = A_2, \quad \mathrm{Ad}(k_0)A_2 = A_1, \quad \mathrm{Ad}(k_0)A_3 = -A_3. \tag{3.13}$$

Proposition 3.1 *The two identifications Ω^{I} and Ω^{II} satisfy:*

$$\Omega^{\mathrm{I}} = \mathrm{Ad}(k_0) \circ \Omega^{\mathrm{II}}$$

on $\mathfrak{k}$.

Proof Since

$$\Omega^{\mathrm{I}} \colon B_1 \mapsto \frac{1}{2}A_1, \quad B_2 \mapsto \frac{1}{2}A_2, \quad B_3 \mapsto \frac{1}{2}A_3,$$
$$\Omega^{\mathrm{II}} \colon B_1 \mapsto \frac{1}{2}A_2, \quad B_2 \mapsto \frac{1}{2}A_1, \quad B_3 \mapsto -\frac{1}{2}A_3,$$

the proposed assertion simply follows from (3.13).

3.3 A Realization of the Subgroup $M = Z_K(\mathfrak{a}_0)$

As K is isomorphic to $SU(2)$, we realize $M = Z_K(\mathfrak{a}_0)$ as a subgroup of $SU(2)$. To do so, first observe that the adjoint action Ad of $SU(2)$ on $\mathfrak{su}(2)$ yields a two-to-one covering map $SU(2) \twoheadrightarrow \mathrm{Ad}(SU(2)) \simeq SO(3)$ of $SO(3)$.

If $\mathrm{Ad}(SU(2))$ is realized as a matrix group with respect to the ordered basis $\{A_2, -A_1, A_3\}$ of $\mathfrak{su}(2)$, then the covering map $SU(2) \twoheadrightarrow \mathrm{Ad}(SU(2))$ corresponds to the isomorphism $\Omega^{\mathrm{I}}\colon \mathfrak{k} \xrightarrow{\sim} \mathfrak{sl}(2, \mathbb{C})$ in (3.7).[1] In this realization, the elements $m_j^{\mathrm{I}} \in SU(2)$ with

$$m_0^{\mathrm{I}} := \begin{pmatrix} 1 & 0 \\ 0 & 1 \end{pmatrix}, \; m_1^{\mathrm{I}} := \begin{pmatrix} \sqrt{-1} & 0 \\ 0 & -\sqrt{-1} \end{pmatrix}, \; m_2^{\mathrm{I}} := \begin{pmatrix} 0 & 1 \\ -1 & 0 \end{pmatrix}, \; m_3^{\mathrm{I}} := \begin{pmatrix} 0 & \sqrt{-1} \\ \sqrt{-1} & 0 \end{pmatrix} \tag{3.14}$$

are mapped to $\pm m_j^{J} \mapsto m_j \in SO(3)$ for $j = 0, 1, 2, 3$, where

$$m_0 = \begin{pmatrix} 1 & 0 & 0 \\ 0 & 1 & 0 \\ 0 & 0 & 1 \end{pmatrix}, \; m_1 = \begin{pmatrix} -1 & 0 & 0 \\ 0 & 1 & 0 \\ 0 & 0 & -1 \end{pmatrix}, \; m_2 = \begin{pmatrix} 1 & 0 & 0 \\ 0 & -1 & 0 \\ 0 & 0 & -1 \end{pmatrix}, \; m_3 = \begin{pmatrix} -1 & 0 & 0 \\ 0 & -1 & 0 \\ 0 & 0 & 1 \end{pmatrix}.$$

Further, since the realization of the covering map $SU(2) \twoheadrightarrow \mathrm{Ad}(SU(2))$ in concern corresponds to the Ω^{I} map, for $Z \in \mathfrak{k}$, we have:

$$\Omega^{\mathrm{I}}(\mathrm{Ad}(m_j)Z) = \mathrm{Ad}(m_j^{\mathrm{I}})\Omega^{\mathrm{I}}(Z) \quad \text{for } j = 0, 1, 2, 3.$$

Similarly, if $\mathrm{Ad}(SU(2))$ is realized as a matrix group with respect to the ordered basis $\{-A_1, A_2, A_3\}$, then the covering map $SU(2) \twoheadrightarrow \mathrm{Ad}(SU(2))$ corresponds to the isomorphism $\Omega^{\mathrm{II}}\colon \mathfrak{k} \xrightarrow{\sim} \mathfrak{sl}(2, \mathbb{C})$ in (3.11). In this realization, the elements

$$m_j^{\mathrm{II}} := k_0^{-1} m_j^{\mathrm{I}} k_0 \tag{3.15}$$

are mapped to $\pm m_j^{\mathrm{II}} \mapsto m_j$. A direct computation shows that m_j^{II} are given as

$$m_0^{\mathrm{II}} = \begin{pmatrix} 1 & 0 \\ 0 & 1 \end{pmatrix}, \; m_1^{\mathrm{II}} = \begin{pmatrix} 0 & 1 \\ -1 & 0 \end{pmatrix}, \; m_2^{\mathrm{II}} = \begin{pmatrix} \sqrt{-1} & 0 \\ 0 & -\sqrt{-1} \end{pmatrix}, \; m_3^{\mathrm{II}} = -\begin{pmatrix} 0 & \sqrt{-1} \\ \sqrt{-1} & 0 \end{pmatrix}. \tag{3.16}$$

By (3.14) and (3.16), we have:

$$m_0^{\mathrm{II}} = m_0^{\mathrm{I}}, \quad m_1^{\mathrm{II}} = m_2^{\mathrm{I}}, \quad m_2^{\mathrm{II}} = m_1^{\mathrm{I}}, \quad m_3^{\mathrm{II}} = -m_3^{\mathrm{I}}. \tag{3.17}$$

[1] The ordered basis $\{A_2, A_1, A_3\}$ in [25, p. 235] should also be $\{A_2, -A_1, A_3\}$.

Since this realization corresponds to the Ω^{II} map, for $Z \in \mathfrak{k}$, we have:

$$\Omega^{\mathrm{II}}(\mathrm{Ad}(m_j)Z) = \mathrm{Ad}(m_j^{\mathrm{II}})\Omega^{\mathrm{II}}(Z) \quad \text{for } j = 0, 1, 2, 3.$$

As $Z_{SO(3)}(\mathfrak{a}_0) = \{m_0, m_1, m_2, m_3\}$, we then realize M as a subgroup of $SU(2)$ as

$$M = \left\{\pm m_0^{\mathrm{I}},\ \pm m_1^{\mathrm{I}},\ \pm m_2^{\mathrm{I}},\ \pm m_3^{\mathrm{I}}\right\}$$
$$\left(= \left\{\pm m_0^{\mathrm{II}},\ \pm m_1^{\mathrm{II}},\ \pm m_2^{\mathrm{II}},\ \pm m_3^{\mathrm{II}}\right\}\right).$$

The subgroup M is isomorphic to the quaternion group Q_8, a noncommutative group of order 8.

3.4 *Irreducible Representations* $\mathrm{Irr}(K)$ *of* K

In order to compute a K-type formula via (2.8), we next discuss the sets $\mathrm{Irr}(K)$ and $\mathrm{Irr}(M)$ of equivalence classes of irreducible representations of K and M, respectively.

We first consider a polynomial realization of $\mathrm{Irr}(K)$. Let $\mathrm{Pol}[t]$ be the space of polynomials of one variable t with complex coefficients and set:

$$\mathrm{Pol}_n[t] := \{p(t) \in \mathrm{Pol}[t] : \deg p(t) \leq n\}.$$

Then we realize the set $\mathrm{Irr}(K)$ of equivalence classes of irreducible representations of $K \simeq SU(2)$ as

$$\mathrm{Irr}(K) \simeq \{(\pi_n, \mathrm{Pol}_n[t]) : n \in \mathbb{Z}_{\geq 0}\}, \tag{3.18}$$

where the representation π_n of $SU(2)$ on $\mathrm{Pol}_n[t]$ is defined as

$$(\pi_n(g)p)(t) := (ct + d)^n p\left(\frac{at + b}{ct + d}\right) \quad \text{for} \quad g = \begin{pmatrix} a & b \\ c & d \end{pmatrix}^{-1}. \tag{3.19}$$

It follows from (3.19) that the elements m_j^{I} defined in (3.14) act on $\mathrm{Pol}_n[t]$ via π_n as

$$m_1^{\mathrm{I}} \colon p(t) \mapsto (\sqrt{-1})^n p(-t); \quad m_2^{\mathrm{I}} \colon p(t) \mapsto t^n p\left(-\frac{1}{t}\right);$$
$$m_3^{\mathrm{I}} \colon p(t) \mapsto (-\sqrt{-1}t)^n p\left(\frac{1}{t}\right). \tag{3.20}$$

By (3.16), the elements m_j^{II} act as

$$m_1^{\mathrm{II}}\colon p(t) \mapsto t^n p\left(-\frac{1}{t}\right); \quad m_2^{\mathrm{II}}\colon p(t) \mapsto (\sqrt{-1})^n p(-t);$$

$$m_3^{\mathrm{II}}\colon p(t) \mapsto (\sqrt{-1}t)^n p\left(\frac{1}{t}\right). \tag{3.21}$$

Let $d\pi_n$ be the differential of the representation π_n. We extend $d\pi_n$ complex-linearly to $\mathfrak{sl}(2,\mathbb{C})$ and also naturally to the universal enveloping algebra $\mathcal{U}(\mathfrak{sl}(2,\mathbb{C}))$. It follows from (3.2) and (3.19) that E_+, E_-, and E_0 act on $\mathrm{Pol}_n[t]$ via $d\pi_n$ as

$$d\pi_n(E_+) = -\frac{d}{dt}, \quad d\pi_n(E_-) = -nt + t^2\frac{d}{dt}, \quad \text{and} \quad d\pi_n(E_0) = -2t\frac{d}{dt} + n. \tag{3.22}$$

3.5 *Irreducible Representations* $\mathrm{Irr}(M)$ *of* M

We next consider $\mathrm{Irr}(M)$ via the isomorphism (3.7). As M is isomorphic to the quaternion group Q_8, the set $\mathrm{Irr}(M)$ consists of four characters and one two-dimensional irreducible representation. For $\varepsilon, \varepsilon' \in \{\pm\}$, we define a character: $\chi^{SO(3)}_{(\varepsilon,\varepsilon')}\colon Z_{SO(3)}(\mathfrak{a}_0) \to \{\pm 1\}$ of $Z_{SO(3)}(\mathfrak{a}_0)$ as

$$\chi^{SO(3)}_{(\varepsilon,\varepsilon')}(\mathrm{diag}(a_1, a_2, a_3)) := |a_1|_\varepsilon\, |a_3|_{\varepsilon'},$$

where $|a|_+ := |a|$ and $|a|_- := a$. Via the character $\chi^{SO(3)}_{(\varepsilon,\varepsilon')}$ of $Z_{SO(3)}(\mathfrak{a}_0)$, we define a character $\chi_{(\varepsilon,\varepsilon')}\colon M \to \{\pm 1\}$ of M as

$$\chi_{(\varepsilon,\varepsilon')}(\pm m_j^{\mathrm{I}}) := \chi^{SO(3)}_{(\varepsilon,\varepsilon')}(m_j) \quad \text{for } j = 0, 1, 2, 3. \tag{3.23}$$

We often abbreviate $\chi_{(\varepsilon,\varepsilon')}$ as $(\varepsilon, \varepsilon')$. The character $(+,+)$, for instance, is the trivial character of M. The set $\mathrm{Irr}(M)$ may then be described as follows:

$$\mathrm{Irr}(M) = \{(+,+), (+,-), (-,+), (-,-), \mathbb{H}\}, \tag{3.24}$$

where $\mathbb{H}$ is the unique genuine two-dimensional representation of $M \simeq Q_8$.

Table 2 Character table for $(\varepsilon, \varepsilon')$ for m_j^J for $J \in \{\mathrm{I}, \mathrm{II}\}$

	$\pm m_0^J$	$\pm m_1^J$	$\pm m_2^J$	$\pm m_3^J$
$(+,+)$	1	1	1	1
$(+,-)$	1	-1	-1	1
$(-,+)$	1	-1	1	-1
$(-,-)$	1	1	-1	-1

We define $(\varepsilon, \varepsilon')(m_j^{\mathrm{II}})$ for $m_j^{\mathrm{II}} = k_0^{-1} m_j^{\mathrm{I}} k_0$ via the following commutative diagram in such a way that we have $(\varepsilon, \varepsilon')(m_j^{\mathrm{II}}) = (\varepsilon, \varepsilon')(m_j^{\mathrm{I}})$ for $j = 0, 1, 2, 3$:

$$\begin{array}{ccc} \mathrm{Pol}_n[t] & \xrightarrow[\sim]{\pi_n(k_0)} & \mathrm{Pol}_n[t] \\ {\scriptstyle \pi_n(m_j^{\mathrm{II}})}\downarrow & \circlearrowleft & \downarrow{\scriptstyle \pi_n(m_j^{\mathrm{I}})} \\ \mathrm{Pol}_n[t] & \xrightarrow[\pi_n(k_0)]{\sim} & \mathrm{Pol}_n[t] \end{array}$$

Table 2 is the character table for $(\varepsilon, \varepsilon')$ with m_j^J for $J \in \{\mathrm{I}, \mathrm{II}\}$.

3.6 Peter–Weyl Theorem for $\mathcal{S}ol_{(u;\lambda)}(\sigma)_K$ with Polynomial Realization

Observe that the action of an element $F \in \mathcal{U}(\mathfrak{k})$ on $\mathrm{Pol}_n[t]$ via $d\pi_n$ is of the form $d\pi_n(\Omega(F))$, where $\Omega \colon \mathfrak{k} \xrightarrow{\sim} \mathfrak{sl}(2, \mathbb{C})$ is an identification of $\mathfrak{k}$ with $\mathfrak{sl}(2, \mathbb{C})$ such as (3.7) and (3.11). To simplify the notation, we write:

$$d\pi_n^{\Omega}(F) = d\pi_n(\Omega(F)) \quad \text{for } F \in \mathcal{U}(\mathfrak{k}). \tag{3.25}$$

Then, for $R(u) \in \mathrm{Diff}_G(I((+,+), \lambda), I(\chi, \nu))$ with $(+,+)$ the trivial character of M, we set:

$$\mathrm{Sol}_{(u)}(n) := \{p(t) \in \mathrm{Pol}_n[t] : d\pi_n^{\Omega}\big(\tau(u^\flat)\big)p(t) = 0\}, \tag{3.26}$$

where $u^\flat \in \mathcal{U}(\mathfrak{k})$ is a compact model of $u \in \mathcal{U}(\bar{\mathfrak{n}})$ (see Definition 2.2). Then the K-type decomposition of $\mathcal{S}ol_{(u;\lambda)}(\sigma)_K$ in (2.8) with the polynomial realization (3.18) becomes:

$$\mathcal{S}ol_{(u;\lambda)}(\sigma)_K \simeq \bigoplus_{n \geq 0} \mathrm{Pol}_n[t] \otimes \mathrm{Hom}_M\left(\mathrm{Sol}_{(u)}(n), \sigma\right). \tag{3.27}$$

We remark that since $K \cap M = M$ and $\mathfrak{m} = \{0\}$, the $(\mathcal{U}(\mathfrak{k} \cap \mathfrak{m}), K \cap M)$-isomorphism (2.3) reduces to an M-isomorphism $\mathcal{U}(\mathfrak{k}) \otimes \mathbb{C}_{(+,+)} \xrightarrow{\sim} \mathcal{U}(\mathfrak{g}) \otimes_{\mathcal{U}(\mathfrak{b})}$

$\mathbb{C}_{(+,+),-\lambda}$. In particular a compact model $u^\flat \in \mathcal{U}(\mathfrak{k})$ of $u \in \mathcal{U}(\bar{\mathfrak{n}})$ is indeed unique in the present situation (see Remark 2.3).

3.7 Recipe for Determining the K-type Formula for $\mathcal{S}ol_{(u;\lambda)}(\sigma)_K$

For later convenience we summarize a recipe for computing the K-type formula for $\mathcal{S}ol_{(u;\lambda)}(\sigma)_K$ via (3.27). Let $R(u) \in \mathrm{Diff}_G(I((+,+),\lambda), I(\chi,\nu))$:

Step 0: Choose an identification $\Omega\colon \mathfrak{k} \simeq \mathfrak{sl}(2,\mathbb{C})$.
Step 1: Find the compact model $u^\flat$ of u (see Definition 2.2).
Step 2: Find the explicit formula of the differential operator $d\pi_n^\Omega(\tau(u^\flat))$.
Step 3: Solve the differential equation $d\pi_n^\Omega(\tau(u^\flat))p(t) = 0$ and classify $n \in \mathbb{Z}_{\geq 0}$ such that $\mathrm{Sol}_{(u)}(n) \neq \{0\}$.
Step 4: For $n \in \mathbb{Z}_{\geq 0}$ with $\mathrm{Sol}_{(u)}(n) \neq \{0\}$, classify the M-representations on $\mathrm{Sol}_{(u)}(n)$.
Step 5: Given $\sigma \in \mathrm{Irr}(M)$, classify $n \in \mathbb{Z}_{\geq 0}$ with $\mathrm{Sol}_{(u)}(n) \neq \{0\}$ such that

$$\mathrm{Hom}_M\left(\mathrm{Sol}_{(u)}(n), \sigma\right) \neq \{0\}.$$

4 Heisenberg Ultrahyperbolic Operator for $\widetilde{SL}(3,\mathbb{R})$

The aim of this section is to discuss a certain second-order differential operator called the Heisenberg ultrahyperbolic operator $R(D_s)$. In this section we apply to $R(D_s)$ the theory described in Sect. 3 for arbitrary intertwining differential operators $R(u)$ for $\widetilde{SL}(3,\mathbb{R})$. We continue the notation and normalizations from Sect. 3.

4.1 Heisenberg Ultrahyperbolic Operator $R(D_s)$

We start with the definition of the Heisenberg ultrahyperbolic operator $R(D_s)$ for $\widetilde{SL}(3,\mathbb{R})$. As $R(D_s)$ being an intertwining differential operator (see Proposition 4.1 below), it is defined for the complex Lie algebra $\mathfrak{g} = \mathfrak{sl}(3,\mathbb{C})$.

Let X and Y be the elements of $\bar{\mathfrak{n}}$ defined in (3.1). For $s \in \mathbb{C}$, we define D_s as

$$D_s := (XY + YX) + s[X,Y] \in \mathcal{U}(\mathfrak{g}).$$

Then the second-order differential operator

$$R(D_s) = R\big((XY + YX) + s[X,Y]\big)$$

is called the *Heisenberg ultrahyperbolic operator* for $\mathfrak{g} = \mathfrak{sl}(3, \mathbb{C})$. For the general definition for $\mathfrak{sl}(m, \mathbb{C})$ with arbitrary rank $m \geq 3$, see [14, Sect. 3].

Recall from Sect. 3.1 that the simple roots $\alpha, \beta \in \Pi$ are realized as $\alpha = \varepsilon_1 - \varepsilon_2$ and $\beta = \varepsilon_2 - \varepsilon_3$. Then, for $s \in \mathbb{C}$, we set:

$$\widetilde{\rho}(s) := \widetilde{\rho} - s\widetilde{\rho}^{\perp} \tag{4.1}$$

with

$$\widetilde{\rho} := \frac{1}{2}(\varepsilon_1 - \varepsilon_3) \quad \text{and} \quad \widetilde{\rho}^{\perp} := \frac{1}{2}(\varepsilon_1 + \varepsilon_3).$$

Remark that as $\widetilde{\rho}^{\perp} \notin \mathfrak{a}^*$, we have $\widetilde{\rho}(s) \notin \mathfrak{a}^*$ unless $s = 0$. For $\sigma \in \mathrm{Irr}(M)$, we write:

$$I(\sigma, \pm\widetilde{\rho}(s)) = I_B(\sigma, \pm\widetilde{\rho}(s)|_{\mathfrak{a}^*}) \tag{4.2}$$

for the parabolically induced representation $I_B(\sigma, \pm\widetilde{\rho}(s)|_{\mathfrak{a}^*})$ defined as in (2.1). Proposition 4.1 shows that the operator $R(D_s)$ is indeed an intertwining differential operator between parabolically induced representations.

Proposition 4.1 *We have:*

$$R(D_s) \in \mathrm{Diff}_G(I((+,+), -\widetilde{\rho}(s)), I((-,-), \widetilde{\rho}(-s))).$$

Consequently, for $\sigma \in \mathrm{Irr}(M)$, *we have:*

$$R(D_s) \otimes \mathrm{id}_\sigma \in \mathrm{Diff}_G(I(\sigma, -\widetilde{\rho}(s)), I((-,-) \otimes \sigma, \widetilde{\rho}(-s))).$$

Proof The first assertion readily follows from [14, Lem. 3.2] and [25, Lem. 6.4]. Lemma 2.1 then concludes the second.

Remark 4.2 Proposition 4.1 in particular shows that the Heisenberg ultrahyperbolic operator $R(D_s)$ cannot be obtained as a residue of the Knapp–Stein intertwining operator.

Our aim is to compute the branching law of the space $\mathcal{S}ol_{(D_s; -\widetilde{\rho}(s))}(\sigma)_K$ of K-finite solutions to $(R(D_s) \otimes \mathrm{id}_\sigma)f = 0$ as a K-representation. The K-type formula in (3.27) with $(u; \lambda) = (D_s; -\widetilde{\rho}(s))$ gives:

$$\mathcal{S}ol_{(D_s; -\widetilde{\rho}(s))}(\sigma)_K \simeq \bigoplus_{n \geq 0} \mathrm{Pol}_n[t] \otimes \mathrm{Hom}_M \left(\mathrm{Sol}_{(D_s)}(n), \sigma\right) \tag{4.3}$$

with

$$\mathrm{Sol}_{(D_s)}(n) = \{p(t) \in \mathrm{Pol}_n[t] : d\pi_n^{\Omega}\left(\tau(D_s^{\flat})\right)p(t) = 0\}.$$

Observe that, for X and Y in (3.1), we have $X, Y \in \mathfrak{g}_0 = \mathfrak{sl}(3, \mathbb{R})$. Thus $\tau(D_s^\flat)$ for $D_s = (XY + YX) + s[X, Y]$ is simply given as

$$\tau(D_s^\flat) = \tau(D_s)^\flat = D_{\bar{s}}^\flat, \tag{4.4}$$

where $\bar{s}$ denotes the complex conjugate of $s \in \mathbb{C}$. Then we write:

$$\mathrm{Sol}(s; n) := \mathrm{Sol}_{(D_{\bar{s}})}(n)$$

so that

$$\begin{aligned}\mathrm{Sol}(s; n) &= \{p(t) \in \mathrm{Pol}_n[t] : d\pi_n^\Omega(\tau(D_{\bar{s}}^\flat))p(t) = 0\} \\ &= \{p(t) \in \mathrm{Pol}_n[t] : d\pi_n^\Omega(D_s^\flat)p(t) = 0\}.\end{aligned} \tag{4.5}$$

Similarly, we put:

$$\mathcal{S}ol(s; \sigma)_K := \mathcal{S}ol_{(D_{\bar{s}}; -\widetilde{\rho}(\bar{s}))}(\sigma)_K.$$

Then the K-type formula (4.3) yields:

$$\mathcal{S}ol(s; \sigma)_K \simeq \bigoplus_{n \geq 0} \mathrm{Pol}_n[t] \otimes \mathrm{Hom}_M\left(\mathrm{Sol}(s; n), \sigma\right). \tag{4.6}$$

Hereafter we consider the branching law of $\mathcal{S}ol(s; \sigma)_K$ instead of $\mathcal{S}ol_{(D_s; -\widetilde{\rho}(s))}(\sigma)_K$.

In order to solve the equation $d\pi_n^\Omega(D_s^\flat)p(t) = 0$ in (4.5), we next find the compact model $D_s^\flat$ of D_s, which is an element of $\mathcal{U}(\mathfrak{k})$ such that

$$D_s^\flat \otimes (\mathbb{1}_{(+,+)} \otimes \mathbb{1}_{\widetilde{\rho}(s)-\rho}) = D_s \otimes (\mathbb{1}_{(+,+)} \otimes \mathbb{1}_{\widetilde{\rho}(s)-\rho})$$

in $\mathcal{U}(\mathfrak{g}) \otimes_{\mathcal{U}(\mathfrak{b})} \mathbb{C}_{(+,+),\,\widetilde{\rho}(s)-\rho}$. For $X, Y \in \bar{\mathfrak{n}}$ in (3.1), define $X^\flat, Y^\flat \in \mathfrak{k}$ as

$$X^\flat := X + \theta(X) \quad \text{and} \quad Y^\flat := Y + \theta(Y),$$

where θ is the Cartan involution defined as $\theta(U) = -U^t$.

Lemma 4.3 *We have:*

$$D_s^\flat = X^\flat Y^\flat + Y^\flat X^\flat + s[X^\flat, Y^\flat] \in \mathcal{U}(\mathfrak{k}). \tag{4.7}$$

Proof One can easily verify that

$$(X^\flat Y^\flat + Y^\flat X^\flat + s[X^\flat, Y^\flat]) \otimes (\mathbb{1}_{(+,+)} \otimes \mathbb{1}_{\widetilde{\rho}(s)-\rho}) = D_s \otimes (\mathbb{1}_{(+,+)} \otimes \mathbb{1}_{\widetilde{\rho}(s)-\rho})$$

in $\mathcal{U}(\mathfrak{g}) \otimes_{\mathcal{U}(\mathfrak{b})} \mathbb{C}_{(+,+),\,\widetilde{\rho}(s)-\rho}$.

4.2 Relationship Between $d\pi_n^{\mathrm{I}}(D_s^\flat)$ and $d\pi_n^{\mathrm{II}}(D_s^\flat)$

In Sect. 3.2, we discussed two identifications of $\mathfrak{k}$ with $\mathfrak{sl}(2, \mathbb{C})$, namely,

$$\Omega^{\mathrm{I}} \colon \mathfrak{k} \xrightarrow{\sim} \mathfrak{sl}(2, \mathbb{C}) \quad \text{and} \quad \Omega^{\mathrm{II}} \colon \mathfrak{k} \xrightarrow{\sim} \mathfrak{sl}(2, \mathbb{C}). \tag{4.8}$$

Thus the compact model $D_s^\flat$ may act on $\mathrm{Pol}_n[t]$ via $d\pi_n$ as

$$d\pi_n(\Omega^{\mathrm{I}}(D_s^\flat)) \quad \text{and} \quad d\pi_n(\Omega^{\mathrm{II}}(D_s^\flat)). \tag{4.9}$$

As for the notation $d\pi_n^{\Omega}(F) = d\pi_n(\Omega(F))$ in (3.25), we abbreviate (4.9) as

$$d\pi_n^J(D_s^\flat) = d\pi_n(\Omega^J(D_s^\flat)) \quad \text{for } J \in \{\mathrm{I}, \mathrm{II}\}.$$

We next discuss a relationship between $d\pi_n^{\mathrm{I}}(D_s^\flat)$ and $d\pi_n^{\mathrm{II}}(D_s^\flat)$.

We begin with the expression of $\Omega^J(D_s^\flat)$ in terms of the $\mathfrak{sl}(2)$-triple $\{E_+, E_-, E_0\}$ in (3.2). First, recall from (3.6) and (3.10) that $\mathfrak{k}$ can be given as

$$\mathfrak{k} = \mathrm{span}_{\mathbb{C}}\{Z_+, Z_-, Z_0\} = \mathrm{span}_{\mathbb{C}}\{W_+, W_-, W_0\},$$

where $\{Z_+, Z_-, Z_0\}$ and $\{W_+, W_-, W_0\}$ are the $\mathfrak{sl}(2)$-triples of $\mathfrak{k}$ defined in (3.5) and (3.9), respectively.

Lemma 4.4 *The compact model $D_s^\flat = X^\flat Y^\flat + Y^\flat X^\flat + s[X^\flat, Y^\flat]$ is expressed as*

$$D_s^\flat = \frac{\sqrt{-1}}{2}\left((Z_+ + Z_-)(Z_+ - Z_-) - (s-1)Z_0\right) \tag{4.10}$$

$$= \frac{1}{2}\left((W_+ + W_-)W_0 - (s-1)(W_+ - W_-)\right). \tag{4.11}$$

Proof A direct computation shows that $X^\flat$ and $Y^\flat$ are expressed in terms of the bases $\{Z_+, Z_-, Z_0\}$ and $\{W_+, W_-, W_0\}$ as

$$X^\flat = \frac{\sqrt{-1}}{2}(Z_+ + Z_-) = -\frac{\sqrt{-1}}{2}(W_+ + W_-),$$

$$Y^\flat = \frac{1}{2}(Z_+ - Z_-) = \frac{\sqrt{-1}}{2}W_0.$$

By substituting these expressions into $D_s^\flat$, one obtains the lemma.

Lemma 4.5 *The elements $\Omega^{\mathrm{I}}(D_s^{\flat})$ and $\Omega^{\mathrm{II}}(D_s^{\flat})$ in $\mathcal{U}(\mathfrak{sl}(2,\mathbb{C}))$ are given as*

$$\Omega^{\mathrm{I}}(D_s^{\flat}) = \frac{\sqrt{-1}}{2}\left((E_+ + E_-)(E_+ - E_-) - (s-1)E_0\right), \tag{4.12}$$

$$\Omega^{\mathrm{II}}(D_s^{\flat}) = \frac{1}{2}\left((E_+ + E_-)E_0 - (s-1)(E_+ - E_-)\right). \tag{4.13}$$

Proof This follows from (3.7), (3.11), and Lemma 4.4.

Recall from (3.12) that

$$k_0 = \frac{1}{\sqrt{2}}\begin{pmatrix} \sqrt{-1} & 1 \\ -1 & -\sqrt{-1} \end{pmatrix} \in SU(2).$$

Proposition 4.6 *The following two conditions on $p(t) \in \mathrm{Pol}_n[t]$ are equivalent:*

(i) $d\pi_n^{\mathrm{II}}(D_s^{\flat})p(t) = 0$;
(ii) $d\pi_n^{\mathrm{I}}(D_s^{\flat})\pi_n(k_0)p(t) = 0$.

Proof It follows from Proposition 3.1 that

$$\Omega^{\mathrm{I}}(D_s^{\flat}) = \mathrm{Ad}(k_0)\Omega^{\mathrm{II}}(D_s^{\flat}). \tag{4.14}$$

Since $d\pi_n^{J}(D_s^{\flat}) = d\pi_n(\Omega^{J}(D_s^{\flat}))$ for $J \in \{\mathrm{I}, \mathrm{II}\}$, the equivalence between (i) and (ii) readily follows from (4.14).

For $J \in \{\mathrm{I}, \mathrm{II}\}$, we write:

$$\mathrm{Sol}_J(s;n) = \{p(t) \in \mathrm{Pol}_n[t] : d\pi_n^{J}(D_s^{\flat})p(t) = 0\}. \tag{4.15}$$

In place of $\mathrm{Sol}(s;n)$ with $\mathrm{Sol}_J(s;n)$, the K-type decomposition (4.6) becomes:

$$\mathcal{S}ol(s;\sigma)_K \simeq \bigoplus_{n\geq 0} \mathrm{Pol}_n[t] \otimes \mathrm{Hom}_M\left(\mathrm{Sol}_J(s;n), \sigma\right). \tag{4.16}$$

It follows from Proposition 4.6 that $\pi_n(k_0)$ yields a linear isomorphism:

$$\pi_n(k_0)\colon \mathrm{Sol}_{\mathrm{II}}(s;n) \xrightarrow{\sim} \mathrm{Sol}_{\mathrm{I}}(s;n). \tag{4.17}$$

Since $m_j^{\mathrm{II}} = k_0^{-1} m_j^{\mathrm{I}} k_0$, by Lemma 2.4, the following diagram commutes:

$$\begin{array}{ccc} \mathrm{Sol}_{\mathrm{II}}(s;n) & \xrightarrow[\sim]{\pi_n(k_0)} & \mathrm{Sol}_{\mathrm{I}}(s;n) \\ {\scriptstyle \pi_n(m_j^{\mathrm{II}})}\downarrow & & \downarrow{\scriptstyle \pi_n(m_j^{\mathrm{I}})} \\ \mathrm{Sol}_{\mathrm{II}}(s;n) & \xrightarrow[\pi_n(k_0)]{\sim} & \mathrm{Sol}_{\mathrm{I}}(s;n) \end{array} \tag{4.18}$$

4.3 Differential Equation $d\pi_n^J(D_s^\flat)f(t) = 0$ for $J \in \{\mathrm{I}, \mathrm{II}\}$

As indicated in the recipe in Sect. 3.7, to determine the K-type decomposition of $\mathcal{S}ol(s; n)_K$, it is crucial to determine the space $\mathrm{Sol}_J(s; n)$ of K-type solutions, for which one needs to find polynomial solutions $p(t) \in \mathrm{Pol}_n[t]$ to differential equations $d\pi_n^J(D_s^\flat)f(t) = 0$ for $J \in \{\mathrm{I}, \mathrm{II}\}$. For this purpose we next investigate these differential equations.

4.3.1 Differential Equation $d\pi_n^{\mathrm{I}}(D_s^\flat)f(t) = 0$

We start with the differential equation $d\pi_n^{\mathrm{I}}(D_s^\flat)f(t) = 0$. The explicit formula of $d\pi_n^{\mathrm{I}}(D_s^\flat)$ is given as follows.

Lemma 4.7 *We have:*

$$-2\sqrt{-1}\, d\pi_n^{\mathrm{I}}(D_s^\flat) = (1-t^4)\frac{d^2}{dt^2} + 2\left((n-1)t^2 + s\right)t\frac{d}{dt} - n\left((n-1)t^2 + s\right). \tag{4.19}$$

Proof Recall from (4.12) that we have

$$-2\sqrt{-1}\,\Omega^{\mathrm{I}}(D_s^\flat) = (E_+ + E_-)(E_+ - E_-) - (s-1)E_0.$$

Now the proposed identity follows from a direct computation with (3.22).

In order to study the differential equation $d\pi_n^{\mathrm{I}}(D_s^\flat)f(t) = 0$, we set:

$$\mathcal{D}_H(a, q; \alpha, \beta, \gamma, \delta; z) := \frac{d^2}{dz^2} + \left(\frac{\gamma}{z} + \frac{\delta}{z-1} + \frac{\varepsilon}{z-a}\right)\frac{d}{dz} + \frac{\alpha\beta z - q}{z(z-1)(z-a)}, \tag{4.20}$$

where $a, q, \alpha, \beta, \gamma, \delta, \varepsilon$ are complex parameters with $a \neq 0, 1$ and $\gamma + \delta + \varepsilon = \alpha + \beta + 1$. Then the differential equation

$$\mathcal{D}_H(a, q; \alpha, \beta, \gamma, \delta; z)f(z) = 0 \tag{4.21}$$

is called *Heun's differential equation*. For a brief account of Eq. (4.21), see Sect. 8.1. The differential equation $d\pi_n^{\mathrm{I}}(D_s^\flat)f(t) = 0$ can be identified with Heun's differential equation as in Lemma 4.8 below. (We are grateful to Hiroyuki Ochiai for pointing it out.)

Lemma 4.8 *We have:*

$$-\frac{\sqrt{-1}d\pi_n^{\mathrm{I}}(D_s^\flat)}{2(1-t^4)t^2} = \mathcal{D}_H(-1, -\frac{ns}{4}; -\frac{n}{2}, -\frac{n-1}{2}, \frac{1}{2}, \frac{1-n-s}{2}; t^2).$$

Proof This follows from a direct computation with a change of variables $z = t^2$ on (4.19).

By Lemma 4.8, to determine the space $\mathrm{Sol}_{\mathrm{I}}(s; n)$ of K-type solutions to the equation $d\pi_n^{\mathrm{I}}(D_s^{\flat})f(t) = 0$, it suffices to find polynomial solutions to the Heun equation:

$$\mathcal{D}_H(-1, -\frac{ns}{4}; -\frac{n}{2}, -\frac{n-1}{2}, \frac{1}{2}, \frac{1-n-s}{2}; t^2)f(t) = 0. \tag{4.22}$$

Let $Hl(a, q; \alpha, \beta, \gamma, \delta; z)$ denote the power series solution $Hl(a, q; \alpha, \beta, \gamma, \delta; z) = \sum_{r=0}^{\infty} c_r z^r$ with $c_0 = 1$ to (4.21) at $z = 0$ [32]. We set:

$$u_{[s;n]}(t) := Hl(-1, -\frac{ns}{4}; -\frac{n}{2}, -\frac{n-1}{2}, \frac{1}{2}, \frac{1-n-s}{2}; t^2), \tag{4.23}$$

$$v_{[s;n]}(t) := tHl(-1, -\frac{(n-2)s}{4}; -\frac{n-1}{2}, -\frac{n-2}{2}, \frac{3}{2}, \frac{1-n-s}{2}; t^2). \tag{4.24}$$

It follows from (8.1) in Sect. 8 that $u_{[s;n]}(t)$ and $v_{[s;n]}(t)$ are linearly independent solutions at $t = 0$ to the Heun equation (4.22). Therefore we have:

$$\mathrm{Sol}_{\mathrm{I}}(s; n) \subset \mathbb{C}u_{[s;n]}(t) \oplus \mathbb{C}v_{[s;n]}(t). \tag{4.25}$$

We shall classify the parameters $(s, n) \in \mathbb{C} \times \mathbb{Z}_{\geq 0}$ such that $u_{[s;n]}(t), v_{[s;n]}(t) \in \mathrm{Sol}_{\mathrm{I}}(s; n)$ in Sect. 6 (see Propositions 6.2 and 6.3).

4.3.2 Differential Equation $d\pi_n^{\mathrm{II}}(D_s^{\flat})f(t) = 0$

We next consider $d\pi_n^{\mathrm{II}}(D_s^{\flat})f(t) = 0$. The equation $d\pi_n^{\mathrm{II}}(D_s^{\flat})f(t) = 0$ can be identified with the hypergeometric equation $\mathcal{D}_F(a, b, c; z)f(z) = 0$ (1.9) as follows.

Lemma 4.9 *We have:*

$$\frac{d\pi_n^{\mathrm{II}}(D_s^{\flat})}{4t} = \mathcal{D}_F(-\frac{n}{2}, -\frac{n+s-1}{4}, \frac{3-n+s}{4}; t^2).$$

Proof First, it follows from (4.13) and (3.22) that $d\pi_n^{\mathrm{II}}(D_s^{\flat})$ is given as

$$2\,d\pi_n^{\mathrm{II}}(D_s^{\flat}) = 2(1-t^2)t\frac{d^2}{dt^2} + ((s+3n-3)t^2 + (s-n+1))\frac{d}{dt} - n(s+n-1)t. \tag{4.26}$$

Then a direct computation with a change of variable $z = t^2$ concludes the lemma.

Similar to the equation $d\pi_n^{\mathrm{I}}(D_s^\flat)f(t) = 0$, Lemma 4.9 shows that, to determine the space $\mathrm{Sol}_{\mathrm{II}}(s; n)$ of K-type solutions to $d\pi_n^{\mathrm{II}}(D_s^\flat)f(t) = 0$, it suffices to find polynomial solutions to the hypergeometric equation:

$$\mathcal{D}_F(-\frac{n}{2}, -\frac{n+s-1}{4}, \frac{3-n+s}{4}; t^2)f(t) = 0. \tag{4.27}$$

We set:

$$a_{[s;n]}(t) := F(-\frac{n}{2}, -\frac{n+s-1}{4}, \frac{3-n+s}{4}; t^2),$$

$$b_{[s;n]}(t) := t^{\frac{1+n-s}{2}} F(-\frac{n+s-1}{4}, -\frac{s-1}{2}, \frac{5+n-s}{4}; t^2),$$

where $F(\alpha, \beta, \gamma; x) \equiv {}_2F_1(\alpha, \beta, \gamma; x)$ is the Gauss hypergeometric series:

$$F(\alpha, \beta, \gamma; x) = \sum_{k=0}^{\infty} \frac{(\alpha, k)(\beta, k)}{(\gamma, k)} \cdot \frac{x^k}{k!}.$$

Here (ℓ, m) stands for the shifted factorial, namely, $(\ell, m) = \frac{\Gamma(\ell+m)}{\Gamma(\ell)}$. Then $a_{[s;n]}(t)$ and $b_{[s;n]}(t)$ form fundamental set of solutions to (4.27) for suitable parameters $(s, n) \in \mathbb{C} \times \mathbb{Z}_{\geq 0}$ for which $a_{[s;n]}(t)$ and $b_{[s;n]}(t)$ are well-defined. Therefore we have:

$$\mathrm{Sol}_{\mathrm{II}}(s; n) \subset \mathbb{C}a_{[s;n]}(t) \oplus \mathbb{C}b_{[s;n]}(t). \tag{4.28}$$

The precise conditions of (s, n) such that $a_{[s;n]}(t), b_{[s;n]}(t) \in \mathrm{Sol}_{\mathrm{II}}(s; n)$ will be investigated in Sect. 5.1.

4.4 Recipe for the K-type Decomposition of $\mathcal{S}ol(s; \sigma)_K$

In Sect. 3.7, we gave a general recipe to determine the K-type formula (3.27) of $\mathcal{S}ol_{(u;\lambda)}(\sigma)_K$. We modify the recipe in such a way that it will fit well for $\mathcal{S}ol(s; \sigma)_K$ with

$$\mathcal{S}ol(s; \sigma)_K \simeq \bigoplus_{n \geq 0} \mathrm{Pol}_n[t] \otimes \mathrm{Hom}_M\left(\mathrm{Sol}_J(s; n), \sigma\right).$$

We first fix an identification:

$$\Omega^J : \mathfrak{k} \xrightarrow{\sim} \mathfrak{sl}(2, \mathbb{C}) \text{ for } J \in \{\mathrm{I}, \mathrm{II}\}.$$

Then the K-type formula of $\mathcal{S}ol(s;\sigma)_K$ may be determined by the following steps:

Step A: Classify $(s,n)\in\mathbb{C}\times\mathbb{Z}_{\geq 0}$ such that $\mathrm{Sol}_J(s;n)\neq\{0\}$. It follows from (4.25) and (4.28) that this is indeed equivalent to classifying $(s,n)\in\mathbb{C}\times\mathbb{Z}_{\geq 0}$ such that

- $u_{[s;n]}(t)\in\mathrm{Pol}_n[t]$ or $v_{[s;n]}(t)\in\mathrm{Pol}_n[t]$ for $J=\mathrm{I}$;
- $a_{[s;n]}(t)\in\mathrm{Pol}_n[t]$ or $b_{[s;n]}(t)\in\mathrm{Pol}_n[t]$ for $J=\mathrm{II}$.

Step B: For $(s,n)\in\mathbb{C}\times\mathbb{Z}_{\geq 0}$ with $\mathrm{Sol}_J(s;n)\neq\{0\}$, classify the M-representations on $\mathrm{Sol}_J(s;n)$.

Step C: Given $\sigma\in\mathrm{Irr}(M)$, classify $(s,n)\in\mathbb{C}\times\mathbb{Z}_{\geq 0}$ with $\mathrm{Sol}_J(s;n)\neq\{0\}$ such that

$$\mathrm{Hom}_M\left(\mathrm{Sol}_J(s;n),\,\sigma\right)\neq\{0\}.$$

In Sects. 5 and 6, we shall proceed with Steps A, B, and C for $J=\mathrm{II}$ and $J=\mathrm{I}$, respectively.

5 Hypergeometric Model $d\pi_n^{\mathrm{II}}(D_s^\flat)f(t)=0$

The aim of this section is to classify the K-type formulas for $\mathcal{S}ol(s;\sigma)_K$ by using the hypergeometric model $d\pi_n^{\mathrm{II}}(D_s^\flat)f(t)=0$. The decomposition formulas are achieved in Theorem 5.16.

5.1 *The Classification of* $\mathrm{Sol}_{\mathrm{II}}(s;n)$

As Step A of the recipe in Sect. 4.4, we first wish to classify $(s,n)\in\mathbb{C}\times\mathbb{Z}_{\geq 0}$ such that $\mathrm{Sol}_{\mathrm{II}}(s;n)\neq\{0\}$, where

$$\mathrm{Sol}_{\mathrm{II}}(s;n)=\{p(t)\in\mathrm{Pol}_n[t]:d\pi_n^{\mathrm{II}}(D_s^\flat)p(t)=0\}.$$

Recall from (4.28) that we have

$$\mathrm{Sol}_{\mathrm{II}}(s;n)\subset\mathbb{C}a_{[s;n]}(t)\oplus\mathbb{C}b_{[s;n]}(t)$$

with

$$a_{[s;n]}(t)=F(-\frac{n}{2},-\frac{n+s-1}{4},\frac{3-n+s}{4};t^2),\tag{5.1}$$

$$b_{[s;n]}(t)=t^{\frac{1+n-s}{2}}F(-\frac{n+s-1}{4},-\frac{s-1}{2},\frac{5+n-s}{4};t^2).\tag{5.2}$$

It thus suffices to classify $(s, n) \in \mathbb{C} \times \mathbb{Z}_{\geq 0}$ such that $a_{[s;n]}(t) \in \mathrm{Pol}_n[t]$ or $b_{[s;n]}(t) \in \mathrm{Pol}_n[t]$.

Given $n \in \mathbb{Z}_{\geq 0}$, we define $I_0^{\pm}$, J_0, I_1, $I_2^{\pm}$, J_2, $I_3 \subset \mathbb{Z}$ as follows:

$$
\begin{aligned}
I_0^{\pm} &:= \{\pm(3+4j) : j = 0, 1, \ldots, \frac{n}{4} - 1\} \quad (n \in 4\mathbb{Z}_{\geq 0});\\
J_0 &:= \{1 + 4j : j = 0, 1, \ldots, \frac{n}{4} - 1\} \quad (n \in 4\mathbb{Z}_{\geq 0});\\
I_1 &:= \{\pm 4j : j = 0, 1, \ldots, \left[\frac{n}{4}\right]\};\\
I_2^{\pm} &:= \{\pm(1+4j) : j = 0, 1, \ldots, \left[\frac{n}{4}\right]\};\\
J_2 &:= \{3 + 4j : j = 0, 1, \ldots, \left[\frac{n}{4}\right] - 1\};\\
I_3 &:= \{\pm(2+4j) : j = 0, 1, \ldots, \left[\frac{n}{4}\right]\}.
\end{aligned}
\tag{5.3}
$$

We start by observing when $a_{[s;n]}(t) \in \mathrm{Pol}_n[t]$. A key observation is that, for $(a, b, c) \in \mathbb{C}$ for which the hypergeometric function $F(a, b, c; z)$ is well-defined, we have:

$$F(a, b, c; z) \in \mathrm{Pol}_n[z] \iff a \in \{0, -1, \ldots, -n\} \text{ or } b \in \{0, -1, \ldots, -n\}.$$

Proposition 5.1 *The following conditions on $(s, n) \in \mathbb{C} \times \mathbb{Z}_{\geq 0}$ are equivalent:*

(i) $a_{[s;n]}(t) \in \mathrm{Pol}_n[t]$.
(ii) *One of the following conditions hold:*

  (a) $n \equiv 0 \pmod 4) : s \in \mathbb{C} \backslash J_0$;
  (b) $n \equiv 1 \pmod 4) : s \in I_1$;
  (c) $n \equiv 2 \pmod 4) : s \in \mathbb{C} \backslash J_2$;
  (d) $n \equiv 3 \pmod 4) : s \in I_3$.

Proof The proposition follows from a careful observation for the parameters of $a_{[s;n]}(t)$ in (5.1). Indeed, suppose that n is even. Then the parameters of $a_{[s;n]}(t)$ imply that $a_{[s;n]}(t) \notin \mathrm{Pol}_n[t]$ if and only if the following conditions are satisfied:

$$\frac{3 - n + s}{4} \in -\mathbb{Z}_{\geq 0}, \quad -\frac{n}{2} < \frac{3 - n + s}{4}, \quad \text{and} \quad -\frac{n + s - 1}{4} < \frac{3 - n + s}{4},$$

which are equivalent to $s \in J_0$ for $n \equiv 0 \pmod 4$ and $s \in J_2$ for $n \equiv 2 \pmod 4$. Now the assertions for (a) and (c) follow from the contrapositive of the arguments.

Next suppose that n is odd. In this case it follows from (5.1) that $a_{[s;n]}(t) \in \mathrm{Pol}_n[t]$ if and only if

$$0 \leq \frac{n + s - 1}{4} \leq \left[\frac{n}{2}\right] \quad \text{and} \quad \frac{n + s - 1}{4} \in \mathbb{Z}_{\geq 0},$$

which are equivalent to $s \in I_1$ for $n \equiv 1 \pmod 4$ and $s \in I_3$ for $n \equiv 3 \pmod 4$. Here we remark that, for $s \in I_1 \cup I_3$, we have $\frac{3-n-s}{4} \notin \mathbb{Z}$. This concludes the proposition.

Suppose that $a_{[s;n]}(t) \in \mathrm{Pol}_n[t]$. If n is even, then generically we have $\deg a_{[s;n]}(t) = n$. Lemma 5.2 classifies the singular parameters of $s \in \mathbb{C}$ for $a_{[s;n]}(t)$ in a sense that $\deg a_{[s;n]}(t) < n$ (see Sect. 5.2.4).

Lemma 5.2 *Suppose that $n \equiv k \pmod 4$ and $s \in \mathbb{C}\backslash J_k$ for $k = 0, 2$. Then the following conditions on $(s, n) \in \mathbb{C} \times \mathbb{Z}_{\geq 0}$ are equivalent:*

(i) $\deg a_{[s;n]}(t) < n$.
(ii) $\deg a_{[s;n]}(t) = \frac{n+s-1}{2}$.
(iii) *One of the following conditions holds:*

  (a) $n \equiv 0 \pmod 4 : s \in I_0^-$;
  (b) $n \equiv 2 \pmod 4 : s \in I_2^-$.

Proof It follows from the parameters of $a_{[s;n]}(t)$ that $\deg a_{[s;n]}(t) < n$ if and only if

$$2 \cdot \frac{n+s-1}{4} < n \quad \text{and} \quad \frac{n+s-1}{4} \in \mathbb{Z}_{\geq 0}, \tag{5.4}$$

which shows the equivalence between (i) and (ii). Moreover, (5.4) is equivalent to

$$s \in \{-n+1+4j : j = 0, 1, 2, \ldots, \frac{n}{2} - 1\}. \tag{5.5}$$

One can readily verify that, under the condition $s \notin J_k$, (5.5) is indeed equivalent to $s \in I_k^-$ for $k = 0, 2$.

We next consider $b_{[s;n]}(t)$ in (5.2).

Proposition 5.3 *The following conditions on $(s, n) \in \mathbb{C} \times \mathbb{Z}_{\geq 0}$ are equivalent:*

(i) $b_{[s;n]}(t) \in \mathrm{Pol}_n[t]$ *with* $b_{[s;n]}(t) \neq a_{[s;n]}(t)$.
(ii) *One of the following conditions holds.*

  (a) $n \equiv 0 \pmod 4 : s \in I_0^+ \cup I_0^- \cup J_0$.
  (b) $n \equiv 1 \pmod 4 : s \in I_1$.
  (c) $n \equiv 2 \pmod 4 : s \in I_2^+ \cup I_2^- \cup J_2$.
  (d) $n \equiv 3 \pmod 4 : s \in I_3$.

Proof Observe that if $b_{[s;n]}(t) \in \mathrm{Pol}[t]$, then the exponent $\frac{1+n-s}{2}$ for $t^{\frac{1+n-s}{2}}$ in (5.2) must satisfy $\frac{1+n-s}{2} \in \mathbb{Z}_{\geq 0}$, which in particular forces $\frac{5+n-s}{4} \notin -\mathbb{Z}_{\geq 0}$. Moreover, if $\frac{1+n-s}{2} = 0$, then

$$b_{[n+1;n]}(t) = F(-\frac{n}{2}, -\frac{n}{2}, 1; t^2) = a_{[n+1;n]}(t).$$

Consequently, we have $b_{[s;n]}(t) \in \mathrm{Pol}_n[t]$ with $b_{[s;n]}(t) \neq a_{[s;n]}(t)$ if and only if either

$$\frac{1+n-s}{2} \in 1+\mathbb{Z}_{\geq 0}, \quad \frac{n+s-1}{4} \in \mathbb{Z}_{\geq 0}, \quad \text{and}$$
$$\frac{1+n-s}{2} + 2 \cdot \frac{n+s-1}{4} \leq n, \tag{5.6}$$

or

$$\frac{1+n-s}{2} \in 1+\mathbb{Z}_{\geq 0}, \quad \frac{s-1}{2} \in \mathbb{Z}_{\geq 0}, \quad \text{and}$$
$$\frac{1+n-s}{2} + 2 \cdot \frac{s-1}{2} \leq n. \tag{5.7}$$

A direct observation shows that the conditions (5.6) and (5.7) are equivalent to

$$s \in (-n+1+4\mathbb{Z}_{\geq 0}) \cap (-\infty, n+1) \tag{5.8}$$

and

$$n \text{ is even and } s \in (1+2\mathbb{Z}_{\geq 0}) \cap (-\infty, n+1), \tag{5.9}$$

respectively. One can directly verify that (5.8) and (5.9) are equivalent to the conditions on $s \in \mathbb{C}$ stated in Proposition 5.3 for $n \equiv k \pmod 4$ for $k = 0, 1, 2, 3$. Indeed, if $n \equiv 0 \pmod 4$, then (5.8) and (5.9) are equivalent to $s \in I_0^- \cup J_0$ and $s \in I_0^+ \cup J_0$, respectively. Since the other three cases can be shown similarly, we omit the proof.

It follows from Propositions 5.1 and 5.3 that if $n \equiv k \pmod 4$ for $k = 0, 2$, then

$$\mathrm{Sol}_{\mathrm{II}}(s;n) = \mathbb{C}a_{[s;n]}(t) \oplus \mathbb{C}b_{[s;n]}(t) \quad \text{for } s \in I_k^+ \cup I_k^-. \tag{5.10}$$

For a later purpose for determining the M-representations on $\mathrm{Sol}_{\mathrm{II}}(s;n)$, for such $n \equiv k \pmod 4$, we define:

$$c^{\pm}_{[s;n]}(t) := a_{[s;n]}(t) \pm C(s;n) b_{[s;n]}(t) \quad \text{for } s \in I_k^-,$$

where

$$C(s;n) := \frac{\left(-\frac{n}{2}, \frac{n+s-1}{4}\right)}{\left(\frac{1-s}{2}, \frac{n+s-1}{4}\right)}. \tag{5.11}$$

Here we have $(\ell, m) = \frac{\Gamma(\ell+m)}{\Gamma(\ell)}$, the shifted factorial. Then, for $n \equiv k \pmod 4$ for $k = 0, 2$, the space $\mathrm{Sol}_{\mathrm{II}}(s; n)$ may be described as

$$\mathrm{Sol}_{\mathrm{II}}(s; n) = \begin{cases} \mathbb{C}a_{[s;n]}(t) \oplus \mathbb{C}b_{[s;n]}(t) & \text{if } s \in I_k^+, \\ \mathbb{C}c^+_{[s;n]}(t) \oplus \mathbb{C}c^-_{[s;n]}(t) & \text{if } s \in I_k^-. \end{cases} \tag{5.12}$$

We then summarize the parameters (s, n) such that $\mathrm{Sol}_{\mathrm{II}}(s; n) \neq \{0\}$ as follows.

Theorem 5.4 *The following conditions on $(s, n) \in \mathbb{C} \times \mathbb{Z}_{\geq 0}$ are equivalent.*

(i) $\mathrm{Sol}_{\mathrm{II}}(s; n) \neq \{0\}$.
(ii) *One of the following conditions is satisfied:*

- $n \equiv 0 \pmod 4 : s \in \mathbb{C}$.
- $n \equiv 1 \pmod 4 : s \in I_1$.
- $n \equiv 2 \pmod 4 : s \in \mathbb{C}$.
- $n \equiv 3 \pmod 4 : s \in I_3$.

Further, for such (s, n), the space $\mathrm{Sol}_{\mathrm{II}}(s; n)$ may be given as follows:

(1) $n \equiv 0 \pmod 4$:

$$\mathrm{Sol}_{\mathrm{II}}(s; n) = \begin{cases} \mathbb{C}a_{[s;n]}(t) & \text{if } s \in \mathbb{C} \backslash (I_0^+ \cup I_0^- \cup J_0), \\ \mathbb{C}a_{[s;n]}(t) \oplus \mathbb{C}b_{[s;n]}(t) & \text{if } s \in I_0^+, \\ \mathbb{C}b_{[s;n]}(t) & \text{if } s \in J_0, \\ \mathbb{C}c^+_{[s;n]}(t) \oplus \mathbb{C}c^-_{[s;n]}(t) & \text{if } s \in I_0^-. \end{cases}$$

(2) $n \equiv 1 \pmod 4$:

$$\mathrm{Sol}_{\mathrm{II}}(s; n) = \mathbb{C}a_{[s;n]}(t) \oplus \mathbb{C}b_{[s;n]}(t) \quad \text{for } s \in I_1.$$

(3) $n \equiv 2 \pmod 4$:

$$\mathrm{Sol}_{\mathrm{II}}(s; n) = \begin{cases} \mathbb{C}a_{[s;n]}(t) & \text{if } s \in \mathbb{C} \backslash (I_2^+ \cup I_2^- \cup J_2), \\ \mathbb{C}a_{[s;n]}(t) \oplus \mathbb{C}b_{[s;n]}(t) & \text{if } s \in I_2^+, \\ \mathbb{C}b_{[s;n]}(t) & \text{if } s \in J_2, \\ \mathbb{C}c^+_{[s;n]}(t) \oplus \mathbb{C}c^-_{[s;n]}(t) & \text{if } s \in I_2^-. \end{cases}$$

(4) $n \equiv 3 \pmod 4$:

$$\mathrm{Sol}_{\mathrm{II}}(s; n) = \mathbb{C}a_{[s;n]}(t) \oplus \mathbb{C}b_{[s;n]}(t) \quad \text{for } s \in I_3.$$

Proof This is a summary of the results in Propositions 5.1 and 5.3 and (5.12).

Remark 5.5 Theorem 5.4 shows that the structure of $\mathrm{Sol}_{\mathrm{II}}(s;n)$ (hypergeometric model) is somewhat complicated. It will be shown in Theorem 6.4 that that of $\mathrm{Sol}_{\mathrm{I}}(s;n)$ (Heun model) is more straightforward.

Remark 5.6 Theorem 5.4 also completely classifies the dimension $\dim_{\mathbb{C}} \mathrm{Sol}_{\mathrm{II}}(s;n)$. When n is even, the dimension $\dim_{\mathbb{C}} \mathrm{Sol}_{\mathrm{II}}(s;n)$ $(= \dim_{\mathbb{C}} \mathrm{Sol}_{\mathrm{I}}(s;n))$ was determined in [12, Thm. 5.13] by factorization formulas of certain tridiagonal determinants (see Theorem 7.5 and Remark 7.6). We shall study such determinants in Sect. 7 from a different point of view.

5.2 *The M-representations on* $\mathrm{Sol}_{\mathrm{II}}(s;n)$

As Step B of the recipe in Sect. 4.4, we next classify the M-representations on $\mathrm{Sol}_{\mathrm{II}}(s;n)$. Here is the classification.

Theorem 5.7 *For each* $(s;n) \in \mathbb{C} \times \mathbb{Z}_{\geq 0}$ *classified in Theorem 5.4, M acts on* $\mathrm{Sol}_{\mathrm{II}}(s;n)$ *as follows:*

(1) $n \equiv 0 \pmod 4$:

$$\mathrm{Sol}_{\mathrm{II}}(s;n) \simeq \begin{cases} (+,+) & \text{if } s \in \mathbb{C}\backslash(I_0^+ \cup I_0^- \cup J_0), \\ (+,+) \oplus (+,-) & \text{if } s \in I_0^+, \\ (+,+) & \text{if } s \in J_0, \\ (+,+) \oplus (-,+) & \text{if } s \in I_0^-. \end{cases}$$

(2) $n \equiv 1 \pmod 4$:

$$\mathrm{Sol}_{\mathrm{II}}(s;n) \simeq \mathbb{H} \quad \text{for } s \in I_1.$$

(3) $n \equiv 2 \pmod 4$:

$$\mathrm{Sol}_{\mathrm{II}}(s;n) \simeq \begin{cases} (-,-) & \text{if } s \in \mathbb{C}\backslash(I_2^+ \cup I_2^- \cup J_2), \\ (-,-) \oplus (-,+) & \text{if } s \in I_2^+, \\ (-,-) & \text{if } s \in J_2, \\ (-,-) \oplus (+,-) & \text{if } s \in I_2^-. \end{cases}$$

(4) $n \equiv 3 \pmod 4$:

$$\mathrm{Sol}_{\mathrm{II}}(s;n) \simeq \mathbb{H} \quad \text{for } s \in I_3.$$

Here the characters $(\varepsilon, \varepsilon')$ *stand for the ones on* $\mathbb{C}a_{[s;n]}(t)$, $\mathbb{C}b_{[s;n]}(t)$, *and* $\mathbb{C}c^{\pm}_{[s;n]}(t)$ *at the same places in Theorem 5.4.*

We prove Theorem 5.7 by considering the following cases separately:

- Case 1: n is odd.
- Case 2: n is even. Let $k \in \{0, 2\}$.
  - Case 2a: $n \equiv k \pmod 4$ and $s \in \mathbb{C} \backslash (I_k^- \cup J_k)$
  - Case 2b: $n \equiv k \pmod 4$ and $s \in I_k^+ \cup J_k$
  - Case 2c: $n \equiv k \pmod 4$ and $s \in I_k^-$

5.2.1 Case 1

We start with the case that $n \equiv k \pmod 4$ and $s \in I_k$ for $k = 1, 3$. In this case, by Theorem 5.4, we have:

$$\mathrm{Sol}_{\mathrm{II}}(s; n) = \mathbb{C}a_{[s;n]}(t) \oplus \mathbb{C}b_{[s;n]}(t).$$

Since the exponent $\frac{1+n-s}{2}$ for $t^{\frac{1+n-s}{2}}$ in (5.17) is odd, $a_{[s;n]}(t)$ and $b_{[s;n]}(t)$ are even and odd functions, respectively.

Observe that, by (3.20), $\pi_n(m_1^{\mathrm{II}})a_{[s;n]}(t)$ is an odd function. As $\mathrm{Sol}_{\mathrm{II}}(s; n)$ is an M-representation, this implies that $\pi_n(m_1^{\mathrm{II}})a_{[s;n]}(t) \in \mathbb{C}b_{[s;n]}(t)$. Similarly, we have: $\pi_n(m_1^{\mathrm{II}})b_{[s;n]}(t) \in \mathbb{C}a_{[s;n]}(t)$.

Lemma 5.8 *Suppose that $n \equiv k \pmod 4$ and $s \in I_k$ for $k = 1, 3$. Then, as an M-representation, we have:*

$$\mathbb{C}a_{[s;n]}(t) \oplus \mathbb{C}b_{[s;n]}(t) \simeq \mathbb{H}.$$

Proof Assume the contrary, that is, there exists a nonzero proper M-invariant subspace $\{0\} \neq V \subsetneq \mathbb{C}a_{[s;n]}(t) \oplus \mathbb{C}b_{[s;n]}(t)$. By the above argument, V is of the form $V = \mathbb{C}(c_1 a_{[s;n]}(t) + c_2 b_{[s;n]}(t))$ for some $c_1, c_2 \in \mathbb{C} \setminus \{0\}$. However, since $a_{[s;n]}(t)$ and $b_{[s;n]}(t)$ are even and odd functions, respectively, it follows from the argument for [25, Prop. 6.11] that m_2^{II} acts on V if and only if $(c_1, c_2) = (0, 0)$, which contradicts the assumption of V. Hence $\mathbb{C}a_{[s;n]}(t) \oplus \mathbb{C}b_{[s;n]}(t)$ is irreducible.

5.2.2 Case 2a

We next consider the characters on $\mathbb{C}a_{[s;n]}(t)$.

Lemma 5.9 *Suppose that $n \equiv k \pmod 4$ and $s \in \mathbb{C} \backslash (I_k^- \cup J_k)$ for $k = 0, 2$. Then M acts on $\mathbb{C}a_{[s;n]}(t)$ as a character.*

Proof We only give a proof for the case $k = 0$, namely, $n \equiv 0 \pmod 4$ and $s \in \mathbb{C}\backslash(I_0^- \cup J_0)$; the other case can be shown similarly. As

$$\mathbb{C}\backslash(I_0^- \cup J_0) = (\mathbb{C}\backslash(I_0^+ \cup I_0^- \cup J_0)) \cup I_0^+,$$

we consider the cases $s \in \mathbb{C}\backslash(I_0^+ \cup I_0^- \cup J_0)$ and $s \in I_0^+$, separately.

First suppose that $s \in \mathbb{C}\backslash(I_0^+ \cup I_0^- \cup J_0)$. By Theorem 5.4, we have $\mathrm{Sol}_{\mathrm{II}}(s; n) = \mathbb{C}a_{[s;n]}(t)$. Since $\mathrm{Sol}_{\mathrm{II}}(s; n)$ is an M-representation by Lemma 2.4, the assertion clearly holds for this case. We next suppose that $s \in I_0^+$. In this case we have $\mathrm{Sol}_{\mathrm{II}}(s; n) = \mathbb{C}a_{[s;n]}(t) \oplus \mathbb{C}b_{[s;n]}(t)$ by Theorem 5.4. As the exponent $\frac{1+n-s}{2}$ for $t^{\frac{1+n-s}{2}}$ of $b_{[s;n]}(t)$ is odd, $a_{[s;n]}(t)$ and $b_{[s;n]}(t)$ are an even and odd function, respectively. The transformation laws (3.21) imply that the action of M on $\mathrm{Pol}_n[t]$ preserves the parities of the polynomials for n even. Hence M acts both on $\mathbb{C}a_{[s;n]}(t)$ and $\mathbb{C}b_{[s;n]}(t)$ as a character.

We next determine the characters on $a_{[s;n]}(t)$ explicitly. It follows from Lemma 5.2 that $a_{[s;n]}(t)$ has degree $\deg a_{[s;n]}(t) = n$ for (s, n) with $n \equiv k \pmod 4$ and $s \in \mathbb{C}\backslash(I_k^- \cup J_k)$ for $k = 0, 2$. Thus, in this case, the hypergeometric polynomial $a_{[s;n]}(t)$ is given as

$$a_{[s;n]}(t) = \sum_{j=0}^{n/2} A_j(s; n)t^{2j} \tag{5.13}$$

with

$$A_j(s; n) = \frac{(-\frac{n}{2}, j)(-\frac{n+s-1}{4}, j)}{(\frac{3-n+s}{4}, j)} \cdot \frac{1}{j!},$$

where (ℓ, m) stands for $(\ell, m) = \frac{\Gamma(\ell+m)}{\Gamma(\ell)}$, the shifted factorial. Recall from (3.16) that we have $M = \{\pm m_j^{\mathrm{II}} : j = 0, 1, 2, 3\}$ with

$$m_0^{\mathrm{II}} = \begin{pmatrix} 1 & 0 \\ 0 & 1 \end{pmatrix}, \; m_1^{\mathrm{II}} = \begin{pmatrix} 0 & 1 \\ -1 & 0 \end{pmatrix}, \; m_2^{\mathrm{II}} = \begin{pmatrix} \sqrt{-1} & 0 \\ 0 & -\sqrt{-1} \end{pmatrix}, \; m_3^{\mathrm{II}} = -\begin{pmatrix} 0 & \sqrt{-1} \\ \sqrt{-1} & 0 \end{pmatrix}.$$

As $m_3^{\mathrm{II}} = m_1^{\mathrm{II}} m_2^{\mathrm{II}}$, it suffices to check the transformation laws of m_1^{II} and m_2^{II} on $a_{[s;n]}(t) = \sum_{j=0}^{n/2} A_j(s; n)t^{2j}$.

Proposition 5.10 *Under the same hypothesis in Lemma 5.9, the character* $(\varepsilon, \varepsilon')$ *on* $\mathbb{C}a_{[s;n]}(t)$ *is given as follows:*

(1) $n \equiv 0 \pmod 4$ *and* $s \in \mathbb{C}\backslash(I_0^- \cup J_0)$:

$$\mathbb{C}a_{[s;n]}(t) \simeq (+, +).$$

(2) $n \equiv 2 \pmod 4$ *and* $s \in \mathbb{C}\backslash(I_2^- \cup J_2)$:

$$\mathbb{C}a_{[s;n]}(t) \simeq (-, -).$$

Proof Since the second assertion can be shown similarly, we only give a proof for the first assertion. Let $n \equiv 0 \pmod 4$ and $s \in \mathbb{C}\backslash(I_0^- \cup J_0)$. We wish to show that both m_1^{II} and m_2^{II} act trivially. First, it is easy to see that the action of m_2^{II} is trivial. Indeed, since $n \equiv 0 \pmod 4$ and $a_{[s;n]}(t)$ is an even function, the transformation law of (3.21) shows that

$$m_2^{\mathrm{II}} \colon a_{[s;n]}(t) \longrightarrow (\sqrt{-1})^n a_{[s;n]}(-t) = a_{[s;n]}(t).$$

In order to show that m_1^{II} also acts trivially, observe that, by (3.21) and (5.13), we have:

$$m_1^{\mathrm{II}} \colon a_{[s;n]}(t) \longrightarrow t^n a_{[s;n]}\left(-\frac{1}{t}\right) = \sum_{j=0}^{n/2} A_{\frac{n}{2}-j}(s;n)t^{2j}.$$

On the other hand, by Lemma 5.9 and Table 2, m_1^{II} acts on $a_{[s;n]}(t)$ by ± 1. Therefore,

$$\sum_{j=0}^{n/2} A_{\frac{n}{2}-j}(s;n)t^{2j} = \pm \sum_{j=0}^{n/2} A_j(s;n)t^{2j}.$$

An easy computation shows $A_{\frac{n}{2}}(s;n) = 1 = A_0(s;n)$, which forces $t^n a_{[s;n]}\left(-\frac{1}{t}\right) = a_{[s;n]}(t)$. Hence m_1^{II} also acts on $\mathbb{C}a_{[s;n]}(t)$ trivially.

5.2.3 Case 2b

Next we consider the characters on $\mathbb{C}b_{[s;n]}(t)$.

Proposition 5.11 *Suppose that* $n \equiv k \pmod 4$ *and* $s \in I_k^+ \cup J_k$ *for* $k = 0, 2$. *Then* M *acts on* $\mathbb{C}b_{[s;n]}(t)$ *as a character as follows:*

(1) $n \equiv 0 \pmod 4$ *and* $s \in I_0^+ \cup J_0$:

$$\mathbb{C}b_{[s;n]}(t) \simeq \begin{cases} (+, -) & \text{if } s \in I_0^+, \\ (+, +) & \text{if } s \in J_0. \end{cases}$$

(2) $n \equiv 2 \pmod 4$ *and* $s \in I_2^+ \cup J_2$:

$$\mathbb{C}b_{[s;n]}(t) \simeq \begin{cases} (-, +) & \text{if } s \in I_2^+, \\ (-, -) & \text{if } s \in J_2. \end{cases}$$

Proof Since the assertions can be shown similarly to Lemma 5.9 and Proposition 5.10, we omit the proof. We remark that it is already shown in the proof of Lemma 5.9 that M acts on $\mathbb{C}b_{[s;n]}(t)$ as a character for (s, n) with $n \equiv 0 \pmod 4$ and $s \in I_0^+$.

5.2.4 Case 2c

Now we consider $\mathbb{C}c^{\pm}_{[s;n]}(t)$ for $n \equiv k \pmod 4$ and $s \in I_k^-$ for $k = 0, 2$, where $\mathbb{C}c^{\pm}_{[s;n]}(t)$ is given as

$$c^{\pm}_{[s;n]}(t) = a_{[s;n]}(t) \pm C(s; n)b_{[s;n]}(t)$$

with $C(s; n)$ in (5.11). As opposed to Case 1, since the exponent $\frac{1+n-s}{2}$ for $t^{\frac{1+n-s}{2}}$ in (5.17) is even, both $a_{[s;n]}(t)$ and $b_{[s;n]}(t)$ are even functions.

Observe that, in this case, $a_{[s;n]}(t)$ has degree $\deg a_{[s;n]}(t) = \frac{n+s-1}{2} < n$ by Lemma 5.2. Therefore, $a_{[s;n]}(t)$ is of the form:

$$a_{[s;n]}(t) = \sum_{j=0}^{(n+s-1)/4} \widetilde{A}_j(s; n)t^{2j}. \tag{5.14}$$

To give an explicit formula of $\widetilde{A}_j(s; n)$, first remark that when $n \equiv 2 \pmod 4$ and $s = -1$, we have $-\frac{n+s-1}{4} = \frac{3-n+s}{4} \in \mathbb{Z}_{\geq 0}$. Namely, for $n = 4\ell + 2$ and $s = -1$, the hypergeometric polynomial $a_{[-1;4\ell+2]}(t)$ is

$$a_{[-1;4\ell+2]}(t) = F(-2\ell - 1, -\ell, -\ell; t^2). \tag{5.15}$$

In this paper we regard (5.15) as

$$a_{[-1;4\ell+2]}(t) = \sum_{j=0}^{\ell} \frac{(-2\ell - 1, j)}{j!} t^{2j}$$

so that $a_{[-1;4\ell+2]}(t)$ and $b_{[-1;4\ell+2]}(t)$ still form a fundamental solution to the hypergeometric equation (4.27). Observe that, for $n = 4\ell + 2$ and $s = -1$, we have:

$$\frac{(-\frac{n}{2}, j)(-\frac{n+s-1}{4}, j)}{(\frac{3-n+s}{4}, j)} = (-2\ell - 1, j).$$

Thus $\widetilde{A}_j(s;n)$ in (5.14) is uniformly given as

$$\widetilde{A}_j(s;n) = \frac{(-\frac{n}{2}, j)(-\frac{n+s-1}{4}, j)}{(\frac{3-n+s}{4}, j)} \cdot \frac{1}{j!}.$$

Since

$$(-\frac{n+s-1}{4}, \frac{n+s-1}{4}) = (-1)^{\frac{n+s-1}{4}} (\frac{n+s-1}{4})!,$$
$$(\frac{3-n+s}{4}, \frac{n+s-1}{4}) = (-1)^{\frac{n+s-1}{4}} (\frac{1-s}{2}, \frac{n+s-1}{4})$$

hold, the expression of $\widetilde{A}_{\frac{n+s-1}{4}}(s;n)$ is simplified as

$$\begin{aligned}\widetilde{A}_{\frac{n+s-1}{4}}(s;n) &= \frac{(-\frac{n}{2}, \frac{n+s-1}{4})(-\frac{n+s-1}{4}, \frac{n+s-1}{4})}{(\frac{3-n+s}{4}, \frac{n+s-1}{4})} \cdot \frac{1}{(\frac{n+s-1}{4})!} \\ &= \frac{(-\frac{n}{2}, \frac{n+s-1}{4})}{(\frac{1-s}{2}, \frac{n+s-1}{4})}.\end{aligned}$$

Then, by (5.11), we have

$$C(s;n) = \widetilde{A}_{\frac{n+s-1}{4}}(s;n). \tag{5.16}$$

It follows from (5.2) that $b_{[s;n]}(t)$ is given as

$$b_{[s;n]}(t) = t^{\frac{1+n-s}{2}} \sum_{j=0}^{(n+s-1)/4} B_j(s;n) t^{2j} \tag{5.17}$$

with

$$B_j(s;n) = \frac{\left(-\frac{n+s-1}{4}, j\right)\left(-\frac{s-1}{2}, j\right)}{\left(\frac{5+n-s}{4}, j\right)} \cdot \frac{1}{j!}.$$

Lemma 5.12 plays a key role to determine the M-representation on $\mathbb{C}c^{\pm}_{[s;n]}(t)$.

Lemma 5.12 *Suppose that* $n \equiv k \pmod 4$ *and* $s \in I_k^-$ *for* $k = 0, 2$. *Then,*

$$\pi_n(m_1^{\mathrm{II}}) a_{[s;n]}(t) = C(s;n) b_{[s;n]}(t).$$

Proof We only show the case of $k = 0$; the other case can be shown similarly, including the exceptional case for $n \equiv 2 \pmod 4$ and $s = -1$. Let $n \equiv 0 \pmod 4$ and $s \in I_0^-$. It follows from (5.10) that we have

$$\mathrm{Sol}_{\mathrm{II}}(s; n) = \mathbb{C}a_{[s;n]}(t) \oplus \mathbb{C}b_{[s;n]}(t).$$

Therefore there exist some constants $c_1, c_2 \in \mathbb{C}$ such that

$$\pi_n(m_1^{\mathrm{II}})a_{[s;n]}(t) = c_1 a_{[s;n]}(t) + c_2 b_{[s;n]}(t). \tag{5.18}$$

On the other hand, by (3.21) and (5.14), we have:

$$\pi_n(m_1^{\mathrm{II}})a_{[s;n]}(t) = t^n a_{[s;n]}\left(-\frac{1}{t}\right) = \sum_{j=0}^{(n+s-1)/4} \widetilde{A}_j(s; n)t^{n-2j}. \tag{5.19}$$

Since

$$n - \frac{n+s-1}{2} > \frac{n+s-1}{2} = \deg a_{[s;n]}(t)$$

for $n \equiv 0 \pmod 4$ and $s \in I_0^-$, all exponents $n - 2j$ for t^{n-2j} in (5.19) is $n - 2j > \deg a_{[s;n]}(t)$. Therefore,

$$\pi_n(m_1^{\mathrm{II}})a_{[s;n]}(t) = c_2 b_{[s;n]}(t).$$

Further, (5.19) and (5.17) show that

$$\widetilde{A}_{\frac{n+s-1}{4}}(s; n) = c_2 B_0(s; n) = c_2.$$

Now (5.16) concludes the assertion.

Proposition 5.13 *Suppose that* $n \equiv k \pmod 4$ *and* $s \in I_k^-$ *for* $k = 0, 2$. *Then* M *acts on* $\mathbb{C}c_{[s;n]}^{\pm}(t)$ *as the following characters:*

(1) $n \equiv 0 \pmod 4$ *and* $s \in I_0^-$.

$$\mathbb{C}c_{[s;n]}^{+}(t) \simeq (+,+) \quad \text{and} \quad \mathbb{C}c_{[s;n]}^{-}(t) \simeq (-,+).$$

(2) $n \equiv 2 \pmod 4$ *and* $s \in I_2^-$.

$$\mathbb{C}c_{[s;n]}^{+}(t) \simeq (-,-) \quad \text{and} \quad \mathbb{C}c_{[s;n]}^{-}(t) \simeq (+,-).$$

Proof As in Lemma 5.12, we only show the case of $k = 0$; the other case can be treated similarly (including the exceptional case for $n \equiv 2 \pmod 4$ and $s = -1$).

Let $n \equiv 0 \pmod 4$ and $s \in I_0^-$. To show the assertion it suffices to only consider m_1^{II}. Indeed, since both $a_{[s;n]}(t)$ and $b_{[s;n]}(t)$ are even functions, by (3.21), m_2^{II} acts on $\mathbb{C}c_{[s;n]}^{\pm}(t)$ trivially.

To consider the action of m_1^{II} on $\mathbb{C}c_{[s;n]}^{\pm}(t)$, observe that, by Lemma 5.12, we have:

$$\pi_n(m_1^{\mathrm{II}})b_{[s;n]}(t) = C(s;n)^{-1}a_{[s;n]}(t).$$

Therefore, for $\varepsilon \in \{+,-\}$, m_1^{II} transforms $c_{[s;n]}^{\varepsilon}(t)$ as

$$m_1^{\mathrm{II}} : c_{[s;n]}^{\varepsilon}(t) \longrightarrow \varepsilon c_{[s;n]}^{\varepsilon}(t).$$

Now Table 2 concludes the assertion.

5.3 *The Classification of* $\mathrm{Hom}_M(\mathrm{Sol}_{\mathrm{II}}(s;n),\sigma)$

As Step C in the recipe in Sect. 4.4, we now classify $(\sigma, s, n) \in \mathrm{Irr}(M) \times \mathbb{C} \times \mathbb{Z}_{\geq 0}$ such that $\mathrm{Hom}_M(\mathrm{Sol}_{\mathrm{II}}(s;n),\sigma) \neq \{0\}$. Given $\sigma \in \mathrm{Irr}(M)$, we define $I^{\pm}((+,-))$, $I^{\pm}((-,+))$, $I(\mathbb{H}) \subset \mathbb{Z} \times \mathbb{Z}_{\geq 0}$ as follows:

$$I^{+}((+,-)) := \{(s,n) \in (3+4\mathbb{Z}_{\geq 0}) \times (4\mathbb{Z}_{\geq 0}) : n > s\},$$

$$I^{-}((+,-)) := \{(s,n) \in -(1+4\mathbb{Z}_{\geq 0}) \times (2+4\mathbb{Z}_{\geq 0}) : n > |s|\},$$

$$I^{+}((-,+)) := \{(s,n) \in (1+4\mathbb{Z}_{\geq 0}) \times (2+4\mathbb{Z}_{\geq 0}) : n > s\},$$

$$I^{-}((-,+)) := \{(s,n) \in -(3+4\mathbb{Z}_{\geq 0}) \times (4\mathbb{Z}_{\geq 0}) : n > |s|\},$$

$$I(\mathbb{H}) := \{(s,n) \in (2\mathbb{Z}) \times (1+2\mathbb{Z}_{\geq 0}) : n > |s| \text{ and } s \equiv n-1 \pmod 4\}. \tag{5.20}$$

Theorem 5.14 *The following conditions on* $(\sigma, s, n) \in \mathrm{Irr}(M) \times \mathbb{C} \times \mathbb{Z}_{\geq 0}$ *are equivalent:*

(i) $\mathrm{Hom}_M(\mathrm{Sol}_{\mathrm{II}}(s;n),\sigma) \neq \{0\}$.
(ii) $\dim_{\mathbb{C}} \mathrm{Hom}_M(\mathrm{Sol}_{\mathrm{II}}(s;n),\sigma) = 1$.
(iii) *One of the following conditions holds:*

- $\sigma = (+,+) : (s,n) \in \mathbb{C} \times 4\mathbb{Z}_{\geq 0}$.
- $\sigma = (-,-) : (s,n) \in \mathbb{C} \times (2+4\mathbb{Z}_{\geq 0})$.
- $\sigma = (+,-) : (s,n) \in I^{+}((+,-)) \cup I^{-}((+,-))$.
- $\sigma = (-,+) : (s,n) \in I^{+}((-,+)) \cup I^{-}((-,+))$.
- $\sigma = \mathbb{H} : \quad (s,n) \in I(\mathbb{H})$.

Further, for such (σ, s, n), *the space* $\mathrm{Hom}_M(\mathrm{Sol}_{\mathrm{II}}(s;n), \sigma)$ *is given as follows:*

(1) $\sigma = (+,+)$: *For* $n \in 4\mathbb{Z}_{\geq 0}$, *we have:*

$$\mathrm{Hom}_M(\mathrm{Sol}_{\mathrm{II}}(s;n), (+,+)) = \begin{cases} \mathbb{C}a_{[s;n]}(t) & \text{if } s \in \mathbb{C}\backslash(I_0^- \cup J_0), \\ \mathbb{C}b_{[s;n]}(t) & \text{if } s \in J_0, \\ \mathbb{C}c^+_{[s;n]}(t) & \text{if } s \in I_0^-. \end{cases}$$

(2) $\sigma = (-,-)$: *For* $n \in 2 + 4\mathbb{Z}_{\geq 0}$, *we have:*

$$\mathrm{Hom}_M(\mathrm{Sol}_{\mathrm{II}}(s;n), (-,-)) = \begin{cases} \mathbb{C}a_{[s;n]}(t) & \text{if } s \in \mathbb{C}\backslash(I_2^- \cup J_2), \\ \mathbb{C}b_{[s;n]}(t) & \text{if } s \in J_2, \\ \mathbb{C}c^+_{[s;n]}(t) & \text{if } s \in I_2^-. \end{cases}$$

(3) $\sigma = (+,-)$: *We have:*

$$\mathrm{Hom}_M(\mathrm{Sol}_{\mathrm{II}}(s;n), (+,-)) = \begin{cases} \mathbb{C}b_{[s;n]}(t) & \text{if } (s,n) \in I^+((+,-)), \\ \mathbb{C}c^-_{[s;n]}(t) & \text{if } (s,n) \in I^-((+,-)). \end{cases}$$

(4) $\sigma = (-,+)$: *We have:*

$$\mathrm{Hom}_M(\mathrm{Sol}_{\mathrm{II}}(s;n), (-,+)) = \begin{cases} \mathbb{C}b_{[s;n]}(t) & \text{if } (s,n) \in I^+((-,+)), \\ \mathbb{C}c^-_{[s;n]}(t) & \text{if } (s,n) \in I^-((-,+)). \end{cases}$$

(5) $\sigma = \mathbb{H}$: *We have*

$$\mathrm{Hom}_M(\mathrm{Sol}_{\mathrm{II}}(s;n), \mathbb{H}) = \mathbb{C}\varphi_{\mathrm{II}}^{(s;n)} \quad \text{for } (s,n) \in I(\mathbb{H}),$$

where $\varphi_{\mathrm{II}}^{(s;n)}$ *is a nonzero* M*-isomorphism*

$$\varphi_{\mathrm{II}}^{(s;n)} : \mathrm{Sol}_{\mathrm{II}}(s;n) \xrightarrow{\sim} \mathbb{H}.$$

Proof The theorem simply follows from Theorems 5.4 and 5.7. Indeed, for instance, suppose that $\sigma = (+,+)$. It then follows from Theorem 5.7 that we have $\mathrm{Hom}_M(\mathrm{Sol}_{\mathrm{II}}(s;n), (+,+)) \neq \{0\}$ if and only if $n \equiv 0 \pmod 4$. Moreover, Theorems 5.4 and 5.7 show that

$$\mathrm{Hom}_M(\mathrm{Sol}_{\mathrm{II}}(s;n), (+,+)) = \begin{cases} \mathbb{C}a_{[s;n]}(t) & \text{if } s \in \mathbb{C}\backslash(I_0^- \cup J_0), \\ \mathbb{C}b_{[s;n]}(t) & \text{if } s \in J_0, \\ \mathbb{C}c^+_{[s;n]}(t) & \text{if } s \in I_0^-. \end{cases}$$

This concludes the assertion for $\sigma = (+,+)$. Similarly, suppose that $\sigma = \mathbb{H}$. Then, by Theorem 5.7, we have $\mathrm{Hom}_M(\mathrm{Sol}_{\mathrm{II}}(s;n), \mathbb{H}) \neq \{0\}$ if and only if $(s,n) \in I(\mathbb{H})$. Furthermore, in this case, $\mathrm{Sol}_{\mathrm{II}}(s;n) \simeq \mathbb{H}$. Therefore, $\mathrm{Hom}_M(\mathrm{Sol}_{\mathrm{II}}(s;n), \mathbb{H})$ is spanned by a nonzero M-isomorphism $\varphi_{\mathrm{II}}^{(s;n)} : \mathrm{Sol}_{\mathrm{II}}(s;n) \xrightarrow{\sim} \mathbb{H}$. Since the other cases can be handled similarly, we omit the proof.

Remark 5.15 It will be shown in Theorem 6.6 that the space $\mathrm{Hom}_M(\mathrm{Sol}_{\mathrm{I}}(s;n), \sigma)$ is simpler than $\mathrm{Hom}_M(\mathrm{Sol}_{\mathrm{II}}(s;n), \sigma)$.

Now we give the K-type formulas for $\mathcal{S}ol(s;\sigma)_K$ for each $\sigma \in \mathrm{Irr}(M)$ in the polynomial realization (3.18) of $\mathrm{Irr}(K)$.

Theorem 5.16 *The following conditions on $(\sigma, s) \in \mathrm{Irr}(M) \times \mathbb{C}$ are equivalent:*

(i) $\mathcal{S}ol(s;\sigma)_K \neq \{0\}$.
(ii) *One of the following conditions holds:*

- $\sigma = (+,+) : s \in \mathbb{C}$.
- $\sigma = (-,-) : s \in \mathbb{C}$.
- $\sigma = (+,-) : s \in 3 + 4\mathbb{Z}$.
- $\sigma = (-,+) : s \in 1 + 4\mathbb{Z}$.
- $\sigma = \mathbb{H} : \quad s \in 2\mathbb{Z}$.

Further, the K-type formulas for $\mathcal{S}ol(s;\sigma)_K$ may be given as follows:

(1) $\sigma = (+,+)$:

$$\mathcal{S}ol(s;(+,+))_K \simeq \bigoplus_{n \geq 0} \mathrm{Pol}_{4n}[t] \qquad \text{for all } s \in \mathbb{C}.$$

(2) $\sigma = (-,-)$:

$$\mathcal{S}ol(s;(-,-))_K \simeq \bigoplus_{n \geq 0} \mathrm{Pol}_{2+4n}[t] \quad \text{for all } s \in \mathbb{C}.$$

(3) $\sigma = (+,-)$:

$$\mathcal{S}ol(s;(+,-))_K \simeq \bigoplus_{n \geq 0} \mathrm{Pol}_{|s|+1+4n}[t] \quad \text{for } s \in 3 + 4\mathbb{Z}.$$

(4) $\sigma = (-,+)$:

$$\mathcal{S}ol(s;(-,+))_K \simeq \bigoplus_{n \geq 0} \mathrm{Pol}_{|s|+1+4n}[t] \quad \text{for } s \in 1 + 4\mathbb{Z}.$$

(5) $\sigma = \mathbb{H}$:

$$\mathcal{S}ol(s; \mathbb{H})_K \simeq \bigoplus_{n \geq 0} \mathrm{Pol}_{|s|+1+4n}[t] \quad \text{for } s \in 2\mathbb{Z}.$$

Proof We only give a proof for $\sigma = (+, -)$; the other cases may be handled similarly. Suppose that $\sigma = (+, -)$. By (4.16) with $J = \mathrm{II}$, we have

$$\mathcal{S}ol(s; (+, -))_K \simeq \bigoplus_{n \geq 0} \mathrm{Pol}_n[t] \otimes \mathrm{Hom}_M \left(\mathrm{Sol}_{\mathrm{II}}(s; n), (+, -)\right).$$

Thus, $\mathcal{S}ol(s; (+, -))_K \neq \{0\}$ if and only if $\mathrm{Hom}_M \left(\mathrm{Sol}_{\mathrm{II}}(s; n), (+, -)\right) \neq \{0\}$ for some $n \in \mathbb{Z}_{\geq 0}$, which is further equivalent to

$$s \in (3 + 4\mathbb{Z}_{\geq 0}) \cup (-(1 + 4\mathbb{Z}_{\geq 0})) = 3 + 4\mathbb{Z}$$

by Theorem 5.14. This shows the assertion between (i) and (ii) for $\sigma = (+, -)$.

To determine the explicit branching law, observe that, by the Frobenius reciprocity, we have

$$\begin{aligned} &\mathrm{Hom}_K(\mathrm{Pol}_n[t], \mathcal{S}ol(s; (+, -))_K) \neq \{0\} \quad \text{if and only if} \\ &\mathrm{Hom}_M(\mathrm{Sol}_{\mathrm{II}}(s; n), (+, -)) \neq \{0\}, \end{aligned}$$

which is equivalent to

$$n \in 4\mathbb{Z}_{\geq 0} \text{ with } n > s \text{ for } s \in 3 + 4\mathbb{Z}_{\geq 0}$$

or

$$n \in 2 + 4\mathbb{Z}_{\geq 0} \text{ with } n > |s| \text{ for } s \in -(1 + 4\mathbb{Z}_{\geq 0})$$

by Theorem 5.14. It also follows from Theorem 5.14 that $\mathcal{S}ol(s; (+, -))_K$ is multiplicity-free. Therefore, we have:

$$\mathcal{S}ol(s; (+, -))_K \simeq \begin{cases} \displaystyle\bigoplus_{\substack{n \equiv 0 \ (\mathrm{mod}\ 4) \\ n > s}} \mathrm{Pol}_n[t] & \text{if } s \in 3 + 4\mathbb{Z}_{\geq 0}, \\ \displaystyle\bigoplus_{\substack{n \equiv 2 \ (\mathrm{mod}\ 4) \\ n > |s|}} \mathrm{Pol}_n[t] & \text{if } s \in -(1 + 4\mathbb{Z}_{\geq 0}), \end{cases}$$

which is equivalent to the proposed formula.

Remark 5.17 The K-type formula for the case of $\sigma = \mathbb{H}$ is obtained also in [36, Thm. 5.3 (2)].

We close this section by giving a proof of Theorem 1.2 in the introduction.

Proof of Theorem 1.2 The assertions simply follow from Theorem 5.16. Indeed, as $\mathcal{S}ol(s;n)_K$ is dense in $\mathcal{S}ol(s;n)$, we have $\mathcal{S}ol(s;n)_K \neq \{0\}$ if and only if $\mathcal{S}ol(s;n) \neq \{0\}$. Then the equivalence $(\pi_n, \mathrm{Pol}_n[t]) \simeq (n/2)$ concludes the theorem.

6 Heun Model $d\pi_n^{\mathrm{I}}(D_s^\flat)f(t) = 0$

The purpose of this short section is to translate the results in Sect. 5 to the Heun model $d\pi_n^{\mathrm{I}}(D_s^\flat)f(t) = 0$. In particular we give a variant of Theorem 5.14 for the space $\mathrm{Hom}_M(\mathrm{Sol}_{\mathrm{I}}(s;n), \sigma)$ in Theorem 6.6. We remark that the results in this section will play a key role to compute certain tridiagonal determinants in Sect. 7.

6.1 Relationships Between $a_{[s;n]}(t)$, $b_{[s;n]}(t)$, $c^{\pm}_{[s;n]}(t)$ and $u_{[s;n]}(t)$, $v_{[s;n]}(t)$

As for the hypergeometric model $d\pi_n^{\mathrm{II}}(D_s^\flat)f(t) = 0$, we start with Step A of the recipe in Sect. 4.4, that is, the classification of $(s,n) \in \mathbb{C} \times \mathbb{Z}_{\geq 0}$ such that $\mathrm{Sol}_{\mathrm{I}}(s;n) \neq \{0\}$. Recall from (4.25) that we have

$$\mathrm{Sol}_{\mathrm{I}}(s;n) \subset \mathbb{C}u_{[s;n]}(t) \oplus \mathbb{C}v_{[s;n]}(t) \tag{6.1}$$

with

$$u_{[s;n]}(t) = Hl(-1, -\frac{ns}{4}; -\frac{n}{2}, -\frac{n-1}{2}, \frac{1}{2}, \frac{1-n-s}{2}; t^2),$$

$$v_{[s;n]}(t) = tHl(-1, -\frac{n-2}{4}s; -\frac{n-1}{2}, -\frac{n-2}{2}, \frac{3}{2}, \frac{1-n-s}{2}; t^2).$$

Thus, equivalently, we wish to classify (s,n) such that $u_{[s;n]}(t) \in \mathrm{Pol}_n[t]$ or $v_{[s;n]}(t) \in \mathrm{Pol}_n[t]$.

To make full use of the results for $\mathrm{Sol}_{\mathrm{II}}(s;n)$, we first transfer elements in $\mathrm{Sol}_{\mathrm{II}}(s;n)$ to $\mathrm{Sol}_{\mathrm{I}}(s;n)$. Recall from (3.12) and (4.17) that

$$k_0 = \frac{1}{\sqrt{2}} \begin{pmatrix} \sqrt{-1} & 1 \\ -1 & -\sqrt{-1} \end{pmatrix} \in SU(2)$$

gives an M-isomorphism

$$\pi_n(k_0)\colon \mathrm{Sol}_{\mathrm{II}}(s;n) \xrightarrow{\sim} \mathrm{Sol}_{\mathrm{I}}(s;n). \tag{6.2}$$

For $p(t), q(t) \in \mathrm{Pol}_n[t]$ with $\pi_n(k_0)p(t) \in \mathbb{C}q(t)$, we write:

$$p(t) \overset{k_0}{\sim} q(t).$$

Then, for $n \in 2\mathbb{Z}_{\geq 0}$, the polynomials $a_{[s;n]}(t)$, $b_{[s;n]}(t)$, $c^{\pm}_{[s;n]}(t) \in \mathrm{Sol}_{\mathrm{II}}(s;n)$ are transferred to $u_{[s;n]}(t)$, $v_{[s;n]}(t) \in \mathrm{Sol}_{\mathrm{I}}(s;n)$ via $\pi_n(k_0)$ as follows.

Lemma 6.1 *Let $n \in 2\mathbb{Z}_{\geq 0}$. We have the following:*

(1) $n \equiv 0 \pmod 4$:

(a) $a_{[s;n]}(t) \overset{k_0}{\sim} u_{[s;n]}(t)$ for $s \in \mathbb{C}\backslash(I_0^- \cup J_0)$.
(b) $b_{[s;n]}(t) \overset{k_0}{\sim} u_{[s;n]}(t)$ *for $s \in J_0$.*
(c) $c^+_{[s;n]}(t) \overset{k_0}{\sim} u_{[s;n]}(t)$ *for $s \in I_0^-$.*
(d) $b_{[s;n]}(t) \overset{k_0}{\sim} v_{[s;n]}(t)$ *for $s \in I_0^+$.*
(e) $c^-_{[s;n]}(t) \overset{k_0}{\sim} v_{[s;n]}(t)$ *for $s \in I_0^-$.*

(2) $n \equiv 2 \pmod 4$:

(a) $a_{[s;n]}(t) \overset{k_0}{\sim} v_{[s;n]}(t)$ *for $s \in \mathbb{C}\backslash(I_2^- \cup J_2)$.*
(b) $b_{[s;n]}(t) \overset{k_0}{\sim} v_{[s;n]}(t)$ *for $s \in J_2$.*
(c) $c^+_{[s;n]}(t) \overset{k_0}{\sim} v_{[s;n]}(t)$ *for $s \in I_2^-$.*
(d) $b_{[s;n]}(t) \overset{k_0}{\sim} u_{[s;n]}(t)$ *for $s \in I_2^+$.*
(e) $c^-_{[s;n]}(t) \overset{k_0}{\sim} u_{[s;n]}(t)$ *for $s \in I_2^-$.*

Proof We only give a proof of (1)(a); the other cases can be shown similarly. Let $n \equiv 0 \pmod 4$ and $s \in \mathbb{C}\backslash(I_0^- \cup J_0)$. It follows from Theorem 5.4 that in this case $a_{[s;n]}(t) \in \mathrm{Sol}_{\mathrm{II}}(s;n)$ and thus $\pi_n(k_0)a_{[s;n]}(t) \in \mathrm{Sol}_{\mathrm{I}}(s;n)$. By (6.1), this implies that $\pi_n(k_0)a_{[s;n]}(t) = c_1 u_{[s;n]}(t) + c_2 v_{[s;n]}(t)$ for some constants $c_1, c_2 \in \mathbb{C}$. We wish to show $c_2 = 0$. To do so, it suffices to show that $\pi_n(k_0)a_{[s;n]}(t)$ is an even function, as $u_{[s;n]}(t)$ and $v_{[s;n]}(t)$ are even and odd functions, respectively. It follows from Theorem 5.7 that m_1^{II} acts on $a_{[s;n]}(t)$ trivially. Since the linear map $\pi_n(k_0)\colon \mathrm{Sol}_{\mathrm{II}}(s;n) \to \mathrm{Sol}_{\mathrm{I}}(s;n)$ is an M-isomorphism, it follows from

(3.15) that m_1^{I} acts on $\pi_n(k_0)a_{[s;n]}(t)$ trivially. Then, by (3.20) and the assumption $n \equiv 0 \pmod 4$, we have:

$$\begin{aligned}\pi_n(k_0)a_{[s;n]}(-t) &= (\sqrt{-1})^n \pi_n(k_0)a_{[s;n]}(-t)\\ &= \pi_n(m_1^{\mathrm{I}})\pi_n(k_0)a_{[s;n]}(t)\\ &= \pi_n(k_0)a_{[s;n]}(t).\end{aligned}$$

Now the assertion follows.

Proposition 6.2 *Given $n \in \mathbb{Z}_{\geq 0}$, the following conditions on $s \in \mathbb{C}$ are equivalent:*

(i) $u_{[s;n]}(t) \in \mathrm{Pol}_n[t]$.
(ii) *One of the following holds:*

  (a) $n \equiv 0 \pmod 4 : s \in \mathbb{C}$.
  (b) $n \equiv 1 \pmod 4 : s \in I_1$.
  (c) $n \equiv 2 \pmod 4 : s \in I_2^+ \cup I_2^-$.
  (d) $n \equiv 3 \pmod 4 : s \in I_3$.

Proof For the case of n even, the assertions simply follow from Theorem 5.4 and Lemma 6.1. For the case of n odd, suppose that $n \equiv 1 \pmod 4$. By Theorem 5.4, we have $\mathrm{Sol}_{\mathrm{II}}(s; n) \neq \{0\}$ if and only if $s \in I_1$. Moreover, for such $s \in I_1$, the dimension $\dim_{\mathbb{C}} \mathrm{Sol}_{\mathrm{II}}(s; n)$ is $\dim_{\mathbb{C}} \mathrm{Sol}_{\mathrm{II}}(s; n) = 2$. Thus $\dim_{\mathbb{C}} \mathrm{Sol}_{\mathrm{I}}(s; n) = 2$ for $s \in I_1$. Now the assertion follows from (6.1). Since the case of $n \equiv 3 \pmod 4$ can be handled similarly, we omit the proof.

Proposition 6.3 *Given $n \in \mathbb{Z}_{\geq 0}$, the following conditions on $s \in \mathbb{C}$ are equivalent:*

(i) $v_{[s;n]}(t) \in \mathrm{Pol}_n[t]$.
(ii) *One of the following holds:*

  (a) $n \equiv 0 \pmod 4 : s \in I_0^+ \cup I_0^-$.
  (b) $n \equiv 1 \pmod 4 : s \in I_1$.
  (c) $n \equiv 2 \pmod 4 : s \in \mathbb{C}$.
  (d) $n \equiv 3 \pmod 4 : s \in I_3$.

Proof Since the proof is similar to the one for Proposition 6.2, we omit the proof.

Theorem 6.4 *The following conditions on $(s, n) \in \mathbb{C} \times \mathbb{Z}_{\geq 0}$ are equivalent.*

(i) $\mathrm{Sol}_{\mathrm{I}}(s; n) \neq \{0\}$.
(ii) *One of the following conditions is satisfied.*

- $n \equiv 0 \pmod 4 : s \in \mathbb{C}$.
- $n \equiv 1 \pmod 4 : s \in I_1$.
- $n \equiv 2 \pmod 4 : s \in \mathbb{C}$.
- $n \equiv 3 \pmod 4 : s \in I_3$.

Moreover, for such (s, n), the space $\mathrm{Sol}_{\mathrm{I}}(s; n)$ may be described as follows:

(1) $n \equiv 0 \pmod 4$:

$$\mathrm{Sol}_{\mathrm{I}}(s; n) = \begin{cases} \mathbb{C}u_{[s;n]}(t) & \text{if } s \in \mathbb{C}\backslash(I_0^+ \cup I_0^-), \\ \mathbb{C}u_{[s;n]}(t) \oplus \mathbb{C}v_{[s;n]}(t) & \text{if } s \in I_0^+, \\ \mathbb{C}u_{[s;n]}(t) \oplus \mathbb{C}v_{[s;n]}(t) & \text{if } s \in I_0^-. \end{cases}$$

(2) $n \equiv 1 \pmod 4$:

$$\mathrm{Sol}_{\mathrm{I}}(s; n) = \mathbb{C}u_{[s;n]}(t) \oplus \mathbb{C}v_{[s;n]}(t) \quad \text{for } s \in I_1.$$

(3) $n \equiv 2 \pmod 4$:

$$\mathrm{Sol}_{\mathrm{I}}(s; n) = \begin{cases} \mathbb{C}v_{[s;n]}(t) & \text{if } s \in \mathbb{C}\backslash(I_2^+ \cup I_2^-), \\ \mathbb{C}u_{[s;n]}(t) \oplus \mathbb{C}v_{[s;n]}(t) & \text{if } s \in I_2^+, \\ \mathbb{C}u_{[s;n]}(t) \oplus \mathbb{C}v_{[s;n]}(t) & \text{if } s \in I_2^-. \end{cases}$$

(4) $n \equiv 3 \pmod 4$:

$$\mathrm{Sol}_{\mathrm{I}}(s; n) = \mathbb{C}u_{[s;n]}(t) \oplus \mathbb{C}v_{[s;n]}(t), \quad \text{for } s \in I_3.$$

Proof This is a summary of the results in Propositions 6.2 and 6.3.

Theorem 6.5 *For $(s, n) \in \mathbb{C} \times \mathbb{Z}_{\geq 0}$ classified in Theorem 6.4, M acts on $\mathrm{Sol}_{\mathrm{I}}(s; n)$ as follows:*

(a) $n \equiv 0 \pmod 4$:

$$\mathrm{Sol}_{\mathrm{I}}(s; n) \simeq \begin{cases} (+, +) & \text{if } s \in \mathbb{C}\backslash(I_0^+ \cup I_0^-), \\ (+, +) \oplus (+, -) & \text{if } s \in I_0^+, \\ (+, +) \oplus (-, +) & \text{if } s \in I_0^-. \end{cases}$$

(b) $n \equiv 1 \pmod 4$:

$$\mathrm{Sol}_{\mathrm{I}}(s; n) \simeq \mathbb{H} \quad \text{for } s \in I_1.$$

(c) $n \equiv 2 \pmod 4$:

$$\mathrm{Sol}_{\mathrm{I}}(s; n) \simeq \begin{cases} (-, -) & \text{if } s \in \mathbb{C}\backslash(I_2^+ \cup I_2^-), \\ (-, +) \oplus (-, -) & \text{if } s \in I_2^+, \\ (+, -) \oplus (-, -) & \text{if } s \in I_2^-. \end{cases}$$

(d) $n \equiv 3 \pmod 4$:

$$\mathrm{Sol}_{\mathrm{I}}(s; n) \simeq \mathbb{H} \quad \text{for } s \in I_3.$$

Here the characters $(\varepsilon, \varepsilon')$ *stand for the ones on* $\mathbb{C}u_{[s;n]}(t)$ *and* $\mathbb{C}v_{[s;n]}(t)$ *at the same places in Theorem 6.4.*

Proof In the case of n odd, the assertions simply follow from Theorem 5.7 via the M-isomorphism $\pi_n(k_0)\colon \mathrm{Sol}_{\mathrm{II}}(s; n) \xrightarrow{\sim} \mathrm{Sol}_{\mathrm{I}}(s; n)$. Similarly, the assertions for n even are drawn by Lemma 6.1, Theorem 6.4, and Propositions 5.10, 5.11, and 5.13.

Recall from (5.20) the subsets $I^{\pm}((+,-))$, $I^{\pm}((-,+))$, $I(\mathbb{H}) \subset \mathbb{Z} \times \mathbb{Z}_{\geq 0}$. We now give the explicit description of $\mathrm{Hom}_M(\mathrm{Sol}_{\mathrm{I}}(s; n), \sigma) \neq \{0\}$.

Theorem 6.6 *The following conditions on* $(\sigma, s, n) \in \mathrm{Irr}(M) \times \mathbb{C} \times \mathbb{Z}_{\geq 0}$ *are equivalent:*

(i) $\mathrm{Hom}_M(\mathrm{Sol}_{\mathrm{I}}(s; n), \sigma) \neq \{0\}$.
(ii) $\dim_{\mathbb{C}} \mathrm{Hom}_M(\mathrm{Sol}_{\mathrm{I}}(s; n), \sigma) = 1$.
(iii) *One of the following conditions holds:*

- $\sigma = (+,+) : (s, n) \in \mathbb{C} \times 4\mathbb{Z}_{\geq 0}$.
- $\sigma = (-,-) : (s, n) \in \mathbb{C} \times (2 + 4\mathbb{Z}_{\geq 0})$.
- $\sigma = (+,-) : (s, n) \in I^{+}((+,-)) \cup I^{-}((+,-))$.
- $\sigma = (-,+) : (s, n) \in I^{+}((-,+)) \cup I^{-}((-,+))$.
- $\sigma = \mathbb{H} : \quad (s, n) \in I(\mathbb{H})$.

Moreover, for such (σ, s, n), *the space* $\mathrm{Hom}_M(\mathrm{Sol}_{\mathrm{I}}(s; n), \sigma)$ *is given as follows:*

(1) $\sigma = (+,+)$: *For* $n \in 4\mathbb{Z}_{\geq 0}$, *we have:*

$$\mathrm{Hom}_M(\mathrm{Sol}_{\mathrm{I}}(s; n), (+,+)) = \mathbb{C}u_{[s;n]}(t) \quad \text{for all } s \in \mathbb{C}.$$

(2) $\sigma = (-,-)$: *For* $n \in 2 + 4\mathbb{Z}_{\geq 0}$, *we have:*

$$\mathrm{Hom}_M(\mathrm{Sol}_{\mathrm{I}}(s; n), (-,-)) = \mathbb{C}v_{[s;n]}(t) \quad \text{for all } s \in \mathbb{C}.$$

(3) $\sigma = (+,-)$: *We have:*

$$\mathrm{Hom}_M(\mathrm{Sol}_{\mathrm{I}}(s; n), (+,-)) = \begin{cases} \mathbb{C}v_{[s;n]}(t) & \text{if } (s, n) \in I^{+}((+,-)), \\ \mathbb{C}u_{[s;n]}(t) & \text{if } (s, n) \in I^{-}((+,-)). \end{cases}$$

(4) $\sigma = (-,+)$: *We have:*

$$\mathrm{Hom}_M(\mathrm{Sol}_{\mathrm{I}}(s; n), (-,+)) = \begin{cases} \mathbb{C}u_{[s;n]}(t) & \text{if } (s, n) \in I^{+}((-,+)), \\ \mathbb{C}v_{[s;n]}(t) & \text{if } (s, n) \in I^{-}((-,+)). \end{cases}$$

(5) $\sigma = \mathbb{H}$: *We have:*

$$\mathrm{Hom}_M(\mathrm{Sol}_{\mathrm{I}}(s;n), \mathbb{H}) = \mathbb{C}\varphi_{\mathrm{I}}^{(s;n)} \quad \text{for } (s,n) \in I(\mathbb{H}),$$

where $\varphi_{\mathrm{I}}^{(s;n)}$ *is a nonzero* M*-isomorphism*

$$\varphi_{\mathrm{I}}^{(s;n)} : \mathrm{Sol}_{\mathrm{I}}(s;n) \xrightarrow{\sim} \mathbb{H}.$$

Proof Since the theorem can be shown similarly to Theorem 5.14, we omit the proof.

7 Sequences $\{P_k(x;y)\}_{k=0}^{\infty}$ and $\{Q_k(x;y)\}_{k=0}^{\infty}$ of Tridiagonal Determinants

The aim of this section is to discuss about two sequences $\{P_k(x;y)\}_{k=0}^{\infty}$ and $\{Q_k(x;y)\}_{k=0}^{\infty}$ of $k \times k$ tridiagonal determinants associated with polynomial solutions to the Heun model $d\pi_n^{\mathrm{I}}(D_s^{\flat})f(t) = 0$. We give factorization formulas for $P_{\left[\frac{n+2}{2}\right]}(x;n)$ and $Q_{\left[\frac{n+1}{2}\right]}(x;n)$ as well as "functional equations" of $\{P_k(x;y)\}_{k=0}^{\infty}$ and $\{Q_k(x;y)\}_{k=0}^{\infty}$. These are achieved in Theorems 7.5 and 7.7 (factorization formulas), and 7.9, 7.11, and 7.13 (functional equations).

7.1 Sequences $\{P_k(x;y)\}_{k=0}^{\infty}$ and $\{Q_k(x;y)\}_{k=0}^{\infty}$ of Tridiagonal Determinants

We start with the definitions of $\{P_k(x;y)\}_{k=0}^{\infty}$ and $\{Q_k(x;y)\}_{k=0}^{\infty}$. Let $a(x)$ and $b(x)$ be two polynomials such that

$$a(x) = 2x(2x-1) \quad \text{and} \quad b(x) = 2x(2x+1). \tag{7.1}$$

For instance, for $x = 1, 2, 3, \ldots$, we have:

$$a(1) = 1 \cdot 2, \quad a(2) = 3 \cdot 4, \quad a(3) = 5 \cdot 6, \ldots,$$
$$b(1) = 2 \cdot 3, \quad b(2) = 4 \cdot 5, \quad b(3) = 6 \cdot 7, \ldots.$$

Similarly, for $x = \frac{1}{2}, \frac{3}{2}, \frac{5}{2}, \ldots$, we have:

$$a\left(\frac{1}{2}\right) = 0 \cdot 1, \quad a\left(\frac{3}{2}\right) = 2 \cdot 3, \quad a\left(\frac{5}{2}\right) = 4 \cdot 5, \ldots,$$

$$b\left(\frac{1}{2}\right) = 1 \cdot 2, \quad b\left(\frac{3}{2}\right) = 3 \cdot 4, \quad b\left(\frac{5}{2}\right) = 5 \cdot 6, \ldots.$$

Clearly, the polynomials $a(x)$ and $b(x)$ satisfy:

$$a(x) = b\left(\frac{2x-1}{2}\right) \quad \text{and} \quad b(x) = a\left(\frac{2x+1}{2}\right). \tag{7.2}$$

We define $k \times k$ tridiagonal determinants $P_k(x; y)$ and $Q_k(x; y)$ in terms of $a(x)$ and $b(x)$ as follows:

(1) $P_k(x; y)$:

- $k = 0 : P_0(x; y) = 1,$
- $k = 1 : P_1(x; y) = yx,$
- $k \geq 2 : P_k(x; y) =$

$$\begin{vmatrix} yx & a(1) & & & & \\ -a\left(\frac{y}{2}\right) & (y-4)x & a(2) & & & \\ & -a\left(\frac{y-2}{2}\right) & (y-8)x & a(3) & & \\ & & \cdots & \cdots & \cdots & \\ & & & -a\left(\frac{y-2k+6}{2}\right) & (y-4k+8)x & a(k-1) \\ & & & & -a\left(\frac{y-2k+4}{2}\right) & (y-4k+4)x \end{vmatrix}.$$

(2) $Q_k(x; y)$:

- $k = 0 : Q_0(x; y) = 1,$
- $k = 1 : Q_1(x; y) = (y-2)x,$
- $k \geq 2 : Q_k(x; y) =$

$$\begin{vmatrix} (y-2)x & b(1) & & & & \\ -b\left(\frac{y-2}{2}\right) & (y-6)x & b(2) & & & \\ & -b\left(\frac{y-4}{2}\right) & (y-10)x & b(3) & & \\ & & \cdots & \cdots & \cdots & \\ & & & -b\left(\frac{y-2k+4}{2}\right) & (y-4k+6)x & b(k-1) \\ & & & & -b\left(\frac{y-2k+2}{2}\right) & (y-4k+2)x \end{vmatrix}.$$

Remark 7.1 In general $P_k(x; y)$ and $Q_k(x; y)$ satisfy the following properties for specific y and k:

1. If $y = n \in 2 + 2\mathbb{Z}_{\geq 0}$, then $P_{\frac{n+2}{2}}(x; n)$ is anti-centrosymmetric:

$$P_{\frac{n+2}{2}}(x;n) = \begin{vmatrix} nx & a(1) & & & & \\ -a\left(\frac{n}{2}\right) & (n-4)x & a(2) & & & \\ & -a\left(\frac{n-2}{2}\right) & (n-8)x & a(3) & & \\ & & \cdots & \cdots & \cdots & \\ & & & -a(2) & -(n-4)x & a\left(\frac{n}{2}\right) \\ & & & & -a(1) & -nx \end{vmatrix}. \tag{7.3}$$

2. If $y = n \in 4 + 2\mathbb{Z}_{\geq 0}$, then $Q_{\frac{n}{2}}(x; n)$ is anti-centrosymmetric:

$$Q_{\frac{n}{2}}(x;n) = \begin{vmatrix} (n-2)x & b(1) & & & & \\ -b\left(\frac{n-2}{2}\right) & (n-6)x & b(2) & & & \\ & -b\left(\frac{n-4}{2}\right) & (n-10)x & b(3) & & \\ & & \cdots & \cdots & \cdots & \\ & & & -b(2) & -(n-6)x & b\left(\frac{n-2}{2}\right) \\ & & & & -b(1) & -(n-2)x \end{vmatrix}. \tag{7.4}$$

3. It follows from (7.2) that for $y = n \in 3 + 2\mathbb{Z}_{\geq 0}$, we have:

$$P_{\frac{n+1}{2}}(x;n) = \begin{vmatrix} nx & a(1) & & & & \\ -b\left(\frac{n-1}{2}\right) & (n-4)x & a(2) & & & \\ & -b\left(\frac{n-3}{2}\right) & (n-8)x & a(3) & & \\ & & \cdots & \cdots & \cdots & \\ & & & -b\,(2) & -(n-6)x & a\left(\frac{n-1}{2}\right) \\ & & & & -b\,(1) & -(n-2)x \end{vmatrix} \tag{7.5}$$

$$Q_{\frac{n+1}{2}}(x;n) = \begin{vmatrix} (n-2)x & b(1) & & & & \\ -a\left(\frac{n-1}{2}\right) & (n-6)x & b(2) & & & \\ & -a\left(\frac{n-3}{2}\right) & (n-10)x & b(3) & & \\ & & \cdots & \cdots & \cdots & \\ & & & -a(2) & -(n-4)x & b\left(\frac{n-1}{2}\right) \\ & & & & -a(1) & -nx \end{vmatrix}. \tag{7.6}$$

We also have $P_1(x;1)=x$ and $Q_1(x;1)=-x$. Therefore,

$$Q_{\frac{n+1}{2}}(x;n)=(-1)^{\frac{n+1}{2}}P_{\frac{n+1}{2}}(x;n)\quad \text{for } n\in 1+2\mathbb{Z}_{\geq 0}. \tag{7.7}$$

The tridiagonal determinants $P_k(x;y)$ and $Q_k(x;y)$ enjoy the following property.

Lemma 7.2 *For $k\in\mathbb{Z}_{\geq 0}$, we have:*

$$P_k(-x;y)=(-1)^kP_k(x;y)\quad\text{and}\quad Q_k(-x;y)=(-1)^kQ_k(x;y).$$

Proof The identities for $k=0,1$ clearly hold by definition. The assertions for $k\geq 2$ follow from the following general property of $k\times k$ tridiagonal determinants:

$$\begin{vmatrix} -a_1 & b_1 & & & & \\ c_1 & -a_2 & b_2 & & & \\ & c_2 & -a_3 & b_3 & & \\ & & \cdots & \cdots & \cdots & \\ & & & c_{k-2} & -a_{k-1} & b_{k-1} \\ & & & & c_{k-1} & -a_k \end{vmatrix} = (-1)^k \begin{vmatrix} a_1 & b_1 & & & & \\ c_1 & a_2 & b_2 & & & \\ & c_2 & a_3 & b_3 & & \\ & & \cdots & \cdots & \cdots & \\ & & & c_{k-2} & a_{k-1} & b_{k-1} \\ & & & & c_{k-1} & a_k \end{vmatrix}. \tag{7.8}$$

From the next subsections y is taken to be $y=n\in\mathbb{Z}_{\geq 0}$, and we shall discuss several properties of $\{P_k(x;n)\}_{k=0}^{\infty}$ and $\{Q_k(x;n)\}_{k=0}^{\infty}$.

7.2 Generating Functions of $\{P_k(x;n)\}_{k=0}^{\infty}$ and $\{Q_k(x;n)\}_{k=0}^{\infty}$

We first give the generating functions of $\{P_k(x;n)\}_{k=0}^{\infty}$ and $\{Q_k(x;n)\}_{k=0}^{\infty}$. Let $u_{[s;n]}(t)$ be the local Heun function defined in (4.23). Write $u_{[s;n]}(t)=\sum_{k=0}^{\infty}U_k(s;n)t^{2k}$ for the power series expansion at $t=0$. It follows from (8.8) with (8.6) in Sect. 8 that each coefficient $U_k(s;n)$ can be given as

$$U_k(s;n)=\begin{vmatrix} E_0^u & -1 & & & & \\ F_1^u & E_1^u & -1 & & & \\ & F_2^u & E_2^u & -1 & & \\ & & \cdots & \cdots & \cdots & \\ & & & F_{k-2}^u & E_{k-2}^u & -1 \\ & & & & F_{k-1}^u & E_{k-1}^u \end{vmatrix} \tag{7.9}$$

with

$$E_0^u=\frac{ns}{a(1)},\quad E_k^u=\frac{(n-4k)s}{a(k+1)},\quad\text{and}\quad F_k^u=\frac{a\left(\frac{n-2k+2}{2}\right)}{a(k+1)}. \tag{7.10}$$

Similarly, Eq. (8.9) with (8.7) in Sect. 8 shows that the coefficients $V_k(s;n)$ of the power series expansion $v_{[s;n]}(t) = \sum_{k=0}^{\infty} V_k(s;n)t^{2k+1}$ of the second solution $v_{[s;n]}(t)$ to the Heun equation (4.22) at $t = 0$ are given as

$$V_k(s;n) = \begin{vmatrix} E_0^v & -1 & & & & \\ F_1^v & E_1^v & -1 & & & \\ & F_2^v & E_2^v & -1 & & \\ & & \cdots & \cdots & \cdots & \\ & & & F_{k-2}^v & E_{k-2}^v & -1 \\ & & & & F_{k-1}^v & E_{k-1}^v \end{vmatrix} \tag{7.11}$$

with

$$E_0^v = \frac{(n-2)s}{b(1)}, \quad E_k^v = \frac{(n-4k-2)s}{b(k+1)}, \quad \text{and} \quad F_k^v = \frac{b\left(\frac{n-2k}{2}\right)}{b(k+1)}. \tag{7.12}$$

Proposition 7.3 shows that $u_{[s;n]}(t)$ and $v_{[s;n]}(t)$ are in fact the "hyperbolic cosine" generating function of $\{P_k(s;n)\}_{k=0}^{\infty}$ and "hyperbolic sine" generating function of $\{Q_k(s;n)\}_{k=0}^{\infty}$, respectively.

Proposition 7.3 *We have:*

$$u_{[s;n]}(t) = \sum_{k=0}^{\infty} P_k(s;n)\frac{t^{2k}}{(2k)!} \quad \text{and} \quad v_{[s;n]}(t) = \sum_{k=0}^{\infty} Q_k(s;n)\frac{t^{2k+1}}{(2k+1)!}.$$

Proof We only show the identity for $u_{[s;n]}(t)$; the assertion for $v_{[s;n]}(t)$ can be shown similarly. We wish to show that $U_k(s;n) = P_k(s;n)/(2k)!$ for all $k \in \mathbb{Z}_{\geq 0}$. By definition, it is clear that $U_0(s;n) = 1 = P_0(s;n)$ and $U_1(s;n) = ns/2 = P_1(s;n)/2!$. We then assume that $k \geq 2$. It follows from (7.10) that $U_k(s;n)$ can be given as

$$U_k(s;n) = \frac{(-1)^k}{\prod_{j=1}^{k} a(j)} P_k(-s;n) = \frac{P_k(s;n)}{\prod_{j=1}^{k} a(j)}.$$

Here, Lemma 7.2 is applied from the second identity to the third. Now the assertion follows from the identity $\prod_{j=1}^{k} a(j) = (2k)!$.

For $R_k(s;n) \in \{P_k(s;n), Q_k(s;n)\}$, we define:

$$\mathrm{Sol}_k\,(R;n) := \{s \in \mathbb{C} : R_k(s;n) = 0\}.$$

We recall from (5.3) the subsets $I_0^{\pm}, I_1, I_2^{\pm}, I_3 \subset \mathbb{Z}$.

Proposition 7.4 *Let $n \in 2\mathbb{Z}_{\geq 0}$. Then $\mathcal{S}\mathrm{ol}_{\frac{n+2}{2}}(P; n)$ and $\mathcal{S}\mathrm{ol}_{\frac{n}{2}}(Q; n)$ are given as follows.*

$$\mathcal{S}\mathrm{ol}_{\frac{n+2}{2}}(P; n) = \begin{cases} \mathbb{C} & if\ n \equiv 0 \pmod 4, \\ I_2^+ \cup I_2^- & if\ n \equiv 2 \pmod 4. \end{cases}$$

$$\mathcal{S}\mathrm{ol}_{\frac{n}{2}}(Q; n) = \begin{cases} I_0^+ \cup I_0^- & if\ n \equiv 0 \pmod 4, \\ \mathbb{C} & if\ n \equiv 2 \pmod 4. \end{cases}$$

Further, for $n \in 1 + 2\mathbb{Z}_{\geq 0}$, the sets $\mathcal{S}\mathrm{ol}_{\frac{n+1}{2}}(P; n)$ and $\mathcal{S}\mathrm{ol}_{\frac{n+1}{2}}(Q; n)$ are given as

$$\mathcal{S}\mathrm{ol}_{\frac{n+1}{2}}(P; n) = \mathcal{S}\mathrm{ol}_{\frac{n+1}{2}}(Q; n) = \begin{cases} I_1 & if\ n \equiv 1 \pmod 4, \\ I_3 & if\ n \equiv 3 \pmod 4. \end{cases}$$

Proof We show the assertion for $\mathcal{S}\mathrm{ol}_{\frac{n+2}{2}}(P; n)$. The other cases can be shown similarly. For $n \in 2\mathbb{Z}_{\geq 0}$, we let

$$S(n) = \begin{cases} \mathbb{C} & \text{if } n \equiv 0 \pmod 4, \\ I_2^+ \cup I_2^- & \text{if } n \equiv 2 \pmod 4. \end{cases}$$

We wish to show $S(n) = \mathcal{S}\mathrm{ol}_{\frac{n+2}{2}}(P; n)$.

First, it follows from Proposition 6.2 that the following conditions on $s \in \mathbb{C}$ are equivalent for $n \in 2\mathbb{Z}_{\geq 0}$:

(i) $u_{[s;n]}(t) \in \mathrm{Pol}_n[t]$;
(ii) $s \in S(n)$.

By Propositions 7.3, this shows that $S(n) \subset \mathcal{S}\mathrm{ol}_{\frac{n+2}{2}}(P; n)$.

If $n \equiv 0 \pmod 4$, we have $S(n) = \mathbb{C} \subset \mathcal{S}\mathrm{ol}_{\frac{n+2}{2}}(P; n) \subset \mathbb{C}$. Thus, $S(n) = \mathcal{S}\mathrm{ol}_{\frac{n+2}{2}}(P; n)$.

If $n \equiv 2 \pmod 4$, then since $S(n) \subset \mathcal{S}\mathrm{ol}_{\frac{n+2}{2}}(P; n)$, we have:

$$P_{\frac{n+2}{2}}(s; n) = 0 \quad \text{for all } s \in I_2^+ \cup I_2^-. \tag{7.13}$$

On the other hand, by (5.3), the cardinality $|I_2^+ \cup I_2^-|$ of $I_2^+ \cup I_2^-$ is given by

$$|I_2^+ \cup I_2^-| = \frac{n+2}{2} = \deg P_{\frac{n+2}{2}}(x; n). \tag{7.14}$$

Now the desired identity $S(n) = \mathcal{S}\mathrm{ol}_{\frac{n+2}{2}}(P; n)$ follows from (7.13) and (7.14).

7.3 Factorization Formulas of $P_{\left[\frac{n+2}{2}\right]}(x;n)$ and $Q_{\left[\frac{n+1}{2}\right]}(x;n)$

We next show the factorization formulas for $P_{\frac{n+2}{2}}(x;n)$ and $Q_{\frac{n}{2}}(x;n)$ for n even, and for $P_{\frac{n+1}{2}}(x;n)$ and $Q_{\frac{n+1}{2}}(x;n)$ for n odd. We remark that any product of the form $\prod_{\ell=j}^{j-1} c_\ell$ with $c_\ell \in \mathrm{Pol}[t]$ (in particular, $c_\ell \in \mathbb{C}$) is regarded as $\prod_{\ell=j}^{j-1} c_\ell = 1$.

7.3.1 Factorization Formulas of $P_{\frac{n+2}{2}}(x;n)$ and $Q_{\frac{n}{2}}(x;n)$ for $n \in 2\mathbb{Z}_{\geq 0}$

We start with $P_{\frac{n+2}{2}}(x;n)$ and $Q_{\frac{n}{2}}(x;n)$ for n even (see (7.3) and (7.4)). For $n \in 2\mathbb{Z}_{\geq 0}$, let α_n and β_n be the products of the coefficients of x on the main diagonal of $P_{\frac{n+2}{2}}(x;n)$ and $Q_{\frac{n}{2}}(x;n)$, respectively. Namely, we have:

$$\alpha_n = \prod_{\ell=0}^{\frac{n}{2}}(n-4\ell) \quad \text{and} \quad \beta_n = \prod_{\ell=0}^{\frac{n-2}{2}}(n-2-4\ell).$$

It is remarked that α_n and β_n may be given as follows:

$$\alpha_n = \begin{cases} 0 & \text{if } n \equiv 0 \pmod 4, \\ (-4)^{\frac{n+2}{4}} \prod_{\ell=0}^{\frac{n-2}{4}}(1+2\ell)^2 & \text{if } n \equiv 2 \pmod 4 \end{cases}$$

and

$$\beta_n = \begin{cases} (-4)^{\frac{n}{4}} \prod_{\ell=0}^{\frac{n-4}{4}}(1+2\ell)^2 & \text{if } n \equiv 0 \pmod 4, \\ 0 & \text{if } n \equiv 2 \pmod 4. \end{cases}$$

Theorem 7.5 *For $n \in 2\mathbb{Z}_{\geq 0}$, the polynomials $P_{\frac{n+2}{2}}(x;n)$ and $Q_{\frac{n}{2}}(x;n)$ are either 0 or factored as follows.*

$$P_{\frac{n+2}{2}}(x;n) = \begin{cases} 0 & if\ n \equiv 0 \pmod 4, \\ \alpha_n \prod_{\ell=0}^{\frac{n-2}{4}}(x^2-(4\ell+1)^2) & if\ n \equiv 2 \pmod 4 \end{cases} \tag{7.15}$$

and

$$Q_{\frac{n}{2}}(x;n) = \begin{cases} \beta_n \prod_{\ell=0}^{\frac{n-4}{4}} (x^2 - (4\ell+3)^2) & if\ n \equiv 0 \pmod 4, \\ 0 & if\ n \equiv 2 \pmod 4. \end{cases} \tag{7.16}$$

Proof We only demonstrate the proof for $P_{\frac{n+2}{2}}(x;n)$; the assertion for $Q_{\frac{n}{2}}(x;n)$ can be shown similarly. It follows from Proposition 7.4 that

$$\mathcal{S}\mathrm{ol}_{\frac{n+2}{2}}(P;n) = \begin{cases} \mathbb{C} & \text{if } n \equiv 0 \pmod 4, \\ I_2^+ \cup I_2^- & \text{if } n \equiv 2 \pmod 4. \end{cases}$$

In particular, as α_n is the product of the coefficients of x in $P_{\frac{n+2}{2}}(x;n)$, we have:

$$P_{\frac{n+2}{2}}(x;n) = \begin{cases} 0 & \text{if } n \equiv 0 \pmod 4, \\ \alpha_n \prod_{s \in I_2^+ \cup I_2^-} (x-s) & \text{if } n \equiv 2 \pmod 4. \end{cases}$$

Now the assertion follows from the identity $\prod_{s \in I_2^+ \cup I_2^-}(x-s) = \prod_{\ell=0}^{\frac{n-2}{4}}(x^2 - (4\ell + 1)^2)$.

Remark 7.6 The factorization formulas (7.15) and (7.16) can also be obtained from [14, Prop. 5.11]. In fact, in [14], the dimension $\dim_{\mathbb{C}} \mathrm{Sol}_{\mathrm{I}}(s;n) (= \dim_{\mathbb{C}} \mathrm{Sol}_{\mathrm{II}}(s;n))$ was determined by using (7.15) and (7.16) (see Remark 5.6).

7.3.2 Factorization Formula of $P_{\frac{n+1}{2}}(x;n)$ for $n \in 1 + 2\mathbb{Z}_{\geq 0}$

We next consider $P_{\frac{n+1}{2}}(x;n)$ and $Q_{\frac{n+1}{2}}(x;n)$ for n odd (see (7.5) and (7.6)). As shown in (7.7), the determinant $Q_{\frac{n+1}{2}}(x;n)$ is given as

$$Q_{\frac{n+1}{2}}(x;n) = (-1)^{\frac{n+1}{2}} P_{\frac{n+1}{2}}(x;n) \quad \text{for } n \in 1 + 2\mathbb{Z}_{\geq 0}.$$

It thus suffices to only consider $P_{\frac{n+1}{2}}(x;n)$. For n odd, let γ_n be the product of the coefficients of x on the main diagonal of $P_{\frac{n+1}{2}}(x;n)$, namely,

$$\gamma_n = \prod_{\ell=0}^{\frac{n-1}{2}} (n - 4\ell).$$

We remark that γ_n may be given as

$$\gamma_n = \begin{cases} (-1)^{\frac{n-1}{4}} \prod_{\ell=0}^{\frac{n-3}{2}} (3+2\ell) & \text{if } n \equiv 1 \pmod 4, \\ (-1)^{\frac{n+1}{4}} \prod_{\ell=0}^{\frac{n-3}{2}} (3+2\ell) & \text{if } n \equiv 3 \pmod 4. \end{cases}$$

Theorem 7.7 *For $n \in 1 + 2\mathbb{Z}_{\geq 0}$, the polynomial $P_{\frac{n+1}{2}}(x; n)$ is factored as*

$$P_{\frac{n+1}{2}}(x; n) = \gamma_n \prod_{\ell=0}^{\frac{n-1}{2}} (x - (n-1) + 4\ell). \tag{7.17}$$

Equivalently, we have:

$$P_{\frac{n+1}{2}}(x; n) = \begin{cases} \gamma_n\, x \prod_{\ell=1}^{\frac{n-1}{4}} (x^2 - (4\ell)^2) & \text{if } n \equiv 1 \pmod 4, \\ \gamma_n \prod_{\ell=0}^{\frac{n-3}{4}} (x^2 - (4\ell+2)^2) & \text{if } n \equiv 3 \pmod 4. \end{cases}$$

Proof Since the argument is similar to that for Theorem 7.5, we omit the proof.

Recall from (1.15) that Sylvester determinant $\mathrm{Sylv}(x; n)$ satisfies the following formula:

$$\mathrm{Sylv}(x; n) = \prod_{\ell=0}^{n} (x - n + 2\ell). \tag{7.18}$$

Equivalently,

$$\mathrm{Sylv}(x; n) = \begin{cases} \prod_{\ell=0}^{\frac{n-1}{2}} (x^2 - (2\ell+1)^2) & \text{if } n \text{ is odd}, \\ x \prod_{\ell=1}^{\frac{n}{2}} (x^2 - (2\ell)^2) & \text{if } n \text{ is even}. \end{cases} \tag{7.19}$$

An observation on (7.19) gives the following expression of $P_{\frac{n+1}{2}}(x; n)$.

Corollary 7.8 For n odd, we have:

$$P_{\frac{n+1}{2}}(x; n) = 2^{n+1} \mathrm{Sylv}(\frac{1}{2}; \frac{n-1}{2}) \mathrm{Sylv}(\frac{x}{2}; \frac{n-1}{2}).$$

Proof By a direct computation, one finds:

$$\prod_{\ell=0}^{\frac{n-1}{2}}(n-4\ell) = 2^{\frac{n+1}{2}}\operatorname{Sylv}(\frac{1}{2};\frac{n-1}{2})$$

and

$$\prod_{\ell=0}^{\frac{n-1}{2}}(x-(n-1)+4\ell) = 2^{\frac{n+1}{2}}\operatorname{Sylv}(\frac{x}{2};\frac{n-1}{2}). \tag{7.20}$$

Now the proposed identity follows from Theorem 7.7 as $\gamma_n = \prod_{\ell=0}^{\frac{n-1}{2}}(n-4\ell)$.

7.4 *Functional Equations of* $\{P_k(x;y)\}_{k=0}^{\infty}$ *and* $\{Q_k(x;y)\}_{k=0}^{\infty}$

We finish this section by showing "functional equations" of $\{P_k(x;y)\}_{k=0}^{\infty}$ and $\{Q_k(x;y)\}_{k=0}^{\infty}$. We consider the following cases, separately:

Case 1: $\{P_k(x;y)\}_{k=0}^{\infty}$ for $y = n \in 2\mathbb{Z}_{\geq 0}$
Case 2: $\{Q_k(x;y)\}_{k=0}^{\infty}$ for $y = n \in 2(1+\mathbb{Z}_{\geq 0})$
Case 3: $\{P_k(x;y)\}_{k=0}^{\infty}$ and $\{Q_k(x;y)\}_{k=0}^{\infty}$ for $y = n \in 1+2\mathbb{Z}_{\geq 0}$

We remark that $\{P_k(x;y)\}_{k=0}^{\infty}$ and $\{Q_k(x;y)\}_{k=0}^{\infty}$ satisfy a *system* of functional equations in Case 3.

For $s \in \mathbb{R}\backslash\{0\}$, we set:

$$\operatorname{sgn}(s) := \begin{cases} +1 & \text{if } s > 0, \\ -1 & \text{if } s < 0. \end{cases}$$

7.4.1 Functional Equation of $\{P_k(x;y)\}_{k=0}^{\infty}$ for $n \in 2\mathbb{Z}_{\geq 0}$

We start with $\{P_k(x;y)\}_{k=0}^{\infty}$ for $n \in 2\mathbb{Z}_{\geq 0}$. Recall from Proposition 7.4 that $\mathcal{S}ol_{\frac{n+2}{2}}(P;n)$ for $n \in 2\mathbb{Z}_{\geq 0}$ is given as

$$\mathcal{S}ol_{\frac{n+2}{2}}(P;n) = \begin{cases} \mathbb{C} & \text{if } n \equiv 0 \pmod 4, \\ I_2^+ \cup I_2^- & \text{if } n \equiv 2 \pmod 4, \end{cases}$$

where

$$I_2^+ \cup I_2^- = \{\pm 1, \pm 5, \pm 9 \ldots, \pm(n-1)\}.$$

We then define a map $\theta_{(P;n)} \colon \mathrm{Sol}_{\frac{n+2}{2}}(P;n) \to \{\pm 1\}$ as

$$\theta_{(P;n)}(s) = \begin{cases} 1 & \text{if } n \equiv 0 \pmod 4, \\ \mathrm{sgn}(s) & \text{if } n \equiv 2 \pmod 4. \end{cases} \tag{7.21}$$

Theorem 7.9 *Let $n \in 2\mathbb{Z}_{\geq 0}$ and $s \in \mathrm{Sol}_{\frac{n+2}{2}}(P;n)$. Then we have $P_k(s;n) = 0$ for $k \geq \frac{n+2}{2}$ and*

$$\frac{P_k(s;n)}{(2k)!} = \theta_{(P;n)}(s) \frac{P_{\frac{n}{2}-k}(s;n)}{(n-2k)!} \tag{7.22}$$

for $k \leq \frac{n}{2}$. Equivalently, for $k \leq \frac{n}{2}$, the following hold:

(1) $n \equiv 0 \pmod 4$: *We have:*

$$\frac{P_k(s;n)}{(2k)!} = \frac{P_{\frac{n}{2}-k}(s;n)}{(n-2k)!} \quad \text{for all } s \in \mathbb{C}. \tag{7.23}$$

(2) $n \equiv 2 \pmod 4$: *For $s \in \{\pm 1, \pm 5, \pm 9 \ldots, \pm(n-1)\}$, we have:*

$$\frac{P_k(s;n)}{(2k)!} = \mathrm{sgn}(s) \frac{P_{\frac{n}{2}-k}(s;n)}{(n-2k)!}. \tag{7.24}$$

Proof Take $s \in \mathrm{Sol}_{\frac{n+2}{2}}(P;n)$. It follows from the equivalence given in the proof of Proposition 7.4 that $P_k(s;n) = 0$ for $k \geq \frac{n+2}{2}$. Thus, to prove the theorem, it suffices to show (7.23) and (7.24) for $k \leq \frac{n}{2}$. We first show that

$$\frac{P_k(s;n)}{(2k)!} = \pm \frac{P_{\frac{n}{2}-k}(s;n)}{(n-2k)!}. \tag{7.25}$$

It follows from Theorem 6.5 and Proposition 7.4 that M acts on $u_{[s;n]}(t)$ as a character $\chi_{(\varepsilon,\varepsilon')}$ in (3.23); in particular, we have:

$$u_{[s;n]}(t) = \chi_{(\varepsilon,\varepsilon')}(m_2^{\mathrm{I}}) \pi_n(m_2^{\mathrm{I}}) u_{[s;n]}(t). \tag{7.26}$$

By (3.20), the identity (7.26) is equivalent to

$$u_{[s;n]}(t) = \chi_{(\varepsilon,\varepsilon')}(m_2^{\mathrm{I}}) t^n u_{[s;n]}\left(-\frac{1}{t}\right). \tag{7.27}$$

As $s \in \mathrm{Sol}_{\frac{n+2}{2}}(P; n)$, by Proposition 7.3, we have

$$u_{[s;n]}(t) = \sum_{k=0}^{\frac{n}{2}} P_k(s; n) \frac{t^{2k}}{(2k)!}.$$

The identity (7.27) thus yields the identity:

$$\sum_{k=0}^{\frac{n}{2}} P_k(s; n) \frac{t^{2k}}{(2k)!} = \chi_{(\varepsilon,\varepsilon')}(m_2^{\mathrm{I}}) \sum_{k=0}^{\frac{n}{2}} P_{\frac{n}{2}-k}(s; n) \frac{t^{2k}}{(n-2k)!}. \tag{7.28}$$

Since $\chi_{(\varepsilon,\varepsilon')}(m_2^{\mathrm{I}}) \in \{\pm 1\}$, the identity (7.25) follows from (7.28).

In order to show (7.23) and (7.24), observe that Theorem 6.5 shows that the character $\chi_{(\varepsilon,\varepsilon')}$ is given as

- $\chi_{(\varepsilon,\varepsilon')} = \chi_{(+,+)}$ for $n \equiv 0 \pmod 4$ and $s \in \mathbb{C}$,
- $\chi_{(\varepsilon,\varepsilon')} = \chi_{(-,+)}$ for $n \equiv 2 \pmod 4$ and $s \in I_2^+$,
- $\chi_{(\varepsilon,\varepsilon')} = \chi_{(+,-)}$ for $n \equiv 2 \pmod 4$ and $s \in I_2^-$.

By Table 2, we have $\chi_{(+,+)}(m_2^{\mathrm{I}}) = \chi_{(-,+)}(m_2^{\mathrm{I}}) = 1$ and $\chi_{(+,-)}(m_2^{\mathrm{I}}) = -1$. Now (7.23) and (7.24) follow from (7.28).

Corollary 7.10 Let $n \in 2\mathbb{Z}_{\geq 0}$. Then the following hold for $P_{\frac{n}{2}}(s; n)$ (see (1.13)).

(1) $n \equiv 0 \pmod 4$: We have

$$P_{\frac{n}{2}}(s; n) = n! \quad \text{for all } s \in \mathbb{C}. \tag{7.29}$$

(2) $n \equiv 2 \pmod 4$: For $s \in \{\pm 1, \pm 5, \pm 9 \ldots, \pm(n-1)\}$, we have

$$P_{\frac{n}{2}}(s; n) = \mathrm{sgn}(s) n!. \tag{7.30}$$

Proof This is simply the case of $k = \frac{n}{2}$ in (7.23) and (7.24).

7.4.2 Functional Equation of $\{Q_k(x; y)\}_{k=0}^{\infty}$ for $n \in 2(1 + \mathbb{Z}_{\geq 0})$

We next consider $\{Q_k(x; y)\}_{k=0}^{\infty}$ for $n \in 2(1 + \mathbb{Z}_{\geq 0})$. Proposition 7.4 shows that

$$\mathrm{Sol}_{\frac{n}{2}}(Q; n) = \begin{cases} I_0^+ \cup I_0^- & \text{if } n \equiv 0 \pmod 4, \\ \mathbb{C} & \text{if } n \equiv 2 \pmod 4, \end{cases}$$

where

$$I_0^+ \cup I_0^- = \{\pm 3, \pm 7, \pm 11, \ldots, \pm(n-1)\}.$$

We then define a map $\theta_{(Q;n)} \colon \mathrm{Sol}_{\frac{n}{2}}(Q; n) \to \{\pm 1\}$ as

$$\theta_{(Q;n)}(s) = \begin{cases} \mathrm{sgn}(s) & \text{if } n \equiv 0 \pmod 4, \\ 1 & \text{if } n \equiv 2 \pmod 4. \end{cases} \tag{7.31}$$

Theorem 7.11 *Let $n \in 2(1 + \mathbb{Z}_{\geq 0})$ and $s \in \mathrm{Sol}_{\frac{n}{2}}(Q; n)$. Then we have $Q_k(s; n) = 0$ for $k \geq \frac{n}{2}$ and*

$$\frac{Q_k(s; n)}{(2k+1)!} = \theta_{(Q;n)}(s) \frac{Q_{\frac{n-2}{2}-k}(s; n)}{(n-2k-1)!} \tag{7.32}$$

for $k \leq \frac{n-2}{2}$. Equivalently, for $k \leq \frac{n}{2} - 1$, the following hold:

(1) $n \equiv 0 \pmod 4$: *For $s \in \{\pm 3, \pm 7, \pm 11, \ldots, \pm(n-1)\}$, we have:*

$$\frac{Q_k(s; n)}{(2k+1)!} = \mathrm{sgn}(s) \frac{Q_{\frac{n-2}{2}-k}(s; n)}{(n-2k-1)!}. \tag{7.33}$$

(2) $n \equiv 2 \pmod 4$: *We have:*

$$\frac{Q_k(s; n)}{(2k+1)!} = \frac{Q_{\frac{n-2}{2}-k}(s; n)}{(n-2k-1)!} \quad \text{for all } s \in \mathbb{C}. \tag{7.34}$$

Proof Since the argument goes similarly to the one for Theorem 7.9, we omit the proof.

Corollary 7.12 Let $n \in 2(1 + \mathbb{Z}_{\geq 0})$. Then the following hold for $Q_{\frac{n-2}{2}}(s; n)$ (see (1.14)):

(1) $n \equiv 0 \pmod 4$: For $s \in \{\pm 3, \pm 7, \pm 11, \ldots, \pm(n-1)\}$, we have:

$$Q_{\frac{n-2}{2}}(s; n) = \mathrm{sgn}(s)(n-1)!. \tag{7.35}$$

(2) $n \equiv 2 \pmod 4$: We have:

$$Q_{\frac{n-2}{2}}(s; n) = (n-1)! \quad \text{for all } s \in \mathbb{C}.$$

Proof This is the case of $k = \frac{n-2}{2}$ in (7.33) and (7.34).

7.4.3 Functional Equations of $\{P_k(x; y)\}_{k=0}^{\infty}$ and $\{Q_k(x; y)\}_{k=0}^{\infty}$ for $n \in 1 + 2\mathbb{Z}_{\geq 0}$

As the last case, we consider the functional equations of $\{P_k(x; y)\}_{k=0}^{\infty}$ and $\{Q_k(x; y)\}_{k=0}^{\infty}$ for $n \in 1 + 2\mathbb{Z}_{\geq 0}$. Proposition 7.4 shows that

$$\mathcal{S}\mathrm{ol}_{\frac{n+1}{2}}(P; n) = \mathcal{S}\mathrm{ol}_{\frac{n+1}{2}}(Q; n) = \begin{cases} I_1 & \text{if } n \equiv 1 \pmod 4, \\ I_3 & \text{if } n \equiv 3 \pmod 4, \end{cases}$$

where

$$I_1 = \{0, \pm 4, \ldots, \pm(n-1)\} \quad \text{and} \quad I_3 = \{\pm 2, \pm 6, \ldots, \pm(n-1)\}.$$

Theorem 7.13 *Let $n \in 1 + 2\mathbb{Z}_{\geq 0}$ and $s \in \mathcal{S}\mathrm{ol}_{\frac{n+1}{2}}(P; n)$. Then we have: $P_k(s; n) = Q_k(s; n) = 0$ for $k \geq \frac{n+1}{2}$ and*

$$\begin{pmatrix} \frac{P_k(s;n)}{(2k)!} \\ \frac{Q_k(s;n)}{(2k+1)!} \end{pmatrix} = \begin{pmatrix} 0 & \frac{P_{\frac{n-1}{2}}(s;n)}{(n-1)!} \\ \frac{Q_{\frac{n-1}{2}}(s;n)}{n!} & 0 \end{pmatrix} \begin{pmatrix} \frac{P_{\frac{n-1}{2}-k}(s;n)}{(n-1-2k)!} \\ \frac{Q_{\frac{n-1}{2}-k}(s;n)}{(n-2k)!} \end{pmatrix}$$

for $k \leq \frac{n-1}{2}$.

In preparation to the proof of Theorem 7.13, we start by showing the following proposition.

Proposition 7.14 *Let $n \in 1 + 2\mathbb{Z}_{\geq 0}$ and $s \in \mathcal{S}\mathrm{ol}_{\frac{n+1}{2}}(P; n)$. Then we have*

$$\frac{P_{\frac{n-1}{2}-k}(s; n)}{(n-1-2k)!} = \frac{P_{\frac{n-1}{2}}(s; n)}{(n-1)!} \cdot \frac{Q_k(s; n)}{(2k+1)!} \tag{7.36}$$

for $k \leq \frac{n-1}{2}$. In particular, the following identity holds:

$$P_{\frac{n-1}{2}}(s; n) \cdot Q_{\frac{n-1}{2}}(s; n) = (n-1)! \cdot n!. \tag{7.37}$$

Proof It follows from (4.23) and (4.24) (or, equivalently, Proposition 7.3) that $u_{[s;n]}(t)$ and $v_{[s;n]}(t)$ are even and odd functions, respectively. Since n is assumed to be odd, the transformed polynomial

$$\pi_n(m_2^{\mathrm{I}}) u_{[s;n]}(t) = t^n u_{[s;n]}\left(-\frac{1}{t}\right) \tag{7.38}$$

is an odd function. Since $\pi_n(m_2^{\mathrm{I}})$ preserves the space $\mathbb{C}u_{[s;n]}(t) \oplus \mathbb{C}v_{[s;n]}(t)$, there exists $c \in \mathbb{C}$ such that

$$\pi_n(m_2^{\mathrm{I}})u_{[s;n]}(t) = c \cdot v_{[s;n]}(t). \tag{7.39}$$

It follows from the proof of Proposition 7.4 that, under the hypothesis, we have

$$u_{[s;n]}(t) = \sum_{k=0}^{\frac{n-1}{2}} \frac{P_k(s;n)}{(2k)!} t^{2k} \quad \text{and} \quad v_{[s;n]}(t) = \sum_{k=0}^{\frac{n-1}{2}} \frac{Q_k(s;n)}{(2k+1)!} t^{2k+1}. \tag{7.40}$$

Then, by (7.38), the identity (7.39) is equivalent to

$$\sum_{k=0}^{\frac{n-1}{2}} \frac{P_{\frac{n-1}{2}-k}(s;n)}{(n-1-2k)!} t^{2k+1} = c \cdot \sum_{k=0}^{\frac{n-1}{2}} \frac{Q_k(s;n)}{(2k+1)!} t^{2k+1}.$$

Thus, for $k = 0, 1, \ldots, \frac{n-1}{2}$, we have:

$$\frac{P_{\frac{n-1}{2}-k}(s;n)}{(n-1-2k)!} = c \cdot \frac{Q_k(s;n)}{(2k+1)!}. \tag{7.41}$$

By substituting $k = 0$ in (7.41), we obtain:

$$\frac{P_{\frac{n-1}{2}}(s;n)}{(n-1)!} = c \cdot 1 = c,$$

which shows (7.36).

To show (7.37), observe that, by substituting $k = \frac{n-1}{2}$ in (7.41), we have:

$$\frac{P_{\frac{n-1}{2}}(s;n)}{(n-1)!} \cdot \frac{Q_{\frac{n-1}{2}}(s;n)}{n!} = 1, \tag{7.42}$$

which is equivalent to (7.37). This concludes the proposition.

Now we prove Theorem 7.13.

Proof of Theorem 7.13 It follows from (7.40) that clearly $P_k(s;n) = Q_k(s;n) = 0$ for $k \geq \frac{n+1}{2}$. Thus, it suffices to assume that $k \leq \frac{n-1}{2}$. One wishes to show that the following identities hold:

$$\frac{P_k(s;n)}{(2k)!} = \frac{P_{\frac{n-1}{2}}(s;n)}{(n-1)!} \cdot \frac{Q_{\frac{n-1}{2}-k}(s;n)}{(n-2k)!}, \tag{7.43}$$

$$\frac{Q_k(s;n)}{(2k+1)!} = \frac{Q_{\frac{n-1}{2}}(s;n)}{n!} \cdot \frac{P_{\frac{n-1}{2}-k}(s;n)}{(n-1-2k)!}. \tag{7.44}$$

To show the first identity (7.43), observe that it follows from (7.36) that, for $k = j$, we have

$$\frac{P_{\frac{n-1}{2}-j}(s;n)}{(n-1-2j)!} = \frac{P_{\frac{n-1}{2}}(s;n)}{(n-1)!} \cdot \frac{Q_j(s;n)}{(2j+1)!}. \tag{7.45}$$

The identity (7.43) is then obtained by substituting $j = \frac{n-1}{2} - k$ in (7.45).

For the second identity (7.44), observe that (7.36) implies that

$$\frac{Q_k(s;n)}{(2k+1)!} = \left(\frac{P_{\frac{n-1}{2}}(s;n)}{(n-1)!}\right)^{-1} \cdot \frac{P_{\frac{n-1}{2}-k}(s;n)}{(n-1-2k)!}.$$

The identity (7.42) now concludes (7.44).

We finish this section by the following corollary of Proposition 7.14.

Corollary 7.15 Under the same hypothesis of Proposition 7.14, we have:

$$\begin{pmatrix} \pi_n(m_2^{\mathrm{I}})u_{[s;n]}(t) \\ \pi_n(m_2^{\mathrm{I}})v_{[s;n]}(t) \end{pmatrix} = \begin{pmatrix} 0 & \frac{P_{\frac{n-1}{2}}(s;n)}{(n-1)!} \\ -\frac{Q_{\frac{n-1}{2}}(s;n)}{n!} & 0 \end{pmatrix} \begin{pmatrix} u_{[s;n]}(t) \\ v_{[s;n]}(t) \end{pmatrix}.$$

Proof We wish to show that the following identities hold:

$$\pi_n(m_2^{\mathrm{I}})u_{[s;n]}(t) = \frac{P_{\frac{n-1}{2}}(s;n)}{(n-1)!} \cdot v_{[s;n]}(t), \tag{7.46}$$

$$\pi_n(m_2^{\mathrm{I}})v_{[s;n]}(t) = -\frac{Q_{\frac{n-1}{2}}(s;n)}{n!} \cdot u_{[s;n]}(t). \tag{7.47}$$

The first identity (7.46) follows from the proof of Proposition 7.14. To show the second identity (7.47), observe that (7.46) with (7.42) implies that

$$v_{[s;n]}(t) = \frac{Q_{\frac{n-1}{2}}(s;n)}{n!} \cdot \pi_n(m_2^{\mathrm{I}})u_{[s;n]}(t). \tag{7.48}$$

Now (7.47) is obtained by applying $\pi_n(m_2^{\mathrm{I}})$ to both sides of (7.48), as $\pi_n(m_2^{\mathrm{I}})^2 = -\mathrm{id}$.

8 Appendix 1: Local Heun Functions

In this appendix we collect several facts and lemmas for the local Heun function $Hl(a, q; \alpha, \beta, \gamma, \delta; z)$ at $z = 0$ that are used in the main part of this paper.

8.1 General Facts

As in (4.20), we set:

$$\mathcal{D}_H(a, q; \alpha, \beta, \gamma, \delta; z) := \frac{d^2}{dz^2} + \left(\frac{\gamma}{z} + \frac{\delta}{z-1} + \frac{\varepsilon}{z-a}\right)\frac{d}{dz} + \frac{\alpha\beta z - q}{z(z-1)(z-a)}$$

with $\gamma + \delta + \varepsilon = \alpha + \beta + 1$. Then the equation

$$\mathcal{D}_H(a, q; \alpha, \beta, \gamma, \delta; z) f(z) = 0$$

is called *Heun's differential equation*. The P-symbol is

$$P\left\{\begin{matrix} 0 & 1 & a & \infty & & \\ 0 & 0 & 0 & \alpha & z & q \\ 1-\gamma & 1-\delta & 1-\varepsilon & \beta & & \end{matrix}\right\}.$$

Let $Hl(a, q; \alpha, \beta, \gamma, \delta; z)$ stand for the local Heun function at $z = 0$ [32]. As in [32], we normalize $Hl(a, q; \alpha, \beta, \gamma, \delta; z)$ so that $Hl(a, q; \alpha, \beta, \gamma, \delta; 0) = 1$. It is known that, for $\gamma \notin \mathbb{Z}$, the functions

$$Hl(a, q; \alpha, \beta, \gamma, \delta; z) \quad \text{and} \quad z^{1-\gamma} Hl(a, q'; \alpha - \gamma + 1, \beta - \gamma + 1, 2 - \gamma, \delta; z) \tag{8.1}$$

with $q' := q + (1 - \gamma)(\varepsilon + a\delta)$ are two linearly independent solutions at $z = 0$ to the Heun equation $\mathcal{D}_H(a, q; \alpha, \beta, \gamma, \delta; z) f(z) = 0$. (See, for instance, [26] and [33, p. 99].)

Remark 8.1 In [33, p. 99], there seems to be a typographical error on the formula:

$$\lambda' = \lambda + (a + b + 1 - c)(1 - c).$$

This should read as

$$\lambda' = \lambda + (a + b + 1 - c + (t - 1)d)(1 - c).$$

Let $Hl(a, q; \alpha, \beta, \gamma, \delta; z) = \sum_{k=0}^{\infty} c_k z^k$ be the power series expansion at $z = 0$. In our normalization we have $c_0 = 1$. Then c_k for $k \geq 1$ satisfy the following recurrence relations (see, for instance, [32, p. 34]).

$$-q + a\gamma c_1 = 0, \tag{8.2}$$

$$P_k c_{k-1} - (Q_k + q)c_k + R_k c_{k+1} = 0, \tag{8.3}$$

where

$$P_k = (k - 1 + \alpha)(k - 1 + \beta),$$
$$Q_k = k[(k - 1 + \gamma)(1 + a) + a\delta + \varepsilon],$$
$$R_k = (k + 1)(k + \gamma)a.$$

Lemma 8.2 shows that each coefficient c_k for $Hl(a, q; \alpha, \beta, \gamma, \delta; z) = \sum_{k=0}^{\infty} c_k z^k$ has a determinant representation.

Lemma 8.2 (See, for instance, [24, Lem. B.1]) *Let $\{a_k\}_{k \in \mathbb{Z}_{\geq 0}}$ be a sequence with $a_0 = 1$ satisfying the following relations:*

- $a_1 = A_0$;
- $a_{k+1} = A_k a_k + B_k a_{k-1}$ $(k \geq 1)$

for some $A_0, A_k, B_k \in \mathbb{C}$. Then a_k can be expressed as

$$a_k = \begin{vmatrix} A_0 & -1 & & & & \\ B_1 & A_1 & -1 & & & \\ & B_2 & A_2 & -1 & & \\ & & \cdots & \cdots & \cdots & \\ & & & B_{k-2} & A_{k-2} & -1 \\ & & & & B_{k-1} & A_{k-1} \end{vmatrix}.$$

8.2 The Local Solutions $u_{[s;n]}(t)$ and $v_{[s;n]}(t)$

Now we consider the following local solutions to (4.22):

$$u_{[s;n]}(t) = Hl(-1, -\frac{ns}{4}; -\frac{n}{2}, -\frac{n-1}{2}, \frac{1}{2}, \frac{1-n-s}{2}; t^2), \tag{8.4}$$

$$v_{[s;n]}(t) = tHl(-1, -\frac{n-2}{4}s; -\frac{n-1}{2}, -\frac{n-2}{2}, \frac{3}{2}, \frac{1-n-s}{2}; t^2). \tag{8.5}$$

Lemma 8.3 *Let* $u_{[s;n]}(t) = \sum_{k=0}^{\infty} U_k(s;n)t^{2k}$ *be the power series expansion of* $u_{[s;n]}(t)$ *at* $t = 0$*. Then the coefficients* $U_k(s;n)$ *for* $k \geq 1$ *satisfy the following recurrence relations:*

$$U_1(s;n) = E_0^u,$$
$$U_{k+1}(s;n) = E_k^u U_k(s;n) + F_k^u U_{k-1}(s;n),$$

where

$$E_0^u = \frac{ns}{2}, \quad E_k^u = \frac{(n-4k)s}{(2k+1)(2k+2)} \quad \text{and} \quad F_k^u = \frac{(n-2k+1)(n-2k+2)}{(2k+1)(2k+2)}. \tag{8.6}$$

Proof This follows from (8.2) and (8.3) for the specific parameters in (8.4).

Lemma 8.4 *Let* $v_{[s;n]}(t) = \sum_{k=0}^{\infty} V_k(s;n)t^{2k+1}$ *be the power series expansion of* $v_{[s;n]}(t)$ *at* $t = 0$*. Then the coefficients* $V_k(s;n)$ $(k \geq 1)$ *satisfy the following recurrence relations:*

$$V_1(s;n) = E_0^v,$$
$$V_{k+1}(s;n) = E_k^v V_k(s;n) + F_k^v V_{k-1}(s;n),$$

where

$$E_0^v = \frac{(n-2)s}{6}, \quad E_k^v = \frac{(n-4k-2)s}{(2k+2)(2k+3)} \quad \text{and} \quad F_k^v = \frac{(n-2k)(n-2k+1)}{(2k+2)(2k+3)}. \tag{8.7}$$

Proof This follows from (8.2) and (8.3) for the specific parameters in (8.5).

It follows from Lemmas 8.2, 8.3, and 8.4 that $U_k(s;n)$ and $V_k(s;n)$ have determinant representations:

$$U_k(s;n) = \begin{vmatrix} E_0^u & -1 & & & & \\ F_1^u & E_1^u & -1 & & & \\ & F_2^u & E_2^u & -1 & & \\ & & \cdots & \cdots & \cdots & \\ & & & F_{k-2}^u & E_{k-2}^u & -1 \\ & & & & F_{k-1}^u & E_{k-1}^u \end{vmatrix} \tag{8.8}$$

and

$$V_k(s;n) = \begin{vmatrix} E_0^v & -1 & & & & \\ F_1^v & E_1^v & -1 & & & \\ & F_2^v & E_2^v & -1 & & \\ & & \cdots & \cdots & \cdots & \\ & & & F_{k-2}^v & E_{k-2}^v & -1 \\ & & & & F_{k-1}^v & E_{k-1}^v \end{vmatrix}. \tag{8.9}$$

9 Appendix 2: Functional Equation of Cayley Continuants and Krawtchouk Polynomials

In this appendix we show Cayley continuants $\{\mathrm{Cay}_k(x;y)\}_{k=0}^{\infty}$ and Krawtchouk polynomials $\{\mathcal{K}_k(x;y)\}_{k=0}^{\infty}$ admit functional equations as in Theorems 7.9 and 7.11. We also briefly discuss a relationship of these polynomials with Jacobi polynomials $P_k^{(\alpha,\beta)}(z)$.

9.1 *Functional Equation of Cayley Continuants* $\{\mathrm{Cay}_k(x;y)\}_{k=0}^{\infty}$

Cayley continuants $\{\mathrm{Cay}_k(x;y)\}_{k=0}^{\infty}$ are $k \times k$ tridiagonal determinants defined as follows:

- $k = 0 : \mathrm{Cay}_0(x;y) = 1$,
- $k = 1 : \mathrm{Cay}_1(x;y) = x$,
- $k \geq 2 : \mathrm{Cay}_k(x;y) = \begin{vmatrix} x & 1 & & & & \\ y & x & 2 & & & \\ & y-1 & x & 3 & & \\ & & \cdots & \cdots & \cdots & \\ & & & y-k+3 & x & k-1 \\ & & & & y-k+2 & x \end{vmatrix}.$

The $(n+1)$th term $\mathrm{Cay}_{n+1}(x;n)$ with $y = n$ is nothing but the Sylvester determinant $\mathrm{Sylv}(x;n)$.

It is known (cf. [17, 30, 37]) that the generating function of the Cayley continuants $\{\mathrm{Cay}_k(x;y)\}_{k=0}^{\infty}$ is given as

$$(1+t)^{\frac{y+x}{2}}(1-t)^{\frac{y-x}{2}} = \sum_{k=0}^{\infty} \mathrm{Cay}_k(x;y)\frac{t^k}{k!}. \tag{9.1}$$

We write:

$$g^C_{[x;n]}(t) := (1+t)^{\frac{n+x}{2}}(1-t)^{\frac{n-x}{2}}$$

and also set

$$\mathcal{S}\mathrm{ol}_k(\mathrm{Cay}; n) = \{s \in \mathbb{C} : \mathrm{Cay}_k(x; n) = 0\}.$$

As $\mathrm{Cay}_{n+1}(x; n) = \mathrm{Sylv}(x; n)$, it follows from (1.16) that

$$\begin{aligned} \mathcal{S}\mathrm{ol}_{n+1}(\mathrm{Cay}; n) &= \{s \in \mathbb{C} : \mathrm{Sylv}(s; n) = 0\} \\ &= \{n - 2\ell : \ell = 0, 1, \ldots, n\} \\ &= \begin{cases} \{0, \pm 2, \pm 4, \ldots, \pm n\} & \text{if } n \text{ is even,} \\ \{\pm 1, \pm 3, \pm 5, \ldots, \pm n\} & \text{if } n \text{ is odd.} \end{cases} \end{aligned} \tag{9.2}$$

Now (9.1) and (9.2) imply the following.

Lemma 9.1 *The following conditions of $s \in \mathbb{C}$ are equivalent:*

(i) $g^C_{[s;n]}(t) \in \mathrm{Pol}_n[t]$.
(ii) $s \in \mathcal{S}\mathrm{ol}_{n+1}(\mathrm{Cay}; n)$.

Proof For the implication (i) ⇒ (ii), observe that if $g^C_{[s;n]}(t) \in \mathrm{Pol}_n[t]$, then it follows from (9.1) that $\mathrm{Cay}_{n+1}(s; n) = 0$, that is, by definition, $s \in \mathcal{S}\mathrm{ol}_{n+1}(\mathrm{Cay}; n)$.

For the other implication (ii) ⇒ (i), suppose that $s \in \mathcal{S}\mathrm{ol}_{n+1}(\mathrm{Cay}; n)$. Then, by (9.2), we have $\frac{n \pm s}{2} \in \mathbb{Z}_{\geq 0}$. Therefore,

$$g^C_{[x;n]}(t) = (1+t)^{\frac{n+s}{2}}(1-t)^{\frac{n-s}{2}} = \pm t^n + (\text{lower order terms}) \in \mathrm{Pol}_n[t].$$

Theorem 9.2 *Let $n \in \mathbb{Z}_{\geq 0}$ and $s \in \mathcal{S}\mathrm{ol}_{n+1}(\mathrm{Cay}; n)$. Then we have $\mathrm{Cay}_k(s; n) = 0$ for $k \geq n+1$ and*

$$\frac{\mathrm{Cay}_k(s; n)}{k!} = (-1)^{\frac{n-s}{2}} \frac{\mathrm{Cay}_{n-k}(s; n)}{(n-k)!} \quad \text{for } k \leq n.$$

In particular we have:

$$\mathrm{Cay}_n(s; n) = (-1)^{\frac{n-s}{2}} n! \quad \text{for } s \in \mathcal{S}\mathrm{ol}_{n+1}(\mathrm{Cay}; n). \tag{9.3}$$

Proof Take $s \in \mathcal{S}\mathrm{ol}_{n+1}(\mathrm{Cay}; n)$. It follows from Lemma 9.1 that, for such s, we have:

$$g^C_{[s;n]}(t) = \sum_{k=0}^{n} \frac{\mathrm{Cay}_k(s; n)}{k!} t^k, \tag{9.4}$$

namely, $\mathrm{Cay}_k(s;n) = 0$ for $k \geq n+1$. Further, (9.4) shows that $t^n g^C_{[s;n]}\left(\frac{1}{t}\right)$ is given as

$$t^n g^C_{[s;n]}\left(\frac{1}{t}\right) = \sum_{k=0}^{n} \frac{\mathrm{Cay}_{n-k}(s;n)}{(n-k)!} t^k.$$

On the other hand, as $g^C_{[x;n]}(t) = (1+t)^{\frac{n+x}{2}}(1-t)^{\frac{n-x}{2}}$, we also have:

$$t^n g^C_{[s;n]}\left(\frac{1}{t}\right) = (-1)^{\frac{n-s}{2}} g^C_{[s;n]}(t).$$

Therefore,

$$\sum_{k=0}^{n} \frac{\mathrm{Cay}_{n-k}(s;n)}{(n-k)!} t^k = (-1)^{\frac{n-s}{2}} \sum_{k=0}^{n} \frac{\mathrm{Cay}_k(s;n)}{k!} t^k,$$

which yields

$$\frac{\mathrm{Cay}_k(x;n)}{k!} = (-1)^{\frac{n-s}{2}} \frac{\mathrm{Cay}_{n-k}(x;n)}{(n-k)!} \quad \text{for all } k \leq n.$$

The second identity is the case $k = n$. This concludes the proposition.

By (9.2), the factorial identity (9.3) can be given more explicitly as follows.

Corollary 9.3 For n even, the values of $\mathrm{Cay}_n(s;n)$ for $s = 0, \pm 2, \ldots, \pm(n-2), \pm n$ are given as follows:

- $n \equiv 0 \pmod 4$:

$$\mathrm{Cay}_n(s;n) = \begin{cases} n! & \text{if } s = 0, \pm 4, \ldots, \pm(n-4), \pm n, \\ -n! & \text{if } s = \pm 2, \pm 6, \ldots, \pm(n-2). \end{cases}$$

- $n \equiv 2 \pmod 4$:

$$\mathrm{Cay}_n(s;n) = \begin{cases} -n! & \text{if } s = 0, \pm 4, \ldots, \pm(n-4), \pm n, \\ n! & \text{if } s = \pm 2, \pm 6, \ldots, \pm(n-2). \end{cases}$$

Similarly, for n odd, the values of $\mathrm{Cay}_n(s; n)$ for $s = \pm 1, \pm 3, \ldots, \pm(n-2), \pm n$ are given as follows:

- $n \equiv 1 \pmod 4$:

$$\mathrm{Cay}_n(s; n) = \begin{cases} n! & \text{if } s = 1, -3, 5, \ldots, -(n-2), n, \\ -n! & \text{if } s = -1, 3, -5, \ldots, n-2, -n. \end{cases}$$

- $n \equiv 3 \pmod 4$:

$$\mathrm{Cay}_n(s; n) = \begin{cases} -n! & \text{if } s = 1, -3, 5, \ldots, n-2, -n, \\ n! & \text{if } s = -1, 3, -5, \ldots, -(n-2), n. \end{cases}$$

9.2 Functional Equation of Krawtchouk Polynomials $\{\mathcal{K}_k(x; y)\}_{k=0}^{\infty}$

For $k \in \mathbb{Z}_{\geq 0}$, we define a polynomial $\mathcal{K}_k(x; y)$ of x and y as

$$\mathcal{K}_k(x; y) = \sum_{j=0}^{k} (-1)^j \binom{x}{j} \binom{y-x}{k-j}, \tag{9.5}$$

where the binomial coefficient $\binom{a}{m}$ is defined as follows.

$$\binom{a}{m} = \begin{cases} \frac{a(a-1)\cdots(a-m+1)}{m!} & \text{if } m \in 1 + \mathbb{Z}_{\geq 0}, \\ 1 & \text{if } m = 0, \\ 0 & \text{otherwise.} \end{cases}$$

Since $\mathcal{K}_k(x; n)$ for $y = n \in \mathbb{Z}_{\geq 0}$ is a Krawtchouk polynomial in the sense of [29, p. 137], we also call $\mathcal{K}_k(x; y)$ a Krawtchouk polynomial in this paper.

Remark 9.4 Symmetric Krawtchouk polynomials (see, for instance, [23, p. 237]) are used to show Sylvester's factorization formula (1.16) in [1]. We remark that the definition of the Krawtchouk polynomial $\mathcal{K}_k(x; n)$ in this paper is different from the one in the cited paper; in particular, $\mathcal{K}_k(x; n)$ is nonsymmetric.

It readily follows from (9.5) that the Krawtchouk polynomials $\{\mathcal{K}_k(x; y)\}_{k=0}^{\infty}$ have the following generating function:

$$(1+t)^{y-x}(1-t)^x = \sum_{k=0}^{\infty} \mathcal{K}_k(x; y) t^k. \tag{9.6}$$

By comparing (9.6) with the generating function (9.1) of the Cayley continuants $\{\mathrm{Cay}_k(x;y)\}_{k=0}^{\infty}$, we have:

$$\mathcal{K}_k(x;y) = \frac{\mathrm{Cay}_k(y-2x;y)}{k!}. \tag{9.7}$$

The factorization formula (9.8) below simply follows from (9.7) and Sylvester's factorization formula (1.16).

Proposition 9.5 *For given $n \in \mathbb{Z}_{\geq 0}$, we have:*

$$\mathcal{K}_{n+1}(x;n) = \frac{(-2)^{n+1}}{(n+1)!}\prod_{\ell=0}^{n}(x-\ell) = (-2)^{n+1}\binom{x}{n+1}. \tag{9.8}$$

As for $\mathcal{S}\mathrm{ol}_k(\mathrm{Cay};n)$, we set:

$$\mathcal{S}\mathrm{ol}_k(\mathcal{K};n) = \{s \in \mathbb{C} : \mathcal{K}_k(s;n) = 0\}.$$

The sequence $\{\mathcal{K}_k(x;y)\}_{k=0}^{\infty}$ of Krawtchouk polynomials enjoys the following functional equation.

Theorem 9.6 *Let $n \in \mathbb{Z}_{\geq 0}$ and $s \in \mathcal{S}\mathrm{ol}_{n+1}(\mathcal{K};n)$. Then we have $\mathcal{K}_k(s;n) = 0$ for $k \geq n+1$ and*

$$\mathcal{K}_k(s;n) = (-1)^s\mathcal{K}_{n-k}(s;n) \quad \text{for } k \leq n.$$

In particular we have

$$\mathcal{K}_n(s;n) = (-1)^s \quad \text{for } s \in \mathcal{S}\mathrm{ol}_{n+1}(\mathcal{K};n). \tag{9.9}$$

Proof Since the proof is the same as that of Theorem 9.2, we omit the proof.

It follows from (9.8) that

$$\begin{aligned}\mathcal{S}\mathrm{ol}_{n+1}(\mathcal{K};n) &= \{s \in \mathbb{C} : \mathcal{K}_{n+1}(s;n) = 0\}\\ &= \{0, 1, 2, \ldots, n\}.\end{aligned}$$

Then the identity (9.9) can be given explicitly as follows:

$$\mathcal{K}_n(s;n) = \begin{cases} 1 & \text{if } s = 0,\ 2,\ 4,\ \ldots,\ n_{\text{even}}, \\ -1 & \text{if } s = 1,\ 3,\ 5,\ \ldots,\ n_{\text{odd}}, \end{cases} \tag{9.10}$$

where n_{odd} and n_{even} are defined as

$$n_{\mathrm{odd}} := \begin{cases} n & \text{if } n \text{ is odd,} \\ n-1 & \text{if } n \text{ is even} \end{cases} \quad \text{and} \quad n_{\mathrm{even}} := \begin{cases} n-1 & \text{if } n \text{ is odd,} \\ n & \text{if } n \text{ is even.} \end{cases}$$

9.3 Relationship with Jacobi Polynomials $P_k^{(\alpha,\beta)}(z)$

For $\alpha, \beta \in \mathbb{C}$, let $P_k^{(\alpha,\beta)}(z)$ denote the Jacobi polynomial:

$$P_k^{(\alpha,\beta)}(z) = \sum_{j=0}^{k} \binom{k+\alpha}{k-j}\binom{k+\beta}{j}\left(\frac{z-1}{2}\right)^j\left(\frac{z+1}{2}\right)^{k-j} \tag{9.11}$$

(cf. [35, p. 68]). The Cayley continuants $\mathrm{Cay}_k(x; y)$ and Krawtchouk polynomials $\mathcal{K}_k(x; y)$ can be expressed by the Jacobi polynomials $P_k^{(\alpha,\beta)}(z)$. Indeed, it follows from (9.11) that the Jacobi polynomials $P_k^{(\alpha,\beta)}(z)$ satisfy the following identity (cf. [4, 27]):

$$\left(1+\frac{z+1}{2}w\right)^{\alpha}\left(1+\frac{z-1}{2}w\right)^{\beta} = \sum_{k=0}^{\infty} P_k^{(\alpha-k,\beta-k)}(z)w^k. \tag{9.12}$$

Let $z = 0$ and $w = 2t$ in (9.12). Then we have:

$$(1+t)^{\alpha}(1-t)^{\beta} = \sum_{k=0}^{\infty} 2^k P_k^{(\alpha-k,\beta-k)}(0)t^k. \tag{9.13}$$

By comparing (9.13) with (9.1) and (9.6), we obtain:

$$\mathrm{Cay}_k(x; y) = k! \cdot 2^k \cdot P_k^{(\frac{y+x}{2}-k,\, \frac{y-x}{2}-k)}(0),$$

$$\mathcal{K}_k(x; y) = 2^k \cdot P_k^{(y-x-k,\, x-k)}(0).$$

Acknowledgments Part of this research was conducted during a visit of the first author at the Department of Mathematics of Aarhus University. He is appreciative of their support and warm hospitality during his stay.

The authors are grateful to Anthony Kable for communication on this paper. They would also like to express their gratitude to Hiroyuki Ochiai for the helpful discussions on the hypergeometric equation and Heun equation, especially for Lemma 4.8. Their deep appreciation also goes to Hiroyoshi Tamori for sending his Ph.D. dissertation, from which they thought of an idea for the hypergeometric model.

Finally, the authors would like to show their gratitude to the anonymous referees for their professional comments and suggestions on a manuscript of this paper.

The first author was partially supported by JSPS Grant-in-Aid for Young Scientists (JP18K13432).

References

1. Askey, R.: Evaluation of Sylvester type determinants using orthogonal polynomials. In: Advances in Analysis, pp. 1–16. World Scientific, Singapore (2005)
2. Barchini, L., Kable, A.C., Zierau, R.: Conformally invariant systems of differential equations and prehomogeneous vector spaces of Heisenberg parabolic type. Publ. Res. Inst. Math. Sci. **44**(3), 749–835 (2008)
3. Barchini, L., Kable, A.C., Zierau, R: Conformally invariant systems of differential operators Adv. Math. **221**(3), 788–811 (2009)
4. Carlitz, L.: A bilinear generating function for the Jacobi polynomials. Boll. Un. Mat. Ital. **18**, 87–89 (1963)
5. Clement, P.A.: A class of triple-diagonal matrices for test purposes. SIAM Rev. **1**, 50–52 (1959)
6. Collingwood, D.H., Shelton, B.: A duality theorem for extensions of induced highest weight modules. Pac. J. Math. **146**(2), 227–237 (1990)
7. Dahl, T.T.: Knapp-Stein Operators and Fourier Transformations, Ph.D. Thesis, Arhus University, 2019
8. Edelman, A., Kostlan, E.: The road from Kac's matrix to Kac's random polynomials. In: Lewis, J. (ed.) Proceedings of the of the Fifth SIAM Conference on Applied Linear Algebra, pp. 503–507. SIAM, Philadelphia (1994)
9. Frahm, J.: Conformally invariant differential operators on Heisenberg groups and minimal representations, Mém. Soc. Math. Fr. (N.S.), 180, vi+139 (2024)
10. Holtz, O.: Evaluation of Sylvester type determinants using block-triangularization. In: Advances in Analysis, pp. 395–405. World Scientific, Singapore (2005)
11. Kable, A.C.: K-finite solutions to conformally invariant systems of differential equations. Tohoku Math. J. **63**(4), 539–559 (2011). Centennial Issue
12. Kable, A.C.: Conformally invariant systems of differential equations on flag manifolds for G_2 and their K-finite solutions. J. Lie Theory **22**(1), 93–136 (2012)
13. Kable, A.C.: The Heisenberg ultrahyperbolic equation: the basic solutions as distributions. Pac. J. Math **258**(1), 165–197 (2012)
14. Kable, A.C.: The Heisenberg ultrahyperbolic equation: K-finite and polynomial solutions. Kyoto J. Math **52**(4), 839–894 (2012)
15. Kable, A.C.: On certain conformally invariant systems of differential equations II: Further study of type A systems. Tsukuba J. Math. **39**, 39–81 (2015)
16. Kable, A.C.: The structure of the space of polynomial solutions to the canonical central systems of differential equations on the block Heisenberg groups: a generalization of a theorem of Korányi . Tohoku Math. J. **70**, 523–545 (2018)
17. Kac, M.: Random walk and the theory of Brownian motion. Am. Math. Monthly **54**, 369–390 (1947)
18. Kobayashi, T., Pevzner, M.: Differential symmetry breaking operators. I. General theory and F-method. Selecta Math. (N.S.) **22**, 801–845 (2016)
19. Kobayashi, T., Ørsted, B.: Analysis on the minimal representation of $O(p, q)$ I. Realization via conformal geometry. Adv. Math. **180**(2), 486–512 (2003)
20. Kobayashi, T., Ørsted, B.: Analysis on the minimal representation of $O(p, q)$ II. Branching laws. Adv. Math. **180**(2), 513–550 (2003)
21. Kobayashi, T., Ørsted, B.: Analysis on the minimal representation of $O(p, q)$ III. Ultrahyperbolic equations on $\mathbb{R}^{p-1,q-1}$. Adv. Math. **180**(2), 551–595 (2003)

22. Kostant, B.: Verma Modules and the Existence of Quasi-Invariant Differential Operators. Noncommutative Harmonic Analysis. Lecture Notes in Mathematics, vol. 466, pp. 101–128. Springer, Berlin (1975)
23. Koekoek, R., Lesky, P.A., Swarttouw, R.F.: Hypergeometric Orthogonal Polynomials and Their q-Analogues. Springer Monographs in Mathematics. Springer, Berlin, Heidelberg (2010)
24. Kristensson, G.: Second Order Differential Equations: Special Functions and Their Classification. Springer, New York (2010)
25. Kubo, T., Ørsted, B.: On the space of K-finite solutions to intertwining differential operators. Represent. Theory **23**, 213–248 (2019)
26. Maier, R.S.: On reducing the Heun equation to the hypergeometric equation. J. Differ. Equ **213**, 171–203 (2005)
27. Milch, P.R.: A probabilistic proof of a formula for Jacobi polynomials by L. Carlitz. Proc. Camb. Philos. Soc. **64**, 695–698 (1968)
28. Muir, T.: The Theory of Determinants in the Historical Order of Development, vol. II, 1911. University of Michigan Library (2006). Reprinted
29. MacWilliams, F.J., Sloane, N.J.A.: The Theory of Error-Correcting Codes. North-Holland Mathematical Library, vol. 16. North-Holland, Amsterdam (1977)
30. Munarini, E., Torri, D.: Cayley continuants. Theor. Comput. Sci. **347**, 353–369 (2005)
31. Rawnsley, J., Sternberg, S.: On representations associated to the minimal nilpotent coadjoint orbit of $SL(3, \mathbf{R})$. Am. J. Math. **104**(6), 1153–1180 (1982)
32. Ronveaux, A. (ed.): Heun's Differential Equations. Oxford Science Publications, Oxford University Press, Oxford (1995). With contributions by Arscott, F.M., Yu. Slavyanov, S., Schmidt, D., Wolf, G., Maroni, P., Duval, A.
33. Slavyanov, S.Y., Lay, W.: Special Functions. A Unified Theory Based on Singularities. Oxford Mathematical Monographs. Oxford Science Publications, Oxford University Press, Oxford (2000). With a foreword by Alfred Seeger
34. Sylvester, J.J.: Théorème sur les déterminants de M. Sylvester. Nouvelles annales de mathématiques: journal des candidats aux écoles polytechnique et normale, 1e série **13**, 305 (1854)
35. Szegö, G.: Orthogonal polynomials, 4th edn. American Mathematical Society Colloquium Publications, vol. 23, xiii+432 pp. American Mathematical Society, Providence (1975)
36. Tamori, H.: Classification of irreducible $(\mathfrak{g}, \mathfrak{k})$-modules associated to the ideals of minimal nilpotent orbits for simple Lie groups of type A. Int. Math. Res. Not. IMRN, rnab356 (2021).
37. Taussky, O., Todd, J.: Another look at a matrix of Mark Kac. Linear Algebra Appl. **150**, 341–360 (1991)
38. Torasso, P.: Quantication géométrique, opérateurs d'entrelacement et représentations unitaires de $(\widetilde{SL})_3(\mathbf{R})$. Acta Math. **150**(3–4), 153–242 (1983)

Gauss–Berezin Integral Operators, Spinors over Orthosymplectic Supergroups, and Lagrangian Super-Grassmannians

Yury A. Neretin

Abstract We obtain explicit formulas for the spinor representation ρ of the real orthosymplectic supergroup $\mathrm{OSp}(2p|2q, \mathbb{R})$ by integral "Gauss–Berezin" operators. Next, we extend ρ to a complex domain and get a representation of a larger semigroup, which is a counterpart of Olshanski subsemigroups in semi-simple Lie groups. Further, we show that ρ can be extended to an operator-valued function on a certain domain in the Lagrangian super-Grassmannian (graphs of elements of the supergroup $\mathrm{OSp}(2p|2q, \mathbb{C})$ are Lagrangian super-subspaces) and show that this function is a "representation" in the following sense: we consider Lagrangian subspaces as linear relations, and composition of two Lagrangian relations in general position corresponds to a product of Gauss–Berezin operators.

1 Introduction

In the present chapter, we consider algebras, superalgebras, and functional spaces over complex numbers $\mathbb{C}$ and, in a few cases, over real numbers $\mathbb{R}$. The transposition of matrices is denoted by $A \mapsto A^t$. The symbol 1_n denotes the unit matrix of size n.

This chapter is an extended variant of preprint https://arxiv.org/abs/0707.0570v3 and a strongly revised version of my earlier preprint [51] exploring other realization of spinors.

Y. A. Neretin (✉)
Department of Mathematics and Scientific Computing, University of Graz, Graz, Austria

High School of Modern Mathematics, MIPT, Moscow, Russia

Faculty of Mechanics and Mathematics, Moscow State University, Moscow, Russia
e-mail: neretin-sbor@yandex.ru; yurii.neretin@univie.ac.at

M. Pevzner, H. Sekiguchi (eds.), *Symmetry in Geometry and Analysis, Volume 2*,
Progress in Mathematics 358, https://doi.org/10.1007/978-981-97-7662-7_9

1.1 Orthosymplectic Spinors

Let $x_1, \dots, x_p$ be real variables, $\xi_1, \dots, \xi_q$ be Grassmann variables, $\xi_i\xi_j = -\xi_j\xi_i$ for all i, j (in particular, $\xi_i^2 = 0$). We consider differential operators

$$1, \quad x_k x_l, \quad x_k\frac{\partial}{\partial x_l}, \quad \frac{\partial}{\partial x_k}\frac{\partial}{\partial x_l}, \quad \xi_m\xi_n, \quad \xi_m\frac{\partial}{\partial \xi_n}, \quad \frac{\partial}{\partial \xi_m}\frac{\partial}{\partial \xi_n}, \tag{1}$$

$$x_k\xi_m, \quad x_k\frac{\partial}{\partial \xi_m}, \quad \xi_m\frac{\partial}{\partial x_l}, \quad \frac{\partial}{\partial x_k}\frac{\partial}{\partial \xi_m}, \tag{2}$$

where $1 \leqslant k, l \leqslant p$, $1 \leqslant m, n \leqslant q$, acting in the space of polynomials in $x_1, \dots, x_p, \xi_1, \dots, \xi_q$. Denote by $\mathcal{R}$ the space of all *complex* linear combinations of such operators. We say that a parity of a nonzero monomial of degree 2 in x_k, $\frac{\partial}{\partial \xi_l}$, ξ_m, and $\frac{\partial}{\partial \xi_n}$ is $\overline{1}$ if it contains either one ξ_k or one $\frac{\partial}{\partial \xi_l}$. Otherwise, parity is $\overline{0}$. So monomials (1) have parity $\overline{0}$ and (2) parity $\overline{1}$. We also say that a nonzero linear combination of monomials of parity $\overline{0}$ (resp. $\overline{1}$) has parity $\overline{0}$ (resp. $\overline{1}$). Let $u, v \in \mathcal{R}$ have parities $p(u)$, $p(v)$. We define the supercommutator of operators u, v by

$$[u, v]_s := uv - (-1)^{p(u)p(v)}vu \tag{3}$$

and extend this operation to the whole $\mathcal{R}$ by bilinearity.

It is easy to see that the space $\mathcal{R}$ is closed with respect to the supercommutator, and therefore, we get a Lie superalgebra; it is isomorphic to a direct sum of the orthosymplectic Lie superalgebra $\mathfrak{osp}(2p|2q)$ and a trivial one-dimensional Lie algebra $\mathbb{C}$.

Let us recall a definition of $\mathfrak{osp}(2p|2q)$. Denote the block $(p+p) \times (p+p)$-matrix $\begin{pmatrix} 0 & 1_p \\ -1_p & 0 \end{pmatrix}$ by J and the block $(q+q) \times (q+q)$-matrix $\begin{pmatrix} 0 & 1_q \\ 1_q & 0 \end{pmatrix}$ by I. The orthosymplectic Lie superalgebra $\mathfrak{osp}(2p|2q)$ consists of complex block $(2p+2q) \times (2p+2q)$-matrices $\begin{pmatrix} A & B \\ C & D \end{pmatrix}$ satisfying the condition[1]

$$\begin{pmatrix} J & 0 \\ 0 & I \end{pmatrix}\begin{pmatrix} A & B \\ C & D \end{pmatrix} + \begin{pmatrix} A^t & C^t \\ -B^t & D^t \end{pmatrix}\begin{pmatrix} J & 0 \\ 0 & I \end{pmatrix} = 0.$$

We say that parity of matrices of the form $\begin{pmatrix} * & 0 \\ 0 & * \end{pmatrix}$ is $\overline{0}$ and parity of $\begin{pmatrix} 0 & * \\ * & 0 \end{pmatrix}$ is $\overline{1}$ and define a supercommutator by formula (3).

[1] J is a canonical form of a non-degenerate skew-symmetric matrix, and I is a canonical form of a nondegenerate complex symmetric matrix of an even order. For our aims, I is more convenient than the unit matrix, which is more natural for general theory.

If $p = 0$, then we get the usual spinor representation of the orthogonal Lie algebra $\mathfrak{o}(2q, \mathbb{C})$ in the Grassmann algebra consisting of "functions" in variables ξ_1, ..., ξ_q. If $q = 0$, then we get a representation of symplectic Lie algebra $\mathfrak{sp}(2q)$ (symplectic spinors or oscillator representation). Spinors and symplectic spinors are distinguished objects of representation theory. Orthosymplectic spinors were considered in numerous works, for instance, Berezin [8] (with the construction mentioned above), Serov [65] (where the spinor representations of the orthosymplectic supergroups $\mathrm{OSp}(2p|r)$ were obtained), and [3, 4, 14, 15, 16, 21, 24, 36, 41, 54].

Remark on Notation In notation $\mathfrak{osp}(2p|2q)$ for the orthosymplectic Lie superalgebra, I firstly write $2p$ corresponding to the "human" (real or complex) variables, and the symplectic Lie algebra $\mathfrak{sp}(2p)$; $2q$ corresponds to Grassmann variables and the orthogonal Lie algebra $\mathfrak{o}(2q)$. In literature, our $\mathfrak{osp}(2p|2q)$ can be denoted by $\mathfrak{osp}(2q|2p)$ or $\mathfrak{spo}(2p|2q)$. ⊠

In this chapter, we write explicit formula for representation of the corresponding global object, which is larger than the real supergroup $\mathrm{OSp}(2p|2q)$. We do not assume that the reader is familiar with super-mathematics and discuss orthosymplectic spinors as a topic of analysis and as a story about integral operators. The text is self-closed; de facto, we use a minimal version of language of super-algebra and super-analysis[2]—and follow DeWitt's [18] way, to consider linear spaces (modules) over Grassmann algebra with infinite number of generators.[3]

1.2 Berezin Formulas

Apparently, first elements of a strange analogy between the spinor representation of the orthogonal groups and the oscillator representation of symplectic groups[4] were observed by K. O. Friedrichs in the early 1950s (see [22]). He considered spinors over symplectic groups $\mathrm{Sp}(2n, \mathbb{R})$ as a kind of a self-obvious object (obtained by an application of the Stone–von Neumann theorem) and initiated a discussion about their extension to the case $n = \infty$, see some historical comments in [52].

[2] Lie superalgebras can have a life of their own, without supergroups, supermanifolds, superintegration, etc.; see [32], [11], Chapter 1. We prefer to discuss supergroups—Lie superalgebras are actually present only in Sect. 10.

[3] Different authors have different points of view to a formalization of "super-analysis." DeWitt's book was an object a justified criticism in [60, 44]. Our work is far from subtleties of analysis on supermanifolds; translation of our results to the more common functorial language is more or less automatical.

[4] *Spinor representation* of orthogonal group is a common term (the representation was discovered by Élie Cartan [13], 1913). The term *oscillator representation* has several synonyms, namely, the Weil representation, the Shale–Weil representation, the Segal–Shale–Weil representation, the harmonic representation, the metaplectic representation, and the symplectic spinors. The term "oscillator representation" was proposed by Irving Segal (who was the first to describe this representation [62], 1959). For further references, see [50, 53].

In the beginning of 1960s, Feliks Berezin obtained explicit formulas [5, 6] for both representations. We briefly recall his results. First of all, let us realize the *real* symplectic group $\mathrm{Sp}(2n, \mathbb{R})$ as the group of *complex* $(n+n) \times (n+n)$ matrices

$$g = \begin{pmatrix} \Phi & \Psi \\ \overline{\Psi} & \overline{\Phi} \end{pmatrix} \tag{4}$$

satisfying the condition

$$g \begin{pmatrix} 0 & 1 \\ -1 & 0 \end{pmatrix} g^t = \begin{pmatrix} 0 & 1 \\ -1 & 0 \end{pmatrix}. \tag{5}$$

Similarly, we realize the *real* orthogonal group $\mathrm{O}(2n, \mathbb{R})$ as the group of *complex* matrices

$$g = \begin{pmatrix} \Phi & \Psi \\ -\overline{\Psi} & \overline{\Phi} \end{pmatrix} \tag{6}$$

satisfying

$$g \begin{pmatrix} 0 & 1 \\ 1 & 0 \end{pmatrix} g^t = \begin{pmatrix} 0 & 1 \\ 1 & 0 \end{pmatrix}. \tag{7}$$

The oscillator (spinor) representation of $\mathrm{Sp}(2n, \mathbb{R})$ (see (4), (5)) is realized by the following integral operators $W(\cdot)$:

$$W \begin{pmatrix} \Phi & \Psi \\ \overline{\Psi} & \overline{\Phi} \end{pmatrix} f(z) = \pm(\det \Phi)^{-1/2} \times$$

$$\times \int_{\mathbb{C}^n} \exp\left\{ \frac{1}{2} \begin{pmatrix} z & \overline{u} \end{pmatrix} \begin{pmatrix} \overline{\Psi}\Phi^{-1} & (\Phi^t)^{-1} \\ \Phi^{-1} & -\Phi^{-1}\Psi \end{pmatrix} \begin{pmatrix} z^t \\ \overline{u}^t \end{pmatrix} \right\} f(u) e^{-|u|^2} du\, d\overline{u} \tag{8}$$

in the space of holomorphic functions on $\mathbb{C}^n$. Here the symbol ^{t} denotes the transposition of matrices; $z = \begin{pmatrix} z_1 & \dots & z_n \end{pmatrix}$, $u = \begin{pmatrix} u_1 & \dots & u_n \end{pmatrix}$ are row vectors,

$$\begin{pmatrix} z & \overline{u} \end{pmatrix} := \begin{pmatrix} z_1 & \dots & z_n & \overline{u}_1 & \dots & \overline{u}_n \end{pmatrix}$$

also is a row vector, and the expression in the curly brackets is a product of a row vector, matrix, and a column vector (i.e., the whole expression is a scalar). We write $W(\cdot)$ in honor of A. Weil.

On the other hand, Berezin obtained formulas for the spinor representation of the group $O(2n, \mathbb{R})$ (we realize $O(2n, \mathbb{R})$ by matrices (6), (7)). Exactly, he wrote "integral operators" of the form

$$\operatorname{spin}\begin{pmatrix} \Phi & \Psi \\ \overline{\Psi} & \overline{\Phi} \end{pmatrix} f(\xi) = \pm(\det \Phi)^{1/2} \times$$
$$\times \int \exp\left\{\frac{1}{2}\begin{pmatrix} \xi & \overline{\eta} \end{pmatrix}\begin{pmatrix} -\overline{\Psi}\Phi^{-1} & (\Phi^t)^{-1} \\ -\Phi^{-1} & -\Phi^{-1}\Psi \end{pmatrix}\begin{pmatrix} \xi^t \\ \overline{\eta}^t \end{pmatrix}\right\} f(\eta) e^{-\eta\overline{\eta}^t} \, d\eta \, d\overline{\eta}, \qquad (9)$$

here $\xi = \begin{pmatrix} \xi_1 & \dots & \xi_n \end{pmatrix}$ and $\eta = \begin{pmatrix} \eta_1 & \dots & \eta_n \end{pmatrix}$ are row-matrices; ξ_j, η_j, and $\overline{\eta}_j$ are anti-commuting variables.[5,6] The integral in the right-hand side is the Berezin integral (see Sect. 2).

In fact, Berezin in his book [6][7] and the subsequent work [7] conjectured that there is the analysis of Grassmann variables parallel to the usual analysis (and these works contain important elements of this analysis). The book includes parallel exposition of the bosonic Fock space (analysis in infinite number of complex variables) and the fermionic Fock space (Grassmann analysis in infinite number of variables); this parallel seemed mysterious. However, the strangest elements of this analogy were formulas (8) and (9). This pushed him at the end of the 1960s to the beginning of the 1970s to the invention of the "super-analysis," which mixes even (complex or real) variables and odd (Grassmann) variables (see [34], Sect. 5).

Apparently, the first case of joining of classical and Grassmann analysis was the paper by Berezin and G. I. Kats [9], 1970, where formal Lie supergroups corresponding to Lie superalgebras were introduced. At least in 1965, Lie superalgebras appeared in algebraic topology (see Milnor, J. Moore [43]; this work was one of the starting points for Berezin and Kats). Starting 1971–1973, Lie superalgebras and supersymmetries were used in the quantum field theory (e.g., [26, 69, 70, 58]; "super" originates from this literature).

[5] Actually, in both cases, Berezin considered $n = \infty$.

[6] Such a formula makes sense only for an open dense subset in $SO(2n, \mathbb{R}) \subset O(2n, \mathbb{R})$; this produces some difficulties below. Spinors over infinite-dimensional symplectic and orthogonal groups were independently introduced by Shale and Stinespring [66, 67]. However, the starting point of super-analysis and standpoint of the present work were Berezin's formulas (8), (9).

[7] On "intellectual history" of this book, its origins, and influence, see [52].

1.3 Aims of This Chapter: Orthosymplectic Spinors

We wish to unite formulas (8)–(9), and to write explicitly the representation of the supergroup[8] $\mathrm{OSp}(2p|2q,\mathbb{R})$, the operators of the representation have the form

$$T(g)f(z,\xi)=\iint \exp\left\{\frac{1}{2}\begin{pmatrix}z & \xi & \overline{u} & \overline{\eta}\end{pmatrix}\cdot\mathfrak{R}(g)\cdot\begin{pmatrix}z^t\\ \xi^t\\ \overline{u}^t\\ \overline{\eta}^t\end{pmatrix}\right\}\times$$
$$\times\, f(u,\eta)\,e^{-u\overline{u}^t-\eta\overline{\eta}^t}\,du\,d\overline{u}\,d\eta\,d\overline{\eta}, \qquad (10)$$

where g ranges in the supergroup $\mathrm{OSp}(2p|2q,\mathbb{R})$ and $\mathfrak{R}(g)$ is a certain block matrix of size $(p+q+p+q)$ composed of elements a supercommutative algebra $\mathcal{A}$ (we prefer to think that $\mathcal{A}$ is the Grassmann algebra with an infinite number of generators $\mathfrak{a}_j$); see Sect. 5.

1.4 Aims of This Chapter: Gauss–Berezin Integral Operators

In [45, 48, 47], it was shown that spinor and oscillator representations are actually representations of categories. We obtain a (non-perfect) super-counterpart of these constructions. Let us explain this in more detail.

First, let us consider the "bosonic" case. Consider a symmetric $(n+n)\times(n+n)$-matrix,

$$S=\begin{pmatrix}A & B\\ B^t & C\end{pmatrix}.$$

Consider a *Gaussian integral operator* [9]

$$\mathfrak{B}_+\begin{bmatrix}A & B\\ B^t & C\end{bmatrix}f(z)=\int_{\mathbb{C}^n}\exp\left\{\frac{1}{2}\begin{pmatrix}z & \overline{u}\end{pmatrix}\begin{pmatrix}A & B\\ B^t & C\end{pmatrix}\begin{pmatrix}z^t\\ \overline{u}^t\end{pmatrix}\right\}f(u)e^{-|u|^2}du\,d\overline{u}. \qquad (11)$$

---

[8] Thus, this chapter returns to the initial point of super-analysis. It seems strange that formula (10) was not written by Berezin himself. The author obtained it after reading the posthumous uncompleted book [8] of Berezin; in my opinion, it contains traces of attempts to do this (see also [10]). In the present chapter, we use tools that were unknown to Berezin. A straightforward extension of [6] leads to cumbersome calculations. Certainly, these difficulties were surmountable. In any case, the present chapter is a kind of a "lost chapter" of Berezin's book [6] and my book [50].

[9] We use the symbol $\mathfrak{B}$ in honor of Berezin.

These operators are more general than (8); it can be shown (this was observed by G. I. Olshanski) that the symmetric matrices $\begin{pmatrix} \overline{\Psi}\Phi^{-1} & (\Phi^t)^{-1} \\ \Phi^{-1} & -\Phi^{-1}\Psi \end{pmatrix}$ in Berezin's formula (8) are unitary.

It can be readily checked that bounded Gaussian operators form a semigroup, which includes the group $\mathrm{Sp}(2n, \mathbb{R})$; its algebraic structure is described below in Sect. 3.

On the other hand, we can introduce *Berezin operators* that are fermionic counterparts of Gaussian operators. Namely, consider a skew-symmetric matrix $\begin{pmatrix} A & B \\ -B^t & C \end{pmatrix}$ and the integral operator

$$\mathfrak{B}_{-}\begin{bmatrix} A & B \\ -B^t & C \end{bmatrix} f(\xi) = \int \exp\left\{\frac{1}{2}\begin{pmatrix} \xi & \overline{\eta} \end{pmatrix}\begin{pmatrix} A & B \\ -B^t & C \end{pmatrix}\begin{pmatrix} \xi^t \\ \overline{\eta}^t \end{pmatrix}\right\} f(\eta) e^{-\eta\overline{\eta}}\, d\eta\, d\overline{\eta}. \tag{12}$$

These operators are more general than (9); it can be shown that in Berezin's formula (9), the skew-symmetric matrix $\begin{pmatrix} -\overline{\Psi}\Phi^{-1} & (\Phi^t)^{-1} \\ -\Phi^{-1} & -\Phi^{-1}\Psi \end{pmatrix}$ is contained in the pseudo-unitary group $\mathrm{U}(n, n)$.

For Berezin operators in general position, the product of Berezin operators has a kernel of the same form. But sometimes this is not the case. It is possible to improve the definition (see Sect. 2) and to obtain a semigroup of Berezin operators.

In this chapter, we introduce "Gauss–Berezin integral operators," which unify bosonic "Gaussian operators" and fermionic "Berezin operators." Our main result is a construction of a canonical bijection between the set of all Gauss–Berezin operators and a certain domain in Lagrangian super-Grassmannian; also we propose a geometric interpretation of products of Gauss–Berezin operators.

Formula (10) for superspinor representation of the supergroup $\mathrm{OSp}(2p|2q)$ is a by-product of the geometric construction.

1.5 Aims of This Chapter: Possible Applications

The spinor and oscillator representations are important at least for the following two reasons.

First, they are a basic tool in representation theory of infinite-dimensional groups (see, e.g., [61, 56, 50]), in particular, for classical groups, the group of diffeomorphisms of the circle, loop groups.[10]

[10] The parabolic induction, which is a main tool for construction of representations of semi-simple Lie groups does not work in these situations.

Second, the Howe duality for spinors is an important topic of classical representation theory (see, for instance, [30, 35, 1]).

Possible applications of our work in similar directions are discussed in Sect. 10.

1.6 Structure of This Chapter

I tried to write a self-contained paper, and no preliminary knowledge of the super-mathematics or representation theory is assumed, but this implies a necessity of various preliminaries, which I provide. Also, we try to minimize the vocabulary of this text.

We start with an exposition of spinor and oscillator representations in Sects. 2 and 3. These sections contain also a discussion of Gaussian integral operators and Berezin (fermionic Gaussian) operators (a detailed exposition of these topics is contained in the books [50] and [53].

In Sects. 6–8, we discuss supergroups $\mathrm{OSp}(2p|2q)$, super-Grassmannians, and superlinear relations. I am trying to use minimally necessary tools; these sections do not contain an introduction to super-science and its basic definitions. For generalities of super-mathematics, see [17, 8, 42, 40, 18, 12, 68, 11].

Section 4 contains a discussion of super-analogue of the Gaussian integral; this is a simple imitation of the well-known formula

$$\int_{\mathbb{R}^n} \exp\Big\{-\frac{1}{2}xAx^t + bx^t\Big\}\,dx = (2\pi)^{n/2}\det(A)^{-1/2}\exp\Big\{-\frac{1}{2}bA^{-1}b^t\Big\}. \tag{13}$$

Apparently, these calculations are written somewhere, but I do not know the references.

Gauss–Berezin integral operators are introduced in Sect. 5. Our main construction is the canonical one-to-one correspondence between superlinear relations and Gauss–Berezin operators, which is obtained in Sect. 9. This immediately produces a representation of real supergroups $\mathrm{OSp}(2p|2q,\mathbb{R})$. For g being in an "open dense" subset in the supergroup, the operators of the representation are of the form (10).

In the last section, we discuss some open problems.

2 A Survey of Orthogonal Spinors: Berezin Operators and Lagrangian Linear Relations

This section is subdivided into three parts.

In Part A, we develop the standard formalism of Grassmann algebras Λ_n and define Berezin operators $\Lambda_n \to \Lambda_m$, which are in some sense morphisms of Grassmann algebras (but not morphisms in the category of algebras).

In Part B, we describe the "geometrical" category **GD**, which is equivalent to the category of Berezin operators. Morphisms of the category **GD** are certainly Lagrangian subspaces.

In Part C, we describe explicitly the correspondence between Lagrangian subspaces and Berezin operators.

In some cases, we present proofs or explanations for a coherent treatment (see [50, Chapter 2]).

A. Grassmann Algebras and Berezin Operators

2.1 Grassmann Variables and Grassmann Algebra

We denote by $\xi_1, \ldots, \xi_n$ Grassmann variables,

$$\xi_i \xi_j = -\xi_j \xi_i,$$

in particular, $\xi_i^2 = 0$. Denote by Λ_n the algebra of polynomials *with complex coefficients* in these variables, evidently, $\dim \Lambda_n = 2^n$. The monomials

$$\xi_{j_1} \xi_{j_2} \ldots \xi_{j_\alpha}, \qquad \text{where } \alpha = 0, 1, \ldots, n \text{ and } j_1 < j_2 < \cdots < j_\alpha, \tag{14}$$

form a basis of Λ_n. Below elements of Grassmann algebra are called *functions*.

2.2 Derivatives

We define *left differentiations* in ξ_j as usual. Exactly, if $f(\xi)$ does not depend on ξ_j, then

$$\frac{\partial}{\partial \xi_j} f(\xi) = 0, \qquad \frac{\partial}{\partial \xi_j} \xi_j f(\xi) = f(\xi).$$

Evidently,

$$\frac{\partial}{\partial \xi_k} \frac{\partial}{\partial \xi_l} = -\frac{\partial}{\partial \xi_l} \frac{\partial}{\partial \xi_k}, \qquad \left(\frac{\partial}{\partial \xi_k}\right)^2 = 0.$$

2.3 Exponentials

Let $f(\xi)$ be an *even* function, i.e., $f(-\xi) = f(\xi)$. We define its exponential as usual,

$$\exp\{f(\xi)\} := \sum_{j=0}^{\infty} \frac{1}{j!} f(\xi)^j.$$

Even functions f and g commute, $fg = gf$; therefore,

$$\exp\{f(\xi) + g(\xi)\} = \exp\{f(\xi)\} \cdot \exp\{g(\xi)\}.$$

2.4 Berezin Integral

Let $\xi_1, \dots, \xi_n$ be Grassmann variables. The *Berezin integral*

$$\int f(\xi)\, d\xi = \int f(\xi_1, \dots, \xi_n)\, d\xi_1 \dots d\xi_n$$

is a linear functional on Λ_n defined by

$$\int \xi_1 \xi_2 \dots \xi_n\, d\xi_1 \dots d\xi_n = 1,$$

integrals of all other monomials are zero.

The following formula for integration by parts holds:

$$\int f(\xi) \cdot \frac{\partial}{\partial \xi_k} g(\xi)\, d\xi = -\int \frac{\partial f(-\xi)}{\partial \xi_k} \cdot g(\xi)\, d\xi.$$

2.5 Integrals with Respect to Odd Gaussian Measure'

Let $\xi_1, \dots, \xi_q$ be as above. Let $\overline{\xi}_1, \dots, \overline{\xi}_q$ be another collection of Grassmann variables,

$$\overline{\xi}_k \overline{\xi}_l = -\overline{\xi}_l \overline{\xi}_k, \qquad \xi_k \overline{\xi}_l = -\overline{\xi}_l \xi_k.$$

We need the following Gaussian expression:

$$\exp\left\{ -\xi\overline{\xi}^t \right\} := \exp\{-\xi_1\overline{\xi}_1 - \xi_2\overline{\xi}_2 - \dots\} =$$
$$= \exp\{\overline{\xi}_1\xi_1\}\exp\{\overline{\xi}_2\xi_2\}\cdots = (1+\overline{\xi}_1\xi_1)(1+\overline{\xi}_2\xi_2)(1+\overline{\xi}_3\xi_3)\dots$$

Denote

$$d\overline{\xi}\,d\xi = d\overline{\xi}_1\,d\xi_1\,d\overline{\xi}_2\,d\xi_2\dots$$

Therefore,

$$\int \Bigl(\prod_{k=1}^{m}(\overline{\xi}_{\alpha_k}\xi_{\alpha_k})\Bigr)\, e^{-\xi\overline{\xi}^t}\,d\overline{\xi}\,d\xi = 1, \tag{15}$$

and the integral is zero for all other monomials. For instance,

$$\int \xi_1\xi_2\overline{\xi}_3\, e^{-\xi\overline{\xi}^t}\,d\overline{\xi}\,d\xi = 0, \qquad \int e^{-\xi\overline{\xi}^t}\,d\overline{\xi}\,d\xi = +1,$$
$$\int \xi_1\overline{\xi}_1\overline{\xi}_{33}\xi_{33}\, e^{-\xi\overline{\xi}^t}\,d\overline{\xi}\,d\xi = -\int (\overline{\xi}_1\xi_1)\,(\overline{\xi}_{33}\xi_{33})\, e^{-\xi\overline{\xi}^t}\,d\overline{\xi}\,d\xi = -1.$$

Evidently,

$$\int f(\overline{\xi})\cdot\frac{\partial}{\partial\xi_k}g(\xi)\,e^{-\xi\overline{\xi}^t}\,d\overline{\xi}\,d\xi = \int \overline{\xi}_k f(-\overline{\xi})\cdot g(\xi)\,e^{-\xi\overline{\xi}^t}\,d\overline{\xi}\,d\xi,$$
$$\int \frac{\partial}{\partial\overline{\xi}_k}f(\overline{\xi})\cdot g(\xi)\,e^{-\xi\overline{\xi}^t}\,d\overline{\xi}\,d\xi = \int f(-\overline{\xi})\cdot\xi_k g(\xi)\,e^{-\xi\overline{\xi}^t}\,d\overline{\xi}\,d\xi.$$

2.6 Integral Operators

Now, consider Grassmann algebras Λ_p and Λ_q consisting of polynomials in Grassmann variables $\xi_1, \dots, \xi_p$ and $\eta_1, \dots, \eta_q$, respectively. For a function $K(\xi, \overline{\eta})$, we define an *integral operator*

$$A_K : \Lambda_q \to \Lambda_p$$

by

$$A_K f(\xi) = \int K(\xi, \overline{\eta}) f(\eta)\, e^{-\eta\overline{\eta}^t}\,d\overline{\eta}\,d\eta.$$

Proposition 2.1 *The map $K \mapsto A_K$ is a one-to-one correspondence of the set of all polynomials $K(\xi, \overline{\eta})$ and the set of all linear maps $\Lambda_q \to \Lambda_p$.*

A proof of Proposition 2.1 is trivial. Indeed, expand

$$K(\xi, \eta) = \sum a_{i_1,\dots,i_l\, j_1,\dots,j_l} \xi_{i_1} \dots \xi_{i_k} \overline{\eta}_{j_1} \dots \overline{\eta}_{j_l}.$$

Then $a_{\dots}$ are the matrix elements[11] of A_K in the standard basis (14).

Proposition 2.2 *If $A : \Lambda_q \to \Lambda_p$ is determined by the kernel $K(\xi, \eta)$ and $B : \Lambda_p \to \Lambda_r$ is determined by the kernel $L(\zeta, \overline{\xi})$, then the kernel of BA is*

$$M(\zeta, \overline{\eta}) = \int L(\zeta, \overline{\xi}) K(\xi, \overline{\eta}) e^{-\xi\overline{\xi}^t} \, d\overline{\xi} \, d\xi.$$

2.7 Berezin Operators in the Narrow Sense

A *Berezin operator* $\Lambda_q \to \Lambda_p$ *in the narrow sense* is an operator of the form

$$\mathfrak{B}\begin{bmatrix} A & B \\ -B^t & C \end{bmatrix} f(\xi) := \int \exp\left\{\frac{1}{2} \begin{pmatrix} \xi & \overline{\eta} \end{pmatrix} \begin{pmatrix} A & B \\ -B^t & C \end{pmatrix} \begin{pmatrix} \xi^t \\ \overline{\eta}^t \end{pmatrix}\right\} f(\eta) \, e^{-\eta\overline{\eta}^t} \, d\overline{\eta} \, d\eta, \tag{16}$$

where $A = -A^t$, $C = -C^t$. Let us explain the notation.

1. $\begin{pmatrix} \xi & \eta \end{pmatrix}$ denotes the row-matrix

$$\begin{pmatrix} \xi & \overline{\eta} \end{pmatrix} := \begin{pmatrix} \xi_1 & \dots & \xi_p & \overline{\eta}_1 & \dots & \overline{\eta}_q \end{pmatrix}.$$

   Respectively, $\begin{pmatrix} \xi^t \\ \overline{\eta}^t \end{pmatrix}$ denotes the transposed column-matrix.

2. The $(p + q) \times (p + q)$-matrix $\begin{pmatrix} A & B \\ -B^t & C \end{pmatrix}$ is skew-symmetric. The whole expression for the kernel has the form

$$\exp\left\{\frac{1}{2} \sum_{k \leqslant p, l \leqslant p} a_{kl} \xi_k \xi_l + \sum_{k \leqslant p, m \leqslant q} b_{km} \xi_k \overline{\eta}_m + \frac{1}{2} \sum_{m \leqslant q, j \leqslant q} c_{mj} \overline{\eta}_m \overline{\eta}_j\right\}.$$

[11] Up to signs.

2.8 Product Formula

Theorem 2.3 *Let*

$$\mathfrak{B}[S_1] = \mathfrak{B}\begin{bmatrix} P & Q \\ -Q^t & R \end{bmatrix} : \Lambda_p \to \Lambda_q, \qquad \mathfrak{B}[S_2] = \mathfrak{B}\begin{bmatrix} K & L \\ -L^t & M \end{bmatrix} : \Lambda_q \to \Lambda_r$$

be Berezin operators in the narrow sense. Assume $\det(1 - MP) \neq 0$. *Then*

$$\mathfrak{B}[S_2]\,\mathfrak{B}[S_1] = \mathrm{Pfaff}\begin{pmatrix} M & 1 \\ -1 & P \end{pmatrix} \mathfrak{B}[S_2 \circ S_1], \tag{17}$$

where

$$\begin{pmatrix} K & L \\ -L^t & M \end{pmatrix} \circ \begin{pmatrix} P & Q \\ -Q^t & R \end{pmatrix} =$$

$$= \begin{pmatrix} K + LP(1 - MP)^{-1}L^t & L(1 - PM)^{-1}Q \\ -Q^t(1 - MP)^{-1}L^t & R - Q^t(1 - MQ)^{-1}MQ \end{pmatrix}. \tag{18}$$

The symbol $\mathrm{Pfaff}(\cdot)$ denotes the Pfaffian (see the next subsection).

A calculation is not difficult (see Sect. 4.9). The formula (18) is clarified in Theorem 2.10.

2.9 Pfaffians and Odd Gaussian Integrals

Let R be a skew-symmetric $2n \times 2n$ matrix. Its *Pfaffian* $\mathrm{Pfaff}(R)$ is defined by the condition

$$\frac{1}{n!}\Big(\frac{1}{2}\sum_{kl} r_{kl}\xi_k\xi_l\Big)^n = \mathrm{Pfaff}(R)\,\xi_1\xi_2\dots\xi_{2n-1}\xi_{2n}$$

(where $\xi_1, \dots, \xi_{2n}$ are Grassmann variables). In other words,

$$\mathrm{Pfaff}(R) = \frac{1}{n!}\int \Big(\frac{1}{2}\sum_{kl} r_{kl}\xi_k\xi_l\Big)^n d\xi = \int \exp\Big\{\frac{1}{2}\xi R \xi^t\Big\}\, d\xi.$$

Recall that

$$\mathrm{Pfaff}(R)^2 = \det R.$$

More generally, let $\xi_1, \ldots, \xi_{2n}, \theta_1, \ldots, \theta_{2n}$ be pairwise anticommuting variables. Then (see [50], Theorem II.4.4)

$$\int \exp\left\{\frac{1}{2}\xi R\xi^t + \sum_{j=1}^{2n}\theta_j\xi_j\right\} d\xi = \mathrm{Pfaff}(R)\exp\left\{\frac{1}{2}\theta R^{-1}\theta^t\right\}. \tag{19}$$

Theorem 2.3 follows from this formula in a straightforward way.

2.10 The Definition of Berezin Operators

Theorem 2.3 suggests an extension of the definition of Berezin operators. Indeed, this theorem is perfect for operators in general position. If $\det(1 - MP) = 0$, then we get an indeterminate of the form $0 \cdot \infty$ in the product formula (17), (18).

For this reason, consider the cone[12] C of all the operators of the form $s \cdot \mathfrak{B}\begin{bmatrix} A & B^t \\ -B^t & C \end{bmatrix}$, where s ranges in $\mathbb{C}$. Certainly, the cone C is not closed. Indeed,

$$\lim_{\varepsilon\to 0} \varepsilon \exp\left\{\frac{1}{\varepsilon}\xi_1\xi_2\right\} = \xi_1\xi_2;$$

$$\lim_{\varepsilon\to 0} \varepsilon^2 \exp\left\{\frac{1}{\varepsilon}(\xi_1\xi_2 + \xi_2\xi_4)\right\} = \xi_1\xi_2\xi_3\xi_4.$$

This suggests the following definition.

Berezin operators are operators $\Lambda_q \to \Lambda_p$, whose kernels have the form (cf. [59])

$$s \cdot \prod_{j=1}^{m}(\xi u_j^t + \overline{\eta} v_j^t) \cdot \exp\left\{\frac{1}{2}\begin{pmatrix}\xi & \overline{\eta}\end{pmatrix}\begin{pmatrix} A & B \\ -B^t & C \end{pmatrix}\begin{pmatrix}\xi^t \\ \overline{\eta}^t\end{pmatrix}\right\}, \tag{20}$$

where

1. $s \in \mathbb{C}$
2. $u_j \in \mathbb{C}^p$ $v_j \in \mathbb{C}^q$ are row-matrices
3. m ranges in the set $\{0, 1, \ldots, p+q\}$
4. $\begin{pmatrix} A & B \\ -B^t & C \end{pmatrix}$ is a skew-symmetric $(p+q) \times (p+q)$-matrix

[12] Below, a *cone* in $\mathbb{C}^N$ is a subset invariant with respect to homotheties $z \mapsto sz$, where $s \in \mathbb{C}$.

2.11 The Space of Berezin Operators

Proposition 2.4

(a) *The cone of Berezin operators is closed in the space of all linear operators* $\Lambda_q \to \Lambda_p$.
(b) *Denote by* $\mathrm{Ber}^{[m]}$ *the set of all Berezin operators with a given number* m *of linear factors in* (20). *Then the closure of* $\mathrm{Ber}^{[m]}$ *is*

$$\mathrm{Ber}^{[m]} \cup \mathrm{Ber}^{[m+2]} \cup \mathrm{Ber}^{[m+4]} \cup \dots$$

Therefore, the cone of all Berezin operators $\Lambda_q \to \Lambda_p$ consists of two components, namely,

$$\bigcup_{m \text{ is even}} \mathrm{Ber}^{[m]} \qquad \text{and} \qquad \bigcup_{m \text{ is odd}} \mathrm{Ber}^{[m]}.$$

They are closures of $\mathrm{Ber}^{[0]}$ and $\mathrm{Ber}^{[1]}$, respectively. Also, kernels $K(\xi, \overline{\eta})$ of Berezin operators satisfy

$$K(-\xi, -\overline{\eta}) = K(\xi, \overline{\eta}), \qquad \text{or} \qquad K(-\xi, -\overline{\eta}) = -K(\xi, \overline{\eta}),$$

respectively.

Remark Kernels $K(\xi, \eta)$ of Berezin operators $\Lambda_q \to \Lambda_p$ are canonically defined polynomials in ξ, η. However, expressions (20) for the kernels of Berezin operators are non-canonical if numbers m of linear factors are $\geqslant 1$. For instance,

$$(\xi_1 + \xi_{33})(\xi_1 + 7\xi_{33}) = 6\xi_1\xi_{33}, \qquad \xi_1 \exp\{\xi_1\xi_2 + \xi_3\xi_4\} = \xi_1 \exp\{\xi_3\xi_4\}.$$

A quadratic form in the exponential in (20) also is not defined canonically. ⊠

2.12 Examples of Berezin Operators

(a) The identity operator is a Berezin operator. Its kernel is $\exp\left\{\sum \xi_j \overline{\eta}_j\right\}$.
(b) More generally, operators with kernels $\exp\left\{\sum_{ij} b_{ij}\xi_i\overline{\eta}_j\right\}$ are operators $\Lambda_q \to \Lambda_p$ defined by the substitution $\eta_i = \sum_j b_{ji}\xi_j$.
(c) The operator with kernel $\xi_1 \exp\{\sum \xi_j\overline{\eta}_j\}$ is the operator $f \mapsto \xi_1 f$.
(e) The operator with kernel $\overline{\eta}_1 \exp\{\sum \xi_j\overline{\eta}_j\}$ is the operator $f \mapsto \frac{\partial}{\partial \xi_1} f$.

(f) The operator in Λ_n determined by the kernel $(\xi_1 + \overline{\eta}_1)(\xi_2 + \overline{\eta}_2)\dots$ is Hodge $\star$-operator. Namely, let $i_1 < i_2 < \dots < i_k$ be a subset in $\{1, 2, \dots, n\}$. Let $j_1 < j_2 < \dots < j_{n-k}$ be the complementary subset. Then

$$\star\big(\xi_1\xi_2\dots\xi_{i_k}\big) = \xi_{j_1}\dots\xi_{j_{n-k}}.$$

(g) The operator with kernel $K(\xi, \eta) = 1$ is the projection to the vector $f(\xi) = 1 \in \Lambda_n$.

(h) The operator $\mathfrak{B}$ with the kernel

$$\exp\left\{\frac{1}{2}a_{ij}\xi_i\xi_j + \sum \xi_j\overline{\eta}_j\right\}$$

is the multiplication operator

$$\mathfrak{B}f(\xi) = \exp\left\{\frac{1}{2}a_{ij}\xi_i\xi_j\right\} f(\xi).$$

(i) A product of Berezin operators is a Berezin operator (see below).

(j) Operators of the spinor representation of $O(2n, \mathbb{C})$ are Berezin operators.

2.13 *Another Definition of Berezin Operators*

Denote by $\mathfrak{D}[\xi_j]$ the following operators in Λ_p

$$\mathfrak{D}[\xi_j]f(\xi) := \left(\xi_j + \frac{\partial}{\partial\xi_j}\right) f(\xi). \tag{21}$$

If g, h do not depend on ξ_j, then

$$\mathfrak{D}[\xi_j]\big(g(\xi) + \xi_j h(\xi)\big) = \xi_j g(\xi) + h(\xi)$$

In the same way, we define the operator $\mathfrak{D}[\eta_j] : \Lambda_q \to \Lambda_q$. Obviously

$$\mathfrak{D}[\xi_i]^2 = 1, \qquad \mathfrak{D}[\xi_i]\,\mathfrak{D}[\xi_j] = -\mathfrak{D}[\xi_j]\,\mathfrak{D}[\xi_i], \quad \text{for } i \neq j. \tag{22}$$

A *Berezin operator* $\Lambda_q \to \Lambda_p$ is any operator that can be represented in the form

$$\mathfrak{D}[\xi_{k_1}]\dots\mathfrak{D}[\xi_{k_\alpha}]\cdot\mathfrak{B}\cdot\mathfrak{D}[\eta_{m_1}]\dots\mathfrak{D}[\eta_{m_\beta}], \tag{23}$$

where

- $\mathfrak{B}$ is a Berezin operator in the narrow sense,
- α ranges in the set $\{0, 1, \dots, p\}$ and β ranges in $\{0, 1, \dots, q\}$

By (22) we can assume $k_1 < \cdots < k_\alpha, m_1 < \cdots < m_\beta$.

Proposition 2.5 *The two definitions of Berezin operators are equivalent.*

Remark Usually, a Berezin operator admits many representations in the form (23). In fact, the space of all Berezin operators is a smooth cone, and formula (23) determines 2^{p+q} coordinate systems on this cone. Any collection of $2^{p+q} - 1$ charts has a non-empty complement. ⊠

2.14 *The Category of Berezin Operators: Groups of Automorphisms*

Theorem 2.6 *Let*

$$\mathfrak{B}_1 : \Lambda_q \to \Lambda_p, \qquad \mathfrak{B}_2 : \Lambda_p \to \Lambda_r$$

be Berezin operators. Then $\mathfrak{B}_2\mathfrak{B}_1 : \Lambda_q \to \Lambda_r$ *is a Berezin operator. If* $\mathfrak{B}$ *is an invertible Berezin operator, then* $\mathfrak{B}^{-1}$ *is a Berezin operator.*

By Theorem 2.6, we get a category whose objects are Grassmann algebras Λ_0, Λ_1, Λ_2, ... and whose morphisms are Berezin operators.

Denote by G_n the group of all invertible Berezin operators $\Lambda_n \to \Lambda_n$. By definition, it contains the group $\mathbb{C}^\times$ of all non-zero scalar operators.

Theorem 2.7 $G_n/\mathbb{C}^\times \simeq \mathrm{O}(2n, \mathbb{C})/\{\pm 1\}$.

Here $\mathrm{O}(2n, \mathbb{C})$ denotes the usual group of orthogonal transformations in $\mathbb{C}^{2n}$.

The map sending each element of $\mathrm{O}(2n, \mathbb{C})/\{\pm 1\}$. Moreover, this isomorphism sending an orthogonal matrix to the corresponding Berezin operator in Λ_n (determined up to a factor) is nothing but the spinor representation of $\mathrm{O}(2n, \mathbb{C})$.

Our next aim is to describe explicitly the category of Berezin operators. In fact, we intend to clarify the strange matrix multiplication (18).

B. Linear Relations and the Category GD

2.15 *Linear Relations*

Let V, W be linear spaces over $\mathbb{C}$. A *linear relation* $P : V \rightrightarrows W$ is a linear subspace $P \subset V \oplus W$.

Remark Let $A : V \to W$ be a linear operator. Its *graph* $\mathrm{graph}(A) \subset V \oplus W$ consists of all vectors $v \oplus Av$. By the definition, $\mathrm{graph}(A)$ is a linear relation,

$$\dim \mathrm{graph}(A) = \dim V.$$

2.16 Product of Linear Relations

Let $P : V \rightrightarrows W$, $Q : W \rightrightarrows Y$ be linear relations. Informally, a product QP of linear relations is a product of many-valued maps. If P takes a vector v to a vector w and Q takes the vector w to a vector y, then QP takes v to y.

Now, we present a formal definition. The *product* QP is a linear relation $QP : V \rightrightarrows Y$ consisting of all $v \oplus y \in V \oplus Y$ such that there exists $w \in W$ satisfying $v \oplus w \in P$, $w \oplus y \in Q$.

In fact, the multiplication of linear relations extends the usual matrix multiplication.

2.17 Imitation of Some Standard Definitions of Matrix Theory

1°. The *kernel* $\ker P$ consists of all $v \in V$ such that $v \oplus 0 \in P$. In other words,

$$\ker P = P \cap (V \oplus 0).$$

2°. The *image* $\mathrm{im}\, P \subset W$ is the projection of P to $0 \oplus W$.
3°. The *domain* $\mathrm{dom}\, P \subset V$ of P is the projection of P to $V \oplus 0$.
4°. The *indefinity* $\mathrm{indef}\, P \subset W$ of P is $P \cap (0 \oplus W)$.

The definitions of a kernel and an image extend the corresponding definitions for linear operators. The definition of a domain extends the usual definition of the domain of an unbounded operator in an infinite-dimensional space. For a linear operator, $\mathrm{indef} = 0$.

2.18 Lagrangian Grassmannian and Orthogonal Groups

First, let us recall some definitions. Let V be a linear space equipped with a non-degenerate symmetric (or skew-symmetric) bilinear form M. A subspace H is *isotropic* with respect to the bilinear form M, if $M(h, h') = 0$ for all $h, h' \in H$. The dimension of an isotropic subspace satisfies

$$\dim H \leqslant \frac{1}{2} \dim V.$$

A *Lagrangian subspace*[13] is an isotropic subspace whose dimension is precisely $\frac{1}{2} \dim V$. By $\mathrm{Lagr}(V)$ we denote the *Lagrangian Grassmannian*, i.e., the space of all Lagrangian subspaces in V.

Consider a space $\mathcal{V}_{2n} = \mathbb{C}^{2n}$ equipped with the symmetric bilinear form $L = L_n$ determined by the matrix $\begin{pmatrix} 0 & 1_n \\ 1_n & 0 \end{pmatrix}$. Denote by $\mathrm{O}(2n, \mathbb{C})$ the group of all linear transformations g of $\mathbb{C}^{2n}$ preserving L, i.e., g must satisfy the condition

$$g \begin{pmatrix} 0 & 1_n \\ 1_n & 0 \end{pmatrix} g^t = \begin{pmatrix} 0 & 1_n \\ 1_n & 0 \end{pmatrix}.$$

Let $m, n = 0, 1, 2, \ldots$. Equip the space $\mathcal{V}_{2n} \oplus \mathcal{V}_{2m}$ with the symmetric bilinear form $L^{\ominus}$ given by

$$L^{\ominus}(v \oplus w, v' \oplus w') = L_n(v, v') - L_m(w, w') \ , \tag{24}$$

where $v, v' \in \mathcal{V}_{2n}$ and $w, w' \in \mathcal{V}_{2m}$.

Observation 2.8 *Let g be an operator in $\mathbb{C}^{2n}$. Then $g \in \mathrm{O}(2n, \mathbb{C})$ if and only if its graph is an $L^{\ominus}$-Lagrangian subspace in $\mathcal{V}_{2n} \oplus \mathcal{V}_{2n}$.*

This is obvious.[14]

2.19 Imitation of Orthogonal Groups: Category GD

Now we define the category **GD**. The objects are the spaces $\mathcal{V}_{2n}$, where $n = 0, 1, 2, \ldots$. There are two types of morphisms $\mathcal{V}_{2n} \to \mathcal{V}_{2m}$:

(a) $L^{\ominus}$-Lagrangian subspaces $P \subset \mathcal{V}_{2n} \oplus \mathcal{V}_{2m}$; we regard them as linear relations.
(b) A distinguished morphism [15] denoted by $\mathsf{null}_{2n,2m}$.

[13] Mostly, the term "Lagrangian subspace" is used for spaces equipped with a skew-symmetric bilinear forms; however, my usage is a usual slang.

[14] I use the term "Observation" for statements that are important for understanding and became trivial or semi-trivial being formulated.

[15] It is not identified with any linear relation.

Now we define a product of morphisms.

– Product of null and any morphism is null.
– Let $P : \mathcal{V}_{2n} \rightrightarrows \mathcal{V}_{2m}$, $Q : \mathcal{V}_{2m} \rightrightarrows \mathcal{V}_{2k}$ be Lagrangian linear relations.
 Assume that

$$\ker Q \cap \operatorname{indef} P = 0 \qquad \text{or, equivalently,} \quad \operatorname{im} P + \operatorname{dom} Q = \mathcal{V}_{2m}. \tag{25}$$

 Then QP is the product of linear relations.
– If the condition (25) is not satisfied, then $QP = \mathsf{null}$.

Theorem 2.9 *The definition is self-consistent, i.e., a product of morphisms is a morphism, and the multiplication is associative.*

At first glance, the appearance of null seems strange; its necessity will be transparent immediately (null corresponds to zero operators in the next theorem). Also, null will be the source of some our difficulties below.

Theorem 2.10 *The category of Berezin operators defined up to scalar factors and the category* **GD** *are equivalent.*

In fact, there is a map that takes each Lagrangian linear relation $P : \mathcal{V}_{2n} \rightrightarrows \mathcal{V}_{2m}$ to a nonzero Berezin operator $\operatorname{spin}(P) : \Lambda_n \to \Lambda_m$ such that for each $R : \mathcal{V}_{2q} \rightrightarrows \mathcal{V}_{2p}$, $Q : \mathcal{V}_{2p} \rightrightarrows \mathcal{V}_{2r}$

$$\operatorname{spin}(Q)\operatorname{spin}(R) = \lambda(Q, R)\operatorname{spin}(QR),$$

where $\lambda(Q, R)$ is a constant. Moreover,

$$\lambda(Q, R) = 0 \text{ if and only if } QR = \mathsf{null}.$$

We describe the correspondence between the category of Berezin operators and the category **GD** explicitly in Theorems 2.16 and 2.17. First, we need some auxiliary facts concerning Lagrangian Grassmannians (for a detailed introduction to Lagrangian Grassmannians, see [2], Sect. 43, [53], Chapter 3).

C. Explicit Correspondence

2.20 Coordinates on the Lagrangian Grassmannian

Recall that $\mathcal{V}_{2n}$ is the space $\mathbb{C}^{2n} = \mathbb{C}^n \oplus \mathbb{C}^n$ equipped with the bilinear form $L = L_n$ with the matrix $\begin{pmatrix} 0 & 1_n \\ 1_n & 0 \end{pmatrix}$. We write $\mathcal{V}_{2n} = \mathbb{C}^{2n}$ as

$$\mathcal{V}_{2n} = \mathcal{V}_n^+ \oplus \mathcal{V}_n^- = \mathbb{C}^n \oplus \mathbb{C}^n,$$

in this decomposition the summands are Lagrangian subspaces.

Lemma 2.11 *Let $H \subset \mathcal{V}_{2n}$ be an n-dimensional subspace such that $H \cap \mathcal{V}_n^- = 0$. Under this condition H is the graph of an operator*

$$T_H : \mathcal{V}_n^+ \to \mathcal{V}_n^- .$$

The following conditions are equivalent:

- *The matrix T_H is skew-symmetric.*
- *The subspace H is Lagrangian.*

Proof We write the bilinear form L as

$$L(v, w) = L(v^+ \oplus v^-, w^+ \oplus w^-) = v^+(w^-)^t + w^+(v^-)^t .$$

By definition, $v \in H$ if and only if $v^- = v^+ T_H$. For $v, w \in P$, we evaluate

$$L(v^+ \oplus v^+ T, w^+ \oplus w^+ T) = v^+(w^+ T_H)^t + v^+ T_H (w^+)^t = v^+(T_H + T_H^t) w^+ .$$

Now the statement becomes obvious. □

Lemma 2.11 defines a coordinate system in $\mathrm{Lagr}(\mathcal{V}_{2n})$. Certainly, this coordinate system does not cover the whole space $\mathrm{Lagr}(\mathcal{V}_{2n})$.

2.21 *Atlas on the Lagrangian Grassmannian*

Denote by $e_1^+, \dots, e_n^+, e_1^-, \dots, e_n^-$ the standard basis in $\mathbb{C}^{2n} = \mathbb{C}^n \oplus \mathbb{C}^n$. Let J be a subset in $\{1, 2, \dots, n\}$. Denote by $\overline{J}$ its complement. We define the subspaces

$$\mathcal{V}_n^+[J] := \big(\oplus_{j \in J} \mathbb{C} e_j^+ \big) \oplus \big(\oplus_{j \notin J} \mathbb{C} e_j^- \big),$$
$$\mathcal{V}_n^-[J] := \big(\oplus_{j \notin J} \mathbb{C} e_j^+ \big) \oplus \big(\oplus_{j \in J} \mathbb{C} e_j^- \big),$$

and then $\mathcal{V}_{2n} = \mathcal{V}_n^+[J] \oplus \mathcal{V}_n^-[J]$. Denote by $\mathcal{M}[J]$ the set of all Lagrangian subspaces H in $\mathcal{V}_{2n}$ such that $H \cap \mathcal{V}_n^-[J] = 0$. Such subspaces are precisely graphs of symmetric operators $\mathcal{V}_n^+[J] \to \mathcal{V}_n^-[J]$.

Proposition 2.12 *The 2^n charts $\mathcal{M}[J]$ cover the whole Lagrangian Grassmannian* $\mathrm{Lagr}(\mathcal{V}_{2n})$.

2.22 *Atlas on the Lagrangian Grassmannian: Elementary Reflections*

We can describe the same maps in a slightly different way. For $i = 1, 2, \dots, n$, define an elementary reflection $\sigma_i : \mathcal{V}_{2n} \to \mathcal{V}_{2n}$ by

$$\sigma_i e_i^+ = e_i^-, \qquad \sigma_i e_i^- = e_i^+, \qquad \sigma_i e_j^\pm = e_j^\pm \qquad \text{for } i \ne j. \tag{26}$$

Observation 2.13

$$\mathcal{M}[J] = \Bigl(\prod_{i \in J} \sigma_i\Bigr) \mathcal{M}[\varnothing]. \tag{27}$$

2.23 *Components of Lagrangian Grassmannian*

Observation 2.14 *The Lagrangian Grassmannian in the space $\mathcal{V}_{2n}$ consists of two connected components.*[16]

We propose two proofs to convince the reader.

1. The group $\mathrm{O}(n, \mathbb{C})$ is dense in $\mathrm{Lagr}(\mathcal{V}_{2n})$. This group has two components.
2. It can be readily checked that $\mathrm{Lagr}(\mathcal{V}_{2n})$ is a homogeneous space

$$\mathrm{Lagr}(\mathcal{V}_{2n}) \simeq \mathrm{O}(2n, \mathbb{C})/\mathrm{GL}(n, \mathbb{C}).$$

The group $\mathrm{O}(2n, \mathbb{C})$ consists of two components, and the group $\mathrm{GL}(n, \mathbb{C})$ is connected. □

Remark Two components of the Lagrangian Grassmannian correspond to two components of the space of Berezin operators. ⊠

2.24 *Coordinates on the Set of Morphisms of* GD

We can apply the reasoning of Sect. 2.20 to $\mathrm{Lagr}(\mathcal{V}_{2n} \oplus \mathcal{V}_{2m})$. Due to the minus in formula (24), we must take care of the signs in Lemma 2.11.

---

[16] Recall that $\mathcal{V}_{2n}$ is equipped with a symmetric bilinear form. The usual Lagrangian Grassmannian discussed in the next section is connected. The orthosymplectic Lagrangian Grassmannian (see Sect. 7) consists of two components.

Lemma 2.15 *Decompose $\mathcal{V}_{2n} \oplus \mathcal{V}_{2m}$ as*[17]

$$\mathcal{V}_{2n} \oplus \mathcal{V}_{2m} = (\mathcal{V}_n^- \oplus \mathcal{V}_m^+) \oplus (\mathcal{V}_n^+ \oplus \mathcal{V}_m^-).$$

Let P be an $(m+n)$-dimensional subspace such that $P \cap (\mathcal{V}_n^+ \oplus \mathcal{V}_m^-) = 0$, i.e., P is the graph of an operator

$$\mathcal{V}_n^- \oplus \mathcal{V}_m^+ \to \mathcal{V}_n^+ \oplus \mathcal{V}_m^-.$$

Then the following conditions are equivalent:

- *$P \in \mathrm{Lagr}(\mathcal{V}_{2n} \oplus \mathcal{V}_{2m})$;*
- *P is the graph of an operator having the form*

$$\begin{pmatrix} A & B \\ B^t & C \end{pmatrix}, \qquad \text{where } A = -A^t, C = -C^t. \tag{28}$$

2.25 Creation-Annihilation Operators

Let $\mathcal{V}_{2n}$ be as above. Decompose

$$\mathcal{V}_{2n} = \mathcal{V}_n^+ \oplus \mathcal{V}_n^-, \qquad \text{where } \mathcal{V}_n^+ := \mathbb{C}^n \oplus 0,\ \mathcal{V}_n^- := 0 \oplus \mathbb{C}^n.$$

Let us write elements of $\mathcal{V}_{2n}$ as

$$v := \begin{pmatrix} v_1^+ & \dots & v_n^+ & v_1^- & \dots & v_n^- \end{pmatrix}.$$

For each $v \in \mathcal{V}_{2n}$, we define a creation–annihilation operator $\widehat{a}(v)$ in Λ_n by

$$\widehat{a}(v) f(\xi) := \Big(\sum_j v_j^+ \xi_j + \sum_j v_j^- \frac{\partial}{\partial \xi_j} \Big) f(\xi).$$

Evidently,

$$\widehat{a}(v)\widehat{a}(w) + \widehat{a}(w)\widehat{a}(v) = L(v, w) \cdot 1.$$

[17] We emphasize that the source space $\mathcal{V}_{2n}$ and the target space $\mathcal{V}_{2m}$ are mixed in the next row.

2.26 A Construction of the Correspondence

Let $\mathfrak{B} : \Lambda_q \to \Lambda_p$ be a nonzero Berezin operator. Consider the subspace $P = P[\mathfrak{B}] \subset \mathcal{V}_{2q} \oplus \mathcal{V}_{2p}$ consisting of $v \oplus w$ such that

$$\widehat{a}(w)\,\mathfrak{B} = \mathfrak{B}\,\widehat{a}(v). \tag{29}$$

Theorem 2.16

(a) *$P[\mathfrak{B}]$ is a morphism of the category* **GD**, *i.e., a Lagrangian subspace.*

(b) *The map $\mathfrak{B} \mapsto P$ is a bijection*

$$\left\{\begin{matrix} Set\ of\ nonzero\ Berezin\ operators\ \Lambda_q \to \Lambda_p \\ defined\ up\ to\ a\ scalar\ factor \end{matrix}\right\} \longleftrightarrow$$

$$\longleftrightarrow \left\{\begin{matrix} The\ set\ of\ non-\ \text{'null'}\ morphisms\ \mathcal{V}_{2q} \to \mathcal{V}_{2p} \\ of\ the\ category\ \mathbf{GD} \end{matrix}\right\}.$$

(c) *Let $\mathfrak{B}_1 : \Lambda_q \to \Lambda_p$, $\mathfrak{B}_2 : \Lambda_p \to \Lambda_r$ be Berezin operators. Then*

$$\mathfrak{B}_2\mathfrak{B}_1 = 0 \quad \textit{if and only if} \quad P[\mathfrak{B}_2]\,P[\mathfrak{B}_1] = \mathsf{null}\,. \tag{30}$$

(d) *Otherwise,*

$$P[\mathfrak{B}_2\mathfrak{B}_1] = P[\mathfrak{B}_2]P[\mathfrak{B}_1].$$

Sketch of Proof

1. First, let us consider a Berezin operator in the narrow sense. We write the Eq. (29)

$$\Big(\sum w_j^+\xi_j + \sum w_j^- \frac{\partial}{\partial \xi_j}\Big)\int \exp\Big\{\frac{1}{2}\,(\xi\ \overline{\eta})\begin{pmatrix} A & B \\ -B^t & C\end{pmatrix}\begin{pmatrix}\xi^t \\ \overline{\eta}^t\end{pmatrix}\Big\} f(\eta)e^{-\eta\overline{\eta}^t}\,d\overline{\eta}\,d\eta =$$

$$= \int \exp\Big\{\frac{1}{2}\,(\xi\ \overline{\eta})\begin{pmatrix} A & B \\ -B^t & C\end{pmatrix}\begin{pmatrix}\xi^t \\ \overline{\eta}^t\end{pmatrix}\Big\}\Big(\sum v_k^+\eta_k + \sum v_k^- \frac{\partial}{\partial \eta_k}\Big) f(\eta)e^{-\eta\overline{\eta}^t}\,d\overline{\eta}\,d\eta,$$

or, equivalently,

$$\Big(\sum w_j^+\xi_j + \sum w_j^- \frac{\partial}{\partial \xi_j} - \sum v_k^-\overline{\eta}_k - \sum v_k^+ \frac{\partial}{\partial \eta_k}\Big)\times$$

$$\times \exp\Big\{\frac{1}{2}\,(\xi\ \overline{\eta})\begin{pmatrix} A & B \\ -B^t & C\end{pmatrix}\begin{pmatrix}\xi^t \\ \overline{\eta}^t\end{pmatrix}\Big\} = 0$$

We differentiate the exponential and get

$$\Big[\sum_j w_j^+ \xi_j + \sum_j w_j^- \Big(\sum_i a_{ji}\xi_i + \sum_m b_{jk}\overline{\eta}_m\Big) - \sum_k v_k^- \overline{\eta}_k - \sum_k v_k^+ \\ + \Big(-\sum_l b_{lk}\xi_l + \sum_m c_{km}\overline{\eta}_m\Big)\Big] \cdot \exp\Big\{\frac{1}{2}\begin{pmatrix}\xi & \overline{\eta}\end{pmatrix}\begin{pmatrix}A & B\\ -B^t & C\end{pmatrix}\begin{pmatrix}\xi^t\\ \overline{\eta}^t\end{pmatrix}\Big\} = 0.$$

Therefore,

$$\begin{pmatrix}v^+ & w^-\end{pmatrix} = \begin{pmatrix}v^- & w^+\end{pmatrix}\begin{pmatrix}A & -B\\ -B^t & -C\end{pmatrix}.$$

By Lemma 2.15, P is a Lagrangian subspace in $\mathcal{V}_{2q} \oplus \mathcal{V}_{2p}$.

2. Next, let $\mathfrak{C}$ be a Berezin operator having the form (23)

$$\mathfrak{C} = \mathfrak{D}[\xi_{k_1}] \dots \mathfrak{D}[\xi_{k_\alpha}] \cdot \mathfrak{B} \cdot \mathfrak{D}[\eta_{m_1}] \dots \mathfrak{D}[\eta_{m_\beta}]. \tag{31}$$

We note that

$$\widetilde{a}(v)\,\mathfrak{D}[\xi_i] = \mathfrak{D}[\xi_i]\,\widetilde{a}(\sigma_i v),$$

where σ_i is an elementary reflection (26). Let the operator $\mathfrak{B}$ satisfy

$$\widehat{a}(w)\,\mathfrak{B} = \mathfrak{B}\,\widehat{a}(v),$$

where $v \oplus w$ ranges in $P[\mathfrak{B}]$. Then the operator $\mathfrak{C}$ satisfies

$$\widehat{a}(\sigma_{k_1} \dots \sigma_{k_\alpha} w)\,\mathfrak{C} = \mathfrak{C}\,\widehat{a}(\sigma_{m_1} \dots \sigma_{m_\beta} v).$$

Therefore, the corresponding linear relation is

$$\sigma_{k_1} \dots \sigma_{k_\alpha} P \sigma_{m_1} \dots \sigma_{m_\beta} \tag{32}$$

and is Lagrangian. This proves (a).

3. By Proposition 2.12, the sets (32) sweep the whole Lagrangian Grassmannian. This proves (b).
4. Let $v \oplus w \in P[\mathfrak{B}_1]$, $w \oplus y \in P[\mathfrak{B}_2]$, i.e.,

$$\widehat{a}(w)\,\mathfrak{B}_1 = \mathfrak{B}_1\,\widehat{a}(v), \qquad \widehat{a}(y)\,\mathfrak{B}_2 = \mathfrak{B}_2\,\widehat{a}(w).$$

Then

$$\mathfrak{B}_2\mathfrak{B}_1\widehat{a}(v) = \mathfrak{B}_2\widehat{a}(w)\mathfrak{B}_1 = \widehat{a}(y)\mathfrak{B}_2\mathfrak{B}_1$$

and this proves (c).

We omit a proof of (d), which is more difficult (see [50, Subsect. II.6.5]). □

2.27 *Explicit Correspondence: Another Description*

The proof above implies also the following theorem (see [50, Theorem II.6.11]).

Theorem 2.17 *Let* P *satisfy Lemma* 2.15. *Then the corresponding Berezin operator* $\mathfrak{B}[P]$ *has the kernel*

$$\exp\left\{\frac{1}{2}\begin{pmatrix}\xi & \overline{\eta}\end{pmatrix}\begin{pmatrix}A & -B\\ B^t & -C\end{pmatrix}\begin{pmatrix}\xi^t\\ \overline{\eta}^t\end{pmatrix}\right\}.$$

3 A Survey of the Oscillator Representation: The Category of Gaussian Integral Operators

In Sects. 3.1–3.2, we define the bosonic Fock space. In Sects. 3.3–3.4, we introduce Gaussian integral operators. The algebraic structure of the category of Gaussian integral operators is described in Sects. 3.6–3.8. For a detailed exposition, see [50, Chapter 4], or [53, Sect. 5.1–5.2].

3.1 *Fock Space*

Denote by $d\lambda(z)$ the Lebesgue measure on $\mathbb{C}^n$ normalized as

$$d\lambda(z) := \pi^{-n} dx_1 \dots dx_n\, dy_1 \dots dy_n, \qquad \text{where } z_j = x_j + iy_j.$$

The bosonic Fock space $\mathbf{F}_n$ is the space of entire functions on $\mathbb{C}^n$ satisfying the condition

$$\int_{\mathbb{C}^n} |f(z)|^2 e^{-|z|^2}\, d\lambda(z) < \infty.$$

We define the inner product in $\mathbf{F}_n$ by the formula

$$\langle f, g\rangle := \int_{\mathbb{C}^n} f(z)\overline{g(z)} e^{-|z|^2}\, d\lambda(z).$$

Theorem 3.1 *The space* $\mathbf{F}_n$ *is complete, i.e., it is a Hilbert space.*

Proposition 3.2 *The monomials $z_1^{k_1} \dots z_n^{k_n}$ are pairwise orthogonal and*

$$\|z_1^{k_1} \dots z_n^{k_n}\|^2 = \prod k_j!$$

3.2 Operators

Theorem 3.3 *For each bounded operator $A : \mathbf{F}_n \to \mathbf{F}_m$, there is a function $K(z, \overline{u})$ on $\mathbb{C}^m \oplus \mathbb{C}^n$ holomorphic in $z \in \mathbb{C}^m$ and antiholomorphic in $u \in \mathbb{C}^n$ such that*

$$Af(z) = \int_{\mathbb{C}^n} K(z, \overline{u}) \, f(u) \, e^{-|u|^2} \, d\lambda(u)$$

(the integral absolutely converges for all f).

3.3 Gaussian Operators

Fix $m, n = 0, 1, 2, \dots$ Let $S = \begin{pmatrix} K & L \\ L^t & M \end{pmatrix}$ be a symmetric $(m+n) \times (m+n)$-matrix, i.e., $S = S^t$. A Gaussian operator

$$\mathfrak{B}[S] = \mathfrak{B}\begin{bmatrix} K & L \\ L^t & M \end{bmatrix} : \mathbf{F}_n \to \mathbf{F}_m$$

is defined by

$$\mathfrak{B}\begin{bmatrix} K & L \\ L^t & M \end{bmatrix} f(z) = \int_{\mathbb{C}^n} \exp\left\{\frac{1}{2}\begin{pmatrix} z & \overline{u} \end{pmatrix}\begin{pmatrix} K & L \\ L^t & M \end{pmatrix}\begin{pmatrix} z^t \\ \overline{u}^t \end{pmatrix}\right\} f(u) \, e^{-|u|^2} \, d\lambda(u),$$

where

$$z := \begin{pmatrix} z_1 & \dots & z_n \end{pmatrix}, \qquad \overline{u} := \begin{pmatrix} \overline{u}_1 & \dots & \overline{u}_n \end{pmatrix}$$

are row-matrices.

For an operator A from the standard Euclidean space $\mathbb{C}^l$ to the standard Euclidean space $\mathbb{C}^k$, denote by $\|A\|$ the operator norm of A,

$$\|A\| = \max_{v \in \mathbb{C}^l, \, \|v\|=1} \|Av\| = \left\{\text{the maximal eigenvalue of } A^* A\right\}^{1/2}.$$

Theorem 3.4 (G. I. Olshanski) *An operator $\mathfrak{B}[S]$ is bounded if and only if*[18]

1. $\|S\| \leqslant 1$,
2. $\|K\| < 1$, $\|M\| < 1$.

3.4 Product Formula

Theorem 3.5 *Let*

$$\mathfrak{B}[S_1] = \mathfrak{B}\begin{bmatrix} P & Q \\ Q^t & R \end{bmatrix} : \mathbf{F}_n \to \mathbf{F}_m, \qquad \mathfrak{B}[S_2] = \mathfrak{B}\begin{bmatrix} K & L \\ L^t & M \end{bmatrix} : \mathbf{F}_m \to \mathbf{F}_k$$

be bounded Gaussian operators. Then their product is

$$\det(1 - MP)^{-1/2}\, \mathfrak{B}[S_2 * S_1],$$

where $S_2 * S_1$ is given by

$$\begin{pmatrix} K & L \\ L^t & M \end{pmatrix} * \begin{pmatrix} P & Q \\ Q^t & R \end{pmatrix} = \begin{pmatrix} K + LP(1-MP)^{-1}L^t & L(1-PM)^{-1}Q \\ Q^t(1-MP)^{-1}L^t & R + Q^t(1-MQ)^{-1}MQ \end{pmatrix}. \tag{33}$$

Theorem 3.6 *Denote by G_n the set of unitary $(n+n)\times(n+n)$ symmetric matrices $\begin{pmatrix} K & L \\ L^t & M \end{pmatrix}$ satisfying $\|K\| < 1$, $\|M\| < 1$. Then G_n is closed with respect to the $*$-multiplication and is isomorphic to the group* $\mathrm{Sp}(2n, \mathbb{R})$.

Observe that formula (33) almost coincides with formula (18). Again, formula (33) hides a product of linear relations.

First, we define analogues of the spaces $\mathcal{V}_{2n}$ from the previous section.

3.5 Complexification of a Linear Space with Bilinear Form

Denote by $\mathcal{W}_{2n}$ the space $\mathbb{C}^{2n} = \mathbb{C}^n \oplus \mathbb{C}^n$. Let us denote its elements by $v = \begin{pmatrix} v^+ & v^- \end{pmatrix}$. We equip $\mathcal{W}_{2n}$ with two forms:

– The skew-symmetric bilinear form

$$\Lambda(v, w) = \Lambda_n(v, w) := \begin{pmatrix} v^+ & v^- \end{pmatrix} \begin{pmatrix} 0 & 1_n \\ -1_n & 0 \end{pmatrix} \begin{pmatrix} (w^+)^t \\ (w^-)^t \end{pmatrix} = v^+(w^-)^t - v^-(w^+)^t; \tag{34}$$

[18] Condition 1 implies $\|K\| \leqslant 1$, $\|M\| \leqslant 1$, while condition 2 strengthens these inequalities.

– The indefinite Hermitian form

$$M(v,w)=M_n(v,w):=\begin{pmatrix}v^+ & v^-\end{pmatrix}\begin{pmatrix}1_n & 0\\ 0 & -1_n\end{pmatrix}\begin{pmatrix}(\overline{v}^+)^t\\ (\overline{w}^-)^t\end{pmatrix}=v^+(\overline{v}^+)^t-w^-(\overline{w}^-)^t. \tag{35}$$

Remark Let us explain the origin of the definition. Consider a real space $\mathbb{R}^{2n}$ equipped with a nondegenerate skew-symmetric bilinear form $\{\cdot,\cdot\}$. Consider the space $\mathbb{C}^{2n}\supset\mathbb{R}^{2n}$. We can extend $\{\cdot,\cdot\}$ to $\mathbb{C}^{2n}$ in the following two ways: First, we can extend it as a bilinear form,

$$\widetilde{\Lambda}(x+iy,x'+iy'):=\{x,x'\}-\{y,y'\}+i\big(\{x,y'\}+\{x',y\}\big).$$

Next, we extend $\{\cdot,\cdot\}$ as a sesquilinear form

$$\widetilde{M}(x+iy,x'+iy'):=\{x,x'\}+\{y,y'\}+i\big(\{y,x'\}-\{x,y'\}\big).$$

This form is anti-Hermitian, $\widetilde{M}(u,z)=-\overline{\widetilde{M}(z,u)}$, it is more convenient to pass to a bilinear form $\Lambda:=i\widetilde{\Lambda}$ and a Hermitian form $M:=i\widetilde{M}$. Thus, we get a space endowed with two forms, skew-symmetric and Hermitian. Denoting by e_l the standard basis in $\mathbb{C}^{2n}$ and passing to a new basis $(e_j+e_{n+j})/\sqrt{2}$, $(e_j-e_{n+j})/\sqrt{2}$, where $j\leqslant n$, we arrive at expressions (34), (35). ⊠

3.6 *The Category* **Sp**

Objects of the category **Sp** are the spaces $\mathcal{W}_{2n}$, where $n=0,1,2,\ldots$.

For $m,n=0,1,2,\ldots$, we equip the direct sum $\mathcal{W}_{2n}\oplus\mathcal{W}_{2m}$ with two forms

$$\begin{aligned}\Lambda^{\ominus}(v\oplus w,v'\oplus w')&:=\Lambda_n(v,v')-\Lambda_m(w,w'),\\ M^{\ominus}(v\oplus w,v'\oplus w')&:=M_n(v,v')-M_m(w,w').\end{aligned}\tag{36}$$

A morphism $\mathcal{W}_{2n}\to\mathcal{W}_{2m}$ is a linear relation $P:\mathcal{W}_{2n}\rightrightarrows\mathcal{W}_{2m}$ satisfying the following conditions.

1. P is $\Lambda^{\ominus}$-Lagrangian.
2. The form $M^{\ominus}$ is non-positive on P.
3. The form $M_{\mathcal{W}_{2n}}$ is strictly negative on $\ker P$, and the form $M_{\mathcal{W}_{2m}}$ is strictly positive on $\operatorname{indef} P$.

A product of morphisms is the usual product of linear relations.[19]

[19] Under the condition 3, for morphisms $P:\mathcal{W}_{2n}\rightrightarrows\mathcal{W}_{2m}$, $Q:\mathcal{W}_{2m}\rightrightarrows\mathcal{W}_{2k}$, we have $\operatorname{im}P\cap\ker Q=0$. For this reason null does not appear in the category **Sp**.

Observation 3.7 *The group of automorphisms of $\mathcal{W}_{2n}$ is the real symplectic group* $\mathrm{Sp}(2n, \mathbb{R})$.

This follows from the remark given in the previous subsection. The group of operators preserving both the forms $\widetilde{\Lambda}$, $\widetilde{M}$ on $\mathbb{C}^{2n}$ preserves also the real subspace $\mathbb{R}^{2n}$.

3.7 Construction of Gaussian Operators from Linear Relations

Recall that $\mathcal{W}_{2n}$ is $\mathbb{C}^{2n} = \mathbb{C}^n \oplus \mathbb{C}^n$. Denote this decomposition by

$$\mathcal{W}_{2n} = \mathcal{W}_n^+ \oplus \mathcal{W}_n^-.$$

Represent a linear relation P as the graph of an operator

$$S = S(P) : \mathcal{W}_m^- \oplus \mathcal{W}_n^+ \to \mathcal{W}_m^+ \oplus \mathcal{W}_n^-.$$

This is possible, because $M^{\ominus}$ is negative semidefinite on the subspace P and is strictly positive on the subspace $\mathcal{W}_m^+ \oplus \mathcal{W}_n^-$; therefore $P \cap (\mathcal{W}_m^+ \oplus \mathcal{W}_n^-) = 0$.

Proposition 3.8 *A matrix S has the form $S(P)$ if and only if it is symmetric and satisfies the Olshanski conditions from Theorem* 3.4.

Theorem 3.9 *For each morphisms*

$$P : \mathcal{W}_{2n} \rightrightarrows \mathcal{W}_{2m}, \qquad Q : \mathcal{W}_{2m} \rightrightarrows \mathcal{W}_{2k},$$

the corresponding Gaussian operators

$$\mathfrak{B}\big[S(P)\big] : \mathbf{F}_n \to \mathbf{F}_m, \qquad \mathfrak{B}\big[S(Q)\big] : \mathbf{F}_m \to \mathbf{F}_k$$

satisfy

$$\mathfrak{B}\big[S(Q)\big]\, \mathfrak{B}\big[S(P)\big] = \lambda(Q, P)\, \mathfrak{B}\big[S(QP)\big],$$

where QP is the product of linear relations and $\lambda(Q, P)$ is a nonzero scalar.

As formulated, the theorem can be proved by direct force.

3.8 Construction of Linear Relations from Gaussian Operators

For

$$\begin{pmatrix} v_1^+ & \dots & v_n^+ & v_1^- & \dots & v_n^- \end{pmatrix} \in \mathcal{V}_{2n},$$

we define the differential operator (a *creation–annihilation operator*)

$$\widehat{a}(v) f(z) = \Bigl(\sum_j v_j^+ z_j + \sum_j v_j^- \frac{\partial}{\partial z_j} \Bigr) f(z).$$

For a given bounded Gaussian operator $\mathfrak{B}[S] : \mathbf{F}_n \to \mathbf{F}_m$ we consider the set P of all $v \oplus w \in \mathcal{W}_{2n} \oplus \mathcal{W}_{2m}$ such that

$$\widehat{a}(w)\, \mathfrak{B}[S] = \mathfrak{B}[S]\, \widehat{a}(v).$$

Theorem 3.10 *The linear relation P is a morphism of the category* **Sp**.

3.9 Details: An Analogue of the Schwartz Space

We define the Schwartz–Fock space $\mathcal{S}\mathbf{F}_n$ as the subspace in $\mathbf{F}_n$ consisting of all

$$f(z) = \sum c_{j_1,\dots,j_n} z^{j_1} \dots z^{j_n}$$

such that for each N

$$\sup_j |c_{j_1,\dots,j_n}| \prod_k j_k!\, {j_k}^N < \infty.$$

Theorem 3.11 *The subspace $\mathcal{S}\mathbf{F}_n$ is a common invariant domain for all Gaussian bounded operators and for all creation–annihilation operators.*

See [53], Subsect. 4.2.4 and Theorem 5.1.5.

3.10 Details: The Olshanski Semigroup $\Gamma\mathrm{Sp}(2n, \mathbb{R})$

The *Olshanski semigroup* $\Gamma\mathrm{Sp}(2n, \mathbb{R})$ is defined as the subsemigroup in $\mathrm{Sp}(2n, \mathbb{C})$ consisting of complex matrices g satisfying the condition

$$g \begin{pmatrix} -1 & 0 \\ 0 & 1 \end{pmatrix} g^* - \begin{pmatrix} -1 & 0 \\ 0 & 1 \end{pmatrix} \leqslant 0,$$

where $g^* = \overline{g}^t$ denotes the adjoint matrix (see [55]).

Equivalently, $g \in \Gamma\mathrm{Sp}(2n, \mathbb{R})$ if and only if

$$M(ug, ug) \leqslant M(u, u) \quad \text{for all } u \in \mathbb{C}^{2n}.$$

The Olshanski semigroup is a subsemigroup in the semigroup of endomorphisms of the object $\mathcal{W}_{2n}$. For details, see [53, Sect. 2.7, 3.5].

4 Gauss–Berezin Integrals

Here we discuss super-analogues of Gaussian integrals. Actually, the final formulas are not used, but their structure is important for us.

Apparently, these integrals are evaluated somewhere, but I do not know a reference. Calculations in the fermionic case are contained in [59].

4.1 Phantom Algebra

Phantom generators $\mathfrak{a}_1, \mathfrak{a}_2, \ldots$ are anticommuting variables,

$$\mathfrak{a}_k \mathfrak{a}_l = -\mathfrak{a}_l \mathfrak{a}_k, \qquad \mathfrak{a}_j^2 = 0.$$

We define a *phantom algebra* $\mathcal{A}$ as the algebra of polynomials in the variables $\mathfrak{a}_j$. *For the sake of simplicity, we assume that the number of variables is infinite*. We also call elements of $\mathcal{A}$ *phantom constants*.

The phantom algebra has a natural $\mathbb{Z}$-grading by degree of monomials,

$$\mathcal{A} = \oplus_{j=0}^{\infty} \mathcal{A}_j.$$

Therefore, $\mathcal{A}$ admits a $\mathbb{Z}_2$-grading, namely,

$$\mathcal{A}_{\text{even}} := \oplus \mathcal{A}_{2j}, \qquad \mathcal{A}_{\text{odd}} := \oplus \mathcal{A}_{2j+1}.$$

We define the automorphism $\mu \mapsto \mu^\sigma$ of $\mathcal{A}$ by the rule

$$\mu^\sigma = \begin{cases} \mu & \text{if } \mu \text{ is even,} \\ -\mu & \text{if } \mu \text{ is odd} \end{cases} \tag{37}$$

(equivalently, $\mathfrak{a}_j^\sigma = -\mathfrak{a}_j$).

The algebra $\mathcal{A}$ is *supercommutative* in the following sense:

$$\mu \in \mathcal{A},\ \nu \in \mathcal{A}_{\text{even}} \quad \Longrightarrow \quad \mu\nu = \nu\mu,$$
$$\mu \in \mathcal{A}_{\text{odd}},\ \nu \in \mathcal{A}_{\text{odd}} \quad \Longrightarrow \quad \mu\nu = -\nu\mu.$$

Also,

$$\mu \in \mathcal{A},\ \nu \in \mathcal{A}_{\text{odd}} \Longrightarrow \quad \nu\mu = \mu^{\sigma}\nu. \tag{38}$$

Next, represent $\mu \in \mathcal{A}$ as $\mu = \sum_{j \geqslant 0} \mu_j$, where $\mu_j \in \mathcal{A}_j$. We define the map

$$\pi_{\downarrow} : \mathcal{A} \to \mathbb{C}$$

by

$$\pi_{\downarrow}(\mu) = \pi_{\downarrow}\Big(\sum\nolimits_{j \geqslant 0} \mu_j\Big) := \mu_0 \in \mathbb{C}.$$

Evidently,

$$\pi_{\downarrow}(\mu_1\mu_2) = \pi_{\downarrow}(\mu_1)\,\pi_{\downarrow}(\mu_2), \qquad \pi_{\downarrow}(\mu_1 + \mu_2) = \pi_{\downarrow}(\mu_1) + \pi_{\downarrow}(\mu_2).$$

Take $\varphi \in \mathcal{A}$ such that $\pi_{\downarrow}(\varphi) = 0$. Then $\varphi^N = 0$ for sufficiently large N. Therefore,

$$(1+\varphi)^{-1} := \sum_{n \geqslant 0} (-\varphi)^n, \tag{39}$$

actually, the sum is finite. In particular, if $\pi_{\downarrow}(\mu) \neq 0$, then μ is invertible.

4.2 A Technical Comment

The aim of this chapter is specific construction, and we use minimal vocabulary necessary for our aims. We regard supergroups $\mathrm{GL}(p|q)$ and $\mathrm{OSp}(2p|2q)$ as groups of matrices over the algebra $\mathcal{A}$. Representations of supergroups are defined in modules over $\mathcal{A}$.

The more common point of view[20] is to consider supergroups as functors from the category of supercommutative algebras to the category of groups. Our

[20] On comparison and criticism of different definitions of super-objects, see, e.g., [44, 60]. Our approach follows DeWitt's book [18]; we use only linear algebra and integral operators (without analysis on manifolds). Wider generality discussed in this subsection is similar to Berezin–Kats [9].

constructions require a restriction of a class of supercommutative algebras.[21] Precisely, algebra $\mathcal{A}$ can be replaced[22] by an arbitrary finitely or countably generated supercommutative algebra $\mathcal{A}'$ over $\mathbb{C}$ satisfying the following properties:

– There is a non-zero homomorphism $\pi_\downarrow : \mathcal{A}' \to \mathbb{C}$, denote $I := \ker \pi_\downarrow$;
– Any finitely generated subalgebra in I is nilpotent.

A pass to algebras equipped with the Krull topology makes situation more flexible. We can consider supercommutative algebras $\mathcal{A}''$ satisfying the following properties:

– There is a non-zero homomorphism $\pi_\downarrow : \mathcal{A}'' \to \mathbb{C}$; denote $I := \ker \pi_\downarrow$.
– I is finitely or countably generated, and $\cap_n I^n = 0$.
– $\mathcal{A}''$ is complete with respect to I-adic topology (Krull topology; see [19, Subsect. 7.7]), i.e., for any sequence $a_n \in \mathcal{A}''/I^n$ such that natural maps $\mathcal{A}''/I^n \to \mathcal{A}''/I^{n-1}$ send a_n to a_{n-1}, there is $a \in \mathcal{A}''$ whose image in each $\mathcal{A}''/I^n$ is a_n.

All our constructions below are functorial with respect to algebras of such types.

Below we use terms *supergroups* and *super-Grassmannians* for objects defined over the algebra $\mathcal{A}$, keeping in mind that this can be translated into the functorial language.

4.3 Berezinian

Let $\begin{pmatrix} P & Q \\ R & T \end{pmatrix}$ be a block $(p+q) \times (p+q)$-matrix; P and T be composed of even phantom constants; and Q and R be composed of odd phantom constants. Then the Berezinian (or Berezin determinant) is[23]

$$\operatorname{ber} \begin{pmatrix} P & Q \\ R & T \end{pmatrix} := (\det P)^{-1} \cdot \det(T - QP^{-1}R).$$

We note that P and $T - QP^{-1}R$ are composed of elements of the commutative algebra $\mathcal{A}_{\text{even}}$; therefore, their determinants are well defined. The Berezinian satisfies the multiplicative property of the usual determinant

$$\operatorname{ber}(A)\operatorname{ber}(B) = \operatorname{ber}(AB).$$

[21] We need exponentials of even elements (42), $\mathbb{C}$-valued integrals (41) and inverses (39), (44). We also must justify the calculation in Theorem 4.3 and the proof of Lemma 7.1.

[22] With some modifications in proofs.

[23] The usual determinant of a block complex matrix $\begin{pmatrix} P & Q \\ R & T \end{pmatrix}$ is $\det P \cdot \det(T - QP^{-1}R)$.

4.4 Functions

We consider three types of variables:

- Real or complex (bosonic) variables, which we denote by x_i and y_j if they are real and z_i and u_j if they are complex
- Grassmann (fermionic) variables, which we denote by ξ_i, η_j, or $\overline{\eta}_j$
- Phantom generators $\mathfrak{a}_j$ as above

Bosonic variables x_l commute with fermionic variables ξ_j and phantom constants $\mu \in \mathcal{A}$. We also assume that fermionic variables ξ_j and phantom generators $\mathfrak{a}_l$ anticommute,

$$\xi_j \mathfrak{a}_l = -\mathfrak{a}_l \xi_j.$$

Fix a collection of bosonic variables $x_1, \dots, x_p$ and a collection of fermionic variables $\xi_1, \dots, \xi_q$. A *function* is a sum of the form

$$f(x, \xi) := \sum_{m \geqslant 0} \sum_{0 < i_1 < \dots < i_k \leqslant q;\ j_1 < \dots < j_m} h_{i_1, \dots, i_k; j_1, \dots, j_m}(x_1, \dots, x_p) \times \\ \times \mathfrak{a}_{j_1} \dots \mathfrak{a}_{j_m} \xi_{i_1} \dots \xi_{i_k}, \qquad (40)$$

where h are smooth functions of $x \in \mathbb{R}^p$. We also write such expressions in the form

$$f(x, \xi) = \sum_{I,J} h_{I,J}(x) \mathfrak{a}^J \xi^I$$

keeping in the mind that I ranges in collections $0 < i_1 < \dots < i_k \leqslant q$ and J in collections $j_1 < \dots < j_m$.

We say that a function f is *even* (respectively *odd*) if it is an even expression in the total collection ξ_i, $\mathfrak{a}_k$. By $f^\sigma(x, \xi)$ we denote the function obtained from $f(x, \xi)$ by the substitution $\mathfrak{a}_j \mapsto -\mathfrak{a}_j$ for all j.

Remark Formally, the fermionic variables and the phantom constants have equal rights in our definition. However, below, their roles are different: ξ_j serve as variables, while elements of $\mathcal{A}$ serve as constants (see (43)). ⊠

Remark There are three ways to understand expressions (40). We consider them as finite sums, but it is possible to consider them as arbitrary formal series in variables $\mathfrak{a}_j$. We also can consider formal series such that each m-th summand depends only on finitely many of $\mathfrak{a}_j$ (considerations below survive in these cases after minor modifications). ⊠

For a given f, we define the function

$$\pi_{\downarrow}(f) := \sum_{I} h_{I,\varnothing}(x)\, \xi^{I} \in C^{\infty}(\mathbb{R}^{p}) \otimes \Lambda_{q}.$$

4.5 Integral

Now, we define the symbols

$$\int f(x,\xi)\, dx \qquad \int f(x,\xi)\, d\xi, \qquad \int f(x,\xi)\, dx\, d\xi.$$

The integration with respect to x is the usual termwise integration in (40),

$$\int_{\mathbb{R}^{p}} f(x,\xi)\, dx := \sum_{I,J} \Big(\int h_{I,J}(x)\, dx \Big) \mathfrak{a}^{J} \xi^{I}. \tag{41}$$

The integration with respect to ξ is the usual termwise Berezin integral,

$$\int f(x,\xi)\, d\xi := \sum_{J} h_{LJ}(x) \mathfrak{a}^{J}, \qquad \text{where } L = \{1, 2, \ldots, q\}.$$

4.6 Exponential

Let $f(x,\xi)$ be an *even* expression in ξ, $\mathfrak{a}$, i.e., $f(x,\xi) = f(x,-\xi)^{\sigma}$. We define its exponential as usual, and it satisfies the usual properties. Namely,

$$\exp\{f(x,\xi)\} := \sum_{n=0}^{\infty} \frac{1}{n!} f(x,\xi)^{n}. \tag{42}$$

Since f_1 and f_2 are even, we have $f_1 f_2 = f_2 f_1$. Therefore, the identity

$$\exp\{f_1 + f_2\} = \exp\{f_1\} \exp\{f_2\}$$

holds.

Observation 4.1 *The series* (42) *converges.*

Indeed,

$$\exp\{f(x,\xi)\} = \exp\{h_{\varnothing\varnothing}(x)\} \prod_{(I,J)\neq(\varnothing,\varnothing)} \exp\{h_{I,J}(x)\mathfrak{a}^I\xi^J\} =$$
$$= \exp\{h_{\varnothing\varnothing}(x)\} \prod_{(I,J)\neq(\varnothing,\varnothing)} (1 + h_{I,J}(x)\mathfrak{a}^I\xi^J).$$

Opening brackets, we get a polynomial in $\mathfrak{a}_j$, ξ_k.

4.7 Gauss–Berezin Integrals: A Special Case

Take p real variables x_i and q Grassmann variables ξ_j. Consider the expression

$$I = \iint \exp\left\{\frac{1}{2}\begin{pmatrix}x & \xi\end{pmatrix}\begin{pmatrix}A & B\\ -B^t & C\end{pmatrix}\begin{pmatrix}x^t\\ \xi^t\end{pmatrix}\right\} dx\, d\xi =$$
$$= \iint \exp\left\{\frac{1}{2}\sum_{ij} a_{ij}x_i x_j + \sum_{ik} b_{ik}x_i\xi_k + \frac{1}{2}\sum_{kl} c_{kl}\xi_k\xi_l\right\} dx\, d\xi. \quad (43)$$

The notation ^{t} denotes the transpose as above; also, $\begin{pmatrix}x & \xi\end{pmatrix}$ is the row-matrix

$$\begin{pmatrix}x & \xi\end{pmatrix} = \begin{pmatrix}x_1 & \dots & x_p & \xi_1 & \dots & \xi_q\end{pmatrix}.$$

Matrices A and C are composed of even phantom constants; A is symmetric, C is skew-symmetric, and B is a matrix composed of odd phantom constants.[24]

Observation 4.2 *The integral converges if and only if the matrix* $\mathrm{Re}\,\pi_\downarrow(A)$ *is negative definite.*

Indeed, the integrand $\exp\{\dots\}$ is a finite sum of the form

$$\exp\left\{\frac{1}{2}\sum_{ij}\pi_\downarrow(a_{ij})x_i x_j\right\} \sum_{i_1<\dots<i_k}\ \sum_{j_1<\dots<j_l} P_{i_1,\dots,i_k;\, j_1,\dots,j_l}(x)\xi_{i_1}\dots\xi_{i_k}\mathfrak{a}_{j_1}\dots\mathfrak{a}_{j_l},$$

where $P_{\dots}(x)$ are polynomials. Under the condition $\mathrm{Re}\,\pi_\downarrow(A) < 0$, a termwise integration is possible. □

[24] The argument of the exponential must be even in ξ, $\mathfrak{a}$. This imposes the constraints of parity for A, B, and C. The symmetry conditions for A, B, and C are the natural conditions for coefficients of a quadratic form in x, ξ.

4.8 Evaluation of the Gauss–Berezin Integral

Let us evaluate integral (43).

Theorem 4.3 *Let* $\operatorname{Re}\pi_\downarrow(A) < 0$. *Then*

$$I = \begin{cases} (2\pi)^{p/2} \det(-A)^{-1/2} \operatorname{Pfaff}(C + B^t A^{-1} B) & \text{if } q \text{ is even}, \\ 0, & \text{otherwise}. \end{cases} \tag{44}$$

Recall that q is the number of Grassmann variables.

Remark The matrix $C + B^t A^{-1} B$ is skew-symmetric and composed of even phantom constants. Therefore, the Pfaffian is well defined. ⊠

Remark Thus, for q, even our expression is a "hybrid" of a Pfaffian and a Berezinian,

$$I^{-2} = -(2\pi)^p \operatorname{ber}\begin{pmatrix} A & B \\ -B^t & C \end{pmatrix}.$$

Similar (but not precisely the same) "hybrid" appeared in [64] and [39]. ⊠

Proof of Theorem 4.3. First, we integrate with respect to x,

$$\exp\left\{\frac{1}{2}\xi C\xi^t\right\} \int_{\mathbb{R}^p} \exp\left\{\frac{1}{2}xAx^t + xB\xi^t\right\} dx =$$

$$= \exp\left\{\frac{1}{2}\xi C\xi^t\right\} \cdot \int_{\mathbb{R}^p} \exp\left\{\frac{1}{2}(x - \xi B^t A^{-1})A(x^t + A^{-1}B\xi^t)\right\}$$

$$\exp\left\{\frac{1}{2}\xi B^t A^{-1} B\xi^t\right\} dx.$$

We substitute

$$y := x - \xi B^t A^{-1}, \tag{45}$$

get

$$\exp\left\{\frac{1}{2}\xi C\xi^t\right\} \cdot \exp\left\{\frac{1}{2}\xi B^t A^{-1} B\xi^t\right\} \int \exp\left\{\frac{1}{2}yAy^t\right\} dy,$$

and arrive at the usual Gaussian integral (13).

Integrating the result, we get

$$\det(-A)^{-1/2}(2\pi)^{p/2} \int \exp\left\{\frac{1}{2}\xi(C + B^t A^{-1} B)\xi^t\right\} d\xi,$$

and arrive at the Pfaffian.

We must justify the substitution (45). Let Φ be a function on $\mathbb{R}^p$ of Schwartz class, let ν be an even expression in $\mathfrak{a}$, ξ, assume that the constant term of ν is 0. Then

$$\int_{\mathbb{R}^p} \Phi(x+\nu)\, dx = \int_{\mathbb{R}^p} \Phi(x)\, dx.$$

Indeed,

$$\Phi(x+\nu) := \sum_{j=0}^{\infty} \frac{1}{j!} \nu^j \frac{d^j}{dx^j} \Phi(x).$$

Actually, the summation is finite. A termwise integration with respect to x gives zero for all $j \neq 0$. □

4.9 Grassmann Gaussian Integral

Observation 4.4 *Let D be a* complex *skew-symmetric matrix of size N, let ξ_k, ζ_k be Grassmann variables. Then the integral*

$$\int \exp\left\{\frac{1}{2}\xi D\xi^t + \xi\zeta^t\right\} d\xi$$

can be represented in the form

$$s \cdot \prod_{j=1}^{m} (\zeta h_j^t) \cdot \exp\left\{\frac{1}{2}\zeta Q\zeta^t\right\},$$

where $s \in \mathbb{C}$, Q is a skew-symmetric matrix, and h_j are row-matrices, $m \leqslant N$, $N-m$ is even.

Indeed, one can find a linear substitution $\xi = \eta S$ such that[25]

$$\xi D\xi^t = \sum_{j=1}^{\gamma} \eta_{2j-1}\eta_{2j1}.$$

Then the integral can be reduced to

$$\det S \int \exp\left\{\sum_{j=1}^{\gamma} \eta_{2j-1}\eta_{2j} + \sum_{k=1}^{N} \eta_k \nu_k\right\} d\eta,$$

[25] In other words, one can reduce a skew-symmetric matrix over $\mathbb{C}$ to a canonical form.

where ν_j are certain linear expressions in ζ_l. So we get

$$\det S \cdot \prod_{j=1}^{\gamma} \int \exp\left\{\eta_{2j-1}\eta_{2j} + \eta_{2j-1}\nu_{2j-1} + \eta_{2j}\nu_{2j}\right\} d\eta_{2j-1}\, d\eta_{2j} \times$$

$$\times \int \exp\Big\{ \sum_{j=2\gamma+1}^{N} \eta_j\nu_j \Big\}\, d\eta_{2\gamma+1} \ldots d\eta_N =$$

$$= \pm \det S \cdot \exp\Big\{ -\sum_{j=1}^{\gamma} \nu_{2j}\nu_{2j+1} \Big\} \prod_{k=2\gamma+1}^{N} \nu_k.$$

Recall that ν_j are certain linear expressions[26] in ζ_l. □.

4.10 More General Gauss–Berezin Integrals

Consider an expression

$$J = \iint \exp\left\{\frac{1}{2}\begin{pmatrix} x & \xi \end{pmatrix}\begin{pmatrix} A & B \\ -B^t & C \end{pmatrix}\begin{pmatrix} x^t \\ \xi^t \end{pmatrix} + xh^t + \xi g^t\right\} dx\, d\xi =$$

$$= \iint \exp\Big\{\frac{1}{2}\sum_{ij} a_{ij}x_i x_j + \sum_{ik} b_{ik}x_i\xi_k + \frac{1}{2}\sum_{kl} c_{kl}\xi_k\xi_l +$$

$$+ \sum_j h_j x_j - \sum_k g_k\xi_k \Big\}\, dx\, d\xi, \qquad (46)$$

here A, B, and C are as above and h^t and g^t are column-vectors $h_j \in \mathcal{A}_{\text{even}}$ and $g_k \in \mathcal{A}_{\text{odd}}$.

We propose two ways to evaluate this integral.

4.11 The First Way to Evaluate Super-Gaussian Integral

Substituting

$$\begin{pmatrix} y & \eta \end{pmatrix} = \begin{pmatrix} x & \xi \end{pmatrix} + \begin{pmatrix} h & g \end{pmatrix}\begin{pmatrix} A & B \\ -B^t & C \end{pmatrix}^{-1},$$

[26] A product of functions ν_m is canonically defined up to a constant factor; equivalently, a linear span of functions ν_m is canonically defined (this sentence is a rephrasing of the Plücker embedding of a Grassmannian into an exterior algebra).

we get

$$\exp\left\{\frac{1}{2}(h\ g)\begin{pmatrix} A & B \\ -B^t & C\end{pmatrix}^{-1}\begin{pmatrix} h^t \\ g^t\end{pmatrix}\right\}\cdot\iint \exp\left\{\frac{1}{2}(y\ \eta)\begin{pmatrix} A & B \\ -B^t & C\end{pmatrix}\begin{pmatrix} y^t \\ \eta^t\end{pmatrix}\right\}dy\,d\eta$$

and arrive at Gauss–Berezin integral (43) evaluated above.

This way is not perfect, because it uses an inversion of a matrix $\begin{pmatrix} A & B \\ -B^t & C\end{pmatrix}$.

Observation 4.5 *A matrix* $\begin{pmatrix} A & B \\ -B^t & C\end{pmatrix}$ *is invertible if and only if* A *and* C *are invertible.*

The necessity is evident; to prove the sufficiency, we note that the matrix

$$T := \begin{pmatrix} A^{-1} & 0 \\ 0 & C^{-1}\end{pmatrix}\begin{pmatrix} A & B \\ -B^t & C\end{pmatrix} - \begin{pmatrix} 1 & 0 \\ 0 & 1\end{pmatrix}$$

is composed of nilpotent elements of $\mathcal{A}$. We write out $(1+T)^{-1} = 1-T+T^2-\dots$, and therefore, our initial matrix $\begin{pmatrix} A & 0 \\ 0 & C\end{pmatrix}(1+T)$ is invertible. □

Matrix A is invertible, because $\operatorname{Re} A < 0$.
But matrix C is skew-symmetric.

- If q is even, then a $q\times q$ skew-symmetric matrix C in general position is invertible. For noninvertible C, we have a chance to remove uncertainty. This way leads to an expression of the form (48) obtained below.
- If q is odd, then C is non-invertible; our approach is not suitable.

4.12 The Second Way to Evaluate Gauss–Berezin Integrals

First, we integrate with respect to x,

$$\exp\left\{\frac{1}{2}\xi C\xi^t + \xi g^t\right\}\int_{\mathbb{R}^p}\exp\left\{\frac{1}{2}xAx^t + xB\xi^t + xh^t\right\}dx =$$
$$= \exp\left\{\frac{1}{2}\xi C\xi^t + \xi g^t\right\}\exp\left\{\frac{1}{2}(h-\xi B^tA^{-1})A(h^t + A^{-1}B\xi^t)\right\}\times$$
$$\times\int_{\mathbb{R}^p}\exp\left\{\frac{1}{2}(x+h-\xi B^t)A(x^t+h^t+B\xi^t)\right\}dx.$$

Substituting $y = x + h - \xi B^t A^{-1}$ and integrating with respect to y, we get

$$(2\pi)^{p/2} \det(-A)^{-1/2} \exp\left\{\frac{1}{2}(h - \xi B^t A^{-1})A(h^t + A^{-1}B\xi^t) + \frac{1}{2}\xi C\xi^t + \xi g^t\right\}.$$

Next, we must integrate with respect to ξ; our integral has a form

$$\int \exp\left\{\frac{1}{2}\xi D\xi^t + \xi r^t\right\} d\xi, \tag{47}$$

where a matrix D is composed of even phantom constants and a vector r is odd.

If D is invertible, we shift the argument again $\eta^t := \xi^t + D^{-1}r^t$ and get

$$\exp\left\{\frac{1}{2}rD^{-1}r^t\right\} \int \exp\left\{\frac{1}{2}\eta D\eta^t\right\} d\eta,$$

the last integral is a Pfaffian. This way is equivalent to the approach discussed in the previous subsection.

Now, consider an arbitrary D. The calculation of Sect. 4.9 does not survive.[27] However, we can write (47) explicitly as follows: for any subset

$$I : i_1 < \cdots < i_{2k}$$

in $\{1, \ldots, q\}$, we consider the complementary subset

$$J : j_1 < \cdots < j_{q-2k}.$$

Define the constant $\sigma(I) = \pm 1$ as follows:

$$(\xi_{i_1}\xi_{i_2}\ldots\xi_{i_{2k}})(\xi_{j_1}\xi_{j_2}\ldots\xi_{j_{q-2k}}) = \sigma(I)\,\xi_1\xi_2\ldots\xi_q.$$

Evidently,

$$\int \exp\left\{\frac{1}{2}\xi D\xi^t + \xi r^t\right\} d\xi =$$

$$= \sum_I \sigma(I)\mathrm{Pfaff}\begin{pmatrix} 0 & d_{i_1 i_2} & \ldots & d_{i_1 i_{2k}} \\ d_{i_2 i_1} & 0 & \ldots & d_{i_2 i_{2k}} \\ \vdots & \vdots & \ddots & \vdots \\ d_{i_{2k} i_1} & d_{i_2 i_{2k}} & \ldots & 0 \end{pmatrix} r_{j_1}\ldots r_{j_{q-2k}}. \tag{48}$$

Recall that $d_{pq} \in \mathcal{A}_{\mathrm{even}}$ and $r \in \mathcal{A}_{\mathrm{odd}}$.

[27] Let D be a skew-symmetric matrix over $\mathcal{A}_{\mathrm{even}}$. If $\pi_\downarrow(D)$ is degenerate, then we cannot reduce D to a normal form.

5 Gauss–Berezin Integral Operators

Here we define super hybrids of Gaussian operators and Berezin operators.

5.1 *Fock–Berezin Spaces*

Fix $p, q = 0, 1, 2, \dots$. Let $z_1, \dots, z_p$ be complex variables, $\xi_1, \dots, \xi_q$ be Grassmann variables. We consider expressions

$$f(z, \xi) =: \sum_{I,J} f_{I,J}(z)\alpha^J \xi^I,$$

where $f_{I,J}$ are entire functions in z and the summation is finite. We define the map $f \mapsto \pi_\downarrow(f)$ as above.

We define the *Fock–Berezin* space $\mathcal{S}\mathbf{F}_{p,q}(\mathcal{A})$ as the space of all functions $f(z, \xi)$ satisfying the condition: *for each I, J, the function $f_{I,J}(z)$ is in the Schwartz–Fock space* $\mathcal{S}\mathbf{F}_p$ (see Sect. 3.9). We say that a sequence $f^{(k)} \in \mathcal{S}\mathbf{F}_{p,q}(\mathcal{A})$ converges [28] to f if

- For all but a finite number of J, all $f_{I,J}^{(k)}$ are zero
- For each I, J, we have a convergence $f_{I,J}^{(k)}(z) \to f_{I,J}(z)$ in $\mathcal{S}\mathbf{F}_p$

Remark We can assume that all the $f_{I,J}$ are in the Hilbert–Fock space $\mathbf{F}_p$. But Gauss–Berezin operators defined below can be unbounded in this space; therefore, this point of view requires descriptions of domains of operators and examination of products of operators. Our definition admits some variations (we chose an open dense subset in $\mathbf{F}_p$, and our choice is volitional). ⊠

5.2 *Another Form of the Gauss–Berezin Integral*

Consider the integral

$$\int \exp\left\{\frac{1}{2}\begin{pmatrix} z & \xi \end{pmatrix}\begin{pmatrix} A & B \\ -B^t & C \end{pmatrix}\begin{pmatrix} z^t \\ \xi^t \end{pmatrix} + z\alpha^t + \xi\beta^t\right\}\times$$

$$\times \exp\left\{\frac{1}{2}\begin{pmatrix} \overline{z} & \overline{\xi} \end{pmatrix}\begin{pmatrix} K & L \\ -L^t & M \end{pmatrix}\begin{pmatrix} \overline{z} \\ \overline{\xi} \end{pmatrix} + \overline{z}\varkappa + \overline{\xi}\lambda^t\right\} \cdot e^{-z\overline{z}^t - \xi\overline{\xi}^t}\, dz\, d\overline{z}\, d\xi\, d\overline{\xi}, \qquad (49)$$

[28] This convergence corresponds to a topology of inductive limit.

where two matrices $\begin{pmatrix} A & B \\ -B^t & C \end{pmatrix}$ and $\begin{pmatrix} K & L \\ -L^t & M \end{pmatrix}$ have the same structure as in Sect. 4.7, row-vectors α and $\varkappa$ are even, and vectors β and λ are odd.

Since $\mathbb{C}^n \simeq \mathbb{R}^{2n}$, this integral is a special case of the Gauss–Berezin integral. We get

$$\text{const} \cdot \exp \left\{ \frac{1}{2} \begin{pmatrix} \alpha & \beta & \varkappa & \lambda \end{pmatrix} \begin{pmatrix} -A & -B & 1 & 0 \\ B^t & -C & 0 & 1 \\ 1 & 0 & -K & -L \\ 0 & -1 & L^t & -M \end{pmatrix}^{-1} \begin{pmatrix} \alpha^t \\ \beta^t \\ \varkappa^t \\ \lambda^t \end{pmatrix} \right\}, \tag{50}$$

where the scalar factor is a hybrid of the Pfaffian and the Berezinian mentioned in Sect. 4.8.

5.3 *Integral Operators*

We write operators $\mathcal{S}\mathbf{F}_{p,q}(\mathcal{A}) \to \mathcal{S}\mathbf{F}_{r,s}(\mathcal{A})$ as

$$Af(z,\xi) = \int K(z,\xi;\overline{u},\overline{\eta})\, f(u,\eta)\, e^{-z\overline{z}^t - \eta\overline{\eta}^t}\, du\, d\overline{u}\, d\overline{\eta}\, d\eta. \tag{51}$$

5.4 *Linear and Antilinear Operators*

We say that an operator $A : \mathcal{S}\mathbf{F}_{p,q}(\mathcal{A}) \to \mathcal{S}\mathbf{F}_{r,s}(\mathcal{A})$ is *linear* if

$$A(f_1 + f_2) = Af_1 + Af_2, \qquad A(\lambda f) = \lambda A f, \text{ where } \lambda \text{ is a phantom constant,}$$

and *antilinear* if

$$A(f_1 + f_2) = Af_1 + Af_2, \qquad A(\lambda f) = \lambda^\sigma A f, \text{ where } \lambda \text{ is a phantom constant,}$$

the automorphism $\lambda \mapsto \lambda^\sigma$. Clearly, operators

$$Af(z,\xi) = \xi_j f(z,\xi), \qquad Bf(z,\xi) = \frac{\partial}{\partial \xi_j} f(z,\xi), \qquad Cf(z,\xi) = \mathfrak{a}_j f(z,\xi)$$

are antilinear.

An integral operator (51) is linear if the kernel $K(z, \xi, \overline{u}, \overline{\eta})$ is an even function in the total collection of all Grassmann variables ξ, $\overline{\eta}$, and $\mathfrak{a}$, i.e.,

$$K(z, \xi, \overline{u}, \eta) = K(z, -\xi, \overline{u}, -\overline{\eta})^{\sigma}.$$

An operator (51) is antilinear if and only if the function K is odd.

Below we meet only linear and antilinear operators.

We define also the (antilinear) operator S of σ-conjugation,

$$\mathsf{S}\, f(z, \xi) = f^{\sigma}(z, \xi). \tag{52}$$

Evidently,

$$\mathsf{S}^2 f = f.$$

5.5 Gauss–Berezin Vectors in the Narrow Sense

A Gauss–Berezin vector (in the narrow sense) is a vector of the form

$$\mathbf{b}\begin{bmatrix} A & B \\ -B^t & C \end{bmatrix} = \lambda \exp\left\{\frac{1}{2}\begin{pmatrix} z & \xi \end{pmatrix}\begin{pmatrix} A & B \\ -B^t & C \end{pmatrix}\begin{pmatrix} z^t \\ \xi^t \end{pmatrix}\right\}, \tag{53}$$

where A, B, and C are as above (see Sect. 4.7).

Observation 5.1 $\mathbf{b}[\cdot] \in \mathcal{S}\mathbf{F}_{p,q}(\mathcal{A})$ *if and only if* $\|\pi_{\downarrow}(A)\| < 1$.

5.6 Gauss–Berezin Operators in the Narrow Sense

A *Gauss–Berezin integral operator in the narrow sense* is an integral operator

$$\mathcal{S}\mathbf{F}_{p,q}(\mathcal{A}) \to \mathcal{S}\mathbf{F}_{r,s}(\mathcal{A}),$$

whose kernel as a function in $z_1, \dots, z_r, \overline{u}_1, \dots, u_p, \xi_1, \dots, \xi_s, \overline{\eta}_1, \overline{\eta}_q$ is a Gauss–Berezin vector. Precisely, a Gauss–Berezin operator has the form

$$\mathfrak{B} f(z, \xi) = \lambda \cdot \iint \exp\left\{\frac{1}{2}\begin{pmatrix} z & \xi & \overline{u} & \overline{\eta} \end{pmatrix}\begin{pmatrix} A_{11} & A_{12} & A_{13} & A_{14} \\ A_{21} & A_{22} & A_{23} & A_{24} \\ A_{31} & A_{32} & A_{33} & A_{34} \\ A_{41} & A_{42} & A_{43} & A_{44} \end{pmatrix}\begin{pmatrix} z^t \\ \xi^t \\ \overline{u}^t \\ \overline{\eta}^t \end{pmatrix}\right\} f(u, \eta) \times$$

$$\times\, e^{-\eta\overline{\eta}^t - u\overline{u}^t}\, du\, d\overline{u}\, d\eta\, d\overline{\eta}, \tag{54}$$

where λ is an even phantom constant and A_{ij} is composed of even phantom constants if $(i + j)$ is even; otherwise, A_{ij} is composed of odd phantom constants. They also satisfy the natural symmetry conditions for a matrix of a quadratic form in the variables z, ξ, $\overline{u}$, and $\overline{\eta}$.

Remark On the other hand, a Gauss–Berezin vector can be regarded as a Gauss–Berezin operator $\mathcal{S}\mathbf{F}_{0,0}(\mathcal{A}) \to \mathcal{S}\mathbf{F}_{p,q}(\mathcal{A})$. ⊠

For Gauss–Berezin operators

$$\mathfrak{B}_1 : \mathcal{S}\mathbf{F}_{p,q}(\mathcal{A}) \to \mathcal{S}\mathbf{F}_{p',q'}(\mathcal{A}), \qquad \mathfrak{B}_2 : \mathcal{S}\mathbf{F}_{p',q'}(\mathcal{A}) \to \mathcal{S}\mathbf{F}_{p'',q''}(\mathcal{A})$$

evaluation of their product is reduced to the Gauss–Berezin integral (49). For operators in general position, we can apply formula (50). Evidently, in this case, the product is a Gauss–Berezin operator again. However, our final Theorem 9.4 avoids this calculation.

Also, considerations of Sect. 2 suggest an extension of the definition of Gauss–Berezin operators.

5.7 General Gauss–Berezin Operators

As above, we define first-order differential operators

$$\mathfrak{D}[\xi_j]f := \Big(\xi_j + \frac{\partial}{\partial \xi_j}\Big) f.$$

If a function f is independent of ξ_j, then

$$\mathfrak{D}[\xi_j]f = \xi_j f, \qquad \mathfrak{D}[\xi_j]\xi_j f = f.$$

Evidently,

$$\mathfrak{D}[\xi_j]^2 = 1, \qquad \mathfrak{D}[\xi_i]\,\mathfrak{D}[\xi_j] = -\mathfrak{D}[\xi_j]\,\mathfrak{D}[\xi_i], \quad , i \neq j.$$

Operators $\mathfrak{D}[\xi_j]$ are antilinear.

A *Gauss–Berezin operator* $\mathcal{S}\mathbf{F}_{p,q}(\mathcal{A}) \to \mathcal{S}\mathbf{F}_{r,s}(\mathcal{A})$ is an operator of the form

$$\mathfrak{C} = \lambda \mathfrak{D}[\xi_{i_1}] \dots \mathfrak{D}[\xi_{i_k}]\,\mathfrak{B}\,\mathfrak{D}[\eta_{m_1}] \dots \mathfrak{D}[\eta_{m_l}] \cdot \mathsf{S}^{k+l}, \tag{55}$$

where operator S is given by (52) and

- $\mathfrak{B}$ is a Gauss –Berezin operator in the narrow sense.
- $i_1 < i_2 < \dots < i_k$, $m_1 < m_2 < \dots < m_l$, and $k, l \geqslant 0$.
- λ is an even invertible phantom constant.

Note that a Gauss–Berezin operator is linear.

Remark We define the set of Gauss–Berezin operators as a union of 2^{p+q} sets. These sets are not disjoint. Actually, we get a supermanifold consisting of two connected components (according to the parity of $k+l$). Each set (55) is open and dense in the corresponding component. This will become obvious below. ⊠

5.8 *Operators $\pi_\downarrow(\mathfrak{B})$ and Boundedness of Gauss–Berezin Operators*

Let $K(z,\xi,\overline{u},\overline{\eta})$ be the kernel of a Gauss–Berezin operator. Then the formula

$$\mathfrak{C}f(z,\xi)=\int\pi_\downarrow\Big(K(z,\xi;\overline{u},\overline{\eta})\Big)\,f(u,\eta)\,e^{-z\overline{z}^t-\eta\overline{\eta}^t}\,du\,d\overline{u}\,d\overline{\eta}\,d\eta$$

determines an integral operator

$$\pi_\downarrow(\mathfrak{C}):\mathbf{F}_p\otimes\Lambda_q\to\mathbf{F}_r\otimes\Lambda_s.$$

Evidently, this operator is a tensor product of a Gaussian operator

$$\pi_\downarrow^+(\mathfrak{C}):\mathbf{F}_p\to\mathbf{F}_r$$

and a Berezin operator

$$\pi_\downarrow^-(\mathfrak{C}):\Lambda_q\to\Lambda_s.$$

For instance, for an operator $\mathfrak{B}$ given by the standard formula (54), we get the Gaussian operator

$$\pi_\downarrow^+(\mathfrak{B})f(z)=\int_{\mathbb{C}^n}\exp\left\{\frac{1}{2}\begin{pmatrix}z&\overline{u}\end{pmatrix}\left[\pi_\downarrow\begin{pmatrix}A_{11}&A_{13}\\A_{31}&A_{33}\end{pmatrix}\right]\begin{pmatrix}z^t\\\overline{u}^t\end{pmatrix}\right\}f(z)\,e^{-z\overline{z}^t}\,dz\,d\overline{z}$$

and the Berezin operator

$$\pi_\downarrow^-(\mathfrak{B})g(\xi)=\int\exp\left\{\frac{1}{2}\begin{pmatrix}\xi&\overline{\eta}\end{pmatrix}\left[\pi_\downarrow\begin{pmatrix}A_{22}&A_{24}\\A_{42}&A_{44}\end{pmatrix}\right]\begin{pmatrix}\xi^t\\\overline{\eta}^t\end{pmatrix}\right\}g(\xi)\,e^{-\xi\overline{\xi}^t}\,d\xi\,d\overline{\xi}.$$

Next,

$$\pi_\downarrow^\pm(\mathfrak{C}_1\mathfrak{C}_2)=\pi_\downarrow^\pm(\mathfrak{C}_1)\,\pi_\downarrow^\pm(\mathfrak{C}_2).\tag{56}$$

Observation 5.2 *The Gauss–Berezin operator* (54) *is bounded in the sense of Fock–Berezin spaces if and only if the operator* $\pi^+_\downarrow(\mathfrak{B})$ *is bounded (i.e., satisfies conditions of Theorem* 3.4).

Proof Only statement "if" requires a proof. A kernel of a Gauss–Berezin operator has the form

$$\sum_{I,J,K} \exp\left\{\frac{1}{2}\begin{pmatrix} z & \overline{u}\end{pmatrix}\left[\pi_\downarrow\begin{pmatrix} A_{11} & A_{13}\\ A_{31} & A_{33}\end{pmatrix}\right]\begin{pmatrix} z^t\\ \overline{u}^t\end{pmatrix}\right\} P_{I,J,K}(z,\overline{u})\mathfrak{a}^I\xi^J\overline{\eta}^K,$$

where $P_{I,J,K}$ is a polynomial in z, $\overline{u}$. Denote the quadratic form in the exponential by $S(z)$, and decompose $P_{I,J,K}(z,\overline{u})$ as a sum of monomials. It is sufficient to show that the following operators are bounded as operators $\mathcal{S}\mathbf{F}_p \to \mathcal{S}\mathbf{F}_r$:

$$\int \prod z_j^{l_j} \exp\{S(z,\overline{u})\} \prod \overline{u}_i^{k_i} f(u)\, e^{-z\overline{z}}\, d\lambda(u).$$

Integrating by parts (see, e.g., [50], Subsect. V.3.5), we arrive at

$$\prod z_j^{l_j} \cdot \int \exp\{S(z,\overline{u})\} \cdot \prod \Big(\frac{\partial}{\partial u_i}\Big)^{k_i} f(u) \cdot e^{-z\overline{z}}\, d\lambda(u).$$

It remains to notice that partial differentiations and multiplications by linear functions are bounded operators in the Fock-Schwartz spaces (see Theorem 3.11). □

5.9 *Products of Gauss–Berezin Operators*

Theorem 5.3 *For each Gauss–Berezin operator*

$$\mathfrak{B}_1 : \mathcal{S}\mathbf{F}_{p,q}(\mathcal{A}) \to \mathcal{S}\mathbf{F}_{p',q'}(\mathcal{A}), \qquad \mathfrak{B}_2 : \mathcal{S}\mathbf{F}_{p',q'}(\mathcal{A}) \to \mathcal{S}\mathbf{F}_{p'',q''}(\mathcal{A}),$$

their product $\mathfrak{B}_2\mathfrak{B}_1$ *is either a Gauss–Berezin operator or*

$$\pi_\downarrow(\mathfrak{B}_2\mathfrak{B}_1 f) = 0 \qquad (57)$$

for all f.

Clearly, the condition (57) also is equivalent to $\pi^-_\downarrow(\mathfrak{B}_2\mathfrak{B}_1 f) = 0$.

For a proof, see Sect. 9. We also present an interpretation of the product in terms of linear relations.

Remark In case (57), the kernel of the product has the form (48), but it is not a Gauss–Berezin operator in our sense. Possibly, this requires to change our definitions. ⊠

5.10 General Gauss–Berezin Vectors

A Gauss–Berezin vector is a vector of the form

$$\mathfrak{D}[\xi_{i_1}] \dots \mathfrak{D}[\xi_{i_k}]\, \mathsf{S}^k \mathfrak{b},$$

where $\mathfrak{b}$ is a Gauss–Berezin vector in the narrow sense.

Remark We write S^k as in (55). However, omitting this factor does not change the definition. ⊠

6 Supergroups OSp(2p|2q)

Here we define a super-analogue of the groups $\mathrm{O}(2n, \mathbb{C})$ and $\mathrm{Sp}(2n, \mathbb{C})$. For a general exposition of supergroups and super-Grassmannians, see books [8, 42, 11, 12].

6.1 Modules $\mathcal{A}^{p|q}$

Let

$$\mathcal{A}^{p|q} := \mathcal{A}^p \oplus \mathcal{A}^q$$

be a direct sum of $(p+q)$ copies of $\mathcal{A}$. We regard elements of $\mathcal{A}^{p|q}$ as row-vectors

$$(v_1, \dots, v_p; w_1, \dots, w_q).$$

We define a structure of $\mathcal{A}$-bimodule on $\mathcal{A}^{p|q}$. The addition in $\mathcal{A}^{p|q}$ is natural. The left multiplication by $\lambda \in \mathcal{A}$ is also natural

$$\lambda \circ (v_1, \dots, v_p; w_1, \dots, w_q) := (\lambda v_1, \dots, \lambda v_p; \lambda w_1, \dots, \lambda w_q).$$

The right multiplications by $\varkappa \in \mathcal{A}$ is

$$(v_1, \dots, v_p; w_1, \dots, w_q) * \varkappa := (v_1 \varkappa, \dots, v_p \varkappa; w_1 \varkappa^\sigma, \dots, w_q \varkappa^\sigma),$$

where σ is the involution of $\mathcal{A}$ defined above.

We define the *even part* of $\mathcal{A}^{p|q}$ as $(\mathcal{A}_{\text{even}})^p \oplus (\mathcal{A}_{\text{odd}})^q$ and the *odd part* as $(\mathcal{A}_{\text{odd}})^p \oplus (\mathcal{A}_{\text{even}})^q$.

6.2 Matrices

Denote by $\mathrm{Mat}(p|q;\mathcal{A})$ the space of $(p+q)\times(p+q)$ matrices over $\mathcal{A}$; we represent such matrices in the block form

$$Q = \begin{pmatrix} A & B \\ C & D \end{pmatrix}.$$

We say that a *matrix Q is even* if all matrix elements of A and D are even and all matrix elements of B and C are odd. *A matrix is odd* if elements of A and D are odd and elements of B and C are even.

A matrix Q acts on the space $\mathcal{A}^{p|q}$ as

$$v \to vQ.$$

Such transformations are compatible with the left $\mathcal{A}$-module structure on $\mathcal{A}^{p|q}$, i.e.,

$$(\lambda \circ v)\, Q = \lambda \circ (vQ) \qquad \text{for any } \lambda \in \mathcal{A},\ v \in \mathcal{A}^{p|q}.$$

However, *even matrices also regard the right $\mathcal{A}$-module structure,*

$$(v * \lambda)\, Q = (vQ) * \lambda \qquad \text{for any } \lambda \in \mathcal{A},\ v \in \mathcal{A}^{p|q}.$$

(we use the rule (38)).

6.3 Super-Transposition

The *supertranspose* of Q is defined by

$$Q^{st} = \begin{pmatrix} A & B \\ C & D \end{pmatrix}^{st} := \begin{cases} \begin{pmatrix} A^t & C^t \\ -B^t & D^t \end{pmatrix} & \text{if } Q \text{ is even,} \\ \begin{pmatrix} A^t & -C^t \\ B^t & D^t \end{pmatrix} & \text{if } Q \text{ is odd,} \end{cases}$$

and

$$(Q_1 + Q_2)^{st} := Q_1^{st} + Q_2^{st}.$$

The following identity holds:

$$(QR)^{st} = \begin{cases} R^{st} Q^{st} & \text{if } Q \text{ or } R \text{ are even,} \\ -R^{st} Q^{st} & \text{if both } R \text{ and } Q \text{ are odd.} \end{cases} \tag{58}$$

Below we use only the first row.

6.4 *The Supergroups* $\mathrm{GL}(p|q; \mathcal{A})$

The group $\mathrm{GL}(p|q; \mathcal{A})$ is the group of *even* invertible matrices in $\mathrm{Mat}(p|q; \mathcal{A})$. The following lemma is trivial.

Lemma 6.1 *An even matrix* $Q \in \mathrm{Mat}(p|q; \mathcal{A})$ *is invertible:*

(a) *If and only if matrices A and D are invertible*
(b) *If and only if matrices* $\pi_\downarrow(A)$ *and* $\pi_\downarrow(D)$ *are invertible.*

Here $\pi_\downarrow(A)$ denotes the matrix composed of elements $\pi_\downarrow(a_{kl})$.
Also, the map $Q \mapsto \pi_\downarrow(Q)$ is a well-defined epimorphism

$$\pi_\downarrow : \mathrm{GL}(p|q; \mathcal{A}) \to \mathrm{GL}(p, \mathbb{C}) \times \mathrm{GL}(q, \mathbb{C})$$

(because $\pi_\downarrow(B) = 0$, $\pi_\downarrow(D) = 0$).

6.5 *The Supergroup* $\mathrm{OSp}(2p|2q; \mathcal{A})$

We define the standard *orthosymplectic form* on $\mathcal{A}^{2p|2q}$ by

$$\mathfrak{s}(u, v) := u J v^{st},$$

where J is a block $(p + p + q + q) \times (p + p + q + q)$ matrix

$$J := \frac{1}{2} \begin{pmatrix} 0 & 1_p & 0 & 0 \\ -1_p & 0 & 0 & 0 \\ 0 & 0 & 0 & 1_q \\ 0 & 0 & 1_q & 0 \end{pmatrix}. \tag{59}$$

The group $\mathrm{OSp}(2p|2q;\mathcal{A})$ is the subgroup in $\mathrm{GL}(2p|2q;\mathcal{A})$ consisting of matrices g satisfying

$$\mathfrak{s}(u,v)=\mathfrak{s}(ug,vg).$$

Equivalently,

$$gJg^{st}=J$$

(for this conclusion, we use (58); since $g\in\mathrm{GL}(p|q;\mathcal{A})$ is even, a sign does not appear).

We also write elements of $\mathrm{OSp}(2p|2q;\mathcal{A})$ as block $(2p+2q)$-matrices $g=\begin{pmatrix}A&B\\C&D\end{pmatrix}$. For such matrix, we have

$$\pi_\downarrow(A)\begin{pmatrix}0&1_p\\-1_p&0\end{pmatrix}\pi_\downarrow(A)^t=\begin{pmatrix}0&1_p\\-1_p&0\end{pmatrix},\qquad \text{i.e., } \pi_\downarrow(A)\in\mathrm{Sp}(2n,\mathbb{C});$$

$$\pi_\downarrow(D)\begin{pmatrix}0&1_q\\1_q&0\end{pmatrix}\pi_\downarrow(D)^t=\begin{pmatrix}0&1_q\\1_q&0\end{pmatrix},\qquad \text{i.e., } \pi_\downarrow(D)\in\mathrm{O}(2n,\mathbb{C}).$$

6.6 *The Super-Olshanski Semigroup* $\Gamma\mathrm{OSp}(2p|2q;\mathcal{A})$

We define the semigroup $\Gamma\mathrm{OSp}(2p|2q;\mathcal{A})$ as a subsemigroup in $\mathrm{OSp}(2p|2q;\mathcal{A})$ consisting of matrices $g=\begin{pmatrix}A&B\\C&D\end{pmatrix}$ such that $\pi_\downarrow(A)$ is contained in the Olshanski semigroup $\Gamma\mathrm{Sp}(2p,\mathbb{R})$ (see Sect. 3.10).

7 Super-Grassmannians

This section is a preparation to the definition of superlinear relations.

7.1 *Super-Grassmannians*

Let $u_1,\dots,u_r$ be even vectors and $v_1,\dots,v_s$ be odd vectors in $\mathcal{A}^{p|q}$. We suppose that

- $\pi_\downarrow(u_j)\in(\mathbb{C}^p\oplus 0)$ are linearly independent.
- $\pi_\downarrow(v_k)\in(0\oplus\mathbb{C}^q)$ are linearly independent.

A supersubspace of superdimension $r|s$ is a left $\mathcal{A}$-module generated by such vectors. Subspaces also are right $\mathcal{A}$-submodules.

We define the *super-Grassmannian* $\mathrm{Gr}^{r|s}_{p|q}(\mathcal{A})$ as the space of all supersubspaces in $\mathcal{A}^{p|q}$ of superdimension $r|s$.

By definition, the map $\pi_\downarrow$ projects $\mathrm{Gr}^{r|s}_{p|q}(\mathcal{A})$ to the product $\mathrm{Gr}^r_p \times \mathrm{Gr}^s_q$ of the usual complex Grassmannians. We denote by $\pi^\pm_\downarrow$ the natural projections

$$\pi^+_\downarrow : \mathrm{Gr}^{r|s}_{p|q}(\mathcal{A}) \to \mathrm{Gr}^r_p, \qquad \pi^-_\downarrow : \mathrm{Gr}^{r|s}_{p|q}(\mathcal{A}) \to \mathrm{Gr}^s_q.$$

7.2 Intersections of Subspaces

Let us examine superdimensions of intersections of subspaces.

Lemma 7.1 *Let L be a subspace of superdimension $r|s$ in $\mathcal{A}^{p|q}$, M be a subspace of superdimension $\rho|\sigma$. Let the following transversality conditions hold:*

$$\pi^+_\downarrow(L) + \pi^+_\downarrow(M) = \mathbb{C}^p, \quad \pi^-_\downarrow(L) + \pi^-_\downarrow(M) = \mathbb{C}^q. \tag{60}$$

Then $L \cap M$ is a subspace, and its superdimension is $(r + \rho - p)|(s + \sigma - q)$.

Remark If the transversality conditions are not satisfied, then incidentally, $L \cap M$ is not a subspace. For instance, consider $\mathcal{A}^{1|1}$ with a basis e_1, e_2 and subspaces

$$L := \mathcal{A}(e_1 + \mathfrak{a}_1 e_2), \qquad M := \mathcal{A} \cdot e_1.$$

Then $L \cap M = \mathcal{A}\mathfrak{a}_1 e_1$ is not a subspace. ⊠

Proof Denote by $I \subset \mathcal{A}$ the ideal spanned by all $\mathfrak{a}_j$, i.e., $\mathcal{A}/I = \mathbb{C}$.

It is easy to see that

$$L + M = \mathcal{A}^{p|q}. \tag{61}$$

Indeed, denote by $e_1, \ldots, e_p, e_{p+1}, \ldots, e_{p+q}$ the standard basis in $\mathbb{C}^p \oplus \mathbb{C}^q$. Then for each k the submodule $L + M$ contains a vector of the form

$$E_k = e_k + \sum_{J \neq \varnothing} \sum_m x_{k,m,J}\, \mathfrak{a}^J e_m,$$

where $x_{k,m,J} \in \mathbb{C}$. Actually, only a finite number of nonzero constants $\mathfrak{a}_j$ are contained in this expression. Without loss of generality, we can assume that this set is $\mathfrak{a}_1, \dots, \mathfrak{a}_N$. Then $L + M$ contains all vectors $\mathfrak{a}_1 \dots \mathfrak{a}_N E_k = \mathfrak{a}_1 \dots \mathfrak{a}_N e_k$ and therefore $L + M \supset \mathfrak{a}_1 \dots \mathfrak{a}_N \cdot \mathbb{C}^p \oplus \mathbb{C}^q$. Next, for each l

$$\mathfrak{a}_1 \dots \mathfrak{a}_{l-1}\mathfrak{a}_{l+1} \dots \mathfrak{a}_N\, E_k - \mathfrak{a}_1 \dots \mathfrak{a}_{l-1}\mathfrak{a}_{l+1} \dots \mathfrak{a}_N\, e_k \;\in\; \mathfrak{a}_1 \dots \mathfrak{a}_N \cdot \mathbb{C}^p \oplus \mathbb{C}^q.$$

Therefore, $\mathfrak{a}_1 \dots \mathfrak{a}_{l-1}\mathfrak{a}_{l+1} \dots \mathfrak{a}_N \cdot \mathbb{C}^p \oplus \mathbb{C}^q \subset L + M$. Repeating this process, we get $\mathbb{C}^p \oplus \mathbb{C}^q \subset L + M$, and this implies (61).

Let $v \in \pi_\downarrow^+(L) \cap \pi_\downarrow^+(M)$. Choose $x \in L$, $y \in M$ such that $\pi_\downarrow(x) = v$, $\pi_\downarrow(y) = v$. Then $x - y \in I \cdot \mathcal{A}^{p|q}$. However, $I \cdot L + I \cdot M = I \cdot \mathcal{A}^{p|q}$; therefore, we can represent

$$x - y = a - b, \qquad \text{where } a \in I \cdot L, b \in I \cdot M.$$

Then

$$(x - a) \in L,\ (y - b) \in M,\ \pi_\downarrow(x - a) = v = \pi_\downarrow(y - b).$$

Thus, for any vector $v \in \pi_\downarrow^+(L) \cap \pi_\downarrow^+(M)$, there is a vector $v^* \in L \cap M$ such that $\pi_\downarrow(v^*) = v$. The same is valid for vectors $w \in \pi_\downarrow^-(L) \cap \pi_\downarrow^-(M)$.

Therefore, $L \cap M$ contains a supersubspace of desired superdimension generated by vectors v^*, w^*. It remains to show that there are no extra vectors in the intersection.

Now, let us vary a phantom algebra $\mathcal{A}$. If $\mathcal{A}$ is an algebra in a finite number of Grassmann constants $\mathfrak{a}_1, \dots, \mathfrak{a}_n$, then this completes a proof, since (61) gives the same superdimension of the intersection over $\mathbb{C}$.

Otherwise, we choose a basis in L and a basis in M. Expressions for basis vectors contain only a finite number of Grassmann constants $\mathfrak{a}_1, \dots, \mathfrak{a}_k$. After this, we apply the same reasoning to algebras $\mathcal{A}[l]$ generated by Grassmann constants $\mathfrak{a}_1, \dots, \mathfrak{a}_l$ for all $l \geqslant k$ and observe that $L \cap M$ does not contain extra vectors. □

7.3 *Atlas on the Super-Grassmannian*

Define an atlas on the super-Grassmannian $\mathrm{Gr}_{p|q}^{r|s}(\mathcal{A})$ as usual. Namely, consider the following complementary subspaces:

$$V_+ := (\mathcal{A}^r \oplus 0) \oplus (\mathcal{A}^s \oplus 0) \qquad V_- := (0 \oplus \mathcal{A}^{p-r}) \oplus (0 \oplus \mathcal{A}^{q-s})$$

in $\mathcal{A}^{p|q}$. Let $S : V_+ \to V_-$ be an even operator. Then its graph is an element of the super-Grassmannian.

Permuting coordinates in $\mathcal{A}^p$ and $\mathcal{A}^q$, we get an atlas that covers the whole super-Grassmannian $\mathrm{Gr}_{p|q}^{r|s}(\mathcal{A})$.

7.4 Lagrangian Super-Grassmannians

Now, equip the space $\mathcal{A}^{2p|2q}$ with the orthosymplectic form $\mathfrak{s}$ as above. We say that a subspace L is *isotropic* if the form $\mathfrak{s}$ is zero on L. A *Lagrangian subspace* L is an isotropic subspace of the maximal possible superdimension, i.e., $\dim L = p|q$.

Observation 7.2 *Let L be a super-Lagrangian subspace. Then*

- $\pi_{\downarrow}^{+}(L)$ *is Lagrangian subspace in* $\mathbb{C}^{2p}$ *with respect to the skew-symmetric bilinear form* $\begin{pmatrix} 0 & 1 \\ -1 & 0 \end{pmatrix}$.
- $\pi_{\downarrow}^{-}(L)$ *is a Lagrangian subspace in* $\mathbb{C}^{2q}$ *with respect to the symmetric bilinear form* $\begin{pmatrix} 0 & 1 \\ 1 & 0 \end{pmatrix}$.

7.5 Coordinates on Lagrangian Super-Grassmannian

Consider the following complementary Lagrangian subspaces:

$$V_+ := (\mathcal{A}^p \oplus 0) \oplus (\mathcal{A}^q \oplus 0), \qquad V_- := (0 \oplus \mathcal{A}^p) \oplus (0 \oplus \mathcal{A}^q). \tag{62}$$

Proposition 7.3 *Consider an even operator* $S : V_+ \to V_-$,

$$S = \begin{pmatrix} A & B \\ C & D \end{pmatrix}.$$

The graph of S is a Lagrangian subspace if and only if

$$A = A^t, \quad D = -D^t, \quad C + B^t = 0. \tag{63}$$

Remark This statement is a super-imitation of Lemma 2.11. ⊠

Proof We write out a vector $h \in \mathcal{A}^{2p|2q}$ as

$$h = (u_+, u_-; v_+, v_-) \in \mathcal{A}^p \oplus \mathcal{A}^p \oplus \mathcal{A}^q \oplus \mathcal{A}^q.$$

Then

$$\mathfrak{s}(h', h) = u'_+ (u_-)^{st} - u'_- (u_+)^{st} + v'_+ (v_-)^{st} + v'_- (v_+)^{st}.$$

Let h be in the graph of S. Then

$$\begin{pmatrix} u_- & v_- \end{pmatrix} = \begin{pmatrix} u_+ & v_+ \end{pmatrix} \begin{pmatrix} A & B \\ C & D \end{pmatrix} = \begin{pmatrix} u_+A + v_+C & u_+B + v_+D \end{pmatrix}$$

and

$$\begin{aligned} \mathfrak{s}(h', h) = u'_+(u_+A + v_+C)^{st} - (u'_+A + v'_+C)u_+^{st} + \\ + v'_+(u_+B + v_+D)^{st} + (u'_+B + v'_+D)v_+^{st}. \end{aligned}$$

Observe that matrices A, B, C, and D are even;[29] for this reason, we write $(u_+A)^{st} = A^{st}(u_+)^{st}$ etc. (see (58)). We arrive at

$$\begin{aligned} u'_+\bigl[A^{st}(u_+)^{st} + C^{st}(v_+)^{st}\bigr] - \bigl[(u'_+A + v'_+C\bigr]u_+^{st} + \\ + v'_+\bigl[B^{st}(u_+)^{st} + D^{st}(v_+)^{st}\bigr] + \bigl[u'_+B + v'_+D\bigr]v_+^{st}. \end{aligned}$$

Next,

$$A^{st} = A^t, \quad B^{st} = -B^t, \quad C^{st} = C^t, \quad D^{st} = D^t.$$

Therefore, we convert our expression to the form

$$u'_+(A - A^t)(u_+)^{st} + v'_+(D + D^t)(v_+)^{st} + (u'_+)(B^t + C)v_+^{st} + (v^+)(B + C^t)u_+^{st}.$$

This expression is zero if and only if the conditions (63) are satisfied. □

7.6 Atlas on the Lagrangian Super-Grassmannian

Now we imitate the construction of Sect. 2.21.

Consider the standard basis in $\mathcal{A}^{2p|2q}$ consisting of vectors, whose coordinates are 0 except one unit. Denote elements of this basis by

$$e_1, \dots, e_p;\ e'_1, \dots, e'_p;\ f_1, \dots, f_q;\ f'_1, \dots, f'_q.$$

In this basis, the matrix of the orthosymplectic form is (59). Consider subsets $I \subset \{1, 2, \dots, p\}$, $J \subset \{1, 2, \dots, q\}$.

[29] Recall that this means that A and D are composed of even phantom constants and C and B of odd phantom constants.

We define

$$V_+[I, J] = \big(\oplus_{i\in I}\mathcal{A}e_i\big) \oplus \big(\oplus_{k\notin I}\mathcal{A}e'_k\big) \oplus \big(\oplus_{j\in J}\mathcal{A}f_j\big) \oplus \big(\oplus_{l\notin J}\mathcal{A}f'_l\big), \tag{64}$$

$$V_-[I, J] = \big(\oplus_{i\notin I}\mathcal{A}e_i\big) \oplus \big(\oplus_{k\in I}\mathcal{A}e'_k\big) \oplus \big(\oplus_{j\notin J}\mathcal{A}f_j\big) \oplus \big(\oplus_{l\in J}\mathcal{A}f'_l\big). \tag{65}$$

We denote by $\mathcal{O}[I, J]$ the set of all the Lagrangian subspaces that are graphs of even operators

$$S : V_+[I, J] \to V_-[I, J].$$

In fact, these operators satisfy the same conditions as in Proposition 7.3 (our initial chart is $\mathcal{O}[\varnothing, \varnothing]$). Thus, we get an atlas on the Lagrangian super-Grassmannian.

7.7 Elementary Reflections

Now we repeat considerations of Sect. 2.22. We define elementary reflections $\sigma[e_i]$, $\sigma[f_j]$ in $\mathcal{A}^{2p|2q}$ by

$$\sigma[e_i]\, e_i^+ = -e_i^-, \quad \sigma[e_i]\, e_i^- = e_i^+,$$
$$\sigma[e_i]\, e_k^\pm = e_k^\pm \ \text{ for } k \neq i, \quad \sigma[e_i]\, f_j^\pm = f_j^\pm,$$

and

$$\sigma[f_j]\, f_j^+ = f_j^-, \quad \sigma[f_j]\, f_j^- = f_j^+, \tag{66}$$

$$\sigma[f_i]\, f_k^\pm = e_k^\pm \ \text{ for } k \neq j, \quad \sigma[f_j]\, e_i^\pm = e_i^\pm, \tag{67}$$

in the first row, we have an extra change of a sign because we want to preserve the symplectic form. Then

$$\mathcal{O}[I, J] = \prod_{i\in I}\sigma[e_i] \cdot \prod_{j\in J}\sigma[f_j] \cdot \mathcal{O}[\varnothing, \varnothing].$$

8 Superlinear Relations

Gauss–Berezin integral operators are enumerated by contractive Lagrangian superlinear relations. These objects are defined in this section.

8.1 Superlinear Relations

We define superlinear relations $P : \mathcal{A}^{p|q} \rightrightarrows \mathcal{A}^{r|s}$ as subspaces in $\mathcal{A}^{p|q} \oplus \mathcal{A}^{r|s}$. Products are defined as above (see Sect. 2.16).

Next, for a superlinear relation, we define complex linear relations

$$\pi_{\downarrow}^{+}(P) : \mathbb{C}^{p} \rightrightarrows \mathbb{C}^{r}, \quad \pi_{\downarrow}^{-}(P) : \mathbb{C}^{q} \rightrightarrows \mathbb{C}^{s}$$

in the natural way. We simply project the super-Grassmannian in $\mathcal{A}^{p|q} \oplus \mathcal{A}^{r|s}$ onto the product of the complex Grassmannians.

8.2 Transversality Conditions

Let V, W, and Y be *complex* linear spaces. We say that linear relations

$$P : V \rightrightarrows W, \;\; Q : W \rightrightarrows Y$$

are *transversal* if

$$\operatorname{im} P + \operatorname{dom} Q = W, \tag{68}$$

$$\operatorname{indef} P \cap \ker Q = 0. \tag{69}$$

We met these conditions in Sect. 2; in what follows, they are even more important.

Theorem 8.1 *If $P : V \rightrightarrows W$, $Q : W \rightrightarrows Y$ are transversal, then*

$$\dim QP = \dim Q + \dim P - \dim W.$$

Proof We rephrase the definition of the product QP as follows (see [50], Prop. II.7.1). Consider the space $V \oplus W \oplus W \oplus Y$ and the following subspaces:

- $P \oplus Q$.
- The subspace H consisting of vectors $v \oplus w \oplus w \oplus y$.
- The subspace $T \subset H$ consisting of vectors $0 \oplus w \oplus w \oplus 0$.

Let us project $(P \oplus Q) \cap H$ on $V \oplus W$ along T. The result is $QP \subset V \oplus W$. By the first transversality condition (68),

$$(P \oplus Q) + H = V \oplus W \oplus W \oplus Y,$$

therefore we know the superdimension of the intersection $S := (P \oplus Q) \cap H$.

By the second condition (69), the projection $H \to V \oplus W$ is injective on S. □

8.3 Transversality for Superlinear Relations

We say that superlinear relations $P : V \rightrightarrows W$ and $Q : W \rightrightarrows Y$ are *transversal* if $\pi_{\downarrow}^{+}(P)$ is transversal to $\pi_{\downarrow}^{+}(Q)$ and $\pi_{\downarrow}^{-}(P)$ is transversal to $\pi_{\downarrow}^{-}(Q)$.

Theorem 8.2 *If $P : V \rightrightarrows W$, $Q : W \rightrightarrows Y$ are transversal superlinear relations, then their product is a superlinear relation and*

$$\dim QP = \dim Q + \dim P - \dim W.$$

Proof We follow the proof of the previous theorem. □

8.4 Lagrangian Superlinear Relations

Consider the spaces $V = \mathcal{A}^{2p|2q}$, $W = \mathcal{A}^{2r|2s}$ endowed with the orthosymplectic forms $\mathfrak{s}_V$, $\mathfrak{s}_W$, respectively. Define the form $\mathfrak{s}^{\ominus}$ on $V \oplus W$ as

$$\mathfrak{s}^{\ominus}(v \oplus w, v' \oplus w') := \mathfrak{s}_V(v, v') - \mathfrak{s}_W(w, w').$$

A Lagrangian superlinear relation $P : V \rightrightarrows W$ is a Lagrangian supersubspace in $V \oplus W$.

Observation 8.3 *Let $g \in \mathrm{OSp}(2p|2q; \mathcal{A})$. Then the graph of g is a Lagrangian superlinear relation $\mathcal{A}^{2p|2q} \rightrightarrows \mathcal{A}^{2p|2q}$.*

Theorem 8.4 *Let $P : V \rightrightarrows W$, $Q : W \rightrightarrows Y$ be transversal Lagrangian superlinear relations. Then $QP : V \rightrightarrows Y$ is a Lagrangian superlinear relation.*

Proof Let $v \oplus w$, $v' \oplus w' \in P$ and $w \oplus y$, $w' \oplus y' \in Q$. By definition,

$$\mathfrak{s}_V(v, v') = \mathfrak{s}_W(w, w') = \mathfrak{s}_Y(y, y'),$$

therefore QP is isotropic. By the virtue of Theorem 8.2, we know $\dim QP$. □

8.5 Components of Lagrangian Super-Grassmannian

As we observed in Sect. 2.23, the orthogonal Lagrangian Grassmannian in the space $\mathbb{C}^{2n}$ consists of two components. The usual symplectic Lagrangian Grassmannian is connected. Therefore, the Lagrangian super-Grassmannian consists of two components.

Below we must distinguish them.

Decompose $V = V_+ \oplus V_-$, $W = W_+ \oplus W_-$ as above (62). We say that the component containing the linear relation

$$(V_+ \oplus W_-) : V \rightrightarrows W$$

is even; the other component is odd.

8.6 Contractive Lagrangian Linear Relations

Now, we again (see Sect. 3) consider the Hermitian form M on $\mathbb{C}^{2p}$; it is defined by a matrix $\begin{pmatrix} 1_p & 0 \\ 0 & -1_p \end{pmatrix}$. Then $\mathbb{C}^{2p}$ becomes an object of the category **Sp**.

We say that a Lagrangian superlinear relation $P : V \rightrightarrows W$ is contractive if $\pi_{\downarrow}^{+}(P)$ is a morphism of the category **Sp**.

8.7 Positive Domain in the Lagrangian Super-Grassmannian

We say that a Lagrangian subspace P in $\mathbb{C}^{2p|2q}$ is positive if the form M defined in the previous subsection is positive on $\pi_{\downarrow}^{+}(P)$.

9 Correspondence Between Lagrangian Superlinear Relations and Gauss–Berezin Operators

Here we prove our main results, namely, Theorems 9.3 and 9.4.

9.1 Creation–Annihilation Operators

Let $V := \mathcal{A}^{2p|2q}$ be a superlinear space endowed with the orthosymplectic bilinear form $\mathfrak{s}$ defined by the matrix (59). For a vector

$$v \oplus w := v_+ \oplus v_- \oplus w_+ \oplus w_- \in \mathcal{A}^{2p|2q},$$

we define the creation-annihilation operator in the Fock–Berezin space $\mathbf{SF}_{p,q}(\mathcal{A})$ by

$$\widehat{a}(v \oplus w) f(z, \xi) = \Big(\sum_i v_+^{(i)} \frac{\partial}{\partial z_i} + \sum_i v_-^{(i)} z_i + \sum_j w_+^{(j)} \frac{\partial}{\partial \xi_i} + \sum_j w_-^{(j)} \xi_j \Big) f(z, \xi).$$

9.2 Supercommutator

We say that a vector $v \oplus w$ is even if v is even and w is odd. It is odd if v is odd and w is even. This corresponds to the definition of even/odd for $(1|0) \times (2p|2q)$ matrices. Let $h = v \oplus w$, $h' = v' \oplus w'$. We define the supercommutator $[\widehat{a}(h), \widehat{a}(h')]_s$ as

$$[\widehat{a}(h), \widehat{a}(h')]_s = \begin{cases} [\widehat{a}(h), \widehat{a}(h')] = \widehat{a}(h)\widehat{a}(h') - \widehat{a}(h')\widehat{a}(h) & \text{if } h \text{ or } h' \text{ is even,} \\ \{\widehat{a}(h), \widehat{a}(h')\} = \widehat{a}(h)\widehat{a}(h') + \widehat{a}(h')\widehat{a}(h) & \text{if } h, h' \text{ are odd.} \end{cases}$$

Then

$$[\widehat{a}(h), \widehat{a}(h')]_s = \mathfrak{s}(h, h') \cdot 1,$$

where 1 denotes the unit operator.

Also, note that an operator $\widehat{a}(h)$ is linear (see Sect. 5.4) if h is even and antilinear if h is odd.

9.3 Annihilators of Gaussian Vectors

Theorem 9.1

(a) *For a Gauss–Berezin vector* $\mathbf{b} \in \mathcal{S}\mathbf{F}_{p,q}(\mathcal{A})$ *consider the set* L *of all vectors* $h \in \mathcal{A}^{2p|2q}$ *such that*

$$\widehat{a}(h)\, \mathbf{b} = 0.$$

Then L *is a positive Lagrangian subspace in* $\mathcal{A}^{2p|2q}$.

(b) *Moreover, the map* $\mathbf{b} \mapsto L$ *is a bijection*

$$\left\{ \begin{matrix} \textit{The set of all Gauss–Berezin vectors} \\ \textit{defined up to an invertible scalar} \end{matrix} \right\} \leftrightarrow \left\{ \begin{matrix} \textit{The positive} \\ \textit{Lagrangian Grassmannian} \end{matrix} \right\}.$$

Before we begin a formal proof, we propose the following (insufficient, but clarifying) argument: Let $h, h' \in L$. If one of them is even, then we write

$$\Big(\widehat{a}(h)\widehat{a}(h') - \widehat{a}(h')\widehat{a}(h)\Big)\, \mathbf{b}.$$

By the definition of L, this is 0. On the other hand, this is $\mathfrak{s}(h, h')\mathbf{b}$. Therefore, $\mathfrak{s}(h, h') = 0$.

If both h, h' are odd, then we write

$$0 = \left(\widehat{a}(h)\widehat{a}(h') + \widehat{a}(h')\widehat{a}(h)\right) \mathbf{b} = \mathfrak{s}(h, h')\mathfrak{b}$$

and arrive at the same result.

Proof First, let $\mathfrak{b}(z, \xi)$ have the standard form (53). We write out

$$\widehat{a}(v \oplus w)\mathfrak{b}(z, \xi) = \Big(\sum_i v_+^{(i)} \frac{\partial}{\partial z_i} + \sum_i v_-^{(i)} z_i + \sum_j w_+^{(j)} \frac{\partial}{\partial \xi_i} + \sum_j w_-^{(j)} \xi_j\Big) \times$$

$$\times \exp\left\{\frac{1}{2} \begin{pmatrix} z & \xi \end{pmatrix} \begin{pmatrix} A & B \\ -B^t & C \end{pmatrix} \begin{pmatrix} z^t \\ \xi^t \end{pmatrix}\right\} =$$

$$= \left(v_+(Ax^t + B\xi^t) + v_- z^t + w_+(-B^t z^t + C\xi^t) + w_-\xi^t\right) \cdot \mathfrak{b}(z, \xi) =$$

$$= \left((v_+ A - w_+ B^t + v_-)z^t + (v_+ B + w_+ D + w_-)\xi^t\right) \cdot \mathfrak{b}(z, \xi).$$

This is zero if and only if

$$\begin{cases} v_- = -(v_+ A - w_+ B^t) \\ w_- = -(v_+ B + w_+ D) \end{cases}.$$

However, this system of equations determines a Lagrangian subspace. The positivity of a Lagrangian subspace is equivalent to $\|\pi_\downarrow(A)\| < 1$ (see, for instance, [53]).

Next, consider an arbitrary Gauss–Berezin vector

$$\mathbf{b}(z, \xi) = \mathfrak{D}[\xi_{i_1}] \dots \mathfrak{D}[\xi_{i_k}] \mathsf{S}^k\ \mathfrak{b}[T], \tag{70}$$

where $\mathfrak{b}[T]$ is a standard Gauss–Berezin vector. We have

$$\widehat{a}(h)\ \mathfrak{D}[\xi_1]\, \mathsf{S} = \mathfrak{D}[\xi_1]\, \mathsf{S}\, \widehat{a}(\sigma[f_1]h), \tag{71}$$

where $\sigma[f_1]$ is an elementary reflection given by (66)–(67).

If h ranges in a Lagrangian subspace, then $\sigma[f_1]h$ also ranges in (another) Lagrangian subspace. Also, a map $\sigma[f_1]$ takes positive subspaces to positive subspaces. Therefore, the statement (a) for vectors

$$\mathfrak{D}[\xi_{i_1}]\, \mathfrak{D}[\xi_{i_2}] \dots \mathfrak{D}[\xi_{i_k}]\, \mathsf{S}^k\ \mathfrak{b}[T] \qquad \text{and} \qquad \mathfrak{D}[\xi_{i_2}] \dots \mathfrak{D}[\xi_{i_k}]\, \mathsf{S}^{k-1}\ \mathfrak{b}[T].$$

is equivalent.

In fact, for fixed $i_1, \dots, i_k$, all vectors of the form (70) correspond to a fixed chart in the Lagrangian super-Grassmannian, namely, to

$$\sigma[f_{i_1}] \cdots \sigma[f_{i_k}] \cdot \mathcal{O}[\varnothing, \varnothing]$$

in notation of Sect. 7.7.

But these charts cover the set of all positive Lagrangian subspaces. □

Theorem 9.2 *For a positive Lagrangian subspace $L \subset \mathcal{A}^{2p|2q}$, consider the system of equations*

$$\widehat{a}(v \oplus w) f(z,\xi) = 0 \quad \textit{for all } v \oplus w \in L, \tag{72}$$

for a function $f(z,\xi)$. All its solutions are of the form $\lambda\mathfrak{b}(z,\xi)$, where $\mathfrak{b}(z,\xi)$ is a Gauss–Berezin vector and λ is a phantom constant.

Proof It suffices to prove the statement for L in the principal chart. Put

$$\varphi(z,\xi) := f(z,\xi)/\mathfrak{b}(z,\xi),$$

i.e.,

$$f(z,\xi) = \mathfrak{b}(z,\xi) \cdot \varphi(z,\xi).$$

By the Leibniz rule,

$$0 = \widehat{a}(v \oplus w)\big(\mathfrak{b}(z,\xi)\varphi(z,\xi)\big) =$$

$$= \Big(\widehat{a}(v \oplus w)\mathfrak{b}(z,\xi)\Big) \cdot \varphi(x,\xi) + \mathfrak{b}(z,\xi) \cdot \Big(\sum_j v_+^{(j)} \frac{\partial}{\partial z_i} + \sum_j w_+^{(j)} \frac{\partial}{\partial \xi_i}\Big).\varphi(z,\xi)$$

The first summand is zero by the definition of $\mathfrak{b}(z,\xi)$. Since v_+, w_+ are arbitrary, we get

$$\frac{\partial}{\partial z_i}\varphi(z,\xi) = 0, \qquad \frac{\partial}{\partial \xi_i}\varphi(z,\xi) = 0.$$

Therefore, $\varphi(z,\xi)$ is a phantom constant. □

9.4 Gauss–Berezin Operators and Superlinear Relations

Let $V = \mathcal{A}^{2p|2q}$, $\widetilde{V} = \mathcal{A}^{2r|2s}$ be (super)spaces endowed with orthosymplectic forms.

Theorem 9.3

(a) *For each contractive Lagrangian superlinear relation $P : V \rightrightarrows \widetilde{V}$, there exists a linear operator*

$$\mathfrak{B}(P) : \mathbf{SF}_{p,q}(\mathcal{A}) \to \mathbf{SF}_{r,s}(\mathcal{A})$$

such that

(1) *The following condition is satisfied:*

$$\widehat{a}(\widetilde{h})\,\mathfrak{B}(P) = \mathfrak{B}(P)\,\widehat{a}(h) \quad \textit{for all } h \oplus \widetilde{h} \in P.$$

(2) *If P is in the even component of Lagrangian super-Grassmannian, then $\mathfrak{B}(P)$ is an integral operator with an even*[30] *kernel. If P is in the odd component, then $\mathfrak{B}(P)\mathsf{S}$ is an integral operator with an odd kernel. Moreover, this operator is unique up to a scalar factor $\in \mathcal{A}_{even}$.*

(b) *Operators $\mathfrak{B}(P)$ are Gauss–Berezin operators, and all Gauss–Berezin operators arise in this way.*

Proof Let us write differential equations for the kernel $K(z, \xi, \overline{y}, \overline{\eta})$ of the operator $\mathfrak{B}(P)$. Denote

$$h = v_+ \oplus v_- \oplus w_+ \oplus w_-, \qquad \widetilde{h} = \widetilde{v}_+ \oplus \widetilde{v}_- \oplus \widetilde{w}_+ \oplus \widetilde{w}_-.$$

Then

$$\widehat{a}(\widetilde{h}) \int K(z, \xi, \overline{y}, \overline{\eta})\, f(y, \eta)\, e^{-y\overline{y}^t - \eta\overline{\eta}^t}\, dy\, d\overline{y}\, d\overline{\eta}\, d\eta =$$
$$= \int K(z, \xi, \overline{y}, \overline{\eta})\, \widehat{a}(h) f(y, \eta)\, e^{-y\overline{y}^t - \eta\overline{\eta}^t}\, dy\, d\overline{y}\, d\overline{\eta} d\eta.$$

Let P be even. Integrating by parts on the right-hand side, we get

$$\Big(\sum_i \widetilde{v}_+^{(i)} \frac{\partial}{\partial z_i} + \sum_i \widetilde{v}_-^{(i)} z_i + \sum_j \widetilde{w}_+^{(j)} \frac{\partial}{\partial \xi_i} + \sum_j \widetilde{w}_-^{(j)} \xi_j\Big) K(z, \xi, \overline{y}, \overline{\eta}) =$$
$$= \Big(\sum_i v_+^{(i)} \overline{y}_i + \sum_i v_-^{(i)} \frac{\partial}{\partial \overline{y}_i} + \sum_j w_+^{(j)} \overline{\eta}_j + \sum_j w_-^{(j)} \frac{\partial}{\partial \overline{\eta}_i}\Big) K(z, \xi, \overline{y}, \overline{\eta}).$$

This system of equations has the form (72) and determines a Gaussian.

Evenness condition was essentially used in this calculation. For instance, for an odd kernel K, we must write $(v_+^{(i)})^\sigma$ instead of $v_+^{(i)}$ on the right-hand side.

Now, let P be odd. Let us try to represent $\mathfrak{B}(P)$ as a product

$$\mathfrak{B}(P) = \mathfrak{C} \cdot \mathfrak{D}[\eta_1] \cdot \mathsf{S}.$$

[30] See Sect. 5.4.

Let L be the kernel of $\mathfrak{C}$.

$$\widehat{a}(\widetilde{h}) \int L(z,\xi,\overline{y},\overline{\eta})\, \mathfrak{D}[\eta_1] \cdot \mathsf{S} \cdot f(y,\eta)\, e^{-y\overline{y}^t - \eta\overline{\eta}^t}\, dy\, d\overline{y}\, d\overline{\eta}\, d\eta =$$

$$= \int L(z,\xi,\overline{y},\overline{\eta})\, \mathfrak{D}[\eta_1] \cdot \mathsf{S} \cdot \widehat{a}(h) f(y,\eta)\, e^{-y\overline{y}^t - \eta\overline{\eta}^t}\, dy\, d\overline{y}\, d\overline{\eta} d\eta.$$

Next, we change the order

$$\mathfrak{D}[\eta_1]\ S\widehat{a}(v) = \widehat{a}(\sigma(f_1)v)\, \mathfrak{D}[\eta_1]\, \mathsf{S},$$

where σ is an elementary reflection of the type (66)–(67).

We again get for L a system of equations determining a Gaussian. □

9.5 *Products of Gauss–Berezin Operators*

Theorem 9.4

(a) *Let $P : \mathcal{A}^{p|q} \rightrightarrows \mathcal{A}^{p'|q'}$, $Q : \mathcal{A}^{p'|q'} \rightrightarrows \mathcal{A}^{p''|q''}$ be contractive Lagrangian relations. Assume that P and Q are transversal. Then*

$$\mathfrak{B}(Q)\, \mathfrak{B}(P) = \lambda \cdot \mathfrak{B}(QP), \tag{73}$$

where $\lambda = \lambda(P, Q)$ is an even invertible phantom constant.

(b) *If P, Q are not transversal, then*

$$\pi_\downarrow\big(\mathfrak{B}(Q)\, \mathfrak{B}(P)\big) = 0.$$

Proof Let $v \oplus w \in P$, $w \oplus y \in Q$. Then

$$\mathfrak{B}(Q)\mathfrak{B}(P)\widehat{a}(v) = \mathfrak{B}(Q)\widehat{a}(w)\mathfrak{B}(P) = \widehat{a}(y)\mathfrak{B}(Q)\mathfrak{B}(P).$$

On the other hand,

$$\widehat{a}(y)\mathfrak{B}(QP) = \mathfrak{B}(QP)\widehat{a}(v).$$

By Theorem 9.3, these relations define a unique operator, and we get (73).

It remains to verify conditions of vanishing of

$$\pi_\downarrow\big(\mathfrak{B}(Q)\, \mathfrak{B}(P)\big) : \mathbf{F}_p \otimes \Lambda_q \to \mathbf{F}_{p''} \otimes \Lambda_{q''}.$$

Here we refer to Sect. 5.8. Our operator is a tensor product of

$$\pi_{\downarrow}^{+}\big(\mathfrak{B}(Q)\big)\,\pi_{\downarrow}^{+}\big(\mathfrak{B}(P)\big) : \mathbf{F}_p \to \mathbf{F}_{p''} \tag{74}$$

and

$$\pi_{\downarrow}^{-}\big(\mathfrak{B}(Q)\big)\,\pi_{\downarrow}^{-}\big(\mathfrak{B}(P)\big) : \Lambda_q \to \Lambda_{q''}. \tag{75}$$

The operator (74) is a product of Gaussian integral operators. By Theorem 3.5, it is nonzero.

The operator (75) is a product of Berezin operators. It is nonzero if and only if $\pi_{\downarrow}^{-}(P)$ and $\pi_{\downarrow}^{-}(Q)$ are transversal; we refer here to Theorem 2.16.c. □

Corollary 9.5 *For an element g of the Olshanski supersemigroup $\Gamma\mathrm{OSp}(2p|2q;\mathcal{A})$ (see Sect. 6.6), denote its graph by* graph(g). *Then $\mathfrak{B}(P(g))$ determines a projective representation of $\Gamma\mathrm{OSp}(2p|2q;\mathcal{A})$ over $\mathcal{A}$,*

$$\mathfrak{B}(\mathrm{graph}(g_1))\,\mathfrak{B}(\mathrm{graph}(g_2)) = \lambda(g_1, g_2)\mathfrak{B}(\mathrm{graph}(g_1 g_2)),$$

where $\lambda(g_1, g_2)$ is an invertible element of $\mathcal{A}$.

Denote by $G(2p|2q;\mathcal{A})$ the group of invertible elements of Olshanski supersemigroup. It is easy to see that $G(2p|2q;\mathcal{A})$ consists of all $g \in \mathrm{OSp}(2p|2q;\mathcal{A})$ such that $\pi_{\downarrow}(g) \in \mathrm{Sp}(2q,\mathbb{R}) \times \mathrm{O}(2p,\mathbb{C})$.[31]

Corollary 9.6 *The map $g \mapsto \mathfrak{B}(\mathrm{graph}(g_1))$ determines a projective representation of the group $G(2p|2ql\mathcal{A})$ over $\mathcal{A}$.*

10 Final Remarks

10.1 Extension of Notion of Gaussian Operators?

Our main result seems incomplete, since Theorem 9.4 describes product of integral operators only if superlinear relations are transversal. However, a product of integral operators can be written explicitly in all cases (see Sect. 4.12). We arrive at the following question:

[31] In $\mathrm{Sp}(2q,\mathbb{R}) \times \mathrm{O}(2p,\mathbb{C})$, we have a product of a real Lie group and a complex Lie group, so it is not a real form (on real forms, see [63, 20]). Moreover, $G(2p|2q;\mathcal{A})$ is not a supergroup in the sense of the usual definitions [9, 18, 12, 42, 11] (because there are no intermediate Lie superalgebras between $\mathfrak{osp}(2p|2q,\mathbb{R})$ and $\mathfrak{osp}(2p|2q,\mathbb{C})$). However, $G(2p|2q;\mathcal{A}')$ depends functorially on algebras $\mathcal{A}'$ described in Sect. 4.2.

Question 10.1

(a) *Is it possible to extend the definitions of Gaussian operators and Lagrangian superlinear relations to make formula* (73) *valid for all* P, Q?
(b) *Is it reasonable to consider the expressions* (48) *as Gaussians?*

10.2 The Infinite-Dimensional Orthosymplectic Supergroup

Recall that constructions of orthogonal and symplectic spinors described in Sects. 2–3 survive well in the infinite-dimensional limit, [50, Chapters IV, VI]. However, there remain gaps between sufficient and necessary conditions for boundedness of Gaussian operators in the bosonic Fock spaces with infinite number of degrees of freedom (see [50, Sect. VI.3-4], [57]) and similar gaps in fermionic case (see [50, Sect. IV.2]). On the other hand, even for finite-dimensional supergroups "unitary representations" are realized by unbounded operators in super-Hilbert spaces (for instance, for orthosymplectic spinors discussed in this work).

Question 10.2 *Find natural topologies in Fock–Berezin space* $\mathbf{F}_{\infty,\infty}(\mathcal{A})$ *with infinite number of variables* $z_1, z_2, \dots, \xi_1, \xi_2, \dots$ *and conditions of boundedness of Gauss–Berezin operators.*

Such topologies can depend on further applications. For instance, in the next subsection, a straightforward repetition of our definition of $\mathbf{SF}_{p,q}$ does not work in that situation.

10.3 The Super-Virasoro Algebras

It is well-known that highest weight representations of the Virasoro algebra appear in a natural way as restrictions of infinite-dimensional Weil representation and infinite-dimensional spinor representation to the Lie algebra of vector fields on the circle (see [61, 46] for more details and [50, Chapter VII]). Let us explain that a similar phenomenon takes place for super-Virasoro algebras and ortho-symplectic spinors.

Consider the circle S^1 with the coordinate $\varphi \in \mathbb{R}/2\pi\mathbb{Z}$, and denote by $C^\infty_+(S^1)$ (resp., $C^\infty_-(S^1)$) the space of smooth complex functions on S^1 such that $f(\varphi+\pi) = f(\varphi)$ (resp., $f(\varphi+\pi) = -f(\varphi)$). Let us denote by $f(\varphi)(d\varphi)^\lambda$ densities of weight λ on the circle, $d\varphi = (d\varphi)^1$, $\frac{d}{d\varphi} = (d\varphi)^{-1}$. Recall that vector fields act on densities by

$$a(\varphi)\frac{d}{d\varphi}\Big(f(\varphi)(d\varphi)^\lambda\Big) = \Big(a(\varphi)f'(\varphi) + \lambda a'(\varphi)f(\varphi)\Big)(d\varphi)^\lambda.$$

Then, $d\varphi = (d\varphi)^1$, $\partial/\partial\varphi = (d\varphi)^{-1}$.

Consider a "super-Witt" Lie superalgebra $\mathfrak{sw}_-$ (see Kirillov [37]), whose elements are

$$a(\varphi)\frac{d}{d\varphi} \oplus b(\varphi)(d\varphi)^{-1/2}, \qquad a \in C_+^\infty(S^1),\ b \in C_-^\infty(S^1), \tag{76}$$

the first summand has parity $\overline{0}$ and the second summand parity $\overline{1}$. The supercommutator is defined as follows:

- Supercommutator of vector fields is the usual commutator.
- Supercommutator of a vector field and a density is the natural action of vector fields on densities of weight $-1/2$.
- The anticommutator of densities $b_1(\varphi)(d\varphi)^{-1/2}$ and $b_2(\varphi)(d\varphi)^{-1/2}$ is the product $2b_1(\varphi)b_2(\varphi)\frac{d}{d\varphi}$.

This supercommutator is a "natural differential geometric operation",[32] i.e., it is invariant with respect to the action of the group of diffeomorphisms r of S^1 satisfying $r(\varphi+\pi) = r(\varphi)+\pi$.

Consider the superlinear space[33] $W := W_{\overline{0}} \oplus W_{\overline{1}}$ consisting of

$$f(\varphi) \oplus g(\varphi)(d\varphi)^{1/2}, \qquad f \in C_+^\infty(S^1)/\mathbb{C},\ g \in C_-^\infty(S^1). \tag{77}$$

We equip W with the orthosymplectic form by

$$\big\{f_1(\varphi) \oplus g_1(\varphi)(d\varphi)^{1/2},\ f_2(\varphi) \oplus g_2(\varphi)(d\varphi)^{1/2}\big\} = \\ = \frac{1}{4}\int_{S^1}(f_1 df_2 - f_2 df_1) + \int_{S^1} g_1 g_2\, d\varphi.$$

Notice that the element $1 \oplus 0$ is contained in the kernel of this form, and we really get a form on the quotient $W := (C_+^\infty(S^1)/\mathbb{C}) \oplus C_-^\infty(S^1)$. Next, we define the inner product on W by

$$\Big\langle \sum p_n e^{2in\varphi} \oplus \sum q_n e^{(2n+1)\varphi}(d\varphi)^{1/2}, \sum p_n' e^{2in\varphi} \oplus \sum q_n' e^{(2n+1)\varphi}(d\varphi)^{1/2}\Big\rangle := \\ = \sum |n|\, p_n \overline{p_n'} + \sum q_n \overline{q_n'}.$$

---

[32] On natural differential operations, see, e.g., [38, 27, 28].

[33] Cf. constructions for Virasoro algebra and the group of diffeomorphisms of circle in [50], Subsect. VII.2.2-2.3, 2.5, VII.3, Sect. VII.3, VIII.6

The superalgebra $\mathfrak{sw}_-$ acts in the space W as follows. Vector fields act in $W_{\overline{0}} \oplus W_{\overline{1}}$ in a natural way.

$$a(\varphi)\frac{d}{d\varphi}\Big(f(\varphi) \oplus g(\varphi)d\varphi^{1/2}\Big) = a(\varphi)f'(\varphi) \oplus \Big(a(\varphi)b'(\varphi) + \tfrac{1}{2}a'(\varphi)b(\varphi)\Big)(d\varphi)^{1/2}.$$

An element $0 \oplus b(\varphi)(d\varphi)^{-1/2}$ acts as

$$f(\varphi) \oplus g(\varphi) \mapsto b(\varphi)g(\varphi) \;\oplus\; b(\varphi)f'(\varphi)(d\varphi)^{1/2} =$$
$$= b(\varphi)(d\varphi)^{-1/2} \cdot g(\varphi)(d\varphi)^{1/2} \;\oplus\; df(\varphi) \cdot b(\varphi)(d\varphi)^{-1/2}.$$

Again, this action is a "natural differential geometric operation."

In this way, we get an embedding of $\mathfrak{sw}_-$ to an infinite-dimensional Lie superalgebra $\mathfrak{osp}(2\infty|2\infty)$. Next, we apply the infinitesimal version of the orthosymplectic spinors setting $p = \infty, q = \infty$ in the notation Sect. 1.1, assuming even variables x_j are complex. We arrive at a *projective* representation of the Lie superalgebra $\mathfrak{sw}_-$ in a tensor product of the bosonic and fermionic Fock spaces with infinite number degrees of freedom. Its restriction to the Lie superalgebra $\mathfrak{sw}_-$ also is a projective representation. The corresponding central extension is the *Neveu–Schwarz super-Virasoro algebra*. For formulas for operators, see, e.g., [25, Subsect. 4.2.2].

Remark The *Ramond super-Virasoro algebra* arises in a similar way: in (76), we assume $a, b \in C^\infty(S^1)$, and get another "super-Witt" algebra $\mathfrak{sw}_+$. Next, in (77), we assume $f, g \in C^\infty(S^1)$. Repeating the construction above, we get an embedding of $\mathfrak{sw}_+$ to an infinite-dimensional Lie superalgebra $\mathfrak{osp}(2\infty|2\infty + 1)$; a construction of the spinor representation for this algebra must be slightly modified (cf. [50], Subsect. III.3.4, VII.3.4). We get a projective representation of $\mathfrak{sw}_+$ or a linear representation of the Ramond Lie superalgebra. ⊠

Question 10.3

(a) *Integrate these representations of the Neveu–Schwarz and Ramond algebras to actions of the corresponding supergroups by Gauss–Berezin operators.*
(b) *Extend highest weight representations of the Ramond and Neveu–Schwarz supergroups to complex semigroups as in* [49], [50], Sect. VII.4-5.

In this context, it is more natural to consider a more general family of super-Virasoro algebras (see [33, 28]).

10.4 "Unitary Representations" of Supergroups

The Howe duality for orthosymplectic spinors exists, and it was a subject of numerous works, e.g., [54, 15, 14, 41, 16]. In particular, this automatically produces many representations of supergroups, which can be extended to Grassmannians.

As far as I know, a notion of *a unitary representation of a Lie supergroup* is commonly recognized. Consider a finite-dimensional *real* Lie superalgebra $\mathfrak{g} = \mathfrak{g}_{\overline{0}} \oplus \mathfrak{g}_{\overline{1}}$, and let $G_{\overline{0}}$ be the Lie group, corresponding to $\mathfrak{g}_{\overline{0}}$. We consider a super-Hilbert space $H = H_{\overline{0}} \oplus H_{\overline{1}}$ and a representation of $\mathfrak{g}$ such that the action of $\mathfrak{g}_{\overline{0}}$ corresponds to a unitary representation of $G_{\overline{0}}$, and for each $X \in \mathfrak{g}_{\overline{1}}$, the operator $\sqrt[4]{-1}X$ is self-adjoint.

Under these conditions, for each $X \in \mathfrak{g}_{\overline{1}} = (\sqrt[4]{-1}X)^2$, the operator $\sqrt{-1}X^2$ is positive (semi)definite. On the other hand, $X^2 = \frac{1}{2}[X, X]_s$ is a generator of $\mathfrak{g}_{\overline{0}}$. In unitary representations, generators of Lie algebras rarely have spectra supported by a semi-axis. This places strong constrains on the Lie algebra $\mathfrak{g}_{\overline{0}}$ and its representation τ in H. In the semi-simple case, this implies (see [55]) that τ is an element of Harish–Chandra [29] highest weight holomorphic series. In particular, few semi-simple real supergroups have non-trivial unitary representations in this sense[34]. This was observed at least in works by Furutsu and Nishiyama [23, 24]. Such "unitary representations" of superalgebras are nice counterparts of the Harish–Chandra holomorphic series and admit explicit classification (Jakobsen [31]), which is a counterpart of the Howe–Enright–Wallach classification of unitary highest weight representations.

Conjecture 10.4 *Any unitary representation of a classical Lie supergroup admits an extension to a certain domain in a certain Grassmanian as in our Theorems* 9.3–9.4.

Acknowledgments The work was supported by grants FWF, projects P19064 and P31591, grant NWO.047.017.015, grant JSPS-RFBR-07.01.91209, and MSHE "Priority 2030" strategic academic leadership program.

I am grateful to D. V. Alekseevski, A. L. Onishchik, and A. S. Losev for explanations of super-algebra and super-analysis. The preliminary variant [51] of this chapter was revised after a discussion with D. Westra, who proposed numerous suggestions to improve the text.

References

1. Adams, J.D.: Discrete spectrum of reductive dual pair $(\mathrm{O}(p, q), \mathrm{Sp}\,(2m))$. Invent. Math. **74**, 449–475 (1983)
2. Arnold, V.I.: Mathematical Methods of Classical Mechanics. Springer, Heidelberg (1978)

[34] Both highest weight and lowest weight representations of a supergroup, say $\mathrm{OSp}(2p|2q; \mathbb{R})$, can be unitary, but they correspond to different $\sqrt[4]{-1}$ (see [23]). For this reason, a tensor product of a highest weight and a lowest weight unitary representation is not unitary. Counterparts of principal series representations for supergroups exist (since there are flag supervarieties and line bundles over them), but they are not unitary. This restricts possible ways for applications of Lie superalgebras to special functions. This is also an obstacle for an extension of Olshanski theory [56] of representations of infinite-dimensional real classical groups to supergroups.
So a picture is poor comparatively to classical representation theory. It remains a minor hope that a notion of unitary representations of supergroups admits some extension.

3. Barbier, S., Claerebout, S., De Bie, H.: A Fock model and the Segal-Bargmann transform for the minimal representation of the orthosymplectic Lie superalgebra osp$(m, 2|2n)$. SIGMA Symmetry Integrability Geom. Methods Appl. **16** , Paper No. 085, 33 pp. (2020)
4. Barbier, S., Frahm, J.: A minimal representation of the orthosymplectic Lie supergroup. Int. Math. Res. Not. IMRN **2021**, 16359–16422 (2021)
5. Berezin, F.A.: Canonical operator transformation in representation of secondary quantization. (Russian) Dokl. Akad. Nauk SSSR **137**, 311–314 (1961); Translated as Soviet Physics Dokl. **6** 212–215 (1961)
6. Berezin, F.A.: The method of second quantization. (Russian) Nauka, Moscow (1965); English transl.: Academic Press, New York (1966)
7. Berezin, F.A.: Automorphisms of Grassmann algebra. Math. Notes, **1**(3), 180–184 (1967)
8. Berezin, F.A.: Introduction to algebra and analysis with anticommuting variables. (Russian), edited by Kirillov, A.A., Palamodov, V.P., Moscow State University, Moscow (1983); English transl.: Introduction to superanalysis. Edited and with a foreword by Kirillov A.A. With an appendix by Ogievetsky V. I. D. Reidel Publishing Co., Dordrecht (1987); Second Russian edition: edited by Leites, D.A. with additions by Leites, D.A., Shander, V.N., Shchepochkina, I.M., MCCME publishers, Moscow (2013)
9. Berezin, F.A., Kats, G.I.: Lie groups with commuting and anticommuting parameters. Math. USSR-Sb. **11**(3), 311–325 (1971)
10. Berezin, F.A., Tolstoy, V.N.: The group with Grassmann structure UOSP(1.2). Comm. Math. Phys. **78**(3), 409–428 (1980/81)
11. Bernstein, I.N., Leites, D.A., Shander, V.N.: Seminar on supersymmetries. (Russian) MCCME Publishers, Moscow (2011)
12. Carmeli, C., Caston, L., Fioresi, R.: Mathematical Foundations of Supersymmetry. EMS Press, New York (2011)
13. Cartan, É.: Les groupes projectifs qui ne laissent invariante aucune multiplicité plane. (French) Bull. Soc. Math. Fr. **41**, 53–96 (1913)
14. Cheng, S.-J., Wang, W.: Howe duality for Lie superalgebras. Compos. Math. **128**(1), 55–94 (2001)
15. Cheng S.-J., Zhang, R.B.: Howe duality and combinatorial character formula for orthosymplectic Lie superalgebras. Adv. Math. **182**(1), 124–172 (2004)
16. Coulembier, K.: The orthosymplectic supergroup in harmonic analysis. J. Lie Theory **23**, 55–83 (2013)
17. Deligne, P., Morgan, J.W.: Notes on supersymmetry (following Joseph Bernstein): In: Deligne, P., et al. (eds.) Quantum fields and strings: a course for mathematicians, vol. 1, pp. 41–97. American Mathematical Society, Providence, RI (1999)
18. DeWitt, B.: Supermanifolds, 2nd edn. Cambridge University Press, Cambridge (1992)
19. Eisenbud, D.: Commutative Algebra. Springer, Berlin (1995)
20. Fioresi, R., Gavarini, F.: Chevalley Supergroups. Mem. Am. Math. Soc. **215**, 1014 (2012)
21. Frappat, L., Sciarrino, A., Sorba, P.: Dictionary on Lie Algebras and Superalgebras. Academic Press, San Diego (2000)
22. Friedrichs, K.O.: Mathematical aspects of the quantum theory of fields. Interscience Publishers, Inc., London (1953)
23. Furutsu, H., Nishiyama, K.: Classification of irreducible super-unitary representations of $\mathfrak{su}(p, q/n)$. Comm. Math. Phys. **141**(3), 475–502 (1991)
24. Furutsu, H., Nishiyama, K.: Realization of irreducible unitary representations of $\mathfrak{osp}(M/N; \mathbb{R})$ on Fock spaces. In: Kawazoe, T., Oshima T., Sano S. (eds.) Representation theory of Lie groups and Lie algebras (Fuji-Kawaguchiko, 1990), pp.1–21. World Scientific, River Edge, NJ (1992)
25. Green, M.B., Schwarz, J.H., Witten, E.: Superstring Theory, vol. 1. Introduction. Cambridge University, Cambridge (1987)

26. Golfand, Yu.A., Likhtman, E.P.: Extension of the algebra of Poincaré group generators and violation of P-invariance. JETP Lett. **13**, 422–455 (1971). Reprinted In: Salam, A., Sezgin, E. (eds.) Supergravities in Diverse Dimensions. Commentary and Reprints, vol. 1, pp. 20–23. World Scientific, Singapore (1988)
27. Grozman, P.: Invariant bilinear differential operators. Commun. Math. **30**(3), 129–188 (2022)
28. Grozman, P., Leites, D., Shchepochkina, I.: Lie superalgebras of string theories. Acta Math. Vietnam. **26**(1), 27–63 (2001)
29. Harish-Chandra: Representations of semisimple Lie groups. IV. Amer. J. Math. **77**, 743–777 (1955)
30. Howe, R.: Transcending classical invariant theory. J. Am. Math. Soc. **37**, 535–552 (1989)
31. Jakobsen, H.P.: The full set of unitarizable highest weight modules of basic classical Lie superalgebras. Mem. Am. Math. Soc. **111**, 116pp. (1994)
32. Kac, V.G.: Lie superalgebras. Adv. Math. **26**(1), 8–96 (1977)
33. Kac, V.G., van de Leur, J.W.: On classification of superconformal algebras. In: Gates, S.J., Jr., Preitschopf, C.R., Siegel, W. (eds.) Strings '88, pp. 77–106. World Scientific, Teaneck, NJ (1989)
34. Karabegov, A., Neretin, Yu., Voronov, Th.: Felix Alexandrovich Berezin and his work. In: Kielanowski, P., Ali, S.T., Odzijewicz, A., Schlichenmaier, M., Voronov, T. (eds.) Geometric methods in physics, pp. 3–33. Birkhäuser/Springer, Basel (2013)
35. Kashiwara, M., Vergne, M.: On the Segal-Shale-Weil representations and harmonic polynomials. Invent. Math. **44**(1), 1–47 (1978)
36. Khudaverdian, H., Voronov, Th.: Thick morphisms of supermanifolds, quantum mechanics, and spinor representation. J. Geom. Phys. **148**, 103540 14 pp. (2020)
37. Kirillov, A.A.: The orbits of the group of diffeomorphisms of the circle, and local Lie superalgebras. Funct. Anal. Appl. **15**(2), 135–137 (1981)
38. Kolář, I., Michor, P.W., Slovák, J.: Natural Operations in Differential Geometry. Springer, Berlin (1993)
39. Lavaud, P.: Superpfaffian. J. Lie Theory **16**(2), 271–296 (2006)
40. Leites, D.A.: Introduction to the theory of supermanifolds. Russ. Math. Surv. **35**(1), 1–64 (1980)
41. Leites, D., Shchepochkina, I.: The Howe duality and Lie superalgebras. In: Duplij, S., Wess, J. (eds.) Noncommutative Structures in Mathematics and Physics (Kiev, 2000), pp. 93–111. Kluwer, Dordrecht (2001)
42. Manin, Yu.I.: Gauge Field Theory and Complex Geometry. Springer, Berlin (1997)
43. Milnor, J.W., Moore, J.C.: On the structure of Hopf algebras. Ann. Math. (2) **81**, 211–264 (1965)
44. Molotkov, V.: Infinite-dimensional and colored supermanifolds. J. Nonlinear Math.Phys. **17**(Suppl. 1), 375–446 (2010)
45. Nazarov, M., Neretin, Yu., Olshanskii, G.: Semi-groupes engendrés par la représentation de Weil du groupe symplectique de dimension infinie. (French) C. R. Acad. Sci. Paris, Sér. I Math. **309**(7), 443–446 (1989)
46. Neretin, Yu.A.: Unitary representations with highest weight of the group of diffeomorphisms of a circle. Funct. Anal. Appl. **17**(3), 235–237 (1983)
47. Neretin, Yu.A.: Spinor representation of an infinite-dimensional orthogonal semigroup and the Virasoro algebra. Funct. Anal. Appl. **23**(3), 196–207 (1990)
48. Neretin, Yu.A.: On a semigroup of operators in the bosonic Fock space. Funct. Anal. Appl. **24**(2), 135–144 (1990)
49. Neretin, Yu.A.: Holomorphic continuations of representations of the group of diffeomorphisms of the circle. Math. USSR, Sb. **67**(1), 75–97 (1990)
50. Neretin, Yu.A.: Categories of Symmetries and Infinite-Dimensional Groups. Oxford University Press, New York (1996)
51. Neretin, Yu.A.: Gauss–Berezin integral operators and spinors over supergroups $\mathrm{OSp}(2p|2q)$, Preprint ESI-1986 (2007). https://www.esi.ac.at/material/scientific-publications/archive

52. Neretin, Yu.A.: "Method of second quantization" of Berezin. View 40 years after. In: Leites, D., Minlos, R.A., Tyutin, I. (eds.) Recollections about Felix Alexandrovich Berezin, the discoverer of supersymmetries, pp. 59–111. MCCME Publishers, Moscow (2008); French translation (La méthode de la seconde quantification de F. A. Berezin regards quarante ans plus tard) In: Anné C., Roubtsov V. (eds.) Les "supermathématiques" et F. A. Berezin, pp. 15–58, Sér. T, Soc. Math. France, Paris (2018)
53. Neretin, Yu.A.: Lectures on Gaussian integral operators and classical groups. EMS Press, Helsinki (2011)
54. Nishiyama, K.: Super dual pairs and highest weight modules of orthosymplectic algebras. Adv. Math. **104**, 66–89 (1994)
55. Olshanski, G.I.: Invariant cones in Lie algebras, Lie semigroups and the holomorphic discrete series. Funct. Anal. Appl. **15**(4), 275–285 (1981)
56. Olshanski, G.I.: Unitary representations of infinite-dimensional pairs (G, K) and the formalism of R. Howe. In: Vershik, A.M., Zhelobenko, D.P. (eds.) Representation of Lie groups and related topics , pp. 269–463. Gordon and Breach, New York (1990)
57. Olshanski, G.I.: The Weil representation and the norms of Gaussian operators. Funct. Anal. Appl. **28**(1), 42–54 (1994)
58. Salam, A., Strathdee , J.A.: Supersymmetry and Nonabelian Gauges. Phys. Lett. B. **51**(4), 353–355 (1974)
59. Sato, M., Miwa, T., Jimbo, M.: Studies on holonomic quantum fields. I. Proc. Japan Acad. Ser. A Math. Sci. **53**(1), 6–10 (1977)
60. Schmitt, T.: Supergeometry and quantum field theory, or: what is a classical configuration? Rev. Math. Phys. **9**(8), 993–1052 (1997)
61. Segal, G.: Unitary representations of some infinite-dimensional groups. Comm. Math. Phys. **80**(3), 301–342 (1981)
62. Segal, I.E.: Foundations of the theory of dynamical systems of infinitely many degrees of freedom (1), Mat.-Fys. Medd. K. Dan. Vidensk. Selsk. **31**(12), 1–39 (1959)
63. Serganova, V.V.: Classification of real simple Lie superalgebras and symmetric superspaces. Funct. Anal. Appl. **17**(3), 200–207 (1983)
64. Sergeev, A.: The invariant polynomials on simple Lie superalgebras. Represent. Theory **3**, 250–280 (1999)
65. Serov, A.A.: The Clifford–Weyl algebra and the spinor group. In: Onishchik, A.L. (ed.) Problems in group theory and homological algebra (Russian), pp. 143–145. Yaroslav Gosudarstvennogo Universiteta, Yaroslavl (1985)
66. Shale, D.: Linear symmetries of free boson fields. Trans. Am. Math. Soc. **103**, 149–167 (1962)
67. Shale, D., Stinespring, W.F.: Spinor representations of infinite orthogonal groups. J. Math. Mech. **14**, 315–322 (1965)
68. Varadarajan, V.S.: Supersymmetry for Mathematicians: An Introduction. American Mathematical Society, Providence, RI (2004)
69. Volkov, D.V., Akulov, V.P.: Is the neutrino a Goldstone particle? Phys. Lett. **B46**, 109–112 (1973)
70. Volkov, D.V., Soroka, V.A.: Higgs effect for Goldstone particles with spin 1/2. JETP Lett. **18**(8), 312–314 (1973)

Toward Gan-Gross-Prasad-Type Conjecture for Discrete Series Representations of Symmetric Spaces

Bent Ørsted and Birgit Speh

It is a pleasure to dedicate this article to Toshiyuki Kobayashi, whose work is fundamental to the understanding of the restriction of representations of reductive groups to subgroups.

Abstract In this chapter, we consider the unitary symmetric spaces of the form $X = U(p,q)/U(1)U(p-1,q)$ and their discrete series representations. Inspired by the work of A. Venkatesh and Y. Sakellaridis on L-groups of p-adic spherical spaces Sakellaridis and Venkatesh (Asterisque 396, 2017) , we formulate and prove natural relative branching laws for the restriction of such discrete series representations to smaller groups of the same type and corresponding unitary symmetric spaces. Using period integrals, some results on the Laplacian of spheres, Schlichtkrull et al. (Sao Paulo J Math Sci 12(2), 295–358, 2018) and results by Kobayashi (Adv Math 388, 38, 2021), we prove an analogue of these conjectures for rank-one unitary symmetric spaces.

1 Introduction

In this chapter, we shall study in detail some branching laws for unitary representations corresponding to a pair of Lie groups $G' \subset G$. We consider a unitary irreducible representation Π of G in a Hilbert space $\mathcal{H}$ and its restriction to the subgroup G'. The branching law gives the decomposition of $\mathcal{H}$ as a direct integral of unitary irreducible representations π of G', possibly with multiplicities. This is the analogue of spectral decomposition of a self-adjoint linear operator on $\mathcal{H}$, and

B. Ørsted
Department of Mathematics, Aarhus University, Aarhus, Denmark
e-mail: orsted@math.au.dk

B. Speh (✉)
Department of Mathematics, Malott Hall, Cornell University, Ithaca, NY, USA
e-mail: speh@math.cornell.edu; bes12@cornell.edu

M. Pevzner, H. Sekiguchi (eds.), *Symmetry in Geometry and Analysis, Volume 2*,
Progress in Mathematics 358, https://doi.org/10.1007/978-981-97-7662-7_10

of special importance is the discrete spectrum. In broad terms, we shall consider families of representations Π as irreducible invariant subspaces of $L^2(G/H)$, the so-called discrete series of a symmetric space G/H (where H is non-compact). Their Langlands parameters are determined by H. Schlichtkrull, [28]. They can also be parametrized using the L-group of the symmetric space (see [19] and [26]). The motivation for this work is to obtain branching laws for discrete series representations of symmetric spaces, which are similar to the branching laws of the Gross-Prasad conjectures for discrete series representations of the orthogonal group (see [8]). We are in particular interested in the representations π in the branching law, which belong to a similar family coming from the discrete spectrum of a symmetric subspace $G'/H' \subset G/H$. This is what we mean by relative branching, and we hope to demonstrate that it is a natural concept, also related to properties of special functions, viewed as spherical functions (or lowest K-types) for unitary representations.

There are essentially three issues to clarify in this context, namely:

- The representations Π appear as irreducible subspaces of the left regular representation $L^2(G/H)$, concretely for $G = U(p,q)$, and H is either $U(1,0) \times U(p-1,q)$ or $U(0,1) \times U(p,q-1)$. They are obtained by cohomological induction from a character of a θ-stable parabolic subgroup. Considered as members of a Arthur-Vogan packet, they also have an associated epsilon character on $\mathbb{Z}_2^3$ [1].
- The subgroup G' in the branching law that we study is $G' = U(p-1,q)$; here, the choice of the imbedding is important, and the parameters for the representations π are again the cohomological parameter, i.e., a character of a θ-stable parabolic subgroup and an epsilon character using the data corresponding to a discrete series representation for a subspace G'/H'.
- There are essentially two qualitatively different situations that we have to consider to determine the branching corresponding to $G' \subset G$. In one case, there is both a continuous spectrum (where purely algebraic methods are difficult to apply) and where we focus on the discrete spectrum. In the other case, the spectrum is purely discrete (and even admissible, i.e., the decomposition is a direct sum of irreducible representations with finite multiplicity).

The representations occur in natural families as a family of representations in the discrete spectrum of spherical spaces G/H, but the representations Π are typically not discrete series representations of G; hence, they are not covered by the prediction of the original Gan-Gross-Prasad conjectures for tempered representations. In special cases, if $|p-q| \leq 1$, they may be covered by the conjectures in [7].

We shall state the parameters of the representations in a way that makes the branching law look simple and natural and use in the proof the idea of branching in stages corresponding to a sequence $G' \subset G \subset G_1$ for a convenient group G_1; we shall also use certain period integrals to illustrate the explicit coupling between a Π and a π in the branching law (representations of G resp. G').

To explain in a little more detail our ideas, consider first compact groups. A classical problem in the representation theory of a connected compact group

Lie group G is the restriction of a representation of G to a (usually connected) subgroup G'. The restriction is a direct sum of irreducible representations of the subgroup. To characterize the representations of G' that occur in the restriction with positive multiplicities i.e., to determine the branching rules, involves complicated combinatorics. In *relative branching*, we consider a pair $G' \subset G$ compact groups, a family $\mathcal{F}$ of representations G with a given geometric realization and a family $\mathcal{F}'$ of representations of G', to have a similar geometric realization. We ask for pairs Π, π with $\Pi \in \mathcal{F}$ and $\pi \in \mathcal{F}'$ with $\mathrm{Hom}_{G'}(\pi, \Pi_{|G'}) \neq 0$. For example, we might have a symmetric subgroup $H \subset G$ and the corresponding family of spherical representations V of G, which are realized in the functions on $X = G/H$. Now we consider the similar family for G'. We consider $H' = G' \cap H$ (assume G' is stable under the symmetry defining H) and $X' = G'/H'$, and representations of G' are realized in the functions on X'. We want to find the multiplicities

$$\dim \mathrm{Hom}_{G'}(V \otimes (V')^*, \mathbb{C})$$

of the H'-spherical representations V' of G' in a given H-spherical representation V of G. A geometric natural way to do this is to study the *period integral*

$$I = \int_{X'} \psi(x')\psi'(x')dx',$$

i.e., the natural pairing of the functions $\psi \in V \subset C^\infty(G/H)$ and $\psi' \in (V')^* \subset C^\infty(G'/H')$. In particular, we may ask for the convergence and the non-vanishing (and exact value) of I for H-invariant spherical functions ψ, respectively, H'-invariant spherical functions ψ'. Even in the compact case, it is an interesting conjecture that the period integral for spherical functions of finite dimensional representations is nonzero if and only if the H'-spherical V' occurs in the branching from V. These questions also make sense for non-compact groups, as we shall see and study in this chapter.

For unitary representations of noncompact reductive Lie groups G, the situation is more complicated since the restriction of an irreducible representation Π to a subgroup G' usually does not decompose into a direct sum of irreducible unitary representations, and we may have both a continuous and also a discrete spectrum in the direct integral decomposition (over the unitary dual of G') of the restriction. Nevertheless, considering $G \times G/G$ as a symmetric space and the discrete series representations as its discrete spectrum, period integrals are important for the restriction of discrete series representations to large subgroups [22].

In a series of papers starting 1991, B. Gross and D. Prasad [8] introduced conjectures for the restriction of discrete series representations of orthogonal groups to orthogonal subgroups. These were generalized by Gan, Gross, and Prasad to the restriction of discrete series representations of unitary groups $U(p, q)$ to unitary subgroups [6] and more recently to Arthur packets. They were proved by Hongyu He for real unitary groups [11] and by R. Beuzart-Plessis in the p-adic case [4]. An important role in these conjectures is played by Vogan packets and also by interlacing relations between the infinitesimal characters of the discrete series representations.

The goal of this chapter is to generalize this circle of ideas as well as the considerations for orthogonal groups in the article "Hidden Symmetries" by T. Kobayashi and one of the authors [18]. In this, we use ideas of A. Venkatesh and Y. Sakellaridis, who define for a spherical space $X = G/H$ a complex Levi subalgebra $\mathfrak{l}_X$ and an L-group [26]. In this chapter, we consider several symmetric spaces G/H for a unitary group $G = U(p,q)$ and $H = G^\sigma$ for an involution σ, and we assume that these symmetric spaces have both the same complex Levi subalgebra $\mathfrak{l}_X = gl(n-2,\mathbb{C}) \oplus \mathbb{C}^2$ and that for both groups, the Lie algebra of its L-group is $sp(2,\mathbb{C})$ (as defined by A. Venkatesh and Y. Sakellaridis). The unitary symmetric spaces satisfying these assumptions are

$$X^+ = G/H^+ = U(p,q)/U(1,0)U(p-1,q)$$

and

$$X^- = G/H^- = U(p,q)/U(0,1)U(p,q-1).$$

Here H^+ is the stabilizer of the first basis vector e_1 and H^- the stabilizer of e_{p+1}. For X^+, the Levi subgroup is

$$L^+ = U(1,0)U(1,0)U(p-2,q),$$

and for X^-, it is

$$L^- = U(p,q-2)U(0,1)U(0,1).$$

To avoid considering special cases, we assume in this chapter

$$p \geq 3 \text{ and } q \geq 3.$$

Under this assumption, all the symmetric spaces we consider have a nonempty discrete L^2 spectrum.

We consider a pair of representations Π^+ and Π^- in the discrete spectrum of $L^2(X^+)$, respectively, $L^2(X^-)$, with the same infinitesimal character

$$(a, \frac{p+q-3}{2}, \frac{p+q-5}{2}, \ldots, -\frac{p+q-5}{2}, -\frac{p+q-3}{2}, -a.)$$

These representations are cohomologically induced from a character of a θ-stable parabolic with Levi subgroup $L^+ = U(1,0)U(1,0)U(p-2,q)$,, respectively, $L^- = U(0,1)U(0,1)U(p,q-2)$. See [23] for the exact description of the θ-stable parabolic and the character. The work of H. Schlichtkrull and E. van den Ban implies that both these representations have multiplicity one in the discrete spectrum of exactly one of the symmetric spaces [3]. Following C. Moeglin and D. Renard, we consider an Arthur packet $A(\phi)$, which contains both of these representations and epsilon characters of the centralizer of the Arthur parameter. For proofs of the results that we use and further details about Arthur parameters of the representations $\Pi^{+/-}$, $\pi^{+/-}$, see [19] sections 3–9, case 3.

Let $G' = U(p-1,q)$ be the subgroup that fixes the basis vector e_p. We have symmetric spaces

$$\begin{aligned} Y^+ &= G'/H^+ \cap U(p-1,q) \\ &= U(p-1,q)/U(1,0)U(p-2,q) \end{aligned}$$

and

$$\begin{aligned} Y^- &= G'/H^- \cap U(p-1,q) \\ &= U(p-1,q)/U(p-1,q-1)U(0,1). \end{aligned}$$

The symmetric spaces Y^+ and Y^- have the same L-group and the same complex Levi subalgebra, and their discrete spectrum is not empty. The representations in the discrete L^2 spectrum are also cohomologically induced from a θ-stable parabolic subgroup, are in the same Arthur packet, and have an epsilon character, which has been determined by C. Moeglin and D. Renard.

We consider in this chapter the relative restriction of representations Π^+, Π^- in the discrete spectrum of the symmetric spaces $L^2(X^+)$, and $L^2(X^-)$ to the discrete spectrum of $L^2(Y^+)$, respectively, $L^2(Y^-)$. More precisely, we consider pairs $\Pi^{+/-}$ and $\pi^{+/-}$ with $\pi^{+/-}$ in the discrete spectrum of $L^2(Y^+)$, respectively, $L^2(Y^-)$, and ask if there are nontrivial symmetry breaking operators in

$$\mathrm{Hom}_{G'}((\Pi^{+/-})^\infty, (\pi^{+/-})^\infty).$$

A *relative branching law* for representations Π and π in the discrete spectrum of two symmetric spaces X, Y with $Y \subset X$ is a rule that describes the pairs (Π, π), so that

$$\mathrm{Hom}_{G'}(\Pi^\infty_{|G'}, \pi^\infty) \neq 0.$$

To simplify the notation, we will in the following omit the superscript ∞.

The goal of this chapter is to find and discuss the relative branching laws for the restriction of representations in the discrete spectrum of $L^2(X^+)$ and $L^2(X^-)$ and those in the discrete spectrum of $L^2(Y^+)$ and $L^2(Y^-)$. To understand the branching, we study both the analytic models of the unitary representations appearing in the discrete spectrum of the symmetric spaces and the corresponding periods integrals.

We follow the conventions of the article by C. Moeglin and D. Renard [19] for the parametrization of cohomological induced representations. Let $a \geq \frac{p+q}{2}$ be a nonnegative integer if $p+q$ is odd and a positive 1/2 integer if $p+q$ is even.

We use the notation Π_a^+ and Π_a^- for the representations of $U(p, q)$ in the discrete spectrum of X^+, respectively X^-, with infinitesimal character

$$(a, \frac{p+q-3}{2}, \frac{p+q-5}{2}, \ldots, -\frac{p+q-5}{2}, -\frac{p+q-3}{2}, -a.)$$

Let b be a non-negative integer if $p+q-1$ is even and a 1/2 integer if $p+q-1$ is odd, and assume that $b \geq \frac{p+q-1}{2}$. We use the notation π_b^+ and π_b^- for the representations of $U(p-1, q)$ in the discrete spectrum of Y^+, respectively Y^-, with infinitesimal character

$$(b, \frac{p+q-4}{2}, \frac{p+q-6}{2}, \ldots, -\frac{p+q-6}{2}, -\frac{p+q-4}{2}, -b).$$

Our first conclusion are as follows: Recall that we are considering in the following symmetry breaking operators defined only on the C^∞ vectors of the representation, i.e., on the Casselman-Wallach realization of the representations.

Theorem 1 *Assume that G and G′ are as discussed above and that representations in the discrete spectrum of the symmetric spaces are parametrized as above.*

1. *Then*

$$dim\ Hom_{G'}(\Pi_a^+{}_{|G'}, \pi_b^+) = 1$$

if and only if $a > b$ *(and zero otherwise) and*

$$dim\ Hom_{G'}(\Pi_a^+{}_{|G'}, \pi_b^-) = 0$$

.

2. *Furthermore*

$$dim\ Hom_{G'}(\Pi_a^-{}_{|G'}, \pi_b^-) = 1$$

if and only if $a < b$ *(and zero otherwise) and*

$$dim\ Hom_{G'}(\Pi_a^-{}_{|G'}, \pi_b^+) = 0.$$

Note the difference here between cases 1 and 2: finitely, many representations in this relative branching in the first case and infinitely many in the second case. This is (as we shall see in the proof) due to non-admissible and admissible branching.

In Sect. 5, we reformulate this theorem using interlacing patterns of infinitesimal characters and epsilon-characters.

We keep the notations and assumptions of the previous theorem. Define

$$A_S(a) = \Pi^+(a) \oplus \Pi^-(a).$$

Similarly, we define for $G' = U(p, q-1)$ the representations π_b^+ and π_b^- and define

$$B'_S(b) = \pi_b^+ \oplus \pi_b^- .$$

Following the ideas in [8], we consider

$$\mathrm{Hom}_{G'}(A_S(a)_{|G'}, B_S(b)).$$

The formulation of conjectures by B. Gross and D. Prasad for discrete series representations of orthogonal groups, our relative branching laws, and the observation that only one of the integers $2a, 2b$ is even inspired the following reformulation of Theorem 1.

Theorem 2 *Assume that $p, q > 3$ and $p \neq q$ and that a and b satisfy the previous assumptions. Using the above notation, we have*

$$\dim \mathrm{Hom}_{G'}(A_S(a)_{|G'}, B_S(b)) = 1.$$

The chapter is organized as follows: In Sect. 2, we introduce our notation. In Sect. 3, we discuss the symmetric spaces, their L-groups, and the representations in the discrete spectrum. In Sect. 4, the concept of relative branching is introduced and illustrated by examples. Relative branching laws are formulated in Sect. 4, as well as some interpretations and a similarity to a branching law to Gross-Prasad branching laws using the epsilon characters of the representations. This is one of our key motivations to state the relative branching law in terms of parameters coming from the theory of spherical spaces. In Sect. 6, we return to the proofs of the branching laws using some classical techniques; in particular, we identify

$$Y^+ = U(p, q)/U(1)U(p-1, q-1)$$

with an orbit of G' on X^+. Flensted-Jensen in his seminal work [5] constructed discrete series for some symmetric spaces G/H, and we shall use his realization of the representations in our special case $L^2(X^+)$. We consider a geometric restriction of the representations in the discrete spectrum of $L^2(X^+)$ to $L^2(Y^+)$ by using the restriction of the K-finite coefficients of the Flensted-Jensen representation to Y^+ and compute the corresponding period integrals. Some of the technical points use recent work of T. Kobayashi [16] on $O(p, q)$, as well as work of H. Schlichtkrull, P. Trapa, and D. Vogan [29] on $U(p, q)$. Note that results in [29], which we use in this chapter, were first obtained by T. Kobayashi in [K1]. In the last part of Sect. 6, we prove the exhaustion of the branching law predicted by the period integrals. In the proof, we consider the sequences

1. $U(p-1, q) \subset O(2p-2, 2q) \times O(2) \subset O(2p, 2q)$
2. $U(p-1, q) \subset U(p, q) \subset O(2p, 2q)$.

We consider restriction in stages along either sequence of groups, and we obtain thus information about the first branching in the second line. In Sect. 7, we discuss branching laws for Π^- [30]. The restriction is an infinite sum of irreducible representations, and we determine the irreducible summands, which are in the discrete spectrum of $L^2(Y^-)$. We determine again K-finite functions in $\Pi^- \subset L^2(X^-)$, which generate the irreducible $(\mathfrak{g}', K')$ modules of these summands, and then show that the restriction of these functions to $Y^- \subset X^-$ is nonzero. The restriction of Π^- to G' was determined using different techniques and in greater generality by Yoshiki Oshima around 2014, and his results show that there are representations in $\Pi^-_{|G'}$, which are not in the discrete spectrum of $L^2(Y^-)$. In Appendix 1, we consider the restriction of discrete series representations of X^+ for $G = U(2, n)$. This case is not covered by in the main part of the article, since p=2. These representations are also a discrete series representation of G. We use here the results of H. He [11]. In Appendix 2, we discuss relative branching and period integrals for compact Riemann symmetric spaces of rank one, extending the discussion also to the quaternionic case.

The second author was introduced to the ideas of Y. Sakellaridis and A. Venkatesh about discrete series representations of spherical spaces in a workshop organized at AIM by A. Venkatesh. She would also like to thank D. Vogan, J. Adams, and D. Renard for helpful discussions about Arthur packets, S. Sahi for many conversations about orthogonal polynomials, and Yoshiki Oshima for sharing his manuscript about branching for a family of cohomologically induced representations of unitary groups. We would like to thank the anonymous referees for helpful suggestions. Much of the joint work was done using the facilities of the virtual AIM research group "Representation theory and non commutative Operator algebras."

Notation $\mathbb{N} = \{0, 1, 2, \ldots, \}$ and $\mathbb{N}_+ = \{1, 2, \ldots, \}$.

2 Notation and Generalities

We introduce in this section the notation and recall some basic results. Consider the Hermitian quadratic form $Q(Z, Z)$ on $\mathbb{C}^{p+q}$ with signature (p, q). We choose the basis $e_1, \ldots, e_p, e_{p+1}, \ldots e_{p+q}$ so that the form is positive on $e_1, \ldots, e_p$ and negative on $e_p, e_{p+1}, \ldots e_{p+q}$ and

$$Q(Z, Z) = z_1\bar{z}_1 + \cdots + z_p\bar{z}_p - z_{p+1}\bar{z}_{p+1} - \cdots - z_{p+q}\bar{z}_{p+q}. \tag{1}$$

We define $G = U(p, q)$ to be the unitary group that preserves the quadratic form Q. Let H^+ be the stabilizer of the vector e_1, and let H^- be the stabilizer of the vector e_{p+1}. Then H^+ is isomorphic to $U(1, 0)U(p - 1, q)$, and H^- is isomorphic to $U(p, q - 1)U(0, 1)$. We consider the unitary symmetric spaces

$$X^+ = G/H^+ = U(p, q)/U(1, 0)U(p - 1, q)$$

and

$$X^- = G/H^- = U(p,q)/U(0,1)U(p,q-1).$$

Let $G' = U(p-1,q)$ be the subgroup that fixes the vector e_p. We consider the symmetric spaces

$$Y^+ = G'/G' \cap H^+ = U(p-1,q)/U(1,0)U(p-2,q)$$

and

$$Y^- = G'/G' \cap H^- = U(p-1,q)/U(0,1)U(p-1,q-1).$$

Consider now the stabilizer $\tilde{G}' \subset G$ of the line through e_p. Then

$$\tilde{G}' = U(p-1,q)U(1,0).$$

It is the fixed-point set of a involution of G. Furthermore,

$$\tilde{H}'^{+} = H^+ \cap \tilde{G}' = U(1,0)U(1,0)U(p-2,q)$$

and

$$\tilde{H}'^{-} = H^- \cap \tilde{G}' = U(1,0)U(p-1,q-1)U(0,1).$$

Thus, we observe

$$\begin{aligned}\tilde{G}'/\tilde{H}'^{+} &= U(p-1,q)U(1,0)/U(1,0)U(p-2,q)U(1,0)\\ &= U(p-1,q)/U(1,0)U(p-2,q)\\ &= G'/H^+ \cap G' = Y^+\end{aligned}$$

and

$$\begin{aligned}\tilde{G}'/\tilde{H}'^{-} &= U(p-1,q)U(1,0)/U(1,0)U(p-1,q-1)U(0,1)\\ &= U(p-1,q)/U(p-1,q-1)U(0,1)\\ &= G'/H^- \cap G' = Y^-.\end{aligned}$$

The maximal compact subgroups of G, G' and H^+ H^- are denoted by K, K', respectively, K_H^+ and K_H^-. The Lie algebras of the groups are denoted by the corresponding lowercase Gothic letters. We use the subscript $_{\mathbb{C}}$ to denote their complexification.

The representations of the larger group G are always denoted by capital Π and those of the smaller group G' by π.

As mentioned in the introduction, to avoid considering special cases, we make in this article the following:

Assumption O

$$p \geq 3 \text{ and } q \geq 3.$$

3 The Symmetric Spaces and Their Discrete Series Representations

In 2017, Y. Sakellaridis and A. Venkatesh published their book about spherical varieties and periods [26], which includes several conjectures of about spherical spaces and representations in their discrete L^2 spectrum. Symmetric spaces X^+ and X^- are spherical varieties, and in this case, conjectures by Y. Sakellaridis and A. Venkatesh have been verified by C. Moeglin and D. Renard [19]. In this section, we recall and summarize those results that are relevant to our article.

In [26], Y. Sakellaridis and A. Venkatesh introduced for a spherical space $X = \hat{G}/\hat{H}$ the concept of a complex Levi subalgebra of the complex Lie algebra of $\hat{\mathfrak{g}}_{\mathbb{C}}$. Symmetric spaces X^+ and X^- are spherical spaces and have the same complex Levi algebra $\mathfrak{l}_X = gl(1)^2 \oplus gl(p+q-2, \mathbb{C})$ [19].

Y. Sakellaridis and A. Venkatesh also defined for a spherical space an L-group. Following the arguments of C. Moeglin, D. Renard, F. Knop, and B. Schalke [13], X^+ and X^- have the same L-group, which we denote by $\check{G}_X$. It is $Sp(2, \mathbb{C}) \times W_{\mathbb{R}}$ (page 11 in [19]). The case $p+q$ odd is problematic, and C. Moeglin and D. Renard use in their proof of the conjecture of Y. Sakellarides and A. Venkatesh the group $\check{G}_X = SO(2, \mathbb{C}) \rtimes W_{\mathbb{R}}$. See Remark 8.3 in [19].

Definition We say that an irreducible representation Π of $U(p,q)$ is a discrete series representation of a symmetric space X for G or equivalently is in the discrete spectrum of a symmetric space X if

$$Hom_G(\Pi, L^2(X)) \neq 0.$$

The work of T. Oshima and T. Matsuki [23] shows that symmetric spaces X^+ and X^- both have discrete series representations. Their Harish-Chandra modules are cohomologically induced from θ-stable parabolic subalgebras $\mathfrak{q}^+$ and $\mathfrak{q}^-$. C. Moeglin and D. Renard show that their Levi subgroups are

$$L^+ = U(1,0)U(1,0)U(p-2,q),$$

respectively,

$$L^- = U(0,1)U(0,1)U(p,q-2).$$

Assume we have (for the standard torus) a positive root system with only one non-compact *simple* root, i.e., that the simple compact roots are $e_i - e_{i+1}$ for $i = 1, \dots, p-1$ and $i = p+1, \dots, p+q-1$ and the simple non-compact root is $e_p - e_{p+1}$. The representations Π^+, respectively, Π^- are parametrized by the characters (of the Levi subgroups)

$$\lambda^+ = (a_1, -a_1, 0, \dots 0)$$

with, respectively,

$$\lambda^- = (0, \dots, 0, a_2, -a_2)$$

with a_i, integers if n is odd and a_i 1/2 integers if n is even. Furthermore, $a_i \geq \frac{p+q-1}{2}$ [19]. We write $\Pi^+_{a_1}$ and $\Pi^-_{a_2}$, but if there is no confusion possible, we drop the subscript. The minimal K-type of $\Pi^+_{a_1}$ is trivial on $U(q)$, and on the other factor, the highest weight is

$$\lambda^+_K = (a_1 - \frac{p+q-1}{2} + q, 0, \dots, 0, -a_1 + \frac{p+q-1}{2} - q)$$

The minimal K-type of $\Pi^-_{a_2}$ is trivial on $U(p)$, and on the other factor, the highest weight is

$$\lambda^-_K = (a_2 - \frac{p+q-1}{2} + q, 0, \dots, 0, -a_2 + \frac{p+q-1}{2} - q)$$

Knapp and Vogan [12]. To simplify the formulas for the highest weights of the K-types, we introduce the notation $a_0 = a - \frac{p+q-1}{2}$. The representations $\Pi^+_{a_1}$, $\Pi^-_{a_2}$ have the same infinitesimal character if $a = a_1 = a_2$. In this case, we use the notation Π^+_a and Π^-_a.

A discrete series representation Π of a symmetric space $X \in \{X^+, X^-\}$ has multiplicity

$$\dim \mathrm{Hom}_G(\Pi, L^2(X)) = 1.$$

See, for example, the survey article by E. van den Ban [3].

An Arthur parameter is a homomorphism $\phi : W_{\mathbb{R}} \times SL(2,\mathbb{C}) \to {}^L U(n)$, so it can be considered as a n (here $n = p+q$) dimensional representation of $W_{\mathbb{R}} \times SL(2,\mathbb{C})$. It is of "bonne parité" if the condition in the first paragraph of 8.3 in [19] or the previous article on Arthur packets of unitary groups by C. Moeglin and D. Renard [20] is satisfied. D. Renard and C. Moeglin verified that the representations Π_a^+, Π_a^- in the discrete spectrum of the symmetric spaces X^+, respectively, X^-, are both in the same Arthur packet $\mathbb{A}(\phi)$ and that the Arthur parameter is of *bonne parité*. For more details, see [19].

In [1], J. Adams, D. Barbasch, and D. Vogan defined for an Arthur parameter ϕ a group

$$\mathcal{A}(\phi) = \text{normalizer } (\phi)/\text{centralizer}(\phi)$$

For a fixed Arthur parameter ϕ, the representations in the Arthur packet of ϕ are parametrized by a characters of $\mathcal{A}(\phi)$. This parametrization depends on the choice of a normalization [1]. In our case, in the notation of C. Moeglin and D. Renard, $\phi = (\chi_a \boxtimes R[1]) \oplus (\chi_0 \boxtimes R[p+q-2]) \oplus (\chi_{-a} \boxtimes R[1]$, and so

$$\mathcal{A}(\phi) = \mathbb{Z}_2 \times \mathbb{Z}_2 \times \mathbb{Z}_2$$

with generators e_1, e_2, e_3. C. Moeglin and D. Renard determine the epsilon characters of the representations as follows:

$$\epsilon(\Pi_a^+) = (1, (-1)^{p+q-2+\frac{(p+q-2)(p+q-3)}{2}}, (-1)^{p+q-1})$$
$$\epsilon(\Pi_a^-) = (-1, (-1)^{p+2q-4+\frac{(p+q-2)(p+q-3)}{2}}, (-1)^{p+q})$$

We observe that not all representations in the Arthur packet $\mathcal{A}(\phi)$ are in the discrete spectrum of a symmetric space. They correspond to the remaining six characters of $\mathcal{A}(\phi)$.

We have the exact sequence of groups

$$Z_{p+q} \to U(1) \times SU(p,q) \to U(p,q)$$

where Z_{p+q} is the set of the $(p+q)$'th roots of unity. We identify a root of unity $\zeta \in U_{p+q}$ with the pair

$$\{(\zeta, \zeta I_{p+q}) \subset U(1) \times SU(p,q)$$

in the center of $U(1) \times SU(p,q)$. U_{p+q} as well as the first factor $U(1)$ act trivially on X^+ and X^-, and thus, both X^+ and X^- are also symmetric spaces for $SU(p,q)$ and have an action of the covering group $U(1) \times SU(p,q)$. The restriction of a representation χ of $U(1)$ to the group Z_{p+q} of $(p+q)$'th roots of unity defines a character χ_1, and an irreducible representation τ of $SU(p,q)$ restricted to the center of $SU(p,q)$ defines a character τ_1. Then the representation $\chi \times \tau$ of $U(1) \times SU(p,q)$ defines a representation of $U(p,q)$ if $\tau_1 = \bar{\chi}_1$. Here $\bar{\chi}$ denotes complex conjugate character. This allows us to lift a representation of $U(p,q)$ to a representation of $U(1) \times SU(p,q)$ via this choice of a character.

Conversely, suppose now that $\check{\Pi}^+$ is a representation in the discrete spectrum of $L^2(SU(p,q)/S(U(1)U(p-1,q))$. To show that it extends to a representation Π of $U(p,q)$, it suffices to show that the center Z_{p+q} of $SU(p,q)$ acts trivially on $\check{\Pi}^+$. Since the center Z_{p+q} acts as a scalar on the irreducible representation $\check{\Pi}^+$ and the center is contained in the maximal compact subgroup, it suffices to show that it acts trivially on the highest weight of the minimal K-type of $\check{\Pi}^+$. Thus, it follows from the formulas in Sect. 6 for the highest weight of the minimal K-types of Π^+ and Π^-.

The discussion of symmetric spaces Y^+ and Y^- and their discrete series representations is exactly analogous to the discussion for X^+ and X^-. We denote the discrete series representations by the letters π^+ and π^-, Since Y^+ and Y^- can also be considered as symmetric spaces of $U(p-1,q)U(1,0)$, we identify π^+ and π^- with $\pi^+ \times$ trivial and $\pi^- \times$ trivial. We have the epsilon characters

$$\epsilon'(\pi_b^+) = (1, (-1)^{p+q-3+\frac{(p+q-3)(p+q-4)}{2}}, (-1)^{p+q-2})$$

$$\epsilon'(\pi_b^-) = (-1, (-1)^{p+2q-5+\frac{(p+q-3)(p+q-4)}{2}}, (-1)^{p+q-1})$$

4 Restriction of Representations and Relative Branching Laws

We introduce in this section *relative branching* and *strong relative branching* for representations in the discrete spectrum of symmetric spaces and discuss some examples.

4.1 Branching

We first discuss the basics about restrictions of infinite dimensional representations and then introduce relative branching and strong relative branching.

Let Π be a unitary representation of G. We say that a unitary representation π of a subgroup G' is a subrepresentation of $\Pi_{|G'}$ if there is a nontrivial G'-equivariant operator *continuous with respect to the unitary topology* so that

$$\mathrm{Hom}_{G'}(\pi, \Pi_{|G'}) \neq 0$$

i.e., if π is a direct summand of the unitary restriction of Π to G'. In this chapter, branching will always refer to unitary branching. A *unitary branching law* is a rule describing the pairs of irreducible unitary representations Π, π so that π is a direct summand of $\Pi_{|G'}$.

Consider groups $G' \subset G$ with symmetric spaces Y, X with $Y \subset X$. Let Π be a unitary representation in the discrete spectrum of X and π a unitary representation in the discrete spectrum of Y. We consider now Casselman-Wallach realizations of representations Π and π of G and G' on the Frechet spaces and consider the G'-equivariant symmetry breaking operators $\mathrm{Hom}_{G'}(\Pi_{|G'}, \pi)$. For a precise definition and discussion of symmetry breaking operators, see [15].

A *relative branching law* for representations in the discrete spectrum of the symmetric spaces X, Y with $Y \subset X$ is a rule that describes the pairs (Π, π) so that

$$\mathrm{Hom}_{G'}(\Pi_{|G'}, \pi) \neq 0.$$

We obtain an important symmetry breaking operator as follows: Consider Π as subrepresentation of $L^2(X)$ and π as subrepresentation of $L^2(Y)$ satisfying the relative branching law. We observe that the restriction $Res_{X \to Y}$ of $C^\infty(X)$-functions to Y induces a symmetry breaking operator from the representation of G on $L^2(X)$ to the representation of G' on $L^2(Y)$. It is defined only on sufficiently smooth vectors (see [21]) and not on the whole representation and is continuous with respect to the Frechet topology.

We say that π is a strong relative restriction of Π if π is a relative restriction of Π and

$$Res_{X \to Y} : C^\infty(X) \cap \Pi \to C^\infty(Y) \cap \pi$$

is a nontrivial symmetry breaking operator. The strong relative restriction law is a set of pairs (Π, π) so that π is a strong relative restriction of Π.

Note that not every symmetry breaking operator is of the above form, and so not every above form and so Π, π may satisfy a relative restriction law, but not a strong relative restriction law.

4.2 Relative Branching for Compact Unitary Symmetric Spaces and Period Integrals

The inclusion of the compact symmetric space Y into the compact symmetric space X induces a push forward of distributions $D(Y)$ into $D(X)$. An interesting distribution from the representation theoretic perspective is integration of a function $F(x)$ in $C^\infty(X)$ (sufficiently regular in order to make the integral converge) and a function $f(y)$ in $C^\infty(Y)$, which transform according to an irreducible finite dimensional representation of G' on $C^\infty(Y)$. The integral

$$F \to \int_Y (f(y)F(y))dy$$

is often referred to as a period integral.

Using the example of the compact group $U(n)$, we discuss in this subsection the differences of branching, relative branching, and strong relative branching using period integrals. We will also consider later other examples of relative branching for representations for non-compact low-dimensional symmetric spaces.

For the group $G = U(n)$, consider the symmetric space (complex projective space)

$$X = U(n)/U(1)U(n-1).$$

The representations of G are parametrized by their highest weight, an n-tuple $(a_1, \ldots, a_n)$ of integers. It defines a representation of $SU(n)$ if the a_i sum up to zero. In particular, a representation has an $U(1)U(n-1)$-fixed vector if the highest weight is $(a, 0, \ldots, 0, -a)$. We restrict the representations to a subgroup $U(n-1)$, which fixes the vector e_n. Then $U(n-1) \cap U(1)U(n-1) = U(1)U(n-2)$, and so get a symmetric space

$$Y = U(n-1)/U(1)U(n-2) \subset X.$$

The classical branching rule for the restriction to a subgroup $U(n-1)$ shows that all representations of $U(n-1)$ with highest weight $(b, 0, \ldots, c)$, which satisfy $0 \leq b \leq a$ and $-a \leq c \leq 0$, are nonzero in the restriction. Representations have a nonzero $U(-1)U(n-2)$ invariant vector iff $c = -b$. So the relative branching is expressed in highest weights.

Lemma 1 (Relative Branching Law for Finite Dimensional Representation of U(n)) *Let Π be a representation in the discrete spectrum of X with highest weight $(a, 0, \ldots, 0, -a)$ and π be a representation in the discrete spectrum of Y with highest weight $(b, 0, \ldots, 0, -b)$. Then*

$$Hom_{G'}(\Pi_{|G'}, \pi) \neq 0$$

iff $a \geq b \geq 0$. In this case, its dimension is one.

The relative branching law is proved using period integrals and Jacobi polynomials in Sect. 6. It states that in this case, the strong relative branching law and the relative branching law are identical. Indeed, what we shall see there is that the strong relative restriction law is obtained simply be restricting a spherical function of $G = U(n)$ to the subgroup $G' = U(n-1)$ and integrating over the subgroup against a spherical function for the subgroup.

4.3 Relative Branching for Discrete Series Representations

A familiar case of relative branching is the restriction of discrete series representations of $U(p, q)$ to $U(p-1, q)$. Here we consider the discrete series representations of $U(p, q)$ as discrete series representations of the symmetric space $X = U(p, q) \times U(p, q)/U(p, q)$, where we embed $U(p, q)$ diagonally into the product. We define the symmetric space Y similarly for $U(p-1, q)$ and consider it as an orbit of $U(p-1, q)$ on X. Then the discrete spectrum of the unitary restriction of Π to $U(p-1, q)$ is isomorphic to its relative discrete spectrum. To see this in general, consider

$$X = G \times G/\text{diagonal } G.$$

Then

$$L^2(X) = \sum \Pi \otimes \Pi^* \oplus \int \Pi \otimes \Pi^*$$

where we symbolically write the discrete part of the spectrum as a sum and the continuous part of the spectrum as an integral. Now the restriction to a reductive $G' \subset G$ will in a similar way decompose as (for every Π)

$$\Pi_{|G'} = \sum \pi_i \oplus \int \pi_j$$

(there could be multiplicities here) and hence the tensor product decomposes as

$$(\Pi \otimes \Pi^*)|_{G'} = (\sum \pi_i \oplus \int \pi_j) \otimes (\sum \pi_i \oplus \int \pi_j)^*.$$

Now we see that the spherical part of (i.e., the relative) branching law amounts to taking the diagonal parts here, and the relative discrete spectrum becomes exactly, for a fixed discrete series Π, the sum

$$\sum \pi_i \otimes {\pi_i}^*$$

which as a discrete part of a branching law is isomorphic to the one for the restriction of the discrete series representation Π restricted to G'. Note here the general principle that the continuous spectrum cannot contribute to the discrete spectrum in this branching—and so the relative branching law is equivalent to the unitary branching law. The unitary branching law for the restriction of discrete series representations of $U(p, q)$ to $U(p-1, q)$ was proved by H. He in 2015. [11]

In the case of representations of a compact group $G = U(p)$ and $G' = U(p-1)$

$$X = G \times G/\text{diagonal } G$$

and

$$L^2(X) = \sum \Pi \otimes \Pi^*,$$

where we sum over all irreducible finite dimensional representations. The analogue of this period integral considered in the previous example can be written as an identity of the characters χ of the representations.

$$I(\chi_\Pi, \chi_\pi) = \int_{G'} \chi_\Pi(x)\chi_\pi(x^{-1})dx.$$

Hence, this counts the multiplicity of π in $\Pi_{|G'}$.

We extend in this chapter the considerations of relative branching to a wider class of discrete series representations. The restriction of a discrete series Π^-, Π^+ of

$$X^- = U(p,q)/U(1)U(p,q-1), \text{ respectively, } X^+ = U(p,q)/U(1)U(p-1,q),$$

to

$$Y^- = U(p-1,q)/U(1)U(p-1,q-1) \text{ respectively,}$$
$$Y^+ = U(p-1,q)/U(1)U(p-2,q-1)$$

$p > 2$, will be discussed in Sects. 6 and 7, and we will prove a relative branching law.

4.4 Period Integrals

Recall that for irreducible representations $\Pi \subset L^2(G/H)$ and $\pi \subset L^2(G'/H')$ in the discrete spectrum, we have

$$\mathrm{Hom}_{G'}(\Pi_{|G'}, \pi) = \mathrm{Hom}_{G'}(\Pi_{|G'} \otimes \pi^\vee, \mathbb{C}),$$

where $\pi^\vee$ is the contragredient representation of π.

We define a G'-invariant linear functional on the C^∞ functions in $\Pi \subset L^2(X)$ by a *period integral*

$$\int_{G'/H'} f(x')f'(x')dx'.$$

Here f is a C^∞-function in $\Pi \subset L^2(X)$, and f' is a C^∞-function in $\pi \subset L^2(Y)$. The technical issue here is if this integral makes sense for the appropriate pairs of functions in the representation spaces. Typically, we consider K-finite, respectively,

K'-finite functions, or smooth vectors (see [21]). In Sect. 6, we will consider the convergence of this integral for pairs (Π^+, π^+) and determine for which pairs it is not zero. We observe that this linear functional depends only on the restriction of the function f to G'/H'. We will use period integrals in Sect. 6 to obtain a strong relative branching law for the restriction of a representation Π_a^+ to G'.

If $X = G/H$ is a compact symmetric space for a compact group G and Y a symmetric space $G'/(H \cap G')$, period integrals can also help obtain strong relative branching laws. If G is a classical and X a rank-one symmetric space, then classical formulas for Jacobi polynomials show that strong relative branching and relative branching coincide. We conjecture that this is always true.

5 The Main Results

In this section, we state main results of the relative branching laws for the restriction of G' of the representations Π_a^+ and Π_a^- in the discrete spectrum of the symmetric spaces X^+, respectively, X^-, and relates our relative branching laws to the formulas and ideas of Gross-Prasad and Gan-Gross Prasad [8, 6, 19]. This allows us to deduce Theorem 2 in the introduction.

5.1 Relative Branching Laws and Interlacing Patterns

We state the relative branching laws for the restriction to G' of the representations Π_a^+ and Π_b^- and deduce Theorem 1 in the introduction. Using interlacing patterns of parameters, we deduce Theorem 2 in the introduction.

Recall that $2a$ is an odd integer if $p+q$ is even and an even integer if $p+q$ is odd. Furthermore, $2b$ is an even integer if $p+q-1$ is odd and an odd integer if is even. We denote the representations by Π_a^+ or sometimes $\Pi_a(\epsilon_1)$, respectively, π_b^+ or $\pi_b(\epsilon_1)$. Recall assumption O in Sect. 2.

Theorem 3 *Assume that G and G' satisfy assumption O and that $a \geq \frac{p+q-1}{2}$ and $b \geq \frac{p+q-2}{2}$. Then*

$$dim\ Hom_{G'}(\Pi_a^+, \pi_b^+) = 1$$

if $a > b \geq \frac{p+q-2}{2}$ and zero otherwise. Furthermore

$$dim\ Hom_{G'}(\Pi_a^+, \pi_b^-) = 0$$

for all $b \geq \frac{p+q-2}{2}$.

We denote the representations by Π_a^- or sometimes $\Pi_a(\epsilon_2)$, respectively, π_b^- or $\pi_b(\epsilon_2)$.

Theorem 4 *Assume that G and G′ satisfy assumption O and that* $a \geq \frac{p+q-1}{2}$ *and* $b \geq \frac{p+q-2}{2}$. *Then*

$$dim\ Hom_{G'}(\Pi_a^-, \pi_b^+) = 0$$

for all $b \geq \frac{p+q-2}{2}$. *Furthermore*

$$dim\ Hom_{G'}(\Pi_a^-, \pi_b^-) = 1$$

iff $b > a \geq \frac{p+q-1}{2}$ *and zero otherwise.*

Theorem 3 and 4 are proved in Sects. 6 and 7.

Combining relative branching laws for Π^+ and Π^- proves the theorem 1 in the introduction.

We simplify the statements of the theorems with following definition (see also the article by C. Moeglin and D. Renard [19])

Definition A pair a, b of half integers has the *property RB,* if

1. $2a = 0 \bmod 2$ and $2b = 1 \bmod 2$ if p+q is odd
   and $2a = 1 \bmod 2$ and $2b = 0 \bmod 2$ if p+q is even
2. $a - \frac{p+q-1}{2} \in \mathbb{N}$ and $b - \frac{p+q-2}{2} \in \mathbb{N}$.

Remarks

1. For a discussion of the first condition RB 1, see the first part of Sect. 3.
2. The condition RB 2 implies that the infinitesimal character of the representations is sufficiently nonsingular and its underlying $(\mathfrak{g}, K)$ module is a cohomologically induced representation in the good range (see [12]).

Ordering the numbers $a, -a, b, -b$ in decreasing order, we have the interlacing patterns

$$P_1(a, b) : (a, b, -b, -a)$$

$$P_2(a, b) : (b, a, -a, -b).$$

Following the ideas of B. Gross and D. Prasad in [8], we reformulate our results. Assume that a, b has the property RB, and define

$$A_S(a) = \Pi_a^+ \oplus \Pi_a^-.$$

Similarly, we define

$$B'_S(b) = \pi_b^+ \oplus \pi_b^-.$$

We consider for $\pi_b \in \{\pi_b^+, \pi_b^-\}$

$$\dim \mathrm{Hom}_{G'}(A_S(a), \pi_b)$$

and for $\Pi_a \in \{\Pi_a^+, \Pi_a^-\}$

$$\dim \mathrm{Hom}_{G'}(\Pi_a, B_S(b)).$$

Thus

Corollary *Assume that $p, q > 3$ and $p \neq q$ and that a, b have property RB.*

1. *There exists exactly one representation $\pi_b \in \{\pi_b^+, \pi_b^-\}$ so that*

$$\dim \mathrm{Hom}_{G'}(A_S(a), \pi_b) \neq 0.$$

2. *There exists exactly one representation $\Pi_a \in \{\Pi_a^+, \Pi_a^-\}$ so that*

$$\dim \mathrm{Hom}_{G'}(\Pi_a, B_S(b)) \neq 0.$$

This implies

Corollary *Assume that $p, q > 3$ and $p \neq q$ and that a and b have the property RB. Then*

$$\dim \mathrm{Hom}_{G'}(A_S(a), B_S(b)) = 1.$$

This corollary is a reformulation of Theorem 2 in the introduction.

5.2 Characters $\mathcal{A}(\phi)$ and Relative Branching

The Gan-Gross-Prasad conjectures for the restriction of discrete series representation discrete series as proved by R. Beuzart-Plessis and H. He are phrased using a character on the group $\mathcal{A}(\phi) = \mathbb{Z}_2^{p+q}$. We show that the interlacing patterns of the infinitesimal characters of the representations can be used to define characters on the group $\mathcal{A}(\phi)$ introduced in Sect. 3. We reformulate the relative branching law relating our relative branching laws to the formulas and ideas of Gross-Prasad and Gan-Gross-Prasad [8, 6, 7, 19].

For an interlacing pattern, we define an ϵ character of $\mathcal{A}(\phi')$ of the group G'. Let $b(>)$ be the number of $a, -a$ in the interlacing pattern larger than b and $b(<)$ the number of $a, -a$ smaller than b in the pattern. Following the ideas of B. Gross and D. Prasad [8], we define

Definition Suppose that a, b have the property RB. Define

$$\check{\epsilon}'(e_1) = (-1)^{1+b(>)}$$
$$\check{\epsilon}'(e_3) = (-1)^{(p+q)-2+1+(-b)(>)}.$$

If a, b have the interlacing pattern P_1, then

$$\check{\epsilon}'(e_1) = 1 \text{ and } \check{\epsilon}'(e_3) = (-1)^{(p+q)-2}.$$

Hence, they are the first and third component of $\epsilon(\pi_b^+)$. (see Sect. 3 for a definition of the characters $\epsilon(\pi_b^+), \epsilon(\pi_b^-)$). If a, b have the interlacing pattern P_2, then

$$\check{\epsilon}'(e_1) = -1 \text{ and } \check{\epsilon}'(e_3) = (-1)^{(p+q)-1}.$$

They are equal to the first and third component of $\epsilon'(\pi_b^-)$. This determines $\check{\epsilon}'$ and thus shows that it is equal to $\epsilon'(\pi_b^+)$, respectively, to $\epsilon'(\pi_b^-)$. So the interlacing patterns determine the representations

$$P_1 \longleftrightarrow \pi_b^+ \tag{2}$$
$$P_2 \longleftrightarrow \pi_b^- . \tag{3}$$

Thus Theorems 1 and 2 can be reformulated as follows:

Theorem 5 *Let a, b be positive numbers with property RB.*

1. *Suppose that a, b satisfy the interlacing pattern P_1 and thus define a pair Π_a^+, π_b^+ of irreducible representations defined in the previous paragraph. Then*

$$\dim Hom_{G'}(\Pi_a^+, \pi_b^+) \neq 0.$$

2. *Suppose that a, b satisfy the interlacing pattern P_2 and thus define a pair Π_a^+, π_b^- of irreducible representations defined in the previous paragraph. Then*

$$\dim Hom_{G'}(\Pi_a^+, \pi_b^-) \neq 0.$$

An equivalent direct formulation avoiding the characters of $\mathcal{A}(\phi)$.

Corollary *Let a, b be positive numbers with the property RB. Then*

1. *a, b satisfy the interlacing pattern P_1 if and only if*

$$\dim Hom_{G'}(\Pi_a^+, \pi_b^+) \neq 0,$$

2. *a, b satisfy the interlacing pattern P_2 if and only if*

$$\dim Hom_{G'}(\Pi_a^-, \pi_b^-) \neq 0.$$

Remarks

1. Although the formulation in the corollary is simpler, the formulation in Theorem 5 suggests generalizations to other unitary symmetric spaces and other groups. A similar approach can be used to reformulate the results in the article "A hidden symmetry of a branching law" [18]. Note that in our case, the dimensions above are one iff they are nonzero.
2. The example in appendix 1 shows that the assumption $p, q > 2$ is necessary. The assumption that $p \neq q$ is not essential, but it is very helpful for the exposition.

6 Relative Branching for Π^+

In this section, we prove the relative branching law in Theorem 3. So we prove if $a \geq \frac{p+q-1}{2}$ and $b \geq \frac{p+q-2}{2}$. Then

$$\dim \mathrm{Hom}_{G'}(\Pi_a^+, \pi_b^+) = 1$$

if $a > b \geq \frac{p+q-2}{2}$ and zero otherwise. Furthermore,

$$\dim \mathrm{Hom}_{G'}(\Pi_a^+, \pi_b^-) = 0$$

for all $b \geq \frac{p+q-2}{2}$.
In the proof, we construct in 6.2 explicit non-vanishing period integrals on the unitary symmetric spaces X^+ and Y^+, thus obtaining a G'-invariant pairing between the discrete series Π_λ^+ and $\pi_{\lambda'}^+$. In the last part, we use branching in stages along the sequences:

- $U(p-1, q) \subset O(2p-2, 2q) \times O(2) \subset O(2p, 2q)$
- $U(p-1, q) \subset U(p, q) \subset O(2p, 2q)$

to complete the proof of a relative branching law for the discrete series of the unitary symmetric space. We will see that all G' invariant linear functionals on $\Pi_a \otimes \pi_b^*$ are obtained by period integrals.

6.1 Background

For the discrete series of an affine symmetric space G/H, there is a remarkable construction of a large part of this due to Flensted-Jensen [5]; this has, as a special case, that of the discrete series of a semisimple Lie group G as done by Harish-Chandra but with a different and independent proof. Using a Poisson transform for a certain dual group and its Riemannian symmetric space, a square-integrable function called ψ_λ on G/H is constructed; this function transforms as an irreducible representation of K, namely, as the corresponding lowest K-type of a discrete series representation of G on G/H, and it generates via the action of the Lie algebra the corresponding Harish-Chandra module. The highest weight μ_λ of the K-type is given explicitly in terms of λ, as is the infinitesimal character of the corresponding representation of G. Here the so-called Flensted-Jensen parameter λ is a linear functional on a certain Cartan subspace of the tangent space at the origin, generalizing the Harish-Chandra parameter for the discrete series in the group case.

Thus, here, we follow the notation of [5], where the parameter λ gives the infinitesimal character of the discrete series representation generated by the function ψ_λ as in formula (8.5) in that chapter. Hopefully, without causing confusion, we use that notation here, so we denote this Flensted-Jensen parameter λ in the formulas below (and in the Appendix). The corresponding lowest K-type $\mu = \mu_\lambda$ will be by an explicit Jacobi polynomial (see also Section 8 of [5]). The main point is that the lowest K-type has a parameter a (a half integer) as we used earlier; in [5], it is called n (an integer) and $n = 2a$ in the formulas that we shall see.

We recall the notation and general setup of the pseudo-Riemannian affine symmetric space $X = G/H$ following M. Flensted-Jensen. (For details, see [5] and [27]). Let K be the maximal compact subgroup $K \subset G$. We have a decomposition $G = (K/K_H)BH$ coming from a Cartan subspace B in the tangent space at the base point for $X = G/H$. If dim B=1 i.e., if rank X =1, then $X = (K/K_H)B$ where $K_H = H \cap K$. In our case, $H = U(1)U(p-1, q)$, and the rank of X is one. Thus, $X = (K/K_H)B$, where $K_H = U(1)U(p-1)U(q)$. So the symmetric space X^+ has coordinates $(U(p)U(q)/(U(1)U(p-1)U(q))\mathbb{R}^+$. We may assume that $G' = K'/K'_{H'}BH'$, i.e., that the two spaces X^+ and Y^+ have the same hyperbolic variable from B.

In 1980, M. Flensted-Jensen [5] discovered a family of irreducible unitary representations in the discrete spectrum $L^2(X^+)$ of some symmetric spaces for *connected semisimple* Lie groups. Although $U(p, q)$ is not semisimple, the considerations in Sect. 3 show that there is a 1-1 correspondence between the representations in the discrete spectrum of $U(p, q)/U(1)U(p-1, q)$ and $SU(p, q)/S(U(1)U(p-1, q))$ and that we can extend the Flensted-Jensen result to representations in the discrete spectrum of $L^2(X^+)$ for $p > 1$. Flensted-Jensen's proof is based on an explicit formula for functions ψ in the lowest K-type of representations in the discrete spectrum of $L^2(X^+)$ [5]. We refer to these functions as Flensted-Jensen functions and the corresponding representations as Flensted-Jensen representations. Flensted-Jensen representations are cohomologically induced from a character of a θ-stable parabolic with Levi subgroup $S(U(1)U(1)U(p-2,q))$, and their Langlands parameters were determined by H. Schlichtkrull [27, 28].

We recall now the explicit construction of M. Flensted-Jensen functions, i.e., the lowest K-types of the Flensted-Jensen representations in the discrete spectrum of $L^2(X^+)$. In this we follow here the notation in [5] and in particular the notation in the last section of [5] where rank-one cases are explained. In particular, we shall follow in subsections the notation there in considering the symmetric space

$$X = SU(p, q+1)/S(U(p, q) \times U(1)),$$

i.e., a switch from the space $SU(p+1, q)/S(U(p, q) \times U(1))$ considered previously by applying an outer isomorphism $SU(p, q+1) \to SU(q+1, p)$. This allows us to cite explicitly formulas of M. Flensted-Jensen. For a representation Π in the discrete spectrum of X, we determine those representations in the discrete spectrum of the restriction of Π to $G' = SU(p, q)$, which are in the discrete spectrum of

$$Y = SU(p, q)/S(U(p, q) \times U(1)).$$

In the proof, we use period integrals

$$I(f, f') = \int_Y f(y')f'(y')dy'$$

for functions $f(x) \in \Pi^+$, $f'(y) \in \pi^+$ and determine whether the integral converges and whether it is nonzero for Flensted-Jensen functions ψ and ψ'. Here $Y = G'/H'$, and $H' = G' \cap H$. The bilinear form $I(\ ,\ )$ is clearly G' invariant and hence gives information about the branching of the representation generated by restricting f from G to G' i.e., the *relative branching* in the sense of finding summands in the branching law of the same type, i.e., Flensted-Jensen representations in the discrete spectrum of Y.

We assume in considerations and calculations of Flensted-Jensen functions and period integrals that

$$G = SU(p, q+1), q > p > 0, \ \ H = S(U(p, q) \times U(1))$$

and H fixing the last coordinate. Note that now $K = S(U(p) \times U(q+1))$ and

$$K/K_H = SU(q+1)/S(U(q) \times U(1)) = P^q(\mathbb{C})$$

the projective space.

6.2 *Period Integrals of Flensted-Jensen Functions*

We will now find the explicit formulas for the Flensted-Jensen functions ψ_λ on X^+ corresponding to the lowest K-type that generates the discrete series representation Π_a^+ in $L^2(X^+)$. Note that by construction, the lowest K-type in this representation is spherical for K/K_H. Then we shall find the corresponding functions on Y^+.

Recall the Jacobi polynomials $P_n^{(\alpha,\beta)}(x)$ of degree n, which are usually normalized by

$$P_n^{(\alpha,\beta)}(1) = \frac{\Gamma(n+\alpha+1)}{\Gamma(n+1)\Gamma(\alpha+1)}.$$

Fact Flensted-Jensen functions on X are of product type, with s the radial variable,

$$\psi_\lambda(kb_s) = (\cosh s)^{-i\lambda-\rho} P_n^{(q-1,0)}(k)$$

for $k \in K,\ b_s \in B$. Here the parameters are (notation as in [5]) $i\lambda = q-p+n\ (n \in 2\mathbb{Z}^+)$, $\rho = p+q$, $\rho_\mathfrak{t} = q$, $\rho - 2\rho_\mathfrak{t} = p-q$, $\mu_\lambda = i\lambda + \rho - 2\rho_\mathfrak{t} = n$.

Jacobi polynomials $P_n^{(\alpha,\beta)}(x)$ with $\alpha = q-1, \beta = 0$ can be considered as spherical functions on the compact Riemannian symmetric space $P^q(\mathbb{C})$ (see [2, 10]). An important classical relation in our situation is

$$P_n^{(\gamma,0)}(x) = \frac{\Gamma(n+1)}{\Gamma(n+\gamma+1)} \sum_{k=0}^{n} \frac{\Gamma(k+\alpha+1)(2k+\alpha+1)}{\Gamma(k+1)} P_k^{(\alpha,0)}(x)$$

where $\gamma = \alpha + 1 = q-1$. This formula is crucial in obtaining the relative branching law for spherical representations from $SU(q+1)$ to $SU(q)$ on the complex projective spaces.

Example Jacobi polynomials corresponding to the projective line $P^1(\mathbb{C})$ are just Legendre polynomials, and spherical representations in this case are representations of $SU(2)$ with integer highest weight a of dimension $d = 2a+1$. We have then for the parameter n in Flensted-Jensen labels $n = 2a$, and the degree of the polynomial is a. In more detail, recall representations in the model where the space is

$$V_m = span\{z^m, z^{m-1}w, z^{m-2}w^2, \ldots, w^m\}$$

i.e., homogeneous polynomials in two variables z, w. The action is the linear action in the variables, and the torus $diag(e^{i\theta}, e^{-i\theta})$ acts with the eigenvalues $\{e^{im\theta}, e^{i(m-2)\theta}, \ldots, e^{-im\theta}\}$ and the representation is spherical for $m \in 2\mathbb{N}$. The spherical vector is $z^n w^n$, $m = 2n$, and the spherical polynomial is then obtained as the matrix coefficient of this vector under the rotations by the angle θ, i.e., it is proportional to

$$(z^n w^n, (\cos\theta \sin\theta z^2 + ((\cos\theta)^2 - (\sin\theta)^2)zw - \cos\theta \sin\theta w^2)^n)$$

using the inner product in V_m. For example, when $n = 2$, we get (a constant times) $3(\cos 2\theta)^2 - 1$, which is the Legendre polynomial of degree 2 in the variable $x = \cos 2\theta$; this is consistent with this being the spherical harmonic on the 2-sphere with the polar angle 2θ corresponding to the double covering $SU(2) \mapsto SO(3)$.

We will now compute the period integrals of Flensted-Jensen functions using branching on the (maximal compact subvarieties) $P^q(\mathbb{C})$ resp. $P^{q-1}(\mathbb{C})$. Consider the restriction to $G' = SU(p, q)$. We have similar formulas for the Flensted-Jensen functions on Y, and thus the corresponding period integral is

$$I(\psi_\lambda, \psi'_{\lambda'}) = \int_0^\infty \int_{P^{q-1}(\mathbb{C})} P_n^{(q-1,0)}(y) P_k^{(q-2,0)}(y) \cdot \\ \cdot (\cosh s)^{-2q-n} (\cosh s)^{-2q+2-k} dy D(s) ds$$

where dy is the invariant measure, and the density

$$D(s) = (\cosh s)^{2q-1} (\sinh s)^{2p-1}.$$

Here $i\lambda' = q - 1 - p + k\ (k \in 2\mathbb{Z}^+)$ is the Flensted-Jensen parameter for G'. The exact value may be calculated in terms of Gamma functions. Note that

$$A(\alpha, \beta) = \int_0^\infty (\sinh t)^\alpha (\cosh t)^{-\beta} dt = \frac{1}{2} B((\alpha - 1)/2, (\beta - \alpha)/2)$$

where the beta function $B(x, y) = \Gamma(x)\Gamma(y)/(\Gamma(x + y)$. So

$$I(\psi_\lambda, \psi'_{\lambda'}) = \\ = \int_0^\infty \int_{P^{q-1}(\mathbb{C})} P_n^{(q-1,0)}(y) P_k^{(q-2,0)}(y) (\cosh s)^{-2q+1-n-k} (\sinh s)^{2p-1} dy ds \\ = A(2p - 1, 2q + n + k - 1) \int_{P^{q-1}(\mathbb{C})} P_n^{(q-1,0)}(y) P_k^{(q-2,0)}(y) dy,$$

and the non-vanishing is governed by the classical formula above. Note also that the convergence is assured by $2p - 2q - n - k < 0\ (n \in \mathbb{N},\ k \in \mathbb{N})$. And observe finally that I is non-vanishing exactly for $0 \leq k \leq n$. (The degree of $P_n(x)$ is n, which here is half the parameter n in the Flensted-Jensen parametrization; see the example below.) Here, the parameters are n, k, and they are exactly the parameters a, b in the previous discussion.

Thus, restricting the spherical function on the left-hand side to the subgroup and integrating against a similar spherical function for the subgroup, we get a nonzero period integral precisely for the representations in Theorem 3. Here we have only treated the period integral, i.e., the explicit coupling between the large representation Π and the small one π for the lowest K-type. However, the same integrals exist and are convergent for all the K-types, resp., K' types in the modules, and it is again clearly invariant. This may be seen using the explicit labeling of the K finite vectors as in, for example, [25], where in formula (3.17), the K-finite basis is given (as spherical harmonics on a product of spheres times a radial dependence essentially (up to a polynomial in $\tanh(s)$) of the form $\cosh(s)^{-(L+p+q-2)}$); in (3.19), the radial decay is seen to be exactly the same as for the lowest K-type; and in (3.20), the labeling of all these is made explicit, consistent with the abstract results (the analogues of the formula by Blattner). This chapter dealt with the case of the real hyperboloids only, i.e., the group is $O(p, q)$, but the same will be the case for our $U(p, q)$, since those representations are restrictions from the real case. Hence, the period integrals give the invariant coupling between the full Harish-Chandra modules in the relevant cases.

Thus, we conclude that there is a relative branching of the discrete series representation Π_λ with lowest K-type ψ_λ in the discrete spectrum of X to the discrete series $\pi_{\lambda'}$ with lowest K-type $\psi'_{\lambda'}$ in the discrete spectrum of Y detected using period integrals containing the parameters (in terms of the degree of the corresponding Jacobi polynomials n, resp., n') $0 \leq n' \leq n$. This confirms part of the relative branching theorem.

6.3 *Branching in Stages*

We conclude the proof of the branching law by proving that a non-vanishing period integral is a necessary and sufficient condition for the relative branching of Π^+ and π^+. We combine results by T. Kobayashi [16] and H. Schlichtkrull, P. Trappa, and D. Vogan [29], respectively, T. Kobayashi [14] , and do the branching law in stages, based on the two sequences:

- $U(p-1, q) \subset O(2p-2, 2q) \times O(2) \subset O(2p, 2q)$
- $U(p-1, q) \subset U(p, q) \subset O(2p, 2q)$.

Step 1 *Setting the stage: Geometric background.*
Consider the real hyperboloid (over the real numbers)

$$\begin{aligned} H_{p,q}(\mathbb{R}) &= O(p, q)/O(p-1, q) \\ &= \{(x_1, x_2, \dots, x_n) \in \mathbb{R}^n \, | x_1^2 + \cdots + x_p^2 - x_{p+1}^2 - \cdots - x_n^2 = 1\} \end{aligned}$$

where $n = p + q$. Similarly we have the complex hyperboloid in complex space

$$
\begin{aligned}
H_{p,q}(\mathbb{C}) &= U(p,q)/U(p-1,q) \\
&= \{(z_1, z_2, \ldots, z_n) \in \mathbb{C}^n \,||z_1|^2 + \cdots + |z_p|^2 \\
&\quad -|z_{p+1}|^2 - \cdots - |z_n|^2 = 1\}
\end{aligned}
$$

again with $n = p + q$. We identify $H_{2p,2q}(\mathbb{R})$ and $H_{p,q}(\mathbb{C})$. We consider $H_{p,q}(\mathbb{C})/U(1)$, i.e., for the projective complex hyperboloid, we identify points on the circles obtained by the action of the center of $U(p,q)$ and conclude that

$$
\begin{aligned}
H_{p,q}(\mathbb{C})/U(1) &= U(p,q)/(U(1) \times U(p-1,q)) \\
&= SU(p,q)/S(U(1) \times U(p-1,q)) = X^+.
\end{aligned}
$$

In this way, we may identify functions on X^+ with U(1) invariant functions $H_{p,q}(\mathbb{C})$ and the discrete spectrum of the symmetric space X^+ with the $U(1)$ invariant part of the discrete spectrum of the real hyperboloid on $H_{2p,2q}(\mathbb{R})$. Here, representations in the discrete spectrum of $H_{2p,2q}(\mathbb{R})$ are denoted by $U(\ell)$ and sometimes to avoid confusions by $U^{SO(2p,2q)}(\ell)$. They are cohomologically induced from a maximal θ-stable parabolic subgroup with Levi subgroup $SO(2,0)SO(2p-2,q)$ and parametrized by an integer $\ell > p+q-1$. Here, $\ell - p - q + 1$ defines a character of $SO(2,0)$. We are restricting representations $U(\ell)$ to U(p-1,q) along both sequences.

Step 2 *From $O(2p,2q)$ to $U(p,q)$.*
We first restrict representations along the second sequence. The restriction of a representation $U(\ell)$ of $SO(2p,2q)$ to $U(p,q)$ is discussed in detail in Theorem 6.1 in [14]. The last section of [29] contains also an account of these results. We now recall the results about the discrete spectrum.
By Theorem 6.1 in [14]

$$
U(\ell)_{|U(p,q)} = \oplus_{x+y=\ell} \Pi^+_{(x,y)}.
$$

Representations $\Pi^+_{(x,y)}$ of U(p,q) are cohomologically induced from a parabolic subgroup with Levi subgroup $U(1,0)U(1,0)U(p-2,q)$. Here, $(x,-y)$ defines characters of $U(1,0)U(1,0)$, which satisfy $x+y=\ell$. Representations $\Pi^+_{(x,y)}$ are in the discrete spectrum of the symmetric space $X^+ = U(p,q)/U(1)U(p-1,q)$ iff $x = y = \ell/2$. Using the notation $\Pi^+_\ell := \Pi^+_{\ell,\ell}$, the relative branching law is $U(\ell), \Pi^+_{\ell/2}$.
From $O(2p,2q)$ to U(p,q). A different argument. A representation χ_s of $U(1)$ defines a line bundle on $U(p,q)/U(1)U(p-1,q)$. T. Kobayashi considers the discrete spectrum of $L^2(U(p,q)/U(1)U(p-1,q), \chi_s)$, and in [14], he

shows that representations $\Pi^-_{(x,y)}$, $x - y = s$ are in the discrete spectrum of $L^2(U(p,q)/U(1)U(p-1,q), \chi_s)$. Thus

$$U(\ell)_{|U(p,q)} \subset \oplus_{|s|<\ell} L^2(U(p,q)/U(1)U(p-1,q), \chi_s).$$

Step 3 *From U(p,q) to U(1)U(p-1,q).*
Observe that $U(1)$ commutes with $U(p-1,q)$; hence, its representation is a character. It also acts by this character on the highest weight vector of the minimal K-type. From [29], we deduce:

1. $U(1)$ acts trivially on discrete spectrum of the restriction of the representation $\Pi^+_{(x,x))}$ to $U(p-1,q)$.
2. If $x \neq y$, $U(1)$ does not act trivially on discrete spectrum of the restriction of the representation $\Pi^+_{(x,y)}$ to $U(p-1,q)$.

From steps 2 and 3 we conclude that the relative branching law (Π^+_ℓ, π) holds for representations in the discrete spectra of X^+ and Y^+ if and only if the relative branching law $(U(\ell), \pi)$ holds for representations in the discrete spectrum of $H_{2p,2q}(\mathbb{R})$ and Y^+.

Step 4 *From $O(2p, 2q)$ to $O(2,0)O(2p-2, 2q)$.*
We consider the restriction along the first sequence. We analyze this using Theorem 1.1 in [16]. The parameters of the representations $U(\ell') \otimes U(\ell")$ in the restriction of $U(\ell)$ to $O(2,0)O(2p-2,2q)$ are sets

$$\Lambda_{++}(\ell) = \{(\ell', \ell") : \ell - \ell' - \ell" - 1 \in 2\mathbb{N}\}$$

$$\Lambda_{+,-}(\ell) = \{(\ell', \ell") : \ell' - \ell" - \ell - 1 \in 2\mathbb{N}\}$$

$$\Lambda_{-,+}(\ell) = \{(\ell', \ell") : \ell" - \ell - \ell' - 1 \in 2\mathbb{N}\}.$$

Observe that

1. The set $\Lambda_{++}(\ell)$ is finite or empty. If $\ell' = 0$ and $U(0)$ the trivial representation, representations in the restriction of $U(\ell)$ are representations of $(O(2)O(2p-2,2q))$. They are discrete series reps of the symmetric space $H_{2p-2,2q}(\mathbb{R}) = O(2p-2,2q)/O(2,0)O(2p-4,2q)$ and $\ell - \ell" - 1 \in 2\mathbb{N}$.
2. The set $\Lambda_{+,-}(\ell)$ is empty if $\ell' = 0$.
3. The set $\Lambda_{-,+}(\ell)$ is also empty. See example 1.2 (2) in [16]

In this case, there is also a continuous spectrum in the restriction. Thus, the relative branching law is $U(\ell), U(\ell")$ with $\ell - \ell" - 1 \in 2\mathbb{N}$.

Step 5 *From $O(2p-2, 2q)$ to U(p-1,q).*
We repeat the argument in Step 2.
We conclude by combining the restrictions of $U(\ell)$ from $SO(2p, 2q)$ to $SO(2,0)SO(2p-2,2q)$ and then from $SO(2p-2,2q)$ to $U(p-1,q)$, where

considering representations $U(\ell")$ in the discrete spectrum of $H_{2p-2,2q}(\mathbb{R})$, we have a relative branching law $U(\ell)$, $\Pi^+_{\ell"}$ with $\ell - \ell" - 1 \in 2\mathbb{N}$. .

Step 6 *Summary and Conclusions.*
We now compare restrictions from $SO(2p, 2q)$ to $SO(2, 0)SO(2p-2, 2q)$ and $SO(2p - 2, 2q)$ to $U(p - 1, q)$ along the first sequence with our results of restricting along the second sequence.

In this first sequence, the restriction of a discrete series Π^O for $H_{2p,2q}(\mathbb{R})$ to $O(2) \times O(2p-2, 2q)$ has both a continuous spectrum and also a discrete spectrum [16]. Note that it is a direct sum with the summands parametrized by characters of $O(2)$. We are interested with representations of $U(p, q)$ and representations of $U(p - 1, q) \times U(1)$ with trivial action of the $U(1)$ in the upper-left corner. This implies that we consider only the summand that is invariant under the subgroup $SO(2)$ of $O(2)O(2p - 2, 2q)$. We observe

1. The direct summand of $U(\ell)_{|O(2)O(2p-2,2q)}$ that is invariant under $O(2)$ contains only finitely many irreducible representations $U(\ell'')$ in its discrete spectrum. Here $\ell - \ell'' - 1$ is an even positive integer.
2. In the continuous spectrum in the branching from $O(2p, 2q)$ to $O(2p - 2, 2q)$, there can be no discrete spectrum for $U(p - 1, q)$.

To see the second observation, suppose that in the continuous spectrum for the Hilbert space one representation had a restriction with a discretely occurring representation, then these would consist of vectors in the Hilbert space, a contradiction, since these only are weakly contained here. Another way to say this is that the direct integral decomposition of a unitary representation is a unique measure class, and hence the discrete part is unique.

Recall that restricting along the second sequence from step 2 and 3 we conclude, that the unitary relative branching law (Π^+_ℓ, π) for representations in the discrete spectra of $L^2(X^+)$, $L^2(Y^+)$ holds if and only if the relative branching law $U(\ell, \pi)$ holds in the discrete spectrum of $L^2(H_{2p,2q}(\mathbb{R}))$ and $L^2(Y^+)$.

Along the first sequence, the first restriction yields the discrete sum over $\pi^{U(p,q)}_{x,y}$, , $x+y = \ell$, $x, y \in \mathbb{Z}$. But since the center of $U(p, q)$ acts by the character of weight $x - y$, the relative part has $x = y$, so we only get one representation. We obtain the (finite, as it turns out) relative discrete spectrum restricting $\pi^{U(p,q)}_{x,x}$ to $U(p - 1, q)$ consisting of the sum of $\pi^{U(p-1,q)}_{a,a}$, $a \leq x$ in agreement with the exhaustion that we wanted. This completes the proof of Theorem 3.

We conclude this section by giving a different perspective on the branching in stages argument, highlighting crucial steps, and including a discussion of the continuous spectrum.

We performed the branching (using [29], Sections 8 and 11, and [16], Theorem 1.1) in stages so that it may be performed in two ways from $O(2p, 2q)$ to $U(p - 1, q)$, namely, along the first sequence, respectively, along the second sequence. In the first sequence, the restriction of a discrete series Π^0 for $H_{2p,2q}(\mathbb{R})$ to $O(2) \times O(2p-2, 2q)$ has both a continuous spectrum and also a discrete spectrum [16]. Note that the discrete spectrum is a direct sum with summands parametrized by characters of $O(2)$. We are interested in representations of $U(p,q)$, respectively, $U(p-1,q)$, which in the case of our relative setting means representations with a trivial action of the center of $U(p,q)$, respectively, $U(p-1,q)$ and representations of $U(p-1,q) \times U(1)$ with trivial action of the $U(1)$ in the upper-left corner. This implies that we consider only the summands that are invariant under the subgroup $SO(2)$ of $O(2)O(2p-2,2q)$. We observe that the direct summand of $\Pi^0_{|O(2)O(2p-2,2q)}$, which is invariant under $O(2)$, contains only finitely many irreducible representations in its discrete spectrum. Furthermore, we observe that in the continuous spectrum of the restriction from $O(2p-2,2q)$ to $U(p-1,q)$, there can be no discrete spectrum of $U(p-1,q)$. To see this observation, suppose that in the continuous spectrum for the Hilbert space, one representation had a restriction with a discretely occurring representation; then these would consist of vectors in the Hilbert space, a contradiction, since these only are weakly contained here.

Remarks

(1) The non-vanishing of the period integral and the above results about branching in stages imply that Theorem 3 is a strong relative branching law.
(2) Using the same results from [16] and [29] for the opposite case of restricting the same representation $\pi^{2p,2q}_{+,\lambda}$ to $O(2p,2q-2) \times O(0,2)$, we find a purely discrete spectrum. (this an admissible restriction), and we find an infinite family of representations in discrete spectrum of $L^2(Y^-)$. Details will be provided in the next section.
(3) As mentioned in Appendix 2, the quaternionic case could have been treated in a similar way by considering $O(4,0) \times O(4p-4,4q) \subset O(4p,4q)$.

7 Relative Branching for Π^-

We prove in this section the main theorem about the relative branching of (Π^-, π^-) for an irreducible representation $\pi^- \subset L^2(G'/H^- \cap G')$ with regular infinitesimal character. The proof in this section uses also restriction in stages and the same ideas as in the previous section. It is also possible to prove it using pseudo dual pairs and the ideas in [30]. This theorem was also proved using different ideas earlier by Y. Oshima. We thank him for informing us about this unpublished result [24].

The representation Π^-_a is cohomologically induced from a θ-stable parabolic subalgebra with Levi subgroup $U(p,q-2)U(0,1)U(0,1)$ and has parameter $(0,\dots 0,a,-a)$, which is trivial on $U(p,q-2)$. We assume that it is in the "good range," i.e., that $a - \frac{p+q-1}{2} \geq 0$. The infinitesimal character of Π^-_a is up to conjugation with an element in the Weyl group

$$(a, \frac{p+q-3}{2}, \dots, -\frac{p+q-3}{2}, -a),$$

and it is the same as the infinitesimal character of Π_a^+. The minimal K-type of Π_a^- is trivial on $U(p)$ and has highest weight

$$(a_0 + p, 0, \ldots, 0, -a_o - p)$$

on $U(q)$ where $a_0 = a - \frac{p+q-1}{2}$. Hence, the K-type is trivial on the center of K and has a $(K \cap K_H)$-fixed vector. The representation Π_a^- is the Langlands subrepresentation of a principal series representation $I(P^-, D, \nu)$ induced from a parabolic subgroup $P^- = M^- A^- N^-$, where A^- is the vector subgroup of $U(p, q-2)$ of dimension $min(q-2, p-2)$. Its centralizer M^- in G is a smaller unitary group, which is non compact for $p, q \geq 3$. (For details, see [12] XI.10).

Recall that the restriction of the representation Π_a^- to $G' \times U(1)$ is admissible, i.e., direct sum of irreducible representations, each occurring with multiplicity one. [17, 30]. We refer to an irreducible representation π^- of G' as *G'-type of* Π^- if there is a character χ_r of U(1) so that $\pi^- \otimes \chi_r$ is isomorphic to a summand of $(\Pi_a^-)_{|G' \otimes U(1)}$. We can use results about K-types to understand the restriction of admissible representations since the restriction of the representation Π_a^- to $G' \times U(1)$ is admissible. This implies that the corresponding $(\mathfrak{g}, K)$ module is admissible [14]. Using pseudo dual pairs, it is proved in Section 4 of [30] that the restriction of Π_a^- to $G_0' = G' \times U(1)$ is a direct sum of representations $\pi_{\mu_{G'}} \otimes \chi_r$. The $(\mathfrak{g}', K')$ module of the representation $\pi_{\mu_{G'}}$ of G' is cohomologically induced from a θ-stable parabolic $\mathfrak{q}'$ with Levi subgroup

$$L' = L \cap G' = U(0, 1)U(0, 1)U(p, q-3)$$

and parameter $\mu_{G'} = (a + 1/2 + l, -a - 1/2 - k, 0, \ldots 0), \quad l, k \in \mathbb{N}$.

Lemma 2 *The G'-types $\pi_{\mu_{G'}}$ of Π_a^- are in the discrete spectrum of $G'/G' \cap H^-$ iff $\chi_0 \otimes \pi_{\mu_{G'}}$ has a nontrivial multiplicity in the restriction of Π_a^- to $U(1) \times G'$. Here, χ_0 denotes the trivial character of $U(1)$.*

Proof Recall that $\pi_{\mu_{G'}}$ is in the discrete spectrum of $L^2(G'/H')$ if k=l. Consider the branching of Π_a^- to $U(1)$. The subgroup $U(1) \subset U(p)$ commutes with G' and hence acts by a scalar on the $(\mathfrak{g}', K')$-modules $A_{\mathfrak{q}'}(\mu_{G'})$. So it suffices to compute this scalar on the minimal K' type. In this case, $K = U(p)U(q)$, and $K' = U(p-1)U(q)$. Using the formulas in [29], we compute the $U(1)$ action on $\mathbf{A}_{\mathfrak{q}'}(\mu_{G'})$ by computing it on the factor $U(p)$ of its lowest K'-type.

Recall that the minimal K' type of a representation in the discrete spectrum of Y^- is representation of $U(p-1) \times U(q)$, which is trivial on the first factor. The highest weight of a representation of $U(p)$ whose restriction to $U(p-1)$ contains an $U(p-1)$-invariant vector is of the form $(r, 0, 0, ...0, -s)$ for positive integers (r, s). By [29], $U(1)$ acts on its $U(p-1)$-invariant vector by the scalar $l = r + s$. Hence, $A_{\mathfrak{q}'}(\mu_{G'})$ is the $(\mathfrak{g}', K')$ module of a representation in the discrete spectrum of Y^- iff $r = s$.

To complete the proof of the relative branching theorem for the restriction of Π_a^- to G', we have to show that $\chi_0 \otimes \pi_{\mu_{G'}}$ has a nontrivial multiplicity in the restriction of Π_a to $U(1)G'$ for $r = s \geq 0$. We can now use K-type calculations or branching by stages. We choose the latter since it complements well the considerations in Sect. 6. Recall the statement of Theorem 4.

Suppose that Π_a^- is an irreducible representation in the discrete spectrum of G/H^- with nonsingular infinitesimal character and parameter $(0, \ldots, 0, a, -a)$, $a \geq 0$.

1. Let π^- be an irreducible infinite dimensional representation with a nonsingular infinitesimal character in the discrete spectrum of $G'/H^- \cap G'$. If

$$\mathrm{Hom}_H((\Pi_a^-)_{|G'}, \pi^-) \neq 0,$$

then π^- has a parameter

$$\mu_{G'} = (0, \ldots 0, a + k + 1/2, -a - k - 1/2), \qquad k \in \mathbb{N}$$

and

$$\dim \mathrm{Hom}_H((\Pi_a^-)_{|G'}, \pi^-) = 1.$$

2. Conversely, if π^- is a representation with a parameter

$$\mu_{G'} = (0, \ldots 0, a + k + 1/2, -a - k - 1/2), \qquad k \in \mathbb{N},$$

then $\chi_0 \otimes \pi^-$ is a direct summand of the representation Π_a^- restricted to $U(1)U(p-1, q)$.

Proof We already proved the first claim. It remains to show that that a representation $\chi_0 \otimes \pi^-$ with parameter

$$\mu_{G'} = (0, \ldots 0, a + k + 1/2, -a - k - 1/2), \qquad k \in \mathbb{N}$$

is a summand of Π^- to $U(1)U(p-1, q)$.

As in the previous section, we proceed by branching of the representation $U^-(\ell)$ along the first sequence and second sequence:

- $U(p-1, q) \subset O(2p-2, 2q) \times O(2) \subset O(2p, 2q)$.
- $U(p-1, q) \subset U(p, q) \subset O(2p, 2q)$.

Here we use the notation of [16]. See Sect. 6 for the definition of the parameter ℓ, or for the exact definition of ℓ, see [16]. We observe that the representation $U^-(\ell)$ of $O(2p, 2q)$ is isomorphic to $U^+(\ell)$ of $O(2q, 2p)$. Consider the restriction of the representation $U^+(\ell)$ of $O(2q, 2p)$ to $O(2q, 2p-2) \times O(0, 2)$, and conclude that it is a direct sum of representations $\chi_r^- \otimes U^+(\ell + k)$, $k \in 2\mathbb{N}$, where χ_r^- is a character of $O(2)$, which is nontrivial on the diagonal matrix (1,-1). (For the exact definition of ℓ, see [16]). So for $r = 0$ using the isomorphism again, we deduce that the restriction of $U^-(\ell)$ to $SO(2p-2, 2q)$ is a direct sum of representations $U^-(\ell+k), k \in 2\mathbb{N}$.

Using the isomorphism again, we see that the discrete spectrum of the restriction of $U^+(\ell)$ of $O(2q, 2p)$ to $U(q, p)$ contains exactly one representation $\Pi^+_{\ell/2,\ell/2}$. This representation is in the discrete spectrum of $L^2(U(q, p)/U(1)U(q-1, p))$. Hence, the representation $\Pi^+_{\ell/2,\ell/2}$ is the only representation in the discrete spectrum of the restriction of $U^-(\ell)$ to $U(p, q-1)$, which is $U(1)$ spherical.

The same argument as in the previous section using the branching of $U^+(\ell)$ along the second sequence completes the proof. □

7.1 Symmetry Breaking by Geometric Restriction

Proposition 1 *Suppose that $b > a$ are positive integers and that the $(\mathfrak{g}, K)$ module of Π^-_a is a subset of $C^\infty_K(X^-) \cap L^2(X^-)$. Let*

$$Res_{G'} : C^\infty_K(X^-) \to C^\infty_{K'}(Y^-)$$

be the restriction. Then the $(\mathfrak{g}', K')$ module of π^-_b is a subrepresentation of $C^\infty_{K'}(Y')$.

Proof $Res_{G'}\Pi^-_a$ is a direct sum of irreducible representation π^-_{a+k} with $k \geq 0$. Recall that this implies that its underlying $(\mathfrak{g}, K)$ module is a direct sum of the corresponding $(\mathfrak{g}', K')$ modules. Recall also that the restriction

$$Res_{G'} : C^\infty_K(X^-) \to C^\infty_{K'}(Y^-)$$

intertwines the G' actions of the K'-finite functions. So the image under the restriction map of the K-finite functions in Π^-_K is a sum of $(\mathfrak{g}', K')$ modules.

To prove the proposition, it suffices to show that if π^-_{a+k} is the $(\mathfrak{g}', K')$ modules of a summand in the restriction of Π^-_a with parameter

$$\mu_{G'} = (0, \ldots 0, a+k+1/2, -a-k-1/2),\ k \in \mathbb{N},$$

then π^-_{a+k} is a submodule of $Res_{G'}(\Pi^- \cap C^\infty_K(X^-))$. To do this, it suffices to show that if f is a K-finite function in Π^-_a generating the $(\mathfrak{g}', K')$ submodule with parameter $\mu_{G'}$, then its restriction to Y is nonzero.

Let δ be the distribution on the functions in $C^\infty(X^-)$ at the identity coset

$$\delta(f) = f(eU(1)U(p, q-1))$$

and δ' the distribution on the functions in $C^\infty(Y^-)$ at the identity coset of Y^-

$$\delta'(f) = f(eU(1)U(p-1, q-1))$$

Since $Y^- \subset X^-$, we get for $f \in \Pi^- \cap C_K^\infty(X^-)$

$$\delta(f) = \delta'(Res_{G'}(f)).$$

Thus, to prove the theorem, it suffices to show that if f_{r,a_0} is a K-finite function in $\Pi_a^- \subset C_K^\infty(X^-)$ in a K-type with highest weight

$$(r, 0, 0, ..., -r) \times (r + a_0, ..., -a_0 - r),$$

which transforms under $U(p-1)$ by the trivial representation, i.e., which is the lowest K-types of a G_0' type of Π^-, then

$$\delta(f_{r,a_0}) = f_{r,a_0}(eH') \neq 0.$$

Recall that $G = KBH^-$ and that B is a one-parameter group. Its centralizer B^G is $U(p-1, q-1)$, and hence,

$$B^G \cap K = U(p-1) \times U(q-1).$$

Furthermore,

$$B^G \cap G' = U(p-1, q-1),$$

and

$$B^G \cap K' = U(p-1) \times U(q-1).$$

Thus, $f_{r,a}$ is a $U(p-1)$ invariant function on $U(p)/U(p-1)$ and an eigenfunction of the Laplacian and thus a harmonic polynomial and has nonzero constant term.

Note that here we may invoke the same ideas about period integrals to see the G'-invariant pairing between the spherical vectors in question, this time seen as Jacobi polynomials on the complex projective space $U(p)/U(1)U(p-1)$. Indeed the K-types listed above exactly parametrize the Flensted-Jensen representations for Y^- in the branching, and the parameter r labels the corresponding summands in the (admissible, multiplicity-free) branching law from G to G' here. The period integral is nonzero by the earlier results on spherical polynomials on $P^{p-1}(\mathbb{C})$ and their pairings with spherical polynomials on the smaller projective space.

Remark As in the Gan-Gross-Prasad conjectures for discrete series representations, the assumption that infinitesimal characters of representations Π^+, Π^- of G are regular is essential in our proofs and in the formulation of the main results in Sect. 5. Nevertheless, a similar result may be true for representations also for with singular infinitesimal characters in the discrete spectrum of rank-one symmetric spaces.

Appendix 1

Although we are making the assumption $p, q \geq 3$, we include a short discussion of the case $p = 2, q > 3$ to illustrate the difficulties in this case.

An interesting example of relative branching is the restriction of a discrete series representation Π_a of

$$X = U(2, n)/U(1)U(1, n)$$

to $G' = U(1, n)$. A discrete series representation of X^+ is also a discrete series representation of $U(2, n)$ since the Levi subgroup of the corresponding θ-stable parabolic subgroup is compact [12]. So we can directly apply the results of H. He to determine representations in the discrete spectrum of its restriction to $U(2, n-1)$ [11]. Following H. He, we assign to the discrete series representation Π_a^+ a sequence

$$(+, -, - \ldots, -, +)$$

of 2+n signs showing that we have two noncompact simple roots for the positive roots defined by the Langlands parameter of Π_a^+. A discrete series representation π of $U(1, n)$ has one sign $\oplus$ and n-1 signs $\ominus$. We have to align them so that the only allowed combinations of two adjacent signs are

$$(\oplus, +), (+\oplus), (-\ominus), (\ominus, -), (+, -), (-, +), (\oplus, \ominus), (\ominus, \oplus)$$

Now, suppose that we are restricting to the group $U(1, n)$. In this case, the only possible patterns for a discrete series representations of $U(1, n)$, which may appear in the restriction, are:

- One $\oplus$ in first place followed by n signs $\ominus$
- n signs $\ominus$ followed by a $\oplus$

These representations are either holomorphic or anti-holomorphic representations.

In the first case, we get one allowed pattern

$$+, \oplus, \ominus, -, \ominus, -, \ldots, \ominus, -, +$$

In the second case, we have also one allowed pattern

$$+, -, \ominus, -, \ldots, \ominus, -, \ominus, \oplus+$$

To determine the infinitesimal characters of the discrete series representations in the discrete spectrum, we have to consider the corresponding interlacing pattern for the infinitesimal characters. We conclude that there are no discrete series representations π in the discrete spectrum of the restriction of Π_a^+ to $U(1, n)$, since there are no allowed interlacing pattern of infinitesimal characters. Thus, the restriction of Π_a^+

to $U(1, n)$ has no discrete spectrum [9], and thus the relative (strong) branching law is $(\Pi_a^+, \emptyset)$.

Now, consider representations Π_a^- in the discrete spectrum of $X^- = U(2,q)/U(1)U(2,q-1)$. They are no longer discrete series representations, but they are cohomologically induced from a θ-stable parabolic with a noncompact Levi subgroup $L^- = U(1)U(1)U(2,q-2)$. The symmetric space

$$Y^- = U(1,q)/U(1)U(1,q-1)$$

is a rank-one symmetric space of the group $U(1,q)$. It has a nonzero discrete spectrum consisting of representations π_b^- for b sufficiently large cohomologically induced from a θ-stable parabolic subgroup with a non-compact Levi subgroup $U(1)U(1)U(1,q-2)$. In addition here may be a finite number of singular representations determined by T. Kobayashi [14]. The arguments in Sect. 7 show that the restriction of Π_a to $U(1,q)$ is a direct sum of irreducible representations and that representations π_b^- are in the discrete spectrum of the restriction of Π_a to $U(1,q)$ for $b > a$.

Appendix 2

Though our main interest is in the case of unitary groups, we may also treat in a similar way all rank-one compact Riemannian symmetric spaces; their spherical polynomials are again Jacobi polynomials $P_n^{(\alpha,\beta)}$. Thus, in the quaternionic case

$$X_c = Sp(n+1)/Sp(1) \times Sp(n) = P^n(\mathbb{H})$$

with spherical polynomials

$$\varphi = P_n^{(2n-1,1)}(x)$$

and also the octonionic case

$$X_c = F_4/Spin(9) = P^2(\mathbb{O})$$

with spherical polynomials

$$\varphi = P_n^{(7,3)}(x)$$

and again $x = \cos(\theta)$ in terms of the polar angle (distance) θ. For the Flensted-Jensen discrete series, we consider in the quaternionic case the subgroup $Sp(n)$ and the corresponding symmetric subspace $X_c' = P^{n-1}(\mathbb{H})$. The quaternionic hyperboloid is now

$$X = Sp(p, q+1)/Sp(p, q) \times Sp(1)$$

with the maximal compact subvariety $P^q(\mathbb{H})$. The Flensted-Jensen function giving the lowest K-type is again of product type

$$\psi_\lambda(kb_s) = (\cosh s)^{-i\lambda-\rho} P_n^{(2q-1,1)}(k)$$

for $k \in K,\ b_s \in B$. Here, the parameters are (notation as in [5])

$$\begin{aligned} i\lambda &= 2q - 2p + 1 + n, (n \in 2\mathbb{Z}^+), \\ \rho &= 2p + 2q + 1, \\ \rho_{\mathfrak{t}} &= q + 1, \\ \rho - 2\rho_{\mathfrak{t}} &= 2p - 2q - 1, \\ \mu_\lambda &= i\lambda + \rho - 2\rho_{\mathfrak{t}} = n. \end{aligned}$$

The n is even corresponding to spherical polynomials on $X_c = P^q(\mathbb{H})$, and the period integral is as before converging and nonzero for $n' \leq n$. Note that since $i\lambda + \rho = 4q + n + 2$, we have the required decay, since the radial density in this case equals (on $X_c = P^q(\mathbb{H})$)

$$D(s) = (\cosh s)^{4q+3}(\sinh s)^{4p-1}.$$

And again, there is a classical identity for these Jacobi polynomials yielding relative branching via period integrals.

References

1. Adams, J., Barbasch, D. and Vogan, D.A.: The Langlands Classification and Irreducible Characters for Real Reductive Groups, vol. 104. Birkhaeuser, Boston (1992)
2. Awonusika, R.O., Taheri, A.: On Jacobi polynomials $(P_k^{\alpha,\beta} : \alpha, \beta > (-1))$ and Maclaurin spectral functions on rank one symmetric spaces. J. Anal. **25**(1), 139–166 (2017)
3. E. van den Ban, P.: The Plancherel theorem for a reductive symmetric space. In: Anker, J.-P., Orsted, B. (eds.) Lie Theory Harmonic Analysis on Symmetric Spaces–General Plancherel Theorems. Progress in Mathematics, vol. 230, pp. 1–97. Birkhauser, Boston (2005)
4. Beuzart-Plessis, R.: Endoscopy and refined local conjecture of Gan-Gross-Prasad for unitary groups. Comp. Math. **151**(7), 1309–1371 (2015)

5. Flensted-Jensen, M.: Discrete series for semisimple symmetric spaces. Ann. Math. **111**(2), 253–311 (1980)
6. Gan, W.T., Gross, B.H., Prasad, D.: Symplectic local root numbers, central critical L-values, and restriction problems in the representation theory of classical groups. Astérisque **346**, 1–109 (2012)
7. Gan, W.T., Gross, B.H., Prasad, D.: Branching laws for classical groups: the non-tempered case. Comp. Math. **156**(11), 2298–2367 (2020)
8. Gross, B.H., Prasad, D.: On the decomposition of a representations of SO_n when restricted to SO_{n-1}. Can. J. Math. **44**, 974–1002 (1992)
9. Gross, B.H., Wallach, N.: Restrictions of small discrete series representations to symmetric subgroups. In: The mathematical legacy of Harish-Chandra: A Celebration of Representation Theory and Harmonic Analysis. Proceedings of Symposia in Pure Mathematics, vol 68 (2000)
10. Heckman, G., Schlichtkrull, H.: Harmonic Analysis and Special Functions on Symmetric Spaces. Perspectives in Mathematics. Academic Press, New York (1994)
11. He, H.: On the Gan-Gross-Prasad Conjecture for unitary groups. Invent. Math. **209**, 837–884 (2017)
12. Knapp, A.W., Vogan, D.: Cohomological Induction and Unitary Representations. Princeton University Press, Princeton (1995). ISBN: 9780691037561
13. Knop, F., Schalke, B.: The dual group of spherical variety. Trans. Moscow **78**, 187–216 (2017)
14. Kobayashi, T.: Discrete decomposability of $A_{\mathfrak{q}}(\lambda)$ with respect to reductive subgroups and its applications. Invent. Math. **117**, 181–205 (1994)
15. Kobayashi, T.: A program for branching problems in the representation theory of real reductive groups. In: Nevins, M., Trapa, P. (eds.) Representations of Reductive Groups: In Honor of David A. Vogan, Jr. on his 60th Birthday, volume 312 of Progress in Mathematics, pp. 277–322. Birkhäuser, Boston (2015)
16. Kobayashi, T.: Branching laws of unitary representations associated to minimal elliptic orbits for indefinite orthogonal group O(p,q). Adv. Math. **388**, 38 pp. (2021)
17. Kobayashi, T., Oshima, Y.: Classification of discretely decomposable $A_{\mathfrak{q}}(\lambda)$ with respect to reductive symmetric pairs. Adv. Math. **231**, 2013–2047 (2012)
18. Kobayashi, T., Speh, B.: A hidden symmetry of a branching law. In: Dobrev, V. (ed.) Lie Theory and Its Applications in Physics. Springer Proceedings in Mathematics & Statistics, vol. 335, pp. 15–28. Springer, Singapore (2020)
19. Moeglin, C., Renard, D.L Séries discrètes des espaces symétriques et paquets d'Arthur. available at arXiv:1906.00725
20. Moeglin, C., Renard, D.: Sur les paqquets d'Arthur des groupes classiques et unitaires non quasi-deployés. In: Relative aspects in Representation Theory, Langlands Functoriality and Automorphic Forms, vol. 2221. Lecture Notes in Mathematics, pp. 341-361. Springer, Berlin (2018)
21. Ørsted, B., Speh, B.: Branching laws for discrete series of some affine symmetric spaces. Pure Appl. Math. Q. **17**(4), 1291–1320 (2021). arXiv:1907.07544
22. Ørsted, B., Vargas, J.: Branching problems in reproducing kernel spaces. Duke Math. J. **169**(18), 3477–3537 (2020)
23. Oshima, T., Matsuki, T.: A description of discrete series representations for semisimple symmetric spaces. Adv. Stud. Pure Math. **4**, 331–390 (1984)
24. Oshima, Y.: Note on the branching laws of $A_{\mathfrak{q}}(\lambda)$ for U(m,n). Private communication
25. Raczka, R., Limic, N., Niederle, J.: Discrete degenerate representations of noncompact rotation groups. I. J. Math. Phys. **7**, 1861 (1966)
26. Sakellaridis, Y., Venkatesh, A.: Periods and harmonic analysis on spherical varieties. Asterisque **396** (2017)
27. Schlichtkrull, H.: Hyperfunctions and Harmonic Analysis on Symmetric Spaces. Progress in Mathematics, vol. 49. Birkhauser, Boston (1984)

28. Schlichtkrull, H.: The Langlands parameters of Flensted-Jensen's discrete series for semisimple symmetric spaces. J. Funct. Anal. **50**(2), 133–150 (1983)
29. Schlichtkrull, H., Trapa, P., Vogan Jr. D.A.: Laplacians on spheres. Sao Paulo J. Math. Sci. **12**(2), 295–358 (2018)
30. Speh, B.: Restriction of some representations of U(p,q) to a symmetric subgroup. In: Representation Theory and Mathematical Physics, a Conference in Honor of G. Zuckerman's 60th Birthday. Contemporary Mathematics, vol. 557 (2011)

Pseudo-dual Pairs and Branching of Discrete Series

Bent Ørsted and Jorge A. Vargas

We are delighted to make this chapter part of a tribute to Toshiyuki Kobayashi, for his indefatigable dedication to representation theory and Lie theory and his many new ideas here.

Abstract For a semisimple Lie group G, we study Discrete Series representations with admissible branching to a symmetric subgroup H. This is done using a canonical associated symmetric subgroup H_0, forming a pseudo-dual pair with H, and a corresponding branching law for this group with respect to its maximal compact subgroup. This is in analogy with either Blattner's or Kostant-Heckman multiplicity formulas and has some resemblance to Frobenius reciprocity. We give several explicit examples and links to Kobayashi-Pevzner theory of symmetry breaking and holographic operators. Our method is well adapted to computer algorithms, such as the Atlas program.

Keywords Admissible restriction · Branching laws · Discrete Series · Multiplicity formulas

MSC2020: Primary 22E45; Secondary 20G05, 17B10

B. Ørsted
Mathematics Department, Aarhus University, Aarhus, Denmark
e-mail: orsted@math.au.dk

J. A. Vargas (✉)
Famaf-CIEM, Ciudad Universitaria, Córdoba, Argentine
e-mail: vargas@famaf.unc.edu.ar

M. Pevzner, H. Sekiguchi (eds.), *Symmetry in Geometry and Analysis, Volume 2*,
Progress in Mathematics 358, https://doi.org/10.1007/978-981-97-7662-7_11

1 Introduction

For a semisimple Lie group G, an irreducible representation (π, V) of G, and closed reductive subgroup $H \subset G$, the problem of decomposing the restriction of π to H has received attention ever since number theory or physics, and other branches of mathematics required a solution. In this chapter, we are concerned with the important particular case of branching representations of the Discrete Series, i.e., those π arising as closed irreducible subspaces of the left regular representation in $L^2(G)$, and breaking the symmetry by a reductive subgroup H. Here, much work has been done. Notable is the paper of Gross-Wallach [9], and the work of Toshiyuki Kobayashi and his school. For further references on the subject, we refer to the overview work of Toshiyuki Kobayashi and references therein. The aim of this note is to compute the decomposition of the restriction of an H-admissible representation π to a symmetric subgroup H (see 3.4.2); in [9], it is derive a duality theorem for Discrete Series representation. Their duality is based on the dual subgroup G^d (this is the dual subgroup that enters the duality introduced by Flensted-Jensen in his study of discrete series for affine symmetric spaces [7]), and roughly speaking, their formula looks like

$$\dim Hom_H(\sigma, \pi_{|H}) = \dim Hom_{\widetilde{K}}(F_\sigma, \tilde{\pi}).$$

Here, π is a irreducible square integrable representation of G, σ is a irreducible representation of H, F_σ is a irreducible representation of a maximal compact subgroup $\widetilde{K}$ of G^d, and $\tilde{\pi}$ is a finite sum of fundamental representations of G^d attached to π. In [23], B. Speh and the first author considered a different duality theorem for restriction to a symmetric subgroup. Let H_0 be the associated subgroup to H (see 3.1); hence, $L := H_0 \cap H$ is a maximal compact subgroup of both H, H_0. Then,

$$\dim Hom_H(\sigma, \pi_{|H}) = \dim Hom_L(\sigma_0, \widetilde{\Pi}). \qquad (\ddagger)$$

Here, π is a certain irreducible representation of G, σ is a irreducible representation of H, σ_0 is the lowest L-type of σ, and $\widetilde{\Pi}$ is a finite sum of irreducible representation of H_0 attached to π. The purpose of this chapter is, for a H-admissible Discrete Series π for G, to show a formula as the above and to provide an explicit isomorphism between the two vector spaces involved in the equality. This is embodied in Theorem 1.

Theorem 1 reduces the branching law in two steps: (1) for the maximal compact subgroup K of G and the lowest K-type of π, branching this under L (maximal compact in H and also in H_0) and (2) branching a Discrete Series of H_0 with respect to L, i.e., finding its L-types with multiplicity. Both of these steps can be implemented in algorithms, as they are available, for example, in the computer program Atlas http://atlas.math.umd.edu.

We would like to point out that T. Kobayashi, T. Kobayashi-Pevzner, and Nakahama have shown a duality formula as (‡) for holomorphic Discrete Series representation π. In order to achieve their result, they have shown an explicit isomorphism between the two vector spaces in the formula. Further, with respect to analyze $res_H(\pi)$, Kobayashi-Oshima have shown a way to compute the irreducible components of $res_H(\pi)$ in the language of Zuckerman modules $A_{\mathfrak{q}}(\lambda)$ [18, 19].

As a consequence of the involved material, we obtain a necessary and sufficient condition for symmetry breaking operators to be represented via normal derivatives. This is presented in Proposition 5.

Another consequence is Proposition 1. That is, for the closure of the linear span of the totality of H_0-translates (resp. H-translates) of the isotypic component associated with the lowest K-type of π, we exhibit its explicit decomposition as a finite sum of Discrete Series representations of H_0 (resp. H).

Our proof is based on the fact that Discrete Series representations are realized in reproducing kernel Hilbert spaces. As a consequence, in Lemma 1, we obtain a general result on the structure of the kernel of a certain restriction map. The proof also relies on the work of Hecht-Schmid [11] and a result of Schmid in [27].

It follows from the work of Kobayashi-Oshima, also, from Tables 1, 2, and 3, that whenever a Discrete Series for G has an admissible restriction to a symmetric subgroup, then, the infinitesimal character of the representation is dominant with respect to either a Borel de Siebenthal system of positive roots or a system of positive roots so that it has two noncompact simple roots, each one having multiplicity one in the highest root. Under the H-admissible hypothesis, the infinitesimal character of each of the irreducible components of $\widetilde{\Pi}$ in formula (‡) has the same property as the infinitesimal character of π. Thus, for most H-admissible Discrete Series, to compute the right-hand side of (‡), we may appeal to the work of the first author and Wolf [26]. Their results let us compute the highest weight of each irreducible factor in the restriction of π to $K_1(\Psi)$. Next, we apply [5, Theorem 5] for the general case.

We may speculate that a formula like (‡) might be true for π, whose underlying Harish-Chandra module is equivalent to a unitarizable Zuckerman module. In this case, the definition of σ_0 would be the subspace spanned by the lowest L-type of σ, and $\widetilde{\Pi}$ would be a Zuckerman module attached to the lowest K-type of π.

The chapter is organized as follows: In Sect. 2, we introduce facts about Discrete Series representation and notation. In Sect. 3, we state the main theorem and begin its proof. As a tool, we obtain information on the kernel of the restriction map.

In Sect. 4, we complete the proof of the main theorem. As a by-product, we obtain information on the kernel of the restriction map, under admissibility hypothesis. We present examples and applications of the main theorem in Sect. 5. This includes lists of multiplicity-free restriction of representations; many of the multiplicity-free representations are non-holomorphic Discrete Series representations. We also dealt with quaternionic and generalized quaternionic representations.

In Sect. 6, we analyze when symmetry breaking operators are represented by means of normal derivatives. Section 7 presents the list of H-admissible Discrete Series and related information.

2 Preliminaries and Some Notation

Let G be an arbitrary matrix, connected semisimple Lie group. Henceforth, we fix a maximal compact subgroup K for G and a maximal torus T for K. Harish-Chandra showed that G admits square integrable irreducible representations if and only if T is a Cartan subgroup of G. For this chapter, we always assume T is a Cartan subgroup of G. Under these hypothesis, Harish-Chandra showed that the set of equivalence classes of irreducible square integrable representations is parameterized by a lattice in $i\mathfrak{t}^\star$. In order to state our results, we need to make explicit this parametrization and set up some notation. As usual, the Lie algebra of a Lie group is denoted by the corresponding lowercase German letter. To avoid notation, the complexification of the Lie algebra of a Lie group is also denoted by the corresponding German letter without any subscript. $V^\star$ denotes the dual space to a vector space V. Let θ be the Cartan involution that corresponds to the subgroup K; the associated Cartan decomposition is denoted by $\mathfrak{g} = \mathfrak{k} + \mathfrak{p}$. Let $\Phi(\mathfrak{g}, \mathfrak{t})$ denote the root system attached to the Cartan subalgebra $\mathfrak{t}$. A root in $\Phi(\mathfrak{g}, \mathfrak{t})$ is *compact* (*noncompact*) whenever the corresponding root space is a subset of $\mathfrak{k}$ ($\mathfrak{p}$). Thus, $\Phi(\mathfrak{g}, \mathfrak{t})$ splits up as the disjoint union of the subset $\Phi_c := \Phi(\mathfrak{k}, \mathfrak{t})$ of compact roots and $\Phi_n := \Phi(\mathfrak{p}, \mathfrak{t})$ the subset of noncompact roots. From now on, we fix a system of positive roots Δ for Φ_c. For this chapter, either the highest weight or the infinitesimal character of an irreducible representation of K is dominant with respect to Δ. The Killing form gives rise to an inner product $(..., ...)$ in $i\mathfrak{t}^\star$. As usual, let $\rho = \rho_G$ denote half of the sum of the roots for some system of positive roots for $\Phi(\mathfrak{g}, \mathfrak{t})$. *A Harish-Chandra parameter* for G is $\lambda \in i\mathfrak{t}^\star$ such that $(\lambda, \alpha) \neq 0$, for every $\alpha \in \Phi(\mathfrak{g}, \mathfrak{t})$, and so that $\lambda + \rho$ lifts to a character of T. To each Harish-Chandra parameter λ, Harish-Chandra associates a unique irreducible square integrable representation $(\pi_\lambda^G, V_\lambda^G)$ of G of infinitesimal character λ. Moreover, he showed the map $\lambda \to (\pi_\lambda^G, V_\lambda^G)$ is a bijection from the set of Harish-Chandra parameters dominant with respect to Δ onto the set of equivalence classes of irreducible square integrable representations for G (see [32, Chap 6]). In short, we will refer to an irreducible square integrable representation as a Discrete Series representation.

Each Harish-Chandra parameter λ gives rise to a system of positive roots

$$\Psi_\lambda = \Psi_{G,\lambda} = \{\alpha \in \Phi(\mathfrak{g}, \mathfrak{t}) : (\lambda, \alpha) > 0\}.$$

From now on, we assume that Harish-Chandra parameters for G are dominant with respect to Δ. Hence, $\Delta \subset \Psi_\lambda$. We write $\rho_n^\lambda = \rho_n = \frac{1}{2}\sum_{\beta \in \Psi_\lambda \cap \Phi_n} \beta$, $(\Psi_\lambda)_n := \Psi_\lambda \cap \Phi_n$. We define $\rho_c = \frac{1}{2}\sum_{\alpha \in \Delta} \alpha$.

We denote by $(\tau, W) := (\pi^K_{\lambda+\rho_n}, V^K_{\lambda+\rho_n})$ the lowest K−type of $\pi_\lambda := \pi_\lambda^G$. The highest weight of $(\pi^K_{\lambda+\rho_n}, V^K_{\lambda+\rho_n})$ is $\lambda + \rho_n - \rho_c$. We recall a theorem of Vogan's thesis [31, 6], which states that (τ, W) determines $(\pi_\lambda, V_\lambda^G)$ up to unitary equivalence. We recall the set of square integrable sections of the vector bundle determined by the principal bundle $K \to G \to G/K$, and the representation (τ, W) of K is isomorphic to the space

$$L^2(G \times_\tau W) \\ := \{f \in L^2(G) \otimes W : f(gk) = \tau(k)^{-1} f(g), g \in G, k \in K\}.$$

Here, the action of G is by left translation $L_x, x \in G$. The inner product on $L^2(G) \otimes W$ is given by

$$(f, g)_{V_\lambda} = \int_G (f(x), g(x))_W dx,$$

where $(..., ...)_W$ is a $K-$invariant inner product on W. Subsequently, L_D (resp. R_D) denotes the left infinitesimal (resp. right infinitesimal) action on functions from G of an element D in universal enveloping algebra $\mathcal{U}(\mathfrak{g})$ for the Lie algebra $\mathfrak{g}$. As usual, Ω_G denotes the Casimir operator for $\mathfrak{g}$. Following Hotta-Parthasarathy [13], Enright-Wallach [6], and Atiyah-Schmid [1], we realize $V_\lambda := V_\lambda^G$ as the space

$$H^2(G, \tau) = \{f \in L^2(G) \otimes W : f(gk) = \tau(k)^{-1} f(g) \\ g \in G, k \in K, R_{\Omega_G} f = [(\lambda, \lambda) - (\rho, \rho)] f\}.$$

We also recall $R_{\Omega_G} = L_{\Omega_G}$ is an elliptic $G-$invariant operator on the vector bundle $W \to G \times_\tau W \to G/K$, and hence, $H^2(G, \tau)$ consists of smooth sections; moreover, point evaluation e_x defined by $H^2(G, \tau) \ni f \mapsto f(x) \in W$ is continuous for each $x \in G$ (cf. [25, Appendix A4]). Therefore, the orthogonal projector P_λ onto $H^2(G, \tau)$ is an integral map (integral operator) represented by the smooth *matrix kernel* or *reproducing kernel* [25, Appendix A1, Appendix A4, Appendix A6]

$$K_\lambda : G \times G \to End_{\mathbb{C}}(W), \tag{1}$$

which satisfies $K_\lambda(\cdot, x)^\star w$ belongs to $H^2(G, \tau)$ for each $x \in G, w \in W$ and

$$(P_\lambda(f)(x), w)_W = \int_G (f(y), K_\lambda(y, x)^\star w)_W dy, \; f \in L^2(G \times_\tau W).$$

For a closed reductive subgroup H, after conjugation by an inner automorphism of G, we may and will assume $L := K \cap H$ is a maximal compact subgroup for H. That is, H is $\theta-$stable. In this chapter, for irreducible square integrable representations (π_λ, V_λ) for G, we would like to analyze its restriction to H. In particular, we study the irreducible $H-$subrepresentations for π_λ. A known fact is that any irreducible $H-$subrepresentation of V_λ is a square integrable representation for H, for a proof (cf. [9]). Thus, owing to the result of Harish-Chandra on the existence of square integrable representations, from now on, we may and will assume H *admits a compact Cartan subgroup*. After conjugation, we may assume $U := H \cap T$ is a maximal torus in $L = H \cap K$. From now on, we set a square integrable representation $V_\mu^H \equiv H^2(H, \sigma) \subset L^2(H \times_\sigma Z)$ of lowest $L-$type $(\pi^L_{\mu+\rho_n^\mu}, V^L_{\mu+\rho_n^\mu}) \equiv: (\sigma, Z)$.

For a representation M and irreducible representation N, $M[N]$ denotes the isotypic component of N, that is, $M[N]$ is the linear span of the irreducible subrepresentations of M equivalent to N. If topology is involved, $M[N]$ is the closure of the linear span.

For a H-admissible representation π, $Spec_H(\pi)$ denotes the set of Harish-Chandra parameters of the irreducible H-subrepresentations of π.

3 Duality Theorem, Explicit Isomorphism

3.1 *Statement and Proof of the Duality Result*

The unexplained notation is as in Sect. 2, our hypotheses are as follows: $(G, H = (G^\sigma)_0)$ is a symmetric pair, and $(\pi_\lambda, V_\lambda^G)$ is a H-admissible, square integrable irreducible representation for G. $K = G^\theta$ is a maximal compact subgroup of G, $H_0 := (G^{\sigma\theta})_0$, and K so that $L = H \cap K = H_0 \cap K$ is a maximal compact subgroup of both H and H_0. By definition, H_0 is the *associated* subgroup to H.

In this section, under our hypothesis, for a H-irreducible factor V_μ^H for $res_H(\pi_\lambda)$, we show an explicit isomorphism from the space

$$Hom_H(V_\mu^H, V_\lambda^G) \text{ onto } Hom_L(V^L_{\mu+\rho_n^\mu}, \pi_\lambda(\mathcal{U}(\mathfrak{h}_0))V_\lambda^G[V^K_{\lambda+\rho_n}]).$$

We also analyze the restriction map $r_0 : H^2(G, \tau) \to L^2(H_0 \times_\tau W)$.

To follow, we present the necessary definitions and facts involved in the main statement.

We consider the linear subspace $\mathcal{L}_\lambda$ spanned by the lowest L-type subspace of each irreducible H-factor of $res_H((L, H^2(G, \tau)))$. That is,

$$\mathcal{L}_\lambda \text{ is the linear span of } \cup_{\mu\in Spec_H(\pi_\lambda)} H^2(G, \tau)[V_\mu^H][V^L_{\mu+\rho_n^\mu}].$$

We recall that our hypothesis yields that the subspace of L-finite vectors in V_λ^G is equal to the subspace of K-finite vectors [16, Prop. 1.6]. Hence, we have $\mathcal{L}_\lambda$ as a subspace of the space of K-finite vectors in $H^2(G, \tau)$.

We also need the subspace

$$\mathcal{U}(\mathfrak{h}_0)W := L_{\mathcal{U}(\mathfrak{h}_0)}H^2(G, \tau)[V^K_{\lambda+\rho_n^\lambda}] \equiv \pi_\lambda(\mathcal{U}(\mathfrak{h}_0))(V_\lambda^G[V^K_{\lambda+\rho_n^\lambda}]).$$

We write $\mathrm{Cl}(\mathcal{U}(\mathfrak{h}_0)W)$ for the closure of $\mathcal{U}(\mathfrak{h}_0)W$. Hence, $\mathrm{Cl}(\mathcal{U}(\mathfrak{h}_0)W)$ is the closure of the left translates by the algebra $\mathcal{U}(\mathfrak{h}_0)$ of the subspace of K-finite vectors

$$H^2(G,\tau)[V^K_{\lambda+\rho_n^\lambda}] = \{K_\lambda(\cdot,e)^\star w : w \in W\} \equiv W.$$

Thus, $\mathcal{U}(\mathfrak{h}_0)W$ consists of analytic vectors for π_λ. Therefore, $\mathrm{Cl}(\mathcal{U}(\mathfrak{h}_0)W)$ is invariant under left translations by H_0. In Proposition 1, we present the decomposition of $\mathcal{U}(\mathfrak{h}_0)W$ as a sum of irreducible representations for H_0.

We point out

The L-module $\mathcal{L}_\lambda$ is equivalent to the underlying L-module in $\mathcal{U}(\mathfrak{h}_0)W$.

This has been proven in [30, (4.5)]. For completeness, we present a proof in Proposition 1.

Under the extra assumption $res_L(\tau)$ is irreducible, we have $\mathcal{U}(\mathfrak{h}_0)W$ as an irreducible $(\mathfrak{h}_0, L)$-module, and in this case, the lowest L-type of $\mathcal{U}(\mathfrak{h}_0)W$ is $(res_L(\tau), W)$. That is, $\mathcal{U}(\mathfrak{h}_0)W$ is equivalent to the underlying Harish-Chandra module for $H^2(H_0, res_L(\tau))$. The Harish-Chandra parameter $\eta_0 \in i\mathfrak{u}^\star$ for $\mathrm{Cl}(\mathcal{U}(\mathfrak{h}_0)W)$ is computed in 3.4.1.

For scalar holomorphic Discrete Series, the classification of symmetric pairs (G, H) such that the equality $\mathcal{U}(\mathfrak{h}_0)W = \mathcal{L}_\lambda$ holds is:

$$(\mathfrak{su}(m,n), \mathfrak{su}(m,l)+\mathfrak{su}(n-l)+\mathfrak{u}(1)), (\mathfrak{so}(2m,2), \mathfrak{u}(m,1)),$$

$(\mathfrak{so}^\star(2n), \mathfrak{u}(1,n-1))$, $(\mathfrak{so}^\star(2n), \mathfrak{so}(2)+\mathfrak{so}^\star(2n-2))$, $(\mathfrak{e}_{6(-14)}, \mathfrak{so}(2,8)+\mathfrak{so}(2))$. [30, (4.6)]. Thus, there exists scalar holomorphic Discrete Series with $\mathcal{U}(\mathfrak{h}_0)W \neq \mathcal{L}_\lambda$.

To follow, we set some more notation. We fix a representative for (τ, W). We write

$$(res_L(\tau), W) = \textstyle\sum_{1\le j\le r} q_j(\sigma_j, Z_j), q_j = \dim Hom_L(Z_j, res_L(W))$$

and the decomposition in isotypic components

$$W = \oplus_{1\le j\le r} W[(\sigma_j, Z_j)] = \oplus_{1\le j\le r} W[\sigma_j].$$

From now on, we fix respective representatives for (σ_j, Z_j) with $Z_j \subset W[(\sigma_j, Z_j)]$.

Henceforth, we denote by

$$\mathbf{H}^2(H_0,\tau) := \sum_j \dim Hom_L(\tau,\sigma_j)\, H^2(H_0,\sigma_j).$$

We think the later module as a linear subspace of

$$\sum_j L^2(H_0 \times_{\sigma_j} W[\sigma_j])_{H_0-disc} \equiv L^2(H_0 \times_\tau W)_{H_0-disc}.$$

Hence, $\mathbf{H}^2(H_0, \tau) \subset L^2(H_0 \times_\tau W)_{H_0-disc}$. We note that when $res_L(\tau)$ is irreducible, then $\mathbf{H}^2(H_0, \tau) = H^2(H_0, res_L(\tau))$.

Owing to both spaces $H^2(H, \sigma)$, $H^2(G, \tau)$ as reproducing kernel spaces, we represent each $T \in Hom_H(H^2(H, \sigma), H^2(G, \tau))$ by a kernel $K_T : H \times G \to Hom_{\mathbb{C}}(Z, W)$ so that $K_T(\cdot, x)^\star w \in H^2(H, \sigma)$ and $(T(g)(x), w)_W = \int_H (g(h), K_T(h, x)^\star w)_Z dh$. Here, $x \in G$, $w \in W$, $g \in H^2(H, \sigma)$. In [25], it is shown K_T is a smooth function, $K_T(h, \cdot)z = K_{T^\star}(\cdot, h)^\star z \in H^2(G, \tau)$, and

$$K_T(e, \cdot)z \in H^2(G, \tau)[V_\mu^H][V^L_{\mu+\rho_n^H}] \tag{2}$$

is a L-finite vector in $H^2(G, \tau)$.

Finally, we recall the restriction map

$$r_0 : H^2(G, \tau) \to L^2(H_0 \times_\tau W), \; r_0(f)(h_0) = f(h_0), h_0 \in H_0,$$

is (L^2, L^2)-continuous [24].

The main result of this section is discussed below.

Theorem 1 *We assume (G, H) is a symmetric pair and $res_H(\pi_\lambda)$ is admissible. We fix a irreducible factor V_μ^H for $res_H(\pi_\lambda)$. Then, the following statements hold:*

(i) The map $r_0 : H^2(G, \tau) \to L^2(H_0 \times_\tau W)$ restricted to $\mathrm{Cl}(\mathcal{U}(\mathfrak{h}_0)W)$ yields a isomorphism between $\mathrm{Cl}(\mathcal{U}(\mathfrak{h}_0)W)$ and $\mathbf{H}^2(H_0, \tau)$.

(ii) For each fixed intertwining L-equivalence

$$D : \mathcal{L}_\lambda[V^L_{\mu+\rho_n^\mu}] = H^2(G, \tau)[V_\mu^H][V^L_{\mu+\rho_n^\mu}] \to (\mathcal{U}(\mathfrak{h}_0)W)[V^L_{\mu+\rho_n^\mu}],$$

the map

$$r_0^D : Hom_H(H^2(H, \sigma), H^2(G, \tau)) \to Hom_L(V^L_{\mu+\rho_n^\mu}, \mathbf{H}^2(H_0, \tau))$$

defined by

$$T \overset{r_0^D}{\longmapsto} (V^L_{\mu+\rho_n^\mu} \ni z \mapsto r_0(D(K_T(e, \cdot)z)) \in \mathbf{H}^2(H_0, \tau))$$

is a linear isomorphism.

Remark 1 When the natural inclusion $H/L \to G/K$ is a holomorphic map, T. Kobayashi, M. Pevzner, and Y. Oshima in [17, 20] have shown a similar dual multiplicity result after replacing the underlying Harish-Chandra module in $\mathbf{H}^2(H_0, \tau)$ by its representation as a Verma module. Also, in the holomorphic setting, Jakobsen-Vergne in [14] has shown the isomorphism $H^2(G, \tau) \equiv \sum_{r\geq 0} H^2(H, \tau_{|L} \otimes S^{(r)}((\mathfrak{h}_0 \cap \mathfrak{p}^+))^\star)$. In [22, 21], we find applications of the result of Kobayashi for their work on decomposing holomorphic Discrete Series. H. Sekiguchi [28] has obtained a similar result of branching laws for singular holomorphic representations.

Remark 2 The proof of Theorem 1 requires to show the map r_0^D is well defined as well as several structure lemmas. Once we verify the map is well defined, we will show injectivity, and Corollary 3, Proposition 4, and linear algebra will give the surjectivity. In Proposition 1, we show (i), and in the same Proposition, we give a proof of the existence of the map D as well as its bijectivity; actually, this result has been shown in [30]. However, we sketch a proof in this note. The surjectivity also depends heavily on a result in [30]; for completeness, we give a proof. We may say that our proof of Theorem 1 is rather long and intricate, involving both linear algebra for finding the multiplicities and analysis of the kernels of the intertwining operators in question to set up the equivalence of the H-morphisms and the L-morphisms. The structure of the branching and corresponding symmetry breaking is however very convenient to apply in concrete situations, and we give several illustrations.

We explicit the inverse map to the bijection r_0^D in Sect. 4.2.

Remark 3 When $\mathcal{L}_\lambda = \mathcal{U}(\mathfrak{h}_0)W$, we may take D equal to the identity map.

Remark 4 A mirror statement to Theorem 1 for symmetry breaking operators is as follows: $Hom_H(H^2(G, \tau), H^2(H, \sigma))$ is isomorphic to $Hom_L(Z, \mathbf{H}^2(H_0, \tau))$ via the map $S \mapsto (z \mapsto (H_0 \ni x \mapsto r_0^D(S^\star)(z)(x) = r_0(D(K_S(\cdot, x)^\star)(z)) \in W))$.

We verify $r_0(D(K_T(e, \cdot)z))(\cdot)$ belongs to $L^2(H_0 \times_\tau W)_{H_0-disc}$.

Indeed, owing to our hypothesis, a result [5] (see [4, Proposition 2.4]) implies π_λ is L-admissible. Hence, [15, Theorem 1.2] implies π_λ is H_0-admissible. Also, [16, Proposition 1.6] shows the subspace of L-finite vectors in $H^2(G, \tau)$ is equal to the subspace of K-finite vectors and $res_{\mathcal{U}(\mathfrak{h}_0)}(H^2(G, \tau)_{K-fin})$ is an admissible, completely algebraically decomposable representation. Thus, the subspace $H^2(G, \tau)[W] \equiv W$ is contained in a finite sum of irreducible $\mathcal{U}(\mathfrak{h}_0)$-factors. Hence, $\mathcal{U}(\mathfrak{h}_0)W$ is a finite sum of irreducible $\mathcal{U}(\mathfrak{h}_0)$ factors. In [9], we find a proof that each irreducible summand for $res_{H_0}(\pi_\lambda)$ is a square integrable representation for H_0; hence, the equivariance and continuity of r_0 yields $r_0(\mathrm{Cl}(\mathcal{U}(\mathfrak{h}_0)W))$ contained in $L^2(H_0 \times_\tau W)_{H_0-disc}$. Section 3.1 shows $K_T(e, \cdot) \in V_\lambda[V_\mu^H][V_{\mu+\rho_n^H}^L]$; hence, $D(K_T(e, \cdot)z)(\cdot) \in \mathcal{U}(\mathfrak{h}_0)W$, and the claim follows.

The map $Z \ni z \mapsto r_0(D(K_T(e, \cdot)z))(\cdot) \in L^2(H_0 \times_\tau W)$ is a L-map.

For this, we recall the equalities

$$K_T(hl, gk) = \tau(k^{-1})K_T(h, g)\sigma(l), k \in K, l \in L, g \in G, h \in H.$$

$$K_T(hh_1, hx) = K_T(h_1, x), h, h_1 \in H, x \in G.$$

Therefore, $K_T(e, hl_1)\sigma(l_2)z = \tau(l_1^{-1})K_T(l_2, h)z = \tau(l_1^{-1})K_T(e, l_2^{-1}h)z$ for $l_1, l_2 \in L, h \in H_0$, and we have shown the claim.

We have enough information to verify the injectivity we have claimed in (i) as well as the injectivity of the map r_0^D. For these, we show a fact valid for an arbitrary reductive pair (G, H) and arbitrary Discrete Series representation.

3.2 Kernel of the Restriction Map

In this section, we show a fact valid for any reductive pair (G, H) and arbitrary representation π_λ. The objects involved here are the restriction map r from $H^2(G, \tau)$ to $L^2(H \times_\tau W)$ and the subspace

$$\mathcal{U}(\mathfrak{h})W := L_{\mathcal{U}(\mathfrak{h})}H^2(G, \tau)[V^K_{\lambda+\rho_n^\lambda}] \equiv \pi_\lambda(\mathcal{U}(\mathfrak{h}))(V_\lambda^G[V^K_{\lambda+\rho_n^\lambda}]). \tag{3}$$

We write $\mathrm{Cl}(\mathcal{U}(\mathfrak{h})W)$ for the closure of $\mathcal{U}(\mathfrak{h})W$. The subspace $\mathrm{Cl}(\mathcal{U}(\mathfrak{h})W)$ is the closure of the left translates by the algebra $\mathcal{U}(\mathfrak{h})$ of the subspace of K-finite vectors

$$\{K_\lambda(\cdot, e)^\star w : w \in W\} = H^2(G, \tau)[W].$$

Thus, $\mathcal{U}(\mathfrak{h})W$ consists of analytic vectors for π_λ. Hence, $\mathrm{Cl}(\mathcal{U}(\mathfrak{h})W)$ is invariant by left translations by H. Therefore, the subspace

$$L_H(H^2(G, \tau)[W]) = \{K_\lambda(\cdot, h)^\star w = L_h(K_\lambda(\cdot, e)^\star w) : w \in W, h \in H\}$$

is contained in $\mathrm{Cl}(\mathcal{U}(\mathfrak{h})W)$. Actually,

$$\mathrm{Cl}(L_H(H^2(G, \tau)[W])) = \mathrm{Cl}(\mathcal{U}(\mathfrak{h})W).$$

The other inclusion follows from that $\mathrm{Cl}(L_H(H^2(G, \tau)[W]))$ is invariant by left translation by H and $\{K_\lambda(\cdot, e)^\star w : w \in W\}$ is contained in the subspace of smooth vectors in $\mathrm{Cl}(L_H(H^2(G, \tau)[W]))$.

The result pointed out in the title of this paragraph is as follows:

Lemma 1 *Let (G, H) be an arbitrary reductive pair and an arbitrary representation $(\pi_\lambda, H^2(G, \tau))$. Then, $Ker(r)$ is equal to the orthogonal subspace to $\mathrm{Cl}(\mathcal{U}(\mathfrak{h})W)$.*

Proof Since, [24], $r : H^2(G, \tau) \to L^2(H \times_\tau W)$ is a continuous map, we have $Ker(r)$ as a closed subspace of $H^2(G, \tau)$. Next, for $f \in H^2(G, \tau)$, it holds the identity

$$(f(x), w)_W = \int_G (f(y), K_\lambda(y, x)^\star w)_W dy, \forall x \in G, \forall w \in W.$$

Thus, $r(f) = 0$ if and only if f is orthogonal to the subspace spanned by $\{K_\lambda(\cdot,h)^\star w : w \in W, h \in H\}$.

Hence, $\mathrm{Cl}(Ker(r)) = (\mathrm{Cl}(L_H(H^2(G,\tau)[W])))^\perp$. Applying considerations after the definition of $\mathrm{Cl}(\mathcal{U}(\mathfrak{h})W)$, we obtain $Ker(r)^\perp = \mathrm{Cl}(\mathcal{U}(\mathfrak{h})W)$. Thus, $Ker(r) = (Ker(r)^\perp)^\perp = \mathrm{Cl}(\mathcal{U}(\mathfrak{h})W)$. □

Corollary 1 *Any irreducible H-discrete factor M for $\mathrm{Cl}(\mathcal{U}(\mathfrak{h})W)$ contains a L-type in $res_L(\tau)$. That is, $M[res_L(\tau)] \neq \{0\}$.*

The corollary follows from that r restricted to $\mathrm{Cl}(\mathcal{U}(\mathfrak{h})W)$ is injective and that Frobenius reciprocity for $L^2(H \times_\tau W)$ holds.

3.3 The Map r_0^D Is Injective

As a consequence of the general fact shown in the previous subsection, we obtain the injectivity in (i) and the map r_0^D is injective.

Corollary 2 *Let (G,H) be a symmetric pair and $H_0 = G^{\sigma\theta}$. Then, the restriction map $r_0 : H^2(G,\tau) \to L^2(H_0 \times_\tau W)$ restricted to the subspace $\mathrm{Cl}(\mathcal{U}(\mathfrak{h}_0)W)$ is injective.*

Corollary 3 *Let (G,H) be a symmetric pair, $H_0 = G^{\sigma\theta}$, and we assume $res_H(\pi_\lambda)$ is H-admissible. Then, the map r_0^D is injective.*

In fact, for $T \in Hom_H(H^2(H,\sigma), H^2(G,\tau))$, if $r_0(D(K_T(e,\cdot)z)) = 0 \forall z \in Z$, then, since $D(K_T(e,x)z) \in \mathcal{U}(\mathfrak{h}_0)W$, the previous corollary implies $D(K_T(e,x)z) = 0 \forall z, x \in G$. Since $K_T(e,\cdot)z \in V_\lambda^G[V_\mu^H][V^L_{\mu+\rho_n^H}]$, and D is injective, we obtain $K_T(e,x)z = 0 \forall z, \forall x$. Lastly, we recall equality $K_T(h,x) = K_T(e,h^{-1}x)$. Hence, we have verified the corollary.

Before we show the surjectivity for the map r_0^D, we would like to comment on other works on the topic of this note.

3.4 Previous Work on Duality Formula and Harish-Chandra Parameters

The setting for this subsection is as follows: (G,H) is a symmetric pair, and $(\pi_\lambda, V_\lambda^G)$ is an irreducible square integrable representation of G and H-admissible. As before, we fix $K, L = H \cap K, T, U = H \cap T$. The following theorem has been shown by Gross and Wallach [9], and a different proof is in [18].

Theorem 2 (Gross-Wallach, T. Kobayashi-Y. Oshima) *We assume (G,H) is a symmetric pair and π_λ^G-is H-admissible; then:*

(a) *$res_H(\pi_\lambda^G)$ is the Hilbert sum of* **inequivalent** *square integrable representations for H, $\pi_{\mu_j}^H$, $j = 1, 2, \ldots$, with respective finite multiplicity $0 < m_j < \infty$.*

(b) *The Harish-Chandra parameters of the totality of discrete factors for $res_H(\pi_\lambda^G)$ belong to a "unique" Weyl Chamber in $i\mathfrak{u}^\star$.*

That is, $V_\lambda^G = \oplus_{1 \le j < \infty} V_\lambda^G[V_{\mu_j}^H] \equiv \oplus_j \; Hom_H(V_{\mu_j}^H, V_\lambda^G) \otimes V_{\mu_j}^H,$

$$\dim Hom_H(V_{\mu_j}^H, V_\lambda^G) = m_j, \qquad \pi_{\mu_j}^H \neq \pi_{\mu_i}^H \text{ iff } i \neq j,$$

and there exists a system of positive roots $\Psi_{H,\lambda} \subset \Phi(\mathfrak{h}, \mathfrak{u})$, such that for all j, $(\alpha, \mu_j) > 0$ for all $\alpha \in \Psi_{H,\lambda}$.

In [30, 18] (see Tables 1, 2, and 3), we find the list of pairs $(\mathfrak{g}, \mathfrak{h})$, as well as systems of positive roots $\Psi_G \subset \Phi(\mathfrak{g}, \mathfrak{t})$, $\Psi_{H,\lambda} \subset \Phi(\mathfrak{h}, \mathfrak{u})$ such that:

- λ dominant with respect to Ψ_G implies $res_H(\pi_\lambda^G)$ is admissible.
- For all μ_j in *a*), we have $(\mu_j, \Psi_{H,\lambda}) > 0$.
- When $U = T$, we have $\Psi_{H,\lambda} = \Psi_\lambda \cap \Phi(\mathfrak{h}, \mathfrak{t})$.

Since (G, H_0) is a symmetric pair, Theorem 2 and its comments apply to (G, H_0) and π_λ. Here, when $U = T$, $\Psi_{H_0,\lambda} = \Psi_\lambda \cap \Phi(\mathfrak{h}_0, \mathfrak{t})$.

From the tables in [30], it follows that any of the system Ψ_λ, $\Psi_{H,\lambda}$, $\Psi_{H_0,\lambda}$ has, at most, two noncompact simple roots, and the sum of the respective multiplicity of each noncompact simple root in the highest root is less than or equal to two.

3.4.1 Computing Harish-Chandra Parameters from Theorem 1

As usual, $\rho_n = \frac{1}{2}\sum_{\beta \in \Psi_\lambda \cap \Phi_n} \beta$, $\rho_n^H = \frac{1}{2}\sum_{\beta \in \Psi_{H,\lambda} \cap \Phi_n} \beta$, $\rho_K = \frac{1}{2}\sum_{\alpha \in \Psi_\lambda \cap \Phi_c} \alpha$, $\rho_L = \frac{1}{2}\sum_{\alpha \in \Psi_{H,\lambda} \cap \Phi_c} \alpha$. We write $res_L(\tau) = res_L(V_{\lambda+\rho_n}^K) = \oplus_{1 \le j \le r} q_j \pi_{\nu_j}^L = \sum_j q_j \sigma_j$, with ν_j dominant with respect to $\Psi_{H,\lambda} \cap \Phi_c$. We recall ν_j is the infinitesimal character (Harish-Chandra parameter) of $\pi_{\nu_j}^L$. Then, the Harish-Chandra parameter for $H^2(H_0, \pi_{\nu_j}^L)$ is $\eta_j = \nu_j - \rho_n^{H_0}$.

According to [27, Lemma 2.22] (see Remark 6), the infinitesimal character of a L-type of $H^2(H_0, \pi_{\nu_j}^L)$ is equal to $\nu_j + B = \eta_j + \rho_n^{H_0} + B$, where B is a sum of roots in $\Psi_{H_0,\lambda} \cap \Phi_n$.

With the isomorphism r_0^D in Theorem 1, we conclude below.

For each subrepresentation $V_{\mu_s}^H$ of $res_H(\pi_\lambda)$, we have $\mu_s + \rho_n^H$ as a L-type of

$$\mathbf{H}^2(H_0, \tau) \equiv \oplus_j q_j \; H^2(H_0, \pi_{\nu_j}^L) \equiv \oplus_j \underbrace{V_{\eta_j}^{H_0} \oplus \cdots \oplus V_{\eta_j}^{H_0}}_{q_j},$$

and the multiplicity of $V_{\mu_s}^H$ is equal to the multiplicity of $V_{\mu_s+\rho_n^H}^L$ in $\mathbf{H}^2(H_0, \tau)$.

3.4.2 Gross-Wallach Multiplicity Formula

To follow, we describe the duality Theorem due to [9]. (G, H) is a symmetric pair. For this paragraph, in order to avoid subindexes, we write $\mathfrak{g} = Lie(G), \mathfrak{h} = Lie(H)$, etc. We recall $\mathfrak{h}_0 = \mathfrak{g}^{\sigma\theta}$. We have decompositions $\mathfrak{g} = \mathfrak{k}+\mathfrak{p} = \mathfrak{h}+\mathfrak{q} = \mathfrak{h}_0+\mathfrak{p}\cap\mathfrak{h}+\mathfrak{q}\cap\mathfrak{k}$. The *dual* real Lie algebra to $\mathfrak{g}$ is $\mathfrak{g}^d = \mathfrak{h}_0+i(\mathfrak{p}\cap\mathfrak{h}+\mathfrak{q}\cap\mathfrak{k})$, and the algebra $\mathfrak{g}^d$ is a real form for $\mathfrak{g}_{\mathbb{C}}$. A maximal compactly embedded subalgebra for $\mathfrak{g}^d$ is $\widetilde{\mathfrak{k}} = \mathfrak{h}\cap\mathfrak{k}+i(\mathfrak{h}\cap\mathfrak{p})$. Let π_λ be a H-admissible Discrete Series for G. One of the main results of [9] attach to π_λ, a finite sum of the underlying Harish-Chandra module of finitely many fundamental representations for G^d, $(\Gamma^{\widetilde{K}}_{H\cap L})^{p_0+q_0}(N(\Lambda))$, so that for each subrepresentation V^H_μ of V^G_λ, we compute the multiplicity $m^{G,H}(\lambda, \mu)$ of V^H_μ by means of Blattner's formula [11] applied to $(\Gamma^{\widetilde{K}}_{H\cap L_1})^{p_0+q_0}(N(\Lambda))$. In more detail, since $Lie(H)_{\mathbb{C}} = Lie(\widetilde{K})_{\mathbb{C}}$, and the center of H is equal to the center of $\widetilde{K}$, for the infinitesimal character μ and the central character χ of V^H_μ, we may associate a finite dimensional irreducible representation $F_{\mu,\chi}$ for $\widetilde{K}$. Then, they show

$$\dim Hom_{\mathfrak{h},H\cap K}(V^H_\mu, V^G_\lambda) = \dim Hom_{\widetilde{K}}(F_{\mu,\chi}, (\Gamma^{\widetilde{K}}_{H\cap L_1})^{p_0+q_0}(N(\Lambda))),$$

$$m^{G,H}(\lambda, \mu) = (-1)^{\frac{1}{2}\dim(H/H\cap L_1)} \sum_{i=1}^{d} \sum_{s\in W_{\widetilde{K}}} \epsilon(s)p(\Lambda_i + \rho_{\widetilde{K}} + s s_{H\cap K}\mu),$$

where $\tau = F^\Lambda = \sum_i M^{\Lambda_i}$ is a sum of irreducible $H \cap L_1$-module and p is the partition function associated with $\Phi(\mathfrak{u}_1/\mathfrak{u}_1 \cap \mathfrak{h}_{\mathbb{C}}, \mathfrak{u})$; here, $\mathfrak{u}_1$ is the nilpotent radical of certain parabolic subalgebra $\mathfrak{q} = \mathfrak{l}_1 + \mathfrak{u}_1$ used to define the $A_{\mathfrak{q}}(\lambda)$ presentation for π_λ. *Explicit example IV* presents the result of [9] for the pair $(SO(2m, 2n), SO(2m, 2n-1))$.

3.4.3 Duflo-Vargas Multiplicity Formula [5]

We keep notation and hypothesis as in the previous paragraph. Then,

$$m^{G,H}(\lambda, \mu) = \pm \sum_{w\in W_K} \epsilon(w)p_{S^H_w}(\mu - q_{\mathfrak{u}}(w\lambda)).$$

Here, $q_{\mathfrak{u}} : \mathfrak{t}^\star \to \mathfrak{u}^\star$ is the restriction map. $p_{S^H_w}$ is the partition function associated with the multiset

$$S^H_w := S^L_w \backslash \Phi(\mathfrak{h}/\mathfrak{l}, \mathfrak{u}), \text{ where, } S^L_w := q_{\mathfrak{u}}(w(\Psi_\lambda)_n) \cup \Delta(\mathfrak{k}/\mathfrak{l}, \mathfrak{u}).$$

We recall for a strict multiset of elements in vector space V the partition function attached to S, roughly speaking, is the function that counts the number of ways of expressing each vector as a non-negative integral linear combinations of elements of S. For a precise definition, see [5] or the proof of Lemma 2.

3.4.4 Harris-He-Olafsson Multiplicity Formula [10]

Notation and hypothesis as in the previous paragraphs. Let

$$r_m : H^2(G,\tau) \to L^2(H \times_{S^m(Ad)\boxtimes\tau} (S^m(\mathfrak{p}\cap\mathfrak{q})^\star \otimes W))$$

be the normal derivative map defined in [24]. Let $\Theta_{\pi_\mu^H}$ denote the Harish-Chandra character of π_μ^H. For f a tempered function in $H^2(G,\tau)$, they define $\phi_{\pi_\lambda,\pi_\mu^H,m}(f) = \Theta_{\pi_\mu^H} \star r_m(f)$. They show

$$m^{G,H}(\lambda,\mu) = \lim_{m\to\infty} \dim \phi_{\pi_\lambda,\pi_\mu^H,m}((H^2(G,\tau)\cap C(G,\tau))[V_{\mu+\rho_n^H}]).$$

3.5 Completion of the Proof of Theorem 1: The Map r_0^D Is Surjective

Item (i) in Theorem 1 is shown in Proposition 1 (c). The existence of the map D is shown in Proposition 1 (e).

To show the surjectivity of r_0^D, we appeal to Theorem 4, [30, Theorem 1], where we show the initial space and the target space are equidimensional, and linear algebra concludes the proof of Theorem 1. Thus, we conclude the proof of Theorem 1 as soon as we complete the proof of Theorem 4 and Proposition 1.

4 Duality Theorem, Proof of Dimension Equality

The purpose of this subsection is to sketch a proof of the equality of dimensions in the duality formula presented in Theorem 1 as well as some consequences. Part of the notation has already been introduced in the previous section. Sometimes, the notation will be explained after it has been used. Unexplained notation is as in [5, 25, 30].

4.1 Dimension Equality Theorem: Statement

The setting is as follows: (G, H) is a symmetric pair, and $(\pi_\lambda, V_\lambda^G) = (L, H^2(G,\tau))$ is a H-admissible irreducible square integrable representation. Then, the Harish-Chandra parameter λ gives rise to systems of positive roots Ψ_λ in $\Phi(\mathfrak{g}, \mathfrak{t})$ and, by means of Ψ_λ in [5], is defined a nontrivial normal connected subgroup $K_1(\Psi_\lambda) =: K_1$ of K; it is shown that the H-admissibility yields $K_1 \subset H$.[1] Thus, $\mathfrak{k} = \mathfrak{k}_1 \oplus \mathfrak{k}_2$, $\mathfrak{l} = \mathfrak{k}_1 \oplus \mathfrak{l} \cap \mathfrak{k}_2$ (as ideals), and $\mathfrak{t} = \mathfrak{t} \cap \mathfrak{k}_1 + \mathfrak{t} \cap \mathfrak{k}_2$, $\mathfrak{u} := \mathfrak{t} \cap \mathfrak{l} = \mathfrak{u} \cap \mathfrak{k}_1 + \mathfrak{u} \cap \mathfrak{k}_2$ is a Cartan subalgebra of $\mathfrak{l}$. Let $q_\mathfrak{u}$ denote restriction map from $\mathfrak{t}^\star$ onto $\mathfrak{u}^\star$. Let K_2 denote the analytic subgroup corresponding to $\mathfrak{k}_2$. We recall $H_0 := (G^{\sigma\theta})_0$, $L = K \cap H = K \cap H_0$. We have $K = K_1 K_2$, $L = K_1(K_2 \cap L)$. We set $\Delta := \Psi_\lambda \cap \Phi(\mathfrak{k}, \mathfrak{t})$. Applying Theorem 2 to both H and H_0, we obtain respective systems of positive roots $\Psi_{H,\lambda}$ in $\Phi(\mathfrak{h}, \mathfrak{u})$, $\Psi_{H_0,\lambda}$ in $\Phi(\mathfrak{h}_0, \mathfrak{u})$. For a list of six-tuples $(G, H, \Psi_\lambda, \Psi_{H,\lambda}, \Psi_{H_0,\lambda}, K_1)$, we refer to [30, Table 1, Table 2, Table 3]. Always, $\Psi_{H,\lambda} \cap \Phi_c(\mathfrak{l}, \mathfrak{u}) = \Psi_{H_0,\lambda} \cap \Phi_c(\mathfrak{l}, \mathfrak{u})$. As usual, either $\Phi_n(\mathfrak{g}, \mathfrak{t})$ or Φ_n denotes the subset of noncompact roots in $\Phi(\mathfrak{g}, \mathfrak{t})$, ρ_n^λ (resp. ρ_n^H, $\rho_n^{H_0}$) denotes one half of the sum of the elements in $\Psi_\lambda \cap \Phi_n(\mathfrak{g}, \mathfrak{t})$ (resp. $\Phi_n \cap \Psi_{H,\lambda}$, $\Phi_n \cap \Psi_{H_0,\lambda}$). When $\mathfrak{u} = \mathfrak{t}$, $\rho_n^\lambda = \rho_n^H + \rho_n^{H_0}$. From now on, the infinitesimal character of an irreducible representation of K (resp L) is dominant with respect to Δ (resp. $\Psi_{H,\lambda} \cap \Phi(\mathfrak{l}, \mathfrak{u})$). The lowest K-type (τ, W) of π_λ decomposes $\pi^K_{\lambda+\rho_n^\lambda} = \pi^{K_1}_{\Lambda_1} \boxtimes \pi^{K_2}_{\Lambda_2}$, with $\pi^{K_s}_{\Lambda_s}$ an irreducible representation for K_s, $s = 1, 2$. We express $\gamma = (\gamma_1, \gamma_2) \in \mathfrak{t}^\star = \mathfrak{t}_1^\star + \mathfrak{t}_2^\star$. Hence, [11, 3], $\Lambda_1 = \lambda_1 + (\rho_n^\lambda)_1$, $\Lambda_2 = \lambda_2 + (\rho_n^\lambda)_2$. Sometimes $(\rho_n^\lambda)_2 \neq 0$. This happens only for $\mathfrak{su}(m, n)$ and some particular systems Ψ_λ (see proof of Lemma 4). Harish-Chandra parameters for the irreducible factors of either $res_H(\pi_\lambda)$ (resp. $res_{H_0}(\pi_\lambda)$)) will always be dominant with respect to $\Psi_{H,\lambda} \cap \Phi(\mathfrak{l}, \mathfrak{u})$ (resp. $\Psi_{H_0,\lambda} \cap \Phi(\mathfrak{l}, \mathfrak{u})$).

In short, we write $\pi_{\Lambda_2} := \pi^{K_2}_{\Lambda_2}$. We write

$$res_{L\cap K_2}(\pi_{\Lambda_2}) = res_{L\cap K_2}(\pi^{K_2}_{\Lambda_2}) = \sum_{\nu_2 \in (\mathfrak{u}\cap\mathfrak{k}_2)^\star} m^{K_2, L\cap K_2}(\Lambda_2, \nu_2)\, \pi^{L\cap K_2}_{\nu_2},$$

as a sum of irreducible representations of $L \cap K_2$.

The set of ν_2 so that $m^{K_2, L\cap K_2}(\Lambda_2, \nu_2) \neq 0$ is denoted by $Spec_{L\cap K_2}(\pi^{K_2}_{\Lambda_2})$. Thus,

$$res_L(\pi^{K_1}_{\Lambda_1} \boxtimes \pi^{K_2}_{\Lambda_2}) = \sum_{\nu_2 \in Spec_{L\cap K_2}(\pi^{K_2}_{\Lambda_2})} m^{K_2, L\cap K_2}(\Lambda_2, \nu_2)\, \pi^{K_1}_{\Lambda_1} \boxtimes \pi^{L\cap K_2}_{\nu_2}, \tag{4}$$

[1] This also follows from tables in [18].

as a sum of irreducible representations of L. Besides, for a Harish-Chandra parameter $\eta = (\eta_1, \eta_2)$ for H_0, we write

$$res_L(\pi^{H_0}_{(\eta_1,\eta_2)}) = \sum_{(\theta_1,\theta_2)\in Spec_L(\pi^{H_0}_{(\eta_1,\eta_2)})} m^{H_0,L}((\eta_1,\eta_2),(\theta_1,\theta_2))\,\pi^L_{(\theta_1,\theta_2)}.$$

The restriction of π_λ to H is expressed by (see 2)

$$res_H(\pi_\lambda) = res_H(\pi^G_\lambda) = \sum_{\mu\in Spec_H(\pi_\lambda)} m^{G,H}(\lambda,\mu)\,\pi^H_\mu.$$

In the above formulas, $m^{\cdot,\cdot}(\cdot,\cdot)$ are non-negative integers and represent multiplicities; for $\nu_2 \in Spec_{L\cap K_2}(\pi^{K_2}_{\Lambda_2})$, ν_2 is dominant with respect to $\Psi_{H,\lambda}\cap\Phi(\mathfrak{k}_2, \mathfrak{u}\cap\mathfrak{k}_2)$, and (Λ_1,ν_2) is $\Psi_{H_0,\lambda}$-dominant (see [30]); in the third formula, (η_1,η_2) is dominant with respect to $\Psi_{H_0,\lambda}$, and (θ_1,θ_2) is dominant with respect to $\Psi_{H_0,\lambda}\cap\Phi_c(\mathfrak{h}_0,\mathfrak{u})$; in the fourth formula, μ is dominant with respect to $\Psi_{H,\lambda}$. Sometimes, for $\mu \in Spec_H(\pi^G_\lambda)$, we replace ρ^μ_n by ρ^H_n.

We make a change of notation:

$$\sigma_j = \pi^{K_1}_{\Lambda_1}\boxtimes\pi^{L\cap K_2}_{\nu_2} \text{ and } q_j = m^{K_2,L\cap K_2}(\Lambda_2,\nu_2).$$

Then, in order to show either the existence of the map D or the surjectivity of the map r^D_0, we need to show:

Theorem 3

$$\begin{aligned}
m^{G,H}(\lambda,\mu) &= \dim Hom_H(H^2(H, V^L_{\mu+\rho^H_n}), H^2(G, V^K_{\lambda+\rho^\lambda_n}))\\
&= \sum_{\nu_2\in Spec_{L\cap K_2}(\pi^{K_2}_{\Lambda_2})} m^{K_2,L\cap K_2}(\Lambda_2,\nu_2)\\
&\qquad\qquad \times \dim Hom_L(V^L_{\mu+\rho^H_n}, H^2(H_0, \pi^{K_1}_{\Lambda_1}\boxtimes\pi^{L\cap K_2}_{\nu_2})).
\end{aligned}$$

A complete proof of the result is in [30]. However, for the sake of completeness and clarity, we would like to sketch a proof. We also present some consequences of the theorem.

Next, we compute the infinitesimal character, $ic(...)$, equivalently the Harish-Chandra parameter, for $H^2(H_0, \pi^{K_1}_{\Lambda_1}\boxtimes\pi^{L\cap K_2}_{\nu_2})$, and restate the previous theorem.

$$\begin{aligned}
&ic(H^2(H_0, \pi^{K_1}_{\Lambda_1}\boxtimes\pi^{L\cap K_2}_{\nu_2})) = (\Lambda_1,\nu_2) - \rho^{H_0}_n\\
&= (\lambda_1 + (\rho^\lambda_n)_1 - (\rho^{H_0}_n)_1, \nu_2 - (\rho^{H_0}_n)_2) = (\lambda_1 + (\rho^H_n)_1, \nu_2 - (\rho^{H_0}_n)_2)\\
&= (\lambda_1, \nu_2 - (\rho^H_n)_2 - (\rho^{H_0}_n)_2) + \rho^H_n = (\lambda_1, \nu_2 - q_{\mathfrak{u}}(\rho^\lambda_n)_2) + \rho^H_n.
\end{aligned}$$

To follow, we state Theorem 3 regardless of the realization of the involved Discrete Series.

Theorem 4 *Duality, dimension formula. The hypothesis is* (G, H) *is a symmetric pair and* π_λ *is a* H*-admissible representation. Then,*

$$
\begin{aligned}
m^{G,H}(\lambda,\mu) &= \dim Hom_H(V_\mu^H, V_\lambda^G)\\
&= \sum_{\nu_2 \in Spec_{L\cap K_2}(\pi_{\Lambda_2}^{K_2})} m^{K_2, L\cap K_2}(\Lambda_2, \nu_2) \dim Hom_L\big(V^L_{\mu+\rho_n^H}, V^{H_0}_{(\lambda_1, \nu_2 - q_{\mathfrak{u}}(\rho_n^\lambda)_2)+\rho_n^H}\big)\\
&= \sum_{\nu_2' \in Spec_{L\cap K_2}(\pi_{\Lambda_2}^{K_2}) - q_{\mathfrak{u}}(\rho_n^\lambda)_2} m^{K_2, L\cap K_2}(\Lambda_2, \nu_2' + q_{\mathfrak{u}}(\rho_n^\lambda)_2)\\
&\qquad \times \dim Hom_L\big(V^L_{\mu+\rho_n^H}, V^{H_0}_{(\lambda_1,\nu_2')+\rho_n^H}\big).
\end{aligned}
$$

Remark In the proof of Lemma 4 we obtain that $(\rho_n^\lambda)_2 \neq 0$ forces $\mathfrak{g} = \mathfrak{su}(m,n)$, $\mathfrak{u} = \mathfrak{t}$, and $(\rho_n^\lambda)_2$ is orthogonal to $(\mathfrak{z}_\mathfrak{k})^\perp$. That is, $(\rho_n^\lambda)_2 \in \mathfrak{z}_\mathfrak{k}^\star$, i.e., for every symmetric pair (G, H), $(\rho_n^\lambda)_2$ determines a central character of $\mathfrak{k}$. It is also verified that since $\Lambda_2 = (\rho_n^\lambda)_2 + \lambda_2$, we have $Spec_{L\cap K_2}(\pi_{\Lambda_2}^{K_2}) = q_{\mathfrak{u}}(\rho_n^\lambda)_2 + Spec_{L\cap K_2}(\pi_{\lambda_2}^{K_2})$, and for $Spec_{L\cap K_2}(\pi_{\Lambda_2}^{K_2}) \ni \nu_2 = q_{\mathfrak{u}}(\rho_n^\lambda)_2 + \nu_2'$, $\nu_2' \in Spec_{L\cap K_2}(\pi_{\lambda_2}^{K_2})$, it holds the equality $m^{K_2, L\cap K_2}(\Lambda_2, \nu_2) = m^{K_2, L\cap K_2}(\lambda_2, \nu_2')$. Finally, Lemma 3 yields the formula in Theorem 4 is equivalent to the formula

$$
m^{G,H}(\lambda,\mu) = \sum_{\nu_2' \in Spec_{L\cap K_2}(\pi_{\lambda_2}^{K_2})} m^{K_2, L\cap K_2}(\lambda_2, \nu_2') \dim Hom_L(V_\mu^L, V^{H_0}_{(\lambda_1,\nu_2')}).
$$

This formula is the one we show in Lemma 2.

The following diagram helps understand the equalities in Theorem 4 and in the next three Lemmas.

$$
\begin{array}{ccc}
Spec_H(V^G_{(\lambda_1,\lambda_2)}) & \xrightarrow{\ \mu\mapsto\mu\ } & \cup_{\nu_2' \in Spec_{L\cap K_2}(\pi_{\lambda_2}^{K_2})} Spec_L(V^{H_0}_{(\lambda_1,\nu_2')}) \\
 & \searrow^{\nu\mapsto\nu+\rho_n^H} & \downarrow{\scriptstyle \nu\mapsto\nu+\rho_n^H} \\
 & & Spec_L(\mathbf{H}^2(H_0,\tau)) = \cup_{\nu_2' \in Spec_{L\cap K_2}(\pi_{\lambda_2}^{K_2})} Spec_L(V^{H_0}_{(\lambda_1,\nu_2')+\rho_n^H})
\end{array}
$$

A consequence of Theorem 4, Lemmas 3, and 2 is:

Corollary 4

$$Spec_H(\pi_\lambda) + \rho_n^H$$
$$= Spec_L(\mathbf{H}^2(H_0, \tau)) = \cup_{\nu_2' \in Spec_{L \cap K_2}(\pi_{\lambda_2}^{K_2})} Spec_L(V^{H_0}_{(\lambda_1, \nu_2') + \rho_n^H}).$$

$$Spec_H(\pi_\lambda) = \cup_{\nu_2' \in Spec_{L \cap K_2}(\pi_{\lambda_2}^{K_2})} Spec_L(V^{H_0}_{(\lambda_1, \nu_2')}).$$

As we pointed out, Theorem 4 follows after we verify the next three Lemmas.

Lemma 2 *The hypothesis is (G, H) is a symmetric space and π_λ is H-admissible. Then*

$$\dim Hom_H(V_\mu^H, V_\lambda^G)$$
$$= \sum_{\nu_2' \in Spec_{L \cap K_2}(\pi_{\lambda_2}^{K_2})} m^{K_2, L \cap K_2}(\lambda_2, \nu_2') \dim Hom_L(V_\mu^L, V^{H_0}_{(\lambda_1, \nu_2')}).$$

Proof (Proof of Lemma 2) With the hypothesis (G, H) is a symmetric pair and π_λ is H-admissible, let us to apply notation and facts in [5, 30] as well as in [12, 3, 9, 18]. The proof is based on an idea in [3] where multiplicities are piled up by means of Dirac delta distributions. That is, let δ_ν denote the Dirac delta distribution at $\nu \in i\mathfrak{u}^\star$. Under our hypothesis, the function $m^{G,H}(\lambda, \mu)$ has polynomial growth in μ; hence, the series $\sum_\mu m^{G,H}(\lambda, \mu)\, \delta_\mu$ converges in the space of distributions in $i\mathfrak{u}^\star$. Since Harish-Chandra parameter is regular, we may and will extend the function $m^{G,H}(\lambda, \cdot)$ to a W_L-skew symmetric function by the rule $m^{G,H}(\lambda, w\mu) = \epsilon(w) m^{G,H}(\lambda, \mu)$, $w \in W_L$. Thus, the series $\sum_{\mu \in HC-param(H)} m^{G,H}(\lambda, \mu)\delta_\mu$ converges in the space of distributions in $i\mathfrak{u}^\star$. Next, for $0 \neq \gamma \in i\mathfrak{u}^\star$, we consider the discrete Heaviside distribution $y_\gamma := \sum_{n \geq 0} \delta_{\frac{\gamma}{2} + n\gamma}$, and for a strict, finite, multiset $S = \{\gamma_1, \ldots, \gamma_r\}$ of elements in $i\mathfrak{u}^\star$, we set

$$y_S := y_{\gamma_1} \star \cdots \star y_{\gamma_r} = \sum_{\mu \in i\mathfrak{u}^\star} p_S(\mu)\delta_\mu.$$

Here, $\star$ is the convolution product in the space of distributions on $i\mathfrak{u}^\star$. p_S is called the *partition* function attached to the set S. Then, in [5], the following equality is presented:

$$\sum_{\mu \in HC-param(H)} m^{G,H}(\lambda, \mu)\, \delta_\mu = \sum_{w \in W_K} \epsilon(w)\, \delta_{q_\mathfrak{u}(w\lambda)} \star y_{S_w^H}.$$

Here, W_S is the Weyl group of the compact connected Lie group S; for a $ad(\mathfrak{u})$-invariant linear subspace R of $\mathfrak{g}_\mathbb{C}$, $\Phi(R, \mathfrak{u})$ denotes the multiset of elements in

$\Phi(\mathfrak{g}, \mathfrak{u})$ such that its root space is contained in R, and $S_w^H = [q_\mathfrak{u}(w(\Psi_\lambda)_n) \cup \Delta(\mathfrak{k}/\mathfrak{l}, \mathfrak{u})]\backslash\Phi(\mathfrak{h}/\mathfrak{l}, \mathfrak{u})$.

Since $K = K_1K_2$, $W_K = W_{K_1} \times W_{K_2}$, we write $W_K \ni w = st, s \in W_{K_1}, t \in W_{K_2}$. We recall the hypothesis yields $K_1 \subset L$. It readily follows: $s\Phi(\mathfrak{h}/\mathfrak{l}, \mathfrak{u}) = \Phi(\mathfrak{h}/\mathfrak{l}, \mathfrak{u})$, $s\Delta(\mathfrak{k}/\mathfrak{l}, \mathfrak{u}) = \Delta(\mathfrak{k}/\mathfrak{l}, \mathfrak{u}), t(\Psi_\lambda)_n = (\Psi_\lambda)_n, t\eta_1 = \eta_1, s\eta_2 = \eta_2$ for $\eta_j \in \mathfrak{k}_j \cap \mathfrak{u}$, $sq_\mathfrak{u}(\cdot) = q_\mathfrak{u}(s\cdot)$. Hence,

$$S_w^H = s([q_\mathfrak{u}((\Psi_\lambda)_n) \cup \Delta(\mathfrak{k}/\mathfrak{l}, \mathfrak{u})]\backslash\Phi(\mathfrak{h}/\mathfrak{l}, \mathfrak{u})) = s(\Psi_n^{H_0}) \cup \Delta(\mathfrak{k}/\mathfrak{l}, \mathfrak{u}).$$

Thus,

$$\begin{aligned}\sum_{w\in W_K} \epsilon(w)\, \delta_{q_\mathfrak{u}(w\lambda)} \star y_{S_w^H} &= \sum_{s,t} \epsilon(st)\delta_{q_\mathfrak{u}(st\lambda)} \star y_{s(\Psi_n^{H_0})\cup\Delta(\mathfrak{k}/\mathfrak{l},\mathfrak{u})} \\ &= \sum_{s,t} \epsilon(st)\delta_{q_\mathfrak{u}(s\lambda_1+t\lambda_2)} \star y_{s(\Psi_n^{H_0})} \star y_{\Delta(\mathfrak{k}/\mathfrak{l},\mathfrak{u})} \\ &= \sum_{s} \epsilon(s)\delta_{(s\lambda_1,0)} \star y_{s(\Psi_n^{H_0})} \star \sum_t \epsilon(t)\delta_{q_\mathfrak{u}(t\lambda_2)} \star y_{\Delta(\mathfrak{k}/\mathfrak{l},\mathfrak{u})}.\end{aligned}$$

Following [12], we write the restriction of $\pi_{\lambda_2}^{K_2}$ to $L \cap K_2$ in the language of Dirac and Heaviside distributions in $i\mathfrak{u}^\star$; hence,

$$\begin{aligned}&\sum_{t\in W_{K_2}} \epsilon(t)\delta_{q_\mathfrak{u}(t\lambda_2)} \star y_{\Delta(\mathfrak{k}_2/(\mathfrak{k}_2\cap\mathfrak{l}),\mathfrak{u})} \\ &\quad= \sum_{\nu_2'\in Spec_{L\cap K_2}(\pi_{\lambda_2}^{K_2})} m^{K_2, L\cap K_2}(\lambda_2, \nu_2') \sum_{w_2\in W_{K_2\cap L}} \epsilon(w_2)\delta_{(0,w_2\nu_2')}.\end{aligned}$$

In the previous formula, we will apply $\Delta(\mathfrak{k}_2/(\mathfrak{k}_2 \cap \mathfrak{l}), \mathfrak{u}) = \Delta(\mathfrak{k}/\mathfrak{l}, \mathfrak{u})$.

We also write in the same language the restriction to L of a Discrete Series $\pi^{H_0}_{(\lambda_1,\nu_2')}$ for H_0. This is

$$\sum_{\nu\in i\mathfrak{u}^\star} m^{H_0,L}((\lambda_1, \nu_2'), \nu)\, \delta_\nu = \sum_{s\in W_{K_1}, t\in W_{K_2\cap L}} \epsilon(st)\delta_{st(\lambda_1,\nu_2')} \star y_{st(\Psi_n^{H_0})}.$$

Putting together the previous equalities, we obtain

$$\begin{aligned}&\sum_\mu m^{G,H}(\lambda, \mu)\, \delta_\mu \\ &\quad= \textstyle\sum_{\nu_2'\in Spec_{L\cap K_2}(\pi_{\lambda_2}^{K_2})} m^{K_2,L\cap K_2}(\lambda_2, \nu_2') \\ &\qquad\times \textstyle\sum_{s\in W_{K_1}, t\in W_{K_2\cap L}} \epsilon(st)\delta_{(st\lambda_1, st\nu_2')} \star y_{st(\Psi_n^{H_0})} \\ &\quad= \textstyle\sum_\nu (\sum_{\nu_2'\in Spec_{L\cap K_2}(\pi_{\lambda_2}^{K_2})} m^{K_2,L\cap K_2}(\lambda_2, \nu_2') m^{H_0,L}((\lambda_1, \nu_2'), \nu))\, \delta_\nu.\end{aligned}$$

Since the family $\{\delta_\nu\}_{\nu \in i\mathfrak{u}^\star}$ is linearly independent, we have shown Lemma 2. □

In order to conclude the proof of the dimension equality, we state and prove a translation invariant property of multiplicity.

Lemma 3 *For a dominant integral $\mu \in i\mathfrak{u}^\star$, it holds:*

$$m^{H_0,L}((\lambda_1, \nu_2') + \rho_n^H, \mu + \rho_n^H) = m^{H_0,L}((\lambda_1, \nu_2'), \mu).$$

Proof We recall that the hypothesis of Lemma 3 is as follows: (G, H) is a symmetric pair, and π_λ is H-admissible. The proof of Lemma 3 is an application of Blattner's multiplicity formula, facts from [11], and observations from [30, Table 1,2,3]. In the next paragraphs, we only consider systems Ψ_λ so that $res_H(\pi_\lambda)$ is admissible. We check the following statements by means of case-by-case analysis and tables in [9] and [30]:

OBS0. Every quaternionic system of positive roots that we are dealing with satisfies the Borel de Siebenthal property, except for the algebra $\mathfrak{su}(2, 2n)$ and systems Ψ_1 (see 5). Its Dynkin diagram is $\bullet$ —— $\circ$ —— $\cdots$ —— $\circ$ —— $\bullet$ ·. Bullet represents noncompact roots, circle compact.

OBS1. Always systems $\Psi_{H,\lambda}$, $\Psi_{H_0,\lambda}$ have the same compact simple roots.

OBS2. When Ψ_λ satisfies the Borel de Siebenthal property, it follows that both systems $\Psi_{H,\lambda}$, $\Psi_{H_0,\lambda}$ satisfy the Borel de Siebenthal property, except for[2] (i) $\mathfrak{g} = \mathfrak{so}(2m, 2n)$, $\mathfrak{h} = \mathfrak{so}(2m, 2) + \mathfrak{so}(2n-2)$, and $\Psi_{H,\lambda}$; and (ii) $\mathfrak{g} = \mathfrak{so}(2m, 2n)$, $\mathfrak{h} = \mathfrak{so}(2m, 2n-2) + \mathfrak{so}(2)$ and $\Psi_{H_0,\lambda}$; for details, see the proof of Remark 5.

OBS3. Ψ_λ satisfies the Borel de Siebenthal property except for two families of algebras: (a) For the algebra $\mathfrak{su}(m, n)$ and the systems $\Psi_a, a = 1, \cdots, m-1$ and $\tilde{\Psi}_b, b = 1, \cdots, n-1$, corresponding systems $\Psi_{H_0,\lambda}$, $\Psi_{H,\lambda}$ do not satisfy the Borel de Siebenthal property. They have two noncompact simple roots. (b) For the algebra $\mathfrak{so}(2m, 2)$, each system $\Psi_\pm$ does not satisfy the Borel de Siebenthal property; however, each associated system $\Psi_{SO(2m,1),\lambda}$, $\Psi_{H_0,\lambda}$ satisfies the Borel de Siebenthal property.

OBS4. For the pair $(\mathfrak{su}(2, 2n), \mathfrak{sp}(1, n))$. Ψ_1 does not satisfy the Borel de Siebenthal property. Here, $\Psi_{H,\lambda} = \Psi_{H_0,\lambda}$, and they have Borel de Siebenthal property.

OBS5. Summing up, both systems $\Psi_{H,\lambda}$, $\Psi_{H_0,\lambda}$ satisfy the Borel de Siebenthal property, except for (i) $(\mathfrak{su}(m, n), \mathfrak{su}(m, k)+\mathfrak{su}(n-k)+\mathfrak{u}(1))$, $(\mathfrak{su}(m, n), \mathfrak{su}(k, n)+\mathfrak{su}(m-k)+\mathfrak{u}(1))$ and the systems $\Psi_a, a = 1, \cdots, m-1$, $\tilde{\Psi}_b, b = 1, \cdots, n-1$; ii) $\mathfrak{g} = \mathfrak{so}(2m, 2n)$, $\mathfrak{h} = \mathfrak{so}(2m, 2)+\mathfrak{so}(2n-2)$ and $\Psi_{H,\lambda}$; and iii) $\mathfrak{g} = \mathfrak{so}(2m, 2n)$, $\mathfrak{h} = \mathfrak{so}(2m, 2n-2) + \mathfrak{so}(2)$ and $\Psi_{H_0,\lambda}$. For details, see the proof of Remark 5.

To continue, we explicit Blattner's formula according to our setting. We recall facts from [11] and finish the proof of Lemma 3 under the assumption $\Psi_{H_0,\lambda}$ that satisfies the property of Borel de Siebenthal. Later on, we consider other systems.

[2] We are indebted to the referee for this observation.

Blattner's multiplicity formula applied to the L-type $V^L_{\mu+\rho_n^H}$ of $V^{H_0}_{(\lambda_1,\nu_2')+\rho_n^H}$ yields

$$\dim Hom_L(V^L_{\mu+\rho_n^H}, V^{H_0}_{(\lambda_1,\nu_2')+\rho_n^H}) \tag{5}$$
$$= \sum_{s\in W_L} \epsilon(s) Q_0(s(\mu+\rho_n^H) - ((\lambda_1,\nu_2') + \rho_n^H + \rho_n^{H_0})).$$

Here, Q_0 is the partition function associated with the set $\Phi_n(\mathfrak{h}_0) \cap \Psi_{H_0,\lambda}$.

We recall a fact that allows to simplify the formula above under our setting.

Fact 1: [11, Statement 4.31]. For a system $\Psi_{H_0,\lambda}$ having the Borel de Siebenthal property, it is shown that in the above sum, if the summand attached to $s \in W_L$ contributes nontrivially, then s belongs to the subgroup $W_U(\Psi_{H_0,\lambda})$ spanned by the reflections about the compact simple roots in $\Psi_{H_0,\lambda}$.

From OBS1 we have $W_U(\Psi_{H_0,\lambda}) = W_U(\Psi_{H,\lambda})$. Owing that either $\Psi_{H_0,\lambda}$ or $\Psi_{H,\lambda}$ has the Borel de Siebenthal property, we apply [11, Lemma 3.3]; hence, $W_U(\Psi_{H,\lambda}) = \{s \in W_L : s(\Psi_{H,\lambda} \cap \Phi_n(\mathfrak{h},\mathfrak{u})) = \Psi_{H,\lambda} \cap \Phi_n(\mathfrak{h},\mathfrak{u})\}$. Thus, for $s \in W_U(\Psi_{H_0,\lambda})$ we have $s\rho_n^H = \rho_n^H$. We apply the equality $s\rho_n^H = \rho_n^H$ in 5, and we obtain

$$\dim Hom_L(V^L_{\mu+\rho_n^H}, V^{H_0}_{(\lambda_1,\nu_2')+\rho_n^H}) = \sum_{s\in W_U(\Psi_{H_0,\lambda})} \epsilon(s) Q_0(s\mu - ((\lambda_1,\nu_2') + \rho_n^{H_0})).$$

Blattner's formula and the previous observations give us that the right-hand side of the above equality is

$$\dim Hom_L(V^L_{\mu}, V^{H_0}_{(\lambda_1,\nu_2')}) = m^{H_0,L}((\lambda_1,\nu_2'),\mu);$$

hence, we have shown Lemma 3 when $\Psi_{H_0,\lambda}$ has the Borel de Siebenthal property.

In order to complete the proof of Lemma 3, owing to OBS5, we are left to consider the pairs $(\mathfrak{su}(m,n), \mathfrak{su}(m,k)+\mathfrak{su}(n-k)+\mathfrak{u}(1))$ and $(\mathfrak{su}(m,n), \mathfrak{su}(k,n)+\mathfrak{su}(m-k)+\mathfrak{u}(1))$ and the systems $\Psi_a, a = 1,\cdots,m-1$, $\tilde{\Psi}_b, b = 1,\cdots,n-1$; $\mathfrak{g} = \mathfrak{so}(2m,2n)$, $\mathfrak{h} = \mathfrak{so}(2m,2) + \mathfrak{so}(2n-2)$ and $\Psi_{H,\lambda}$; $\mathfrak{g} = \mathfrak{so}(2m,2n)$, $\mathfrak{h} = \mathfrak{so}(2m,2n-2) + \mathfrak{so}(2)$ and $\Psi_{H_0,\lambda}$. For details, see the proof of Remark 5.

The previous reasoning says we are left to extend Fact 1, [11, Statement (4.31)], for the pairs $(\mathfrak{su}(m,n), \mathfrak{su}(m,k)+\mathfrak{su}(n-k)+\mathfrak{u}(1))$ (resp. $(\mathfrak{su}(m,n), \mathfrak{su}(k,n)+\mathfrak{su}(m-k)+\mathfrak{u}(1))$) and the systems $(\Psi_a)_{a=1,\cdots,m-1}$ (resp. $(\tilde{\Psi}_b)_{b=1,\cdots,n-1}$); $\mathfrak{g} = \mathfrak{so}(2m,2n)$, $\mathfrak{h} = \mathfrak{so}(2m,2n-2) + \mathfrak{so}(2)$ and $\Psi_{H_0,\lambda}$. For details, see the proof of Remark 5. Under this setting, we first verify:

Remark 5 If $w \in W_L$ and $Q_0(w\mu - (\lambda + \rho_n^{H_0})) \neq 0$, then $w \in W_U(\Psi_{H_0,\lambda})$. □

To show Remark 5, we follow [11]. To begin with, we consider the pairs associated with $\mathfrak{g} = \mathfrak{su}(m,n)$. We fix as Cartan subalgebra $\mathfrak{t}$ of $\mathfrak{su}(m,n)$ the set of diagonal matrices in $\mathfrak{su}(m,n)$. For certain orthogonal basis $\epsilon_1,\ldots,\epsilon_m,\delta_1,\ldots,\delta_n$

of the dual vector space to the subspace of diagonal matrices in $\mathfrak{gl}(m+n,\mathbb{C})$, we may and will choose $\Delta = \{\epsilon_r - \epsilon_s, \delta_p - \delta_q, 1 \le r < s \le m, 1 \le p < q \le n\}$; the set of noncompact roots is $\Phi_n = \{\pm(\epsilon_r - \delta_q)\}$. We recall the positive roots systems for $\Phi(\mathfrak{g},\mathfrak{t})$ containing Δ are in a bijective correspondence with the totality of lexicographic orders for the basis $\epsilon_1,\dots,\epsilon_m,\delta_1,\dots,\delta_n$ which contains the "suborder" $\epsilon_1 > \cdots > \epsilon_m$, $\delta_1 > \cdots > \delta_n$. The two holomorphic systems correspond to the orders $\epsilon_1 > \cdots > \epsilon_m > \delta_1 > \cdots > \delta_n$; $\delta_1 > \cdots > \delta_n > \epsilon_1 > \cdots > \epsilon_m$. We fix $1 \le a \le m-1$, and let Ψ_a denote the set of positive roots associated with the order $\epsilon_1 > \cdots > \epsilon_a > \delta_1 > \cdots > \delta_n > \epsilon_{a+1} > \cdots > \epsilon_m$. We fix $1 \le b \le n-1$, and let $\tilde{\Psi}_b$ denote the set of positive roots associated with the order $\delta_1 > \cdots > \delta_b > \epsilon_1 > \cdots > \epsilon_m > \delta_{b+1} > \cdots > \delta_n$. Since $\mathfrak{h} = \mathfrak{su}(m,k)+\mathfrak{u}(n-k)$, $\mathfrak{h}_0 = \mathfrak{su}(m,n-k)+\mathfrak{u}(k)$. The root systems for $(\mathfrak{h},\mathfrak{t})$ and $(\mathfrak{h}_0,\mathfrak{t})$, respectively, are

$$\Phi(\mathfrak{h},\mathfrak{t}) = \{\pm(\epsilon_r - \epsilon_s), \pm(\delta_p - \delta_q), \pm(\epsilon_i - \delta_j), 1 \le r < s \le m,$$
$$1 \le p < q \le k,\ or,\ k+1 \le p < q \le n, 1 \le i \le m, 1 \le j \le k\}.$$

$$\Phi(\mathfrak{h}_0,\mathfrak{t}) = \{\pm(\epsilon_r - \epsilon_s), \pm(\delta_p - \delta_q), \pm(\epsilon_i - \delta_j), 1 \le r < s \le m,$$
$$1 \le p < q \le k\ or\ k+1 \le p < q \le n, 1 \le i \le m, k+1 \le j \le n\}.$$

Systems $\Psi_{H,\lambda} = \Psi_\lambda \cap \Phi(\mathfrak{h},\mathfrak{t})$, $\Psi_{H_0,\lambda} = \Psi_\lambda \cap \Phi(\mathfrak{h}_0,\mathfrak{t})$, which correspond to Ψ_a, are systems associated with the respective lexicographic orders

$$\epsilon_1 > \cdots > \epsilon_a > \delta_1 > \cdots > \delta_k > \epsilon_{a+1} > \cdots > \epsilon_m,\ \delta_{k+1} > \cdots > \delta_n.$$

$$\epsilon_1 > \cdots > \epsilon_a > \delta_{k+1} > \cdots > \delta_n > \epsilon_{a+1} > \cdots > \epsilon_m,\ \delta_1 > \cdots > \delta_k.$$

Without loss of generality, and in order to simplify notation, we may and will assume $\mathfrak{h}_0 = \mathfrak{su}(m,n)$, $\Psi_{H_0,\lambda} = \Psi_a$, and we show Remark 5 for $\mathfrak{su}(m,n)$ and Ψ_a. Q denotes the partition function for $\Psi_a \cap \Phi_n$.

The subroot system spanned by the compact simple roots in Ψ_a is

$\Phi_U = \{\epsilon_i - \epsilon_j, 1 \le i \le a, 1 \le j \le a \text{ or } a+1 \le i \le m, a+1 \le j \le m\} \cup \{\delta_i - \delta_j, 1 \le i \ne j \le n\}$.

$\Psi_a \cap \Phi_c \backslash \Phi_U = \{\epsilon_i - \epsilon_j, 1 \le i \le a, a+1 \le j \le m\}$.

$\Psi_a \cap \Phi_n = \{\epsilon_i - \delta_j, \delta_j - \epsilon_r, 1 \le i \le a, a+1 \le r \le m, 1 \le j \le n\}$.

$2\rho_n^{H_0} = n(\epsilon_1 + \cdots + \epsilon_a) - n(\epsilon_{a+1} + \cdots + \epsilon_m) - (a - (m-a))(\delta_1 + \cdots + \delta_n)$.

A finite sum of noncompact roots in Ψ_a is equal to

$B = \sum_{1 \le j \le a} A_j \epsilon_j - \sum_{a+1 \le i \le m} B_i \epsilon_i + \sum_r C_r \delta_r$ with A_j, B_i as non-negative numbers.

Let $w \in W_L$ so that $Q(w\mu - (\lambda + \rho_n^{H_0})) \ne 0$. Hence, $\mu = w^{-1}(\lambda + \rho_n^{H_0} + B)$, with B a sum of roots in $\Psi_a \cap \Phi_n$. Thus, w^{-1} is the unique element in W_L that takes $\lambda + \rho_n^{H_0} + B$ to the Weyl chamber determined by $\Psi_a \cap \Phi_c$.

Let $w_1 \in W_U(\Psi_a)$ so that $w_1(\lambda + \rho_n + B)$ is $\Psi_a \cap \Phi_U$-dominant. Next we verify $w_1(\lambda + \rho_n + B)$ is $\Psi_a \cap \Phi_c$-dominant. For this, we fix $\alpha \in \Psi_a \cap \Phi_c \backslash \Phi_U$

and check $(w_1(\lambda + \rho_n + B), \alpha) > 0$. $\alpha = \epsilon_i - \epsilon_j, i \leq a < j$, and $w_1 \in W_U(\Psi_a)$; hence, $w_1^{-1}(\alpha) = \epsilon_r - \epsilon_s, r \leq a < s$ belongs to Ψ_a. Thus, $(w_1(\lambda + \rho_n + B), \alpha) = (\lambda + \rho_n + B, w_1^{-1}\alpha) = (\lambda, w_1^{-1}\alpha) + (\rho_n, w_1^{-1}\alpha) + (B, w_1^{-1}\alpha) = (\lambda, w_1^{-1}\alpha) + n - (-n) + A_r + B_s$; the first summand is positive because λ is Ψ_a-dominant, while the third and fourth are non-negative. Therefore, $w^{-1} = w_1$, and we have shown Remark 5 for $\mathfrak{g} = \mathfrak{su}(m, n)$.

To follow, we fix $\mathfrak{g} = \mathfrak{so}(2m, 2n), n > 1, \mathfrak{h} = \mathfrak{so}(2m, 2n-2) + \mathfrak{so}(2), \mathfrak{h}_0 = \mathfrak{so}(2m, 2) + \mathfrak{so}(2n-2)$; then, for certain orthogonal basis $\epsilon_1, \ldots, \epsilon_m, \delta_1, \ldots, \delta_n$ of the dual vector space to $\mathfrak{t}$, the system of positive roots $\Psi_\lambda = \{\epsilon_i \pm \epsilon_j, \delta_r \pm \delta_s, \epsilon_a \pm \delta_b, 1 \leq i < j \leq m, 1 \leq r < s \leq n, 1 \leq a \leq m, 1 \leq b \leq n\}$ so that $K_1(\Psi_\lambda) = SO(2m) \subset H$. Since $n > 1$, this is a Borel de Siebenthal system.

$\Psi_{H,\lambda} = \{\epsilon_i \pm \epsilon_j, \delta_r \pm \delta_s, \epsilon_a \pm \delta_b, 1 \leq i < j \leq m, 1 \leq r < s \leq n-1, 1 \leq a \leq m, 1 \leq b \leq n-1\}$.

For $n > 2$, $\Psi_{H,\lambda}$ is Borel de Siebenthal; for $n = 2$, $\Psi_{H,\lambda}$ is not Borel de Siebenthal.

$\Psi_{H_0,\lambda} = \{\epsilon_i \pm \epsilon_j, \delta_r \pm \delta_s, \epsilon_a \pm \delta_n, 1 \leq i < j \leq m, 1 \leq r < s \leq n-1, 1 \leq a \leq m\}$. Since $n > 1$, $\Psi_{H_0,\lambda}$ is not a Borel de Siebenthal system. The simple roots for $\Psi_{H_0,\lambda}$ are $\epsilon_1 - \epsilon_2, \cdots, \epsilon_{m-1} - \epsilon_m, \epsilon_m \pm \delta_n, \delta_1 - \delta_2, \cdots, \delta_{n-3} - \delta_{n-2}, \delta_{n-2} \pm \delta_{n-1}$.

$\Phi_U = \{\epsilon_i - \epsilon_j, \delta_r \pm \delta_s, 1 \leq i < j \leq m, 1 \leq r < s \leq n-1\}$.

$W_L = span\{S_{\epsilon_i \pm \epsilon_j}, S_{\delta_r \pm \delta_s}, 1 \leq i < j \leq m, 1 \leq r < s \leq n-1\}$.

$W_U(\Psi_{H_0,\lambda}) = span\{S_{\epsilon_i - \epsilon_j}, S_{\delta_r \pm \delta_s}, 1 \leq i < j \leq m, 1 \leq r < s \leq n-1\}$.

$\Psi_\lambda \cap \Phi_c \backslash \Phi_U = \{\epsilon_i + \epsilon_j, 1 \leq i < j \leq m\}$.

Q_0 is the partition function associated with $\Psi_{H_0,\lambda} \cap \Phi_n = \{\epsilon_i \pm \delta_n, 1 \leq i \leq m\}$.

$\rho_n^{H_0} = \epsilon_1 + \cdots + \epsilon_m$.

A finite sum of noncompact roots B in $\Psi_{H_0,\lambda}$ is equal to
$B = \sum_{j=1}^{j=m} (a_j + b_j)\epsilon_j + (\sum_{1 \leq j \leq m} (a_j - b_j))\delta_n, a_j \geq 0, b_j \geq 0 \forall j$.

Let $w \in W_L$ so that $Q_0(w\mu - (\lambda + \rho_n)) \neq 0$. Hence, $\mu = w^{-1}(\lambda + \rho_n + B)$, with B a sum of roots in $\Psi_\lambda \cap \Phi_n$. Thus, w^{-1} is the unique element in W_L that takes $\lambda + \rho_n + B$ to the Weyl chamber determined by $\Psi_\lambda \cap \Phi_c$.

Let $w_1 \in W_U(\Psi_{H_0,\lambda})$ so that $w_1(\lambda + \rho_n + B)$ is $\Psi_{H_0,\lambda} \cap \Phi_U$-dominant. Next we verify $w_1(\lambda + \rho_n + B)$ is $\Psi_{H_0,\lambda} \cap \Phi_c$-dominant. For this, we fix $\alpha \in \Psi_{H_0,\lambda} \cap \Phi_c \backslash \Phi_U$ and check $(w_1(\lambda + \rho_n + B), \alpha) > 0$. $\alpha = \epsilon_i + \epsilon_j$, and $w_1 \in W_U(\Psi_{H_0,\lambda})$; hence, $w_1^{-1}(\alpha) = \epsilon_r + \epsilon_s$ belongs to $\Psi_{H_0,\lambda}$. Thus, $(w_1(\lambda + \rho_n + B), \alpha) = (\lambda + \rho_n + B, w_1^{-1}\alpha) = (\lambda, w_1^{-1}\alpha) + (\rho_n, w_1^{-1}\alpha) + (B, w_1^{-1}\alpha)$; the first summand is positive because λ is $\Psi_{H_0,\lambda}$-dominant, while the second and third are non-negative. Therefore, $w^{-1} = w_1$, and we have shown Remark 5 for $\mathfrak{g} = \mathfrak{so}(2m, 2n)$.

Hence, we have concluded the proof of Lemma 3. □

Lemma 4[3] *We recall $\Lambda_2 = \lambda_2 + (\rho_n^\lambda)_2$. We claim:*

$$m^{K_2, L\cap K_2}(\Lambda_2, \nu_2' + q_{\mathfrak{u}}(\rho_n^\lambda)_2) = m^{K_2, L\cap K_2}(\lambda_2, \nu_2').$$

$$Spec_{L\cap K_2}(\pi_{\Lambda_2}^{K_2}) = Spec_{L\cap K_2}(\pi_{\lambda_2}^{K_2}) + q_{\mathfrak{u}}(\rho_n^\lambda)_2.$$

[3] We thank the referee for improving this Lemma.

In fact, when Ψ_λ is holomorphic, ρ_n^λ is in $\mathfrak{z}_\mathfrak{k} = \mathfrak{k}_1$; hence, $(\rho_n^\lambda)_2 = 0$. In [30], it is shown that when K is semisimple, $(\rho_n^\lambda)_2 = 0$. Actually, this is owing to the fact that simple roots for $\Psi_\lambda \cap \Phi(\mathfrak{k}_2, \mathfrak{t}_2)$ are simple roots for Ψ_λ and that ρ_n^λ is orthogonal to every compact simple root for Ψ_λ. For other triples $(\mathfrak{g}, \Psi, \mathfrak{h})$, the previous considerations together with that $(\rho_n^\lambda)_2$ is orthogonal to $\mathfrak{k}_1$ yield that $(\rho_n^\lambda)_2$ belongs to the dual of the center of $\mathfrak{l} \cap \mathfrak{k}_2$. Thus, we must analyze the triples $(\mathfrak{g}, \Psi, \mathfrak{h})$ so that $\mathfrak{z}_{\mathfrak{l}\cap\mathfrak{k}_2} \neq \{0\}$ and compute $(\rho_n^G)_2 = (\rho_n^\lambda)_2$. From Tables 1, 2, and 3, we deduce we are left to analyze $(\mathfrak{su}(m, n), \Psi_a, \mathfrak{h})$, $(\mathfrak{so}(2m, 2), \Psi_\pm, \mathfrak{h})$. Here, $\mathfrak{k}_2(\Psi_a) = \mathfrak{z}_k + \{0\} \times \mathfrak{su}(n)$, $\mathfrak{k}_2(\Psi_\pm) = \mathfrak{z}_k$. For $\mathfrak{so}(2m, 2)$ we follow the notation in 4.1.3, then $\mathfrak{t}_1^\star = span(\epsilon_1, \ldots, \epsilon_m)$, $\mathfrak{t}_2^\star = span(\delta_1)$, and $\rho_n^{\Psi_\pm} = c(\epsilon_1 + \cdots + \epsilon_m) \in \mathfrak{t}_1^\star$. Hence, $(\rho_n^\lambda)_2 = 0$. For $\mathfrak{su}(m, n)$ we follow the notation in Lemma 3. It readily follows that for $1 \leq a < m$, $\rho_n^{\Psi_a} = \frac{n}{m}\big((m-a)(\epsilon_1 + \cdots + \epsilon_a) - a(\epsilon_{a+1} + \cdots + \epsilon_m)\big) + \frac{2a-m}{2m}\big(n(\epsilon_1 + \cdots + \epsilon_m) - m(\delta_1 + \cdots + \delta_n)\big)$. The first summand is in $(\mathfrak{t} \cap \mathfrak{su}(m))^\star$, and the second summand belongs to $\mathfrak{z}_\mathfrak{k}^\star$, whence, $(\rho_n^{\Psi_a})_2 = 0$ if and only if $2a = m$. Hence, for $(\mathfrak{su}(2, 2n), \Psi_1, \mathfrak{h} = \mathfrak{sp}(1, n))$, we have $q_\mathfrak{u}((\rho_n^{\Psi_1})_2) = 0$. Conclusion, when $\rho_n^\lambda)_2 \neq 0$, $(\rho_n^{\Psi_a})_2$ determines a character of the center of $\mathfrak{k} \subset \mathfrak{k}_2 \cap \mathfrak{l}$. Hence, $\pi_{\Lambda_2}^{K_2}$ is equal to $\pi_{\lambda_2}^{K_2}$ times a central character of K. Thus, the Lemma follows.

As a corollary, we obtain: $(\rho_n^\lambda)_2 \neq 0$ forces $\mathfrak{g} = \mathfrak{su}(m, n)$, $\mathfrak{h} = \mathfrak{s}(\mathfrak{u}(m, k) + \mathfrak{u}(n - k))$, $m > 1$, $n > 1$, $1 \leq k \leq n-1$, $\mathfrak{u} = \mathfrak{t}$. Thus, we always have $q_\mathfrak{u}((\rho_n^\lambda)_2) = (\rho_n^\lambda)_2$.

4.1.1 Conclusion Proof of Theorem 4

We just put together Lemmas 3, 2, and 4; hence, we obtain the equalities we were searching for. This concludes the proof of Theorem 4.

4.1.2 Existence of D

To follow, we show the existence of the isomorphism D in Theorem 4 (ii) and derive the decomposition into irreducible factors of the semisimple $\mathfrak{h}_0$-module $\mathcal{U}(\mathfrak{h}_0)W$. In the meantime, we also consider some particular cases of Theorem 4. Before we proceed, we comment on the structure of the representation τ.

4.1.3 Representations π_λ so that $res_L(\tau)$ Is Irreducible

Under our H-admissibility hypothesis of π_λ, we analyze the cases so that the representation $res_L(\tau)$ is irreducible. The next structure statements are verified in [30]. To begin with, we recall the decomposition $K = K_1 Z_K K_2$ (this is not a direct product; Z_K connected center of K) and the direct product $K = K_1 K_2$; we also recall that actually, either K_1 or K_2 depends on Ψ_λ. When π_λ is a holomorphic representation $K_1 = Z_K$ and $\mathfrak{k}_2 = [\mathfrak{k}, \mathfrak{k}]$; when Z_K is nontrivial and π_λ is

Table 1 Case $U = T$, Ψ_λ non-holomorphic

G	H	H_0	Ψ_λ	K_1
$\mathfrak{su}(m,n)$	$\mathfrak{su}(m,k) \oplus \mathfrak{su}(n-k) \oplus \mathfrak{u}(1)$	$\mathfrak{su}(m,n-k) \oplus \mathfrak{su}(k) \oplus \mathfrak{u}(1)$	Ψ_a	$\mathfrak{su}(m)$
$\mathfrak{su}(m,n)$	$\mathfrak{su}(k,n) \oplus \mathfrak{su}(m-k) \oplus \mathfrak{u}(1)$	$\mathfrak{su}(m-k,n) \oplus \mathfrak{su}(k) \oplus \mathfrak{u}(1)$	$\tilde{\Psi}_b$	$\mathfrak{su}(n)$
$\mathfrak{so}(2m,2n), m > 2$	$\mathfrak{so}(2m,2k) \oplus \mathfrak{so}(2n-2k)$	$\mathfrak{so}(2m,2n-2k) \oplus \mathfrak{so}(2k)$	$\Psi_\pm$	$\mathfrak{so}(2m)$
$\mathfrak{so}(4,2n)$	$\mathfrak{so}(4,2k) \oplus \mathfrak{so}(2n-2k)$	$\mathfrak{so}(4,2n-2k) \oplus \mathfrak{so}(2k)$	$\Psi_\pm$	$\mathfrak{su}_2(\alpha_{max})$
$\mathfrak{so}(2m,2n+1), m > 2$	$\mathfrak{so}(2m,k) \oplus \mathfrak{so}(2n+1-k)$	$\mathfrak{so}(2m,2n+1-k) \oplus \mathfrak{so}(k)$	$\Psi_\pm$	$\mathfrak{so}(2m)$
$\mathfrak{so}(4,2n+1)$	$\mathfrak{so}(4,k) \oplus \mathfrak{so}(2n+1-k)$	$\mathfrak{so}(4,2n+1-k) \oplus \mathfrak{so}(k)$	$\Psi_\pm$	$\mathfrak{su}_2(\alpha_{max})$
$\mathfrak{so}(4,2n), n > 2$	$\mathfrak{u}(2,n)_1$	$w\mathfrak{u}(2,n)_1$	$\Psi_{1\,-1}$	$\mathfrak{su}_2(\alpha_{max})$
$\mathfrak{so}(4,2n), n > 2$	$\mathfrak{u}(2,n)_2$	$w\mathfrak{u}(2,n)_2$	$\Psi_{1\,1}$	$\mathfrak{su}_2(\alpha_{max})$
$\mathfrak{so}(4,4)$	$\mathfrak{u}(2,2)_{11}$	$w\mathfrak{u}(2,2)_{11}$	$\Psi_{1\,-1},\ w_{\epsilon,\delta}\Psi_{1\,-1}$	$\mathfrak{su}_2(\alpha_{max})$
$\mathfrak{so}(4,4)$	$\mathfrak{u}(2,2)_{12}$	$w\mathfrak{u}(2,2)_{12}$	$\Psi_{1\,-1},\ w_{\epsilon,\delta}\Psi_{1\,1}$	$\mathfrak{su}_2(\alpha_{max})$
$\mathfrak{so}(4,4)$	$\mathfrak{u}(2,2)_{21}$	$w\mathfrak{u}(2,2)_{21}$	$\Psi_{1\,1},\ w_{\epsilon,\delta}\Psi_{1\,-1}$	$\mathfrak{su}_2(\alpha_{max})$
$\mathfrak{so}(4,4)$	$\mathfrak{u}(2,2)_{22}$	$w\mathfrak{u}(2,2)_{22}$	$\Psi_{1\,1},\ w_{\epsilon,\delta}\Psi_{1\,1}$	$\mathfrak{su}_2(\alpha_{max})$
$\mathfrak{sp}(m,n)$	$\mathfrak{sp}(m,k) \oplus \mathfrak{sp}(n-k)$	$\mathfrak{sp}(m,n-k) \oplus \mathfrak{sp}(k)$	Ψ_+	$\mathfrak{sp}(m)$
$\mathfrak{f}_{4(4)}$	$\mathfrak{sp}(1,2) \oplus \mathfrak{su}(2)$	$\mathfrak{so}(5,4)$	Ψ_{BS}	$\mathfrak{su}_2(\alpha_{max})$
$\mathfrak{e}_{6(2)}$	$\mathfrak{so}(6,4) \oplus \mathfrak{so}(2)$	$\mathfrak{su}(4,2) \oplus \mathfrak{su}(2)$	Ψ_{BS}	$\mathfrak{su}_2(\alpha_{max})$
$\mathfrak{e}_{7(-5)}$	$\mathfrak{so}(8,4) \oplus \mathfrak{su}(2)$	$\mathfrak{so}(8,4) \oplus \mathfrak{su}(2)$	Ψ_{BS}	$\mathfrak{su}_2(\alpha_{max})$
$\mathfrak{e}_{7(-5)}$	$\mathfrak{su}(6,2)$	$\mathfrak{e}_{6(2)} \oplus \mathfrak{so}(2)$	Ψ_{BS}	$\mathfrak{su}_2(\alpha_{max})$
$\mathfrak{e}_{8(-24)}$	$\mathfrak{so}(12,4)$	$\mathfrak{e}_{7(-5)} \oplus \mathfrak{su}(2)$	Ψ_{BS}	$\mathfrak{su}_2(\alpha_{max})$

Table 2 Case $U \neq T$, Ψ_λ non-holomorphic

G	H	H_0	Ψ_λ	K_1
$\mathfrak{su}(2, 2n),\ n > 2$	$\mathfrak{sp}(1, n)$	$\mathfrak{sp}(1, n)$	Ψ_1	$\mathfrak{su}_2(\alpha_{max})$
$\mathfrak{su}(2, 2)$	$\mathfrak{sp}(1, 1)$	$\mathfrak{sp}(1, 1)$	Ψ_1	$\mathfrak{su}_2(\alpha_{max})$
$\mathfrak{su}(2, 2)$	$\mathfrak{sp}(1, 1)$	$\mathfrak{sp}(1, 1)$	$\tilde{\Psi}_1$	$\mathfrak{su}_2(\alpha_{max})$
$\mathfrak{so}(2m, 2n), m > 2$	$\mathfrak{so}(2m, 2k+1) + \mathfrak{so}(2n-2k-1)$	$\mathfrak{so}(2m, 2n-2k-1) + \mathfrak{so}(2k+1)$	$\Psi_\pm$	$\mathfrak{so}(2m)$
$\mathfrak{so}(4, 2n)$,	$\mathfrak{so}(4, 2k+1) + \mathfrak{so}(2n-2k-1)$	$\mathfrak{so}(4, 2n-2k-1) + \mathfrak{so}(2k+1)$	$\Psi_\pm$	$\mathfrak{su}_2(\alpha_{max})$
$\mathfrak{so}(2m, 2), m > 2$	$\mathfrak{so}(2m, 1)$	$\mathfrak{so}(2m, 1)$	$\Psi_\pm$	$\mathfrak{so}(2m)$
$\mathfrak{so}(4, 2)$,	$\mathfrak{so}(4, 1)$	$\mathfrak{so}(4, 1)$	$\Psi_\pm$	$\mathfrak{su}_2(\alpha_{max})$
$\mathfrak{e}_{6(2)}$	$\mathfrak{f}_{4(4)}$	$\mathfrak{sp}(3, 1)$	Ψ_{BS}	$\mathfrak{su}_2(\alpha_{max})$

Table 3 π_λ^G holomorphic Discrete Series. The last two lines show the unique holomorphic pairs so that $U \neq T$

G	H (a)	H_0 (b)
$\mathfrak{su}(m,n), m \neq n$	$\mathfrak{su}(k,l) + \mathfrak{su}(m-k,n-l) + \mathfrak{u}(1)$	$\mathfrak{su}(k,n-l) + \mathfrak{su}(m-k,l) + \mathfrak{u}(1)$
$\mathfrak{su}(n,n)$	$\mathfrak{su}(k,l) + \mathfrak{su}(n-k,n-l) + \mathfrak{u}(1)$	$\mathfrak{su}(k,n-l) + \mathfrak{su}(n-k,l) + \mathfrak{u}(1)$
$\mathfrak{so}(2,2n)$	$\mathfrak{so}(2,2k) + \mathfrak{so}(2n-2k)$	$\mathfrak{so}(2,2n-2k) + \mathfrak{so}(2k)$
$\mathfrak{so}(2,2n)$	$\mathfrak{u}(1,n)$	$\mathfrak{u}(1,n)$
$\mathfrak{so}(2,2n+1)$	$\mathfrak{so}(2,k) + \mathfrak{so}(2n+1-k)$	$\mathfrak{so}(2,2n+1-k) + \mathfrak{so}(k)$
$\mathfrak{so}^\star(2n)$	$\mathfrak{u}(m,n-m)$	$\mathfrak{so}^\star(2m) + \mathfrak{so}^\star(2n-2m)$
$\mathfrak{sp}(n,\mathbb{R})$	$\mathfrak{u}(m,n-m)$	$\mathfrak{sp}(m,\mathbb{R}) + \mathfrak{sp}(n-m,\mathbb{R})$
$\mathfrak{e}_{6(-14)}$	$\mathfrak{so}(2,8) + \mathfrak{so}(2)$	$\mathfrak{so}(2,8) + \mathfrak{so}(2)$
$\mathfrak{e}_{6(-14)}$	$\mathfrak{su}(2,4) + \mathfrak{su}(2)$	$\mathfrak{su}(2,4) + \mathfrak{su}(2)$
$\mathfrak{e}_{6(-14)}$	$\mathfrak{so}^\star(10) + \mathfrak{so}(2)$	$\mathfrak{su}(5,1) + \mathfrak{sl}(2,\mathbb{R})$
$\mathfrak{e}_{7(-25)}$	$\mathfrak{so}^\star(12) + \mathfrak{su}(2)$	$\mathfrak{su}(6,2)$
$\mathfrak{e}_{7(-25)}$	$\mathfrak{so}(2,10) + \mathfrak{sl}(2,\mathbb{R})$	$\mathfrak{e}_{6(-14)} + \mathfrak{so}(2)$
$\mathfrak{su}(n,n)$	$\mathfrak{so}^\star(2n)$	$\mathfrak{sp}(n,\mathbb{R})$
$\mathfrak{so}(2,2n)$	$\mathfrak{so}(2,2k+1) + \mathfrak{so}(2n-2k-1)$	$\mathfrak{so}(2,2n-2k-1) + \mathfrak{so}(2k+1)$

not a holomorphic representation, we have $Z_K \subset K_2$; for $\mathfrak{g} = \mathfrak{su}(m,n)$, $\mathfrak{h} = \mathfrak{su}(m,k) + \mathfrak{su}(n-k) + \mathfrak{z}_L$, we have $\mathbf{T} \equiv Z_K \subset Z_L \equiv \mathbf{T}^2$. Here, $Z_K \subset L$, and $\tau_{|L}$ irreducible, forces $\tau = \pi_{\Lambda_1}^{SU(m)} \boxtimes \pi_\chi^{Z_K} \boxtimes \pi_{\rho_{SU(n)}}^{SU(n)}$; for $\mathfrak{g} \not\equiv \mathfrak{su}(m,n)$ and G/K a Hermitian symmetric space, we have to consider the next two examples.

For both cases we have $K_2 = Z_K(K_2)_{ss}$ and $Z_K \not\subseteq L$.

(1) When $\mathfrak{g} = \mathfrak{so}(2m,2)$, $\mathfrak{h} = \mathfrak{so}(2m,1)$ and $\Psi_\lambda = \Psi_\pm$, then $\mathfrak{k}_1 = \mathfrak{so}(2m)$, $\mathfrak{k}_2 = \mathfrak{z}_K$, and obviously, $res_L(\tau)$ is always an irreducible representation. Here, $\pi_{\Lambda_2}^{K_2}$ is a one-dimensional representation.
(2) When $\mathfrak{g} = \mathfrak{su}(2,2n)$, $\mathfrak{h} = \mathfrak{sp}(1,n)$, $\Psi_\lambda = \Psi_1$, then $\mathfrak{k}_1 = \mathfrak{su}_2(\alpha_{max})$, $\mathfrak{k}_2 = \mathfrak{su}(2n) + \mathfrak{z}_\mathfrak{k}$, $\mathfrak{l} \equiv \mathfrak{su}_2(\alpha_{max}) + \mathfrak{sp}(n)$. Examples of $\tau_{|L}$ irreducible are[4] $\tau = \pi_{b\Lambda_1}^{K_1} \boxtimes \pi_\chi^{Z_K} \boxtimes \pi_{\rho_{SU(2n)}+a\widetilde{\Lambda}_1}^{SU(2n)}$, $\tau = \pi_{b\Lambda_1}^{K_1} \boxtimes \pi_\chi^{Z_K} \boxtimes \pi_{\rho_{SU(2n)}+a\widetilde{\Lambda}_{2n-1}}^{SU(2n)}$, $a \geq 0, b \geq 2$, Λ_1 (resp. $\widetilde{\Lambda}_1$) is the highest weight of the first fundamental representation of $\mathfrak{su}_2$ (resp. $\mathfrak{su}(2n)$), $\widetilde{\Lambda}_{2n-1}$ is the highest weight of the dual representation to the first fundamental representation for $\mathfrak{su}(2n)$.

For $\mathfrak{g} = \mathfrak{so}(2m,2n)$, $\mathfrak{h} = \mathfrak{so}(2m,2n-1)$, $n > 1$, $\Psi_\lambda \cap \Phi_n = \{\epsilon_i \pm \delta_j\}$, $\mathfrak{k}_1 = \mathfrak{so}(2m)$, and if λ is so that $\lambda + \rho_n^\lambda = ic(\tau) = (\sum c_i\epsilon_i, k(\delta_1 + \cdots + \delta_{n-1} \pm \delta_n)) + \rho_K$, then $res_L(\tau)$ is irreducible, and $\pi_{\Lambda_2}^{K_2} = \pi_{k(\delta_1+\cdots+\delta_{n-1}\pm\delta_n)+\rho_{K_2}}^{K_2}$ is not a one-dimensional representation for $k > 0$. It follows from the classical branching laws that these are the unique $\tau's$ such that $res_L(\tau)$ is irreducible.

We do not know the pairs (π_λ, τ) so that π_λ is H-admissible and $res_L(\tau)$ is irreducible. We believe that for $\mathfrak{g} \not\equiv \mathfrak{so}(2m,2n)$ or $\mathfrak{g} \not\equiv \mathfrak{su}(m,n)$, we could conclude

[4] This was pointed out by the referee.

that τ is the tensor product of an irreducible representation of K_1 times a one-dimensional representation of K_2. That is, $\tau \equiv \pi_{\phi_1}^{K_1} \boxtimes \pi_{\rho_{K_2}}^{K_2} \otimes \pi_{\chi}^{Z_{K_2}}$.

In Sect. 5.3, we show that whenever a symmetric pair (G, H) so that some Discrete Series for G is H-admissible, then there exists H-admissible Discrete Series for G so that its lowest K-type restricted to L is irreducible.

4.1.4 Analysis of $\mathcal{U}(\mathfrak{h}_0)W$, $\mathcal{L}_\lambda$, Existence of D, Case $\tau_{|_L}$ Is Irreducible

As before, our hypothesis is as follows: (G, H) is a symmetric space, and π_λ^G is H-admissible. For this paragraph, we add the hypothesis $\tau_{|_L} = res_L(\tau)$ is irreducible. We recall that $\mathcal{U}(\mathfrak{h}_0)W = L_{\mathcal{U}(\mathfrak{h}_0)}(H^2(G, \tau)[W])$, $\mathcal{L}_\lambda = \oplus_{\mu \in Spec_H(\pi_\lambda)} H^2(G, \tau)[V_\mu^H][V_{\mu+\rho_n^\mu}^L]$. We claim:

(a) If a H-irreducible discrete factor of V_λ^G contains a copy of $\tau_{|L}$, then $\tau_{|L}$ is the lowest L-type of such factor.
(b) The multiplicity of $res_L(\tau)$ in $H^2(G, \tau)$ is one.
(c) $\mathrm{Cl}(\mathcal{U}(\mathfrak{h}_0)W)$ is H_0-equivalent to $H^2(H_0, \tau)$.
(d) $\mathcal{L}_\lambda$ is L-equivalent to $H^2(H_0, \tau)_{L-fin}$.
(e) $\mathcal{L}_\lambda$ is L-equivalent to $\mathcal{U}(\mathfrak{h}_0)W$. Thus, D exists.

We rely on:

Remark 6

(1) Two Discrete Series for H are equivalent if and only if their respective lowest L-types are equivalent [31].
(2) For any Discrete Series π_λ, the highest weight (resp. infinitesimal character) of any K-type is equal to the highest weight of the lowest K-type (resp. the infinitesimal character of the lowest K-type) plus a sum of noncompact roots in Ψ_λ [27, Lemma 2.22].

From now on $ic(\phi)$ denotes the infinitesimal character (Harish-Chandra parameter) of the representation ϕ.

Let V_μ^H be a discrete factor for $res_H(\pi_\lambda)$ so that $\tau_{|_L}$ is a L-type. Then, Theorem 4 implies $V_{\mu+\rho_n^H}^L$ is a L-type for $H^2(H_0, \tau)$. Hence, after we apply Remark 6, we obtain

$\mu + \rho_n^H + B_1 = ic(\tau_{|_L})$ with B_1 a sum of roots in $\Psi_{H,\lambda} \cap \Phi_n$.

$\mu + \rho_n^H = ic(\tau_{|_L}) + B_0$ with B_0 a sum of roots in $\Psi_{H_0,\lambda} \cap \Phi_n$.

Thus, $B_0 + B_1 = 0$, whence $B_0 = B_1 = 0$ and $\mu + \rho_n^H = ic(\tau_{|_L})$, and we have verified (a).

Due to H-admissibility hypothesis, we have $\mathcal{U}(\mathfrak{h})W$ is a finite sum of irreducible underlying modules of Discrete Series for H. Now, Corollary 1 to Lemma 1 yields that a copy of a V_μ^H contained in $\mathcal{U}(\mathfrak{h})W$ contains a copy of $V_\lambda^G[W]$. Thus, (a) implies $\tau_{|_L}$ is the lowest L-type of such V_μ^H. Hence, $H^2(H, \tau)$ is nonzero. Now, Theorem 4 together with the fact that the lowest L-type of a Discrete Series has multiplicity one yields that $\dim Hom_H(H^2(H,\tau), V_\lambda^G) = 1$. Also, we obtain $\dim Hom_{H_0}(H^2(H_0,\tau), V_\lambda^G) = 1$. Thus, whenever $\tau_{|L}$ occurs in $res_L(V_\lambda^G)$, $\tau_{|L}$ is realized in $V_\lambda^G[W]$. In other words, the isotypic component $V_\lambda^G[\tau_{|_L}] \subset V_\lambda^G[W]$. Hence, (b) holds.

Owing our hypothesis, we may write $\mathcal{U}(\mathfrak{h}_0)W = N_1 + ... + N_k$, with each N_j being the underlying Harish-Chandra module of a irreducible square integrable representation for H_0. Since Lemma 1 shows r_0 is injective in $\mathcal{U}(\mathfrak{h}_0)W$, $r_0(\mathrm{Cl}(N_j))$ is a Discrete Series in $L^2(H_0 \times_{res_L(\tau)} W)$; hence, Frobenius reciprocity implies $\tau_{|_L}$ is a L-type for N_j. Hence, (b) and (a) yield $\mathcal{U}(\mathfrak{h}_0)W$ is $\mathfrak{h}_0$-irreducible, and (c) follows.

By definition, the subspace $\mathcal{L}_\lambda$ is the linear span of the subspaces $V_\lambda^G[V_\mu^H][V^L_{\mu+\rho_n^H}]$ with $\mu \in Spec_H(\pi_\lambda)$. Since

$$\dim Hom_L(V^L_{\mu+\rho_n^H}, V_\lambda^G[V_\mu^H][V^L_{\mu+\rho_n^H}]) = \dim Hom_H(V_\mu^H, V_\lambda^G)$$

$$= \dim Hom_L(V^L_{\mu+\rho_n^H}, H^2(H_0,\tau)) = \dim Hom_L(V^L_{\mu+\rho_n^H}, H^2(H_0,\tau)[V^L_{\mu+\rho_n^H}]),$$

and both L-modules are isotypical, (d) follows. Finally, (e) follows from (c) and (d).

Whenever $\pi^{K_2}_{\Lambda_2}$ is the trivial representation, Theorem 4 and Lemma 3 justify

$$\dim Hom_H(V_\mu^H, V_\lambda^G) = \dim Hom_L(V^L_{\mu+\rho_n^H}, V^{H_0}_{(\lambda_1,\rho_{K_2\cap L})+\rho_n^H})$$

$$= \dim Hom_L(V^L_{\mu+\rho_n^H}, H^2(H_0,\tau)) = \dim Hom_L(V^L_\mu, V^{H_0}_{(\lambda_1,\rho_{K_2\cap L})}),$$

the infinitesimal character of $H^2(H_0,\tau)$ is $(\lambda_1, \rho_{K_2\cap L}) + q_{\mathrm{u}}(\rho_n^\lambda) - \rho_n^{H_0} = (\lambda_1, \rho_{K_2\cap L}) + \rho_n^H$. Thus, $H^2(H_0,\tau) \equiv V^{H_0}_{(\lambda_1,\rho_{K_2\cap L})+\rho_n^H}$.

4.1.5 Analysis of $\mathcal{U}(\mathfrak{h}_0)W$, $\mathcal{L}_\lambda$, Existence of D, for General (τ, W)

We recall that by definition, $\mathcal{L}_\lambda = \oplus_{\mu\in Spec_H(\pi_\lambda)} H^2(G,\tau)[V_\mu^H][V^L_{\mu+\rho_n^\mu}]$, $\mathcal{U}(\mathfrak{h}_0)W = L_{\mathcal{U}(\mathfrak{h}_0)}(H^2(G,\tau)[W])$.

Proposition 1 *The hypothesis is as follows: (G, H) is a symmetric pair and π_λ a H-admissible square integrable representation of lowest K-type (τ, W). We write $res_L(\tau) = q_1\sigma_1 + \cdots + q_r\sigma_r$, with $(\sigma_j, Z_j) \in \widehat{L}$, $q_j > 0$. Then,*

(a) If a H-irreducible discrete factor for $res_H(\pi_\lambda)$ contains a copy of σ_j, then σ_j is the lowest L-type of such factor.
(b) The multiplicity of σ_j in $res_L(H^2(G,\tau))$ is equal to q_j.
(c) $r_0 : \mathrm{Cl}(\mathcal{U}(\mathfrak{h}_0)W) \to \mathbf{H}^2(H_0,\tau)$ is a H_0-equivalence.
(d) $\mathcal{L}_\lambda$ is L-equivalent to $\mathbf{H}^2(H_0,\tau)_{L-fin}$.
(e) $\mathcal{L}_\lambda$ is L-equivalent to $\mathcal{U}(\mathfrak{h}_0)W$. Therefore, D exists.

Proof Let V_μ^H be a discrete factor for $res_H(\pi_\lambda)$ so that some irreducible factor of $\tau_{|L}$ is a L-type. Then, Theorem 4 implies $V^L_{\mu+\rho_n^H}$ is a L-type for $\mathbf{H}^2(H_0, \tau) = \oplus_j q_j H^2(H_0, \sigma_j)$. Let's say $V^L_{\mu+\rho_n^H}$ is a subrepresentation of $H^2(H_0, \sigma_i)$. We recall $ic(\phi)$ denotes the infinitesimal character (Harish-Chandra parameter) of the representation ϕ. Hence, after we apply Remark 6, we obtain

$\mu + \rho_n^H + B_1 = ic(\sigma_j)$ with B_1 a sum of roots in $\Psi_{H,\lambda} \cap \Phi_n$.

$\mu + \rho_n^H = ic(\sigma_i) + B_0$ with B_0 a sum of roots in $\Psi_{H_0,\lambda} \cap \Phi_n$.

Thus, $B_0 + B_1 = ic(\sigma_j) - ic(\sigma_i)$. Now, since $\mathfrak{k} = \mathfrak{k}_1 + \mathfrak{k}_2$, $\mathfrak{k}_1 \subset \mathfrak{l}$, $\tau = \pi^{K_1}_{\Lambda_1} \boxtimes \pi^{K_2}_{\Lambda_2}$, we may write $\sigma_s = \pi^{K_1}_{\Lambda_1} \boxtimes \phi_s$, with $\phi_s \in \widehat{L \cap K_2}$; hence, $ic(\sigma_j) - ic(\sigma_i) = ic(\phi_j) - ic(\phi_i)$. Since each ϕ_t is an irreducible factor of $res_{L\cap K_2}(\pi^{K_2}_{\Lambda_2})$, we have $ic(\phi_j) - ic(\phi_i)$ is equal to the difference of two sum of roots in $\Phi(\mathfrak{k}_2, \mathfrak{t} \cap \mathfrak{k}_2)$. The hypothesis forces that the simple roots for $\Psi_\lambda \cap \Phi(\mathfrak{k}_2, \mathfrak{t} \cap \mathfrak{k}_2)$ are compact simple roots for Ψ_λ (see [5]); hence, $ic(\sigma_j) - ic(\sigma_i)$ is a linear combination of compact simple roots for Ψ_λ. On the other hand, $B_0 + B_1$ is a sum of noncompact roots in Ψ_λ. Now $B_0 + B_1$ cannot be a linear combination of compact simple roots, unless $B_0 = B_1 = 0$. Thus, $ic(\sigma_i) = ic(\sigma_j)$, and $Z_j \equiv V^L_{\sigma_j}$ is the lowest L-type of V_μ^H; we have verified (a).

Due to H-admissibility hypothesis, $\mathcal{U}(\mathfrak{h})W$ is a finite sum of irreducible underlying Harish-Chandra modules of Discrete Series for H. Thus, a copy of certain V_μ^H contained in $\mathcal{U}(\mathfrak{h})W$ contains $W[\sigma_j]$. Hence, σ_j is the lowest L-type of such V_μ^H. Hence, $H^2(H, \sigma_j)$ is nonzero, and it is equivalent to a subrepresentation of $\mathrm{Cl}(\mathcal{U}(\mathfrak{h})W)$.

We claim, for $i \neq j$, no σ_j is a L-type of $\mathrm{Cl}(\mathcal{U}(\mathfrak{h})W)[H^2(H, \sigma_i)]$.

Indeed, if σ_j were a L-type in $\mathrm{Cl}(\mathcal{U}(\mathfrak{h})W)[H^2(H, \sigma_i)]$, then, σ_j would be a L-type of a Discrete Series of lowest L-type equal to σ_i; according to (a), this forces $i = j$, a contradiction. Now, we compute the multiplicity of $H^2(H, \sigma_j)$ in $H^2(G, \tau)$. For this, we apply Theorem 4. Thus, $\dim Hom_H(V_\lambda^G, H^2(H, \sigma_j)) = \sum_i q_i \dim Hom_L(\sigma_j, H^2(H, \sigma_i)) = q_j$.

In order to realize the isotypic component corresponding to $H^2(H, \sigma_j)$, we write $V_\lambda^G[W][\sigma_j] = R_1 + \cdots + R_{q_j}$ an explicit sum of L-irreducible modules. Then, owing to (a), $L_{\mathcal{U}(\mathfrak{h})}(R_r)$ contains a copy N_r of $H^2(H, \sigma_j)$, and R_r is the lowest L-type of N_r. Therefore, the multiplicity computation yields $H^2(G, \tau)[H^2(H, \sigma_j)] = N_1 + \cdots + N_{q_j}$. Hence, (b) holds. A corollary of this computation is:

$$Hom_H(H^2(H, \sigma_j), (\mathrm{Cl}(\mathcal{U}(\mathfrak{h})W))^\perp) = \{0\}.$$

Verification of (c). After we recall Lemma 1, $r_0 : \mathrm{Cl}(\mathcal{U}(\mathfrak{h}_0)W) \to L^2(H_0 \times_\tau W)$ is injective, and we apply to the algebra $\mathfrak{h} := \mathfrak{h}_0$ the statement (b) together with the computation to show (b); we make the choice of the $q_j's$ subspaces Z_j as a lowest L-type subspace of $W[Z_j]$. Thus, the image via r_0 of $\mathcal{U}(\mathfrak{h}_0)Z_j$ is a subspace of $L^2(H_0 \times_{\sigma_j} Z_j)$. Since Hotta-Parthasarathy [13], Atiyah-Schmid [1], and Enright-

Wallach [6] have shown $H^2(H_0, \sigma_j)$ has multiplicity one in $L^2(H_0 \times_{\sigma_j} Z_j)$, we obtain the image of r_0 is equal to $\mathbf{H}^2(H_0, \tau)$.

The proof of (d) and (e) are word by word as the one for 4.1.4. □

Corollary 5 *The multiplicity of $H^2(H, \sigma_j)$ in $res_H(H^2(G, \tau))$ is equal to*

$$q_j = \dim Hom_L(\sigma_j, H^2(G, \tau)).$$

Corollary 6 *For each σ_j, $\mathcal{L}_\lambda[Z_j] = \mathrm{Cl}(\mathcal{U}(\mathfrak{h}_0)W)[Z_j] = H^2(G, \tau)[W][Z_j] =$ "W"$[Z_j]$. Thus, we may fix $D = I_{\text{“}W\text{”}[Z_j]} : \mathcal{L}_\lambda[Z_j] \to \mathrm{Cl}(\mathcal{U}(\mathfrak{h}_0)W)[Z_j]$.*

4.2 Explicit Inverse Map to r_0^D

We consider three cases: $res_L(\tau)$ is irreducible, $res_L(\tau)$ is multiplicity-free, and general case. Formally, they are quite alike; however, for us, it has been illuminating to consider the three cases. As a by-product, we obtain information on compositions $r^\star r$, $r_0^\star r_0$; a functional equation that must be satisfied by the kernel of a holographic operator; for some particular discrete factor $H^2(H, \sigma)$ of $res_H(\pi_\lambda)$, the reproducing kernel for $H^2(G, \tau)$ is an extension of the reproducing kernel for $H^2(H, \sigma)$ as well as that the holographic operator from $H^2(H, \sigma)$ into $H^2(G, \tau)$ is just a plain extension of functions.

4.2.1 Case (τ, W) Restricted to L Is Irreducible

In Tables 1, 2, and 3, we show the list of the triples (G, H, π_λ) such that (G, H) is a symmetric pair and π_λ is H-admissible. In Sect. 5.3, we show that if there exists (G, H, π_λ) so that π_λ is H-admissible, then there exists a H-admissible $\pi_{\lambda'}$ so that its lowest K-type restricted to L is irreducible and λ' is dominant with respect to Ψ_λ. We denote by η_0 the Harish-Chandra parameter for $H^2(H_0, \tau) \equiv \mathrm{Cl}(\mathcal{U}(\mathfrak{h}_0)W)$.

We set $d(\pi)$ for the formal degree of an irreducible square integrable representation π and define $c = d(\pi_\lambda) \dim W / d(\pi_{\eta_0}^{H_0})$. Next, we show the following.

Proposition 2 *We assume the setting as well as the hypothesis in Theorem 1, and further (τ, W) restricted to L is irreducible.*
Let $T_0 \in Hom_L(Z, H^2(H_0, \tau))$, then the kernel K_T corresponding to $T := (r_0^D)^{-1}(T_0) \in Hom_H(H^2(H, \sigma), H^2(G, \tau))$ is

$$K_T(h, x)z = (D^{-1}[\int_{H_0} \frac{1}{c} K_\lambda(h_0, \cdot)(T_0(z)(h_0))dh_0])(h^{-1}x).$$

Proof We systematically apply Theorem 1. Under our assumptions, $\mathbf{H}^2(H_0, \tau)$ is an irreducible representation, and $\mathbf{H}^2(H_0, \tau) = H^2(H_0, \tau)$;

$\mathrm{Cl}(\mathcal{U}(\mathfrak{h}_0)(H^2(G,\tau)[W]))$ is H_0-irreducible. We define $\tilde{r}_0 := rest(r_0) : \mathrm{Cl}(\mathcal{U}(\mathfrak{h}_0)H^2(G,\tau)[W]) \to H^2(H_0,\tau)$ as an isomorphism. To follow, we notice the inverse of $\tilde{r}_0$ is, up to a constant, equal to $r_0^\star$ restricted to $H^2(H_0,\tau)$. This is because functional analysis yields equalities $\mathrm{Cl}(Im(r_0^\star)) = ker(r_0)^\perp = \mathrm{Cl}(\mathcal{U}(\mathfrak{h}_0)W)$ and $Ker(r_0^\star) = Im(r_0)^\perp = H^2(H_0,\tau)^\perp$. Thus, Schur's lemma applied to irreducible modules $H^2(H_0,\tau), \mathrm{Cl}(\mathcal{U}(\mathfrak{h}_0)W)$ implies there exists nonzero constants b, d so that $(\tilde{r}_0 r_0^\star)_{|_{H^2(H_0,\tau)}} = bI_{H^2(H_0,\tau)}, r_0^\star \tilde{r}_0 = dI_{\mathrm{Cl}(\mathcal{U}(\mathfrak{h}_0)W)}$. Hence, the inverse to $\tilde{r}_0$ follows. In Sect. 4.2.2, we show $b = d = d(\pi_\lambda)\dim W/d(\pi_{\eta_0}^{H_0}) = c$.

For $x \in G$, $f \in H^2(G,\tau)$, the identity $f(x) = \int_G K_\lambda(y,x)f(y)dy$ holds. Thus, $r_0(f)(p) = f(p) = \int_G K_\lambda(y,p)f(y)dy$, for $p \in H_0$, $f \in H^2(G,\tau)$, and we obtain

$$K_{r_0}(x,h_0) = K_\lambda(x,h_0), \quad K_{r_0^\star}(h_0,x) = K_{r_0}(x,h_0)^\star = K_\lambda(h_0,x).$$

Hence, for $g \in H^2(H_0,\tau)$, we have

$$\tilde{r}_0^{-1}(g)(x) = \frac{1}{c}\int_{H_0} K_{r_0^\star}(h_0,x)g(h_0)dh_0 = \frac{1}{c}\int_{H_0} K_\lambda(h_0,x)g(h_0)dh_0.$$

Therefore, for $T_0 \in Hom_L(Z, H^2(H_0,\tau))$, the kernel K_T of the element T in $Hom_H(H^2(H,\sigma), H^2(G,\tau))$ such that $r_0^D(T) = T_0$ satisfies for $z \in Z$

$$D^{-1}([r_0^{-1}(T_0(z)(\cdot))])(\cdot) = K_T(e,\cdot)z \in V_\lambda^G[H^2(H,\sigma)][Z] \subset H^2(G,\tau).$$

More explicitly, after we recall $K_T(e,h^{-1}x) = K_T(h,x)$,

$$K_T(h,x)z = (D^{-1}[\int_{H_0} \frac{1}{c}K_\lambda(h_0,\cdot)(T_0(z)(h_0))dh_0])(h^{-1}x).$$

□

Corollary 7 *For any T in $Hom_H(H^2(H,\sigma), H^2(G,\tau))$, we have*

$$K_T(h,x)z = (D^{-1}[\int_{H_0} \frac{1}{c}K_\lambda(h_0,\cdot)(r_0(D(K_T(e,\cdot)z))(h_0))dh_0])(h^{-1}x).$$

Corollary 8 *When D is the identity map, we obtain*

$$\begin{aligned} K_T(h,x)z &= \int_{H_0} \frac{1}{c}K_\lambda(h_0,h^{-1}x)(T_0(z)(h_0))dh_0 \\ &= \int_{H_0} \tfrac{1}{c}K_\lambda(hh_0,x)K_T(e,h_0)z\,dh_0. \end{aligned}$$

The equality in the conclusion of Proposition 2 is equivalent to

$$D(K_T(e,\cdot))(y) = \int_{H_0} \frac{1}{c} K_\lambda(h_0, y) D(K_T(e,\cdot))(h_0) dh_0,\ y \in G.$$

Hence, we have derived a formula that let us to recover the kernel K_T (resp. $D(K_T(e,\cdot))(\cdot)$) from $K_T(e,\cdot)$ (resp. $D(K_T(e,\cdot))(\cdot)$) restricted to H_0!

Remark 7 We notice

$$r_0^\star r_0(f)(y) = \int_{H_0} K_\lambda(h_0, y) f(h_0) dh_0, \quad f \in H^2(G,\tau),\ y \in G. \tag{6}$$

Since we are assuming $\tau_{|L}$ is irreducible, $\mathrm{Cl}(\mathcal{U}(\mathfrak{h}_0)W)$ is irreducible; hence, Lemma 1 let us obtain a scalar multiple of $r_0^\star r_0$ to be the orthogonal projector onto the irreducible factor $\mathrm{Cl}(\mathcal{U}(\mathfrak{h}_0)W)$.

Hence, the orthogonal projector onto $\mathrm{Cl}(\mathcal{U}(\mathfrak{h}_0)W)$ is given by $\frac{d(\pi_{\eta_0}^{H_0})}{d(\pi_\lambda)\dim W} r_0^\star r_0$.

Thus, the kernel K_{λ,η_0} of the orthogonal projector onto $\mathrm{Cl}(\mathcal{U}(\mathfrak{h}_0)W)$ is

$$K_{\lambda,\eta_0}(x,y) := \frac{d(\pi_{\eta_0}^{H_0})}{d(\pi_\lambda)\dim W} \int_{H_0} K_\lambda(p,y) K_\lambda(x,p) dp.$$

Doing $H := H_0$, we obtain a similar result for the kernel of the orthogonal projector onto $\mathrm{Cl}(\mathcal{U}(\mathfrak{h})W)$.

The equality $(r_0 r_0^\star)_{|H^2(H_0,\tau)} = c I_{H^2(H_0,\tau)}$ yields the first claim in:

Proposition 3 *Assume $res_L(\tau)$ is irreducible. Then:*

(a) For every $g \in H^2(H_0, \tau_{|L})$ (resp. $g \in H^2(H, \tau_{|L})$), the function $r_0^\star(g)$ (resp. $r^\star(g)$) is an extension of a scalar multiple of g.
(b) The kernel K_λ^G is an extension of a scalar multiple of $K_{\tau_{|L}}^H$.

When we restrict holomorphic Discrete Series, this fact naturally happens (see [22], [25, Example 10.1], and references therein).

Proof Let $r : H^2(G,\tau) \to L^2(H \times_\tau W)$ be the restriction map. The duality H, H_0 and Theorem 1 applied to $H := H_0$ imply $H^2(H,\tau) = r(\mathrm{Cl}(\mathcal{U}(\mathfrak{h})W))$, as well as that there exists, up to a constant, a unique $T \in Hom_H(H^2(H,\tau), H^2(G,\tau)) \equiv Hom_L(W, H^2(H_0,\tau)) \equiv \mathbb{C}$. It follows from the proof of Proposition 2 that, up to a constant, $T = r^\star$ restricted to $H^2(H,\tau)$. After we apply the equality $T(K_\mu^H(\cdot,e)^\star z)(x) = K_T(e,x)z$ (see [24]), we obtain

$$r^\star(K_\mu^H(\cdot,e)^\star z)(y) = K_\lambda(y,e)^\star z.$$

Also, Schur's lemma implies $rr^\star$ restricted to $H^2(H,\tau)$ is a constant times the identity map. Thus, for $h \in H$, we have $rr^\star(K_\mu^H(\cdot,e)^\star w)(h) = qK_\mu^H(h,e)^\star w$. For the value of q, see 4.2.2. Putting together, we obtain,

$$K_\lambda(h,e)^\star z = r(K_\lambda(\cdot,e)^\star z)(h) = qK_\mu^H(h,e)^\star z.$$

Hence, for $h, h_1 \in H$, we have

$$K_\lambda(h_1,h)^\star z = K_\lambda(h^{-1}h_1,e)^\star z = qK_\mu^H(h^{-1}h_1,e)^\star z = qK_\mu^H(h_1,h)^\star z$$

as we have claimed. □

By the same token, after we set $H := H_0$, we obtain:

For $res_L(\tau)$ *irreducible,* $(\sigma, Z) = (res_L(\tau), W)$*, and* $V_{\eta_0}^{H_0} = H^2(H_0,\sigma)$*, the kernel* K_λ *extends a scalar multiple of* $K_{\eta_0}^{H_0}$. Actually, $r_0(K_\lambda(\cdot,e)^\star w) = cK_{\eta_0}^{H_0}(\cdot,e)^\star w$.

Remark 8 We would like to point out that the equality

$$r^\star(K_\mu^H(\cdot,e)^\star(z))(y) = K_\lambda(y,e)^\star z$$

implies $res_H(\pi_\lambda)$ is H-algebraically discretely decomposable. Indeed, we apply a theorem shown by Kobayashi [16, Lemma 1.5], which says that when $(V_\lambda^G)_{K-fin}$ contains an irreducible $(\mathfrak{h}, L)$ irreducible submodule, then V_λ^G is discretely decomposable. We know $K_\lambda(y,e)^\star z$ is a K-finite vector; the equality implies $K_\lambda(y,e)^\star z$ is $\mathfrak{z}(\mathcal{U}(\mathfrak{h}))$-finite. Hence, owing to Harish-Chandra [32, Corollary 3.4.7 and Theorem 4.2.1], $H^2(G,\tau)_{K-fin}$ contains a nontrivial irreducible $(\mathfrak{h}, L)$-module, and the fact shown by Kobayashi applies.

4.2.2 Value of $b = d = c$ when $res_L(\tau)$ Is Irreducible

We show $b = d = d(\pi_\lambda)\dim W/d(\pi_{\eta_0}^{H_0}) = c$. In fact, the constant b, d satisfies $(r_0^\star r_0)_{\mathcal{U}(\mathfrak{h}_0)W} = dI_{\mathcal{U}(\mathfrak{h}_0)W}$ and $(r_0 r_0^\star)_{|H^2(H_0,\tau)} = bI_{H^2(H_0,\tau)}$. Now, it readily follows $b = d$. To evaluate $r_0^\star r_0$ at $K_\lambda(\cdot,e)^\star w$, for $h_1 \in H_0$, we compute, for $h_1 \in H_0$,

$$bK_\lambda(h_1,e)^\star w = r_0^\star r_0(K_\lambda(\cdot,e)^\star w)(h_1) = \int_{H_0} K_\lambda(h_0,h_1)K_\lambda(h_0,e)^\star dh_0 w$$
$$= d(\pi_\lambda)^2 \int_{H_0} \Phi(h_1^{-1}h_0)\Phi(h_0)^\star dh_0 w.$$

Here, Φ is the spherical function attached to the lowest K-type of π_λ. Since we are assuming $res_L(\tau)$ is an irreducible representation, $\mathcal{U}(\mathfrak{h}_0)W$ is an irreducible $(\mathfrak{h}_0, L)$-module, and it is equivalent to the underlying Harish-Chandra module for $H^2(H_0, res_L(\tau))$. Thus, the restriction of Φ to H_0 is the spherical function attached to the lowest L-type of the irreducible square integrable representation $\mathrm{Cl}(\mathcal{U}(\mathfrak{h}_0)W) \equiv H^2(H_0, res_L(\tau))$. We fix a orthonormal basis $\{w_i\}$ for $\mathcal{U}(\mathfrak{h}_0)W[W]$. We recall

$\Phi(x)w = P_W\pi(x)P_Ww = \sum_{1\leq i\leq \dim W}(\pi(x)w, w_i)_{L^2}w_i,$

$\Phi(x^{-1}) = \Phi(x)^\star.$

For $h_1 \in H_0$, we compute, and to justify steps, we appeal to the invariance of Haar measure and to the orthogonality relations for matrix coefficients of irreducible square integrable representations, and we recall $d(\pi_{\eta_0}^{H_0})$ denotes the formal degree for $H^2(H_0, res_L(\tau))$.

$$\begin{aligned}
\int_{H_0}\Phi(h_1^{-1}h)\Phi(h)^\star w dh &= \sum_{i,j}\int_{H_0}(\pi(h_1^{-1}h)w_j, w_i)_{L^2}(\pi(h^{-1})w, w_j)_{L^2}w_i dh\\
&= \sum_{i,j}\int_{H_0}(\pi(h)w_j, \pi(h_1)w_i)_{L^2}\overline{(\pi(h)w_j, w)_{L^2}}w_i dh\\
&= 1/d(\pi_{\eta_0}^{H_0})\sum_{i,j}(w_j, w_j)_{L^2}\overline{(\pi(h_1)w_i, w)_{L^2}}w_i\\
&= \dim W/d(\pi_{\eta_0}^{H_0})\sum_i(\pi(h_1^{-1})w, w_i)_{L^2}w_i\\
&= \dim W/d(\pi_{\eta_0}^{H_0})\Phi(h_1)^\star w.
\end{aligned}$$

Thus,

$$\begin{aligned}
r_0^\star r_0(K_\lambda(\cdot, e)^\star w)(h_1) &= d(\pi_\lambda)^2\int_{H_0}\Phi(h_1^{-1}h)\Phi(h)^\star w dh\\
&= \frac{d(\pi_\lambda)^2 \dim W}{d(\pi_{\eta_0}^{H_0})d(\pi_\lambda)}K_\lambda(h_1, e)^\star w.
\end{aligned}$$

The functions $K_\lambda(\cdot, e)^\star w, r_0^\star r_0(K_\lambda(\cdot, e)^\star w)(\cdot)$ belong to $\mathrm{Cl}(\mathcal{U}(\mathfrak{h}_0)W)$, the injectivity of r_0 on $\mathrm{Cl}(\mathcal{U}(\mathfrak{h}_0)W)$, forces, for every $x \in G$

$$r_0^\star r_0(K_\lambda(\cdot, e)^\star w)(x) = d(\pi_\lambda)\dim W/d(\pi_{\eta_0}^{H_0})K_\lambda(x, e)^\star w.$$

Hence, we have computed $b = d = c$.

4.2.3 Analysis of r_0^D for Arbitrary (τ, W), (σ, Z)

We recall the decomposition $W = \sum_{\nu_2'\in Spec_{L\cap K_2}(\pi_{\Lambda_2}^{K_2})} W[\pi_{\Lambda_1}^{K_1}\boxtimes\pi_{\nu_2'}^{L\cap K_2}]$.

A consequence of Proposition 1 is that $r_0^\star$ maps $\mathbf{H}^2(H_0, W[\pi_{\Lambda_1}^{K_1}\boxtimes\pi_{\nu_2}^{L\cap K_2}])$ into $\mathrm{Cl}(\mathcal{U}(\mathfrak{h}_0)W[\pi_{\Lambda_1}^{K_1}\boxtimes\pi_{\nu_2}^{L\cap K_2}])$. In consequence, $r_0 r_0^\star$ restricted to $\mathbf{H}^2(H_0, W[\pi_{\Lambda_1}^{K_1}\boxtimes$

$\pi_{\nu_2}^{L\cap K_2}]$) is a bijective H_0-endomorphism C_j. Hence, the inverse map of r_0 restricted to $\mathrm{Cl}(\mathcal{U}(\mathfrak{h}_0)W[\pi_{\Lambda_1}^{K_1}\boxtimes\pi_{\nu_2}^{L\cap K_2}])$ is $r_0^\star C_j^{-1}$. Since $H^2(H_0,\pi_{\Lambda_1}^{K_1}\boxtimes\pi_{\nu_2}^{L\cap K_2})$ has a unique lowest L-type, we conclude C_j is determined by an element of $Hom_L(\pi_{\Lambda_1}^{K_1}\boxtimes\pi_{\nu_2}^{L\cap K_2}, H^2(H_0,\pi_{\Lambda_1}^{K_1}\boxtimes\pi_{\nu_2}^{L\cap K_2})[\pi_{\Lambda_1}^{K_1}\boxtimes\pi_{\nu_2}^{L\cap K_2}])$. Since for $D\in\mathcal{U}(\mathfrak{h}_0)$, $w\in W$ we have $C_j(L_D w)=L_D C_j(w)$, we obtain C_j, a zero-order differential operator on the underlying Harish-Chandra module of $H^2(H_0,\pi_{\Lambda_1}^{K_1}\boxtimes\pi_{\nu_2}^{L\cap K_2})$. Summing up, we have that the inverse to $r_0:\mathrm{Cl}(\mathcal{U}(\mathfrak{h}_0)W)\to\mathbf{H}^2(H_0,\tau)$ is the function $r_0^\star(\oplus_j C_j^{-1})$. For $T\in Hom_H(H^2(H,\sigma),H^2(G,\tau))$ and $T_0\in Hom_L(Z,\mathbf{H}^2(H_0,\tau))$ so that $r_0^D(T)=T_0$, we obtain the equalities

$$K_T(e,x)z=(D^{-1}[\int_{H_0}K_\lambda(h_0,\cdot)((\oplus_j C_j^{-1})T_0(z))(h_0)dh_0])(x).$$

$$K_T(h,x)z=(D^{-1}[\int_{H_0}K_\lambda(h_0,\cdot)(\oplus_j C_j^{-1})(r_0(D(K_T(e,\cdot)z))(\cdot))(h_0)dh_0])(h^{-1}x).$$

When D is the identity, the formula simplifies as the one in the second Corollary to Proposition 2.

4.2.4 Eigenvalues of $r_0^\star r_0$

For general case, we recall $r_0^\star r_0$ intertwines the action of H_0. Moreover, Proposition 1 and its Corollary give that for each L-isotypic component $Z_1\subseteq W$ of $res_L(\tau)$, we have $\mathcal{U}(\mathfrak{h}_0)W[Z_1]=Z_1$. Thus, each isotypic component of $res_L((\mathcal{U}(\mathfrak{h}_0)W)[W])$ is invariant by $r_0^\star r_0$; in consequence, $r_0^\star r_0$ leaves invariant the subspace "W" $=H^2(G,\tau)[W]=\{K_\lambda(\cdot,e)^\star w, w\in W\}$. Since $Ker(r_0)=(\mathrm{Cl}(\mathcal{U}(\mathfrak{h}_0)W))^\perp$, $r_0^\star r_0$ is determined by the values it takes on "W". Now, we assume $res_L(\tau)$ is a multiplicity-free representation. We write $Z_1^\perp=Z_2\oplus\cdots\oplus Z_q$, where Z_j are L-invariant and L-irreducible. Thus, Proposition 1 implies $\mathrm{Cl}(\mathcal{U}(\mathfrak{h}_0)W)=\mathrm{Cl}(\mathcal{U}(\mathfrak{h}_0)Z_1)\oplus\cdots\oplus\mathrm{Cl}(\mathcal{U}(\mathfrak{h}_0)Z_q)$. This is an orthogonal decomposition, each summand is irreducible, and no irreducible factor is equivalent to the other. For $1\le i\le q$, let η_i denote the Harish-Chandra parameter for $\mathrm{Cl}(\mathcal{U}(\mathfrak{h}_0)Z_i)$.

Proposition 4 *When $res_L(\tau)$ is a multiplicity-free representation, the linear operator $r_0^\star r_0$ on $\mathrm{Cl}(\mathcal{U}(\mathfrak{h}_0)Z_i)$ is equal to $\frac{d(\pi_\lambda)\dim Z_i}{d(\pi_{\eta_i}^{H_0})}$ times the identity map.*

Proof For the subspace $\mathrm{Cl}(\mathcal{U}(\mathfrak{h}_0)W)[W]$, we choose a $L^2(G)$-orthonormal basis $\{w_j\}_{1\le j\le \dim W}$ equal to the union of respective $L^2(G)$-orthonormal basis for $\mathrm{Cl}(\mathcal{U}(\mathfrak{h}_0)Z_i)[Z_i]$. Next, we compute and freely make use of the notation in 4.2.2. Owing to our multiplicity-free hypothesis, we have that $r_0^\star r_0$ restricted to $\mathrm{Cl}(\mathcal{U}(\mathfrak{h}_0)Z_i)$ is equal to a constant d_i times the identity map. Hence, on $w\in\mathrm{Cl}(\mathcal{U}(\mathfrak{h}_0)Z_i)[Z_i]$, we have $d_i w=d(\pi_\lambda)^2\int_{H_0}\Phi(h_0)\Phi(h_0)^\star w dh_0$.

Now, $\Phi(h_0) = (a_{ij}) = ((\pi_\lambda(h_0)w_j, w_i)_{L^2(G)})$, Hence, the pq-coefficient of the product $\Phi(h_0)\Phi(h_0)^\star$ is equal to

$\sum_{1\leq j\leq \dim W}(\pi_\lambda(h_0)w_j, w_p)_{L^2(G)}\overline{(\pi_\lambda(h_0)w_j, w_q)}_{L^2(G)}$.

Let I_i denote the set of indexes j so that $w_j \in Z_i$. Thus, $\{1, \ldots, \dim W\}$ is equal to the disjoint union $\cup_{1\leq i\leq q} I_i$. A consequence of Proposition 1 is the $L^2(G)$-orthogonality of the subspaces $\mathrm{Cl}(\mathcal{U}(\mathfrak{h}_0)Z_j)$; hence, for $t \in I_a, q \in I_d$ and $a \neq d$, we have $(\pi_\lambda(h_0)w_q, w_t)_{L^2(G)} = 0$. Therefore, with the previous observation and the disjointness of the sets I_r, let us obtain that for $i \neq d$, $p \in I_i, q \in I_d$, each summand in

$$\sum_{1\leq j\leq \dim W}\int_{H_0}(\pi_\lambda(h_0)w_j, w_p)_{L^2(G)}\overline{(\pi_\lambda(h_0)w_j, w_q)}_{L^2(G)}dh_0$$

is equal to zero.

For $p, q \in I_i$, we apply the previous computation and the orthogonality relations to the irreducible representation $\mathrm{Cl}(\mathcal{U}(\mathfrak{h}_0)Z_i)$. We obtain

$\sum_{1\leq j\leq \dim W}\int_{H_0}(\pi_\lambda(h_0)w_j, w_p)_{L^2(G)}\overline{(\pi_\lambda(h_0)w_j, w_q)}_{L^2(G)}dh_0$

$= \sum_{j\in I_i}\int_{H_0}(\pi_\lambda(h_0)w_j, w_p)_{L^2(G)}\overline{(\pi_\lambda(h_0)w_j, w_q)}_{L^2(G)}dh_0$

$= \sum_{j\in I_i}\frac{1}{d(\pi_{\eta_i}^{H_0})}(w_j, w_j)_{L^2(G)}(w_q, w_p)_{L^2(G)} = \frac{\dim Z_i}{d(\pi_{\eta_i}^{H_0})}\delta_{pq}$.

Thus, we have shown Proposition 4. □

Remark 9 Even, when $res_L(\tau)$ is not multiplicity-free, the conclusion in Proposition 4 holds. In fact, let us denote the L-isotypic component of $res_L(\tau)$ again by Z_i. Now, the proof goes as the one for Proposition 4 till we need to compute

$= \sum_{j\in I_i}\int_{H_0}(\pi_\lambda(h_0)w_j, w_p)_{L^2(G)}\overline{(\pi_\lambda(h_0)w_j, w_q)}_{L^2(G)}dh_0$.

For this, we decompose "Z_i" $= \sum_s Z_{i,s}$ as a $L^2(G)$-orthogonal sum of irreducible L-modules, and we choose the orthonormal basis for "Z_i" as a union of orthonormal basis for each $Z_{i,s}$. Then, we have the $L^2(G)$-orthogonal decomposition $\mathrm{Cl}(\mathcal{U}(\mathfrak{h}_0)Z_i) = \sum_s \mathrm{Cl}(\mathcal{U}(\mathfrak{h}_0)Z_{i,s})$. Now, the proof $res_L(\tau)$, which is multiplicity-free.

5 Examples

We present three types of examples. The first is *multiplicity-free representations.* A simple consequence of the duality theorem is that it readily follows examples of symmetric pair (G, H) and square integrable representation π_λ^G so that $res_H(\pi_\lambda)$ is H-admissible and the multiplicity of each irreducible factor is one. This is equivalent with determining when the representation $res_L(\mathbf{H}^2(H_0, \tau))$ is multiplicity-free. The second is *explicit examples*. Here, we compute the Harish-Chandra parameters of irreducible factors for some $res_H(H^2(G, \tau))$. The third is *existence of representations so that its lowest K-types restricted to L is an irreducible representation.*

In order to present the examples, we need information on certain families of representations.

5.1 Multiplicity-Free Representations

In this section, we generalize the work of T. Kobayashi and his coworkers in the setting of Hermitian symmetric spaces and holomorphic Discrete Series.

Before we present the examples, we would like to comment:

(a) Assume a Discrete Series π_λ has an admissible restriction to a subgroup H. Then, any Discrete Series $\pi_{\lambda'}$ for λ' dominant with respect to Ψ_λ is H-admissible [15].
(b) If $res_H(\pi_\lambda)$ is H-admissible and a multiplicity-free representation, then the restriction to L of the lowest K-type for π_λ is multiplicity-free. This follows from the duality theorem.
(c) In the next paragraphs, we will list families $\mathcal{F}$ of Harish-Chandra parameters of Discrete Series for G so that each representation in the family has a multiplicity-free restriction to H. It may happen that $\mathcal{F}$ is the whole set of Harish-Chandra parameters on a Weyl chamber or $\mathcal{F}$ is a proper subset of a Weyl Chamber. Information on $\mathcal{F}$ for holomorphic representations is in [18, 19].
(d) Every irreducible $(\mathfrak{g}, K)$-module for either $\mathfrak{g} \equiv \mathfrak{su}(n, 1)$ or $\mathfrak{g} \equiv \mathfrak{so}(n, 1)$, restricted to K, is a multiplicity-free representation.

5.1.1 Holomorphic Representations

For G so that G/K is a Hermitian symmetric space, it has been shown by Harish-Chandra that G admits Discrete Series representations with one-dimensional lowest K-type. Here, we further assume that the smooth imbedding $H/L \to G/K$ is holomorphic; equivalently, the center of K is contained in L, and π_λ is a holomorphic representation. Under this hypothesis, it was shown by Kobayashi [17] that a holomorphic Discrete Series for G has a multiplicity-free restriction to the subgroup H whenever it is a scalar holomorphic Discrete Series. Moreover, [17, Theorem 8.8] computes the Harish-Chandra parameter of each irreducible factor. Also, from the work of Kobayashi and Nakahama, we find a description of the restriction to H of arbitrary holomorphic Discrete Series representations. As a consequence, we find restrictions that are not multiplicity-free.

In [19], we find a complete list of the pairs $(\mathfrak{g}, \mathfrak{h})$ so that $H/L \to G/K$ is a holomorphic embedding. The list in [17] can be constructed below.

Also, Theorem 1 let us verify that the following pairs $(\mathfrak{g}, \mathfrak{h})$ so that $res_H(\pi_\lambda)$ is multiplicity-free for any holomorphic π_λ. For this, we list the associated $\mathfrak{h}_0$.
$(\mathfrak{su}(m, n), \mathfrak{s}(\mathfrak{u}(m-1, n) + \mathfrak{u}(1)))$, $\mathfrak{h}_0 = \mathfrak{su}(1, n) + \mathfrak{su}(m-1) + \mathfrak{u}(1)$.
$(\mathfrak{su}(m, n), \mathfrak{s}(\mathfrak{u}(m, n-1) + \mathfrak{u}(1)))$, $\mathfrak{h}_0 = \mathfrak{su}(n-1) + \mathfrak{su}(m, 1) + \mathfrak{u}(1)$.

The list is correct, owing to the fact that any Discrete Series for $SU(n, 1)$ restricted to K is a multiplicity-free representation.

5.1.2 Quaternionic Real Forms, Quaternionic Representations

In [9], the authors considered and classified quaternionic real forms, and also, they made a careful study of quaternionic representations. To follow, we bring out the essential facts for us. From [9], we read that the list of Lie algebra of quaternionic groups is $\mathfrak{su}(2, n)$, $\mathfrak{so}(4, n)$, $\mathfrak{sp}(1, n)$, $\mathfrak{e}_{6(2)}$, $\mathfrak{e}_{7(-5)}$, $\mathfrak{e}_{8(-24)}$, $\mathfrak{f}_{4(4)}$, and $\mathfrak{g}_{2(2)}$. For each quaternionic real form G, there exists a system of positive roots $\Psi \subset \Phi(\mathfrak{g}, \mathfrak{t})$ so that the maximal root α_{max} in Ψ is compact, α_{max} is orthogonal to all compact simple roots, and α_{max} is not orthogonal to each noncompact simple roots. Hence, $\mathfrak{k}_1(\Psi) \equiv \mathfrak{su}_2(\alpha_{max})$. The system Ψ is not unique. We refer to such a system of positive roots *a quaternionic system.*

Let us recall that a *quaternionic representation* is a Discrete Series for a quaternionic real form G so that its Harish-Chandra parameter is dominant with respect to a quaternionic system of positive roots and so that its lowest K-type is equivalent to a irreducible representation for $K_1(\Psi)$ times the trivial representation for K_2. A fact shown in [9] is given a quaternionic system of positive roots, for all but finitely many representations (τ, W) equivalent to the tensor of a nontrivial representation for $K_1(\Psi)$ times the trivial representation of K_2, the following holds: τ is the lowest K-type of a quaternionic (unique) irreducible square integrable representation $H^2(G, \tau)$. We define a *generalized quaternionic representation* to be a Discrete Series representation π_λ so that its Harish-Chandra parameter is dominant with respect to a quaternionic system of positive roots.

From Tables 1 and 2, we readily read the pairs $(\mathfrak{g}, \mathfrak{h})$ so that $\mathfrak{g}$ is a quaternionic Lie algebra, and hence, we have a list of generalized quaternionic representations of G with admissible restriction to H.

Let (G, H) denote a symmetric pair so that a quaternionic representation $(\pi_\lambda, H^2(G, \tau))$ is H-admissible. Then, from [30, 5, 4] we have $\mathfrak{k}_1(\Psi_\lambda) \equiv \mathfrak{su}_2(\alpha_{max}) \subset \mathfrak{l}$ and π_λ is L-admissible. In consequence, [16], π_λ is H_0-admissible. By definition, for a quaternionic representation π_λ, we have $\tau_{|_L}$ is irreducible; hence, $\mathbf{H}^2(H_0, \tau)$ is irreducible. Moreover, after checking on [30] or Tables 1 and 2, the list of systems $\Psi_{H_0,\lambda}$, it follows that $H^2(H_0, \tau)$ is again a quaternionic representation. Finally, in order to present a list of quaternionic representations with multiplicity-free restriction to H, we recall that it follows from the duality theorem that $res_H(H^2(G, \tau))$ is multiplicity-free if and only if $res_L(H^2(H_0, \tau))$ is a multiplicity-free representation and that on [9, Page 88], it is shown that a quaternionic representation for H_0 is L-multiplicity-free if and only if $\mathfrak{h}_0 = \mathfrak{sp}(n, 1), n \geq 1$.

To follow, we list pairs $(\mathfrak{g}, \mathfrak{h})$ where multiplicity-free restriction holds for all quaternionic representations.

$(\mathfrak{su}(2, 2n), \mathfrak{sp}(1, n))$, $\mathfrak{h}_0 = \mathfrak{sp}(1, n), n \geq 1$.

$(\mathfrak{so}(4, n), \mathfrak{so}(4, n-1))$, $\mathfrak{h}_0 = \mathfrak{so}(4, 1) + \mathfrak{so}(n-1)$ (n even or odd).

$(\mathfrak{sp}(1, n), \mathfrak{sp}(1, k) + \mathfrak{sp}(n-k)), \mathfrak{h}_0 = \mathfrak{sp}(1, n-k) + \mathfrak{sp}(k)$.
$(\mathfrak{f}_{4(4)}, \mathfrak{so}(5, 4)), \mathfrak{h}_0 = \mathfrak{sp}(1, 2) \oplus \mathfrak{su}(2)$.
$(\mathfrak{e}_{6(2)}, \mathfrak{f}_{4(4)}), \mathfrak{h}_0 = \mathfrak{sp}(3, 1)$.
A special pair is:
$(\mathfrak{su}(2, 2), \mathfrak{sp}(1, 1)), \mathfrak{h}_0 = \mathfrak{sp}(1, 1)$.
Here, multiplicity free holds for any π_λ so that λ is dominant with respect to a system of positive roots that defines a quaternionic structure on G/K. For details, see [30, Table 2] or Explicit example II.

5.1.3 More Examples of Multiplicity-Free Restriction

Next, we list pairs $(\mathfrak{g}, \mathfrak{h})$ and systems of positive roots $\Psi \subset \Phi(\mathfrak{g}, \mathfrak{t})$ so that $\pi_{\lambda'}$ is H-admissible and multiplicity-free for every element λ' dominant with respect to Ψ. We follow either Tables 1, 2, and 3 or [19]. For each $(\mathfrak{g}, \mathfrak{h})$, we list the corresponding $\mathfrak{h}_0$.
$(\mathfrak{su}(m, n), \mathfrak{su}(m, n-1) + \mathfrak{u}(1)), \Psi_a, \tilde{\Psi}_b, \mathfrak{h}_0 = \mathfrak{su}(m, 1) + \mathfrak{su}(n-1) + \mathfrak{u}(1)$.
$(\mathfrak{so}(2m, 2n+1), \mathfrak{so}(2m, 2n)), \Psi_\pm, \mathfrak{h}_0 = \mathfrak{so}(2m, 1) + \mathfrak{so}(2n)$.
$(\mathfrak{so}(2m, 2), \mathfrak{so}(2m, 1)), \Psi_\pm, \mathfrak{h}_0 = \mathfrak{so}(2m, 1)$.
$(\mathfrak{so}(2m, 2n), \mathfrak{so}(2m, 2n-1)), n > 1, \Psi_\pm, \mathfrak{h}_0 = \mathfrak{so}(2m, 1) + \mathfrak{so}(2n-1)$.

5.2 *Explicit Examples*

5.2.1 Quaternionic Representations for $Sp(1, b)$

For further use, we present an intrinsic description for the $Sp(1) \times Sp(b)$-types of a quaternionic representation for $Sp(1, b)$; a proof of the statements is in [8]. Quaternionic representations for $Sp(1, b)$ are representations of lowest $Sp(1) \times Sp(b)$-type $S^n(\mathbb{C}^2) \boxtimes \mathbb{C}, n \geq 1$. We label the simple roots for the quaternionic system of positive roots Ψ as in [9], $\beta_1, \ldots, \beta_{b+1}$; the long root is β_{b+1}, β_1 is adjacent to just one simple root, and the maximal root β_{max} is adjacent to $-\beta_1$. Let $\Lambda_1, \ldots, \Lambda_{d+1}$ be the associated fundamental weights. Thus, $\Lambda_1 = \frac{\beta_{max}}{2}$. Let $\tilde{\Lambda}_1, \ldots, \tilde{\Lambda}_b$ denote the fundamental weights for "$\Psi \cap \Phi(\mathfrak{sp}(b))$". The irreducible $L = Sp(1) \times Sp(b)$-factors of

$$H^2(Sp(1, b), \pi^{Sp(1)}_{n\frac{\beta_{max}}{2}} \boxtimes \pi^{Sp(b)}_{\rho_{Sp(b)}}) = H^2(Sp(1, b), S^{n-1}(\mathbb{C}^2) \boxtimes \mathbb{C})$$

are

$$\{S^{n-1+m}(\mathbb{C}^2) \boxtimes S^m(\mathbb{C}^{2b}) \equiv \pi^{Sp(1)}_{(n+m)\frac{\beta_{max}}{2}} \boxtimes \pi^{Sp(b)}_{m\tilde{\Lambda}_1 + \rho_{Sp(b)}}, \ m \geq 0\}.$$

The multiplicity of each L-type in $H^2(Sp(1, b), S^{n-1}(\mathbb{C}^2) \boxtimes \mathbb{C})$ is one.

5.2.2 Explicit Example I

We develop this example in detail. We restrict quaternionic representations for $Sp(1,d)$ to $Sp(1,k) \times Sp(d-k)$. For this, we need to review definitions and facts in [8, 19, 30]. The group $G := Sp(1,d)$ is a subgroup of $GL(\mathbb{C}^{2+2d})$. A maximal compact subgroup of $Sp(1,d)$ is the usual immersion of $Sp(1) \times Sp(d)$. Actually, $Sp(1,d)$ is a quaternionic real form for $Sp(\mathbb{C}^{1+d})$. $Sp(1,d)$ has a compact Cartan subgroup T, and there exists an orthogonal basis $\delta, \epsilon_1, \dots, \epsilon_d$ for $i\mathfrak{t}^\star$ so that

$\Phi(\mathfrak{sp}(d+1,\mathbb{C}), \mathfrak{t}) = \{\pm 2\delta, \pm 2\epsilon_1, \dots, \pm 2\epsilon_d, \pm(\epsilon_i \pm \epsilon_j), 1 \le i < j \le d, \pm(\delta \pm \epsilon_s), 1 \le s \le d\}$.

We fix $1 \le k < d$. We consider the usual immersion of $H := Sp(1,k) \times Sp(d-k)$ into $Sp(1,d)$. Thus,

$\Phi(\mathfrak{h}, \mathfrak{t}) := \{\pm 2\delta, \pm 2\epsilon_1, \dots, \pm 2\epsilon_d, \pm(\epsilon_i \pm \epsilon_j), 1 \le i < j \le k,$
$\text{or, } k+1 \le i < j \le d, \pm(\delta \pm \epsilon_s), 1 \le s \le k\}$.

Then, H_0 is isomorphic to $Sp(1, d-k) \times Sp(k)$. We have

$\Phi(\mathfrak{h}_0, \mathfrak{t}) := \{\pm 2\delta, \pm 2\epsilon_1, \dots, \pm 2\epsilon_d, \pm(\epsilon_i \pm \epsilon_j), k+1 \le i < j \le d,$
$\text{or, } 1 \le i < j \le k, \pm(\delta \pm \epsilon_s), k+1 \le s \le d\}$.

From now on, we fix the quaternionic system of positive roots

$$\Psi := \{2\delta, 2\epsilon_1, \dots, 2\epsilon_d, (\epsilon_i \pm \epsilon_j), 1 \le i < j \le d, (\delta \pm \epsilon_s), 1 \le s \le d\}.$$

Then, $\alpha_{max} = 2\delta$, and $\rho_n^\Psi = d\delta$. The Harish-Chandra parameter λ of a quaternionic representation π_λ is dominant with respect to Ψ. Hence, $\Psi_\lambda = \Psi$. The systems in Theorem 2 are $\Psi_{H,\lambda} = \Phi(\mathfrak{h}, \mathfrak{t}) \cap \Psi$ and $\Psi_{H_0,\lambda} = \Phi(\mathfrak{h}_0, \mathfrak{t}) \cap \Psi$. Also, [5], $\Phi(\mathfrak{k}_1 := \mathfrak{k}_1(\Psi), \mathfrak{t}_1 := \mathfrak{t} \cap \mathfrak{k}_1) = \{\pm 2\delta\}$, $\Phi(\mathfrak{k}_2 := \mathfrak{k}_2(\Psi), \mathfrak{t}_2 := \mathfrak{t} \cap \mathfrak{k}_2) = \{\pm 2\epsilon_1, \dots, \pm 2\epsilon_d, \pm(\epsilon_i \pm \epsilon_j), 1 \le i < j \le d\}$. Thus, $K_1(\Psi) \equiv SU_2(2\delta) \equiv Sp(1) \subset H$, and $K_2 \equiv Sp(d)$. Hence, for a Harish-Chandra parameter $\lambda = (\lambda_1, \lambda_2), \lambda_j \in i\mathfrak{t}_j^\star$ dominant with respect to Ψ, the representation π_λ is H-admissible.

The lowest K-type of a generalized quaternionic representation π_λ is the representation $\tau = \pi^K_{\lambda+\rho_n^\lambda} = \pi^{K_1}_{\lambda_1+d\delta} \boxtimes \pi^{K_2}_{\lambda_2}$. Since $\rho_{K_2} = d\epsilon_1 + (d-1)\epsilon_2 + \cdots + \epsilon_d$, for $n \ge d+1$, the functional $\mathfrak{t}^\star \ni \lambda_n := n\delta + \rho_{K_2}$ is a Harish-Chandra parameter dominant with respect to Ψ, and the lowest K-type τ_n of π_{λ_n} is $\pi^{K_1}_{(n+d)\delta} \boxtimes \pi^{K_2}_{\rho_{K_2}}$. That is, $\pi^K_{\lambda_n+\rho_n^\lambda}$ is equal to a irreducible representation of $K_1 \equiv Sp(1) = SU_2(2\delta)$ times the trivial representation of $K_2 \equiv Sp(d)$. The family $(\pi_{\lambda_n})_n$ exhausts, up to equivalence, the set of quaternionic representations for $Sp(1,d)$. Now, $\mathbf{H}^2(H_0, \tau_n)$ is the irreducible representation of lowest L-type equal to the irreducible representation $\pi^{K_1}_{(n+d)\delta}$ of K_1 times the trivial representation of $K_2 \cap L$. Actually, $\mathbf{H}^2(H_0, \pi^{K_1}_{(n+d)\delta} \boxtimes \pi^{K_2}_{\rho_{K_2}})$ is a realization of the quaternionic representation $H^2(Sp(1, n-k), \pi^{Sp(1)}_{(n+d)\delta} \boxtimes \pi^{Sp(n-k)}_{\rho_{Sp(n-k)}})$ for $Sp(1, d-k)$ times the trivial representation of $Sp(k)$. In [8, Proposition 6.3], it is shown that the representation $H^2(Sp(1, n-k), \pi^{Sp(1)}_{(n+d)\delta} \boxtimes \pi^{Sp(n-k)}_{\rho_{Sp(n-k)}})$ restricted to L is a multiplicity-free representation, and also, they list the highest weights of the totality of L-irreducible factors. To follow, we explicit such a computation. For this, we recall 5.2.1 and notice $b = d - k$; $\Lambda_1 = \delta$, $\beta_{max} = 2\delta$, $\tilde{\Lambda}_1 = \epsilon_1$; as $Sp(1)$-module, $S^p(\mathbb{C}^2) \equiv \pi^{SU_2(2\delta)}_{(p+1)\delta}$; for $p \geq 1$, as $Sp(p)$-module $S^m(\mathbb{C}^{2p}) \equiv \pi^{Sp(p)}_{m\epsilon_1 + \rho_{Sp(p)}}$. Then, the irreducible $L = Sp(1) \times Sp(d-k) \times Sp(k)$-factors of

$$\mathbf{H}^2(H_0, \pi^{K_1}_{(n+d)\delta} \boxtimes \pi^{K_2}_{\rho_{K_2}}) \equiv H^2(Sp(1, d-k), \pi^{Sp(1)}_{(n+d)\delta} \boxtimes \pi^{Sp(d-k)}_{\rho_{Sp(d-k)}}) \boxtimes \mathbb{C}$$

are multiplicity-free, and it is the set of inequivalent representations

$$\{S^{n+d-1+m}(\mathbb{C}^2) \boxtimes S^m(\mathbb{C}^{2(d-k)}) \boxtimes \mathbb{C}$$
$$\equiv \pi^{Sp(1)}_{(n+d+m)\delta} \boxtimes \pi^{Sp(d-k)}_{m\epsilon_{k+1} + \rho_{Sp(d-k)}} \boxtimes \pi^{Sp(k)}_{\rho_{Sp(k)}}, \ m \geq 0\}.$$

Here, $\rho_{Sp(d-k)} = (d-k)\epsilon_{k+1} + (d-k-1)\epsilon_{k+2} + \cdots + \epsilon_d$, and $\rho_{Sp(k)} = k\epsilon_1 + (k-1)\epsilon_2 + \cdots + \epsilon_k$.

We compute $\Psi_{H,\lambda} = \{2\delta, 2\epsilon_1, \ldots, 2\epsilon_d, (\epsilon_i \pm \epsilon_j), 1 \leq i < j \leq k \,\text{or}\, k+1 \leq i < j \leq d, (\delta \pm \epsilon_s), 1 \leq s \leq k\}$. $\rho^\mu_n = \rho^H_n = k\delta$. Now, from Theorem 1, we have $Spec_H(\pi_\lambda) + \rho^H_n = Spec_L(\mathbf{H}^2(H_0, \tau))$; hence, we conclude below.

The representation $res_{Sp(1,k) \times Sp(d-k)}(\pi^{Sp(1,d)}_{\lambda_n})$ is a multiplicity-free representation, and the totality of Harish-Chandra parameters of the $Sp(1,k) \times Sp(d-k)$-irreducible factors is the set

$$\{(n+d+m)\delta + m\epsilon_{k+1} + \rho_{Sp(k)} + \rho_{Sp(d-k)}\} - \rho^H_n =$$
$$(n+d+m-k)\delta + m\epsilon_{k+1} + (d-k)\epsilon_{k+1} + \cdots + \epsilon_d + k\epsilon_1 + \cdots + \epsilon_k, m \geq 0\}.$$

Hence, $res_{Sp(1,k) \times Sp(d-k)}(\pi^{Sp(1,d)}_{\lambda_n})$ is equivalent to the Hilbert sum

$$\oplus_{m \geq 0} V^{Sp(1,k) \times Sp(d-k)}_{(n+d+m-k)\delta + m\epsilon_{k+1} + \rho_{Sp(k)} + \rho_{Sp(d-k)}}$$
$$\equiv \oplus_{m \geq 0} H^2(Sp(1,k) \times Sp(d-k), \pi^{Sp(1) \times Sp(k) \times Sp(d-k)}_{(n+d+m)\delta + m\epsilon_{k+1} + \rho_{Sp(k)} + \rho_{Sp(d-k)}}).$$

A awkward point of our decomposition is that it does not provide an explicit description of the H-isotypic components for $res_H(V^G_\lambda)$.

5.2.3 Explicit Example II

We restrict from $Spin(2m, 2)$, $m \geq 2$, to $Spin(2m, 1)$. We notice the isomorphism between $(Spin(4,2), Spin(4,1))$ and the pair $(SU(2,2), Sp(1,1))$. In this setting, $K = Spin(2m) \times Z_K$, $L = Spin(2m)$, and $Z_K \equiv \mathbb{T}$. Obviously, we may conclude that any irreducible representation of K is irreducible when restricted to L. In this case, $H_0 \equiv Spin(2m, 1)$, and (for $m = 2$, $H_0 \equiv Sp(1,1)$) and $\mathbf{H}^2(H_0, \tau)$ is irreducible. Therefore, the duality theorem together with that of any irreducible representation for $Spin(2m, 1)$ is $L = Spin(2m)$-multiplicity-free [29, page 11]; we obtain

Any $Spin(2m, 1)$ – *admissible representation* $\pi_\lambda^{Spin(2m,2)}$ *is multiplicity free.*

For $(Spin(2m, 2), Spin(2m, 1))$ in [30, Table 2], [19], it is verified that any π_λ, with λ dominant with respect to one of the systems $\Psi_\pm$ (see proof of 4), has admissible restriction to $Spin(2m, 1)$ and no other π_λ has admissible restriction to $Spin(2m, 1)$.

In [30, 15, 16, Table 2], it is verified that any square integrable representation π_λ with λ dominant with respect to a quaternionic system for $SU(2, 2)$ has admissible restriction to $Sp(1, 1)$. As in 5.2.2, we may compute the Harish-Chandra parameters for the irreducible components of $res_{Sp(1,1)}(\pi_\lambda^{SU(2,2)})$.

5.2.4 Explicit Example III

$(\mathfrak{e}_{6(2)}, \mathfrak{f}_{4(4)})$. We fix a compact Cartan subgroup $T \subset K$ so that $U := T \cap H$ is a compact Cartan subgroup of $L = K \cap H$. Then, there exists a quaternionic and Borel de Siebenthal positive root system Ψ_{BS} for $\Phi(\mathfrak{e}_6, \mathfrak{t})$ so that after we write the simple roots as in Bourbaki (see [8, 30]), the compact simple roots are $\alpha_1, \alpha_3, \alpha_4, \alpha_5, \alpha_6$ (they determine the A_5-Dynkin sub-diagram), and α_2 is noncompact. α_2 is adjacent to $-\alpha_{max}$ and to α_4. In [30], it is verified that Ψ_{BS} is the unique system of positive roots such that $\mathfrak{k}_1(\Psi_{BS}) = \mathfrak{su}_2(\alpha_{max})$.

The automorphism σ of $\mathfrak{g}$ acts on the simple roots as follows:

$$\sigma(\alpha_2) = \alpha_2, \ \sigma(\alpha_1) = \alpha_6, \ \sigma(\alpha_3) = \alpha_5, \ \sigma(\alpha_4) = \alpha_4.$$

Hence, $\sigma(\Psi_{BS}) = \Psi_{BS}$. Let $h_2 \in i\mathfrak{t}^\star$ be so that $\alpha_j(h_2) = \delta_{j2}$ for $j = 1, \ldots, 6$. Then, $h_2 = \frac{2H_{\alpha_{max}}}{(\alpha_{max}, \alpha_{max})}$ and $\theta = Ad(exp(\pi i h_2))$. A straightforward computation yields the following: $\mathfrak{k} \equiv \mathfrak{su}_2(\alpha_{max}) + \mathfrak{su}(6)$, $\mathfrak{l} \equiv \mathfrak{su}_2(\alpha_{max}) + \mathfrak{sp}(3)$; the fix point subalgebra for $\theta\sigma$ is isomorphic to $\mathfrak{sp}(1, 3)$. Thus, the pair $(\mathfrak{e}_{6(2)}, \mathfrak{sp}(1, 3))$ is the associated pair to $(\mathfrak{e}_{6(2)}, \mathfrak{f}_{4(4)})$. Let $q_\mathfrak{u}$ denote the restriction map from $\mathfrak{t}^\star$ to $\mathfrak{u}^\star$. Then, for λ dominant with respect to Ψ_{BS}, the simple roots for $\Psi_{H,\lambda} = \Psi_{\mathfrak{f}_{4(4)},\lambda}$, respectively, $\Psi_{\mathfrak{sp}(1,3),\lambda}$,, are

$$\alpha_2, \ \alpha_4, \ q_\mathfrak{u}(\alpha_3) = q_\mathfrak{u}(\alpha_5), \ q_\mathfrak{u}(\alpha_1) = q_\mathfrak{u}(\alpha_6).$$

$$\beta_1 = q_\mathfrak{u}(\alpha_2 + \alpha_4 + \alpha_5) = q_\mathfrak{u}(\alpha_2 + \alpha_4 + \alpha_3), \ \beta_2 = q_\mathfrak{u}(\alpha_1) = q_\mathfrak{u}(\alpha_6),$$

$$\beta_3 = q_\mathfrak{u}(\alpha_3) = q_\mathfrak{u}(\alpha_5), \ \beta_4 = \alpha_4.$$

The fundamental weight $\tilde{\Lambda}_1$ associated with β_1 is equal to $\frac{1}{2}\beta_{max}$. Hence, $\tilde{\Lambda}_1 = \beta_1 + \beta_2 + \beta_3 + \frac{1}{2}\beta_4 = \alpha_2 + \frac{3}{2}\alpha_4 + \alpha_3 + \alpha_5 + \frac{1}{2}(\alpha_1 + \alpha_6)$.

Thus, from the duality theorem, for the quaternionic representation

$$H^2(E_{6(2)}, \pi^{SU_2(\alpha_{max})\times SU(6)}_{n\frac{\alpha_{max}}{2}+\rho_{SU(6)}})$$

the set of Harish-Chandra parameters of the irreducible $F_{4(4)}$-factors is equal to: $-\rho_n^H$ *plus the set of infinitesimal characters of the $L \equiv SU_2(\alpha_{max}) \times Sp(3)$-irreducible factors for*

$$res_{SU_2(\alpha_{max})\times Sp(3)}(H^2(Sp(1,3), \pi^{SU_2(\alpha_{max})\times Sp(3)}_{n\frac{\alpha_{max}}{2}+\rho_{Sp(3)}})).$$

Here, $-\rho_n^H = -d_H\frac{\alpha_{max}}{2}$, $d_H = d_{\mathfrak{f}_{4(4)}} = 7$ (see [9]).
Therefore, from the computation in 5.2.1, we obtain

$$res_{F_{4(4)}}(\pi^{E_{6(2)}}_{n\frac{\alpha_{max}}{2}+\rho_{SU(6)}-\rho_n^G}) = \oplus_{m\geq 0} V^{F_{4(4)}}_{(n-7+m)\frac{\alpha_{max}}{2}+m\tilde{\Lambda}_1+\rho_{Sp(3)}}.$$

Here, $\rho_{Sp(3)} = 3\beta_2 + 5\beta_3 + 3\beta_4 = \frac{3}{2}(\alpha_1+\alpha_6) + \frac{5}{2}(\alpha_3+\alpha_5) + 3\alpha_4$.

5.2.5 Comments on Admissible Restriction of Quaternionic Representations

As usual, (G, H) is a symmetric pair, H is not a compact group, and $(\pi_\lambda, H^2(G,\tau))$ is an H-admissible, non-holomorphic, square integrable representation. We further assume G/K as well as H/L holds a quaternionic structure and the inclusion $H/L \hookrightarrow G/K$ respects the respective quaternionic structures. Then, from Tables 1, 2, and 3, it follows:

(a) λ is dominant with respect to a quaternionic system of positive roots. That is, π_λ is a generalized quaternionic representation.
(b) H_0/L has a quaternionic structure.
(c) Each system $\Psi_{H,\lambda}$ and $\Psi_{H_0,\lambda}$ is a quaternionic system.
(d) The representation $\mathbf{H}^2(H_0,\tau)$ is a sum of generalized quaternionic representations.
(e) When π_λ is quaternionic, then the representation $\mathbf{H}^2(H_0,\tau)$ is equal to $H^2(H_0, res_L(\tau))$; hence, it is quaternionic. Moreover, [8] computed the highest weight and the respective multiplicity of each of its L-irreducible factors.
(f) Thus, the duality theorem 1 together with (a)–(e) let us compute the Harish-Chandra parameters of the irreducible H-factors for a quaternionic representation π_λ. Actually, the computation of the Harish-Chandra parameters is quite similar to the computation in *Explicit example I, Explicit example III*.

In the following paragraph, we consider particular quaternionic symmetric pairs. One pair is $(\mathfrak{f}_{4(4)}, \mathfrak{so}(5,4))$. Here, $\mathfrak{h}_0 \equiv \mathfrak{sp}(1,2) + \mathfrak{su}(2)$. Thus, for any Harish-Chandra parameter λ dominant with respect to the quaternionic system of positive roots, we have that π_λ restricted to $SO(5,4)$ is an admissible representation, and the duality theorem allows us compute either multiplicities or Harish-Chandra parameters of the restriction. Moreover, since quaternionic Discrete Series for $Sp(1,2) \times SU(2)$ are multiplicity-free, [8], we have that quaternionic Discrete Series for $\mathfrak{f}_{4(4)}$, restricted to $SO(5,4)$, are multiplicity-free. It seems that it can be deduced from the branching rules for the pair $(Sp(3), Sp(1) \times Sp(2))$ that a generalized quaternionic representation, $res_{SO(5,4)}(\pi_\lambda)$ is multiplicity-free if and only π_λ is quaternionic.

For the pair $(\mathfrak{f}_{4(4)}, \mathfrak{so}(5,4))$, if we attempt to deduce our decomposition result from the work of [9], we have to consider the group of Lie algebra $\mathfrak{g}^d \equiv \mathfrak{f}_{4(-20)}$, its maximal compactly embedded subalgebra is isomorphic to $\mathfrak{so}(9)$, a simple Lie algebra, hence no Discrete Series for G^d has an admissible restriction to H_0 (see [18, 5]). Thus, it is not clear to us how to deduce our Duality result from the Duality Theorem in [8].

For the pairs $(\mathfrak{e}_{6(2)}, \mathfrak{so}(6,4) + \mathfrak{so}(2))$, $(\mathfrak{e}_{7(-5)}, \mathfrak{e}_{6(2)} + \mathfrak{so}(2))$, for each G, generalized quaternionic representations do exist, and they are H-admissible. For these pairs, the respective $\mathfrak{h}_0$ are as follows: $\mathfrak{su}(2,4) + \mathfrak{su}(2)$, $\mathfrak{su}(6,2)$. In these two cases, the Maplesoft, developed by Silva-Vergne [2], allows to compute L-Harish-Chandra parameters and respective multiplicity for each Discrete Series for $H_0 \equiv SU(p,q) \times SU(r)$; hence, the duality formula yields the Harish-Chandra Parameters for $res_H(\pi_\lambda)$ and their multiplicity.

5.2.6 Explicit Example IV

The pair $(SO(2m,n), SO(2m,n-1))$ is considered in [9]. We recall their result and sketch how to derive the result from our duality theorem. We only consider the case $\mathfrak{g} = \mathfrak{so}(2m, 2n+1)$. Here, $\mathfrak{k} = \mathfrak{so}(2m) + \mathfrak{so}(2n+1)$, $\mathfrak{h} = \mathfrak{so}(2m,2n)$, $\mathfrak{h}_0 = \mathfrak{so}(2m,1) + \mathfrak{so}(2n)$, and $\mathfrak{l} = \mathfrak{so}(2m) + \mathfrak{so}(2n)$. We fix a Cartan subalgebra $\mathfrak{t} \subset \mathfrak{l} \subset \mathfrak{k}$. Then, there exists an orthogonal basis $\epsilon_1, \dots, \epsilon_m, \delta_1, \dots, \delta_n$ for $i\mathfrak{t}^\star$ so that

$$\Delta = \{(\epsilon_i \pm \epsilon_j), 1 \le i < j \le m, (\delta_r \pm \delta_s), 1 \le r < s \le n\} \cup \{\delta_j\}_{1 \le j \le n}.$$

$$\Phi_n = \{\pm(\epsilon_r \pm \delta_s), r = 1, \dots, m, s = 1, \dots, n\} \cup \{\pm\epsilon_j, j = 1, \dots, m\}.$$

Systems of positive roots Ψ_λ so that π_λ^G is an admissible representation of H are systems $\Psi_\pm$ associated with the lexicographic orders $\epsilon_1 > \cdots > \epsilon_m > \delta_1 > \cdots > \delta_n$, $\epsilon_1 > \cdots > \epsilon_{m-1} > -\epsilon_m > \delta_1 > \cdots > \delta_{n-1} > -\delta_n$. Here, for $m \ge 3$, $\mathfrak{k}_1(\Psi_\pm) = \mathfrak{so}(2m)$. For $m = 2$, $\mathfrak{k}_1(\Psi_\pm) = \mathfrak{su}_2(\epsilon_1 \pm \epsilon_2)$. Then,

$\Psi_{H,+} = \{(\epsilon_i \pm \epsilon_j), 1 \le i < j \le m, (\delta_r \pm \delta_s), 1 \le r < s \le n\} \cup \{(\epsilon_r \pm \delta_s), r = 1, \dots, m, s = 1, \dots, n\}$,

$\Psi_{H_0,+} = \{(\epsilon_i \pm \epsilon_j), 1 \le i < j \le m, (\delta_r \pm \delta_s), 1 \le r < s \le n\} \cup \{\epsilon_j, j = 1, \dots, m\}$.

$\mathfrak{g}^d = \mathfrak{so}(2m+2n, 1)$. Thus, from either our duality theorem or from [9], we infer that whenever $res_H(\pi_\lambda)$ is H-admissible, then, $res_H(\pi_\lambda)$ is a multiplicity-free representation. Hence, we are left to compute the Harish-Chandra parameters for $res_{SO(2m,2n)}(H^2(SO(2m,2n+1),\pi_{\Lambda_1}^{SO(2m)}\boxtimes\pi_{\Lambda_2}^{SO(2n+1)}))$. For this, according to the duality theorem, we have to compute the infinitesimal characters of each irreducible factor of the underlying L-module in

$$\mathbf{H}^2(H_0,\tau)=\sum_{\nu\in Spec_{SO(2n)}(\pi_{\Lambda_2}^{SO(2n+1)})} H^2\big(SO(2m,1),\pi_{\Lambda_1}^{SO(2m)}\big)\boxtimes V_\nu^{SO(2n)}.$$

The branching rules for $res_{SO(2m)}(H^2(SO(2m,1),\pi_{\Lambda_1}^{SO(2m)}))$ are found in [29] and other references, while the branching rule for $res_{SO(2n)}(\pi_{\Lambda_2}^{SO(2n+1)})$ can be found in [29]. From both computations, we deduce [9, Proposition 3], for $\lambda = \sum_{1\le i\le m}\lambda_i\epsilon_i+\sum_{1\le j\le n}\lambda_{m+j}\delta_j$; then V_μ^H is a H-subrepresentation of $H^2(G,\tau)\equiv V_\lambda^{SO(2m,2n+1)}$ $(\mu=\sum_{1\le i\le m}\mu_i\epsilon_i+\sum_{1\le j\le n}\mu_{j+m}\delta_j)$ if and only if

$$\mu_1>\lambda_1>\cdots>\mu_m>\lambda_m,\ \lambda_{m+1}>\mu_{m+1}>\ldots\lambda_{m+n}>|\mu_{m+n}|.$$

5.3 Existence of Discrete Series Whose Lowest *K*-Type Restricted to $K_1(\Psi)$ Is Irreducible

Let G a semisimple Lie group that admits square integrable representations. This hypothesis allows to fix a compact Cartan subgroup $T\subset K$ of G. Duflo and Vargas [5] defined for each system of positive roots $\Psi\subset\Phi(\mathfrak{g},\mathfrak{t})$ a normal subgroup $K_1(\Psi)\subset K$ so that for a symmetric pair (G,H), with H a θ-invariant subgroup, the following holds: for any Harish-Chandra parameter dominant with respect to Ψ, the representation $res_H(\pi_\lambda)$ is H-admissible if and only if $K_1(\Psi)$ is a subgroup of H. For a holomorphic system Ψ, $K_1(\Psi)$ is equal to the center of K; for a quaternionic system of positive roots, $K_1(\Psi)\equiv SU_2(\alpha_{max})$. Either for the holomorphic family or for a quaternionic real forms we find that among the H-admissible Discrete Series for G, there are many examples of the following nature: the lowest K-type of π_λ is equal to a irreducible representation of $K_1(\Psi)$ tensor with the trivial representation for K_2, [9]. To follow, under the general setting at the beginning of this paragraph, we verify.

For each system of positive roots $\Psi\subset\Phi(\mathfrak{g},\mathfrak{t})$, there exist Discrete Series with Harish-Chandra parameter dominant with respect to Ψ and so that its lowest K-type is equal to a irreducible representation of $K_1(\Psi)$ tensor with the trivial representation for $K_2(\Psi)$.

We may assume $K_1(\Psi)$ is a proper subgroup of K. Then, when $K_1(\Psi)=Z_K$, Harish-Chandra showed there exists such a representation. For G a quaternionic real form, Ψ a quaternionic system of positive roots, $K_1(\Psi)=SU_2(\alpha_{max})$, then, in [9], we find a proof of the statement. From tables in [5, 30], we are left to consider the triples $(G,K,K_1(\Psi))$ so that their respective Lie algebras are in the triples

$(\mathfrak{su}(m,n), \mathfrak{su}(m)+\mathfrak{su}(n)+\mathfrak{u}(1), \mathfrak{su}(m)), m>2,$
$(\mathfrak{sp}(m,n), \mathfrak{sp}(m)+\mathfrak{sp}(n), \mathfrak{sp}(m)).$
$(\mathfrak{so}(2m,n), \mathfrak{so}(2m)+\mathfrak{so}(n), \mathfrak{so}(2m)).$

We analyze the second triple of the list. To follow, G is so that its Lie algebra is $\mathfrak{sp}(m,n)$, $n \geq 2, m > 1$. We want to produce Discrete Series representations so that the lowest K-type restricted to $K_1(\Psi)$ is still irreducible. Here, $\mathfrak{k} = \mathfrak{sp}(m)+\mathfrak{sp}(n)$. We fix maximal torus $T \subset K$ and describe the root system as in [30]. For the system of positive roots $\Psi := \{\epsilon_i \pm \epsilon_j, i<j, \delta_r \pm \delta_s, r<s, \epsilon_a \pm \delta_b, 2\epsilon_a, 2\delta_b, 1 \leq a,i,j \leq m, 1 \leq b,r,s \leq n\}$, we have $K_1(\Psi) = K_1 \equiv Sp(m)$, $K_2(\Psi) = K_2 \equiv Sp(n)$. Obviously, there exists a system of positive roots $\tilde{\Psi}$ so that $K_1(\tilde{\Psi}) \equiv Sp(n)$, $K_2(\tilde{\Psi}) \equiv Sp(m)$. For any other system of positive roots in $\Phi(\mathfrak{g},\mathfrak{t})$, we have that the associated subgroup K_1 is equal to K. It readily follows that $\lambda := \sum_{1\leq j\leq m} a_j\epsilon_j + \rho_{K_2}$ is a Ψ-dominant Harish-Chandra parameter when the coefficients a_j are all integers so that $a_1 > \cdots > a_m >> 0$. Since ρ_n^λ belongs to $span_{\mathbb{C}}\{\epsilon_1,\ldots,\epsilon_m\}$, it follows that the lowest K type of π_λ is equivalent to a irreducible representation for $Sp(m)$ times the trivial representation for $Sp(n)$. With the same proof, it is verified that the statement holds for the third triple. For the first triple, we further assume $G = SU(p,q)$. Thus, K is the product of two simply connected subgroups times a one-dimensional torus Z_K; we notice $\rho_n^{\Psi_a} = \rho_{\mathfrak{g}}^\lambda - \rho_K$; hence, $\rho_n^{\Psi_a}$ lifts to a character of K. Thus, as in the case $\mathfrak{sp}(m,n)$, we obtain π_λ with λ dominant with respect to Ψ_a so that its lowest $K = SU(p)SU(q)Z_K$-type is the tensor product of an irreducible representation for $SU(p)Z_K$ times the trivial representation for $SU(q)$. Since $\rho_n^{\Psi_a}$ lifts to a character of K, after some computation, the claim follows.

6 Symmetry Breaking Operators and Normal Derivatives

For this subsection, (G,H) is a symmetric pair, and π_λ is a square integrable representation. Our aim is to generalize a result in [22, Theorem 5.1]. Kobayashi and Pevzner [20] considered symmetry breaking operators expressed by means of normal derivatives. They obtain results for holomorphic embedding of a rank-one symmetric pairs. As before, $H_0 = G^{\sigma\theta}$ is the associated subgroup. We recall $\mathfrak{h}\cap\mathfrak{p}$ is orthogonal to $\mathfrak{h}_0\cap\mathfrak{p}$ and that $\mathfrak{h}\cap\mathfrak{p} \equiv T_{eL}(H/L)$, and $\mathfrak{h}_0\cap\mathfrak{p} \equiv T_{eL}(H_0/L)$. Hence, for $X \in \mathfrak{h}_0\cap\mathfrak{p}$, more generally for $X \in \mathcal{U}(\mathfrak{h}_0)$, we say R_X is a normal derivative to H/L differential operator, in short, *normal derivative*. Other ingredients necessary for the next proposition are subspaces $\mathcal{L}_\lambda$ and $\mathcal{U}(\mathfrak{h}_0)W$. The latter subspace is contained in the subspace of K-finite vectors, whereas according to a well-known conjecture, whenever $res_H(\pi_\lambda)$ is not discretely decomposable, the former subspace is disjoint to the subspace of G-smooth vectors. When $res_H(\pi_\lambda)$ is H-admissible, $\mathcal{L}_\lambda$ is contained in the subspace of K-finite vectors. However, it might not be equal to $\mathcal{U}(\mathfrak{h}_0)W$ as we have pointed out. The next proposition and its converse dealt with consequences of the equality $\mathcal{L}_\lambda = \mathcal{U}(\mathfrak{h}_0)W$.

Proposition 5 *We assume (G, H) is a symmetric pair. We also assume there exists a irreducible representation (σ, Z) of L so that $H^2(H, \sigma)$ is a irreducible factor of $H^2(G, \tau)$ and $H^2(G, \tau)[H^2(H, \sigma)][Z] = \mathcal{L}_\lambda[Z] = \mathcal{U}(\mathfrak{h}_0)W[Z] = L_{\mathcal{U}(\mathfrak{h}_0)}(H^2(G, \tau)[W])[Z]$. Then, $res_H(\pi_\lambda)$ is H-admissible. Moreover, any symmetry breaking operator from $H^2(G, \tau)$ into $H^2(H, \sigma)$ is represented by a normal derivative differential operator.*

We show a converse to Proposition 5 in 6.1.

Proof We begin by recalling that $H^2(G, \tau)[W] = \{K_\lambda(\cdot, e)^\star w, w \in W\}$ is a subspace of $H^2(G, \tau)_{K-fin}$; hence, $L_{\mathcal{U}(\mathfrak{h}_0)}(H^2(G, \tau)[W])[Z]$ is a subspace of $H^2(G, \tau)_{K-fin}$. Owing to our hypothesis, we then have $\mathcal{L}_\lambda[Z]$ as a subspace of $H^2(G, \tau)_{K-fin}$. Next, we quote a result of Harish-Chandra: a $\mathcal{U}(\mathfrak{h})-$finitely generated, $\mathfrak{z}(\mathcal{U}(\mathfrak{h}))-$finite, module has a finite composition series. Thus, $H^2(G, \tau)_{K-fin}$ contains an irreducible $(\mathfrak{h}, L)$-submodule. For a proof (cf. [32, Corollary 3.4.7 and Theorem 4.2.1]). Now, in [15, Lemma 1.5], we find a proof of the following: if a $(\mathfrak{g}, K)-$module contains an irreducible $(\mathfrak{h}, L)-$submodule, then the $(\mathfrak{g}, K)-$module is $\mathfrak{h}-$algebraically decomposable. Thus, $res_H(\pi_\lambda)$ is algebraically discretely decomposable. In [16, Theorem 4.2], it is shown that under the hypothesis (G, H) is a symmetric pair, for Discrete Series, $\mathfrak{h}$-algebraically discrete decomposable is equivalent to H-admissibility; hence, $res_H(\pi_\lambda)$ is H-admissible. Let $S : H^2(G, \tau) \to H^2(H, \sigma) = V_\mu^H$ be a continuous intertwining linear map. Then, we have shown in 2, for $z \in Z$, $K_S(\cdot, e)^\star z \in H^2(G, \tau)[V_\mu^H][Z]$. We fix an orthonormal basis $\{z_p\}$, $p = 1, \ldots, \dim Z$ for Z. The hypothesis

$$H^2(G, \tau)[V_\mu^H][Z] = L_{\mathcal{U}(\mathfrak{h}_0)}(H^2(G, \tau)[W])[Z]$$

implies that for each p, there exist $D_p \in \mathcal{U}(\mathfrak{h}_0)$ and $w_p \in W$ so that $K_S(\cdot, e)^\star z_p = L_{D_p} K_\lambda(\cdot, e)^\star w_p$. Next, we fix $f_1 \in H^2(G, \tau)^\infty$, $h \in H$ and set $f := L_{h^{-1}}(f_1)$, and then $f(e) = f_1(h)$, and we recall $D^\star$ (resp. $\check{D}$) is the formal adjoint of $D \in \mathcal{U}(\mathfrak{g})$, (resp. is the image under the anti homomorphism of $\mathcal{U}(\mathfrak{g})$ that extends minus the identity of $\mathfrak{g}$). We have,

$$\begin{aligned} (S(f)(e), z_p)_Z &= \int_G (f(y), K_S(y, e)^\star z_p)_W dy \\ &= \int_G (L_{D_p^\star} f(y), K_\lambda(y, e)^\star w_p)_W dy \\ &= (L_{D_p^\star} f(e), w_p)_W \\ &= (R_{\check{D}_p^\star} f(e), w_p)_W. \end{aligned} \tag{7}$$

Thus, for each $z \in Z$ and f_1 smooth vector, we obtain

$$(S(f_1)(h), z)_Z = \sum_p (S(f_1)(h), (z, z_p)_Z z_p)_Z = \sum_p (R_{\check{D}_p^\star} f_1(h), w_p)_W (z_p, z)_Z.$$

As in [25, Proof of Lemma 2], we conclude that for any $f \in H^2(G,\tau)$

$$S(f)(h) = \sum_{1\leq p\leq \dim Z} (R_{\check{D}_p^\star} f(h), w_p)_W\ z_p. \tag{8}$$

Since $D_p \in \mathcal{U}(\mathfrak{h}_0)$, such an expression of $S(f)$ is a representation in terms of normal derivatives. □

6.1 Converse to Proposition 5

We want to show that if every element in $Hom_H(H^2(G,\tau), H^2(H,\sigma))$ has an expression as differential operator by means of "normal derivatives," then the equality $\mathcal{L}_\lambda[Z] = H^2(G,\tau)[H^2(H,\sigma)][Z] = \mathcal{U}(\mathfrak{h}_0)W[Z]$ holds.

In fact, the hypothesis $S(f)(h) = \sum_{1\leq p\leq \dim Z}(R_{\check{D}_p^\star} f(h), w_p)_W\ z_p$, $D_p \in \mathcal{U}(\mathfrak{h}_0)$, yields $K_S(\cdot,e)^\star z = L_{D_z} K_\lambda(\cdot,e)^\star w_z$, $D_z \in \mathcal{U}(\mathfrak{h}_0)$, $w_z \in W$. The fact that (σ, Z) has multiplicity one in $H^2(H,\sigma)$ gives

$$\dim Hom_H(H^2(G,\tau), H^2(H,\sigma)) = \dim Hom_L(Z, H^2(G,\tau)[H^2(H,\sigma)][Z]).$$

Hence, functions

$$\{K_S(\cdot,e)^\star z, z \in Z, S \in Hom_H(H^2(G,\tau), H^2(H,\sigma))\}$$

span $H^2(G,\tau)[H^2(H,\sigma)][Z]$. Therefore, $H^2(G,\tau)[H^2(H,\sigma)][Z]$ is contained in $\mathcal{U}(\mathfrak{h}_0)W[Z] = L_{\mathcal{U}(\mathfrak{h}_0)}H^2(G,\tau)[W][Z]$. Owing to Theorem 1, both spaces have the same dimension; hence, the equality holds.

The pairs so that Proposition 5[5] holds for scalar holomorphic Discrete Series are $(\mathfrak{su}(m,n), \mathfrak{su}(m,l)+\mathfrak{su}(n-l)+\mathfrak{u}(1))$, $(\mathfrak{so}(2m,2), \mathfrak{u}(m,1))$, $(\mathfrak{so}^\star(2n), \mathfrak{u}(1,n-1))$, $(\mathfrak{so}^\star(2n), \mathfrak{so}(2)+\mathfrak{so}^\star(2n-2))$, $(\mathfrak{e}_{6(-14)}, \mathfrak{so}(2,8)+\mathfrak{so}(2))$. See [30, (4.6)].

6.2 Comments on the Interplay Among Subspaces, $\mathcal{L}_\lambda$, $\mathcal{U}(\mathfrak{h}_0)W$, $H^2(G,\tau)_{K-fin}$, and Symmetry Breaking Operators

It readily follows that the subspace $\mathcal{L}_\lambda[Z] = V_\lambda^G[H^2(H,\sigma)][Z]$ is equal to the closure of the linear span of

$\mathcal{K}_{Sy}(G,H) := \{K_{S^\star}(e,\cdot)z = K_S(\cdot,e)^\star z, z \in Z, S \in Hom_H(V_\lambda^G, V_\mu^H)\}$.

[5] In the work in progress, we have shown the proposition holds for $(\mathfrak{sp}(m,1), \mathfrak{sp}(m-1,1)+\mathfrak{sp}(1))$ and a quaternionic representation.

(1) $H^2(G,\tau)_{K-fin} \cap \mathcal{L}_\lambda[Z]$ is equal to the linear span of the elements in $\mathcal{K}_{Sy}(G,H)$ so that the corresponding symmetry breaking operator is represented by a differential operator. See [25, Lemma 4.2].
(2) $\mathcal{U}(\mathfrak{h}_0)W \cap \mathcal{L}_\lambda[Z]$ is equal to the linear span corresponding to elements K_S in $\mathcal{K}_{Sy}(G,H)$ so that S is represented by a normal derivative differential operator. This is shown in Proposition 5 and its converse.
(3) The set of symmetry breaking operators represented by a differential operator is not the null space if and only if $res_H(\pi_\lambda)$ is H-discretely decomposable. See [25, Theorem 4.3] and the proof of Proposition 5.
(4) We believe that from Nakahama's thesis, it is possible to construct examples of $V_\lambda^G[H^2(H,\sigma)][Z] \cap \mathcal{U}(\mathfrak{h}_0)W[Z] \neq \{0\}$, so that the equality $V_\lambda^G[H^2(H,\sigma)][Z] = \mathcal{U}(\mathfrak{h}_0)W[Z]$ does not hold! That is, there are symmetry breaking operators represented by plain differential operators and some of the operators are not represented by normal derivative operators.

6.3 A Functional Equation for Symmetry Breaking Operators

Notation is as in Theorem 1. We assume (G,H) is a symmetric pair and $res_H(\pi_\lambda)$ is admissible. The objects involved in the equation are the following: $H_0 = G^{\sigma\theta}$, $Z = V^L_{\mu+\rho_n^H}$ the lowest L-type for V_μ^H, $\mathcal{L}_\lambda = \sum_\mu H^2(G,\tau)[V_\mu^H][V^L_{\mu+\rho_n^H}]$, $\mathcal{U}(\mathfrak{h}_0)W = L_{\mathcal{U}(\mathfrak{h}_0)}H^2(G,\tau)[W]$, L-isomorphism $D : \mathcal{L}_\lambda[Z] \to \mathcal{U}(\mathfrak{h}_0)W[Z]$, a H-equivariant continuous linear map $S : H^2(G,\tau) \to H^2(H,\sigma)$, the kernel $K_S : G \times H \to Hom_{\mathbb{C}}(W,Z)$ corresponding to S, 2 implies $K_S(\cdot,e)^\star z \in \mathcal{L}_\lambda[Z]$, finally, we recall $K_\lambda : G \times G \to Hom_{\mathbb{C}}(W,W)$ the kernel associated with the orthogonal projector onto $H^2(G,\tau)$. Then,

Proposition 6 *For $z \in Z$, $y \in G$ we have*

$$D(K_S(e,\cdot)^\star(z))(y) = \frac{1}{c}\int_{H_0} K_\lambda(h_0,y) D(K_S(e,\cdot)^\star(z))(h_0) dh_0.$$

Here, $c = d(\pi_\lambda)\dim W/d(\pi_{\eta_0}^{H_0})$. When D is the identity map, the functional equation turns into

$$K_S(x,h) = \frac{1}{c}\int_{H_0} K_S(h_0,e) K_\lambda(x,hh_0) dh_0$$

The functional equation follows from Proposition 2 applied to $T := S^\star$. The second equation follows after we compute the adjoint of the first equation.

We note that as in the case of holographic operators, a symmetry breaking operator can be recovered from its restriction to H_0.

We also note that [22] has shown a different functional equation for K_S for scalar holomorphic Discrete Series and holomorphic embedding $H/L \to G/K$.

7 Tables

For an arbitrary symmetric pair (G, H), whenever π_λ^G is an admissible representation of H, we define,

$$K_1 = \begin{cases} Z_K & \text{if } \Psi_\lambda \text{ holomorphic} \\ K_1(\Psi_\lambda) & \text{otherwise} \end{cases}$$

In the next tables, we present the 5-tuple satisfying the following: (G, H) is a symmetric pair, H_0 is the associated group to H, Ψ_λ is a system of positive roots such that π_λ^G is an admissible representation of H, and $K_1 = Z_1(\Psi_\lambda)K_1(\Psi_\lambda)$. Actually, instead of writing Lie groups, we write their respective Lie algebras. Each table is in part a reproduction of tables in [18, 30]. The tables can also be computed by means of techniques presented in [5]. Note that each table is "symmetric" when we replace H by H_0. As usual, α_{max} denotes the highest root in Ψ_λ. Unexplained notation is as in [30].

8 Partial List of Symbols and Definitions

- (τ, W), (σ, Z), $L^2(G \times_\tau W)$, $L^2(H \times_\sigma Z)$ (cf. Sect. 2).
- $H^2(G, \tau) = V_\lambda = V_\lambda^G$, $H^2(H, \sigma) = V_\mu^H, \pi_\mu^H, \pi_\nu^K$. (cf. Sect. 2).
- $\pi_\lambda = \pi_\lambda^G$, $d_\lambda = d(\pi_\lambda)$ formal degree of π_λ, P_λ, P_μ, K_λ, K_μ, (cf. Sect. 2).
- P_X orthogonal projector onto subspace X.
- $\Phi(x) = P_W\pi(x)P_W$ spherical function attached to the lowest K-type W of π_λ.
- $K_\lambda(y, x) = d(\pi_\lambda)\Phi(x^{-1}y)$.
- $M_{K-fin}(resp. M^\infty)$ $K-$finite vectors in M (resp. smooth vectors in M).
- dg, dh Haar measures on G and H.
- A unitary representation is *square integrable*, equivalently a *Discrete Series* representation, (resp. *integrable*) if some nonzero matrix coefficient is square integrable (resp. integrable) with respect to Haar measure on the group in question.
- $\Theta_{\pi_\mu^H}(...)$ Harish-Chandra character of the representation π_μ^H.
- For a module M and a simple submodule N, $M[N]$ denotes the *isotypic component* of N in M. That is, $M[N]$ is the sum of all irreducible submodules isomorphic to N. If topology is involved, we define $M[N]$ to be the closure of $M[N]$.

- M_{H-disc} is the closure of the linear subspace spanned by the totality of $H-$irreducible submodules. $M_{disc} := M_{G-disc}$
- A representation M is $H-$*discretely decomposable* if $M_{H-disc} = M$.
- A representation is $H-$*admissible* if it is $H-$discretely decomposable, and each isotypic component is equal to a finite sum of $H-$irreducible representations.
- $\mathcal{U}(\mathfrak{g})$ (resp. $\mathfrak{z}(\mathcal{U}(\mathfrak{g})) = \mathfrak{z}_{\mathfrak{g}}$) universal enveloping algebra of the Lie algebra $\mathfrak{g}$(resp. center of universal enveloping algebra).
- $\mathrm{Cl}(X)$ =closure of the set X.
- I_X identity function on set X.
- $\mathbb{T}$ one-dimensional torus.
- Z_S identity connected component of the center of the group S.
  $S^{(r)}(V)$ the rth-symmetric power of the vector space V.

Acknowledgments The authors would like to thank T. Kobayashi for much insight and inspiration on the problems considered here. Also, we thank Michel Duflo, Birgit Speh, Yosihiki Oshima, and Jan Frahm for conversations on the subject. Part of the research in this chapter was carried out within the online research community on Representation Theory and Noncommutative Geometry sponsored by the American Institute of Mathematics. Also, some of the results in this note were the subject of a talk in the "Conference in Honor of Prof. Toshiyuki Kobayashi" to celebrate his 60th birthday. The authors thank the organizers for the facilities to present and participate in such a wonderful meeting via Zoom. Finally, we thank the referees for their true expert and careful advice in improving the chapter.

References

1. Atiyah, M., Schmid, W.: A geometric construction of the discrete series for semisimple Lie groups. Invent. Math. **42**, 1–62 (1977)
2. Baldoni Silva, M., Vergne, M.: Discrete series representations and K multiplicities for $U(p, q)$. User's guide, arXiv:1008.5360 (2010)
3. Duflo, M., Heckman, G., Vergne, M.: Projection d'orbites, formule de Kirillov et formule de Blattner. In: Harmonic Analysis on Lie Groups and Symmetric Spaces (Kleebach, 1983). Mémoires de la Société Mathématique de France (N.S.), vol. 15 (1984), pp. 65–128
4. Duflo, M., Galina, E., Vargas, J.: Square integrable representations of reductive Lie groups with admissible restriction to $SL_2(R)$. J. Lie Theory **27**(4), 1033–1056 (2017)
5. Duflo, M., Vargas, J.: Branching laws for square integrable representations. Proc. Japan Acad. Ser. A Math. Sci. **86**(3), 49–54 (2010)
6. Enright, Thomas J., Wallach, Nolan R.: The fundamental series of representations of a real semisimple Lie algebra. Acta Math. **140**, 1–32 (1978)
7. Flensted-Jensen, M.: Discrete series for semisimple symmetric spaces. Ann. Math. (2) **111**(2), 253–311 (1980)
8. Gross, B., Wallach, N.: On quaternionic discrete series representations, and their continuation. J. Reine Angew **481**, 73–123 (1996)
9. Gross, B., Wallach, N.: Restriction of small discrete series representations to symmetric subgroups. The mathematical legacy of Harish-Chandra (Baltimore, MD, 1998), Proceedings of Symposia in Pure Mathematics, vol. 68 (2000), pp. 255–272
10. Harris, B., He, H., Olafsson, G.: The continuous spectrum in Discrete Series branching laws. Int. J. Math. **24**(7), Article ID 1350049, 29 p. (2013)

11. Hecht, H., Schmid, W.: A proof of Blattner's conjecture. Invent. Math. **31**, 129–154 (1976). https://doi.org/10.1007/BF01404112
12. Heckman, G.: Projections of orbits and asymptotic behavior of multiplicities for compact connected lie groups. Invent. Math. **67**, 333–356 (1982)
13. Hotta, R., Parthasarathy, R.: Multiplicity formulae for discrete series. Lnvent. Math. **26**, 133–178 (1974)
14. Jakobsen, H., Vergne M.: Restrictions and expansions of holomorphic representations. J. Funct. Anal. **34**, 29–53 (1979)
15. Kobayashi, T.: Discrete decomposability of the restriction of $A_{\mathfrak{q}}(\lambda)$ with respect to reductive subgroups and its applications. Invent. Math. **117**, 181–205 (1994)
16. Kobayashi, T.: Discrete decomposability of the restriction of $A_{\mathfrak{q}}(\lambda)$ with respect to reductive subgroups, III. Restriction of Harish-Chandra modules and associated varieties. Invent. Math. **131**, 229–256 (1998)
17. Kobayashi, T.: Multiplicity-free theorems of the restrictions of unitary highest weight modules with respect to reductive symmetric pairs. In: Representation Theory and Automorphic Forms, vol. 255. Progress in Mathematics. Birkhäuser, Basel (2007), pp. 45–109
18. Kobayashi, T., Oshima, Y.: Classification of discretely decomposable $A_{\mathfrak{q}}(\lambda)$ with respect to reductive symmetric pairs. Adv. Math. **231**(3-4), 2013–2047 (2012)
19. Kobayashi, T., Oshima, Y.: Classification of symmetric pairs with discretely decomposable restriction of $(\mathfrak{g}, K)$-modules. J. für die reine und angewandte Mathematik (Crelle's Journal) **703**, 201–223 (2015)
20. Kobayashi, T., Pevzner, M.: Differential symmetry breaking operators. II: Rankin-Cohen operators for symmetric pairs. Sel. Math. New Ser. **22**(2), 847–911 (2016)
21. Labriet, Q.: A geometrical point of view for branching problems for holomorphic discrete series of conformal Lie groups. Int. J. Math. **33**(10–11) (2022). https://doi.org/10.1142/S0129167X22500690
22. Nakahama, R.: Construction of intertwining operators between holomorphic discrete series representations. SIGMA **15**, 101–202 (2019)
23. Ørsted, B., Speh, B.: Branching laws for some unitary representations of SL(4, R), symmetry, integrability and geometry: methods and applications. SIGMA **4**, 017–036 (2008)
24. Ørsted, B., Vargas, J.: Restriction of Discrete Series representations (Discrete spectrum). Duke Math. J. **123**, 609–631 (2004)
25. Ørsted, B., Vargas, J.: Branching problems in reproducing kernel spaces. Duke Math. J. **169**(18), 3477–3537 (2020)
26. Ørsted, B., Wolf, J.: Geometry of the Borel-de Siebenthal discrete series. J. Lie Theory. **20**, 175–212 (2010)
27. Schmid, W.: Some properties of square-integrable representations of semisimple lie groups. Ann. Math. 2nd Ser. **102**, 535–564 (1975)
28. Sekiguchi, H.: Branching rules of singular unitary representations with respect to symmetric pairs (A_{2n-1}, d_n). Int. J. Math. **24**(4), Article ID 1350011, 25 p. (2013)
29. Thieleker, E.: On the quasisimple irreducible representations of the Lorentz groups. Trans. Am. Math. Soc. **179**, 465–505 (1973)
30. Vargas, J.: Associated symmetric pair and multiplicities of admissible restriction of discrete series. Int. J. Math. **27**, 12 (2016). https://doi.org/10.1142/S0129167X16501007
31. Vogan, D.: The algebraic structure of the representations of semisimple Lie groups I. Ann. Math. **109**, 1–60 (1979)
32. Wallach, N.: Real Reductive Groups I. Academic Press, New York (1988)

Integral Transformations of Hypergeometric Functions with Several Variables

Toshio Oshima

Abstract As a generalization of Riemann-Liouville integral, we introduce integral transformations of convergent power series, which can be applied to hypergeometric functions with several variables.

1 Introduction

Suppose a function $\phi(x)$ satisfies a linear ordinary differential equation on $\mathbb{P}^1$. Then, the Riemann-Liouville integral (1) of $\phi(x)$ induces a middle convolution of the differential equation defined by Katz [Ka]. The multiplication of $\phi(x)$ by a simple function $(x-c)^\lambda$ induces an addition of the differential equation, which is also important. For example, any rigid irreducible linear Fuchsian differential equation is constructed by successive applications of middle convolutions and additions from the trivial equation $u' = 0$. Hence, we have an integral representation of its solution, which is shown first by Katz [Ka] in the case of Fuchsian systems of the first order and by the author [O1] in the case of single differential equations of higher orders. Here the equation is called rigid if it is free from accessory parameters, namely, the equation is globally determined by the local structure at the singular points. Applying these transformations to linear ordinary differential equations on $\mathbb{P}^1$, we study many fundamental problems on their solutions in [O1].

The rigid Fuchsian ordinary differential equation on $\mathbb{P}^1$ can be extended to a Knizhnik-Zamolodchikov-type equation (KZ equation in short, cf. [KZ]) regarding the singular points as new variables. Haraoka [Ha] shows this by extending middle convolutions on KZ equations, and its generalization for equations with irregular singularities is given by the author [O4, O5]. Then these transformations are also useful to hypergeometric functions with several variables including Appell's hypergeometric functions (cf. [O3]).

T. Oshima (✉)
Josai University, Chiyodaku, Tokyo, Japan
e-mail: oshima@ms.u-tokyo.ac.jp

M. Pevzner, H. Sekiguchi (eds.), *Symmetry in Geometry and Analysis, Volume 2*,
Progress in Mathematics 358, https://doi.org/10.1007/978-981-97-7662-7_12

These transformations do not give an integral representation of Appell's hypergeometric series F_4 but K. Aomoto gives an integral representation of F_4, which is written in [O1, §13.10.2]. We define integral transformations on the space of convergent power series of several variables in Sect. 2 and study hypergeometric functions with several variables, which extend a brief study of Appell's hypergeometric functions in [O1, §13.10]. The transformations are invertible, and they are generalizations of Riemann-Liouville integrals of functions with a single variable. On the space of hypergeometric functions with several variables, we have important transformations such as the multiplications of suitable functions and coordinate transformations. Note that a holonomic Fuchsian differential equation of several variables has solutions of convergent power series times simple functions $x_1^{\lambda_1}\cdots x_n^{\lambda_n}$ at its normally crossing singular points (cf. [KO]). We study a combination of the integral transformations and multiplications by these simple functions.

In Sect. 3, we show that the transformations defined in Sect. 2 give integral representations of Appell's or Lauricella's hypergeometric series and certain Horn's hypergeometric series with irregular singularities.

In Sect. 4, we show that our study gives a result related to the connection problem on the solutions, which will be discussed in [MO] and related to the study by Matsubara [Ma, §3].

In Sect. 5, a combination of the integral transformations with coordinate transformations defined by products of powers of coordinate functions parametrized by $GL(n,\mathbb{Z})$. The transformations given in Sect. 5 are related to A-hypergeometric series introduced by Gel'fand, Kapranov, and Zelevinsky [GKZ]. The transformation

$$\sum_{m=0}^{\infty}\sum_{n=0}^{\infty} c_{m,n}x^m y^n \mapsto \sum_{m=0}^{\infty}\sum_{n=0}^{\infty} c_{m,n}\frac{(\alpha)_{p_1m+q_1n}(\beta)_{p_2m+q_2n}}{(\gamma)_{(p_1+p_2)m+(q_1+q_2)n}}x^m y^n$$

of convergent power series is an example. Here, p_1, p_2, q_1, and q_2 are non-negative integers with $\begin{pmatrix} p_1 & p_2 \\ q_1 & q_2 \end{pmatrix} \in GL(2,\mathbb{Z})$, and we put $(a)_k = a(a+1)\cdots(a+k-1)$.

In Sect. 6, we study the transformation of differential equations corresponding to the transformations of their solutions.

In Sect. 7, we study the transformation given in Sect. 5, which keeps the space of KZ equations

$$\mathscr{M}: \begin{cases} \dfrac{\partial u}{\partial x} = \dfrac{A_{01}}{x-y}u + \dfrac{A_{02}}{x-1}u + \dfrac{A_{03}}{x}u, \\ \dfrac{\partial u}{\partial y} = \dfrac{A_{01}}{y-x}u + \dfrac{A_{12}}{y-1}u + \dfrac{A_{13}}{y}u \end{cases}$$

and give the induced transformations of the residue matrices $A_{i,j}$ defining the equations. The transformation is reduced to a coordinate transformation corresponding to the coordinate symmetries described in [O3, §6] and a middle convolution of the

KZ equation. Hence, we apply the result in [Ha, O3] to them. The hypergeometric series

$$\sum_{m=0}^{\infty}\sum_{n=0}^{\infty}\frac{\prod_{i=1}^{p}(\alpha_i)_m\prod_{j=1}^{q}(\beta_j)_n\prod_{k=1}^{r}(\gamma_k)_{m+n}}{\prod_{i=1}^{p}(1-\alpha_i')_m\prod_{j=1}^{q}(1-\beta_j')_n\prod_{k=1}^{r}(1-\gamma_k')_{m+n}}x^m y^n$$

with $\alpha_1' = \beta_1' = 0$ is a typical example satisfying a KZ equation, which is a generalization of Appell's F_1.

To the KZ equation, we show Theorem 3, which gives an interesting correspondence between *simple* solutions along a line (cf. Definition 6) and *simple* solutions at a singular point where three singular lines meet.

In Sect. 8, we restrict our transformations in Sect. 7 to certain ordinary differential equations of Shlesinger canonical form. The transformation may be interesting since it may change the index of rigidity defined by [Ka].

Several applications of the results in this chapter will be given in other papers.

2 Integral Transformations

The Riemann-Liouville transform $I_c^\mu\phi$ of a function $\phi(x)$ is defined by

$$(I_{x,c}^\mu\phi)(x) = (I_c^\mu\phi)(x) := \frac{1}{\Gamma(\mu)}\int_c^x \phi(t)(x-t)^{\mu-1}\mathrm{d}t. \tag{1}$$

Here c is usually a singular point of an integrable function $\phi(x)$. Since

$$\begin{aligned}\int_0^x \phi(t)(x-t)^{\mu-1}\mathrm{d}t &= \int_0^1 \phi(xs)(x-xs)^{\mu-1}x\mathrm{d}s \qquad (t=xs)\\ &= x^\mu\int_0^1\phi(sx)(1-s)^{\mu-1}\mathrm{d}s,\end{aligned}$$

the transformation

$$(K_x^\mu\phi)(x) := \frac{1}{\Gamma(\mu)}\int_0^1 \phi(tx)(1-t)^{\mu-1}\mathrm{d}t \tag{2}$$

of a function $\phi(x)$ satisfies

$$K_x^\mu = x^{-\mu}I_0^\mu, \tag{3}$$

$$K_x^\mu x^\alpha = \frac{\Gamma(\alpha+1)}{\Gamma(\alpha+\mu+1)}x^\alpha. \tag{4}$$

Definition 1 We extend the integral transform K_x^μ to a function $\phi(x)$ of several variables $x = (x_1, \dots, x_n)$ by

$$K_x^\mu \phi(x) := \frac{1}{\Gamma(\mu)} \int_{\substack{t_1>0,\dots,t_n>0 \\ t_1+\dots+t_n<1}} (1 - t_1 - \dots - t_n)^{\mu-1} \phi(t_1 x_1, \dots, t_n x_n) \mathrm{d}t_1 \cdots \mathrm{d}t_n.$$

Note that

$$\begin{aligned}
\int_0^{1-s} t^\alpha (1 - s - t)^{\mu-1} \mathrm{d}t &= \int_0^{1-s} t^\alpha (1-s)^{\mu-1} (1 - \tfrac{t}{1-s})^{\mu-1} \mathrm{d}t \\
&= (1-s)^{\alpha+\mu} \int_0^1 t^\alpha (1-t)^{\mu-1} \mathrm{d}t \\
&= \frac{\Gamma(\alpha+1)\Gamma(\mu)}{\Gamma(\alpha+\mu+1)} (1-s)^{\alpha+\mu}
\end{aligned}$$

and hence

$$\begin{aligned}
&\int_{\substack{t_1>0,\dots,t_n>0 \\ t_1+\dots+t_n<1}} t^{\alpha_1} \cdots t^{\alpha_n} (1 - t_1 - \dots - t_n)^{\mu-1} \mathrm{d}t \\
&= \int_0^1 t_1^{\alpha_1} \mathrm{d}t_1 \int_0^{1-t_1} t_2^{\alpha_2} \mathrm{d}t_2 \cdots \int_0^{1-t_1-\dots-t_{n-1}} t_n^{\alpha_n} (1 - t_1 - \dots - t_n)^{\mu-1} \mathrm{d}t_n \\
&= \frac{\Gamma(\mu)\Gamma(\alpha_n+1)}{\Gamma(\alpha_n+\mu+1)} \int_0^1 t_1^{\alpha_1} \mathrm{d}t_1 \cdots \int_0^{1-t_1-\dots-t_{n-2}} t_{n-1}^{\alpha_{n-1}} \\
&\quad \times (1 - t_1 - \dots - t_{n-1})^{\alpha_n+\mu} \mathrm{d}t_{n-1} \\
&= \frac{\Gamma(\mu)\Gamma(\alpha_n+1)}{\Gamma(\alpha_n+\mu+1)} \times \frac{\Gamma(\alpha_n+\mu+1)\Gamma(\alpha_{n-1}+1)}{\Gamma(\alpha_{n-1}+\alpha_n+\mu+2)} \times \cdots \\
&\quad \cdots \times \frac{\Gamma(\alpha_2+\dots+\alpha_n+\mu+n-1)\Gamma(\alpha_1+1)}{\Gamma(\alpha_1+\dots+\alpha_n+\mu+n)} \\
&= \frac{\Gamma(\mu)\Gamma(\alpha_1+1)\cdots\Gamma(\alpha_n+1)}{\Gamma(\alpha_1+\dots+\alpha_n+\mu+n)}.
\end{aligned}$$

Therefore we have

$$K_x^\mu x^{\boldsymbol{\alpha}} = \frac{\Gamma(\boldsymbol{\alpha}+1)}{\Gamma(|\boldsymbol{\alpha}+1|+\mu)} x^{\boldsymbol{\alpha}} \tag{5}$$

and

$$
\begin{aligned}
&K_x^{\mu}\phi(x_1)x_2^{\alpha_2-1}\cdots x_n^{\alpha_n-1} \\
&= \frac{\Gamma(\alpha_2)\cdots\Gamma(\alpha_n)}{\Gamma(\alpha_2+\cdots+\alpha_n+\mu)}(I_0^{\alpha_2+\cdots+\alpha_n+\mu}\phi)(x_1)\cdot x_2^{\alpha_2-1}\cdots x_n^{\alpha_n-1}.
\end{aligned} \tag{6}
$$

Here and hereafter, we use the notation

$$
\begin{aligned}
&\mathbb{N} = \{0, 1, 2, , \ldots\}, \\
&\mathbf{m} \geq 0 \Leftrightarrow m_1 \geq 0, \ldots, m_n \geq 0, \\
&|\boldsymbol{\alpha}| = \alpha_1 + \cdots + \alpha_n,\ \boldsymbol{\alpha} + c = (\alpha_1 + c, \ldots, \alpha_n + c),\ \mathbf{m}! = m_1!\cdots m_n!, \\
&x^{\boldsymbol{\alpha}} = \mathbf{x}^{\boldsymbol{\alpha}} = x_1^{\alpha_1}\cdots x_n^{\alpha_n},\ (c-\mathbf{x})^{\boldsymbol{\alpha}} = (c-x_1)^{\alpha_1}\cdots(c-x_n)^{\alpha_n}, \\
&\Gamma(\boldsymbol{\alpha}) = \Gamma(\alpha_1)\cdots\Gamma(\alpha_n),\ (\boldsymbol{\alpha})_{\mathbf{m}} = \frac{\Gamma(\boldsymbol{\alpha}+\mathbf{m})}{\Gamma(\boldsymbol{\alpha})}
\end{aligned} \tag{7}
$$

for $\boldsymbol{\alpha} = (\alpha_1, \ldots, \alpha_n) \in \mathbb{C}^n$, $\mathbf{m} = (m_1, \ldots, m_n) \in \mathbb{N}^n$ and the variable $x = (x_1, \ldots, x_n)$.

The arguments in this section are valid when $\operatorname{Re}\alpha_1 > 0, \ldots, \operatorname{Re}\alpha_n > 0$ and $\operatorname{Re}\mu > 0$, but the right-hand side of (5) is meromorphic for $\boldsymbol{\alpha}$ and μ, and we define $K_x^{\mu}x^{\boldsymbol{\alpha}}$ by the analytic continuation with respect to these parameters.

We will define the inverse L_x^{μ} of K_x^{μ}. Suppose $0 \leq \operatorname{Re} s < 1$ and $0 < c < 1 - \operatorname{Re} s$. Then

$$
\begin{aligned}
&\int_{c-i\infty}^{c+i\infty} t^{-\alpha}(1-s-t)^{-\tau}\tfrac{dt}{t} = (1-s)^{-\tau}\int_{c-i\infty}^{c+i\infty} t^{-\alpha}(1-\tfrac{t}{1-s})^{-\tau}\tfrac{dt}{t} \\
&= (1-s)^{-\alpha-\tau}\int_{\frac{c-i\infty}{1-s}}^{\frac{c+i\infty}{1-s}} t^{-\alpha}(1-t)^{-\tau}\tfrac{dt}{t} \\
&= (1-s)^{-\alpha-\tau}\int_{c-i\infty}^{c+i\infty} t^{-\alpha}(1-t)^{-\tau}\tfrac{dt}{t} \\
&= (1-s)^{-\alpha-\tau}(-e^{-\tau\pi i}+e^{\tau\pi i})\int_1^{\infty} t^{-\alpha}(t-1)^{-\tau}\tfrac{dt}{t} \\
&= (1-s)^{-\alpha-\tau}\cdot 2i\sin\tau\pi\int_0^1 (\tfrac{1}{u})^{-\alpha}(\tfrac{1}{u}-1)^{-\tau}\tfrac{du}{u} \quad (u=\tfrac{1}{t}) \\
&= \frac{2\pi i(1-s)^{-\alpha-\tau}}{\Gamma(\tau)\Gamma(1-\tau)}\int_0^1 u^{\alpha+\tau-1}(1-u)^{-\tau}du \\
&= 2\pi i\frac{\Gamma(\alpha+\tau)}{\Gamma(\tau)\Gamma(\alpha+1)}(1-s)^{-\alpha-\tau}.
\end{aligned}
$$

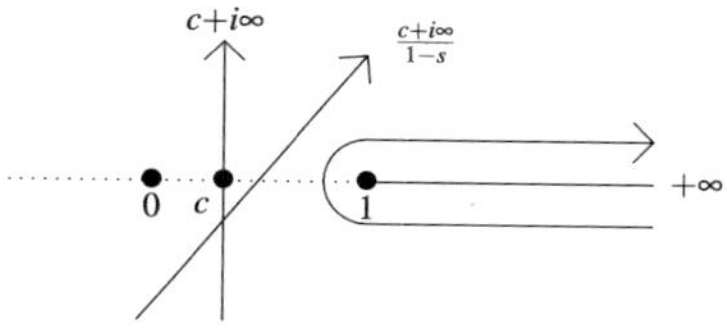

Here the path of the integration of the above first line is $(-\infty, \infty) \ni s \mapsto c + is$, and we also use the path $1 \quad +\infty$ of the integration in the above.

Thus, we have

$$\begin{aligned}
&\int_{\frac{1}{n+1}-i\infty}^{\frac{1}{n+1}+i\infty}\cdots\int_{\frac{1}{n+1}-i\infty}^{\frac{1}{n+1}+i\infty} t^{-\boldsymbol{\alpha}}(1-t_1-\cdots-t_n)^{-\tau}\tfrac{dt_1}{t_1}\cdots\tfrac{dt_n}{t_n}\\
&=(2\pi i)^n\frac{\Gamma(\alpha_n+\tau)}{\Gamma(\tau)\Gamma(\alpha_n+1)}\frac{\Gamma(\alpha_{n-1}+\alpha_n+\tau)}{\Gamma(\alpha_n+\tau)\Gamma(\alpha_{n-1}+1)}\cdots\\
&\quad\frac{\Gamma(|\boldsymbol{\alpha}|+\tau)}{\Gamma(\alpha_2+\cdots+\alpha_n+\tau)\Gamma(\alpha_1+1)}\\
&=(2\pi i)^n\frac{\Gamma(|\boldsymbol{\alpha}+1|+\tau-n)}{\Gamma(\boldsymbol{\alpha}+1)\Gamma(\tau)}.
\end{aligned}$$

Definition 2 We define the transformation

$$\begin{aligned}
(L_x^\mu\phi)(x) &:= \frac{\Gamma(\mu+n)}{(2\pi i)^n}\int_{\frac{1}{n+1}-i\infty}^{\frac{1}{n+1}+i\infty}\cdots\int_{\frac{1}{n+1}-i\infty}^{\frac{1}{n+1}+i\infty}\phi(\tfrac{x_1}{t_1},\dots,\tfrac{x_n}{t_n})\\
&\quad\times(1-|\mathbf{t}|)^{-\mu-n}\tfrac{dt_1}{t_1}\cdots\tfrac{dt_n}{t_n}\\
&=\frac{\Gamma(\mu+n)}{(2\pi i)^n}\int_{\left|\frac{2s_1}{n+1}-1\right|=1}\cdots\int_{\left|\frac{2s_n}{n+1}-1\right|=1}\phi(s_1x_1,\dots,s_nx_n)\\
&\quad\times(1-\tfrac{1}{s_1}-\cdots-\tfrac{1}{s_n})^{-\mu-n}\tfrac{ds_1}{s_1}\cdots\tfrac{ds_n}{s_n}.
\end{aligned}$$

In the above, we mainly consider the case when $\phi(x)=x^{\boldsymbol{\lambda}}\varphi(x)$ with a convergent power series $\varphi(x)$. Then we have

$$L_x^\mu x^{\boldsymbol{\alpha}}=\frac{\Gamma(|\boldsymbol{\alpha}+1|+\mu)}{\Gamma(\boldsymbol{\alpha}+1)}x^{\boldsymbol{\alpha}}. \tag{8}$$

When $n=1$, we have

$$\begin{aligned}
(L_x^\mu\phi)(x) &= \frac{\Gamma(\mu+1)}{2\pi i}\int_{\frac12-i\infty}^{\frac12+i\infty}\phi(\tfrac{x}{t})(1-t)^{-\mu-1}\tfrac{dt}{t}\\
&=\frac{\sin(\mu+1)\pi\cdot\Gamma(\mu+1)}{\pi}\int_1^\infty\phi(\tfrac{x}{t})(1-t)^{-\mu-1}\tfrac{dt}{t}\\
&=\frac{1}{\Gamma(-\mu)}\int_0^1 s^\mu\phi(s)(x-s)^{-\mu-1}ds\\
&=I_0^{-\mu}(x^\mu\phi).
\end{aligned}$$

In general, we have

$$L_x^{\mu}(\phi(x_1)x_2^{\alpha_2-1}\cdots x_n^{\alpha_n-1}) = \frac{\Gamma(\alpha_2+\cdots+\alpha_n+\mu)}{\Gamma(\alpha_2)\cdots\Gamma(\alpha_n)} \times (I_{x_1,0}^{-\alpha_2-\cdots-\alpha_n-\mu} x_1^{\alpha_2+\cdots+\alpha_n+\mu}\phi(x_1))x_2^{\alpha_2-1}\cdots x_n^{\alpha_n-1}. \tag{9}$$

Definition 3 We define two transformations

$$K_x^{\mu,\boldsymbol{\lambda}} := x^{1-\boldsymbol{\lambda}} K_x^{\mu} x^{\boldsymbol{\lambda}-1} \text{ and } L_x^{\mu,\boldsymbol{\lambda}} := x^{1-\boldsymbol{\lambda}} L_x^{\mu} x^{\boldsymbol{\lambda}-1} \tag{10}$$

which act on the ring $\mathscr{O}_0$ of convergent power series of $x = (x_1, \ldots, x_n)$.

We have

$$K_x^{\mu,\boldsymbol{\lambda}} x^{\boldsymbol{\alpha}} = \frac{\Gamma(\boldsymbol{\lambda}+\boldsymbol{\alpha})}{\Gamma(|\boldsymbol{\lambda}+\boldsymbol{\alpha}|+\mu)} x^{\boldsymbol{\alpha}} \text{ and } L_x^{\mu,\boldsymbol{\lambda}} x^{\boldsymbol{\alpha}} = \frac{\Gamma(|\boldsymbol{\lambda}+\boldsymbol{\alpha}|+\mu)}{\Gamma(\boldsymbol{\lambda}+\boldsymbol{\alpha})} x^{\boldsymbol{\alpha}}.$$

Theorem 1 *Putting*

$$u(x) = \sum_{\mathbf{m}\geq 0} c_{\mathbf{m}} x^{\mathbf{m}} = \sum_{m_1=0}^{\infty}\cdots\sum_{m_n=0}^{\infty} c_{\mathbf{m}} x^{\mathbf{m}} \in \mathscr{O}_0 \quad (c_{\mathbf{m}} \in \mathbb{C}),$$

$$K_x^{\mu,\boldsymbol{\lambda}} u(x) = \sum_{\mathbf{m}} c_{\mathbf{m}}^K x^{\mathbf{m}} \text{ and } L_x^{\mu,\boldsymbol{\lambda}} u(x) = \sum_{\mathbf{m}} c_{\mathbf{m}}^L x^{\mathbf{m}} \quad (c_{\mathbf{m}}^K,\ c_{\mathbf{m}}^L \in \mathbb{C}),$$

we have

$$c_{\mathbf{m}}^K = \frac{\Gamma(\boldsymbol{\lambda})}{\Gamma(|\boldsymbol{\lambda}|+\mu)} \frac{(\boldsymbol{\lambda})_{\mathbf{m}}}{(|\boldsymbol{\lambda}|+\mu)_{|\mathbf{m}|}} c_{\mathbf{m}}, \tag{11}$$

$$c_{\mathbf{m}}^L = \frac{\Gamma(|\boldsymbol{\lambda}|+\mu)}{\Gamma(\boldsymbol{\lambda})} \frac{(|\boldsymbol{\lambda}|+\mu)_{|\mathbf{m}|}}{(\boldsymbol{\lambda})_{\mathbf{m}}} c_{\mathbf{m}}. \tag{12}$$

By the analytic continuation with respect to the parameters the transformations $K_x^{\mu,\boldsymbol{\lambda}}$ and $L_x^{\mu,\boldsymbol{\lambda}}$ are well-defined if

$$\lambda_\nu \notin \mathbb{Z}_{\leq 0} \quad (\nu = 1, \ldots, n) \tag{13}$$

and

$$|\boldsymbol{\lambda}| + \mu \notin \mathbb{Z}_{\leq 0}, \tag{14}$$

respectively. Namely, we may consider that $K_x^{\mu,\boldsymbol{\lambda}}$ and $L_x^{\mu,\boldsymbol{\lambda}}$ are defined by Theorem 1 by using (11) and (12). Hence, if (13) and (14) are valid, $K_x^{\mu,\boldsymbol{\lambda}}$ and $L_x^{\mu,\boldsymbol{\lambda}}$ are bijective on $\mathscr{O}_0$, and the map $K_x^{\mu,\boldsymbol{\lambda}} \circ L_x^{\mu,\boldsymbol{\lambda}}$ is the identity map.

Proposition 1 *By Eqs.* (6) *and* (9), *we have*

$$K_x^{\mu,\boldsymbol{\lambda}}\phi(x_1)=\frac{\Gamma(\lambda_2)\cdots\Gamma(\lambda_n)}{\Gamma(\lambda_2+\cdots+\lambda_n+\mu)}x_1^{-|\boldsymbol{\lambda}|-\mu+1}I_{x_1,0}^{|\boldsymbol{\lambda}|-\lambda_1+\mu}x_1^{\lambda_1-1}\phi(x_1), \tag{15}$$

$$L_x^{\mu,\boldsymbol{\lambda}}\phi(x_1)=\frac{\Gamma(\lambda_2+\cdots+\lambda_n+\mu)}{\Gamma(\lambda_2)\cdots\Gamma(\lambda_n)}x_1^{1-\lambda_1}I_{x_1,0}^{-|\boldsymbol{\lambda}|+\lambda_1-\mu}x_1^{|\boldsymbol{\lambda}|+\mu-1}\phi(x_1). \tag{16}$$

3 Some Hypergeometric Functions

Under notation (7), we note that

$$(1-|\mathbf{x}|)^{-\lambda}=\sum_{\mathbf{m}\ge 0}\frac{(\lambda)_{|\mathbf{m}|}}{\mathbf{m}!}x^{\mathbf{m}}\ \text{ and }\ e^{|\mathbf{x}|}=\sum_{\mathbf{m}\ge 0}\frac{\mathbf{x}^{\mathbf{m}}}{\mathbf{m}!}.$$

Lauricella's hypergeometric series (cf. [La, Er]) and their integral representations are given as follows (cf. Theorem 1):

$$\begin{aligned}F_A(\lambda_0,\boldsymbol{\mu},\boldsymbol{\lambda};\mathbf{x})&:=\sum_{\mathbf{m}\ge 0}\frac{(\lambda_0)_{|\mathbf{m}|}(\boldsymbol{\mu})_{\mathbf{m}}}{(\boldsymbol{\lambda})_{\mathbf{m}}\mathbf{m}!}\mathbf{x}^{\mathbf{m}}\\&=\frac{\Gamma(\boldsymbol{\lambda})}{\Gamma(\boldsymbol{\mu})}K_{x_1}^{\lambda_1-\mu_1,\mu_1}\cdots K_{x_n}^{\lambda_n-\mu_n,\mu_n}(1-|\mathbf{x}|)^{-\lambda_0}\\&=\frac{\Gamma(\boldsymbol{\lambda})}{\Gamma(\lambda_0)}L_x^{\lambda_0-|\boldsymbol{\lambda}|,\boldsymbol{\lambda}}(1-\mathbf{x})^{-\boldsymbol{\mu}},\end{aligned} \tag{17}$$

$$F_B(\boldsymbol{\lambda},\boldsymbol{\lambda}',\mu;\mathbf{x}):=\sum_{\mathbf{m}\ge 0}\frac{(\boldsymbol{\lambda})_{\mathbf{m}}(\boldsymbol{\lambda}')_{\mathbf{m}}}{(\mu)_{|\mathbf{m}|}\mathbf{m}!}\mathbf{x}^{m}=\frac{\Gamma(\mu)}{\Gamma(\boldsymbol{\lambda})}K_x^{\mu-|\boldsymbol{\lambda}|,\boldsymbol{\lambda}}(1-\mathbf{x})^{-\boldsymbol{\lambda}'}, \tag{18}$$

$$F_C(\mu,\lambda_0,\boldsymbol{\lambda};\mathbf{x}):=\sum_{\mathbf{m}\ge 0}\frac{(\mu)_{|\mathbf{m}|}(\lambda_0)_{|\mathbf{m}|}}{(\boldsymbol{\lambda})_{\mathbf{m}}\mathbf{m}!}\mathbf{x}^{\mathbf{m}}=\frac{\Gamma(\boldsymbol{\lambda})}{\Gamma(\mu)}L_x^{\mu-|\boldsymbol{\lambda}|,\boldsymbol{\lambda}}(1-|\mathbf{x}|)^{-\lambda_0}, \tag{19}$$

$$F_D(\lambda_0,\boldsymbol{\lambda},\mu;\mathbf{x}):=\sum_{\mathbf{m}\ge 0}\frac{(\lambda_0)_{|\mathbf{m}|}(\boldsymbol{\lambda})_{\mathbf{m}}}{(\mu)_{|\mathbf{m}|}\mathbf{m}!}\mathbf{x}^{\mathbf{m}}=\frac{\Gamma(\mu)}{\Gamma(\boldsymbol{\lambda})}K_x^{\mu-|\boldsymbol{\lambda}|,\boldsymbol{\lambda}}(1-|\mathbf{x}|)^{-\lambda_0}. \tag{20}$$

When $n = 2$, namely, the number of variables equals 2, functions F_D, F_A, F_B, and F_C are Appell's hypergeometric series F_1, F_2, F_3, and F_4 (cf. [AK]), respectively. Moreover, we give examples of confluent Horn's series (cf. [Ho, Er]):

$$\begin{aligned}\Phi_2(\beta, \beta'; \gamma; x, y) &:= \sum_{m=0}^{\infty}\sum_{n=0}^{\infty} \frac{(\beta)_m(\beta')_n}{(\gamma)_{m+n} m! n!} x^m y^n \\ &= \frac{\Gamma(\gamma)}{\Gamma(\beta)\Gamma(\beta')} K_{x,y}^{\gamma-\beta-\beta',\beta,\beta'} e^{x+y},\end{aligned} \tag{21}$$

$$\begin{aligned}\Psi_1(\alpha; \beta; \gamma, \gamma'; x, y) &:= \sum_{m=0}^{\infty}\sum_{n=0}^{\infty} \frac{(\alpha)_{m+n}(\beta)_m}{(\gamma)_m(\gamma')_n m! n!} x^m y^n \\ &= \frac{\Gamma(\gamma)\Gamma(\gamma')}{\Gamma(\alpha)} L_{x,y}^{\alpha-\gamma-\gamma',\gamma,\gamma'} (1-x)^{-\beta} e^{y},\end{aligned} \tag{22}$$

$$\begin{aligned}\Psi_2(\alpha; \gamma', \gamma'; x, y) &:= \sum_{m=0}^{\infty}\sum_{n=0}^{\infty} \frac{(\alpha)_{m+n}}{(\gamma)_m(\gamma')_n m! n!} x^m y^n \\ &= \frac{\Gamma(\gamma)\Gamma(\gamma')}{\Gamma(\alpha)} L_{x,y}^{\alpha-\gamma-\gamma',\gamma,\gamma'} e^{x+y}.\end{aligned} \tag{23}$$

When $n = 1$, (17), (18), (19), and (20) are reduced to an integral representation of Gauss hypergeometric series. In fact, putting $(\lambda_0, \boldsymbol{\mu}, \boldsymbol{\lambda}) = (\alpha, \beta, \gamma)$ in (17), we have

$$\begin{aligned}F(\alpha, \beta, \gamma; x) &= \sum_{m=0}^{\infty} \frac{(\alpha)_m(\beta)_m}{(\gamma)_m m!} x^m \\ &= \frac{\Gamma(\gamma)}{\Gamma(\alpha)} K_x^{\gamma-\alpha,\alpha} (1-x)^{-\beta} \\ &= \frac{\Gamma(\gamma)}{\Gamma(\alpha)\Gamma(\gamma-\alpha)} \int_0^1 t^{\alpha-1}(1-t)^{\gamma-\alpha-1}(1-tx)^{-\beta} \mathrm{d}t\end{aligned} \tag{24}$$

and Kummer function is

$$\begin{aligned}{}_1F_1(\alpha; \gamma; x) &:= \sum_{n=0}^{\infty} \frac{(\alpha)_n}{(\gamma)_n} \frac{x^n}{n!} \\ &= \frac{\Gamma(\gamma)}{\Gamma(\alpha)} K_x^{\gamma-\alpha,\alpha} e^x = \frac{\Gamma(\gamma)}{\Gamma(\alpha)\Gamma(\gamma-\alpha)} \int_0^1 t^{\alpha-1}(1-t)^{\gamma-\alpha-1} e^{tx} \mathrm{d}t.\end{aligned} \tag{25}$$

4 A Connection Problem

Integral representations of hypergeometric functions are useful for the study of global structure of the functions. Rigid linear ordinary differential equations on the Riemann sphere with regular or unramified irregular irregularities are reduced to the trivial equation by successive applications of middle convolutions and additions. These transformations correspond to the transformations of their solutions defined by Riemann-Liouville integrals and multiplications by elementary functions such as $(x-c)^\lambda$ or $e^{r(x)}$ with rational functions $r(x)$ of x.

In [O1, Chapter 12] and [O6], analyzing the asymptotic behavior of the Riemann-Liouville integral when variable x tends to a singular point of the function, we get the change of connection coefficients and Stokes coefficients under the integral transformations and finally such coefficients of the hypergeometric function we are interested in.

In this section, a generalization of this way of study is shown in the case of several variables, which will be explained by using F_1. Since

$$F_1(a,b,b',c;x,y)=\frac{\Gamma(c)}{\Gamma(b)\Gamma(b')}K_{x,y}^{c-b-b',b,b'}(1-x-y)^{-a},$$

Proposition 1 implies

$$F_1(a,b,b',c;x,0)=\frac{\Gamma(c)}{\Gamma(b)}x^{1-c}I_0^{c-b}x^{b-1}(1-x)^{-a}.$$

Equalities (3) and (24) show $F_1(a,b,b',c;x,0)=F(b,a,c;x)$, but first we do not use this fact.

We pursue the changes of the Riemann scheme under the procedure given by Proposition 1 (cf. [O1, Chapter 5]). They are the change when we apply I_0^{c-b} to $x^{b-1}(1-x)^{-a}$ and the change when we multiply the resulting function by $\frac{\Gamma(c)}{\Gamma(b)}x^{1-c}$, which are

$$\begin{aligned}
&x^{b-1}(1-x)^{-a}:\begin{Bmatrix} x=0 & 1 & \infty \\ \underline{b-1} & -a & a-b+1\end{Bmatrix}\\
&\xrightarrow{I_0^{c-b}}\begin{Bmatrix} x=0 & 1 & \infty \\ 0 & 0 & b-c+1 \\ \underline{c-1} & c-a-b & a-c+1\end{Bmatrix}\\
&\xrightarrow{\times\frac{\Gamma(c)}{\Gamma(b)}x^{1-c}}\begin{Bmatrix} x=0 & 1 & \infty \\ 1-c & 0 & b \\ \underline{0} & c-a-b & a\end{Bmatrix}.
\end{aligned}\tag{26}$$

Then, $F_1(a,b,b',c;x,0)$ is characterized as the holomorphic function in a neighborhood of 0 with the Riemann scheme (26) and $F_1(a,b,b',c;0,0)=1$.

Since $F_1(a, b, b', c; x, y)$ satisfies a system of differential equations with singularities $x = 0, 1, \infty$, $y = 0, 1, \infty$ and $x = y$ in $\mathbb{P}^1 \times \mathbb{P}^1$, we have a connection relation

$$F_1(a, b, b', c; x, y) = (-\tfrac{1}{x})^a C_a f_1^a(x, y) + (-\tfrac{1}{x})^b C_b f_1^b(x, y) \tag{27}$$

in a neighborhood of $(-\infty, 0) \times \{0\}$ in $\mathbb{P}^1 \times \mathbb{P}^1$. Here $f_1^a(x, y)$ and $f_1^b(x, y)$ are holomorphic in a neighborhood of $[-\infty, 0) \times \{0\}$ and $f_1^a(-\infty, 0) = f_1^b(-\infty, 0) = 1$, and connection coefficients C_a and C_b are given by those of $F_1(a, b, b', c; x, 0)$, namely,

$$C_a = \frac{\Gamma(c)\Gamma(b-a)}{\Gamma(b)\Gamma(c-a)} \quad \text{and} \quad C_b = \frac{\Gamma(c)\Gamma(a-b)}{\Gamma(a)\Gamma(c-b)}. \tag{28}$$

Remark 1 We note that the general expression [O1, (0.25)] of the connection coefficient of Gauss hypergeometric function is simple and easy to be specialized (cf. [O2]). Moreover, the connection formula

$$\begin{aligned} F(a, b, c; x) &= (-\tfrac{1}{x})^a C_a F(a, a-c+1, a-b+1; \tfrac{1}{x}) \\ &\quad + (-\tfrac{1}{x})^b C_b F(b, b-c+1, b-a+1; \tfrac{1}{x}) \end{aligned}$$

follows from

$$\begin{aligned} \left\{\begin{matrix} x=0 & 1 & \infty & \\ 0 & 0 & a; & x \\ 1-c & c-a-b & b & \end{matrix}\right\} &= \left\{\begin{matrix} x=0 & 1 & \infty & \\ a & 0 & 0; & \frac{1}{x} \\ b & c-a-b & 1-c & \end{matrix}\right\} \\ &= (-x)^a \left\{\begin{matrix} x=0 & 1 & \infty & \\ 0 & 0 & a; & \frac{1}{x} \\ 1-(a-b+1) & c-a-b & a-c+1 & \end{matrix}\right\}. \end{aligned}$$

We explicitly calculate (27) in the following way:

$$\begin{aligned} F_1(a, b, b', c; x, y) &= \sum_{m=0}^{\infty}\sum_{n=0}^{\infty} \frac{(a)_{m+n}(b)_m(b')_n}{(c)_{m+n}m!n!} x^m y^n \\ &= \sum_{n=0}^{\infty} \frac{(a)_n(b')_n}{(c)_n n!} y^n \sum_{m=0}^{\infty} \frac{(a+n)_m(b)_m}{(c+n)_m m!} x^m = \sum_{n=0}^{\infty} \frac{(a)_n(b')_n}{(c)_n n!} y^n \\ &\quad \times F(a+n, b, c+n; x) \\ &= \sum_{n=0}^{\infty} \frac{(a)_n(b')_n}{(c)_n n!} \end{aligned}$$

$$
\begin{aligned}
&\Big((-x)^{-a-n}\frac{\Gamma(c+n)\Gamma(b-a-n)}{\Gamma(b)\Gamma(c-a)}F(a+n,a-c+1,a-b+n+1;\tfrac{1}{x})\\
&\quad+(-x)^{-b}\frac{\Gamma(c+n)\Gamma(a-b+n)}{\Gamma(a+n)\Gamma(c-b+n)}F(b,b-c-n+1,b-a-n+1;\tfrac{1}{x})\Big)\\
&=(-\tfrac{1}{x})^{a}F_1^a+(-\tfrac{1}{x})^{b}F_1^b,\\
F_1^a&=\sum\frac{\Gamma(c)(a)_n(b')_n\Gamma(b-a-n)(a+n)_m(a-c+1)_m}{\Gamma(b)\Gamma(c-a)(c)_n(a-b+n+1)_m m!n!}(\tfrac{1}{x})^m(-\tfrac{y}{x})^n\\
&=\sum\frac{\Gamma(c)(a)_{m+n}(b')_n\Gamma(a-b+1)\Gamma(b-a)(a-c+1)_m}{\Gamma(b)\Gamma(c-a)\Gamma(a-b+n+1)(a-b+n+1)_m m!n!}(\tfrac{1}{x})^m(\tfrac{y}{x})^n\\
&=\sum\frac{\Gamma(c)\Gamma(a-b+1)\Gamma(b-a)(a)_{m+n}(b')_n(a-c+1)_m}{\Gamma(b)\Gamma(c-a)\Gamma(a-b+1+m+n)m!n!}(\tfrac{1}{x})^m(\tfrac{y}{x})^n\\
&=\sum\frac{\Gamma(c)\Gamma(b-a)(a)_{m+n}(b')_n(a-c+1)_m}{\Gamma(b)\Gamma(c-a)(a-b+1)_{m+n}m!n!}(\tfrac{1}{x})^m(\tfrac{y}{x})^n\\
&=\sum\frac{\Gamma(c)\Gamma(b-a)(a)_{m+n}(a-c+1)_m(b')_n}{\Gamma(b)\Gamma(c-a)(a-b+1)_{m+n}m!n!}(\tfrac{1}{x})^m(\tfrac{y}{x})^n\\
&=\frac{\Gamma(c)\Gamma(b-a)}{\Gamma(b)\Gamma(c-a)}F_1(a,a-c+1,b',a-b+1;\tfrac{1}{x},\tfrac{y}{x}),\\
F_1^b&=\sum\frac{\Gamma(c)(b')_n\Gamma(a-b+n)(b)_m(b-c-n+1)_m}{\Gamma(a)\Gamma(c-b+n)(b-a-n+1)_m m!n!}(\tfrac{1}{x})^m y^n\\
&=\sum\frac{\Gamma(c)(b)_m(b')_n\Gamma(a-b+n)\Gamma(b-c-n+1)(b-c-n+1)_m\,(-\tfrac{1}{x})^m(-y)^n}{\Gamma(a)\Gamma(c-b)\Gamma(b-c+1)(a-b+n-1)\cdots(a-b+n-m)m!n!}\\
&=\sum\frac{\Gamma(c)(b)_m(b')_n\Gamma(a-b+n-m)\Gamma(b-c-n+1+m)(-\tfrac{1}{x})^m(-y)^n}{\Gamma(a)\Gamma(c-b)\Gamma(b-c+1)m!n!}\\
&=\sum\frac{\Gamma(c)(b)_m(b')_n\Gamma(a-b)(a-b)_{n-m}(b-c+1)_{m-n}}{\Gamma(a)\Gamma(c-b)m!n!}(-\tfrac{1}{x})^m(-y)^n\\
&=\frac{\Gamma(c)\Gamma(a-b)}{\Gamma(a)\Gamma(c-b)}G_2(b,b',a-b,b-c+1;-\tfrac{1}{x},-y).
\end{aligned}
$$

Here

$$
G_2(\alpha,\beta,\gamma,\delta;x,y):=\sum_{m=0}^{\infty}\sum_{n=0}^{\infty}\frac{(\alpha)_m(\beta)_n(\gamma)_{n-m}(\delta)_{m-n}}{m!n!}x^m y^n \tag{29}
$$

and we have

$$
\begin{aligned}
f_1^a(x,y) &= F_1(a, a-c+1, b', a-b+1; \tfrac{1}{x}, \tfrac{y}{x}),\\
f_1^b(x,y) &= G_2(b, b', a-b, b-c+1; -\tfrac{1}{x}, -y)
\end{aligned} \tag{30}
$$

in (27).

Remark 2 The argument above is justified since $F_1(a; b, b'; c; x, y)$ satisfies a differential equation, which has regular singularities along the hypersurface defined by $y = 0$ in $\{(x, y) \in \mathbb{C}^2 \mid \operatorname{Re} x < 0\}$ or Kummer's formula

$$
F(\alpha, \beta, \gamma; x) = (1-x)^{-\alpha} F(\alpha, \gamma-\beta, \gamma; \tfrac{x}{x-1}).
$$

We give an answer to a part of connection problem of Appell's F_1, which satisfies a KZ equation of rank 3, and the equation allows the coordinate transformations on $(\mathbb{P}^1)^5$ corresponding to permutations of five coordinates. By the action of this transformation, we get Kummer type formula for F_1 and solve the connection problem for F_1. Note that the singularity of the origin is not of the normally crossing type, but, for example, map $(x, y) \mapsto (\frac{1}{x}, \frac{y}{x})$ is one of the coordinate transformations, and the blowing up of the origin naturally corresponds to this coordinate transformation. This enables us to get all the analytic continuation of F_1 in $\mathbb{P}^1 \times \mathbb{P}^1$ in terms of F_1 and G_2 as is given in the above special case (cf. Sect. 7). Another independent solution f_1^c of the equation at $(\infty, 0)$ is characterized by the fact that the analytic continuation of f_1^c in a suitable neighborhood of $(-\infty, 0) \times \{0\}$ is a scalar multiple of f_1^c and then f_1^c is expressed by using F_1 as in the case of $(-\frac{1}{x})^a f_1^a$. This will be discussed in [MO] with more general examples.

5 More Transformations

In this section, we examine transformations of power series obtained by a suitable class of coordinate transformations and transformations $K_x^{\mu,\boldsymbol{\lambda}}$ and $L_x^{\mu,\boldsymbol{\lambda}}$. For a coordinate transformation $x \mapsto R(x)$ of $\mathbb{C}^n$, we put

$$
(T_{x\to R(x)}\phi) = \phi(R(x))
$$

for functions $\phi(x)$.

Definition 4 Choose a subset of indices $\{i_1, \dots, i_k\} \subset \{1, \dots, n\}$. Then we put $\mathbf{y} = (x_{i_1}, \dots, x_{i_k})$. For $\mu \in \mathbb{C}$ and $\boldsymbol{\lambda} \in \mathbb{C}^k$, we define

$$
\begin{aligned}
K_{\mathbf{y},x\to R(x)}^{\mu,\boldsymbol{\lambda}} &:= T_{x\to R(x)}^{-1} \circ K_{\mathbf{y}}^{\mu,\boldsymbol{\lambda}} \circ T_{x\to R(x)},\\
L_{\mathbf{y},x\to R(x)}^{\mu,\boldsymbol{\lambda}} &:= T_{x\to R(x)}^{-1} \circ L_{\mathbf{y}}^{\mu,\boldsymbol{\lambda}} \circ T_{x\to R(x)}.
\end{aligned}
$$

Let $\mathbf{p} = \Big(p_{i,j}\Big)_{\substack{1\le i\le n\\ 1\le j\le n}} \in GL(n,\mathbb{Z})$. We denote

$$
\begin{aligned}
x^{\mathbf{p}} = \mathbf{x}^{\mathbf{p}} &= (x^{p_{*,1}}, \dots, x^{p_{*,n}}) = \Big(\prod_{\nu=1}^{n} x_\nu^{p_{\nu,1}}, \dots, \prod_{\nu=1}^{n} x_\nu^{p_{\nu,n}}\Big),\\
\mathbf{pm} &= (p_{1,*}\mathbf{m}, \dots, p_{n,*}\mathbf{m}) = \Big(\sum_{\nu=1}^{n} p_{1,\nu}m_\nu, \dots, \sum_{\nu=1}^{n} p_{n,\nu}m_\nu\Big)
\end{aligned}
\tag{31}
$$

with $\mathbf{m} = (m_1, \dots, m_n) \in \mathbb{Z}^n$. Then $T_{x\to x^{\mathbf{p}}}^{-1} = T_{x\to x^{\mathbf{p}^{-1}}}$

We examine the transformations $K^{\mu,\boldsymbol{\lambda}}_{\mathbf{y},x\to x^{\mathbf{p}}}$ and $L^{\mu,\boldsymbol{\lambda}}_{\mathbf{y},x\to x^{\mathbf{p}}}$ under the assumption

$$
p_{i_\nu,j} \ge 0 \qquad (1\le \nu\le k,\ 1\le j\le n). \tag{32}
$$

Since

$$
\big(T_{x\to x^{\mathbf{p}}}^{-1} T_{x\to(t_1x_1,\dots,t_nx_n)} T_{x\to x^{\mathbf{p}}}\phi\big)(x) = \phi\Big(x_1\prod_{\nu=1}^{n} t_\nu^{p_{\nu,1}}, \dots, x_n\prod_{\nu=1}^{n} t_\nu^{p_{\nu,n}}\Big),
$$

we have

$$
\begin{aligned}
&\big(K^{\mu,\boldsymbol{\lambda}}_{(x_{i_1},\dots,x_{i_k}),x\to x^{\mathbf{p}}}\phi\big)(x)\\
&\quad= \frac{1}{\Gamma(\mu)}\int_{\substack{t_1>0,\dots t_k>0\\ t_1+\dots+t_k<1}} \mathbf{t}^{\boldsymbol{\lambda}-1}(1-|\mathbf{t}|)^{\mu-1}\phi\Big(x_1\prod_{\nu=1}^{k} t_\nu^{p_{i_\nu,1}}, \dots, x_n\prod_{\nu=1}^{k} t_\nu^{p_{i_\nu,n}}\Big)\mathrm{d}t,
\end{aligned}
\tag{33}
$$

$$
\begin{aligned}
\big(L^{\mu,\boldsymbol{\lambda}}_{(x_{i_1},\dots,x_{i_k}),x\to x^{\mathbf{p}}}\phi\big)(x) &= \frac{\Gamma(\mu+k)}{(2\pi i)^k}\int_{c-i\infty}^{c+i\infty}\cdots\int_{c-i\infty}^{c+i\infty} \mathbf{t}^{\boldsymbol{\lambda}-1}(1-|\mathbf{t}|)^{-\mu-k}\\
&\quad\times\phi\Big(\frac{x_1}{\prod_{\nu=1}^{k} t_\nu^{p_{i_\nu,1}}}, \dots, \frac{x_n}{\prod_{\nu=1}^{k} t_\nu^{p_{i_\nu,n}}}\Big)\frac{\mathrm{d}t_1}{t_1}\cdots\frac{\mathrm{d}t_k}{t_k} \quad \text{with } c = \tfrac{1}{k+1}.
\end{aligned}
\tag{34}
$$

We note that (32) assures that these transformations are defined on $\mathscr{O}_0$.

Proposition 2 *Denoting*

$$
(\mathbf{pm})_{i_1,\dots,i_k} = \Big(\sum_{\nu=1}^{n} p_{i_1,\nu}m_\nu, \dots, \sum_{\nu=1}^{n} p_{i_k,\nu}m_\nu\Big),
$$

we have

$$K^{\mu,\boldsymbol{\lambda}}_{(x_{i_1},\dots,x_{i_k}),x\to x^{\mathbf{p}}}x^{\mathbf{m}} = \frac{\Gamma(\boldsymbol{\lambda}+(\mathbf{pm})_{i_1,\dots,i_k})}{\Gamma(|\boldsymbol{\lambda}+(\mathbf{pm})_{i_1,\dots,i_k}|+\mu)}x^{\mathbf{m}}, \tag{35}$$

$$L^{\mu,\boldsymbol{\lambda}}_{(x_{i_1},\dots,x_{i_k}),x\to x^{\mathbf{p}}}x^{\mathbf{m}} = \frac{\Gamma(|\boldsymbol{\lambda}+(\mathbf{pm})_{i_1,\dots,i_k}|+\mu)}{\Gamma(\boldsymbol{\lambda}+(\mathbf{pm})_{i_1,\dots,i_k})}x^{\mathbf{m}}. \tag{36}$$

We give some examples hereafter in this section.

Let $(p_1,\dots,p_n)$ be a non-zero vector of non-negative integers. Suppose the greatest common divisor of $p_1,\dots,p_n$ equals 1. Then there exists $\mathbf{p}=\big(p_{i,j}\big)\in GL(n,\mathbb{Z})$ with $p_{1,j}=p_j$ and

$$K^{\mu,\lambda}_{x_1,x\to\mathbf{x}^{\mathbf{p}}}x^{\mathbf{m}} = \frac{\Gamma(\lambda)}{\Gamma(\lambda+\mu)}\frac{(\lambda)_{p_1m_1+\cdots+p_nm_n}}{(\lambda+\mu)_{p_1m_1+\cdots+p_nm_n}}x^{\mathbf{m}}.$$

In particular we have

$$K^{\mu,\lambda}_{x_1,x\to(x_1,\frac{x_1}{x_2},\dots,\frac{x_1}{x_n})}x^{\mathbf{m}} = \frac{\Gamma(\lambda)}{\Gamma(\lambda+\mu)}\frac{(\lambda)_{m_1+\cdots+m_n}}{(\lambda+\mu)_{m_1+\cdots+m_n}}x^{\mathbf{m}}, \tag{37}$$

$$F_D(\lambda_0,\boldsymbol{\lambda},\mu;\mathbf{x}) = \frac{\Gamma(\mu)}{\Gamma(\lambda_0)}K^{\mu-\lambda_0,\lambda_0}_{x_1,x\to(x_1,\frac{x_1}{x_2},\dots,\frac{x_1}{x_n})}(1-\mathbf{x})^{-\boldsymbol{\lambda}} \quad \text{(cf. (20))}. \tag{38}$$

Here we note that the coordinate transformation $x\mapsto(x_1,\frac{x_1}{x_2},\dots,\frac{x_1}{x_n})$ gives a transformation of KZ equations of n variables (cf. [O3, §6]).

Let $\mathbf{p}=\begin{pmatrix}p_1 & p_2\\ q_1 & q_2\end{pmatrix}\in GL(2,\mathbb{Z})$ with $p_1,\ p_2,\ q_1,\ q_2\ge 0$. Put $\tilde{\mathbf{p}}=\mathbf{p}\otimes I_{n-2}\in GL(n,\mathbb{Z})$. Then

$$K^{\mu,(\lambda_1,\lambda_2)}_{(x_1,x_2),x\to x^{\tilde{\mathbf{p}}}}x^{\mathbf{m}} = \frac{\Gamma(\lambda_1)\Gamma(\lambda_2)}{\Gamma(\lambda_1+\lambda_2+\mu)}\frac{(\lambda_1)_{p_1m_1+p_2m_2}(\lambda_2)_{q_1m_1+q_2m_2}}{(\lambda_1+\lambda_2+\mu)_{(p_1+q_1)m_1+(p_2+q_2)m_2}}x^{\mathbf{m}},$$

$$L^{\mu,(\lambda_1,\lambda_2)}_{(x_1,x_2),x\to x^{\tilde{\mathbf{p}}}}x^{\mathbf{m}} = \frac{\Gamma(\lambda_1+\lambda_2+\mu)}{\Gamma(\lambda_1)\Gamma(\lambda_2)}\frac{(\lambda_1+\lambda_2+\mu)_{(p_1+q_1)m_1+(p_2+q_2)m_2}}{(\lambda_1)_{p_1m_1+p_2m_2}(\lambda_2)_{q_1m_1+q_2m_2}}x^{\mathbf{m}}.$$

Successive applications of these transformations to $(1-|\mathbf{x}|)^{-\lambda}$ or $(1-\mathbf{x})^{-\boldsymbol{\lambda}}$ or $e^{|\mathbf{x}|},\dots,$ we have many examples of integral representations of power series whose coefficients of $\frac{x^{\mathbf{m}}}{\mathbf{m}!}$ are expressed by the quotient of products of the form $(\lambda)_{p_1m_1+\cdots+p_nm_n}$.

The series

$$\phi(x,y)=\sum_{m=0}^{\infty}\sum_{n=0}^{\infty}\frac{\prod_{\nu=1}^{K}(a_\nu)_{m+n}\prod_{\nu=1}^{M}(b_\nu)_m\prod_{\nu=1}^{N}(c_\nu)_n}{\prod_{\nu=1}^{K'}(a'_\nu)_{m+n}\prod_{\nu=1}^{M'}(b'_\nu)_m\prod_{\nu=1}^{N'}(c'_\nu)_n}\frac{x^m}{m!}\frac{y^n}{n!} \tag{39}$$

with the condition

$$(K+M)-(K'+M')=(K+N)-(K'+N')=1$$

is an example. Then Appell's hypergeometric functions F_1, F_2, F_3, and F_4 correspond to $(K, M, N; K', N', N') = (1,1,1;1,0,0)$, $(1,1,1;0,1,1)$, $(0,2,2;1,0,0)$, and $(2,0,0;0,1,1)$, respectively. In general $\phi(x, y)$ may have several integral expressions as in the case of F_1 and F_2. Series (39) with $M = M' + 1$, $N = N' + 1$, and $K = K$ is a generalization of Appell's F_1, which will be given in Sect. 7 as an example.

Series

$$\begin{aligned} &K_x^{\gamma_2-\beta_2,\beta_2}\cdot(1-x)^{\alpha_2}\cdot K_x^{\gamma_1-\beta_1,\beta_1}(1-x)^{-\alpha_1} \\ &=\frac{\Gamma(\beta_1)\Gamma(\beta_2)}{\Gamma(\gamma_1)\Gamma(\gamma_2)}\sum_{m=0}^{\infty}\sum_{n=0}^{\infty}\frac{(\alpha_1)_m(\alpha_2)_m(\beta_1)_m(\beta_2)_{m+n}}{(\gamma_1)_m(\gamma_2)_{m+n}m!n!}x^{m+n} \end{aligned} \tag{40}$$

of $x \in \mathbb{C}$ satisfies a Fuchsian differential equation with the spectral type $211, 211, 211$ (cf. [O1, §13.7.5] and Remark 7) and the coefficients of x^k is not simple.

6 Differential Equations

In this section, we examine the differential equations satisfied by our invertible integral transformations of a function $u(x)$ in terms of the differential equation satisfied by $u(x)$. We denote by $W[x]$ the ring of differential operators with polynomial coefficients and put $W(x) = \mathbb{C}[x] \otimes W[x]$. Then $W[x]$ is called a Weyl algebra.

First, we review the related results in [O1]. The integral transformation $u \mapsto I_c^\mu u$ given by (1) satisfies

$$I_c^{-\mu}\circ I_c^\mu = \mathrm{id}, \tag{41}$$

$$I_c^\mu\circ\partial=\partial\circ I_c^\mu \text{ and } I_c^\mu\circ\vartheta=(\vartheta-\mu)\circ I_c^\mu \tag{42}$$

under the notation

$$\partial=\tfrac{d}{dx},\quad \vartheta=x\partial. \tag{43}$$

Hence, for an ordinary differential operator $P \in W[x]$, we define the middle convolution $\mathrm{mc}_\mu(P)$ of P by

$$\mathrm{mc}_\mu(P):=\partial^{-m}\sum_{i,j}a_{i,j}\partial^i(\vartheta-\mu)^j\in W[x]. \tag{44}$$

Here we first choose a positive integer k so that

$$\partial^k P = \sum_{i\geq 0,\ j\geq 0} a_{i,j}\partial^i\vartheta^j \quad ([a_{i,j},x]=[a_{i,j},\partial]=0), \tag{45}$$

and then we choose the maximal positive integer m so that $\mathrm{mc}_\mu(P) \in W[x]$. The number k can be taken to be the degree of P with respect to x. Then we have

$$Pu = 0 \ \Rightarrow\ (\mathrm{mc}_\mu(P))I_c^\mu u = 0. \tag{46}$$

The transformation $u \mapsto f(x)u$ of $u(x)$ defined by a suitable function $f(x)$ induces an automorphism $\mathrm{Ad}(f)$ of $W(x)$. Namely, $\mathrm{Ad}(f)$ is called an addition and defined by

$$\mathrm{Ad}(f)\partial = \partial - \tfrac{\partial(f)}{f} \ \text{ and } \ \mathrm{Ad}(f)x = x. \tag{47}$$

Hence $\frac{\partial(f)}{f}$ should be a rational function. Then $f(x)$ can be a function $(x-c)^\lambda$ or $f(x) = e^{r(x)}$ with a rational function $r(x)$.

There is another transformation $\mathrm{R}P$ of $P \in W(x) \setminus \{0\}$ where we define $\mathrm{R}P = r(x)P$ with $r(x) \in \mathbb{C}[x] \setminus \{0\}$ so that $r(x)P \in W[x]$ has the minimal degree with respect to x. Then $\mathrm{R}P$ is called the *reduced representative* of P. When we consider $\mathrm{mc}_\mu(P)$, we usually replace P by $\mathrm{R}P$.

Let $Pu = 0$ be a rigid Fuchsian differential equation on $\mathbb{P}^1$. Then it is proved in [O1] that P is obtained by successive applications of $\mathrm{Ad}(f)$ and $\mathrm{mc}_\mu \circ \mathrm{R}$ to ∂, and hence, we have an integral representation of the solution to this equation and moreover its expansion into a power series.

In a similar way, the author [O1, §13.10] examines Appell's hypergeometric functions using the integral transformation

$$\begin{aligned} J_x^\mu(u)(x) &:= \int_\Delta (1-s_1x_1-\cdots-s_nx_n)^\mu u(s_1,\ldots,s_n)\mathrm{d}s \\ &= \frac{1}{x_1\cdots x_n}\int_{\Delta'}(1-t_1-\cdots-t_n)^\mu u(\tfrac{t_1}{x_1},\ldots,\tfrac{t_n}{x_n})\mathrm{d}t \quad (t_j = s_jx_j) \end{aligned} \tag{48}$$

with certain regions Δ and Δ' of integrations and get integral representations of Appell's hypergeometric functions. For example, we put $u(x) = x_1^{\beta-1}x_2^{\beta'-1}(1-x_1-x_2)^{\gamma-\beta-\beta'-1}$ and $\Delta = \{(s_1,s_2) \mid s_1 \geq 0,\ s_2 \geq 0,\ 1-s_1-s_2 \geq 0\}$ to get Appell's F_1 and put $u(x) = x_1^{\lambda_1-1}(1-x_1)^{\lambda_2-1}x_2^{\lambda_1'-1}(1-x_2)^{\lambda_2'-1}$ and $\Delta = \{(s_1,s_2) \mid 0 \leq s_1 \leq 1,\ 0 \leq s_2 \leq 1\}$ to get Appell's F_2.

We show there the commuting relations

$$\begin{aligned} J_x^{\mu} \circ \vartheta_j &= (-1 - \vartheta_j) \circ J_x^{\mu}, \\ J_x^{\mu} \circ \partial_j &= x_j(\mu - \vartheta_1 - \cdots - \vartheta_n) \circ J_x^{\mu}, \end{aligned} \tag{49}$$

which correspond to (42) and imply the following proposition, and then we get the differential equations satisfied by Appell's hypergeometric functions.

In general, we have the following proposition:

Proposition 3 ([O1, Proposition 13.2]) *For a differential operator*

$$P = \sum_{\boldsymbol{\alpha},\boldsymbol{\beta}\in\mathbb{N}^n} c_{\boldsymbol{\alpha},\boldsymbol{\beta}} \partial^{\boldsymbol{\alpha}} \vartheta^{\boldsymbol{\beta}}$$

we have

$$J_x^{\mu}\bigl(Pu(x)\bigr) = J_x^{\mu}(P) J_x^{\mu}\bigl(u(x)\bigr),$$

$$J_x^{\mu}(P) := \sum_{\boldsymbol{\alpha},\boldsymbol{\beta}\in\mathbb{N}^n} c_{\boldsymbol{\alpha},\boldsymbol{\beta}} \Bigl(\prod_{k=1}^{n} \bigl(x_k(\mu - \vartheta_1 - \cdots - \vartheta_n)\bigr)^{\alpha_k} \Bigr) (-\boldsymbol{\vartheta} - 1)^{\boldsymbol{\beta}}.$$

Here the sums are finite and we use the notation.

$$\partial_x = \tfrac{\partial}{\partial x},\ \partial_y = \tfrac{\partial}{\partial y},\ \vartheta_x = x\partial_x,\ \vartheta_y = y\partial_y,\ \partial_i = \tfrac{\partial}{\partial x_i},\ \vartheta_i = x_i\partial_i. \tag{50}$$

Comparing the definition of integral transformations, we have the following:

Proposition 4 *The integral transformations defined in Sect. 2 is expressed by J_x^{μ} as follows:*

$$\begin{aligned} K_x^{\mu} &= \frac{1}{\Gamma(\mu)} T_{x\to(\frac{1}{x_1},\ldots,\frac{1}{x_n})} \circ x_1 \cdots x_n \cdot J_x^{\mu-1} \\ &\quad \textit{with}\quad \Delta' = \bigl\{(t_1,\ldots,t_n) \mid t_1 > 0, \ldots, t_n > 0,\ t_1 + \cdots + t_n < 1\bigr\} \end{aligned} \tag{51}$$

and

$$\begin{aligned} L_x^{\mu} &= \frac{\Gamma(\mu+n)}{(2\pi i)^n} J_x^{-\mu-n} \circ T_{x\to(\frac{1}{x_1},\ldots,\frac{1}{x_n})} \circ x_1 \cdots x_n \\ &\quad \textit{with}\quad \Delta' = \bigl\{(t_1,\ldots,t_n) \mid \operatorname{Re} t_1 = \cdots = \operatorname{Re} t_n = \tfrac{1}{n+1}\bigr\}. \end{aligned} \tag{52}$$

For $\mathbf{p} \in GL(n, \mathbb{Z})$ we put $\mathbf{q} = \mathbf{p}^{-1}$. Then

$$T_{x \to x^{\mathbf{p}}}(x_j) = x^{p_{*,j}} = \prod_{\nu=1}^{n} x_\nu^{p_{\nu,j}} \text{ and } T_{x \to x^{\mathbf{p}}}(\partial_i) = \sum_{j=1}^{n} q_{i,j} \frac{x_j}{x^{p_{j,i}}} \partial_j.$$

In particular

$$T_{x \to (\frac{1}{x_1}, \ldots, \frac{1}{x_n})}(x_j) = \tfrac{1}{x_j}, \quad T_{x \to (\frac{1}{x_1}, \ldots, \frac{1}{x_n})}(\partial_j) = -x_j^2 \partial_j$$

and

$$T_{x \to (\frac{1}{x_1}, \ldots, \frac{1}{x_n})} \circ x_1 \ldots x_n = \frac{1}{x_1 \cdots x_n} \circ T_{x \to (\frac{1}{x_1}, \ldots, \frac{1}{x_n})}$$

and thus we have the following lemma:

Lemma 1 *Defining*

$$\begin{aligned} \widetilde{u}(x_1, \ldots, x_n) &:= \tfrac{1}{x_1 \cdots x_n} u(\tfrac{1}{x_1}, \ldots, \tfrac{1}{x_n}), \\ \widetilde{P} = P^{\sim} &:= \sum a_{\boldsymbol{\alpha}}(\tfrac{1}{x_1}, \ldots, \tfrac{1}{x_n}) \prod_\nu (-x_\nu^2 \partial_\nu - x_\nu)^{\beta_\nu} \\ \textit{for} \quad P &= \sum a_{\boldsymbol{\alpha}}(x) \partial^{\boldsymbol{\alpha}} \in W(x), \end{aligned} \tag{53}$$

we have

$$\widetilde{Pu} = \widetilde{P}\widetilde{u}, \tag{54}$$

$$\widetilde{x_j} = \tfrac{1}{x_j}, \ \widetilde{\partial_j} = -x_j^2 \partial_j - x_j = -x_j(\vartheta_j + 1), \ \widetilde{\vartheta}_j = -\vartheta_j - 1. \tag{55}$$

Hence, Proposition 3, Proposition 4, and Lemma 1 show

$$K_x^\mu \circ \vartheta_j = (\widetilde{-1 - \vartheta_j}) \circ K_x^\mu = \vartheta_j \circ K_x^\mu, \tag{56}$$

$$\begin{aligned} K_x^\mu \circ \partial_j &= \big(x_j(\mu - 1 - \vartheta_1 - \cdots - \vartheta_n)\big)^{\sim} \circ K_x^\mu \\ &= \tfrac{1}{x_j}(\vartheta_1 + \cdots + \vartheta_n + \mu + n - 1) \circ K_x^\mu. \end{aligned} \tag{57}$$

Similarly, we have $L_x^\mu \circ \vartheta_i = \vartheta_i \circ L_x^\mu$ and

$$L_x^\mu \circ x_j(\vartheta_j + 1) = (x_j(\vartheta_1 + \cdots + \vartheta_n + \mu + n)) \circ L_x^\mu. \tag{58}$$

These relations can be checked by applying them to $x^{\boldsymbol{\alpha}}$. For example, it follows from (8) that

$$\begin{aligned}
L_x^\mu \circ x_j(\vartheta_j+1)x^{\boldsymbol{\alpha}} &= L_x^\mu(\alpha_j+1)x_jx^{\boldsymbol{\alpha}}\\
&= (\alpha_j+1)\frac{\Gamma(|\boldsymbol{\alpha}|+\mu+n+1)}{\Gamma(\alpha_1+1)\cdots\Gamma(\alpha_j+2)\cdots\Gamma(\alpha_n+1)}x_jx^{\boldsymbol{\alpha}}\\
&= \frac{(|\boldsymbol{\alpha}|+\mu+n)\Gamma(|\boldsymbol{\alpha}|+\mu+n)}{\Gamma(\boldsymbol{\alpha}+1)}x_jx^{\boldsymbol{\alpha}}\\
&= x_j(\vartheta_1+\cdots+\vartheta_n+\mu+n)L_x^\mu(x^{\boldsymbol{\alpha}}).
\end{aligned}$$

We also note that (56) is directly given by

$$(\vartheta_i K_x^\mu u)(x) = \frac{1}{\Gamma(\mu)}\int_0^1 (1-|\mathbf{t}|)^{\mu-1}t_ix_i(\partial_i u)(tx)\mathrm{d}t = (K_x^\mu\vartheta_i u)(x)$$

and the equality

$$\tfrac{\partial}{\partial t_i}\big((1-|\mathbf{t}|)^{\mu-1}u(tx)\big) = -(\mu-1)(1-|\mathbf{t}|)^{\mu-2}u(tx) + (1-t)^{\mu-1}x_i(\partial_i u)(tx)$$

shows

$$x_iK_x^\mu\partial_i = (\mu-1)K_x^{\mu-1}$$

and therefore

$$\begin{aligned}
\mu\int_0^1(1-|\mathbf{t}|)^{\mu-1}u(tx)\mathrm{d}t &= x_i\int_0^1(1-|\mathbf{t}|)^{\mu}(\partial_i u)(tx)\mathrm{d}t\\
&= x_i\int_0^1(1-|\mathbf{t}|)^{\mu-1}(1-t_1-\cdots-t_n)(\partial_i u)(tx)\mathrm{d}t,\\
x_iK_x^\mu\partial_i u &= \mu K_x^\mu u + \sum_{\nu=1}^n\frac{1}{\Gamma(\mu)}\frac{x_i}{x_\nu}\int_0^1(1-|\mathbf{t}|)^{\mu-1}(x_\nu\partial_i u)(tx)\mathrm{d}t\\
&= \mu K_x^\mu u + \sum_{\nu=1}^n\frac{x_i}{x_\nu}K_x^\mu\partial_i x_\nu u - K_x^\mu u\\
&= (\mu-1)K_x^\mu u + \sum_{\nu=1}^n K_x^\mu\partial_\nu x_\nu u\\
&= (\mu+n-1)K_x^\mu u + \sum_{\nu=1}^n\vartheta_\nu K_x^\mu u,
\end{aligned}$$

which implies (57).

Thus, we have the following theorem:

Theorem 2 *Suppose $u(x)$ satisfies $Pu(x) = 0$ with a certain $P \in W(x)$.*

(i) *Putting*

$$Q = \mathrm{R}P = \sum_{\boldsymbol{\alpha},\,\boldsymbol{\beta} \in \mathbb{N}^n} a_{\boldsymbol{\alpha},\boldsymbol{\beta}} x^{\boldsymbol{\alpha}} \partial^{\boldsymbol{\beta}}, \tag{59}$$

we choose $\boldsymbol{\gamma} \in \mathbb{Z}^n$ so that

$$\partial^{\boldsymbol{\gamma}} Q = \sum_{\boldsymbol{\alpha},\,\boldsymbol{\beta} \in \mathbb{N}^n} c_{\boldsymbol{\alpha},\boldsymbol{\beta}} \partial^{\boldsymbol{\alpha}} \vartheta^{\boldsymbol{\beta}} \quad (c_{\boldsymbol{\alpha},\boldsymbol{\beta}} \in \mathbb{C}). \tag{60}$$

Then we have $K_x^{\mu}(\partial^{\boldsymbol{\gamma}} Q) K_x^{\mu} u(x) = 0$ with

$$K_x^{\mu}\Big(\sum c_{\boldsymbol{\alpha},\boldsymbol{\beta}} \partial^{\boldsymbol{\alpha}} \vartheta^{\boldsymbol{\beta}}\Big) := \mathrm{R} \sum c_{\boldsymbol{\alpha},\boldsymbol{\beta}} \Big(\prod_{k=1}^{n} \big(\tfrac{1}{x_k}(\vartheta_1 + \cdots + \vartheta_n + \mu + n - 1)\big)^{\alpha_k} \Big) \vartheta^{\boldsymbol{\beta}}. \tag{61}$$

(ii) *Putting*

$$Q = \mathrm{R}\tilde{P}, \tag{62}$$

we choose $\boldsymbol{\gamma} \in \mathbb{Z}^n$ so that (60) *holds. Then we have $L_x^{\mu}(\partial^{\boldsymbol{\gamma}} Q) L_x^{\mu} u(x) = 0$ with*

$$L_x^{\mu}\Big(\sum c_{\boldsymbol{\alpha},\boldsymbol{\beta}} \partial^{\boldsymbol{\alpha}} \vartheta^{\boldsymbol{\beta}}\Big) := \mathrm{R} \sum c_{\boldsymbol{\alpha},\boldsymbol{\beta}} \Big(\prod_{k=1}^{n} \big(x_k(\mu - \vartheta_1 - \cdots - \vartheta_n)\big)^{\alpha_k} \Big) (-\boldsymbol{\vartheta} - 1)^{\boldsymbol{\beta}}. \tag{63}$$

Remark 3

(i) In Theorem 2 i), $\boldsymbol{\gamma} = (\gamma_1, \ldots, \gamma_n)$ can be taken by

$$\gamma_j = \max\{0, \alpha_j - \beta_j \,|\, a_{(\alpha_1,\ldots,\alpha_n),(\beta_1,\ldots,\beta_n)} \neq 0\} \qquad (1 \le j \le n).$$

(ii) If $P \in W[x, y]$ in Theorem 2, it is clear that the theorem is valid under the assumption $c_{\boldsymbol{\alpha},\boldsymbol{\beta}} \in W[y]$.

(iii) Suppose $u(x) \in \mathscr{O}_0$ satisfies $P_1 P_2 u(x) = 0$ with $P_1,\ P_2 \in W[x]$ and $\{u \in \mathscr{O}_0 \mid P_1 u = 0\} = \{0\}$. Then $P_2 u(x) = 0$.

(iv) Without the assumption (32), we can define transformations $K^{\mu,\boldsymbol{\lambda}}_{x,x\to x^{\mathbf{P}}}$ and $L^{\mu,\boldsymbol{\lambda}}_{x,x\to x^{\mathbf{P}}}$ on $\mathscr{O}_0$ by (35) and (36). Even in this case, the results in this section are clearly valid.

We will calculate some examples. By the integral expression

$$F_1(\lambda_0, \lambda_1, \lambda_2, \mu; x, y) = \frac{\Gamma(\mu)}{\Gamma(\lambda_1)\Gamma(\lambda_2)} K_{x,y}^{\mu-\lambda_1-\lambda_2,\lambda_1,\lambda_2}(1-x-y)^{-\lambda_0}$$
$$= C_1 \int_{\substack{s>0,\, t>0\\ s+t<1}} (1-s-t)^{\mu-\lambda_1-\lambda_2-1} s^{\lambda_1-1} t^{\lambda_2-1} (1-sx-ty)^{-\lambda_0} \mathrm{d}s\mathrm{d}t$$

corresponding to (20) and (33), we calculate the system of differential equations satisfied by $F_1(\lambda_0, \lambda_1, \lambda_2, \mu; x, y)$ as follows. Putting

$$h := x^{\lambda_1-1} y^{\lambda_2-1} (1-x-y)^{-\lambda_0},$$

we have

$$\begin{aligned}
\mathrm{Ad}(h)\partial_x &= \partial_x - \tfrac{\lambda_1-1}{x} - \tfrac{\lambda_0}{1-x-y},\\
\mathrm{Ad}(h)\partial_y &= \partial_y - \tfrac{\lambda_2-1}{y} - \tfrac{\lambda_0}{1-x-y},\\
\mathrm{Ad}(h)(\vartheta_x + \vartheta_y) &= \vartheta_x + \vartheta_y - \tfrac{\lambda_0}{1-x-y} - (\lambda_1 + \lambda_2 - \lambda_0 - 2),\\
\mathrm{Ad}(h)(\vartheta_x + \vartheta_y - \partial_x) &= \vartheta_x + \vartheta_y - \partial_x + \tfrac{\lambda_1-1}{x} - (\lambda_1 + \lambda_2 - \lambda_0 - 2).
\end{aligned}$$

Hence, we put

$$\begin{aligned}
Q &:= \mathrm{R}\,\mathrm{Ad}(h)(\vartheta_x + \vartheta_y - \partial_x)\\
&= (\vartheta_x + 1)(\vartheta_x + \vartheta_y - \lambda_1 - \lambda_2 + \lambda_0 + 2) - \partial_x\vartheta_x + \lambda_1 - 1.
\end{aligned}$$

and we have

$$\begin{aligned}
K_x^{\mu-\lambda_1-\lambda_2}(Q) &= x(\vartheta_x + 1)(\vartheta_x + \vartheta_y - \lambda_1 - \lambda_2 + \lambda_0 + 2)\\
&\quad - (\vartheta_x + \vartheta_y + \mu - \lambda_1 - \lambda_2 + 1)(\vartheta_x - \lambda_1 + 1),\\
\mathrm{Ad}(x^{1-\lambda_1} y^{1-\lambda_2}) K_x^{\mu}(Q) &= x(\vartheta_x + \lambda_1)(\vartheta_x + \vartheta_y + \lambda_0) - (\vartheta_x + \vartheta_y + \mu - 1)\vartheta_x\\
&= x\big((\vartheta_x + \lambda_1)(\vartheta_x + \vartheta_y + \lambda_0) - \partial_x(\vartheta_x + \vartheta_y + \mu - 1)\big).
\end{aligned}$$

Hence, $F_1(\lambda_0, \lambda_1, \lambda_2, \mu; x, y)$ is a solution of the system

$$\begin{cases}
(\vartheta_x + \lambda_1)(\vartheta_x + \vartheta_y + \lambda_0) - \partial_x(\vartheta_x + \vartheta_y + \mu - 1)u_1 = 0,\\
(\vartheta_y + \lambda_2)(\vartheta_x + \vartheta_y + \lambda_0) - \partial_y(\vartheta_x + \vartheta_y + \mu - 1)u_1 = 0.
\end{cases} \tag{64}$$

Next, we consider the integral representation

$$F_1(\lambda_0, \lambda_1, \lambda_2, \mu; x, y) = \frac{\Gamma(\mu)}{\Gamma(\lambda_0)} K^{\mu-\lambda_0,\lambda_0}_{x,(x,y)\to(x,\frac{x}{y})} (1-x)^{-\lambda_1}(1-y)^{-\lambda_2}$$
$$= C_1' \int_0^1 t^{\lambda_0-1}(1-t)^{\mu-\lambda_0}(1-tx)^{-\lambda_1}(1-ty)^{-\lambda_2}dt$$

corresponding to (38). Since

$$T_{(x,y)\mapsto(x,\frac{x}{y})} : \partial_x \mapsto \tfrac{1}{x}(\vartheta_x + \vartheta_y),\ \partial_y \mapsto -\tfrac{y}{x}\vartheta_y,\ \vartheta_x \mapsto \vartheta_x + \vartheta_y,\ \vartheta_y \mapsto -\vartheta_y,$$

we have

$$\begin{aligned}
\partial_x &\xrightarrow{\mathrm{Ad}((1-x)^{-\lambda_1}(1-y)^{-\lambda_2})} \partial_x - \tfrac{\lambda_1}{1-x} \xrightarrow{T_{(x,y)\to(x,\frac{x}{y})}} \tfrac{1}{x}(\vartheta_x+\vartheta_y) - \tfrac{\lambda_1}{1-x}\\
&\xrightarrow{\mathrm{Ad}(x^{\lambda_0-1})} \tfrac{1}{x}(\vartheta_x+\vartheta_y) - \tfrac{\lambda_0-1}{x} - \tfrac{\lambda_1}{1-x}\\
&\xrightarrow{\mathrm{R}} (1-x)(\vartheta_x+\vartheta_y-\lambda_0+1) - \lambda_1 x\\
&\xrightarrow{\partial_x} (\partial_x - \vartheta_x - 1)(\vartheta_x+\vartheta_y-\lambda_0+1) - \lambda_1(\vartheta_x+1)\\
&\xrightarrow{K_x^{\mu-\lambda_0}} (\partial_x + \tfrac{\mu-\lambda_0}{x} - \vartheta_x - 1)(\vartheta_x+\vartheta_y-\lambda_0+1) - \lambda_1(\vartheta_x+1)\\
&\xrightarrow{\mathrm{Ad}(x^{1-\lambda_0})} (\partial_x + \tfrac{\mu-1}{x} - \vartheta_x - \lambda_0)(\vartheta_x+\vartheta_y) - \lambda_1(\vartheta_x-\lambda_0)\\
&\xrightarrow{T_{(x,y)\to(x,\frac{x}{y})}} (\tfrac{1}{x}(\vartheta_x+\vartheta_y) + \tfrac{\mu-1}{x} - \vartheta_x - \vartheta_y - \lambda_0)\vartheta_x - \lambda_1(\vartheta_x-\lambda_0)\\
&= \partial_x(\vartheta_x+\vartheta_y+\mu-1) - (\vartheta_x+\lambda_1)(\vartheta_x+\vartheta_y+\lambda_0).
\end{aligned}$$

Thus, we also get the system (64) characterizing $F_1(\lambda_0, \lambda_1, \lambda_2, \mu; x, y)$.

We have similar calculations for other Appell's hypergeometric series as follows:

$$F_2(\lambda_0; \mu_1, \mu_2; \lambda_1, \lambda_2; x, y) = \frac{\Gamma(\lambda_1)\Gamma(\lambda_2)}{\Gamma(\mu_1)\Gamma(\mu_2)} K_x^{\lambda_1-\mu_1,\mu_1} K_y^{\lambda_2-\mu_2,\mu_2}(1-x-y)^{-\lambda_0}$$
$$= C_2 \int_0^1\int_0^1 s^{\mu_1}t^{\mu_2}(1-s)^{\lambda_1-\mu_1-1}(1-t)^{\lambda_2-\mu_2-1}(1-sx-ty)^{-\lambda_0}\tfrac{ds}{s}\tfrac{dt}{t},$$

$$\begin{aligned}
\partial_x &\xrightarrow{\mathrm{Ad}(x^{\mu_1-1}y^{\mu_2-1}(1-x)^{\lambda_1-\mu_1-1}(1-y)^{\lambda_2-\mu_2-1})} \partial_x - \tfrac{\mu_1-1}{x} + \tfrac{\lambda_1-\mu_1-1}{1-x}\\
&\xrightarrow{\mathrm{R}} x(1-x)\partial_x + (\lambda_1-2)x - (\mu_1-1)\\
&\xrightarrow{\partial_x} \partial_x x(-\vartheta_x+\lambda_1-2) + \partial_x(\vartheta_x-\mu_1+1)
\end{aligned}$$

$$\xrightarrow{J_{x,y}^{-\lambda_0}} -\vartheta_x(\vartheta_x+1+\lambda_1-2)+x(-\lambda_0-\vartheta_x-\vartheta_y)(-1-\vartheta_x-\mu_1+1)$$
$$= x\big((\vartheta_x+\mu_1)(\vartheta_x+\vartheta_y+\lambda_0)-\partial_x(\vartheta_x+\lambda_1-1)\big)$$

and

$$F_3(\lambda_1,\lambda_2;\lambda_1',\lambda_2';\mu;x,y)=\frac{\Gamma(\mu)}{\Gamma(\lambda_1)\Gamma(\lambda_2)}K_{x,y}^{\mu-\lambda_1-\lambda_2}(1-x)^{-\lambda_1'}(1-y)^{-\lambda_2'}$$
$$= C_3\int_{\substack{s>0,t>0\\ s+t<1}} s^{\lambda_1}t^{\lambda_2}(1-s-t)^{\mu-1}(1-sx)^{-\lambda_1'}(1-ty)^{-\lambda_2'}\tfrac{ds}{s}\tfrac{dt}{t},$$
$$\partial_x \xrightarrow{\mathrm{R\,Ad}(x^{\lambda_1-1}y^{\lambda_2-1}(1-x)^{-\lambda_1'}(1-y)^{-\lambda_2'})} x(1-x)\partial_x+(\lambda_1-\lambda_1'-1)x-(\lambda_1-1)$$
$$\xrightarrow{\partial_x} \partial_x x(-\vartheta_x+\lambda_1-\lambda_1'-1)+\tfrac{1}{x}(\vartheta_x+\vartheta_y+\mu-\lambda_1'-\lambda_2-1)(\vartheta_x-\lambda_1+1)$$
$$\xrightarrow{\mathrm{Ad}(x^{1-\lambda_1}y^{1-\lambda_2})} -(\vartheta_x+\lambda_1)(\vartheta_x+\lambda_1')+\partial_x(\vartheta_x+\vartheta_y+\mu-1)$$

and

$$F_4(\mu,\lambda_0;\lambda_1,\lambda_2;x,y)=\frac{\Gamma(\lambda_1)\Gamma(\lambda_2)}{\Gamma(\mu)}L_{x,y}^{\mu-\lambda_1-\lambda_2,\lambda_1,\lambda_2}(1-x-y)^{-\lambda_0}$$
$$= C_4\int_{\frac13-i\infty}^{\frac13+i\infty} s^{\lambda_1}t^{\lambda_2}(1-s-t)^{\lambda_1+\lambda_2-\mu-2}(1-sx-ty)^{-\lambda_0}\tfrac{ds}{s}\tfrac{dt}{t},$$
$$\partial_x \xrightarrow{\mathrm{Ad}((1-x-y)^{-\lambda_0})} \partial_x-\tfrac{\lambda_0}{1-x-y}$$
$$\partial_x-\vartheta_x-\vartheta_y-\lambda_0 \qquad \big((\partial_x-\vartheta_x-\vartheta_y-\lambda_0)(1-x-y)^{-\lambda_0}=0\big)$$
$$\xrightarrow{\mathrm{Ad}(x^{\lambda_1-1}y^{\lambda_2-1})} \partial_x-\tfrac{\lambda_1-1}{x}-(\vartheta_x+\vartheta_y+\lambda_0-\lambda_1+1-\lambda_2+1)$$
$$\xrightarrow{\frac{1}{xy}T_{(x,y)\mapsto(\frac1x,\frac1y)}} -x(\vartheta_x+1)-(\lambda_1-1)-(\lambda_1-1)x+\vartheta_x+\vartheta_y-\lambda_0+\lambda_1+\lambda_2$$
$$\xrightarrow{\partial_x} -\partial_x x(\vartheta_x+\lambda_1)+\partial_x(\vartheta_x+\vartheta_y-\lambda_0+\lambda_1+\lambda_2)$$
$$\xrightarrow{\mathrm{Ad}(L_{x,y}^{\mu-\lambda_1-\lambda_2})} \vartheta_x(-\vartheta_x-1+\lambda_1)$$
$$+x(\lambda_1+\lambda_2-\mu-2-\vartheta_x-\vartheta_y)(-\vartheta_x-\vartheta_y-\lambda_0+\lambda_1+\lambda_2$$
$$\xrightarrow{\partial_x} -\partial_x x(\vartheta_x+\lambda_1)+\partial_x(\vartheta_x+\vartheta_y-\lambda_0+\lambda_1+\lambda_2)$$
$$\xrightarrow{L_{x,y}^{\mu-\lambda_1-\lambda_2}} \vartheta_x(-\vartheta_x-1-\lambda_1)$$

$$+x(\lambda_1+\lambda_2-\mu-2-\vartheta_x-\vartheta_y)(-\vartheta_x-\vartheta_y-\lambda_0+\lambda_1+\lambda_2-2)$$

$$\xrightarrow{\mathrm{Ad}(x^{1-\lambda_1}y^{1-\lambda_2})} x\Big((\vartheta_x+\vartheta_y+\mu)(\vartheta_x+\vartheta_y+\lambda_0)-\partial_x(\vartheta_x+\lambda_1-1)\Big).$$

Here, C_1, C_1', C_2, C_3, and C_4 are constants easily obtained from the integral formula in Sect. 2 with putting $x=y=0$. Hence,

$$\begin{cases} u_2 = F_2(\lambda_0;\mu_1,\mu_2;\lambda_1,\lambda_2;x,y),\\ u_3 = F_3(\lambda_1,\lambda_2;\lambda_1',\lambda_2';\mu;x,y),\\ u_4 = F_4(\mu,\lambda_0;\lambda_1,\lambda_2;x,y) \end{cases} \tag{65}$$

are solutions of the system

$$\begin{cases} \big((\vartheta_x+\mu_1)(\vartheta_x+\vartheta_y+\lambda_0)-\partial_x(\vartheta_x+\lambda_1-1)\big)u_2=0,\\ \big((\vartheta_y+\mu_2)(\vartheta_x+\vartheta_y+\lambda_0)-\partial_y(\vartheta_y+\lambda_2-1)\big)u_2=0, \end{cases} \tag{66}$$

$$\begin{cases} \big((\vartheta_x+\lambda_1)(\vartheta_x+\lambda_1')-\partial_x(\vartheta_x+\vartheta_y+\mu-1)\big)u_3=0,\\ \big((\vartheta_y+\lambda_2)(\vartheta_x+\lambda_2')-\partial_y(\vartheta_x+\vartheta_y+\mu-1)\big)u_3=0, \end{cases} \tag{67}$$

$$\begin{cases} \big((\vartheta_x+\vartheta_y+\mu)(\vartheta_x+\vartheta_y+\lambda_0)-\partial_x(\vartheta_x+\lambda_1-1)\big)u_4=0,\\ \big((\vartheta_x+\vartheta_y+\mu)(\vartheta_x+\vartheta_y+\lambda_0)-\partial_y(\vartheta_y+\lambda_2-1)\big)u_4=0. \end{cases} \tag{68}$$

Remark 4 The above systems are directly obtained from the adjacent relations of the coefficients of Appell's hypergeometric series. Here we get them by the transformations of systems of differential equations corresponding to integral transformations of functions discussed in this chapter so that it can be applied to general cases.

7 KZ Equations

A Pfaffian system

$$du = \sum_{0\le i<j\le q} A_{i,j}\frac{d(x_i-x_j)}{x_i-x_j}u \tag{69}$$

with an unknown N vector u and constant square matrices $A_{i,j}$ of size N is called a KZ (Knizhnik-Zamolodchikov-type) equation of rank N (cf. [KZ]), which equals the system of equations

$$\mathcal{M}:\ \frac{\partial u}{\partial x_i} = \sum_{\substack{0\le\nu\le q\\ \nu\ne i}} \frac{A_{i,\nu}}{x_i-x_\nu}u \qquad (i=0,\dots,q) \tag{70}$$

with denoting $A_{j,i} = A_{i,j}$. The matrix $A_{i,j}$ is called the *residue matrix* of $\mathscr{M}$ at $x_i = x_j$. Here we always assume the *integrability condition*

$$\begin{cases} [A_{i,j}, A_{k,\ell}] = 0 & (\forall\{i,j,k,\ell\} \subset \{0,\dots,q\}), \\ [A_{i,j}, A_{i,k} + A_{j,k}] = 0 & (\forall\{i,j,k\} \subset \{0,\dots,q\}), \end{cases} \tag{71}$$

which follows from the condition $\mathrm{dd}u = 0$. Here $i,\ j,\ k,\ \ell$ are mutually different indices:

Definition 5 Using the notation

$$A_{i,i} = A_{\emptyset} = A_i = 0, \quad A_{i,j} = A_{j,i} \quad (i,\ j \in \{0,1,\dots,q+1\}),$$

$$A_{i,q+1} := -\sum_{\nu=0}^{n} A_{i,\nu},$$

$$A_{i_1,i_2,\dots,i_k} := \sum_{1\le\nu<\nu'\le k} A_{i_\nu,i_{\nu'}} \quad (\{i_1,\dots,i_k\} \subset \{0,\dots,q+1\}),$$

we have

$$[A_I, A_J] = 0 \quad \text{if } I \cap J = \emptyset \text{ or } I \subset J \text{ with } I,\ J \subset \{0,\dots,q+1\}. \tag{72}$$

The matrix $A_{i,j}$ is called the residue matrix of $\mathscr{M}$ at $x_i = x_j$, and x_{q+1} corresponds to ∞ in $P^1_{\mathbb{C}}$.

We note that any rigid irreducible Fuchsian system

$$\mathscr{N} : \frac{\mathrm{d}u}{\mathrm{d}x} = \sum_{i=1}^{q} \frac{B_i}{x - x_i} u \tag{73}$$

can be extended to KZ equation $\mathscr{M}$ with $x = x_0$ and $B_i = A_{0,i}$, which follows from the result by Haraoka [Ha] extending a middle convolution on KZ equations.

We assume that $\mathscr{M}$ is irreducible at a generic value of the holomorphic parameter contained in $\mathscr{M}$. Then it is shown in [O3, §1] that $A_{0,\dots,q}$ is a scalar matrix κI_N with $\kappa \in \mathbb{C}$ and by the gage transformation $u \mapsto (x_{q-1} - x_q)^{-\kappa} u$, we may assume that $\mathscr{M}$ is *homogeneous*, which means

$$A_{i_0,\dots,i_q} = 0 \quad (0 \le i_0 < i_1 < \cdots < i_q \le q+1). \tag{74}$$

Then the symmetric group $\mathfrak{S}_{q+2}$, which is identified with the permutation group of the set of indices $\{0, 1, \dots, q+1\}$, naturally acts on the space of the homogeneous KZ equations (cf. [O3, §6]):

$$\begin{array}{ccccccc} x_0 & x_1 & x_2 & \cdots & x_{q-2} & x_{q-1} & x_q & x_{q+1} \\ \circ & \circ & \circ & \cdots & \circ & \circ & \circ & \circ \\ x & y_1 & y_2 & & y_{q-2} & 1 & 0 & \infty \end{array}$$

$$\begin{array}{lll} (0,1) & : x \leftrightarrow y_1, & \\ (i,i+1) & : y_i \leftrightarrow y_{i+1} & (1 \le i \le q-3), \\ (q-2,q-1) & : (x,y_1,\dots,y_{q-1},y_{q-2}) \leftrightarrow \left(\frac{x}{y_{q-2}}, \frac{y_1}{y_{q-2}}, \dots, \frac{y_{q-1}}{y_{q-2}}, \frac{1}{y_{q-2}}\right), & \\ (q-1,q) & : (x,y_1,\dots,y_{q-1},y_{q-2}) \leftrightarrow (1-x, 1-y_1, \dots, 1-y_{q-1}, 1-y_{q-2}), & \\ (q,q+1) & : (x,y_1,\dots,y_{q-1},y_{q-2}) \leftrightarrow \left(\frac{1}{x}, \frac{1}{y_1}, \dots, \frac{1}{y_{q-1}}, \frac{1}{y_{q-2}}\right). & \end{array}$$

Here we put $(x_0, \dots, x_{q+1}) = (x, y_1, \dots, y_{q-1}, 1, 0, \infty)$ by a transformation $\mathbb{P}^1 \ni x \mapsto ax + b$, which keeps residue matrices $A_{i,j}$.

For simplicity, we assume $n = q - 1 = 2$ and put $(x_0, x_1, x_2, x_3, x_4) = (x, y, 1, 0, \infty)$. Then (74) means

$$A_{01} + A_{01} + A_{03} + A_{12} + A_{13} + A_{23} = 0 \tag{75}$$

and the five residue matrices A_{01}, A_{01}, A_{03}, A_{12}, and A_{13} uniquely determine the other five residue matrices A_{23} and A_{i4} with $0 \le i \le 3$, and the action of $\mathfrak{S}_5$ is generated by the 4 involutions

$$\begin{array}{rcl} (x_0, x_1, x_2, x_3, x_4) & \to & (x, y, 1, 0, \infty), \\ x_0 \leftrightarrow x_1 & \to & (x, y) \leftrightarrow (y, x), \\ x_1 \leftrightarrow x_2 & \to & (x, y) \leftrightarrow (\frac{x}{y}, \frac{1}{y}), \\ x_2 \leftrightarrow x_3 & \to & (x, y) \leftrightarrow (1-x, 1-y), \\ x_3 \leftrightarrow x_4 & \to & (x, y) \leftrightarrow (\frac{1}{x}, \frac{1}{y}). \end{array}$$

In particular, the KZ system is determined by the equation

$$\mathscr{M} : \begin{cases} \dfrac{\partial u}{\partial x} = \dfrac{A_{01}}{x-y}u + \dfrac{A_{02}}{x-1}u + \dfrac{A_{03}}{x}u, \\ \dfrac{\partial u}{\partial y} = \dfrac{A_{01}}{y-x}u + \dfrac{A_{12}}{y-1}u + \dfrac{A_{13}}{y}u. \end{cases} \tag{76}$$

Remark 5 Coordinate transformations corresponding to the involutions

$$\begin{array}{lcl} (x_0, x_1, x_2, x_3, x_4) \leftrightarrow (x_2, x_1, x_0, x_4, x_3) & \to & (x, y) \leftrightarrow (x, \frac{x}{y}) \\ (x_0, x_1, x_2, x_3, x_4) \leftrightarrow (x_0, x_2, x_1, x_4, x_3) & \to & (x, y) \leftrightarrow (\frac{y}{x}, y) \end{array}$$

give the local coordinates of the blowing up of the singularities of Eq. (76) at the origin:

$$\begin{array}{cccc} \mathfrak{S}_5 \ni x_0 \leftrightarrow x_1 & x_1 \leftrightarrow x_2 & x_2 \leftrightarrow x_3 & x_3 \leftrightarrow x_4 \\ (x,y) \mapsto (y,x) & (\frac{x}{y},\frac{1}{y}) & (1-x,1-y) & (\frac{1}{x},\frac{1}{y}) \end{array}$$

$$\begin{aligned} (x,y) &\leftrightarrow (x,\tfrac{x}{y}) \\ \{|x|<\varepsilon,\ |y|<C|x|\} &\leftrightarrow \{|x|<\varepsilon,\ |y|>C^{-1}\} \\ x=y=0 &\leftrightarrow x=0 \end{aligned}$$

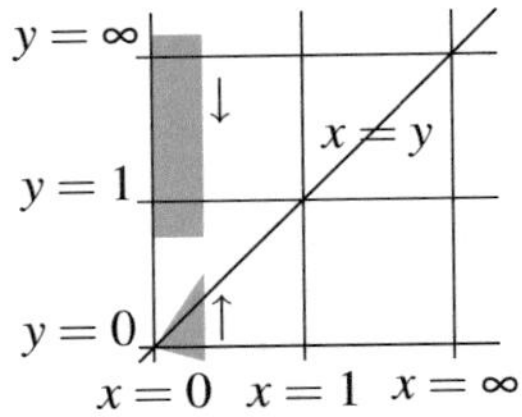

$y=\infty$, $y=1$, $y=0$, $x=y$, $x=0$ $x=1$ $x=\infty$

Now we review the result in [DR, DR2, Ha] by using the transformations defined in this chapter. The convolution of KZ equation (70) corresponds to the transformation defined by

$$(\tilde{\mathrm{mc}}_\mu u)(x,y) := \begin{pmatrix} I_{x,0}^{\mu+1}\frac{u(x,y)}{x-y} \\ I_{x,0}^{\mu+1}\frac{u(x,y)}{y} \\ I_{x,0}^{\mu+1}\frac{u(x,y)}{x} \end{pmatrix} = \begin{pmatrix} \frac{1}{\Gamma(\mu+1)}\int_0^x (1-t)^\mu \frac{u(t,y)}{t-y}\mathrm{d}t \\ \frac{1}{\Gamma(\mu+1)}\int_0^x (1-t)^\mu \frac{u(t,y)}{y}\mathrm{d}t \\ \frac{1}{\Gamma(\mu+1)}\int_0^x (1-t)^\mu \frac{u(t,y)}{t}\mathrm{d}t \end{pmatrix}.$$

We put $\tilde{K}_x^\mu = x^{-\mu}\circ \tilde{\mathrm{mc}}_\mu$ and $\tilde{K}_x^{\mu,\lambda} = x^{-\lambda}\circ \tilde{K}_x^\mu \circ x^\lambda$. Then

$$(\tilde{K}_x^{\mu,\lambda}u)(x,y) = \begin{pmatrix} K_x^{\mu+1,\lambda}\frac{xu(x,y)}{x-y} \\ K_x^{\mu+1,\lambda}\frac{xu(x,y)}{x-1} \\ K_x^{\mu+1,\lambda}u(x,y) \end{pmatrix}. \tag{77}$$

Putting $\tilde{u} = \tilde{K}_x^{\mu,\lambda}u$ for a solution u of KZ equation (70), we have the KZ equation

$$\frac{\partial \tilde{u}}{\partial x_i} = \sum_{\substack{0\le\nu\le 3\\ \nu\ne i}} \frac{\tilde{A}_{i,\nu}}{x_i - x_\nu}\tilde{u} \tag{78}$$

satisfied by $\tilde{u}$.

Since this equation is reducible in general, we consider the reduced equation

$$\bar{\mathscr{M}} : \frac{\partial \bar{u}}{\partial x_i} = \sum_{\substack{0\le\nu\le 3\\ \nu\ne i}} \frac{\bar{A}_{i,\nu}}{x_i - x_\nu}\bar{u}. \tag{79}$$

The residue matrices $\tilde{A}_{i,j}$ and $\bar{A}_{i,j}$ are obtained from the results in [DR, DR2, Ha]:

$$\tilde{A}_{01} = \begin{pmatrix} \mu + A_{01} & A_{02} & A_{03} + \lambda \\ 0 & 0 & 0 \\ 0 & 0 & 0 \end{pmatrix}, \qquad \tilde{A}_{02} = \begin{pmatrix} 0 & 0 & 0 \\ A_{01} & \mu + A_{02} & A_{03} + \lambda \\ 0 & 0 & 0 \end{pmatrix},$$
$$\tilde{A}_{03} = \begin{pmatrix} -\mu - \lambda & 0 & 0 \\ 0 & -\mu - \lambda & 0 \\ A_{01} & A_{02} & A_{03} \end{pmatrix}, \qquad \tilde{A}_{04} = \begin{pmatrix} -A_{01} + \lambda & -A_{02} & -A_{03} - \lambda \\ -A_{01} & -A_{02} + \lambda & -A_{03} - \lambda \\ -A_{01} & -A_{02} & -A_{03} \end{pmatrix},$$
$$\tilde{A}_{12} = \begin{pmatrix} A_{12} + A_{02} & -A_{02} & 0 \\ -A_{01} & A_{12} + A_{01} & 0 \\ 0 & 0 & A_{12} \end{pmatrix}, \quad \tilde{A}_{13} = \begin{pmatrix} A_{13} + A_{03} + \lambda & 0 & -A_{03} - \lambda \\ 0 & A_{13} & 0 \\ -A_{01} & 0 & A_{01} + A_{13} \end{pmatrix},$$
$$\tilde{A}_{14} = \begin{pmatrix} A_{23} - \mu - \lambda & 0 & 0 \\ A_{01} & A_{02} + A_{03} + A_{23} & 0 \\ A_{01} & 0 & A_{02} + A_{03} + A_{23} \end{pmatrix}, \tag{80}$$
$$\tilde{A}_{23} = \begin{pmatrix} A_{23} & 0 & 0 \\ 0 & A_{03} + A_{23} + \lambda & -A_{03} - \lambda \\ 0 & -A_{02} & A_{02} + A_{23} \end{pmatrix},$$
$$\tilde{A}_{24} = \begin{pmatrix} A_{01} + A_{13} + A_{03} & A_{02} & 0 \\ 0 & A_{13} - \mu - \lambda & 0 \\ 0 & A_{02} & A_{01} + A_{13} + A_{03} \end{pmatrix},$$
$$\tilde{A}_{34} = \begin{pmatrix} A_{12} + A_{01} + A_{02} + \mu & 0 & A_{03} + \lambda \\ 0 & A_{12} + A_{01} + A_{02} + \mu & A_{03} + \lambda \\ 0 & 0 & A_{12} \end{pmatrix}.$$

Here we denote $A_{01} = A_{0,1}$, etc. for simplicity.

Then the subspace

$$\begin{aligned} \mathscr{L} &:= \begin{pmatrix} \ker A_{01} \\ \ker A_{02} \\ \ker A_{03} + \lambda \end{pmatrix} + \ker(\tilde{A}_{04} - \mu - \lambda) \\ &= \begin{pmatrix} \ker A_y \\ \ker A_1 \\ \ker A_0 + \lambda \end{pmatrix} + \ker \begin{pmatrix} A_y + \mu & A_1 & A_0 + \lambda \\ A_y & A_1 + \mu & A_0 + \lambda \\ A_y & A_1 & A_0 + \mu + \lambda \end{pmatrix} \end{aligned} \tag{81}$$

of $\mathbb{C}^{3N}$ satisfies $\tilde{A}_{i,j}\mathscr{L} \subset \mathscr{L}$. We define $\bar{A}_{i,j}$ the square matrices of size $3N - \dim \mathscr{L}$, which correspond to linear transformations induced by $\tilde{A}_{i,j}$, respectively, on the quotient space $\mathbb{C}^{3N}/\mathscr{L}$.

It is known that if Eq. (73) is irreducible, then the corresponding ordinary differential equation defined by $\bar{A}_{0,1}$, $\bar{A}_{0,2}$, and $\bar{A}_{0,3}$ is irreducible (cf. [DR]), and so is the equation

$$\begin{cases} \dfrac{\partial \bar{u}}{\partial x} = \dfrac{\bar{A}_{01}}{x-y}\bar{u} + \dfrac{\bar{A}_{02}}{x-1}\bar{u} + \dfrac{\bar{A}_{03}}{x}\bar{u}, \\ \dfrac{\partial \bar{u}}{\partial y} = \dfrac{\bar{A}_{01}}{y-x}\bar{u} + \dfrac{\bar{A}_{12}}{y-1}\bar{u} + \dfrac{\bar{A}_{13}}{y}\bar{u}. \end{cases} \tag{82}$$

Note that if λ and μ are generic, we have

$$\mathscr{L} = \begin{pmatrix} \ker A_{01} \\ \ker A_{02} \\ 0 \end{pmatrix}.$$

Next, we examine transformations

$$\begin{aligned} \tilde{K}_y^{\mu,\lambda} &:= T_{(x,y)\mapsto(y,x)} \circ \tilde{K}_x^{\mu,\lambda} \circ T_{(x,y)\mapsto(y,x)}, \\ \tilde{K}_{x,y}^{\mu,\lambda} &:= T_{(x,y)\mapsto(x,\frac{x}{y})} \circ \tilde{K}_x^{\mu,\lambda} \circ T_{(x,y)\mapsto(x,\frac{x}{y})}. \end{aligned} \tag{83}$$

Note that $(x, y) \mapsto (y, x)$ and $(x, y) \mapsto (x, \frac{x}{y})$ correspond to $(x_0, x_1, x_2, x_3, x_4) \mapsto (x_1, x_0, x_2, x_3, x_4)$ and $(x_0, x_1, x_2, x_3, x_4) \mapsto (x_2, x_1, x_0, x_4, x_3)$, respectively. Hence, KZ equations satisfied by $\tilde{K}_y^{\mu,\lambda}u$ and $\tilde{K}_{x,y}^{\mu,\lambda}u$ are easily obtained from their definition and the equation satisfied by $\tilde{K}_x^{\mu,\lambda}u$. We consider the equation satisfied by $\tilde{K}_{x,y}^{\mu,\lambda}u$.

Putting

$$\tilde{u}(x,y) = (\tilde{K}_{x,y}^{\mu,\lambda}u)(x,y) = \begin{pmatrix} T_{(x,y)\mapsto(x,\frac{x}{y})} K_x^{\mu+1,\lambda} \frac{x}{x-y} u(x,\frac{x}{y}) \\ T_{(x,y)\mapsto(x,\frac{x}{y})} K_x^{\mu+1,\lambda} \frac{x}{x-1} u(x,\frac{x}{y}) \\ T_{(x,y)\mapsto(x,\frac{x}{y})} K_x^{\mu+1,\lambda} u(x,\frac{x}{y}) \end{pmatrix}, \tag{84}$$

residue matrices of KZ equation satisfied by $\tilde{u}(x, y)$ are given by

$$\tilde{A}_{01} = \begin{pmatrix} A_{01}+A_{02} & -A_{02} & 0 \\ -A_{12} & A_{01}+A_{12} & 0 \\ 0 & 0 & A_{01} \end{pmatrix}, \quad \tilde{A}_{02} = \begin{pmatrix} 0 & 0 & 0 \\ A_{12} & A_{02}+\mu & A_{24}+\lambda \\ 0 & 0 & 0 \end{pmatrix},$$

$$\tilde{A}_{03} = \begin{pmatrix} A_{03} & A_{02} & 0 \\ 0 & A_{14}-\mu-\lambda & 0 \\ 0 & A_{02} & A_{03} \end{pmatrix}, \quad \tilde{A}_{04} = \begin{pmatrix} A_{04}+A_{24} & 0 & 0 \\ 0 & A_{04}+A_{24}+\lambda & -A_{24}-\lambda \\ 0 & -A_{02} & A_{02}+A_{04} \end{pmatrix},$$

$$\tilde{A}_{12}=\begin{pmatrix}A_{12}+\mu & A_{02} & A_{24}+\lambda\\ 0&0&0\\ 0&0&0\end{pmatrix},\quad \tilde{A}_{13}=\begin{pmatrix}A_{04}-\mu-\lambda & 0 & 0\\ A_{12} & A_{13} & 0\\ A_{12} & 0 & A_{13}\end{pmatrix},$$
$$\tilde{A}_{14}=\begin{pmatrix}A_{14}+A_{24}+\lambda & 0 & -A_{24}-\lambda\\ 0 & A_{14}+A_{24} & 0\\ -A_{12} & 0 & A_{12}+A_{14}\end{pmatrix},$$
$$\tilde{A}_{23}=\begin{pmatrix}-A_{12}+\lambda & -A_{02} & -A_{24}-\lambda\\ -A_{12} & -A_{02}+\lambda & -A_{24}-\lambda\\ -A_{12} & -A_{02} & -A_{24}\end{pmatrix},\quad \tilde{A}_{24}=\begin{pmatrix}-\mu-\lambda & 0 & 0\\ 0 & -\mu-\lambda & 0\\ A_{12} & A_{02} & A_{24}\end{pmatrix}, \tag{85}$$
$$\tilde{A}_{34}=\begin{pmatrix}A_{01}+A_{02}+A_{12}+\mu & 0 & A_{24}+\lambda\\ 0 & A_{01}+A_{02}+A_{12}+\mu & A_{24}+\lambda\\ 0 & 0 & A_{01}\end{pmatrix}$$

and the invariant subspace to define the required residue matrices $\bar{A}_{i,j}$ is

$$\begin{aligned}\mathscr{L}&=\begin{pmatrix}\ker A_{12}\\ \ker A_{02}\\ \ker A_{24}+\lambda\end{pmatrix}+\ker(\tilde{A}_{23}-\mu-\lambda)\\ &=\begin{pmatrix}\ker B_1\\ \ker A_1\\ \ker A_{24}+\lambda\end{pmatrix}+\ker\begin{pmatrix}B_1+\mu & A_1 & A_{24}+\lambda\\ B_1 & A_1+\mu & A_{24}+\lambda\\ B_1 & A_1 & A_{24}+\mu+\lambda\end{pmatrix}\subset\mathbb{C}^{3N}.\end{aligned} \tag{86}$$

Lastly, in this section, we give an example of hypergeometric series characterized by a KZ equation. Namely, applying

$$\prod_{i=2}^{p}\tilde{K}_x^{-\alpha_i'-\alpha_i,\alpha_i}\prod_{j=2}^{q}\tilde{K}_y^{-\beta_j'-\beta_j,\beta_j}\prod_{r=1}^{r}\tilde{K}_{x,y}^{-\gamma_k'-\gamma_k,\gamma_k} \tag{87}$$

to a solution of the equation $\mathrm{d}u=\alpha_1 u\frac{\mathrm{d}x}{x-1}+\beta_1 u\frac{\mathrm{d}y}{y-1}$, we get KZ equation (76) with the generalized Riemann scheme (see [O3, §4] for its definition)

$$\left\{\begin{matrix}A_{01} & A_{02} & A_{03} & A_{04} & A_{12}\\ [0]_{pq+(p+q-1)r} & [0]_{pr+(p+r-1)q} & [\alpha_i']_{q+r} & [\alpha_i]_{q+r} & [0]_{qr+(q+r-1)p}\\ [-\alpha''-\beta'']_r & [-\alpha''-\gamma'']_q & \beta_j+\gamma_k' & \beta_j'+\gamma_k & [-\beta''-\gamma'']_p\end{matrix}\right.$$
$$\left.\begin{matrix}A_{13} & A_{23} & A_{14} & A_{24} & A_{34}\\ [\beta_j']_{p+r} & [\gamma_k]_{p+q} & [\beta_j]_{p+r} & [\gamma_k']_{p+q} & [0]_{pq+qr+rp-(p+q+r)+1}\\ \alpha_i+\gamma_k' & \alpha_i+\beta_j & \alpha_i'+\gamma_k & \alpha_i'+\beta_j' & [-\alpha''-\beta''-\gamma'']_2\\ &&&& [-\alpha''-\beta'']_{r-1}\\ &&&& [-\beta''-\gamma'']_{p-1}\\ &&&& [-\alpha''-\gamma'']_{q-1}\end{matrix}\right\}, \tag{88}$$

$$
\begin{aligned}
&\alpha_i'' := \alpha_i + \alpha_i',\ \beta_j'' := \beta_j + \beta_j',\ \gamma_k'' := \gamma_k + \gamma_k',\ \alpha_1' = \beta_1' = 0,\\
&\alpha'' = \sum_{i=1}^{p} \alpha_i'',\ \beta'' = \sum_{j=1}^{q} \beta_j'',\ \gamma'' = \sum_{k=1}^{r} \gamma_k'',\\
&1 \le i \le p,\ 1 \le j \le q,\ 1 \le k \le r \quad (p \ge 1,\ q \ge 1,\ r \ge 1).
\end{aligned}
\tag{89}
$$

Here, for example, the eigenvalues of the square matrix A_{01} of size $R = pq+qr+rp$ are 0 with multiplicity $pq + (p+q-1)r$ and $-\alpha'' - \beta''$ with multiplicity r. If the parameters $\alpha_i, \beta_j, \gamma_k, \alpha_i', \beta_j'$, and γ_k' are generic, matrices $A_{i,j}$ are semi-simple, and the KZ equation is irreducible.

Note the hypergeometric series

$$
\begin{aligned}
\phi(x, y) = \sum_{m=0}^{\infty}\sum_{n=0}^{\infty} \frac{\prod_{i=1}^{p}(\alpha_i)_m \prod_{j=1}^{q}(\beta_j)_n \prod_{k=1}^{r}(\gamma_k)_{m+n}}{\prod_{i=1}^{p}(1-\alpha_i')_m \prod_{j=1}^{q}(1-\beta_j')_n \prod_{k=1}^{r}(1-\gamma_k')_{m+n}} x^m y^n\\
\text{with } \alpha_1' = \beta_1' = 0
\end{aligned}
\tag{90}
$$

is a component of a solution of this KZ equation (cf. (77)).

The Riemann scheme (88) is obtained by [O3, Theorem 7.1] and (87). The precise argument and a further study of the hypergeometric series (90) will be given in [MO].

The index of the rigidity of this KZ equation with respect to x equals

$$
\begin{aligned}
\mathrm{Idx}_x \mathscr{M} &= (R-q)^2 + q^2 + (R-r)^2 + r^2 + 2(p(q+r)^2 + qr) - 2R^2\\
&= 2 - 2(q-1)(r-1)(q+r+1)
\end{aligned}
\tag{91}
$$

and hence, the ordinary differential equation with respect to the variable x is rigid if and only if $r = 1$ or $q = 1$.

If $p = q = r = 1$, the corresponding KZ equation (78) is given by (85) with

$$
\begin{aligned}
&(A_{01}, A_{02}, A_{03}, A_{04}, A_{12}, A_{13}, A_{14}, A_{23}, A_{24}, A_{34}, \lambda, \mu)\\
&= (0, \alpha_1, 0, -\alpha_1, \beta_1, 0, -\beta_1, -\alpha_1 - \beta_1, \alpha_1 + \beta_1, \gamma_1, -\gamma_1 - \gamma_1')
\end{aligned}
$$

and it follows from (84) that the equation has a solution with the last component

$$
\phi(x, y) = F_1(\gamma_1, \alpha_1, \beta_1, 1 - \gamma_1'; x, y).
$$

We define a simple local solution to (76) at the origin, which includes the solution we have just considered.

Definition 6 We define that a local solution to Eq. (76) near the origin has a *simple monodromy* if the analytic continuation of the solution in a neighborhood of the origin spans one-dimensional space. We simply call the solution a *simple* solution at the origin. We also define that a local solution of the equation to (76) near the line $x = 0$ has a simple monodromy and call it a simple solution along $x = 0$ if the analytic continuation of the solution in a neighborhood of $x = 0$ spans one-dimensional space.

By correspondence between Eqs. (76) and (70) with $q = 3$ and moreover a transformation by $\mathfrak{S}_5$, we define a local solution at $x_i = x_j = x_k$ and a local solution at $x_s = x_t$ to Eq. (70) with $q = 3$ when $\{i, j, k, s, t\} = \{0, 1, 2, 3, 4\}$.

Here, for example, the path of the analytic continuation in the latter case, namely, along $x = 0$, is in $\{(x, y) \in \mathbb{P}^1 \times \mathbb{P}^1 \mid |x| < \epsilon,\ 0 < |x| < \epsilon|y|\}$ with a small positive number ϵ.

Then we have the following theorem:

Theorem 3 *Suppose* $\{i, j, k, s, t\} = \{0, 1, 2, 3, 4\}$ *as above. To Eq.* (70) *with* $q = 3$, *there is a one-to-one correspondence between a simple solution at* $x_i = x_j = x_k$ *and a simple solution along* $x_s = x_t$.

Proof The coordinate $(x, \frac{x}{y})$ is a local coordinate of a blowing up of the singularities of Eq. (70) around the origin. Then the origin corresponds to the line $x = 0$. This coordinate transformation corresponds to the map $(x_0, x_1, x_2, x_3, x_4) \mapsto (x_2, x_1, x_0, x_4, x_3)$, which is explained in Remark 5. Since $x_2 = x_4$ corresponds to $x_0 = x_3$, we have the theorem when $(i, j, k, s, t) = (0, 1, 3, 2, 4)$. Note that the coordinate $(\frac{y}{x}, y)$ gives the same conclusion. Then the symmetry $\mathfrak{S}_5$ proves the theorem. □

This theorem says that an eigenvalue of A_{24} with free multiplicity corresponds to a simple solution at $x_0 = x_1 = x_3$, which is the origin in (x, y) coordinate. Hence, at the origin, we have pq independent simple solutions of the KZ equation with the Riemann scheme (88) if the parameters are generic.

Remark 6 The space of local solutions at a normally crossing singular point defined by $x_i = x_j$ and $x_s = x_t$ under the above notation is spanned by simple solutions if the parameters are generic (cf. [KO]).

Remark 7 The transformation

$$\mathscr{O}_0 \ni u = \sum c_{m,n} x^m y^n \mapsto (1-x)^{-a}(1-y)^{-b} u = \sum (a)_i (b)_j c_{m,n} x^{i+m} y^{j+n}$$

induces the transformation $(A_{02}, A_{12}) \mapsto (A_{02} + a, A_{12} + b)$ of Eq. (76). Then coefficients of the resulting power series may be complicated (cf. (40)).

Remark 8 The transformation of residue matrices induced by $\tilde{K}_x^\lambda$, $\tilde{K}_y^\lambda$, and the $\tilde{K}_{x,y}^\lambda$ and the calculation of Riemann scheme (88) for given p, q, r, etc. are supported by functions `m2mc` and `mc2grs` in the library [O7] of the computer algebra `Risa/Asir`. For example, by the commands

```
R=os_md.mc2grs(0,["K",[4,3,2]]);
os_md.mc2grs(R,"get"|dviout=1,div=5);
```

we get (88) on a display in the case $(p, q, r) = (4, 3, 2)$. This is enabled by using a PDF file output by functions in the library under the computer algebra. Here div=5 indicates to divide the Riemann scheme by five columns into two parts as in (88), and R is a list of simultaneous eigenspace decomposition at 15 normally crossing singularities of the corresponding KZ equation. The algorithm for the calculation is given by [O3, Theorem 7.1]. Moreover, by the command

```
os_md.mc2grs(R,"rest"|dviout=1);
```

we get the Riemann scheme of the induced equations on ten singular hypersurfaces corresponding to eigenvalues of ten residue matrices. If "spct" is indicated in place of "rest", a table of spectral types with respect to variables x_i for $i = 0, \dots, 4$ and indices of rigidities are displayed. If "dviout=-1" is indicated in place of dviout=1, the result is given by a TEX source in place of displaying the result on a screen. If "|dviout=1" is not indicated, the result is given in a format recognized by Risa/Asir.

8 Fuchsian Ordinary Differential Equations

In this section, we consider a Fuchsian differential equation

$$\mathscr{N} : \frac{du}{dx} = \frac{A_y}{x-y}u + \frac{A_1}{x-1}u + \frac{A_0}{x}u \tag{92}$$

with regular singularities at $x = 0, 1, y$, and ∞. Here, residue matrices A_y, A_1, and A_0 are constant square matrices of size N. If (92) is irreducible and rigid or

$$\dim\bigl(Z(A_y) \cap Z(A_1) \cap Z(A_0)\bigr) = 1 \tag{93}$$

and

$$\dim Z(A_y) + \dim Z(A_1) + \dim Z(A_0) + \dim Z(A_y + A_1 + A_0) - 2N^2 = 2, \tag{94}$$

Equation (92) is constructed by successive applications of middle convolutions and additions to the trivial equation $u' = 0$. Here $Z(A)$ denotes the space of the centralizer in $M(N, \mathbb{C})$ for $A \in M(N, \mathbb{C})$, and the left-hand side of (94) is the index of the rigidity of the equation. We note that middle convolutions can be replaced by the transformation of equations induced by $\tilde{K}_x^{\mu}$ with additions.

We assume that Eq. (92) can be extended to the compatible equation

$$\frac{\partial u}{\partial y} = \frac{A_y}{x-y}u + \frac{B_1}{y-1}u + \frac{B_0}{y}u. \tag{95}$$

If Eq. (92) is rigid, it extends to the compatible equation, and moreover, if the equation satisfies (93), matrices B_0 and B_1 are uniquely determined by (92) up to the difference of scalar matrices. We can apply the transformations induced by $\tilde{K}_y^{\mu}$ and $\tilde{K}_{x,y}^{\mu}$ to (92). Note that these transformations may change the index of rigidity as was shown in the last section.

Transformations induced by $\tilde{K}_x^{\mu}$, $\tilde{K}_y^{\mu}$, and $\tilde{K}_{x,y}^{\mu}$ are given by (96), (97), and (98), respectively, with calculating the induced matrices of the residue matrices on $\mathbb{C}^{3N}/\mathscr{L}$.

$$\begin{aligned}
\frac{\mathrm{d}\tilde{u}}{\mathrm{d}x} &= \frac{\begin{pmatrix} A_y+\mu & A_1 & A_0 \\ 0 & 0 & 0 \\ 0 & 0 & 0 \end{pmatrix}}{x-y}\tilde{u} + \frac{\begin{pmatrix} 0 & 0 & 0 \\ A_y & A_1+\mu & A_0 \\ 0 & 0 & 0 \end{pmatrix}}{x-1}\tilde{u} + \frac{\begin{pmatrix} -\mu & 0 & 0 \\ 0 & -\mu & 0 \\ A_y & A_1 & A_0 \end{pmatrix}}{x}\tilde{u}, \\
\mathscr{L} &= \begin{pmatrix} \ker A_y \\ \ker A_1 \\ \ker A_0 \end{pmatrix} + \ker \begin{pmatrix} A_y+\mu & A_1 & A_0 \\ A_y & A_1+\mu & A_0 \\ A_y & A_1 & A_0+\mu \end{pmatrix},
\end{aligned} \tag{96}$$

$$\begin{aligned}
\frac{\mathrm{d}\tilde{u}}{\mathrm{d}x} &= \frac{\begin{pmatrix} A_y+\mu & B_1 & B_0 \\ 0 & 0 & 0 \\ 0 & 0 & 0 \end{pmatrix}}{x-y}\tilde{u} + \frac{\begin{pmatrix} A_1+B_1 & -B_1 & 0 \\ -A_y & A_1+A_y & 0 \\ 0 & 0 & A_1 \end{pmatrix}}{x-1}\tilde{u} + \frac{\begin{pmatrix} A_0+B_0 & 0 & -B_0 \\ 0 & A_0 & 0 \\ -A_y & 0 & A_0+A_y \end{pmatrix}}{x}\tilde{u}, \\
\mathscr{L} &= \begin{pmatrix} \ker A_y \\ \ker B_1 \\ \ker B_0 \end{pmatrix} + \ker \begin{pmatrix} A_y+\mu & B_1 & B_0 \\ B_y & A_1+\mu & B_0 \\ B_y & A_1 & B_0+\mu \end{pmatrix},
\end{aligned} \tag{97}$$

$$\begin{aligned}
\frac{\mathrm{d}\tilde{u}}{\mathrm{d}x} &= \frac{\begin{pmatrix} A_y+A_1 & -A_1 & 0 \\ -B_1 & A_y+B_1 & 0 \\ 0 & 0 & A_y \end{pmatrix}}{x-y}\tilde{u} + \frac{\begin{pmatrix} 0 & 0 & 0 \\ B_1 & A_1+\mu & A_{24} \\ 0 & 0 & 0 \end{pmatrix}}{x-1}\tilde{u} + \frac{\begin{pmatrix} A_0 & A_1 & 0 \\ 0 & A_{14}-\mu & 0 \\ 0 & A_1 & A_0 \end{pmatrix}}{x}\tilde{u}, \\
\mathscr{L} &= \begin{pmatrix} \ker B_1 \\ \ker A_1 \\ \ker A_{24} \end{pmatrix} + \ker \begin{pmatrix} B_1+\mu & A_1 & A_{24} \\ B_1 & A_1+\mu & A_{24} \\ B_1 & A_1 & A_{24}+\mu \end{pmatrix}.
\end{aligned} \tag{98}$$

Here

$$\begin{aligned}
A_{14} &= -A_y - B_0 - B_1, \\
A_{24} &= -(A_{02} + A_{12} + A_{23}) = (A_{01} + A_{02} + A_{03} + A_{12} + A_{13}) - A_{02} - A_{12} \\
&= A_{01} + A_{03} + A_{13} = A_y + A_0 + B_0.
\end{aligned}$$

Acknowledgments This work was supported by Grants-in-Aid for Scientific Research (C), No. 18K03341, Japan Society for the Promotion of Science.

References

[AK] Appell, K., Kampé de Fériet, J.: Fonctions hypergéométriques et hypersphériques polynomes d'Hermite. Gauthier-Villars (1926)

[DR] Dettweiler, M., Reiter, S.: An algorithm of Katz and its applications to the inverse Galois problems. J. Symbolic Comput. **30**, 761–798 (2000)

[DR2] Dettweiler, M., Reiter, S.: Middle convolution of Fuchsian systems and the construction of rigid differential systems. J. Algebra **318**, 1–24 (2007)

[Er] Erdélyi, A., Magnus, W., Oberhettinger, F., Tricomi, F.G.: Higher Transcendental Functions, vol. 3. McGraw-Hill Book, New York (1953)

[GKZ] Gel'fand, I.M., Kpranov, M.M., Zelevinsky, A.V.: Generalized Euler integrals and A-hypergeometric functions. Adv. Math. **84**, 255–271 (1990)

[Ha] Haraoka, Y.: Middle convolution for completely integrable systems with logarithmic singularities along hyperplane arrangements. Adv. Stud. Pure Math. **62**, 109–136 (2012)

[Ho] Horn J.: Über die Convergenz der hypergeometrischen Reihen zweier und dreier Veränderlichen. Math. Ann. **34**, 544–600 (1889)

[KO] Kashiwara, M., Oshima, T.: Systems of differential equations with regular singularities and their boundary value problems. Annal. Math. **106**, 145–200 (1977)

[Ka] Katz, N.M.: Rigid Local Systems. Annals of Mathematics Studies, vol. 139. Princeton University Press, Princeton (1995). https://doi.org/10.1515/9781400882595

[KZ] Knizhnik, K., Zamolodchikov, A.: Current algebra and Wess-Zumino model in 2 dimensions. Nucl. Phys. B **247**, 83–103 (1984)

[La] Lauricella, G.: Sulle funzioni ipergeometriche a piu variabili. Rendiconti del Circolo Matematico di Palermo. **7**, 111–158 (1983)

[Ma] Matsubara-Hao, S.-J.: Global analysis of GG systems. Int. Math. Res. Not. **2022**(19), 14923–14963 (2022). https://doi.org/10.1093/imrn/rnab144

[MO] Matsubara-Hao, S.-J., Oshima, T.: Generalized hypergeometric functions with several variables. Indag. Math. ArXiv.2311.18611. In press

[O1] Oshima, T.: Fractional Calculus of Weyl Algebra and Fuchsian Differential Equations. MSJ Memoirs, vol. 28. Mathematical Society of Japan, Tokyo (2012)

[O2] Oshima, T.: An Elementary Approach to the Gauss Hypergeometric Function. Josai Mathematical Monographs, vol. 6, pp. 3–23 (2013). https://doi.org/10.20566/1344777-06-3

[O3] Oshima, T.: Transformation of KZ type equations. Microlocal Analysis and Singular Perturbation Theory. RIMS Kôkyûroku Bessatsu, vol. B61, pp. 141–162 (2017)

[O4] Oshima, T.: Confluence and versal unfolding of Pfaffian systems. Josai Math. Monogr. **12**, 117–151 (2020). https://doi.org/10.20566/13447777-12-117

[O5] Oshima, T.: Versal unfolding of irregular singularities of a linear differential equation on the Riemann sphere. Publ. RIMS Kyoto Univ. **57**, 893–920 (2021). https://doi.org/10.4171/PRIMS/57-3-6

[O6] Oshima, T.: Riemann-Liouville transform and linear differential equations on the Riemann sphere. In: Recent Trends in Formal and Analytic Solutions of Differential Equations. Contemporary Mathematics, vol. 782, 57–91. American Mathematical Society (2023)

[O7] Oshima, T.: `os_muldif.rr`, a library of computer algebra `Risa/Asir`. (2008–2023). http://www.ms.u-tokyo.ac.jp/~oshima/

Printed in the United States
by Baker & Taylor Publisher Services